Catalysis for Fuels

Cape Town, South Africa

24–26 January 2017

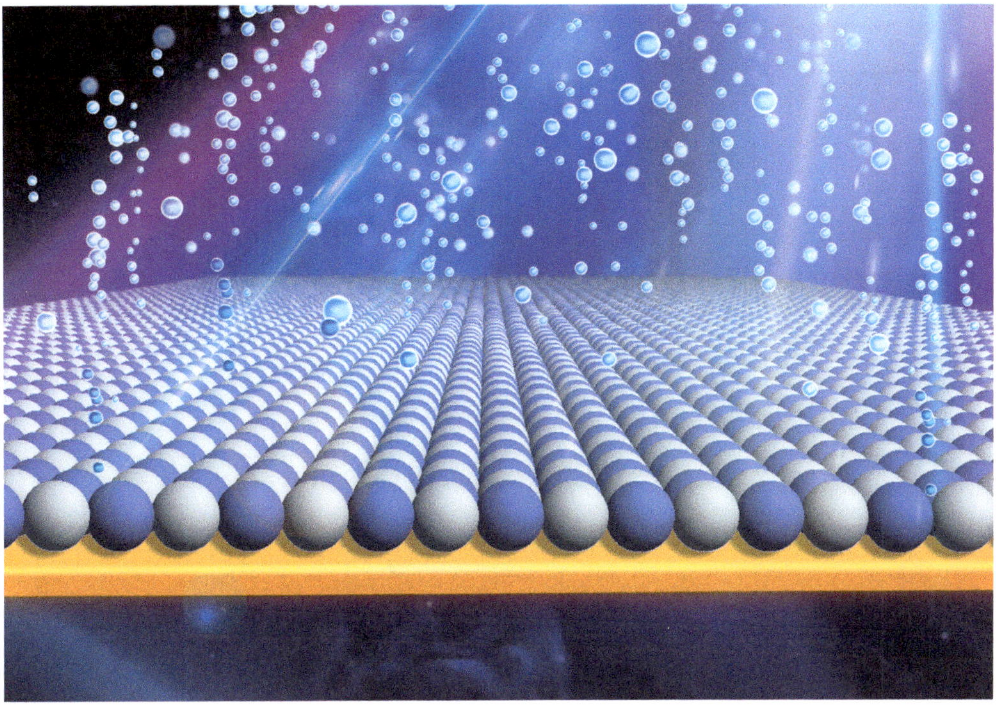

FARADAY DISCUSSIONS

Volume 197, 2017

ROYAL SOCIETY OF CHEMISTRY

The Faraday Division of the Royal Society of Chemistry, previously the Faraday Society, was founded in 1903 to promote the study of sciences lying between chemistry, physics and biology.

Faraday Discussions (Print ISSN 1359-6640, Electronic ISSN 1364-5498) is published 10 times a year by the Royal Society of Chemistry, Thomas Graham House, Science Park, Milton Road, Cambridge, UK CB4 0WF.

Volume 197 ISBN 13: 978-1-78262-945-0

2017 annual subscription price: print+electronic £1010 US $1782; electronic only £962, US $1693. Customers in Canada will be subject to a surcharge to cover GST. Customers in the EU subscribing to the electronic version only will be charged VAT.

All orders, with cheques made payable to the Royal Society of Chemistry, should be sent to the Royal Society of Chemistry Order Department, Royal Society of Chemistry, Thomas Graham House, Science Park, Milton Road, Cambridge, CB4 0WF, UK Tel +44 (0)1223 432398; E-mail **orders@rsc.org**

If you take an institutional subscription to any Royal Society of Chemistry journal you are entitled to free, site-wide web access to that journal. You can arrange access via Internet Protocol (IP) address at **www.rsc.org/ip**

Customers should make payments by cheque in sterling payable on a UK clearing bank or in US dollars payable on a US clearing bank.

Printed in the UK

MIX
Paper from responsible sources
FSC www.fsc.org FSC® C013604

Faraday Discussions are unique international discussion meetings that focus on rapidly developing areas of chemistry and its interfaces with other scientific disciplines.

Catalysis for Fuels

Faraday Discussions

www.rsc.org/faraday_d

A General Discussion on Catalysis for Fuels was held in Cape Town, South Africa on the 24th, 25th and 26th of January 2017.

CONTENTS

ISSN 1359-6640; ISBN 978-1-78262-945-0

Cover
See Domen *et al.*, *Faraday Discuss.*, 2017, **197**, 491–504.

Particulate photocatalyst sheets for Z-scheme water splitting: advantages over powder suspension and photoelectrochemical systems and future challenges.

Image reproduced by permission of Kazunari Domen from *Faraday Discuss.*, 2017, **197**, 491–504.

INTRODUCTORY LECTURE

PAPERS AND DISCUSSIONS

CONCLUDING REMARKS

ADDITIONAL INFORMATION

Faraday Discussions

PAPER

Catalytic routes to fuels from C_1 and oxygenate molecules

Shuai Wang, Iker Agirrezabal-Telleria, Aditya Bhan,† Dante Simonetti,‡ Kazuhiro Takanabe§ and Enrique Iglesia*

Received 16th January 2017, Accepted 20th January 2017
DOI: 10.1039/c7fd00018a

This account illustrates concepts in chemical kinetics underpinned by the formalism of transition state theory using catalytic processes that enable the synthesis of molecules suitable as fuels from C_1 and oxygenate reactants. Such feedstocks provide an essential bridge towards a carbon-free energy future, but their volatility and low energy density require the formation of new C–C bonds and the removal of oxygen. These transformations are described here through recent advances in our understanding of the mechanisms and site requirements in catalysis by surfaces, with emphasis on enabling concepts that tackle ubiquitous reactivity and selectivity challenges. The hurdles in forming the first C–C bond from C_1 molecules are illustrated by the oxidative coupling of methane, in which surface O-atoms form OH radicals from O_2 and H_2O molecules. These gaseous OH species act as strong H-abstractors and activate C–H bonds with earlier transition states than oxide surfaces, thus rendering activation rates less sensitive to the weaker C–H bonds in larger alkane products than in CH_4 reactants. Anhydrous carbonylation of dimethyl ether forms a single C–C bond on protons residing within inorganic voids that preferentially stabilize the kinetically-relevant transition state through van der Waals interactions that compensate for the weak CO nucleophile. Similar solvation effects, but by intrapore liquids instead of inorganic hosts, also become evident as alkenes condense within MCM-41 channels containing isolated Ni^{2+} active sites during dimerization reactions. Intrapore liquids preferentially stabilize transition states for C–C bond formation and product desorption, leading to unprecedented reactivity and site stability at sub-ambient temperatures and to 1-alkene dimer selectivities previously achieved only on organometallic systems with co-catalysts or activators. C_1 homologation selectively forms C_4 and C_7 chains with a specific backbone (isobutane, triptane) on solid acids, because of methylative growth and

Department of Chemical Engineering, University of California at Berkeley, Chemical Sciences Division, E. O. Lawrence Berkeley National Laboratory, Berkeley, CA 94720, USA. E-mail: iglesia@berkeley.edu

† Current address: Department of Chemical Engineering and Materials Science, University of Minnesota, Minneapolis, MN 55455.

‡ Current address: Department of Chemical and Biomolecular Engineering, University of California at Los Angeles, Los Angeles, CA, 90095.

§ Current address: King Abdullah University of Science and Technology, KAUST Catalysis Center, Physical Science and Engineering Division, Thuwal, 23955, Saudi Arabia.

hydride transfer rates that reflect the stability of their carbenium ion transition states and are unperturbed by side reactions at low temperatures. Aldol condensation of carbonyl compounds and ketonization of carboxylic acids form new C–C bonds concurrently with O-removal. These reactions involve analogous elementary steps and occur on acid–base site pairs on TiO_2 and ZrO_2 catalysts. Condensations are limited by α-H abstraction to form enolates *via* concerted interactions with predominantly unoccupied acid–base pairs. Ketonization is mediated instead by C–C bond formation between hydroxy-enolates and monodentate carboxylates on site pairs nearly saturated by carboxylates. Both reactions are rendered practical through bifunctional strategies, in which H_2 and a Cu catalyst function scavenge unreactive intermediates, prevent sequential reactions and concomitant deactivation, and remove thermodynamic bottlenecks. Alkanal–alkene Prins condensations on solid acids occur concurrently with alkene dimerization and form molecules with new C–C bonds as skeletal isomers unattainable by other routes. Their respective transition states are of similar size, leading to selectivities that cannot sense the presence of a confining host. Prins condensation reactions benefit from weaker acid sites because their transition states are less charged than those for oligomerization and consequently less sensitive to conjugate anions that become less stable as acids weaken.

1. Background and context

The role of catalysis, and specifically of processes enabled by heterogeneous catalysts, in the conversion and upgrading of molecules intended for use as fuels and energy carriers has been remarkable in breadth and impact[1] throughout nearly a century of symbiotic relationships between concepts and technologies that have evolved side-by-side. In fact, together with the engineering and design of chemical processes, the science and practice of heterogeneous catalysis have been critical in the evolution of modern-day fuels by enabling their improved performance and environmental compatibility; catalysis and its fundamental concepts and tools will remain so as we evolve from crude oils to transitional sources of fuels and feedstocks, such as natural gas[2] and biomass,[3] and ultimately to carbon-free energy carriers derived from geothermal, wind, and solar resources.

Even a cursory review of such a storied history would be impossible; it is not the intent of this brief and modest account. Instead, the discourse focuses on conceptual advances in upgrading streams derived from natural gas – C_1 molecules that lack C–C bonds – or from oxygenates of plausible biogenic origins. We have chosen, as a matter of comfort and uniformity, illustrative case studies from our research group. The emphasis on concepts in each section is intentional, without any intent to distract the reader from the realities of heterogeneous catalysis as a discipline with a historical record of formidable practical impact.

Low volatility and high energy density are primary characteristics of transportation fuels, imparted in current refining strategies by conversion and upgrading units. C_1 molecules, typified by CH_4, the molecule with the highest energy density per C-atom, lack C–C bonds. Oxygenates, in turn, lack the required energy density, and O-removal forms carbon chains that are too short to meet volatility requirements, thus requiring the concurrent formation of new C–C bonds during O-removal. Appropriately, therefore, what follows focuses on the formation of new C–C bonds (homologation, carbonylation, oligomerization) and deoxygenation steps that concurrently form C–C bonds (aldol and Prins condensations, ketonization) as the enabling catalytic routes.

2. Catalysis for the conversion and upgrading of diverse feedstocks without C–C bonds and the challenges inherent in activation of C_1 molecules

The recurring interest in unconventional carbon sources as precursors to chemicals and transportation fuels periodically returns the catalysis field to the challenges of converting molecules, such as methane, methanol, and dimethyl ether, which lack C–C bonds. Such C_1 species, and methane specifically, bring forth formidable thermodynamic and kinetic challenges in directing the catalytic route towards products of higher value and practical use as fuels or chemicals. Such inherent hurdles and bottlenecks, with roots in thermodynamics but transferred into the realm of dynamics through linear free energy relations, lead to the significant complexity and cost that characterize the commercial deployment of C_1 conversion to more valuable molecules. Here, we discuss a conceptual framework to provide guidance for addressing these challenges in the form of some examples. In doing so, the reader will notice that we evolve the products of C_1 conversion from C_2 (oxidative coupling, carbonylation) to C_4 (dimerization) and ultimately C_4–C_7 (homologation) chains. The manuscript excludes, as a result of space constraints, any discussion of C_1 polymerization processes known as the Fischer–Tropsch synthesis and specifically of our recent studies of the unique reactivity and selectivity conferred upon catalytic surfaces by the presence of dense adlayers of chemisorbed CO.

2.1. Oxidative coupling of methane: selectivity conferred by strong H-abstractors formed *in situ* from O_2–H_2O mixtures

The kinetic consequences of ubiquitous thermodynamic hurdles in activating strong bonds preferentially over weaker ones, the inherent limitations of catalysis in circumventing them, and the pre-eminent requirement for processes that stress simplicity and inexpensive oxidants, such as O_2, are illustrated by the extensive studies of the process known as the oxidative coupling of methane (OCM). Here, we discuss how H-abstractors stronger than O_2, but formed *in situ* in the form of OH radicals from H_2O–O_2 mixtures, are used to moderate the kinetic preference for abstracting H-atoms from C–H bonds in C_2H_6 products instead of CH_4 reactants; these homogeneous activation routes also circumvent surface routes that favor unselective oxidation of even the stronger bonds in C_2H_4 because of their strong π-interactions with oxide surfaces.

OCM leads to the direct formation of a C–C bond from C_1 molecules *via* the coupling of two hydrocarbon radicals; it is favored under CH_4-rich conditions that tend to minimize the CO_x formation that prevails at combustion-type conditions. OCM involves coupled heterogeneous–homogeneous reactions, in which a C–H bond in CH_4 is activated *via* H-abstraction with O_2-derived surface species to form H_2O and gaseous CH_3 radicals that undergo subsequent homogeneous coupling reactions to form C_2H_6.[4] The subsequent transformation of C_2H_6, which contains weaker C–H bonds than those in CH_4, proceeds *via* H-abstraction steps similar to those for CH_4 to form reactive species that ultimately form CO_x and C_2H_4. Such pathways, described schematically in Scheme 1, limit attainable C_2 yields,[5,6] because the conversion of C_2 products is more facile than that of CH_4 on oxide surfaces at the temperatures required to form the required CH_3 radical intermediates.

A detailed kinetic and isotopic study of OCM primary and secondary reactions on $Mn/Na_2WO_4/SiO_2$ [7] provided a rigorous mechanistic interpretation of the effects of H_2O in enhancing rates through parallel homogeneous activation routes involving OH radicals; such radicals were formed at oxide surfaces *via* reactions of H_2O (added or formed during OCM as CH_4 conversion increased) with O_2-derived surface O* species, as depicted schematically in Scheme 1[8,9] and demonstrated experimentally from the higher C_2 yields attainable in the presence of the H_2O formed as the reaction proceeds (Fig. 1). Such parallel OH-mediated routes contribute significantly to C_2 formation rates and selectivities on such alkali-promoted catalyst compositions, which tend to show lower reactivity for C–H bond activation than Li–MgO and Sr–La_2O_3 compositions, on which surface-mediated routes prevail at their lower operating temperatures.

Isotopic tracer studies using mixtures of $^{13}CH_4/O_2$ with $^{12}C_2H_6$ or $^{12}C_2H_4$ in the presence or absence of added H_2O showed that H_2O increases the activation rates for all C–H bonds (k_1, k_2, and k_3), but to a different extent, thus leading to lower k_3/k_1 ratios (Fig. 2), consistent with the higher attainable C_2 product yields with added H_2O (Fig. 1).[8,9] H-abstraction by OH radicals occurs less selectively for the weaker C–H bonds in C_2H_6 than for the stronger CH_4 bonds than in drier environments, in which surface O* species carry out the required H-abstractions. In the absence of added H_2O, C–H bonds in C_2H_4 were more reactive than in CH_4 in spite of their greater strength (Fig. 2), but gave rate constant ratios slightly below unity when H_2O was added, as expected from stronger C–H bonds in C_2H_4 than CH_4. Such kinetic models illustrate the essential effects of H_2O in enabling higher C_{2+} yields, as shown in Fig. 1. The blue data points in Fig. 1 correspond to these ratios of rate constants for C–H activation by surface O* species (dry), while the red data points consider the additional contributions to such ratios by C–H activation mediated by OH radicals in the gas phase (wet).

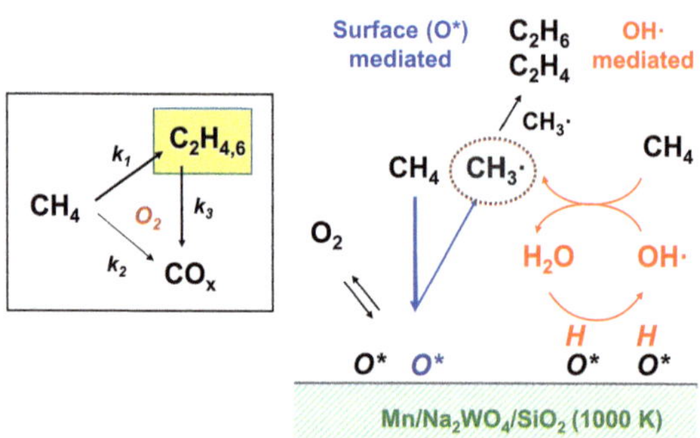

Scheme 1 Simplified scheme for the OCM reaction network and pseudo-first-order rate constants for each step (left); representative scheme for OCM reaction steps involving CH_3 radical generation *via* surface O* mediated (dry) and OH radical gas-phase mediated (wet) pathways on $Mn/Na_2WO_4/SiO_2$ (right).

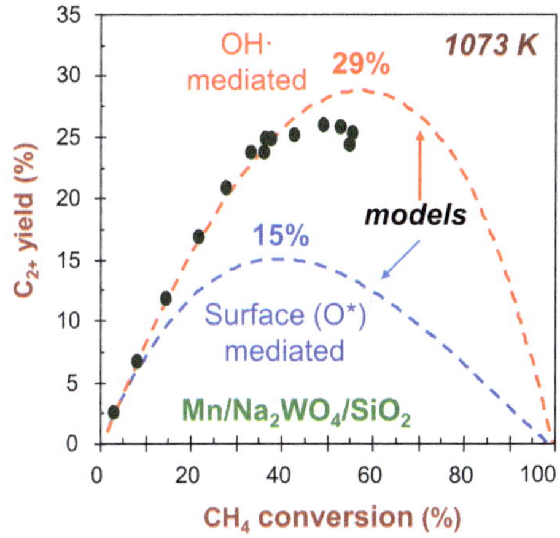

Fig. 1 Measured C_{2+} yields as a function of CH_4 conversion (data points) and model predictions from measured rate constants in this figure and the reaction steps in Scheme 1. Blue curve for surface O*-mediated (dry) route and red curve for combined paths mediated by surface O* (dry) and OH radicals (wet) routes (0.02 g, 1073 K, 4.0 kPa CH_4, 1.3 kPa O_2 (initial), 1.5 kPa O_2 (added)). Adapted with permission from ref. 8 (Copyright 2008 Wiley-VCH).

These data suggest the involvement of parallel C–H activation routes (Scheme 1) involving pathways in which H-abstraction is mediated by either oxygen species on surfaces or by OH radicals, formed *via* H_2O/O_2 equilibration on catalyst surfaces, with specific involvement of Na or K cations.[10] These coupled homogeneous-catalytic sequences are consistent with the kinetic effects of O_2, CH_4, and H_2O on the rates of primary and secondary reactions and with the observed H/D isotope effects. These isotopic data also show that C–H activation is irreversible and kinetically-relevant and that O_2 dissociation is quasi-equilibrated for surface routes, but becomes irreversible as O_2/H_2O ratios increase.

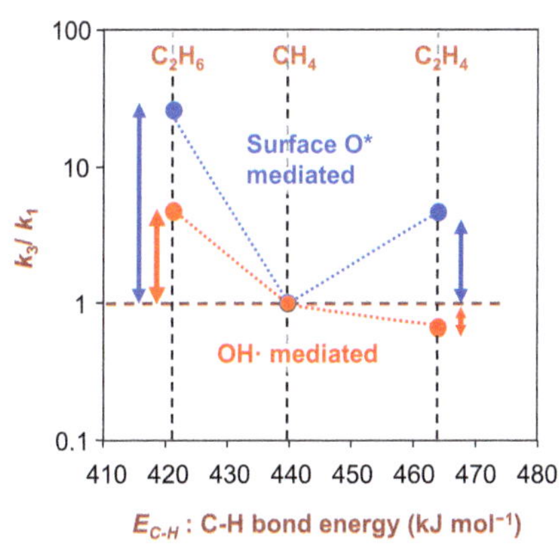

Fig. 2 The abscissa shows the C–H bond energies for CH_4, C_2H_6, and C_2H_4. The lower measured k_3/k_1 rate constants for C–H bond activation in ethene or ethane (k_3) and methane (k_1) (as defined in Scheme 1) *via* surface O* (dry) and gas-phase OH radical gas-phase (wet) pathways lead to the higher C_{2+} yields achieved with added H_2O (Fig. 1).

OCM rates and C_{2+} yields are higher when H_2O is present because these homogeneous OH-mediated pathways involve a much stronger H-abstractor than $O_2(g)$ or surface O* species. Indeed, the ratio of pseudo-first-order rate constants for the activation of CH_4 to that for C_2H_6 increased monotonically and roughly exponentially as the enthalpy of the H-abstraction step increases for homogeneous H-abstraction reactions (Fig. 3),[8] thus rendering OH radicals less selective for weaker C–H bonds than lattice O-atoms. Such trends reflect C–H activation transition states that occur earlier along the reaction coordinate as steps become more exothermic, leading to reactant-like structures in which C–H bonds are less fully cleaved and to activation barriers that are therefore less sensitive to the strength of the C–H bonds. Soft abstractors and low temperatures are often invoked as promising strategies to improve attainable C_2 yields in OCM; such strategies are not only incompatible with each other, but in fact counterproductive, because they strengthen the sensitivity of activation barriers to C–H bond strength. These concepts have been recently confirmed for the selective dehydrogenation of C_2H_6 with O_2/H_2O co-reactants.[11]

2.2. Anhydrous carbonylation of dimethyl ether: confinement of transition states within inorganic pockets of molecular dimensions and the selective formation of a single C–C bond

The intended formation of C–C bonds from stable CH_4 molecules *via* the OCM routes in Section 2.1 illustrates the inherent challenges in the selective formation of the first C–C bond from C_1 molecules. The carbonylation of C_1 species forms such C–C bonds, but becomes thermodynamically possible only for oxygen-containing C_1 molecules, such as CH_3OH or CH_3OCH_3 (dimethyl ether, DME).

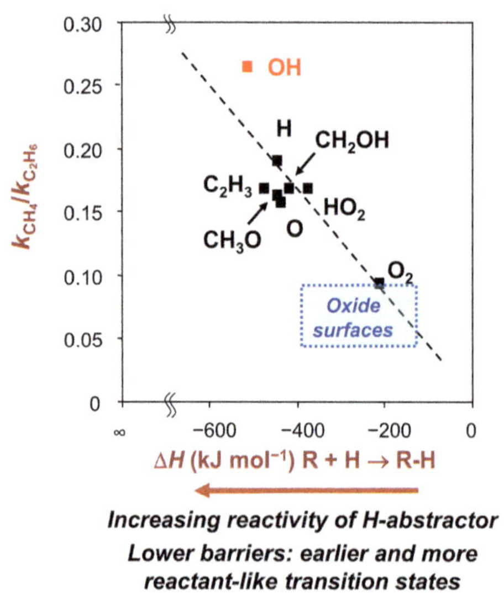

Fig. 3 Rate constant ratios for H-abstraction from CH_4 and C_2H_6 (1073 K, $k_{CH_4}/k_{C_2H_6}$) for different abstractors (R) as a function of the reaction enthalpy (ΔH for R + H = R–H). Data points are for homogeneous reactions of each R radical; the rectangle indicates range of H-abstraction properties of O*. Adapted with permission from ref. 8 (Copyright 2008 Wiley-VCH).

In fact, such carbonylation reactions provide the only known route to form one single C–C bond with nearly perfect selectivity from two C_1 species. Current processes require the use of halides to form methylating species capable of reacting with a very weak nucleophile such as CO and organometallic complexes of Rh or Ir as catalysts, as implemented in the large-scale synthesis of acetic acid from CH_3OH–CO reactants.[12] These processes also require the exhaustive removal of H_2 from CO-containing streams to prevent deleterious side reactions that detract from selectivity and catalyst stability and lead to costly purification of product streams.

Efforts to catalyze CO reactions with CH_3OH on acid-forms of zeotypes led to very low rates and fast deactivation, symptomatic of competing hydrolysis reactions of bound methyls with H_2O molecules formed *via* fast CH_3OH homologation and dehydration reactions and of the need, unrecognized at the time, for voids of specific size in order to compensate for the weak nucleophilic character of CO. These hurdles were surmounted by the use of DME as the substrate, which allows carbonylation to occur under essentially anhydrous conditions with the consequent formation of methyl acetate (MeOAc), and by the discovery that side pockets in MOR structures and the smaller channels in two-dimensional networks in FER zeolites, both consisting of structures formed by inorganic rings containing eight O-atoms (eight-membered ring, 8-MR), provide effective van der Waals contacts with the transition state that mediates the formation of the C–C bond in the kinetically-relevant step.[13–15] Such catalytic routes allow the direct use of synthesis gas to form a product (methyl acetate) that can be converted, if so warranted by markets, to acetic acid *via* reactions with H_2O or by coupling of CH_3OH dehydration (to DME) with MeOAc hydrolysis. MeOAc forms with >99% selectivity on these catalysts at moderate temperatures (<550 K); it can be used to form ethanol *via* hydrogenolysis and ethene *via* ethanol dehydration, thus providing the only known routes for the formation of molecules containing a single C–C bond from C_1 feedstocks.

DME carbonylation involves a kinetically-relevant step consisting of backside attack by CO on DME-derived methyls bound at Al sites on the acid-forms of zeotypes (Fig. 4). This mechanism is consistent with kinetic and isotopic data and with infrared spectra, as well as with the stoichiometric formation of CH_3 groups from DME and their subsequent reaction with CO and scavenging of acyls formed by another DME molecule;[13–15] it is also in agreement with theoretical treatments that followed these experimental demonstrations and mechanistic interpretations.[16,17] This catalytic cycle (Fig. 4A) forms an acyl intermediate that cannot desorb without an intervening methoxylation event, in which acyls react with DME to desorb MeOAc and to re-form a bound CH_3. This step avoids the recurrence of the slow formation of such CH_3 species from DME, which, under anhydrous conditions, needs to occur only once as an initiation step in a long propagation cycle, in which CO backside attack (Fig. 4B) and methoxylation provide the propagation elementary steps. Turnover rates (per H^+) were highest on MOR structures (Fig. 4C), but varied significantly among MOR samples prepared or treated using different protocols. Rates were lower on FER samples and essentially undetectable on aluminosilicates lacking 8-MR structures (BEA, MFI, FAU, SiO_2–Al_2O_3) (Fig. 4C).

Such inferences about the required structural motifs led to the use of spectroscopic and titration methods to determine the location of protons within

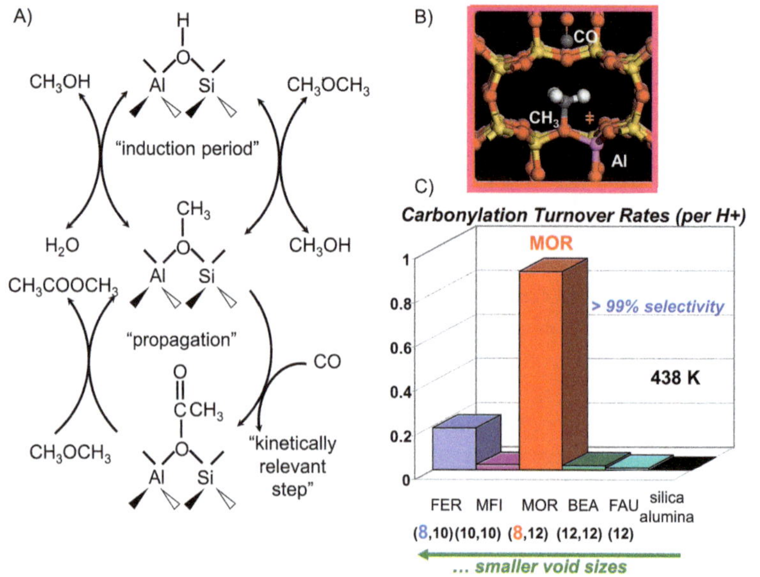

Fig. 4 (A) A proposed set of elementary steps describing the induction period and propagation reaction sequences in the carbonylation of dimethyl ether on proton form zeolites [adapted from Cheung *et al.*[14]]; (B) the transition state resembles a methyl carbenium ion stabilized *via* van der Waals interactions with the anionic zeolite framework and the carbon monoxide nucleophile [adapted from Neurock[16]]; (C) turnover rates (per proton) for dimethyl ether carbonylation (438 K, 0.93 MPa CO, 20 kPa DME, 50 kPa Ar) for H-form zeolites and amorphous silica alumina.

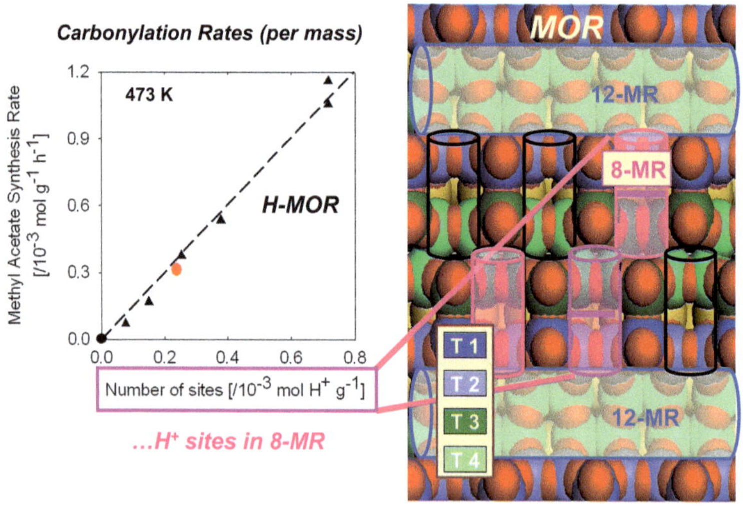

Fig. 5 DME carbonylation rates per unit mass (438 K, 3.34 $cm^3\ s^{-1}\ g^{-1}$, 0.93 MPa CO, 20 kPa DME, 50 kPa Ar) plotted against the number of H^+ sites per unit mass in 8-MR channels of Na-exchanged H-MOR (▲; Si/Al = 10, Zeolyst; Na/Al = 0.0–0.55) and Co-exchanged MOR (●; Co/Al = 0.65). Adapted from ref. 13.

different local environments in each given framework, as specifically in MOR, which exhibits the highest reactivity and in which protons reside at four crystallographically distinct locations that are located within two (8-MR and 12-MR) very different types of environments (Fig. 5). The use of pyridine as an organic titrant that can access only the 12-MR protons led to undetectable changes in rates, consistent with the kinetic irrelevance of H^+ species at 12-MR locations. The deconvolution of location-specific O–H infrared bands led to an accurate assessment of the fraction of protons at 8-MR and 12-MR locations. The rate data in Fig. 5 are plotted against the number of H^+ within 8-MR MOR structures for samples prepared or treated using different protocols or in which H^+ was exchanged for Na^+ to different extents. The excellent correlation between carbonylation rates and protons at 8-MR Al sites indicates that the relevant transition state structures are preferentially stabilized (relative to their bound CH_3 precursors) at 8-MR motifs that prevail in MOR and FER frameworks, while protons at Al sites within larger voids (10-MR, 12-MR motifs) exhibit much lower reactivity and lead to unselective homologation reactions. The specificity of such confinement effects for CH_3 reactions with CO was unprecedented at the time for inorganic solids, which predominantly influence reactivity by size exclusion of bulkier transition states or by restricting access and egress of certain molecules to the confining structures containing the active sites.

These stabilization effects depend on effective van der Waals contacts between the transition state and the confining voids and on such contacts being more effective for the backside CO-attack transition state than for its bound methyl precursors. Such enthalpic benefits of confinement prevail over the consequent entropic penalties because of the pre-eminent role of enthalpy in determining Gibbs free energies of activation at the moderate temperatures of DME carbonylation. As a result the best fit becomes a tight fit and the resulting solvation by the inorganic host preferentially stabilizes specific guests *via* dispersive forces. This preeminent role of dispersion becomes evident from carbonylation transition state enthalpies derived from density functional theory (DFT), which are not influenced by location when functionals that neglect such forces are used, but become favored at 8-MR locations when using either higher levels of theory or functionals that can account for non-specific induced dipole interactions.[16]

2.3. Dimerization and solvation by a liquid phase: selectivity for the formation of one C–C bond between ethene molecules and site stability conferred by intrapore liquids

The catalytic routes in the previous two sections have formed a single C–C bond from C_1 molecules, with different selectivities, at very different temperatures, and exploiting unrelated concepts and chemistries. This section shows that the chemistry required to form a single C–C bond between two ethene molecules (to form 1-butene) benefits from the preferential solvation of the C–C bond formation and product desorption transition states by an extended intrapore ethene liquid phase within MCM-41 channels that contain active Ni sites;[18] such a liquid phase forms *via* capillary condensation at high pressures and sub-ambient temperatures (240–260 K).[18] In this example, the host structure for the transition state guest is the mobile liquid solvent, in contrast with the quasi-rigid walls of microporous aluminosilicates in the DME carbonylation example of Section 2.2.

Heterogeneous ethene dimerization catalysts based on active Ni structures dispersed on mesoporous or microporous oxides typically require higher temperatures (380–500 K) to achieve rates similar to those observed at lower temperatures on homogeneous catalysts, but benefit from much less costly product-catalyst separation and from their ability to operate without activators or co-catalysts and to be regenerated without degradation, in contrast with homogeneous and MOF-based catalysts (Table 1). The higher temperatures required for oxide-based catalysts lead, however, to low 1-butene selectivities and to C_4 yield losses resulting from sequential oligomerization reactions; the latter reactions form larger chains that bind strongly to Ni active sites, thus requiring frequent thermal treatments for their removal *via* oxidation reactions or *via* the desorption of strongly-bound oligomers.

The operation of Ni-MCM-41 catalysts at pressures and temperatures that lead to the filling of MCM-41 mesopores with liquid ethene leads to a sharp decrease in their first-order deactivation rate constants (k_d, Fig. 6) and to a concomitant significant increase in catalyst life over those observed at higher temperatures in gas-filled mesopores (half-life reported as $1/k_d$; Table 1). Such stabilization becomes evident at relative ethene pressures of about 0.6 (defined as the ratio of ethene pressure to its saturation pressure at each temperature) (Fig. 6), which corresponds to that required for condensation within the 1.0–2.2 nm channels in MCM-41 (as determined from their pore-filling isotherms) where the active Ni sites are likely to reside. These liquid-filled channels lead to the conversion of ethene to 1-butene (1-C_4) with very high selectivity (98% butenes in products; 93% 1-butene in butenes) and with unprecedented turnover rates (>10 s^{-1}; per Ni atom), even at sub-ambient temperatures (243–258 K), without exhibiting detectable deactivation or requiring the aluminoxane activators essential for turnovers in homogeneous and Ni–MOF catalysts (Table 1).

Such unique reactivity, selectivity, and stability are imparted by the intrapore liquid phase, containing ethene and butene products, which preferentially

Table 1 Ethene dimerization rates, and C_4 and 1-butene selectivities for Ni-based catalysts

Catalyst	Ethene dimerization							
Support	Ni (% wt)	Pressure (MPa)	Temp. (K)	Rate (g_{butene} g_{cat}^{-1} h^{-1})	Rate (per Ni atom) (g_{butene} mol_{Ni}^{-1} h^{-1})	C_4 in products (%)	1-Butene in C_4 (%)	$1/k_d$ (h)
MOF[20a]	10.0	1.5	283	205	10 400	94 (95)	95	12
Ni–phosphine[19b]	—	6.0	298	—	26 600	9	50	—
FAU[22c]	3.7	1.5	388	10	530	53	—	65
MCM-41 [21,23d]	2.0	3.5	423	67	15 700	58	20	91
MCM-41 [18e]	0.2	3.0	258	43	38 900	98 (97)	92 (91)	630
	5.0	3.0	258	135	8700	98 (96)	93 (92)	612

[a] *n*-Heptane solvent; Et_2AlCl activator (Al/Ni = 70). [b] Homogeneous α-nitroketonate Ni(II) complex; toluene solvent; MAO activator (Al/Ni = 100). [c] Ni-exchanged Na–FAU zeolite (Si/Al = 3); no solvent, full conversion. [d] Ni-exchanged Al/MCM-41 (Si/Al = 9); *n*-heptane solvent. [e] Ni-exchanged Al/MCM-41 (Si/Al = 40); no solvent at 5% (or 40%) conversion.

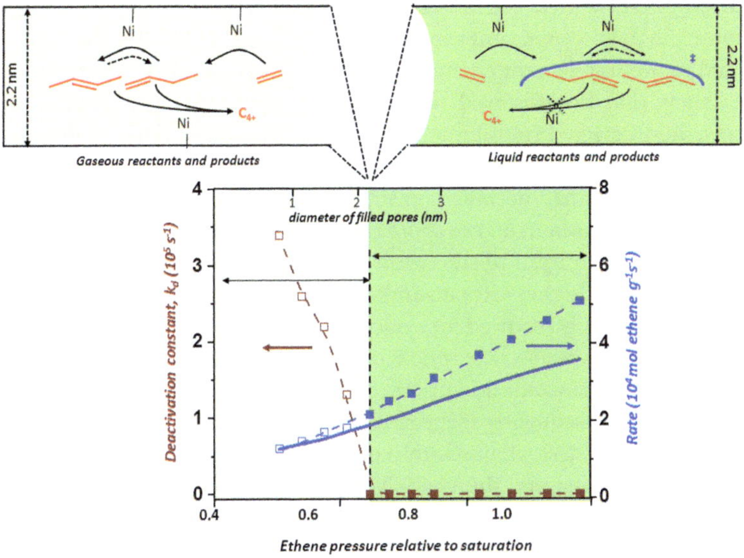

Fig. 6 Ethene dimerization first-order deactivation constant (k_d, data symbols, left), trend for deactivation constant (dashed, left), rates per catalyst mass (data symbols, right), trend for rates (dashed, right) and rates corrected by ethene fugacity (solid line, right) as a function of ethene pressure relative to saturation on 5.0% wt Ni-MCM-41 at 243 K. Symbols (□) and (■) denote conditions leading to intrapore gas and liquid, respectively. The top *x*-axis denotes the required MCM-41 pore diameter (2.2 nm) that leads to liquid reactants and products in pores containing exchanged Ni sites. The presence of intrapore liquids solvates the transition state that mediates the desorption of 1-butene and enables the transition from unstable to stable rates.

solvates the transition states that mediate the desorption of the primary 1-butene dimers before their isomerization or subsequent growth. Similarly, intrapore liquids derived from intrapore condensation of propene during its dimerization also lead to high rates and selectivities to hexenes and to stable operation conditions leading to capillary condensation (0.3 MPa, 260 K). These transition states occur late along the reaction coordinate, because of the endothermic nature of desorption processes; their product-like nature, resembling a desorbing alkene significantly interacting with the surrounding liquid phase, allows their preferential stabilization over strongly-bound transition states, which occur later along the reaction coordinate and mediate subsequent isomerization and oligomerization reactions of bound 1-alkenes formed as primary products in alkene dimerization elementary steps.

The remarkable stability of these materials in the presence of intrapore liquids allows the unequivocal attribution of reactivity to isolated Ni cations based on rates (per mass) that reach a constant value at a Ni^{2+}/H^{+} exchange stoichiometry of unity, consistent with the plausible involvement of $\{Ni^{2+}\text{–}(OH)^{-}\}^{+}$ species as active sites. Such stability also allows the mechanistic interpretation of dimerization rates in terms of elementary steps and of selectivity in terms of subsequent reactions that can occur during one surface sojourn at Ni sites or after 1-butene or its isomers readsorb along the catalyst bed with increasing residence time. Dimerization turnovers are limited by the reaction of a bound ethene, present at low coverages at Ni sites, with an unbound ethene molecule, leading to rates that are approximately second order in ethene pressure. Transition state formalisms required to accurately treat chemical dynamics in thermodynamically non-ideal media show that the observed rate dependence on ethene fugacity, which is stronger than expected from bimolecular transition states, reflects the solvation effects by intrapore liquids, which lead to the kinetically-detectable stabilization of the C–C bond formation transition state. These intrapore hydrocarbon liquids, which are present without an extrapore continuous liquid flow in the catalyst bed, avoid potential leaching of Ni sites while leading to dimerization rates that are, in fact, significantly higher at sub-ambient temperatures than at the higher temperatures typically required for oxide-based Ni catalysts (Table 1). Solvation by liquids also increases the desorption of surface 1-butene, as a primary dimerization product, thus minimizing its conversion to regioisomers or to larger oligomers, and consequently increasing the selectivity to 1-butene regioisomers within butenes and to C_4 chains within products. These phenomena highlight, as in carbonylation catalysis, the enabling role of solvents, whether as inorganic hosts or extended liquids, in modifying reactivity and selectivity by stabilizing specific transition states, a subject examined again in Section 3.3 in the context of alkene–alkanal Prins condensations on acid-forms of zeotypes.

2.4. Methylative homologation of alkenes and alkanes using C_1 oxygenates: formation of highly branched alkanes *via* the orderly methylation and timely hydride transfer inherent in "undisturbed" carbenium ion reactions

The catalytic strategies described in the previous three sections, as well as CH_3OH/DME homologation routes at high temperatures (>600 K),[24] lead to C_1-derived aliphatic and aromatic molecules with additional C–C bonds. Metathesis can then shuffle C–C bonds among alkenes to form chains of intermediate length,[25] including *via* tandem dimerization-metathesis strategies that can form alkenes with odd carbon numbers (*e.g.* propene from ethene).[26] Alkenes and alkanes can also grow into larger chains *via* oligomerization and alkylation processes on solid acids.[27] These strategies allow some modest control over the chain length and backbone structure in products, but the required temperatures typically lead to secondary skeletal isomerization and β-scission reactions; these secondary reactions lead to product slates diverse in chain length and backbone structure.

The remarkable ability of some metal halides in liquid methanol (*e.g.* Zn and In iodides) to form supra-equilibrium concentrations of triptane (2,2,3-trimethylbutane) and isobutane at low temperatures (400–500 K)[28,29] led us to consider the use of solid acids at these lower temperatures but at higher DME pressures (60–250 kPa) than typically used in homologation of C_1 oxygenates *via* processes known as methanol-to-hydrocarbon chemistries.[24] In doing so, we uncovered the previously unrecognized ability of solid acids at these temperatures to also form triptane and isobutane with unprecedented selectivity,[30–33] while also demonstrating the likely involvement of protons in the metal halide systems, where protons can form *via in situ* hydrolysis of halides by the H_2O formed in C_1 homologation steps. The resulting mechanistic insights, gained through extensive kinetic and isotopic tracer studies, provided a unique vantage point into the "rules" for methylation and hydride transfer reactions of bound alkoxide intermediates.[31] It also opened an unexpected opportunity to introduce unreactive light alkanes into homologation cycles, *via* chemistry that became known as methylative homologation of alkanes,[32] thus fulfilling the hydrogen requirements for alkane formation from CH_3OH and DME and allowing the recycle of the isobutane products into the methylation–hydrogen transfer catalytic cycle.

In fact, the unique selectivities to isobutane and triptane reflect the preference of alkoxides to methylate at specific locations and to desorb *via* H-transfer when they reach a specific length and skeletal backbone structure along the methylation path. Preferential methylation occurs at positions that preserve a four-carbon backbone along the chain growth path merely because of the stability of the carbenium ion transition states that mediate methylation events (Scheme 2). Slow skeletal isomerization and β-scission reactions of these four-carbon backbone species prevent their deviation from the path to triptane, while fast isomerization and β-scission occur for any chains that grow beyond triptane with the consequent formation of isobutyl chains that desorb as isobutane. Facile hydrogen transfer to *tert*-butoxides and 2,3,3-trimethyl *sec*-butoxides relative to the methylation of isobutene and triptene causes these species to preferentially exit the chain growth sequence as isobutane and triptane.

These reactions are shown to occur on all solids containing Brønsted acid sites (Fig. 7)[33] and are mediated by carbenium ion transition states that favor methylation (instead of hydrogen transfer, isomerization, or cracking) when alkoxides have secondary and/or tertiary carbons adjacent to one another, but which favor termination (instead of methylation) when alkoxides contain tertiary carbons vicinal to primary carbons (Fig. 8).[31] These desorption events (*via* H-transfer)

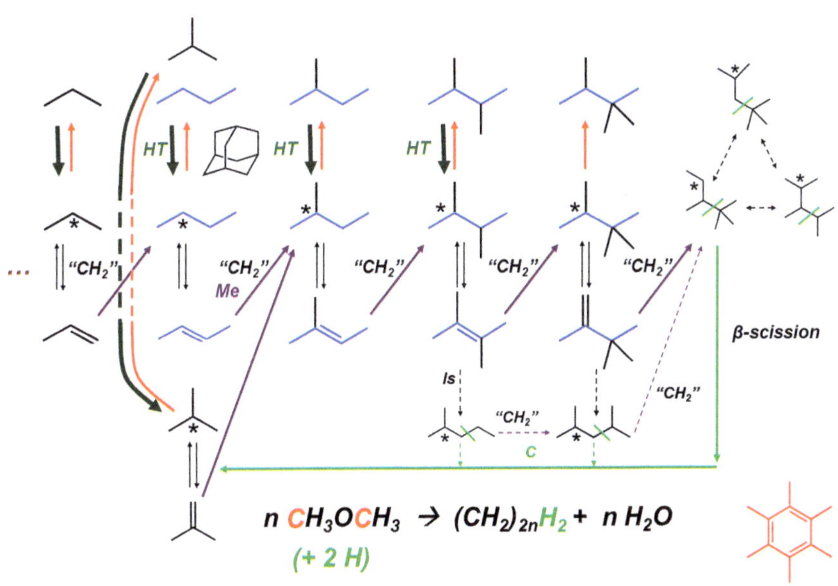

Scheme 2 Chain growth sequence for the homologation of dimethyl ether, including re-incorporation of alkanes *via* dehydrogenation (using adamantane as a hydrogen transfer co-catalyst). HT, Me, Is, and C labels represent hydrogen transfer, methylation, isomerization, and cracking reactions, respectively. 'CH_2' represents surface methylating species derived from dimethyl ether. Reprinted from ref. 32 with permission from Elsevier.

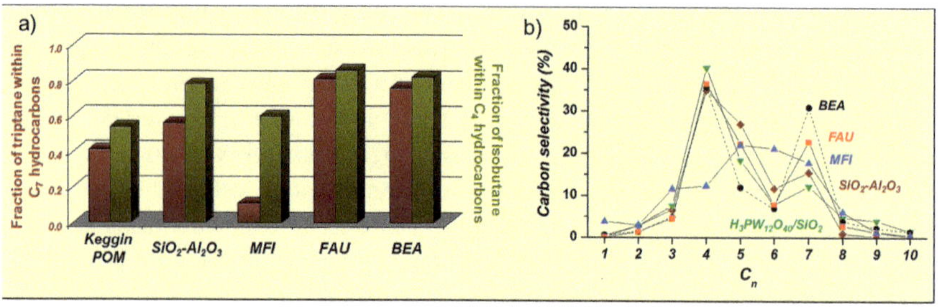

Fig. 7 (a) Selectivities to triptane within C_7 products (red bars) and isobutane within C_4 products (green bars) for dimethyl ether homologation at 473 K on zeolites (BEA, FAU, MFI) and amorphous solid acids (5 wt% $H_3PW_{12}O_{40}/SiO_2$ and SiO_2–Al_2O_3). (b) Product selectivities (carbon basis) for DME homologation at 473 K on BEA (●), FAU (■), MFI (▲), 5 wt% $H_3PW_{12}O_{40}/SiO_2$ (▼), and SiO_2–Al_2O_3 (◆). Reprinted from ref. 33 with permission from Elsevier.

involve transition states that cannot readily form in 10-MR zeolites (*e.g.* MFI), which must therefore desorb fragments of alkoxide intermediates as alkenes *via* β-scission in order to complete catalytic turnovers. Therefore, such 10-MR aluminosilicates, which are typically used in C_1 homologation processes at higher temperatures,[24] cannot form products with the skeleton structure and size dictated by the undisturbed carbenium ion chemistry that preferentially forms isobutane and triptane. In contrast, hydrogen transfer readily occurs, often facilitated by confinement within voids, on aluminosilicates with 12-MR voids and channels (FAU, BEA) or mesoporous networks (amorphous SiO_2–Al_2O_3); such unique selectivities are evident for both stronger acids (phosphotungstic Keggin polyoxometalates dispersed on SiO_2) and for the significantly weaker aluminosilicate acids (Fig. 7).[33]

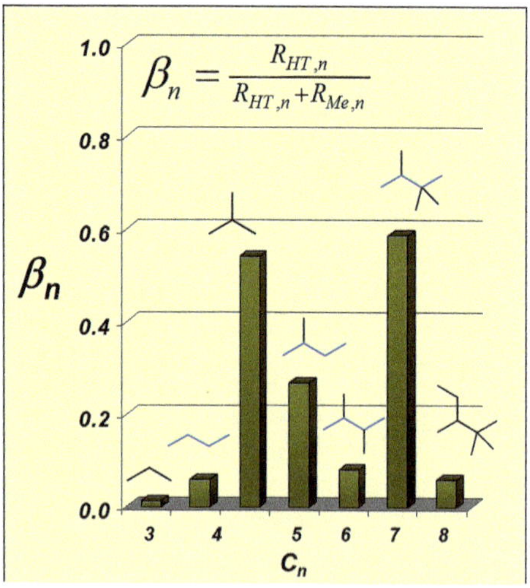

Fig. 8 Termination probabilities (β; defined as the ratio of the rate of hydrogen transfer to the sum of the rates of hydrogen transfer and methylation) of molecules formed along the chain growth path to triptane.[31]

Competitive reactions between ^{13}C-labeled DME and unlabeled alkenes were used to measure the rates and selectivities of methylative growth and of H-transfer for alkene-derived alkoxides of specific chain length and backbone.[31] These studies demonstrated that C_1-species add to alkoxides at locations that retain a four-carbon backbone, because such routes avoid cyclic transition states with primary carbenium-ion character. The specific isomers formed *via* methylation exhibit much lower rates for skeletal isomerization (and β-scission) than for H-transfer or subsequent methylation, thus mitigating backbone disruptions along the path to triptane (Fig. 9 and 10). Less frequent isomerization events (and infrequent methylation at other positions) form backbones that deviate from the four-carbon chain required for triptane formation. These species, however, rapidly grow to C_{8+} hydrocarbons because isomers that lack the four-carbon backbone form alkoxides with secondary-tertiary structures that prefer to methylate instead of terminating *via* hydrogen transfer (Fig. 10). Chain termination probabilities (β_n, defined in Fig. 8) reached a maximum value for triptyl species and high values for *tert*-butoxides. The C_8 species resulting from methylation of triptane (and other C_7 isomers) exhibit very low β values (Fig. 8), because they undergo facile β-scission to form *tert*-butoxides that desorb *via* rapid H-transfer to form isobutane. Thus, any chain growth "mistakes" along the path to triptyl backbones are resolved kinetically *via* the formation of isobutane, which, as we show below, can be re-inserted into the methylation cycle *via* its own involvement in H-transfer. Such pathways lead to chains that preferentially terminate as triptane and isobutane because tertiary carbenium ions that mediate hydrogen

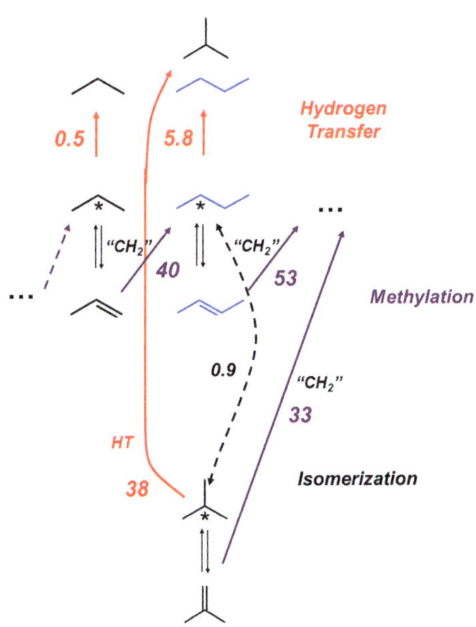

Fig. 9 Rates (μmol [mol$_{Al}$ s]$^{-1}$) of methylation, hydrogen transfer, and skeletal isomerization for C_3 and C_4 species formed during homologation reactions of dimethyl ether at low temperatures. Rates were determined from co-reactions of ^{13}C-labeled dimethyl ether (78 kPa) and unlabeled propene, *trans*-2-butene, or isobutene (0.5 kPa) at 473 K on H-BEA (Si/Al = 12.5).[31]

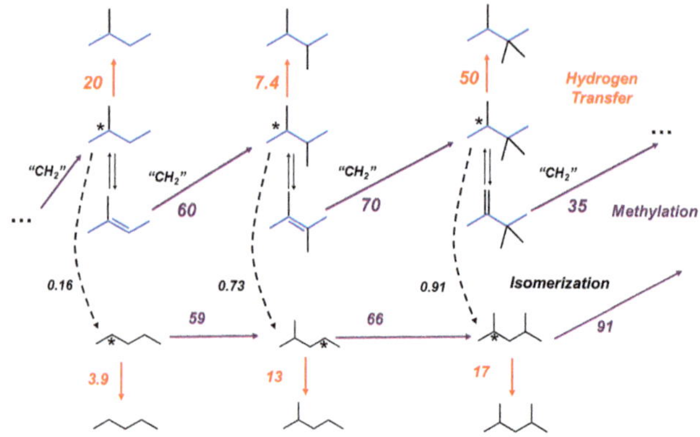

Fig. 10 Rates (μmol [mol$_{Al}$ s]$^{-1}$) of methylation, hydrogen transfer, and skeletal isomerization for C_5–C_7 species formed during homologation reactions of dimethyl ether at low temperatures. Rates were determined from co-reactions of ^{13}C-labeled dimethyl ether (78 kPa) and unlabeled 2-methyl-2-butene, 2-pentene, 2,3-dimethyl-2-butene, 2-methyl-2-pentene, triptene, or 2,4-dimethyl-2-pentene (0.5 kPa) at 473 K on H-BEA (Si/Al = 12.5).[31]

transfer in triptyl and isobutyl species are much more stable than the primary carbenium ions required for their respective methylation reactions.

The stoichiometry of DME or CH_3OH homologation reactions leads to the formation of alkanes that contain more H-atoms than those remaining after oxygen rejection as H_2O from DME (or CH_3OH), thus requiring the stoichiometric formation of unsaturated co-products, predominantly hexamethylbenzene (Scheme 2). The methylative homologation of alkanes with DME/CH_3OH leads to the orderly addition of methyl groups to an alkane-derived alkoxide to form larger alkanes, without requiring any addition or rejection of H-atoms.[32] The incorporation of added alkanes into C_1-homologation chains occurs on all solid acids, as shown by the reactions of unlabeled alkanes with different backbones to ^{13}C-labeled DME, which lead to the alkane-derived ^{12}C fractions shown in Fig. 11 for H-BEA. Such alkanes form alkoxides *via* a concerted H-transfer step to growing alkoxide chains, thus enabling alkoxide desorption as alkanes.

The presence of H-donor alkanes containing tertiary C–H bonds leads to higher rates of H-transfer, and the rates of H-transfer (from all H-donors) are markedly enhanced by using adamantane, which acts as a hydride transfer co-catalyst (Fig. 11). Adamantane dissects the H-transfer process, which typically occurs *via* the concerted transfer from a donor to an acceptor, into two separate steps: H-transfer from adamantane to alkoxides (to give an alkane and an adamantyl cation) and then from alkanes to adamantyl cations (to re-form adamantane and a new alkane-derived alkoxide). These sequential H-transfer steps mediated by adamantane lead to lower barriers than the concerted direct transfer because of the involvement of particularly stable tertiary adamantyl cations in each of the sequential H-transfer steps. Such methylative homologation routes allow the upgrading of light alkanes to highly-branched larger alkanes, the return of isobutane formed through the kinetic resolution of over-methylation mistakes in DME homologation cycles, and the substantial elimination of the hexamethylbenzene stoichiometric requirements.

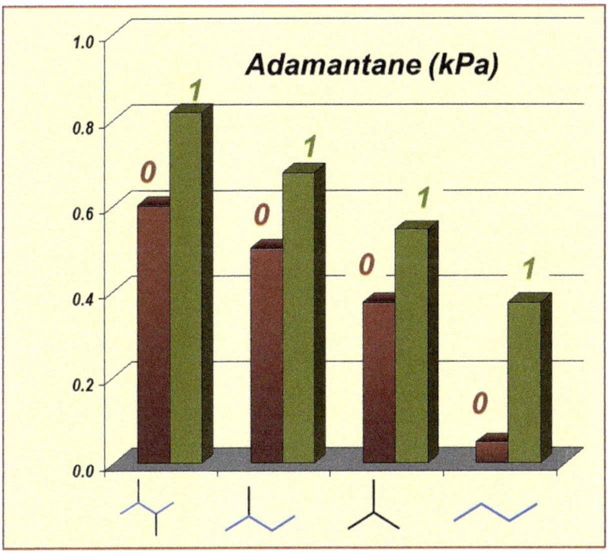

Fig. 11 Fraction of ^{12}C-atoms in triptane products from co-reaction of ^{13}C-labeled dimethyl ether with unlabeled alkanes (2,3-dimethylbutane, isopentane, isobutane, or *n*-butane) at 473 K on H–BEA in the presence (green bars) and absence (red bars) of adamantane hydride transfer co-catalysts (1 kPa).

3. Catalysis for oxygen removal and chain growth reactions of oxygenates: elementary steps and site requirements in aldol condensation, ketonization, and Prins condensation on acid–base site pairs and Brønsted acid sites at oxide surfaces

The examples discussed in this section address reactions of oxygenates through which we seek to increase their energy density *via* O-removal as H_2O or CO_2, while preventing the consequent increase in volatility through the concurrent formation of new C–C bonds. The specific reactions of emphasis are known as aldol condensation,[34–37] ketonization,[38–40] and Prins condensation;[41–43] they are depicted along with some mechanistic details in Scheme 3 and discussed here in the context of conceptual frameworks that we have chosen in order to tackle practical hurdles in controlling reactivity, selectivity, or stability on oxide catalysts that represent the state-of-the-art for these transformations.

In the studies cited above, aldol condensations involve reactions of mixtures of alkanols and alkanals (or alkanones) formed *via* the equilibration of alkanol/H_2 feeds on the hydrogenation–dehydrogenation Cu function present as a physical mixture with the oxide-based condensation and esterification catalysts. The carbonyl compounds in these mixtures form enolate intermediates *via* C–H activation at the α-position to their C=O group (Section 3.1). Subsequent nucleophilic attack at the electrophilic C-atom in another carbonyl molecule

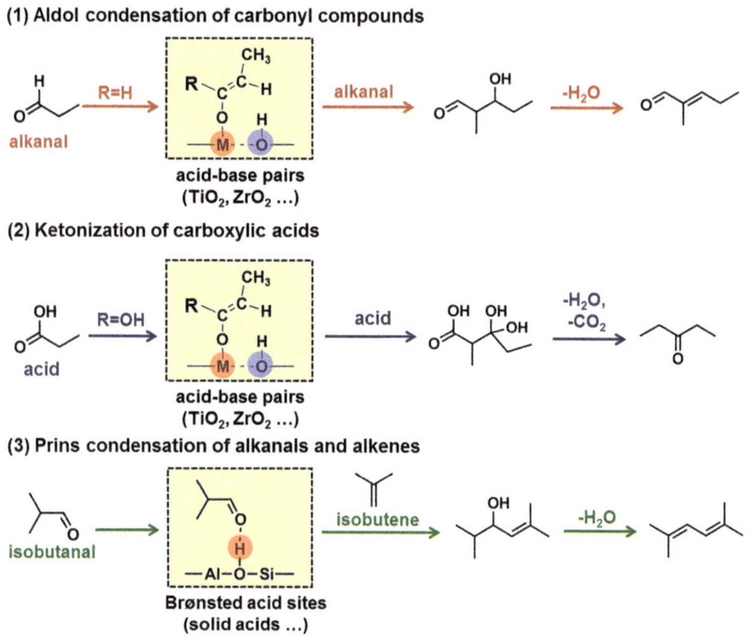

Scheme 3 C–C coupling reactions for oxygenates on Lewis acid–base pairs of metal oxides and on Brønsted acid sites of solid acids.

leads to the formation of aldol species that subsequently dehydrate to α,β-unsaturated molecules and ultimately hydrogenate to equilibrated alkanal/alkanone/alkanol mixtures. Ketonization, in contrast, involves reactions of carboxylic acids through α-C–H activation to form 1-hydroxy-enolates that nucleophilically attack the electrophilic C-atom in another acid to form 3,3-dihydroxy-carboxylic acids, *via* steps reminiscent of those involved in aldol condensations (Section 3.2). The 3,3-dihydroxy-carboxylic acids that form from 1-hydroxy-enolates are unstable; they undergo facile dehydration and decarboxylation to form alkanones, with the loss of one C-atom from the two carboxylic acid reactants (Section 3.2). Prins condensations involve nucleophilic attack by the terminal C-atom in the C=C bond in alkenes at the electrophilic carbonyl C-atom in alkanals to form α,β-unsaturated alkanols (Section 3.3); subsequent dehydration of the α,β-unsaturated alkanols forms dienes with specific skeletal structures unattainable in other C–C formation reactions (*e.g.* alkene oligomerization[44,45] and aldol condensation of carbonyl compounds[35–37]).

All three types of reactions achieve the intended loss of O-atoms and a concomitant increase in product molecular weight and lead to the formation of distinct skeletal structures in products. The loss of oxygen and the increase in molecular weight are evident from the data shown in Fig. 12 for reactions of equilibrated alkanal/alkanone/alkanol mixtures on anatase TiO_2 catalysts used as physical mixtures with Cu/SiO_2 co-catalysts. The O/C ratio in the reactor effluent decreases (Fig. 12a) and its average molecular weight increases (Fig. 12b) as the conversion of the equilibrated pool increases with increasing reactor residence time as a result of condensation and esterification events.

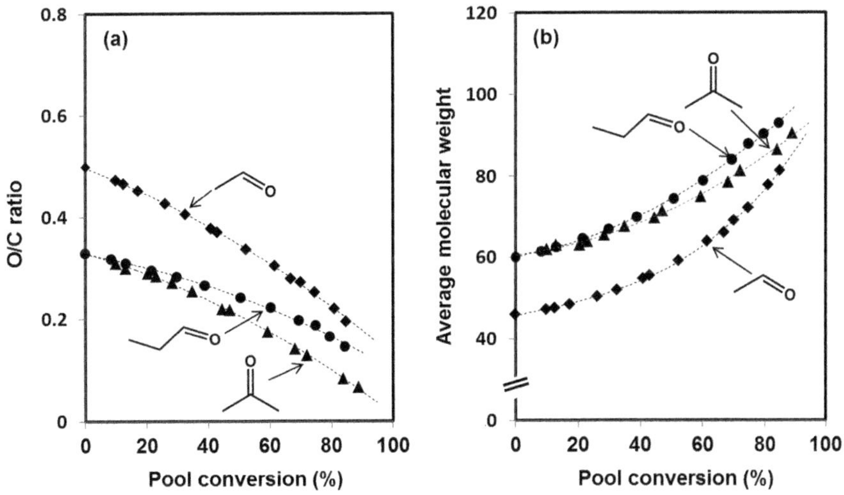

Fig. 12 (a) O/C ratio and (b) average molecular weight of the molecules in the reactor effluent as a function of alkanol–alkanal (alkanone) pool conversion (varied through changes in residence time) for acetaldehyde (◆), propanal (●) and acetone (▲) reactants formed *in situ* from the respective alkanol (TiO_2(P25) + 20% wt Cu/SiO_2 (1 : 1 mass), 523 K, 0.8 kPa ethanol, 1-propanol, or 2-propanol, 40 kPa H_2).[36,37] Dashed curves indicate trends.

These catalytic chemistries and the prevailing reactivity, selectivity, and catalyst stability hurdles that have obstructed their deployment in large-scale practice are addressed here through several strategies. These strategies are the result of mechanistic inquiries about the identity and kinetic relevance of plausible intermediates and elementary steps, the role of binding and confinement effects on the stability of the relevant transition states, and the kinetic coupling made possible by the presence of acid–base site pairs and a hydrogenation–dehydrogenation function within molecular diffusion distances. These strategies exploit:

(i) The scavenging of intermediates that lead to losses in yield or in available active sites through kinetic cascades in which a metal function catalyzes hydrogenation–dehydrogenation events and acid–base pairs promote C–C bond formation and deoxygenation from alkanal/alkanone/alkanol mixtures (aldol condensation, Section 3.1) or from carboxylic acids (ketonization, Section 3.2);

(ii) The stabilization of C–C bond formation transition states *via* concerted interactions with acid–base pairs at metal oxide surfaces; these concepts are based on the early studies of Tanabe about bond-making and bond-breaking at metal–oxygen (M–O) sites exhibiting moderate acid and base strength[46,47] (Sections 3.1 and 3.2); such balanced properties lead to high condensation turnover rates and selectivities without requiring strong basic sites that are inhibited by the H_2O and CO_2 molecules formed as condensation by-products;

(iii) The inherent and previously unrecognized mechanistic connections between condensation and ketonization reactions, involving common enolate-like intermediates, the required Ti–O and Zr–O acid–base pairs of balanced and moderate strength, and the need to scavenge unreactive intermediates (unsaturated alkanals and alkanones in aldol condensation and bidentate carboxylates in ketonization) through bifunctional cascade strategies (Sections 3.1 and 3.2);

(iv) The preferential stabilization of alkene–alkanal Prins condensation transition states on weaker Brønsted acids over those that mediate alkene oligomerization because their smaller charge decreases the enthalpic penalty to form the conjugate anion, thus dampening the effects of acid strength, as well as the reactivity enhancements brought forth by confinement of transition states within inorganic voids of molecular dimensions, which stabilize the two transition states to the same extent because of their similar size (Section 3.3).

In all three types of reactions, a seamless connection between theory and experiment is made possible by accurate site counts, by reactivity and activation free energy comparisons based on turnover rates, by the evidence for mechanistic proposals gathered from spectroscopic and isotopic methods, and by the use of theoretical methods that rigorously account for van der Waals interactions and of structural models that consider the complex but known structures of the acid–base pairs and the Brønsted acid sites that catalyze these reactions.

3.1. Aldol condensation and esterification of oxygenate reactants mediated by concerted H-abstraction transition states on acid–base pairs of intermediate strength

Aldol condensations of alkanals and alkanones form new C–C bonds and remove O-atoms *via* activation of α-C–H bonds (step 3, Scheme 4), nucleophilic attack of carbonyls by enolates (step 5, Scheme 4), and dehydration of aldols to α,β-unsaturated carbonyl compounds (step 7, Scheme 4). These reactions lead to the decrease in O-content and an increase in product molecular weight evident from the data in Fig. 12. Oxides, such as TiO_2 and ZrO_2, catalyze such reactions, but their unfavorable thermodynamics limit rates and favor the accumulation of unsaturated intermediates that lead to loss of selectivity *via* subsequent chain

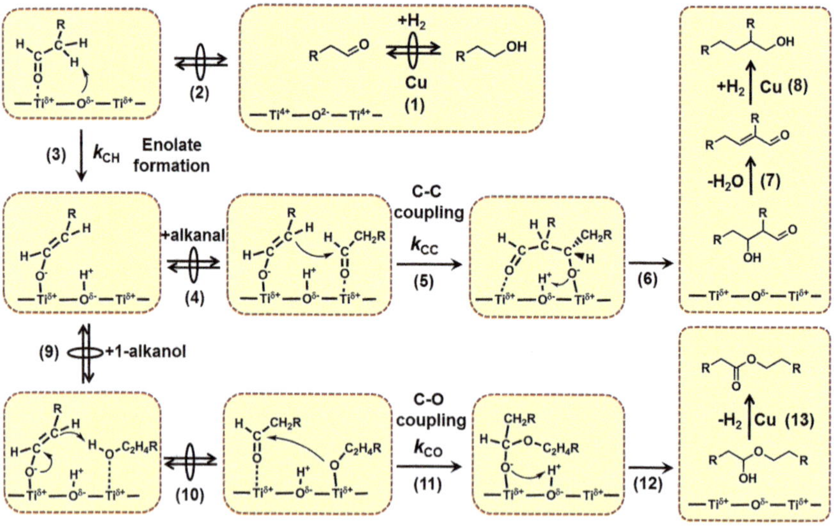

Scheme 4 Aldol condensation and esterification pathways for C_2–C_5 alkanol–alkanal reactants (CH_2RCH_2OH–$CH_2RCH{=}O$, R = H or alkyl) on bifunctional TiO_2 + Cu/SiO_2 mixtures.[36,37] Quasi-equilibrated steps are noted by a circle over double arrows. Reaction numbers used throughout the text are in parentheses.

growth and the formation of residues that block active sites and lead to fast deactivation.[48]

A metal function, such as Cu, Pt, or Pd, even as physical mixtures with metal oxides, and the presence of H_2 leads to the facile hydrogenation of these unsaturated products and intermediates (step 8, Scheme 4), thus removing thermodynamic bottlenecks and preventing side reactions and the otherwise prevalent loss of selectivity and active sites. Such a metal function also rapidly interconverts alkanols and their dehydrogenation products (alkanals, alkanones), rendering all such species equivalent as condensation (and esterification) reactants (step 1, Scheme 4). The use of equilibrated 1-alkanol–alkanal mixtures and of a metal function also leads to esterification turnovers, made possible *via* the dehydrogenation of hemiacetals (step 13, Scheme 4) formed as unstable initial products of C–O bond formation in esterification, *via* reactions between 1-alkanols and alkanal-derived enolates (steps 9–12, Scheme 4).

Kinetic, isotopic, and theoretical assessments of the aldol condensation and esterification rates and of the mechanism of C_2–C_5 alkanols on bifunctional anatase TiO_2 mixtures with Cu/SiO_2 [36,37] have shown that the formation of enolates from alkanals/alkanones (step 3, Scheme 4) is the sole kinetically-relevant step for all reactants, consistent with measured and DFT-derived H/D kinetic isotope effects (2.3 and 2.4, respectively, ethanal-d_6, 523 K) and with the similar activation free energies obtained from theory and experiment. The C–C bonds form *via* subsequent coupling of enolates with alkanals or alkanones with the elimination of H_2O to give the α,β-unsaturated carbonyl products (steps 5–7, Scheme 4). New C–O bonds can also form from reactions of alkanal-derived enolates with 1-alkanols, a step that proceeds *via* sequential H-transfer and alkanal–alkoxide coupling steps to form unstable hemiacetals (steps 10–12, Scheme 4). The formation of C–O and C–C bonds (steps 5 and 11, Scheme 4) from a common enolate precursor determines the fate of the latter species, but not their formation rates, and thus the esterification/condensation rate ratios. Neither step (C–C or C–O bond formation) is limited by reactions involving C–H or O–H bonds, in agreement with the small measured (1.4) and DFT-derived (1.2) kinetic isotope effects (ethanol-d_6, 523 K).

Site titration experiments during acetone condensation demonstrated the involvement of acid–base pairs of intermediate strength on both TiO_2 and ZrO_2. On anatase TiO_2, enolate formation rates were unaffected by CO_2 and reversibly inhibited by pyridine and propanoic acid titrants (Fig. 13),[36] consistent with concerted involvement of both centers in Ti–O pairs with balanced acid and base strength. The use of propanoic acid as a strongly-bound titrant during alkanal or alkanone reactions allowed direct measurements of the number of site pairs; these numbers (3.7 nm^{-2}) are consistent with the atomic density of low-index anatase surfaces and with the kinetic irrelevance of any minority defect sites, often invoked to account for the reactive properties of TiO_2 surfaces. Such site counts allowed rates to be expressed as turnover rates and measured activation free energies, allowing the accurate benchmarking of measured and DFT-derived activation free energies.

The DFT-derived activation free energies ($\Delta G^{\ddagger}_{CH}$) for enolate formation from C_2–C_5 oxygenates are in excellent agreement with the measured values (Fig. 14a), as is also the case for the difference in free energies between the esterification and condensation branching reactions from a given enolate ($\Delta G^{\ddagger}_{CO-CC}$, Fig. 14b). The

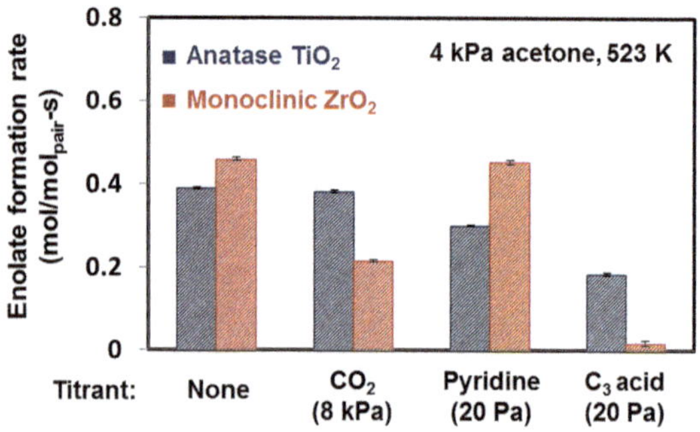

Fig. 13 Effects of CO_2, pyridine, and propanoic acid introductions on enolate formation rates of acetone on anatase TiO_2 and monoclinic ZrO_2 (oxide + 20% wt Cu/SiO_2 (1 : 1 mass), 523 K, 4 kPa acetone, 4 kPa CO_2, 20 Pa pyridine, 20 Pa propanoic acid, 40 kPa H_2).[36,37]

stronger α-C–H bonds in alkanones lead to lower reactivity (higher $\Delta G^{\ddagger}_{CH}$) than for alkanals, while longer and more branched alkyls at the α-C position cause steric hindrance and higher $\Delta G^{\ddagger}_{CH}$ values for enolate formation (Fig. 14a). DFT-derived free energy differences between C–O and C–C bond coupling transition states ($\Delta G^{\ddagger}_{CO-CC}$) also agree with the measured values (Fig. 14b), indicating that alkyl substitution effects are stronger for condensation than esterification transition states (steps 5 and 11, Scheme 4). This reflects their requirement for more precise orbital alignment and the enforcement of such alignment at locations more proximate to the alkyl attachment point for condensation than for esterification transition states.[37]

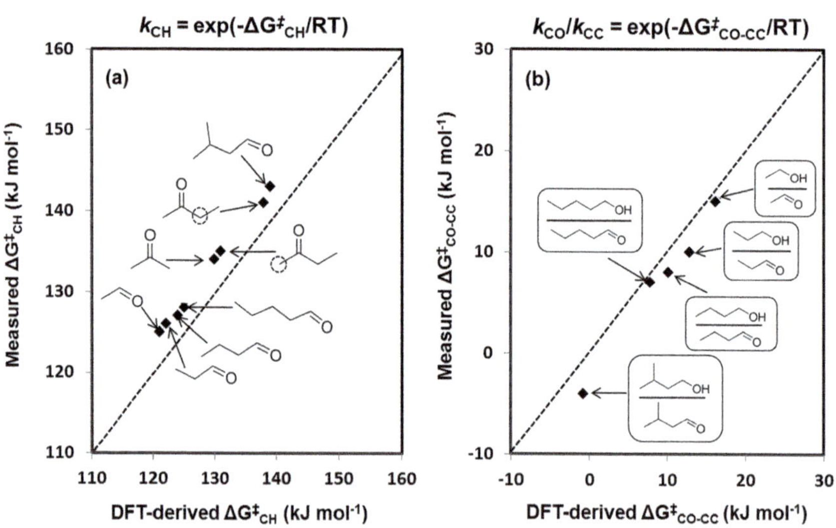

Fig. 14 Measured and DFT-derived (a) $\Delta G^{\ddagger}_{CH}$ and (b) $\Delta G^{\ddagger}_{CO-CC}$ for C_2–C_5 alkanals and alkanones on anatase TiO_2 surfaces (Ti_{5c}–O_{3c}–Ti_{5c} structures of an anatase $Ti_5O_{19}H_{18}$ cluster used as an acid–base pair model,[36] B3LYP + D3BJ, 6-311G(d,p) for C, H, and O atoms, LANL2DZ for Ti atoms; 523 K, 1 bar). Dashed lines represent the parity lines. Adapted with permission from ref. 37. Copyright 2016 American Chemical Society.

These mechanistic conclusions and their consequences for reactivity and selectivity apply to a broad range of oxygenate reactants and remain unchanged on Zr–O pairs at monoclinic ZrO_2 surfaces, which exhibit enolate formation turnover rates (Fig. 13), activation free energies, and isotope effects similar to those on anatase TiO_2.[37] Enolate formation rates on monoclinic ZrO_2 were reversibly inhibited by CO_2, but not by pyridine (Fig. 13), because acid–base pairs on monoclinic ZrO_2 contain stronger basic sites and weaker acid sites than anatase TiO_2. Yet, such site pairs provide the balance of acid–base strength required to stabilize the concerted α-C–H bond activation transition states that mediate enolate formation. These conclusions are consistent with DFT-derived interaction energies of H^+ and OH^- (as probes of the strength of acid and basic sites) on site pairs prevalent on low-index surfaces of ZrO_2 and TiO_2.[37]

3.2. Ketonization of carboxylic acids mediated by C–C coupling transition states on acid–base pairs of intermediate strength

The replacement of a H-atom at the carbonyl C-atom in alkanals by an OH group to give carboxylic acids transforms their C–C bond formation chemistries from aldol condensation into ketonization routes. Ketonization converts carboxylic acids into equimolar mixtures of alkanones, CO_2, and H_2O, while forming a new C–C bond and removing three of the four O-atoms in the two acid reactants (Scheme 3(2)). As in the case of aldol condensation (Section 3.1), ketonization requires the cleavage of the α-C–H bond in one of the acid reactants before the formation of the new C–C bond at the activated C-atom *via* reaction with another acid molecule. These reactions are catalyzed by metal oxides, such as TiO_2 and ZrO_2, which also catalyze aldol condensations, suggestive of a mechanistic link between these two types of chemical transformations.

The demonstrated involvement of enolates in aldol condensation on TiO_2 and ZrO_2 (Section 3.1) makes it plausible that similar enolate-like species (1-hydroxy-enolates) are also involved in ketonization catalytic sequences (step 2, Scheme 5). These species would then participate in nucleophilic attack with another acid to form 3,3-dihydroxy-carboxylic acids with the new C–C bond (step 3, Scheme 5). Such molecules readily decompose *via* dehydration and decarboxylation reactions to form alkanones (steps 4–6, Scheme 5), in contrast with the H_2O elimination step that forms condensation products from aldols without the loss of a C-atom (step 7, Scheme 4). These similarities in mechanisms and the common involvement of Lewis acid–base pairs, supported by theoretical treatments,[40] allow us to extend the use of the titration methods developed to measure the number of active sites during aldol condensation reactions to their ketonization counterparts.[36,37]

Acetic acid ketonization turnover rates (1.0 kPa; 523 K) are slightly higher on monoclinic ZrO_2 (0.51 ks^{-1}, Fig. 15) than on tetragonal ZrO_2 or anatase TiO_2 (0.44 and 0.25 ks^{-1}), but much lower on rutile TiO_2 (0.024 ks^{-1}). These reactivity trends are observed in parallel with deactivation rate constants that are larger on the more active oxides (Fig. 15). Infrared spectra during ketonization on anatase TiO_2 [40] show that bidentate acetates (*AcO*), bound to two vicinal Ti–O pairs with their two O-atoms each interacting with one Ti center and the dissociated proton bound onto an O site, form *via* slow dissociation on acid–base pairs (step 8, Scheme 5) and act as unreactive spectator species during steady-state catalysis. Cu co-catalysts, present as physical mixtures with these oxides and added H_2

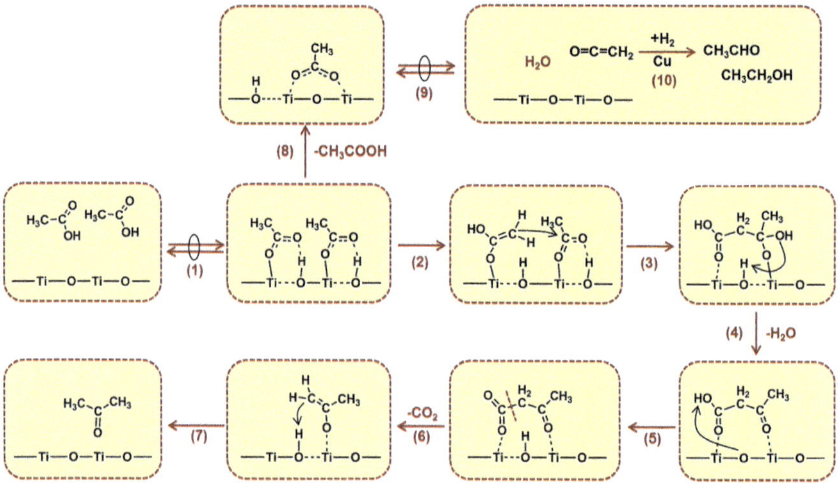

Scheme 5 Ketonization pathway for acetic acid on bifunctional TiO_2 + Cu/SiO_2 mixtures. Quasi-equilibrated steps are noted by a circle over double arrows. Reaction numbers used throughout the text are in parentheses.

markedly decreased deactivation rates and led to higher steady-state ketonization rates.[40] These protecting effects reflect the scavenging of ketene, present at trace concentrations *via* adsorption–desorption equilibrium with *AcO* (step 9, Scheme 5), to form acetaldehyde and ethanol (step 10, Scheme 5), thus providing a route to lower steady-state coverages of unreactive *AcO* and thus lower deactivation rates during ketonization.

Combined kinetic, spectroscopic, and DFT assessments of reactions of C_2–C_4 carboxylic acids on anatase TiO_2 [40] have shown that monodentate carboxylates, the reactive intermediates in 1-hydroxy-enolate formation (step 2, Scheme 5), bind strongly at Ti–O pairs *via* concerted interactions of its two carbonyl O-atoms with the Ti center and the abstracted H-atom (bound at the basic O-atom in the site pair) (*e.g.* AcO* for acetic acid; step 1, Scheme 5); such binding properties lead to

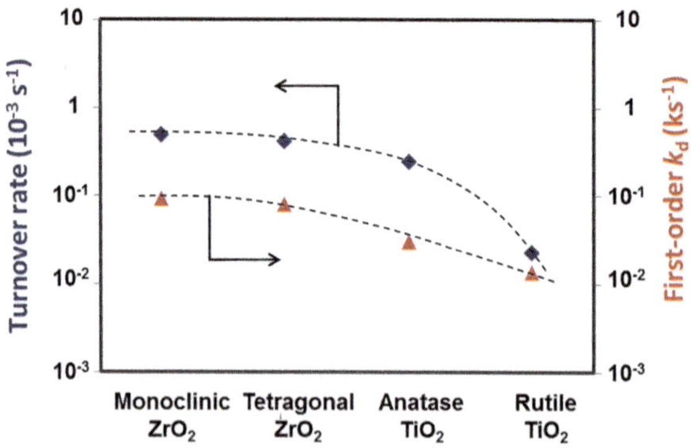

Fig. 15 Turnover rates and first-order deactivation constants (k_d) for acetic acid ketonization on TiO_2 and ZrO_2 catalysts (523 K, 1.0 kPa acetic acid).[40] Dashed lines indicate trends.

densely-covered surfaces at all practical ketonization conditions. The weak measured H/D kinetic isotope effects (acetic acid-d_4, 1.1, 523 K) and the nearly saturated monolayers of monodentate carboxylates evident in infrared spectra during catalysis, indicate that the nucleophilic attack of co-adsorbed acids by 1-hydroxy-enolates (step 3, Scheme 5) is the sole kinetically-relevant step in ketonization catalytic cycles. These conclusions are consistent with measured kinetic effects of acid pressure, with the effects of alkyl substituents on ketonization turnover rates, and with DFT-derived activation free energy barriers using periodic slab models.[40]

Ketonization rates of C_2–C_4 carboxylic acids on anatase TiO_2 become independent of acid pressures even at pressures (~1 kPa) well below those typical of catalytic practice (>0.1 MPa). These data indicate that ketonization occurs on oxide surfaces nearly saturated with monodentate carboxylate intermediates.[40] Zero-order ketonization turnover rates reflect free energy differences ($\Delta G^{\ddagger}_{CC}$) between the C–C coupling transition state and two monodentate carboxylate species (Scheme 6). Acetic acid turnover rates on monoclinic ZrO_2 give $\Delta G^{\ddagger}_{CC}$ values that are smaller than on anatase TiO_2 (136 *vs.* 166 kJ mol^{-1}, 523 K, 0.1 MPa acetic acid[40]), consistent with DFT-derived values on the low-index exposed surfaces of these oxides (140 *vs.* 160 kJ mol^{-1}). These $\Delta G^{\ddagger}_{CC}$ differences, indicative of the stability of the C–C bond formation transition state relative to the monodentate carboxylate precursors, show that ketonization reactions are favored on monoclinic ZrO_2 surfaces over anatase TiO_2 surfaces. In contrast, aldol condensation turnover rates, which reflect the stability of the enolate formation transition state relative to a gaseous carbonyl reactant, are similar on the acid–base site pairs present on these two surfaces (Section 3.1). The different consequences of the properties of acid–base site pairs for these two reactions reflect the specific roles of the Lewis acid and base centers on their respective transition states and precursors, as discussed next.

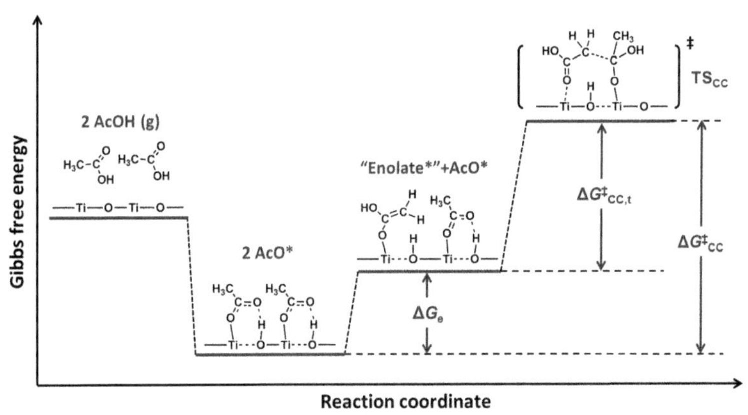

Scheme 6 Schematic reaction coordinate diagram for ketonization of acetic acid on anatase TiO_2. $\Delta G^{\ddagger}_{CC}$ represents the experimentally accessible apparent free energy barrier for the C–C bond formation, and equals to the sum of the free energy difference between a bound 1-hydroxyl-enolate and a monodentate carboxylate (ΔG_e) and the free energy difference between the C–C coupling transition state and its 1-hydroxyl-enolate–carboxylate pair precursor ($\Delta G^{\ddagger}_{CC,t}$).

Ketonization activation free energy ($\Delta G^{\ddagger}_{CC}$) differences between Zr–O and Ti–O pairs can reflect differences in either the free energy of formation of 1-hydroxy-enolates from AcO* on M–O site pairs (M = Ti, Zr; ΔG_e, Scheme 6) or that of the formation of the C–C coupling transition state from its 1-hydroxy-enolate–AcO* precursors ($\Delta G^{\ddagger}_{CC,t}$, Scheme 6). Their relative contributions depend on the coverages and reactivity of the 1-hydroxy-enolates in forming the C–C bond with a vicinal AcO*. DFT treatments show that monoclinic ZrO_2 surfaces are more reactive than those of anatase TiO_2 because they form higher 1-hydroxy-enolate concentrations at Zr centers (ΔG_e: 44 *vs.* 111 kJ mol^{-1} for monoclinic ZrO_2 and anatase TiO_2), as a consequence of stronger basicity,[37] which disfavors the reprotonation of 1-hydroxy-enolates to form AcO*.[38] In contrast, the activation of α-C–H bonds in carbonyl compounds to form enolates on acid–base pairs involves concerted transition states. In such structures, the α-H-atom is abstracted by the basic O site and the incipiently-formed enolate moiety is stabilized by a Lewis acid center. The stronger basic sites on ZrO_2 (relative to TiO_2) compensate for the weaker Zr centers in the concerted stabilization of enolate moieties, leading to the similar observed condensation reactivity of TiO_2 and ZrO_2 surfaces.

Aldol condensation and ketonization both require acid–base pairs and involve nucleophilic attack of enolate-like species at the carbonyl C-atom of another molecule to form a C–C bond. The weak adsorption of alkanals and alkanones leads to acid–base pairs that remain essentially uncovered during condensation turnovers; as a result such turnovers are limited by enolate formation. The dissociation of carboxylic acids, in contrast, forms strongly-bound monodentate carboxylates, which saturate surfaces at all conditions of ketonization catalysis. Ketonization turnovers become limited by C–C coupling of (carboxylate-derived) 1-hydroxy-enolates with another acid on such surfaces. The acid–base balance required for enolate formation and the stronger basic moieties that increase enolate coverages during ketonization illustrate how catalytic sequences, intrinsically connected *via* analogous elementary rearrangements, lead to different acid–base requirements as a result of the identity of the kinetically-relevant step.

3.3. Alkanal–alkene Prins condensation mediated by C–C coupling transition states on Brønsted solid acids: effects of acid strength and of confinement within voids of molecular size

The activation of alkanals (or alkanones) and carboxylic acids on acid–base site pairs forms enolate-like species that act as strong nucleophiles in the subsequent formation of a new C–C bond (Sections 3.1 and 3.2). In contrast, protonation of carbonyl compounds on Brønsted acids renders these species strongly electrophilic, leading to their facile ability to react even with weak nucleophiles, such as alkenes, to form new C–C bonds. These latter routes are discussed in this section in the context of alkanal–alkene Prins condensation reactions. Prins condensations form C–C bonds between the carbonyl C-atoms in alkanals and the terminal C-atoms at C=C bonds in alkenes, with the concurrent removal of O-atoms *via* dehydration steps that convert the α,β-unsaturated alkanol formed to dienes (Scheme 3(3)). Specifically, isobutanal–isobutene reactions selectively form 2,5-dimethyl-hexadiene regioisomers (2,5-DMH); their skeletal structures render such species convenient as precursors to *p*-xylene from isobutene and isobutanal co-reactants that can be derived from isobutanol, a product of biomass fermentations.[42]

Isobutanal–isobutene Prins condensation occurs on solid acids, in parallel with isobutene oligomerization to 2,4,4-trimethyl-pentene regioisomers (2,4,4-TMP; steps 1′–5′, Scheme 7) and with secondary isomerization, cyclization, and β-scission of 2,5-DMH primary condensation products. Non-coordinating 2,6-di-*tert*-butyl-pyridine titrants fully suppressed reactivity for Prins condensation and oligomerization on all acids;[43] thus, both reactions occur exclusively on Brønsted acid sites, without contributions from Lewis acids or acid–base site pairs. Turnover rates (per proton) for Prins condensation (r_{Prins}) and oligomerization (r_{oligo}) both increased as the strength of the acid sites increased from amorphous aluminosilicates to niobic acid and heteropolytungstic acids ($H_3PW_{12}O_{40}$) (Fig. 16), but r_{Prins}/r_{oligo} ratios were significantly higher on the weaker acids (Fig. 16). Kinetic data and theoretical treatments indicate that isobutanal–isobutene Prins condensation rates are limited by C–C bond formation between a H-bonded isobutanal and a gaseous isobutene molecule (step 3, Scheme 7); parallel isobutene oligomerization reactions are limited by C–C bond formation between an isobutene-derived alkoxide and a gaseous isobutene molecule (step 4′, Scheme 7).

Measured second-order rate constants for Prins condensation (k_{Prins}) and oligomerization (k_{oligo}) were both smaller on Al-MCM-41 (12 ± 2 and 1.2 ± 0.3 ks^{-1} kPa^{-2} per H^+; 473 K) than on stronger $H_3PW_{12}O_{40}$ acids (25 ± 1 and 12 ± 1 ks^{-1} kPa^{-2} per H^+),[43] while k_{Prins}/k_{oligo} ratios were much higher on Al-MCM-41 (10.3 ± 0.2; 473 K) than on $H_3PW_{12}O_{40}$ (2.0 ± 0.1). These trends are consistent with theoretical estimates based on aluminosilicate slabs and Keggin $H_3PW_{12}O_{40}$ cluster models, which confirm the lower reactivity but higher Prins selectivity of weaker acids. The C–C formation transition states for Prins condensation (TS_{Prins};

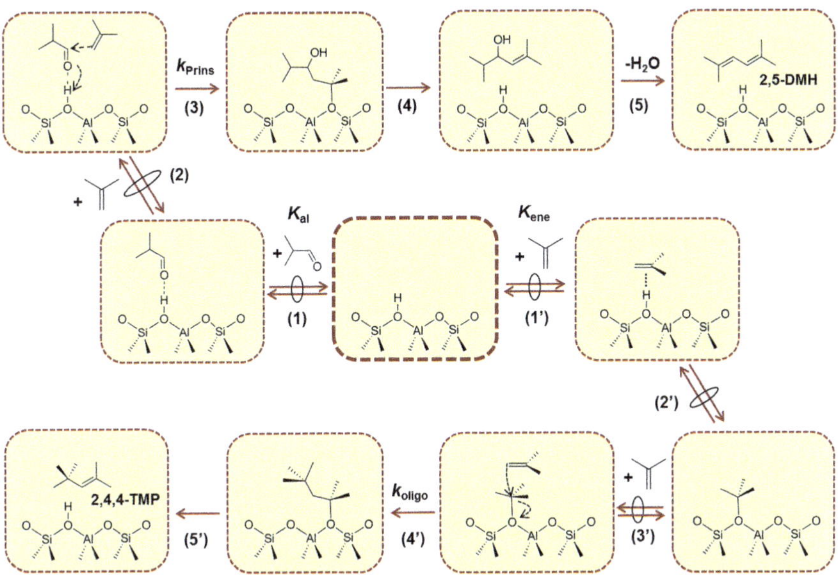

Scheme 7 Reaction pathways for isobutanal–isobutene Prins condensation and isobutene oligomerization on Brønsted solid acids (shown for aluminosilicates as an illustrative example). Quasi-equilibrated steps are noted by a circle over double arrows. Reaction numbers used throughout the text are in parentheses ((1–5) for Prins condensation; (1′–5′) for oligomerization).

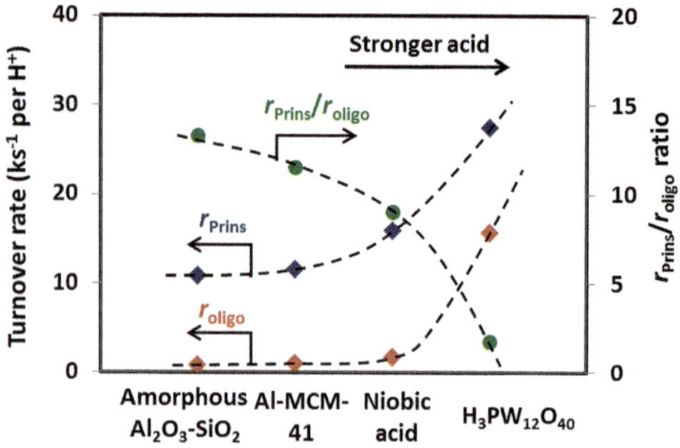

Fig. 16 Turnover rates of Prins condensation (r_{Prins}) and isobutene oligomerization (r_{oligo}) and their rate ratios on solid acids (473 K, 2.0 kPa isobutanal, 1.0 isobutene).[43]

step 3, Scheme 7) and oligomerization (TS_{oligo}; step 4′, Scheme 7) are positively-charged, but TS_{Prins} structures become full ion-pairs only on stronger acids, while TS_{oligo} structures are present as full ion-pairs, irrespective of acid strength. As a result, both k_{Prins} and k_{oligo} increase with acid strength, but the TS_{oligo} structure preferentially benefits from the greater stability of the conjugate anions in the stronger acids, leading to higher 2,5-DMH selectivities, reflected on the k_{Prins}/k_{oligo} ratios, on weaker acids.

The TS_{Prins} and TS_{oligo} structures on aluminosilicates are similar in size (van der Waals volume 0.153 *vs.* 0.144 nm^3) and are therefore stabilized to the same extent by van der Waals contacts with the walls of the confining voids. Consequently, their similar size and the similar acid strength of sites in aluminosilicates, irrespective of framework structure,[49] lead to similar k_{Prins}/k_{oligo} ratios (15–20, 473 K) on all mesoporous and microporous aluminosilicates.[50] These TS_{Prins} and TS_{oligo} structures, however, are significantly larger than their respective precursors (H-bonded isobutanal and *tert*-butoxide, respectively), causing both r_{Prins} and r_{oligo} values to sense the size (and shape) of the confining voids. Measured first-order Prins condensation rate constants (k_{Prins}/K_{al}; K_{al}, isobutanal adsorption constant) reflect the Gibbs free energy of the TS_{Prins} relative to a H-bonded isobutanal and a gaseous isobutene molecule. These k_{Prins}/K_{al} values increase with increasing void size (from TON to BEA to FAU, Fig. 17), but then decrease for even larger voids (Al-MCM-41), because of less effective van der Waals contacts between the transition state and the mesoporous voids. Protons at the cage-like intersections formed by MFI channels gave slightly larger k_{Prins}/K_{al} values (~two-fold) than those within cylindrical channels in BEA (Fig. 17), in spite of their similar diameters, as a result of the quasi-spherical shape of MFI intersections, which provide a better fit for the geometric features of TS_{Prins} structures.

The conversion of primary 2,5-DMH products to skeletal (2,4-DMH, 3,4-DMH) or cyclic (1,4-dimethyl-cyclohexene) isomers is favored for voids of intermediate size (MFI, BEA),[50] suggesting that the required transition states are preferentially stabilized relative to TS_{Prins}. MFI structures showed a unique propensity to form 1,4-dimethyl-cyclohexene, apparently as a result of cage-channel undulations

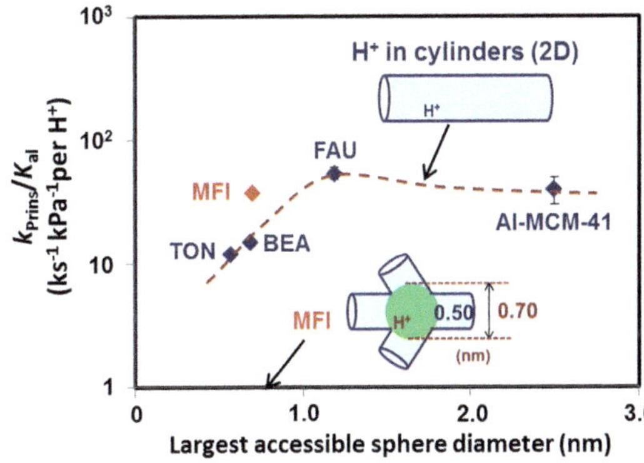

Fig. 17 Measured k_{Prins}/K_{al} values as a function of the largest accessible sphere diameter of aluminosilicate voids at 473 K.[50]

along the MFI structure, which allow the formation of 2,5-DMH and its skeletal isomers at cage-like locations, but impede their egress *via* diffusion through the interconnecting apertures. Their intra-cage cyclization can form faster diffusing species (*e.g.*, 1,4-dimethyl-cyclohexene); such molecules exhibit similar cross sectional areas but much shorter lengths than the acyclic dimethyl isomers.[50]

4. Conclusions

The examples discussed here represent a small subset within the theme of catalysts for fuels; they intend to illustrate the ubiquitous challenges in converting atypical carbon feedstocks, which lack the required volatility or energy density, into liquid fuels *via* the formation of new C–C bonds and the removal of excess oxygen. The formidable hurdles imposed by these requirements are addressed by the preferential stabilization of specific transition states, conferred through solvation within inorganic voids or by intrapore liquids, by acid sites or acid–base site pairs of specific acid/base strength and geometry, by the *in situ* generation of co-reactants that shift transition states along the reaction coordinate, by cascade strategies that protect active sites and circumvent thermodynamic bottlenecks, and by enforcing carbenium-ion chemistries unperturbed by side reactions. These hurdles and their intended solutions are framed here in the context of their underpinnings in chemical kinetics concepts, specifically the formalism of transition state theory, enabled by a seamless combination of kinetic, spectroscopic, isotopic, and theoretical methods. These foundational strategies transcend, in our view, the specific problems at hand, as we hope the diverse problems tackled have been able to demonstrate.

Acknowledgements

The work described here represents contributions also from Patricia Cheung and John Ahn, in addition to those of the co-authors, who also acknowledge collaborations with Jay Labinger, John Bercaw, and Glenn Sunley in the C_1 homologation and carbonylation studies. The financial support by BP, p.l.c. for all the examples shown and by the Marie Sklodowska Curie Fellowship Program, and the

Vermeulen Chair Endowment Fund at the University of California-Berkeley for the dimerization studies are acknowledged with thanks.

References

1 B. C. Gates, J. R. Katzer and G. C. A. Schuit, *Chemistry of Catalytic Processes*, McGraw-Hill, 1979; C. H. Bartholomew and R. J. Farrauto, *Fundamentals of Industrial Catalytic Processes*, John Wiley & Sons, 2011; J. N. Armor, *Catal. Today*, 2011, **163**, 3; C. Martínez and A. Corma, *Coord. Chem. Rev.*, 2011, **255**, 1559; W. Vermieren and J. P. Gilson, *Top. Catal.*, 2009, **52**, 1131.
2 A. I. Olivos-Suarez, A. Szecsenyi, E. J. M. Hensen, J. Ruiz-Martinez, E. V. Pidko and J. Gascon, *ACS Catal.*, 2016, **6**, 2695; C. Mesters, *Annu. Rev. Chem. Biomol. Eng.*, 2016, 7, 223.
3 G. W. Huber, S. Iborra and A. Corma, *Chem. Rev.*, 2006, **106**, 4044; D. M. Alonso, J. Q. Bond and J. A. Dumesic, *Green Chem.*, 2010, **12**, 1493; M. A. Mellner, J. M. Gallo, D. M. Alonso and J. A. Dumesic, *ACS Catal.*, 2015, **5**, 3354; J. S. Luterbacher, D. M. Alonso and J. A. Dumesic, *Green Chem.*, 2014, **16**, 4816; G. W. Huber and A. Corma, *Angew. Chem., Int. Ed.*, 2007, **46**, 2.
4 J. H. Lunsford, *Catal. Today*, 2000, **63**, 165.
5 J. A. Labinger and K. C. Ott, *J. Phys. Chem.*, 1987, **91**, 2682; J. A. Labinger, *Catal. Lett.*, 1988, **1**, 371.
6 C. A. Mims, R. Mauti, A. M. Dean and K. D. Rose, *J. Phys. Chem.*, 1994, **98**, 13357; S. C. Reyes, E. Iglesia and C. P. Kelkar, *Chem. Eng. Sci.*, 1993, **48**, 2643; S. C. Reyes, C. P. Kelkar and E. Iglesia, *Catal. Lett.*, 1993, **19**, 13372.
7 A. Palermo, J. P. H. Vazquez, A. F. Lee, M. S. Tikhov and R. M. Lambert, *J. Catal.*, 1998, **177**, 259; S. Pak, P. Qiu and J. H. Lunsford, *J. Catal.*, 1998, **79**, 222.
8 K. Takanabe and E. Iglesia, *Angew. Chem., Int. Ed.*, 2008, **47**, 7689.
9 K. Takanabe and E. Iglesia, *J. Phys. Chem. C*, 2009, **113**, 10131.
10 Y. Liang, Z. Li, M. Nourdine, S. Shahid and K. Takanabe, *ChemCatChem*, 2014, **6**, 1245.
11 K. Takanabe and S. Shahid, *AIChE J.*, 2017, **63**, 105.
12 G. J. Sunley and D. J. Watson, *Catal. Today*, 2000, **58**, 293.
13 A. Bhan, A. Allian, G. Sunley, D. Law and E. Iglesia, *J. Am. Chem. Soc.*, 2007, **129**, 4919.
14 P. Cheung, A. Bhan, G. L. Sunley, D. Law and E. Iglesia, *J. Catal.*, 2007, **245**, 110.
15 A. Bhan and E. Iglesia, *Acc. Chem. Res.*, 2008, **41**, 559.
16 M. Neurock, *Ind. Eng. Chem. Res.*, 2010, **49**, 10183.
17 M. Boronat, C. Martinez and A. Corma, *Phys. Chem. Chem. Phys.*, 2011, **13**, 2603.
18 I. Agirrezabal-Telleria and E. Iglesia, *J. Catal.*, 2017, submitted.
19 J. C. Mol, *J. Mol. Catal. A: Chem.*, 2004, **213**, 39.
20 J. Canivet, S. Aguado, Y. Schuurman and D. Farrusseng, *J. Am. Chem. Soc.*, 2013, **135**, 4195.
21 R. D. Andrei, M. I. Popa, F. Fajula and V. Hulea, *J. Catal.*, 2015, **323**, 76.
22 J. Heveling, A. Van Der Beek and M. De Pender, *Appl. Catal.*, 1988, **42**, 325.
23 A. Lacarriere, J. Robin, D. Swierczynski, A. Finiels, F. Fajula, F. Luck and V. Hulea, *ChemSusChem*, 2012, **5**, 1787.
24 M. Stöcker, *Microporous Mesoporous Mater.*, 1999, **29**, 3–48; C. D. Chang and A. J. Silvestri, *J. Catal.*, 1977, **47**, 249; C. D. Chang, *Catal. Rev.: Sci. Eng.*, 1983, **25**, 1; S. Kolbe, *Acta Chem. Scand., Ser. A*, 1986, **40**, 711.

25 K. Amakawa, S. Wrabetz, J. Kröhnert, G. Tzolova-Müller, R. Schlögl and A. Trunschke, *J. Am. Chem. Soc.*, 2012, **134**, 11462; K. Ding, A. Gulec, A. M. Johnson, T. L. Drake, W. Wu, Y. Lin, E. Weitz, L. D. Marks and P. C. Stair, *ACS Catal.*, 2016, **6**, 5740; A. Salameh, C. Coperet, J. M. Basset, V. P. W. Bohm and M. Roper, *Adv. Synth. Catal.*, 2007, **349**, 238; S. Lwin and I. E. Wachs, *ACS Catal.*, 2014, **4**, 2505.
26 R. D. Andrei, M. I. Popa, F. Fajula, C. Cammarano, A. A. Khudhair, K. Bouchmella, P. H. Mutin and V. Hulea, *ACS Catal.*, 2015, **5**, 2774; L. Li, R. D. Palcheva and K. J. Jens, *Top. Catal.*, 2013, **56**, 783.
27 S. A. Tabak, F. J. Krambeck and W. E. Garwood, *AIChE J.*, 1986, **32**, 1526; R. J. Quann, L. A. Green, S. A. Tabak and F. J. Krambeck, *Ind. Eng. Chem. Res.*, 1988, **27**, 565; A. Corma and A. Martínez, *Catal. Rev.: Sci. Eng.*, 1993, **35**, 483; L. F. Albright, *Ind. Eng. Chem. Res.*, 2009, **48**, 1409; G. S. Nivarthy, A. Feller, K. Seshan and J. A. Lercher, *Microporous Mesoporous Mater.*, 2000, **75**, 35.
28 J. E. Bercaw, P. L. Diaconescu, R. H. Grubbs, R. D. Kay, S. Kitching, J. A. Labinger, X. Li, P. Mehrkhodavandi, G. E. Morris, G. J. Sunley and P. Vagner, *J. Org. Chem.*, 2006, **71**, 8907; J. E. Bercaw, P. L. Diaconescu, R. H. Grubbs, N. Hazari, R. D. Kay, J. A. Labinger, P. Mehrkhodavandi, G. E. Morris, G. J. Sunley, P. Vagner and J. E. Bercaw, *Inorg. Chem.*, 2007, **46**, 11371.
29 N. Hazari, J. Labinger, D. Simonetti and E. Iglesia, *Acc. Chem. Res.*, 2012, **45**, 653.
30 J. H. Ahn, B. Temel and E. Iglesia, *Angew. Chem.*, 2009, **48**, 3814.
31 D. A. Simonetti, J. H. Ahn and E. Iglesia, *J. Catal.*, 2011, **277**, 173.
32 D. A. Simonetti, J. H. Ahn and E. Iglesia, *ChemCatChem*, 2011, **3**, 704.
33 D. Simonetti, R. T. Carr and E. Iglesia, *J. Catal.*, 2012, **285**, 19.
34 M. Sad, M. Neurock and E. Iglesia, *J. Am. Chem. Soc.*, 2011, **133**, 20384.
35 D. A. Simonetti and J. A. Dumesic, *Catal. Rev.: Sci. Eng.*, 2009, **51**, 441.
36 S. Wang, K. Goulas and E. Iglesia, *J. Catal.*, 2016, **340**, 302.
37 S. Wang and E. Iglesia, *J. Phys. Chem. C*, 2016, **120**, 21589.
38 T. N. Pham, T. Sooknoi, S. P. Crossley and D. E. Resasco, *ACS Catal.*, 2013, **3**, 2456.
39 A. Pulido, B. Oliver-Tomas, M. Renz, M. Boronat and A. Corma, *ChemSusChem*, 2013, **6**, 141.
40 S. Wang and E. Iglesia, *J. Catal.*, 2017, **345**, 183.
41 T. Yamaguchi and C. Nishimichi, *Catal. Today*, 1993, **16**, 555.
42 T. J. Taylor; J. D. Taylor; M. W. Peters and D. E. Henton, *U.S. Pat.* 8,742,187, 2014.
43 S. Wang and E. Iglesia, *ACS Catal.*, 2016, **6**, 7664.
44 M. L. Sarazen, E. Doskocil and E. Iglesia, *ACS Catal.*, 2016, **6**, 7059.
45 M. L. Sarazen, E. Doskocil and E. Iglesia, *J. Catal.*, 2016, **344**, 553.
46 K. Tanabe, in *Acid–base bifunctional catalysis*, ed. J. Fraissard and L. Petrakis, Kluwer Academic Publishers, Dordrecht, 1994, p. 353.
47 K. Tanabe, *Acid–Base Catalysis*, Kodansha-VCH, 1989.
48 J. E. Rekoske and M. A. Barteau, *Ind. Eng. Chem. Res.*, 2011, **50**, 41.
49 A. J. Jones and E. Iglesia, *ACS Catal.*, 2015, **5**, 5741.
50 S. Wang and E. Iglesia, *ACS Catal.*, submitted.

Faraday Discussions

PAPER

Micro-kinetic simulations of the catalytic decomposition of hydrazine on the Cu(111) surface†

Saeedeh S. Tafreshi,[a] Alberto Roldan[b] and Nora H. de Leeuw*[ab]

Received 7th September 2016, Accepted 19th October 2016
DOI: 10.1039/c6fd00186f

Hydrazine (N_2H_4) is produced at industrial scale from the partial oxidation of ammonia or urea. The hydrogen content (12.5 wt%) and price of hydrazine make it a good source of hydrogen fuel, which is also easily transportable in the hydrate form, thus enabling the production of H_2 *in situ*. N_2H_4 is currently used as a monopropellant thruster to control and adjust the orbits and altitudes of spacecrafts and satellites; with similar procedures applicable in new carbon-free technologies for power generators, *e.g.* proton-exchange membrane fuel cells. The N_2H_4 decomposition is usually catalysed by the expensive Ir/Al_2O_3 material, but a more affordable catalyst is needed to scale-up the process whilst retaining reaction control. Using a complementary range of computational tools, including newly developed micro-kinetic simulations, we have derived and analysed the N_2H_4 decomposition mechanism on the Cu(111) surface, where the energetic terms of all states have been corrected by entropic terms. The simulated temperature-programmed reactions have shown how the pre-adsorbed N_2H_4 coverage and heating rate affect the evolution of products, including NH_3, N_2 and H_2. The batch reactor simulations have revealed that for the scenario of an ideal Cu terrace, a slow but constant production of H_2 occurs, 5.4% at a temperature of 350 K, while the discharged NH_3 can be recycled into N_2H_4. These results show that Cu(111) is not suitable for hydrogen production from hydrazine. However, real catalysts are multi-faceted and present defects, where previous work has shown a more favourable N_2H_4 decomposition mechanism, and, perhaps, the decomposition of NH_3 improves the production of hydrogen. As such, further investigation is needed to develop a general picture.

1. Introduction

Hydrazine (N_2H_4) decomposition by heterogeneous catalysis is employed in a proton-exchange membrane fuel cell (PEMFC), due to its hydrogen content of 12.5 wt%. Moreover, since hydrazine hydrate is liquid under mild conditions and its

[a]*Department of Chemistry, University College London, 20 Gordon Street, London, WC1H 0AJ, UK*
[b]*School of Chemistry, Cardiff University, Main Building, Park Place, Cardiff, CF10 3AT, UK. E-mail: DeLeeuwN@cardiff.ac.uk; Tel: +44 (0)2920870658*

† Electronic supplementary information (ESI) available. See DOI: 10.1039/c6fd00186f

decomposition over catalysts at room temperature is exothermic without the need for added energy, it is ideal for portable applications such as space vehicles and satellites.[1–5] Hydrazine is also used in a monopropellant thruster to control and adjust the orbits and altitudes of spacecrafts and satellites, based on the production of much larger volumes of nitrogen, hydrogen and ammonia gases from hydrazine.[2] The most important catalyst used in the hydrazine decomposition reaction is Ir/Al_2O_3 with a very high loading of iridium (20–40%).[2,6–8] However, owing to the high price and limited resources of iridium, considerable research has been focused on the development of active but cheaper and readily available alternative catalysts for hydrazine decomposition. Al-Haydari *et al.* showed that hydrazine molecules adsorb molecularly on a Cu film at 243 K with 20% of the adsorption being reversible; further dissociative adsorption continues above 303 K, producing mainly NH_3 with N_2 and H_2 gaseous by-products.[9] As for the production of hydrogen, we aim to investigate the catalytic activity of Cu towards hydrazine decomposition for the production of hydrogen, and hinder the use of hydrocarbon steam reforming at industrial scale.[10]

Micro-kinetic simulations are convenient tools to approach a catalytic process from the atomic level to reactor conditions. In recent years, there has been increasing interest in the development of micro-kinetic models for various industrially relevant processes, such as ammonia synthesis,[11–14] oxidation,[15,16] and decomposition,[14,17] methanol synthesis[18] and decomposition,[19–21] ethylene oxidation[22], and the water gas shift reaction.[23–25] To construct a reliable micro-kinetic model, it is necessary to investigate all the relevant processes involved, such as adsorption, desorption, and surface reactions. Density functional theory (DFT) is commonly used to determine the energy profile along the reaction pathway that will be employed in the micro-kinetic modelling.[26,27]

We have successfully investigated the dissociative adsorption of hydrazine (N_2H_4) on the planar and stepped Cu(111) surfaces by first-principles calculations,[28] where the thermodynamic and kinetic potential energy surface (PES) showed that intermolecular dehydrogenation of hydrazine to produce NH_3 and N_2 is the favoured route among the explored reaction network.[28] Based on the identified mechanism, we have established a micro-kinetic model to simulate a batch reactor where hydrazine is in contact with the planar Cu(111) surface, using 52 elementary reactions including adsorption, desorption, and reactions on the surface. We have evaluated the effects of temperature, initial N_2H_4 coverage and heating rate on a temperature-programmed reaction (TPR), as well as the selectivity towards the formation of NH_3, N_2, and H_2, resulting in an excellent agreement with the experimental results.

2. Computational methods

In the heterogeneous catalytic system, the constant rate of each surface elementary step is commonly computed using the transition-state theory (TST) approximations of Eyring[29] and Evans and Polanyi.[30] Although TST is widely applicable and provides a useful description of the chemical reaction rates, it has limitations. For example, rather than surmount the reaction energy barrier, the reactants could quantum mechanically tunnel across the barrier, even though their energy is considerably less than the energy needed to go over the barrier. This could be important where the energy barrier is low, as the probability of tunnelling increases with a decreasing energy barrier.[31] TST also fails in its description of reactions at high temperature, where the motion of molecules becomes more complex and collisions may lead to transition states far away from the lowest energy saddle point on the potential energy surface.[32] TST assumes that each intermediate is long-lived in each elementary step to reach a Boltzmann distribution of energy, and it thus fails in situations where intermediates are very short-lived.[33] TST also assumes that the transition states can evolve only to products, while in some cases they may return to the reactants. However, this theory still remains very useful in calculating the thermodynamic properties of the transition state and the reaction rates. More information about TST and rate constants is supplied in the ESI.†

Based on the mechanisms investigated in our previous work on the decomposition of N_2H_4 on the Cu(111) surface, the micro-kinetic model constitutes the 52 reactions summarized in Table 1. Following the approach of a previous study,[33] in this model, the lateral adsorbate–adsorbate interactions are negligible. Surface diffusion is also neglected, assuming that its energy barrier is much smaller than any reaction barrier in the decomposition process. Non-limiting mass transfer is also assumed.

Upon defining the rate equations, we used numerical methods to solve the set of differential equations, which describe the relationship between the species coverages and time.

All of the thermodynamic and kinetic parameters were extracted from calculations based on density functional theory (DFT)[34–36] using the VASP code.[37–40] The total energy calculations were performed using the Perdew–Burke–Ernzerhof (PBE)[41] form of the generalized gradient approximation (GGA), whereas the projector augmented wave (PAW) method was used to consider the effect of the inner cores on the valence density.[42,43] To improve the description of the long-range interactions, and following our previous work on the Cu–hydrazine system,[27] we employed the DFT-D2 method of Grimme as implemented in VASP,[44] which has been shown to improve accuracy on several systems.[27,34,45] Plane wave basis sets were used with an energy cut-off at 600 eV, which gave bulk energies converged to within 0.001 eV per atom. This high value for the cut-off energy ensured that no Pulay stresses occurred within the cell during relaxations. A $5 \times 5 \times 1$ Monkhorst–Pack grid[46] of K-points was used to sample the Brillouin zone for surfaces. The slabs were modelled with a 2×2 supercell, $p(4 \times 4)$, with an area of 88.37 Å^2 and a vacuum layer of 20 Å between slabs. The adsorbate and the top three out of four layers of the slabs were allowed to relax during structural optimisation, in line with previous studies.[47,48] Different slab thicknesses were tested until convergence was achieved.

A combination of two techniques were used to identify transition state (TS) structures: the climbing image nudged elastic band (NEB) method[49,50] and the improved dimer method (IDM),[51] which we verified by a single imaginary frequency associated with the reaction coordinate.

We have calculated the adsorption (E_{ads}) and desorption ($E_{des} = -E_{ads}$) energies for the species using eqn (1);

$$E_{ads} = E^{surf}_{molecule} - (E^{surf} + E^{gas}_{molecule}) \tag{1}$$

where $E^{surf}_{molecule}$ is the total energy of the species adsorbed on a relaxed Cu(111) surface, and E^{surf} and $E^{gas}_{molecule}$ are the energies of the naked surface and isolated gas-phase molecules, respectively. Within this definition, a negative E_{ads} value

Table 1 Calculated ZPE corrected reaction (E_r^{ZPE}) and barrier (E_a^{ZPE}) energies for the reaction pathways considered. The pre-exponential factors (A_0) and reaction rate constants (k) at 300 and 650 K are also included. The E_r^{ZPE} of the adsorption and desorption processes are the corresponding E_{ads}^{ZPE} and E_{des}^{ZPE}, which for each species were calculated relative to the gas-phase species. Note that "*" and (X*) indicate a free site and the adsorbed species on the surface, respectively. The units of A_0 and k for first order and second order reactions are s^{-1} and $ML^{-1}\ s^{-1}$, respectively

	Reactions	E_r^{ZPE} (eV)	E_a^{ZPE} (eV)	A_0 300 K	A_0 650 K	k 300 K	k 650 K
Adsorption–desorption							
R0	$N_2H_4 + * \rightarrow N_2H_4*$	−0.94	—	1.04×10^4	7.06×10^3	1.21×10^{-2}	4.91×10^{-4}
R1	$N_2H_4* \rightarrow N_2H_4 + *$	0.94	—	2.31×10^{11}	1.75×10^{10}	5.33×10^4	2.71×10^{11}
R2	$NH_3* \rightarrow NH_3 + *$	0.78	—	4.98×10^{11}	9.19×10^{10}	2.01×10^8	9.10×10^{11}
R3	$NH_3 + * \rightarrow NH_3*$	−0.78	—	1.43×10^4	9.69×10^3	2.68	7.15×10^{-2}
R4	$N_2* \rightarrow N_2 + *$	0.11	—	6.72×10^{10}	6.93×10^9	8.26×10^{12}	8.10×10^{13}
R5	$N_2 + * \rightarrow N_2*$	−0.11	—	1.11×10^4	7.55×10^3	6.10×10^{-1}	8.83×10^{-2}
R6	$H* + H* \rightarrow H_2 + 2*$	0.45	1.08	5.81×10^{12}	6.01×10^{12}	1.04×10^8	1.02×10^{14}
R7	$H_2 + 2* \rightarrow H* + H*$	−0.45	0.65	4.16×10^4	2.82×10^4	9.83×10^2	1.42×10^2
N_2H_x (x = 1–4) dehydrogenation							
R8	$N_2H_4* \rightarrow N_2H_3* + H*$	0.16	1.30	8.33×10^{12}	2.14×10^{13}	1.30×10^{-9}	2.54×10^3
R9	$N_2H_3* + H* \rightarrow N_2H_4*$	−0.16	1.14	3.12×10^{13}	1.06×10^{14}	6.13×10^{-6}	9.15×10^5
R10	$N_2H_3* \rightarrow NNH_2* + H*$	0.55	1.26	2.54×10^{13}	1.14×10^{14}	8.37×10^{-8}	1.78×10^5
R11	$NNH_2* + H* \rightarrow N_2H_3*$	−0.55	0.71	1.45×10^{13}	6.00×10^{13}	1.23×10^7	2.84×10^{11}
R12	$N_2H_3* \rightarrow NHNH* + H*$	0.64	1.35	1.47×10^{13}	4.26×10^{13}	7.84×10^{-10}	4.75×10^3
R13	$NHNH* + H* \rightarrow N_2H_3*$	−0.64	0.71	5.52×10^{12}	9.61×10^{12}	4.74	1.95×10^7
R14	$NNH_2* \rightarrow NNH* + H*$	0.45	1.12	9.77×10^{12}	2.12×10^{13}	2.46×10^{-6}	7.04×10^4
R15	$NNH* + H* \rightarrow NNH_2*$	−0.45	0.67	1.81×10^{13}	6.00×10^{13}	8.64×10^4	2.35×10^{10}
R16	$NHNH* \rightarrow NNH* + H*$	0.46	1.47	2.03×10^{13}	4.88×10^{13}	3.3×10^{-5}	6.33×10^5
R17	$NNH* + H* \rightarrow NHNH*$	−0.46	1.01	1.75×10^{13}	4.51×10^{13}	8.12×10^3	4.60×10^9

Table 1 (*Contd.*)

	Reactions	E_r^{ZPE} (eV)	E_a^{ZPE} (eV)	A_0 300 K	A_0 650 K	k 300 K	k 650 K
R18	$NNH^* \rightarrow N_2^* + H^*$	−1.62	0.17	1.54×10^{13}	4.16×10^{13}	4.64×10^{10}	5.81×10^{12}
R19	$N_2^* + H^* \rightarrow NNH^*$	1.62	1.79	2.13×10^{12}	3.51×10^{12}	9.17×10^{-19}	1.44×10^{-2}
N_2H_x (x = 1–4) N–N decoupling							
R20	$N_2H_4^* \rightarrow NH_2^* + NH_2^*$	−0.98	0.69	1.68×10^{13}	5.74×10^{13}	1.22×10^{2}	1.11×10^{9}
R21	$NH_2^* + NH_2^* \rightarrow N_2H_4^*$	0.98	1.67	3.72×10^{13}	8.93×10^{13}	1.90×10^{-14}	6.54×10
R22	$N_2H_3^* \rightarrow NH_2^* + NH^*$	−0.39	0.90	1.10×10^{13}	3.54×10^{13}	9.1×10^{-3}	7.86×10^{6}
R23	$NH_2^* + NH^* \rightarrow N_2H_3^*$	0.39	1.29	1.17×10^{13}	2.59×10^{13}	3.68×10^{-9}	4.40×10^{3}
R24	$NNH_2^* \rightarrow NH_2^* + N^*$	0.12	1.35	2.87×10^{12}	4.42×10^{12}	3.58×10^{-11}	5.54×10
R25	$NH_2^* + N^* \rightarrow NNH_2^*$	−0.12	1.23	1.40×10^{13}	3.64×10^{13}	1.07×10^{-7}	3.48×10^{4}
R26	$NHNH^* \rightarrow NH^* + NH^*$	−0.38	0.79	2.42×10^{12}	3.05×10^{12}	9.97×10^{-2}	7.09×10^{5}
R27	$NH^* + NH^* \rightarrow NHNH^*$	0.38	1.17	8.86×10^{12}	1.43×10^{13}	6.14×10^{-7}	1.86×10^{4}
R28	$NNH^* \rightarrow NH^* + N^*$	0.15	1.42	1.47×10^{12}	1.62×10^{12}	4.48×10^{-13}	1.83
R29	$NH^* + N^* \rightarrow NNH^*$	−0.15	1.27	7.53×10^{12}	1.28×10^{13}	4.08×10^{-9}	1.69×10^{3}
NH_x (x = 1–3) dehydrogenation							
R30	$NH_3^* \rightarrow NH_2^* + H^*$	0.60	1.41	3.26×10^{12}	9.05×10^{12}	3.27×10^{-12}	6.85×10
R31	$NH_2^* + H^* \rightarrow NH_3^*$	−0.60	0.81	1.04×10^{13}	2.57×10^{13}	6.00×10^{-1}	2.99×10^{7}
R32	$NH_2^* \rightarrow NH^* + H^*$	0.56	1.40	3.12×10^{12}	4.64×10^{12}	6.60×10^{-6}	1.52×10^{4}
R33	$NH^* + H^* \rightarrow NH_2^*$	−0.56	0.84	1.08×10^{13}	2.99×10^{13}	2.14×10^{-1}	2.43×10^{7}
R34	$NH^* \rightarrow N^* + H^*$	1.34	1.79	6.31×10^{12}	1.22×10^{13}	5.16×10^{-20}	1.70×10^{-2}
R35	$N^* + H^* \rightarrow NH^*$	−1.34	0.45	6.89×10^{12}	1.44×10^{13}	2.62×10^{3}	6.53×10^{8}
Interaction of NH_2 molecules							
R36	$2NH_2^* \rightarrow NH^* + NH_3^*$	0.00	0.45	1.54×10^{13}	2.58×10^{13}	1.08×10^{7}	4.69×10^{10}
R37	$NH^* + NH_3^* \rightarrow 2NH_2^*$	0.00	0.45	1.19×10^{13}	2.88×10^{13}	6.20×10^{5}	1.97×10^{10}

Table 1 (*Contd.*)

Reactions	E_r^{ZPE} (eV)	E_a^{ZPE} (eV)	A_0 300 K	A_0 650 K	k 300 K	k 650 K
Interaction of NH_2 with N_2H_x (x = 1–4)						
R38 $N_2H_4^* + NH_2^* \rightarrow N_2H_3^* + NH_3^*$	−0.36	0.42	2.42×10^{12}	2.73×10^{12}	4.81×10^{4}	2.38×10^{8}
R39 $N_2H_3^* + NH_3^* \rightarrow N_2H_4^* + NH_2^*$	0.36	0.78	2.72×10^{12}	4.52×10^{12}	3.17×10^{-2}	8.22×10^{5}
R40 $N_2H_3^* + NH_2^* \rightarrow NHNH^* + NH_3^*$	0.08	0.64	5.64×10^{12}	9.39×10^{12}	1.10×10^{2}	7.79×10^{7}
R41 $NHNH^* + NH_3^* \rightarrow N_2H_3^* + NH_2^*$	−0.08	0.56	1.04×10^{12}	8.98×10^{11}	3.73×10	2.06×10^{6}
R42 $N_2H_3^* + NH_2^* \rightarrow NNH_2^* + NH_3^*$	−0.13	0.53	1.37×10^{13}	3.14×10^{13}	5.58×10^{3}	2.35×10^{9}
R43 $NNH_2^* + NH_3^* \rightarrow N_2H_3^* + NH_2^*$	0.13	0.66	1.98×10^{12}	2.46×10^{12}	5.18	3.44×10^{6}
R44 $NHNH^* + NH_2^* \rightarrow NNH^* + NH_3^*$	−0.23	0.25	1.53×10^{13}	2.39×10^{13}	2.11×10^{9}	4.60×10^{11}
R45 $NNH^* + NH_3^* \rightarrow NHNH^* + NH_2^*$	0.23	0.48	4.44×10^{12}	5.50×10^{12}	2.41×10^{4}	4.00×10^{8}
R46 $NNH_2^* + NH_2^* \rightarrow NNH^* + NH_3^*$	0.12	0.30	2.55×10^{13}	5.38×10^{13}	1.05×10^{9}	1.06×10^{12}
R47 $NNH^* + NH_3^* \rightarrow NNH_2^* + NH_2^*$	−0.12	0.18	7.72×10^{12}	1.60×10^{13}	1.21×10^{10}	8.63×10^{11}
R48 $NNH^* + NH_2^* \rightarrow N_2^* + NH_3^*$	−2.09	0.08	1.60×10^{13}	3.11×10^{13}	4.38×10^{12}	2.54×10^{13}
R49 $N_2^* + NH_3^* \rightarrow NNH^* + NH_2^*$	2.09	2.17	1.00×10^{12}	1.43×10^{12}	1.25×10^{-25}	3.26×10^{-6}
N_2 dissociation						
R50 $N_2^* \rightarrow N^* + N^*$	3.20	4.69	6.71×10^{11}	7.03×10^{11}	1.09×10^{-68}	1.53×10^{-26}
R51 $N^* + N^* \rightarrow N_2^*$	−3.20	1.49	3.01×10^{13}	9.02×10^{13}	1.08×10^{-11}	1.51×10^{3}

means a release of energy during adsorption. The reaction energies (E_r) were estimated by the difference in energy between the final and initial states; hence, a negative E_r indicates an exothermic process. The forward activation barrier (E_a) was defined as the energy difference between the transition state (TS) and the initial state. We have also considered the effect of temperature on E_{ads}, E_r and E_a; see the ESI for details.†

3. Results and discussion

The 52 elementary steps for hydrazine decomposition summarised in Table 1 include the adsorption and desorption of reactants and products, N–N decoupling, dehydrogenation of N_2H_x (x = 1–4) and NH_x (x = 1–3), and intermolecular interactions on the surface. We have also included in Table 1 the reaction (E_r^{ZPE}) and barrier (E_a^{ZPE}) corrected with the zero-point energy, the pre-exponential factors (A_0), and the reaction rate constants (k) of each elementary step at 300 and 650 K. We have represented the reaction rate constants as a function of the temperature in Fig. S3 of the ESI.† Calculation of the reaction rate constants help us to determine the reaction rate as a function of the temperature, from which we can conclude which step is the rate-limiting reaction. The adsorption sticking coefficients (S_0) of N_2H_4, NH_3, N_2, and H_2 on the Cu(111) surface at 300 and 650 K are also provided in Table 2, where similar results have been reported for NH_3 on Ru(0001),[52] N_2 on Fe(100) and Fe(111)[53], and H_2 on low-index Cu surfaces.[54] We have calculated two different micro-kinetic models. In the first part of our work, we have modelled a temperature-programmed reaction (TPR) where, starting from pre-adsorbed N_2H_4, the temperature increased at different rates from 100 to 500 K, while any gas was extracted to avoid the re-adsorption of gases (R0, R3, R5 and R7). In the second section, we have explored the catalytic activity of copper surfaces towards N_2H_4 dissociation in a batch reactor under varying conditions, starting from a situation where the naked Cu surface is in contact with a given pressure of N_2H_4. The rate equations of the elementary reactions and corresponding differential equations are listed in the ESI.†

3.1. Temperature programmed reaction simulation

Fig. 1 shows the simulated TPR spectra of N_2H_4, NH_3, N_2 and H_2 gases from different initial N_2H_4 coverages. The TPR plots show desorption of species from the surface as the temperature increases. The desorption peaks therefore show the temperature at which the molecules have the highest desorption rate. As Fig. 1 shows, different initial N_2H_4 coverages do not change the temperature of the maximum desorption rate. Fig. 1 in conjunction with Fig. 2 relates the pressures of desorbed gases and the coverages of the most abundant species on the surface, *i.e.* N_2H_4, N_2H_3 and NH, as a function of temperature. Fig. 1a indicates that N_2H_4 desorption takes place at around 213 K, similarly to the desorption peak on a Rh foil surface.[55] As Fig. 2b shows, N_2H_4 disappears completely from the surface at around 220 K by desorbing or converting to species such as N_2H_3 and NH_3, which

Table 2 Calculated sticking coefficients (S_0) of N_2H_4, NH_3, N_2, and H_2 adsorption on the Cu(111) surface at 300 and 650 K. Note that "*" and (X*) indicate a free site and the adsorbed species on the surface, respectively

	Reactions	S_0 this work, 300 K	S_0 this work, 650 K	S_0 other works
R0	$N_2H_4 + * \rightarrow N_2H_4*$	1.17×10^{-6}	6.96×10^{-8}	—
R3	$NH_3 + * \rightarrow NH_3*$	1.88×10^{-4}	7.38×10^{-6}	$\leq 2 \times 10^{-4}$, 300–500 K (ref. 52)
R5	$N_2 + * \rightarrow N_2*$	5.5×10^{-5}	1.2×10^{-5}	1.0×10^{-6} to 1.0×10^{-7}, 500 K (ref. 53)
R7	$H_2 + * \rightarrow 2H*$	2.36×10^{-2}	5.03×10^{-3}	1.0×10^{-5} to 5.0×10^{-2}, 190 K (ref. 54)

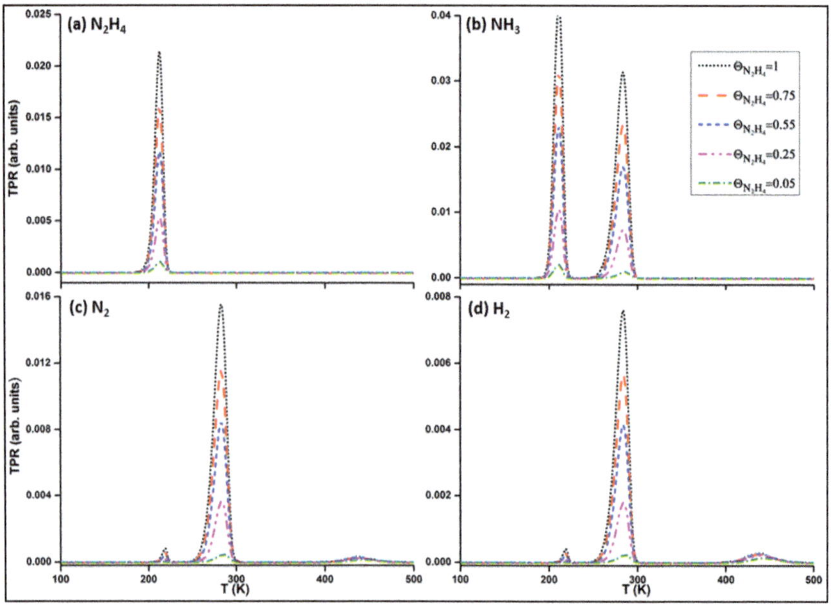

Fig. 1 Simulated TPR spectra for N_2H_4, NH_3, N_2, and H_2 desorption from the Cu(111) surface, starting from adsorbed N_2H_4 at different initial coverages at a reaction time of 1 s with a 1 K min^{-1} heating rate.

desorb quickly from the surface ($k_2 \sim 10^8$ s^{-1}). NH_3 starts to desorb at around 190 K which agrees well with the results of the thermal desorption spectroscopy (TDS) study of NH_3 adsorption on Cu(100), where it desorbs at 185 ± 10 K.[56] The high amount of N_2H_3 on the surface between 200–300 K (Fig. 2b) and NH_3 desorbed from the surface, *i.e.* the first peak of the NH_3 TPR at 211 K (Fig. 1b), indicate that an inter-molecular dehydrogenation mechanism is taking place. N_2H_4 produces NH_2 from N–N decoupling (R38), which, at this low temperature, is feasible from kinetic and thermodynamic points of view. This reaction is the most favoured step in the temperature range of 200–265 K, where the N_2H_3 coverage increases.

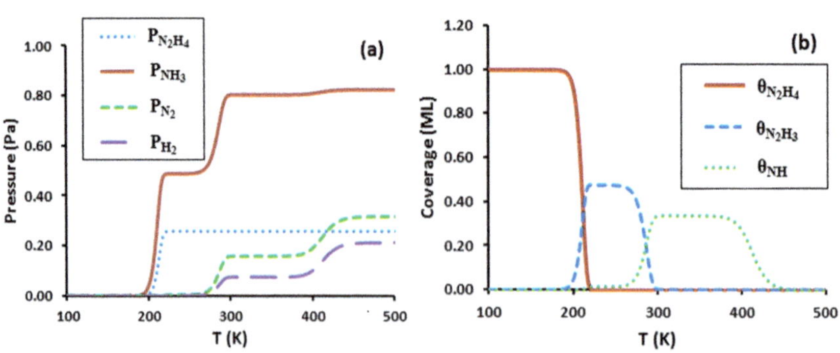

Fig. 2 (a) The partial pressure of desorbed N_2H_4, NH_3, N_2 and H_2 gases and (b) surface coverage of N_2H_4, N_2H_3 and NH as a function of temperature with an initial N_2H_4 full coverage in the TPR simulation at a reaction time of 1 s with a 1 K min^{-1} heating rate.

However, at higher temperatures, the N_2H_3 coverage decreases by reacting with NH_2 intermediates, losing hydrogen atoms (R40, R42, R44, R46 and R48) and resulting in the formation of other species, *i.e.* the NH_3 peak at 284 K (Fig. 1b).

The N_2 and H_2 have three desorption peaks, two smaller peaks at around 219 and 440 K and a maximum desorption peak at 284 K. The N_2 and H_2 desorption peaks at ~219 K are due to the recombination of H and N ad-atoms on the surface, produced by the decomposition of intermediates. The produced N_2 and H_2 desorb from the surface due to the small desorption energies of 0.11 and 0.45 eV, respectively, which are in agreement with measurements on single crystals and polycrystalline Cu, where the heat of adsorption of N_2 on Cu(110) was determined to be 0.088 eV using helium scattering,[57] while the adsorption energy of H_2 lies between 0.39 and 0.48 eV on various forms of unsupported Cu.[54,58–64] The maximum N_2 and H_2 desorption peaks at around 284 K, which appear at the same temperature as the second NH_3 desorption peak, are due to reactions of intermolecular dehydrogenation, resulting in the production of NNH. This leads to the reaction between NNH and NH_2 (R48), as well as NNH decomposition (R18). The smaller peaks at higher temperatures correspond to the recombination of atomic H and N following the decomposition reactions of NH to N and H on the surface (R34). NH is stable on the surface until 450 K, as shown in Fig. 2b, in line with other studies of hydrazine dissociation on Ni[65] and Rh.[66]

Fig. 3 shows the spectra of the gases produced during the TPR, at an initial condition of full coverage of N_2H_4, for a temperature range from 100 to 600 K and with three heating rates of 1, 5 and 8 K min^{-1}. As the heating rate increases, so does the temperature at which the desorption rate is at its maximum (peak

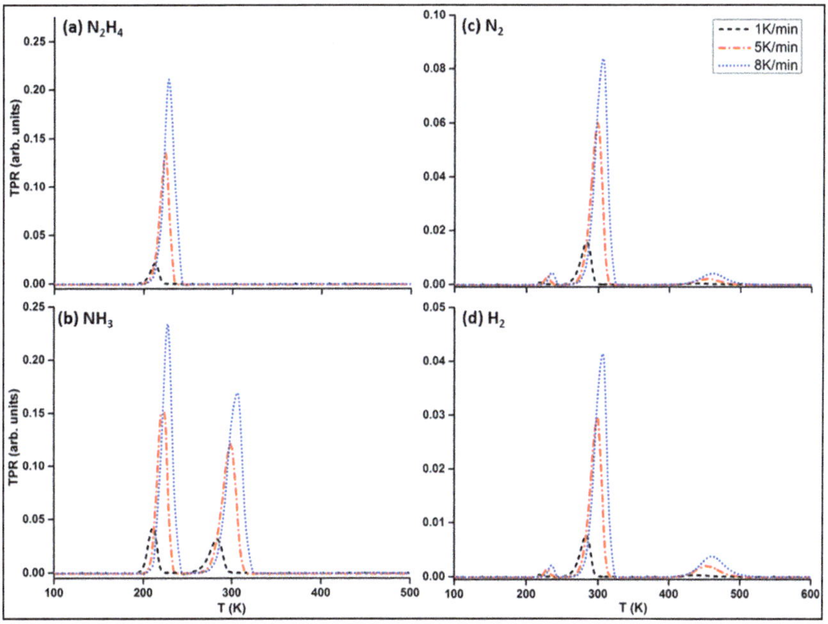

Fig. 3 Simulated TPR curves for N_2H_4, NH_3, N_2 and H_2 desorption from the Cu(111) surface, starting from an initial N_2H_4 coverage of 1 ML (full coverage) at a reaction time of 1 s for different heating rates.

temperature), whereas the intensity of the peak increases as well, in agreement with an experimental TPR study of hydrazine decomposition on an Al_2O_3-supported Ir catalyst.[67] These changes in peak temperature and intensity are related to a sudden variation of the pressures and the derivative slope. The higher the heating rate, the more abrupt are the changes in pressure and the higher the slope of $\frac{\delta P}{\delta T}$. Note that the abrupt desorption of N_2H_4 reduces the time for further reaction on the surface to produce NH_3, N_2 and H_2, indicating that a slower heating rate helps to increase the yield of NH_3, N_2 and H_2, Fig. 4.

The composition of the exhaled gas resulting from the hydrazine interaction with the Cu(111) surface at different temperatures is given in Table 3 and is in agreement with the experimental report by Al-Haydari *et al.*[9] NH_3 is the main gaseous product of hydrazine decomposition, whereas H_2 is the least present. According to the experiment, the temperature at which hydrazine starts to decompose is 303 K when NH_3, N_2 and some H_2 desorb from the Cu film.[9] In our simulation, as shown in Fig. 3, hydrazine starts to decompose at ~190 K, when the first NH_3 can be observed, while N_2 and H_2 desorb at a higher temperature of 219 K. This discrepancy between the experiment and simulation may be due to the adsorption of gases into the liquid hydrazine, as well as the use of a Cu film without a well-defined Cu structure, which may include defects. Our previous works have shown that the introduction of defects on the surface provides more favourable sites for stronger hydrazine adsorption, resulting in higher temperatures for decomposition and desorption.[34,35] Moreover, the heating rate and the reaction time of the experimental study were not reported.

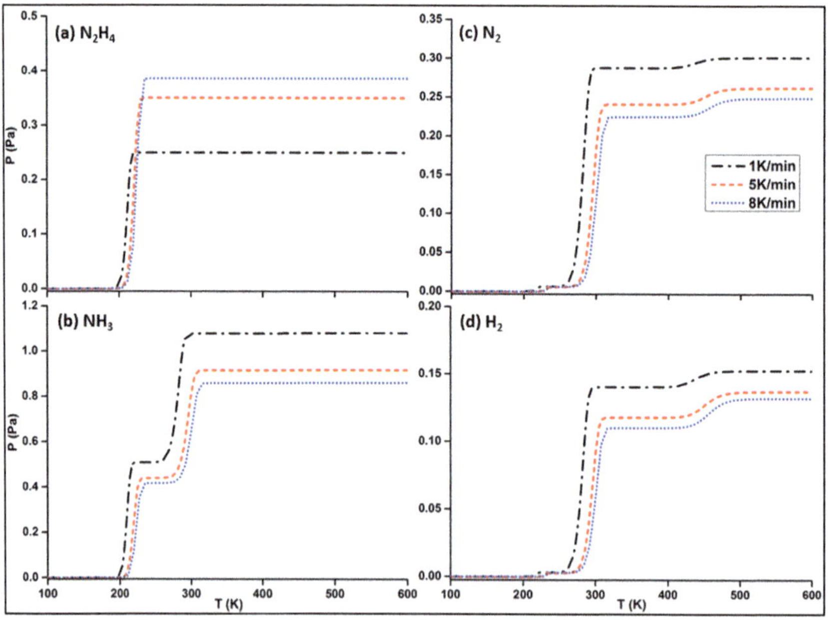

Fig. 4 N_2H_4, NH_3, N_2 and H_2 evolution from Cu(111) surface as a function of temperature for an initial N_2H_4 coverage of 1 ML (full coverage) in the TPR simulation at a reaction time of 1 s for different heating rates.

Table 3 Percentage composition of gaseous products throughout hydrazine decomposition on the Cu(111) surface from the initial N_2H_4 full coverage in the TPR simulation at a reaction time of 1 s with a 1 K min^{-1} heating rate at different temperatures, in comparison with experimental reports[9]

	NH_3 (%)		N_2 (%)		H_2 (%)	
T (K)	This work	Ref. 9	This work	Ref. 9	This work	Ref. 9
303	77.28	75.60	15.33	18.75	7.38	5.62
333	77.28	72.63	15.34	22.93	7.39	4.58
363	77.23	69.49	15.36	27.23	7.41	3.25
393	75.25	69.56	16.37	27.34	8.38	3.10

3.2. Batch reactor simulation

The micro-kinetic simulation of a batch reactor, discussed in this section, considers all elementary steps in Table 1 and starts from a situation where the naked Cu surface is exposed to N_2H_4 gas. We have carried out the micro-kinetic simulations of hydrazine decomposition at a small initial N_2H_4 pressure of 6 Pa in

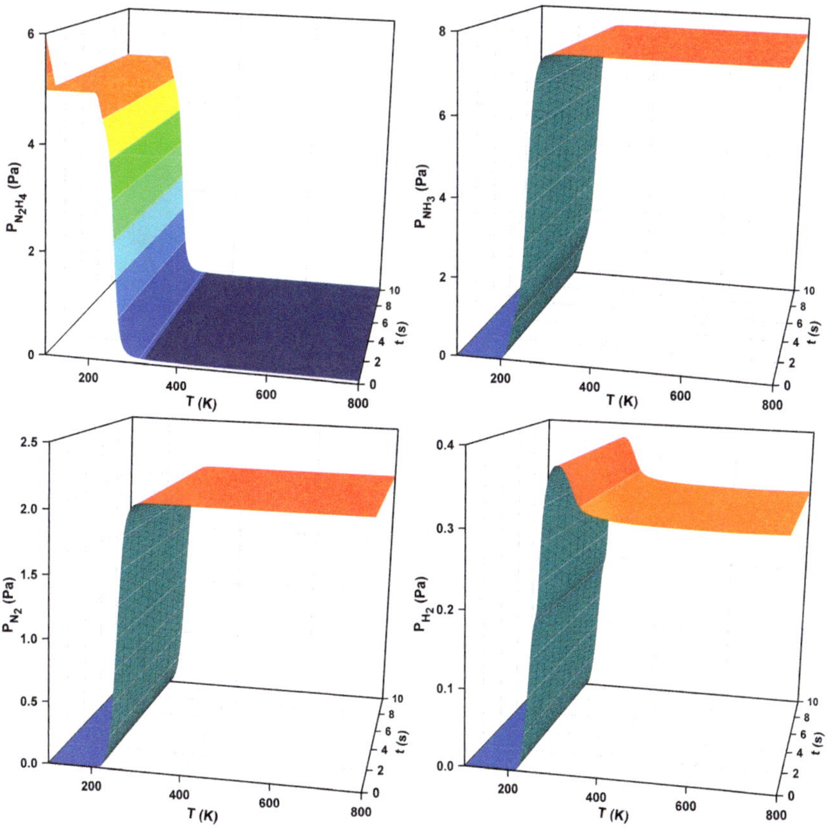

Fig. 5 N_2H_4, NH_3, N_2 and H_2 evolution from the Cu(111) surface as a function of temperature and time for an initial N_2H_4 pressure of 6 Pa with a 1 K min^{-1} heating rate in the batch reactor simulation.

the temperature range of 100–800 K. The corresponding differential equations are listed in the ESI.†

We have represented the pressure of N_2H_4, NH_3, N_2 and H_2 as a function of the temperature and the reaction time in 3D plots (Fig. 5). They show that at 100 K N_2H_4 reaches full coverage of the exposed surface. The decomposition of the N_2H_4 molecules on the surface starts at around 200 K and the N_2H_4 molecules still in the gas phase can occupy the resulting empty sites on the surface. N_2H_4 pressure reaches an equilibrium state at around 300 K, when its pressure decreases below 0.1 Pa. The NH_3 starts to desorb from the Cu(111) surface at around 200 K, while N_2 and H_2 appear in the gas phase at the higher temperature of 220 K. NH_3 and N_2 desorption reach equilibrium at around 300 K, with pressures of 7.5 and 2.1 Pa respectively, and any increase in the temperature does not further affect their desorption. The H_2 pressure increases with increasing temperature up to 300 K (0.38 Pa), and then decreases to 0.33 Pa at 370 K, due to the adsorption and reaction with N ad-atoms producing NH_3.

To understand the N_2H_4 overall decomposition mechanisms, the coverage of some prominent intermediates is plotted in Fig. 6 showing how the coverages of N_2H_4, N_2H_3, NH_3, H and NH change with temperature at 1 s after the surface was covered fully with adsorbed N_2H_4.

The observed trends for the coverages of N_2H_4, N_2H_3 and NH in the batch reactor simulation are the same as those in the TPR simulation, although the coverages of the species are different owing to the adsorption of species on the surface. The hydrazine N–N decoupling (R20) is again the most preferred reaction mechanism among the ones studied due to a relatively low energy barrier. The produced NH_2 intermediate subtracts one of the hydrogens of a co-adsorbed hydrazine molecule, resulting in the production of NH_3 and N_2H_3 molecules on the Cu(111) surface (R38), which is an exothermic reaction ($E_r = -0.36$ eV) with a relatively low energy barrier of 0.42 eV. Fig. 6a shows that as soon as the coverage of N_2H_4 molecules on the surface starts to decrease, the coverage of the N_2H_3 intermediate increases until all N_2H_4 is converted to N_2H_3 and NH_3 at 230 K, when almost the entire surface is covered with N_2H_3 and some NH_3 molecules. The coverage of NH_3, Fig. 6b, decreases quickly when the temperature reaches 215 K due to a relatively small desorption energy of 0.78 eV, which compares well with the experimental desorption energy on Cu(001) of 0.72 ± 0.07 eV.[56] N_2H_3 is stable

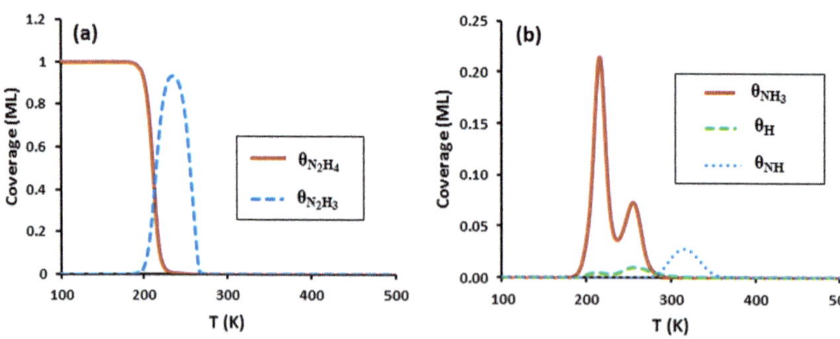

Fig. 6 The surface coverage of (a) N_2H_4 and N_2H_3, (b) NH_3, H and NH as a function of temperature with an initial N_2H_4 pressure of 6 Pa, in the batch reactor simulation at a reaction time of 1 s with a 1 K min^{-1} heating rate.

on the surface in the temperature range of 190–265 K and NH_3 has obtained enough energy to desorb from the surface. The NH_2 present on the surface, obtained from the H_2N–NH_2 bond breaking, reacts with N_2H_3 (R40 and R42) in an intermolecular dehydrogenation mechanism, causing a decrease in the N_2H_3 coverage at around 230 K and also leading to other species on the surface.

The existence of H atoms on the surface in the temperature range of 200–350 K (Fig. 6b) indicates the dehydrogenation of adsorbed species with energy barriers higher than 1.2 eV (R8, R10, R12, R14 and R16). The production of NH from the dehydrogenation of NH_3 and NH_2 on the surface (R30 and R32) takes place at temperatures higher than 270 K, because of an even higher energy barrier of 1.40 eV, which results in the observed NH peak in Fig. 6b centred at 315 K. From here, the NH coverage starts to decrease *via* decomposition to N and H atoms on the surface (R34), with an energy barrier of 1.79 eV. The fact that NH is stable in the temperature range of 270–370 K is in line with the experimental work by Gland *et al.*, where they showed that NH is stable during hydrazine decomposition until 365 K on Ni(111)[65] as well as on Ru,[68] Rh,[66,69] Ir,[70] W,[71] and Mo[72] surfaces.

The associative desorption of hydrogen, produced by dehydrogenation, results in the peak centred at ~300 K, shown in Fig. 5. The highly endothermic reaction between N_2 and H on the surface (R19), with an energy barrier of 1.79 eV, could only occur at high temperatures, and leads to a decrease of H_2 pressure at around 300 K, shown in Fig. 5, due to dissociative adsorption of H_2 in order to provide enough H atoms on the surface.

We have also examined the N_2H_4 conversion and NH_3, N_2 and H_2 selectivities at a fixed temperature of 350 K as a function of the initial N_2H_4 pressure, varied from 10^{-6} to 100 Pa, shown in Fig. 7a. The system reaches the highest NH_3 selectivity of 81.9% for an initial N_2H_4 pressure of 0.1 Pa, the N_2 selectivity reaches its maximum (36.3%) at $P_{N_2H_4} = 100$ Pa, and the maximum H_2 selectivity (5.4%) occurs for an initial N_2H_4 pressure of 0.001 Pa. N_2H_4 conversion is complete for $P_{N_2H_4} \leq 10$ Pa.

Fig. 7b plots the N_2H_4 conversion and NH_3, N_2 and H_2 selectivities with temperature. The N_2H_4 conversion reaches its maximum at ~330 K (98.5%), while NH_3, N_2 and H_2 selectivities converge to 64.5%, 35.1% and 2.6%, respectively, at ~265 K.

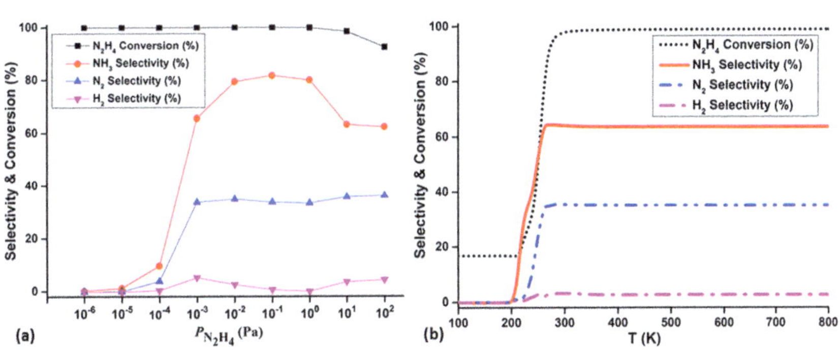

Fig. 7 N_2H_4 conversion and NH_3, N_2 and H_2 selectivities as a function of (a) initial N_2H_4 pressure at 350 K, (b) temperature with an initial N_2H_4 pressure of 6 Pa in the batch reactor simulation at a reaction time of 1 s with a 1 K min^{-1} heating rate.

4. Conclusions

A micro-kinetic model based on results from density functional theory calculations was established, taking into account adsorption, desorption and reaction processes of reactants, intermediates and products, involved in the N_2H_4 decomposition on the Cu(111) surface. Two simulation models have been considered; the first model started from a situation of pre-adsorbed N_2H_4 and considered the constant removal of gases from the reactor, *i.e.* simulating TPR spectra. In the second model, we have considered the naked Cu(111) surface in contact with N_2H_4, where all the gaseous products from the N_2H_4 decomposition are allowed to adsorb and desorb freely until the system reaches equilibrium in the batch reactor. The simulated TPR shows gas desorption peaks depending on the heating rate and the initial N_2H_4 coverage. The simulations of the batch reactor show NH_3 being the major gaseous product on the extended surfaces, in agreement with experiments. The representation of the coverages of the intermediates with temperature shows that N_2H_3 and NH are the most stable intermediates on the surface during N_2H_4 decomposition in the 190–265 and 270–370 K temperature ranges, respectively. Temperature and initial N_2H_4 pressures affect the N_2H_4 conversion and the NH_3, N_2 and H_2 selectivities. The highest NH_3, N_2 and H_2 selectivities obtained in the simulation at 350 K are 81.9%, 36.3% and 5.4%, respectively, while an initial N_2H_4 pressure of 6 Pa gives a conversion of 98.5% at 330 K with NH_3, N_2 and H_2 selectivities of 64.5%, 35.1% and 2.6%, respectively. These results show that Cu(111) is not suitable for hydrogen production from hydrazine, as the dominant product is ammonia. It is known, however, that low coordinated metals are more active and may stabilise intermediates favouring the NH_3 decomposition pathway and therefore increase the production of H_2. Future studies of N_2H_4 decomposition processes on surfaces such as the (001) and (011), and on common defects, *e.g.* steps, will provide a general picture of the feasible production of H_2 from a cheap compound such as N_2H_3 on an abundant Cu catalyst. This study, whose findings are in line with available experiments, validates the proposed mechanisms and shows that micro-kinetic simulations are an effective tool to predict yields and selectivities from DFT results under a wide range of temperature and pressure conditions. This method can now be further rolled out to alternative systems, including metal and non-metal systems.

Acknowledgements

S. S. T acknowledges University College London and the UCL Industrial Doctorate Centre in Molecular Modelling and Material Science for an Overseas Research Scholarship. N. H. d. L. acknowledges the Royal Society for an Industry Fellowship and A. R. is grateful to the Ramsay Memorial Trust and University College London for the award of a Ramsay Fellowship. *Via* our membership of the UK's HPC Materials Chemistry Consortium, which is funded by EPSRC (EP/L000202), this work made use of the facilities of HECToR and ARCHER, the UK's national high-performance computing service, which is funded by the Office of Science and Technology through EPSRC's High End Computing Programme, as well as the UCL Legion High Performance Computing facility (Legion@UCL), and associated support services.

References

1 M. Y. Zheng, R. H. Cheng, X. W. Chen, N. Li, L. Li, X. D. Wang and T. Zhang, *Int. J. Hydrogen Energy*, 2005, **30**, 1081–1089.
2 E. W. Schmidt, *Hydrazine and Its Derivatives: Preparation, Properties, Applications*, Wiley, New York, 2nd edn, 2001.
3 S. G. Pakdehi, M. Salimi and M. Rasoolzadeh, *Researches and Applications in Mechanical Engineering*, 2014, **3**, 21–25.
4 J. Block and G. Schulzek, *J. Catal.*, 1973, **30**, 327–329.
5 M. Grunze, *Surf. Sci.*, 1979, **81**, 603–625.
6 S. Mary, C. Kappenstein, S. Balcon, S. Rossignol and E. Gengembre, *Appl. Catal., A*, 1999, **182**, 317–325.
7 S. Balcon, S. Mary, C. Kappenstein and E. Gengembre, *Appl. Catal., A*, 2000, **196**, 179–190.
8 J. P. Contour and G. Pannetie, *J. Catal.*, 1972, **24**, 434–445.
9 Y. K. Alhaydari, J. M. Saleh and M. H. Matloob, *J. Phys. Chem.*, 1985, **89**, 3286–3290.
10 A. Haryanto, S. Fernando, N. Murali and S. Adhikari, *Energy Fuels*, 2005, **19**, 2098–2106.
11 P. Stoltze, *Phys. Scr.*, 1987, **36**, 824–864.
12 K. Honkala, A. Hellman, I. N. Remediakis, A. Logadottir, A. Carlsson, S. Dahl, C. H. Christensen and J. K. Norskov, *Science*, 2005, **307**, 555–558.
13 A. Hellman, K. Honkala, I. N. Remediakis, A. Logadottir, A. Carlsson, S. Dahl, C. H. Christensen and J. K. Norskov, *Surf. Sci.*, 2006, **600**, 4264–4268.
14 A. Hellman, K. Honkala, I. N. Remediakis, A. Logadottir, A. Carlsson, S. Dahl, C. H. Christensen and J. K. Norskov, *Surf. Sci.*, 2009, **603**, 1731–1739.
15 C. C. Wang, J. Y. Wu, T. L. M. Pham and J. C. Jiang, *ACS Catal.*, 2014, **4**, 639–648.
16 G. Novell-Leruth, J. M. Ricart and J. Perez-Ramirez, *J. Phys. Chem. C*, 2008, **112**, 13554–13562.
17 K. Rasim, M. Bobeth, W. Pompe and N. Seriani, *J. Mol. Catal. A: Chem.*, 2010, **325**, 15–24.
18 T. S. Askgaard, J. K. Norskov, C. V. Ovesen and P. Stoltze, *J. Catal.*, 1995, **156**, 229–242.
19 J. Greeley and M. Mavrikakis, *J. Am. Chem. Soc.*, 2004, **126**, 3910–3919.
20 S. Kandoi, J. Greeley, M. A. Sanchez-Castillo, S. T. Evans, A. A. Gokhale, J. A. Dumesic and M. Mavrikakis, *Top. Catal.*, 2006, **37**, 17–28.
21 A. A. Gokhale, S. Kandoi, J. P. Greeley, M. Mavrikakis and J. A. Dumesic, *Chem. Eng. Sci.*, 2004, **59**, 4679–4691.
22 C. Stegelmann, N. C. Schiodt, C. T. Campbell and P. Stoltze, *J. Catal.*, 2004, **221**, 630–649.
23 A. A. Gokhale, J. A. Dumesic and M. Mavrikakis, *J. Am. Chem. Soc.*, 2008, **130**, 1402–1414.
24 L. C. Grabow, A. A. Gokhale, S. T. Evans, J. A. Dumesic and M. Mavrikakis, *J. Phys. Chem. C*, 2008, **112**, 4608–4617.
25 R. J. Madon, D. Braden, S. Kandoi, P. Nagel, M. Mavrikakis and J. A. Dumesic, *J. Catal.*, 2011, **281**, 1–11.

26 A. Roldán, G. Novell, J. M. Ricart and F. Illas, *J. Phys. Chem. C*, 2010, **114**, 5101–5106.

27 N. Y. Dzade, A. Roldan and N. H. de Leeuw, *J. Chem. Phys.*, 2013, **139**, 124708–124708.

28 S. S. Tafreshi, A. Roldan and N. H. de Leeuw, *Phys. Chem. Chem. Phys.*, 2015, **17**, 21533–21546.

29 H. Eyring, *J. Chem. Phys.*, 1935, **3**, 107–115.

30 M. G. Evans and M. Polanyi, *Trans. Faraday Soc.*, 1935, **31**, 0875–0893.

31 D. S. Sholl and J. A. Steckel, *Density functional theory: a practical introduction*, Wiley, Hoboken, N.J., 2009.

32 J. Pineda and S. D. Schwartz, *Philos. Trans. R. Soc., B*, 2006, **361**, 1433–1438.

33 E. V. Anslyn and D. A. Doughtery, Transition State Theory and Related Topics, *Modern Physical Organic Chemistry*, University Science Books, 2006.

34 S. S. Tafreshi, A. Roldan, N. Y. Dzade and N. H. de Leeuw, *Surf. Sci.*, 2014, **622**, 1–8.

35 S. S. Tafreshi, A. Roldan and N. H. de Leeuw, *J. Phys. Chem. C*, 2014, **118**, 26103–26114.

36 S. S. Tafreshi, A. Roldan and N. H. de Leeuw, *Surf. Sci.*, 2015, **637–638**, 140–148.

37 G. Kresse and J. Furthmuller, *Phys. Rev. B*, 1996, **54**, 11169–11186.

38 G. Kresse and J. Furthmuller, *Comput. Mater. Sci.*, 1996, **6**, 15–50.

39 G. Kresse and J. Hafner, *Phys. Rev. B*, 1993, **47**, 558–561.

40 G. Kresse and J. Hafner, *Phys. Rev. B*, 1994, **49**, 14251–14269.

41 J. P. Perdew, K. Burke and M. Ernzerhof, *Phys. Rev. Lett.*, 1996, **77**, 3865–3868.

42 G. Kresse and D. Joubert, *Phys. Rev. B*, 1999, **59**, 1758–1775.

43 P. E. Blochl, *Phys. Rev. B*, 1994, **50**, 17953–17979.

44 S. Grimme, *J. Comput. Chem.*, 2006, **27**, 1787–1799.

45 S. Irrera, A. Roldan, G. Portalone and N. H. De Leeuw, *J. Phys. Chem. C*, 2013, **117**, 3949–3957.

46 H. J. Monkhorst and J. D. Pack, *Phys. Rev. B*, 1976, **13**, 5188–5192.

47 R. Koitz, A. P. Seitsonen, M. Iannuzzi and J. Hutter, *Nanoscale*, 2013, **5**, 5589–5595.

48 M. Gajdos and J. Hafner, *Surf. Sci.*, 2005, **590**, 117–126.

49 G. Mills and H. Jonsson, *Phys. Rev. Lett.*, 1994, **72**, 1124–1127.

50 G. Mills, H. Jonsson and G. K. Schenter, *Surf. Sci.*, 1995, **324**, 305–337.

51 A. Heyden, A. T. Bell and F. J. Keil, *J. Chem. Phys.*, 2005, **123**, 224101–224115.

52 L. R. Danielson, M. J. Dresser, E. E. Donaldson and J. T. Dickinson, *Surf. Sci.*, 1978, **71**, 599–614.

53 F. Bozso, G. Ertl, M. Grunze and M. Weiss, *J. Catal.*, 1977, **49**, 18–41.

54 G. Anger, A. Winkler and K. D. Rendulic, *Surf. Sci.*, 1989, **220**, 1–17.

55 J. Prasad and J. L. Gland, *Langmuir*, 1991, **7**, 722–726.

56 I. Hussla, H. Seki, T. J. Chuang, Z. W. Gortel, H. J. Kreuzer and P. Piercy, *Phys. Rev. B*, 1985, **32**, 3489–3501.

57 A. Marmier, C. Ramseyer, P. N. M. Hoang, C. Girardet, J. Goerge, P. Zeppenfeld, M. Buchel, R. David and G. Comsa, *Surf. Sci.*, 1997, **383**, 321–339.

58 M. J. Sandoval and A. T. Bell, *J. Catal.*, 1993, **144**, 227–237.

59 P. B. Rasmussen, P. M. Holmblad, H. Christoffersen, P. A. Taylor and I. Chorkendorff, *Surf. Sci.*, 1993, **287**, 79–83.

60 J. M. Campbell and C. T. Campbell, *Surf. Sci.*, 1991, **259**, 1–17.

61 C. T. Rettner, D. J. Auerbach and H. A. Michelsen, *Phys. Rev. Lett.*, 1992, **68**, 2547–2550.
62 B. E. Hayden, D. Lackey and J. Schott, *Surf. Sci.*, 1990, **239**, 119–126.
63 C. Astaldi, A. Bianco, S. Modesti and E. Tosatti, *Phys. Rev. Lett.*, 1992, **68**, 90–93.
64 J. Harris, *Surf. Sci.*, 1989, **221**, 335–345.
65 J. L. Gland, G. B. Fisher and G. E. Mitchell, *Chem. Phys. Lett.*, 1985, **119**, 89–92.
66 J. Prasad and J. L. Gland, *Langmuir*, 1991, **7**, 722–726.
67 J. L. Falconer and H. Wise, *J. Catal.*, 1976, **43**, 220–233.
68 H. Rauscher, K. L. Kostov and D. Menzel, *Chem. Phys.*, 1993, **177**, 473–496.
69 W. M. Daniel and J. M. White, *Surf. Sci.*, 1986, **171**, 289–302.
70 B. J. Wood and H. Wise, *J. Catal.*, 1975, **39**, 471–480.
71 R. C. Cosser and F. C. Tompkins, *Trans. Faraday Soc.*, 1971, **67**, 526–544.
72 R. C. A. Contaminard and F. C. Tompkins, *Trans. Faraday Soc.*, 1971, **67**, 545–555.

Faraday Discussions

ROYAL SOCIETY OF CHEMISTRY

PAPER

Theoretical insights into the sites and mechanisms for base catalyzed esterification and aldol condensation reactions over Cu†

Matthew Neurock,*[ab] Zhiyuan Tao,[b] Ashwin Chemburkar,[a] David D. Hibbitts[bc] and Enrique Iglesia[d]

Received 25th October 2016, Accepted 21st November 2016
DOI: 10.1039/c6fd00226a

Condensation and esterification are important catalytic routes in the conversion of polyols and oxygenates derived from biomass to fuels and chemical intermediates. Previous experimental studies show that alkanal, alkanol and hydrogen mixtures equilibrate over Cu/SiO_2 and form surface alkoxides and alkanals that subsequently promote condensation and esterification reactions. First-principle density functional theory (DFT) calculations were carried out herein to elucidate the elementary paths and the corresponding energetics for the interconversion of propanal + H_2 to propanol and the subsequent C–C and C–O bond formation paths involved in aldol condensation and esterification of these mixtures over model Cu surfaces. Propanal and hydrogen readily equilibrate with propanol *via* C–H and O–H addition steps to form surface propoxide intermediates and equilibrated propanal/propanol mixtures. Surface propoxides readily form *via* low energy paths involving a hydrogen addition to the electrophilic carbon center of the carbonyl of propanal or *via* a proton transfer from an adsorbed propanol to a vicinal propanal. The resulting propoxide withdraws electron density from the surface and behaves as a base catalyzing the activation of propanal and subsequent esterification and condensation reactions. These basic propoxides can readily abstract the acidic C_α–H of propanal to produce the $CH_3CH^{(-)}CH_2O^*$ enolate, thus initiating aldol condensation. The enolate can subsequently react with a second adsorbed propanal to form a C–C bond and a β-alkoxide alkanal intermediate. The β-alkoxide alkanal can subsequently undergo facile hydride transfer to form the 2-formyl-3-pentanone intermediate that decarbonylates to give the 3-pentanone product. Cu is unique in that it rapidly catalyzes the decarbonylation of the C_{2n} intermediates to form C_{2n-1} 3-pentanone as the major product with very small yields of C_{2n} products. This is

[a]Department of Chemical Engineering and Material Science, University of Minnesota, Minneapolis, MN, 55455, USA. E-mail: mneurock@umn.edu
[b]Department of Chemical Engineering, University of Virginia, Charlottesville, VA, 22904, USA
[c]Department of Chemical Engineering, University of Florida, Gainesville, FL, 32611, USA
[d]Department of Chemical and Biomolecular Engineering, University of California, Berkeley, CA 94720, USA
† Electronic supplementary information (ESI) available. See DOI: 10.1039/c6fd00226a

likely due to the absence of Brønsted acid sites, present on metal oxide catalysts, that rapidly catalyze dehydration of the hemiacetal or hemiacetalate over decarbonylation. The basic surface propoxide that forms on Cu can also attack the carbonyl of a surface propanal to form propyl propionate. Theoretical results indicate that the rates for both aldol condensation and esterification are controlled by reactions between surface propoxide and propanal intermediates. In the condensation reaction, the alkoxide abstracts the weakly acidic hydrogen of the C_α–H of the adsorbed alkanal to form the surface enolate whereas in the esterification reaction the alkoxide nucleophilically attacks the carbonyl group of a vicinal bound alkanal. As both condensation and esterification involve reactions between the same two species in the rate-limiting step, they result in the same rate expression which is consistent with experimental results. The theoretical results indicate that the barriers between condensation and esterification are within 3 kJ mol^{-1} of one another with esterification being slightly more favored. Experimental results also report small differences in the activation barriers but suggest that condensation is slightly preferred.

1. Introduction

Aldol condensation and Guerbet coupling as well as other alkanal/alkanol coupling reactions provide attractive routes for the removal of oxygen and hydrocarbon chain growth, which are important steps for the conversion of oxygenates to fuels as well as chemicals. Aldol condensation reactions are currently used industrially in the synthesis of 2-ethyl hexanal, methyl isobutyl ketone and Guerbet alcohols[2] and are also being considered for the conversion of bio-alcohols to fuels.[3,4] Aldol condensation and Guerbet reactions typically proceed *via* coupling of aldehydes and alcohols, respectively, to form β-hydroxy alkanals and alkanones that dehydrate to form α,β-unsaturated carbonyl intermediates on acid or base catalysts.[5] Base-catalyzed aldol condensation mechanisms are thought to proceed *via* the abstraction of the weakly acidic hydrogen at the α-C of an alkanal, thus resulting in the formation of an enolate that subsequently attacks the electrophilic carbonyl group of a vicinal alkanal to form a new C–C bond. These reactions can be carried out over a range of different homogeneous catalysts including Cu(II), Co(II), Mn(II), and Zn(II)[6] together with base promoters (NaOH, NaOR, and Na_2CO_3) or basic ligands,[7–10] as well as on heterogeneous base catalysts, including MgO,[11–14] Mg–AlO_x, ZnO,[15,16] phosphates,[17] and base-promoted mesoporous silicas, such as diamino-functionalized MCM-41 (ref. 18) and hydrotalcites.[19,20]

Previous kinetic studies have shown that while C_n alcohols react *via* aldol condensation paths to form C_{2n-1} alkanones over CuO/ZnO/Al_2O_3 and Cu/ZnO catalysts,[21] it is the basic sites on the ZnO support that are actually responsible for C–C bond formation. In a recent study, Sad *et al.*[1] demonstrated for the first time that both aldol condensation as well as esterification can proceed over non-basic catalysts, specifically monofunctional Cu/SiO_2 catalysts. Propanol (C_3H_7OH) was found to rapidly equilibrate with propanal (C_3H_6O) and H_2 to form an equilibrated propanol–propanal–H_2 reactant pool that subsequently reacts *via* the three predominant paths shown in Scheme 1: (1) dehydration of propanol to form propene and propane *via* subsequent hydrogenation steps; (2) aldol condensation followed by decarbonylation and dehydrogenation to form 3-pentanone ($C_5H_{10}O$)

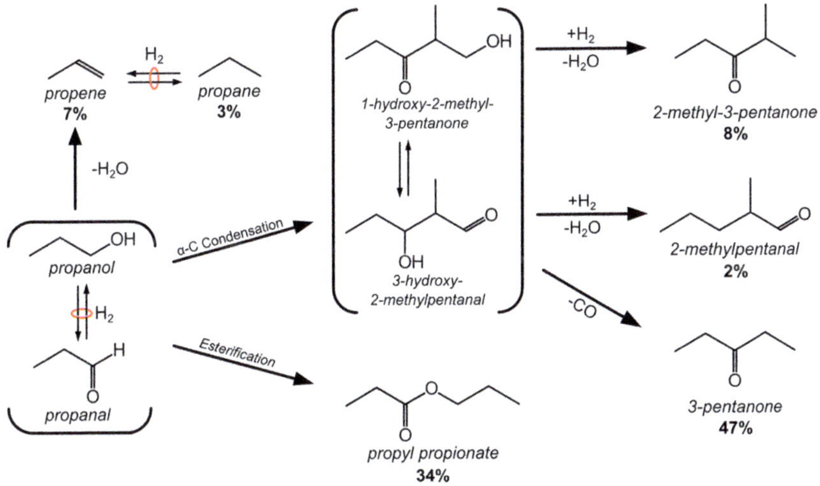

Scheme 1 Reaction paths for the conversion of propanol/propanal/H_2 over Cu/SiO_2, adapted from Sad *et al.*[1] Selectivities for six major products are given based on the following conditions: 5 wt% Cu/SiO_2, 5.6% dispersion, 503 K, 2160 g cat.-ks per mol propanol, 0.64 kPa propanol, 80 kPa H_2, balance He, 3% conversion.

or dehydration paths to form minor 2-methyl 3-pentanone and 2-methyl pentanal ($C_6H_{12}O$) products; and (3) esterification to form propyl propionate ($C_6H_{12}O_2$). The reported selectivities to form the propene, propane, 2-methyl-3-pentanone, 2-methyl pentanal, 3-pentanone, and propyl propionate at 3% conversion (shown in Scheme 1) are 7%, 3%, 8%, 2%, 47%, and 34%, respectively. Detailed kinetic analyses suggested that both esterification and condensation reactions proceed *via* the *in situ* formation of surface alkoxides on Cu that act as bases that catalyze C–C and C–O bond-forming reactions.

Further kinetic analyses showed that the rate equations for esterification and condensation follow the exact same functional form with the rate of condensation being ~2 times faster than the rate of esterification. This suggests that they proceed *via* a similar kinetically-relevant step that precedes a kinetic branch point unaffected by H_2 and alkanal/alkanol pressures, or that they have distinct kinetically-relevant steps that share the same reactant precursors.[1] Experimental results were used together with simple estimates from gas phase molecular calculations to suggest that condensation proceeds by the reaction of a surface enolate with an adsorbed propanal to form a hemi-acetalate (β-alkoxide alkanal) that decarbonylates to form 3-pentanone. In contrast, propyl propionate was proposed to be formed *via* a Cu-catalyzed esterification of two aldehydes, similar to the classic base-catalyzed esterification routes proposed by Cannizaro and Tishchenko as reported in the literature.[22,23]

First principle density functional theoretical (DFT) calculations were carried out herein to provide insights into the elementary steps, elucidate the mechanisms and establish the kinetics for Cu-catalyzed alcohol dehydrogenation, esterification and condensation. The simulation results indicate that propoxide surface intermediates are readily formed on Cu *via* the hydrogenation and dehydrogenation of the propanal and propanol, respectively, and act as a base, co-catalyzing esterification and condensation pathways. Adsorbed alkanals can also catalyze promote hydride transfer from the adsorbed alkoxide intermediates to remove condensation and esterification products from the surface and to regenerate the catalytic propoxides.

2. Computational methods

All of the calculations reported herein were carried out using non-local gradient corrected periodic, plane-wave density functional theory (DFT) calculations using the Vienna *ab initio* Simulation Program (VASP).[24] The calculations were carried out over model Cu(111) surfaces to mimic the coordinatively-saturated sites that comprise >95% of the most active large Cu particles. The influence of edge and corner sites on the activation of C–H and O–H bonds of propanol were also examined by carrying out calculations on a Cu(110) surface and on a 201 atom cuboctahedral Cu cluster (Cu_{201}). The details for all of the simulations are reported in the ESI.†

The adsorption energies for all of the reactant, intermediate and product molecules were calculated as:

$$\Delta E_{\text{ads}} = E_{\text{surf+ads}} - E_{\text{surf}} - E_{\text{ads}} \tag{1}$$

where $E_{\text{surf+ads}}$, E_{surf}, and E_{ads} are the energies of the surface–adsorbate complex, the bare metal surface, and the adsorbate in vacuum, respectively. The activation barriers and reaction energies were calculated as:

$$A^* + B^* \rightarrow TS^* \rightarrow C^* + D^* \tag{2}$$

$$\Delta E_{\text{act}} = E_{\text{TS*}} + \delta E_{\text{surf}} - E_{\text{A*}} - E_{\text{B*}} \tag{3}$$

$$\Delta E_{\text{rxn}} = E_{\text{C*}} + E_{\text{D*}} - E_{\text{A*}} - E_{\text{B*}} \tag{4}$$

where $E_{\text{i*}}$ and $E_{\text{TS*}}$ refer to the energies of adsorbed intermediate i (A*, B*, C* or D*) and the transition state (TS*), respectively. δE_{surf} refers to the change in energy required to bring the "infinitely" separated species on the surface together into the reactant state where they sit adjacent to one another, thus accounting for any attractive or repulsive interactions.

The intrinsic activation barriers for the elementary steps reported herein all refer to the direct energy difference between the transition state and the bound reactant(s) state along the elementary step reaction coordinate ($E_{\text{TS*}} - E_{\text{A*}} - E_{\text{B*}}$ in eqn (3)). In order to compare with experimental results, we also calculate apparent activation energies where the transition state energies are referenced to the most abundant species present on the surface under reaction conditions rather than from the elementary step reactant state. Experimental results suggest that the surfaces are covered under reaction conditions. We therefore use adsorbed propoxide and the gas phase alkanal as the zero energy reference state to calculate apparent barriers for most of the systems discussed herein. The reaction entropies and free energies of activation were also calculated for the rate controlling esterification and condensation steps as discussed in ESI S.1.†

3. Results and discussion

3.1 Conversion between propanol and propanal

3.1.1 Equilibrium of propanal and hydrogen to propanol. As was discussed earlier, mixtures of propanal and hydrogen along with propanol readily react and equilibrate over Cu/SiO_2. DFT calculations were used here to examine the elementary steps in the hydrogenation of propanal to propanol over Cu(111), as well at the reverse reactions for the dehydrogenation of propanol to propanal and hydrogen. Propanal was calculated to preferentially adsorb atop in an η_1 configuration resulting in an adsorption energy of -28 kJ mol^{-1} which is consistent with the adsorption site and energy reported from ultrahigh vacuum experimental studies carried out on Cu(111)[25,26] and previous theoretical studies[27–29] Hydrogen readily dissociates over the Cu(111) surface to form two H* surface species with an overall energy of -32 kJ mol^{-1}, consistent with experimentally reported adsorption energies of ~40 kJ mol^{-1}.[30] The adsorbed propanal can subsequently hydrogenate by the initial addition of H* to the oxygen or the carbon of its carbonyl to form the hydroxypropyl ($CH_3CH_2CH^*OH$) or propoxide ($CH_3CH_2O^*$) intermediate, respectively, as shown in Scheme 2. The paths which proceed *via* propoxide and hydroxypropyl intermediates are known more generally in the literature as the alkoxide and hydroxyalkyl paths.

The propoxide path which is shown in Fig. 1 (unfilled symbols) proceeds by the initial addition of hydrogen to the carbon of the bound carbonyl to form the propoxide resulting in a barrier of 63 kJ mol^{-1} when taken with respect to the adsorbed propanal and a single H*, which mimics what would be measured on the covered surfaces found experimentally. The overall reaction energy for this step is highly exothermic with an overall energy of -52 kJ mol^{-1}. The propoxide intermediate subsequently hydrogenates to form adsorbed propanol. The intrinsic activation energy and the overall reaction energy to hydrogenate the propoxide to form propanol were calculated to be +99 kJ mol^{-1} and -21 kJ mol^{-1}, respectively. The propanol that forms desorbs from the surface with an energy of +29 kJ mol^{-1}. The overall potential energy surface for this path shown in Fig. 1 (unfilled symbols) suggests that propoxide forms and builds up on the surface as its rate of removal appears to be controlled by the high barrier for the second hydrogen addition step.

In the hydroxypropyl path which is shown in Fig. 1 (filled symbols), the barrier to add hydrogen first to the oxygen of the carbonyl rather than the carbon was

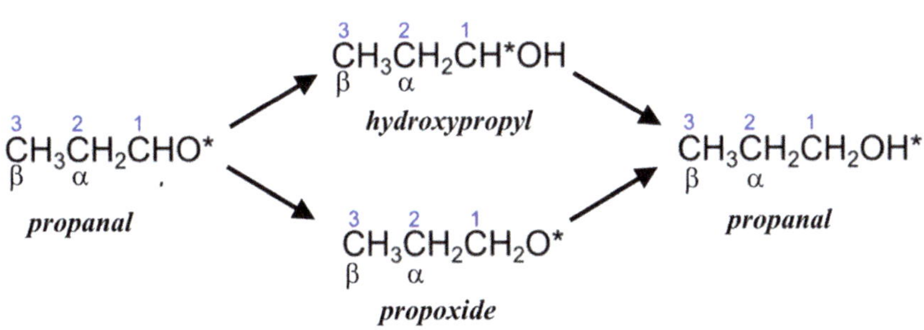

Scheme 2 Routes for the dehydrogenation of propanol to form propanal.

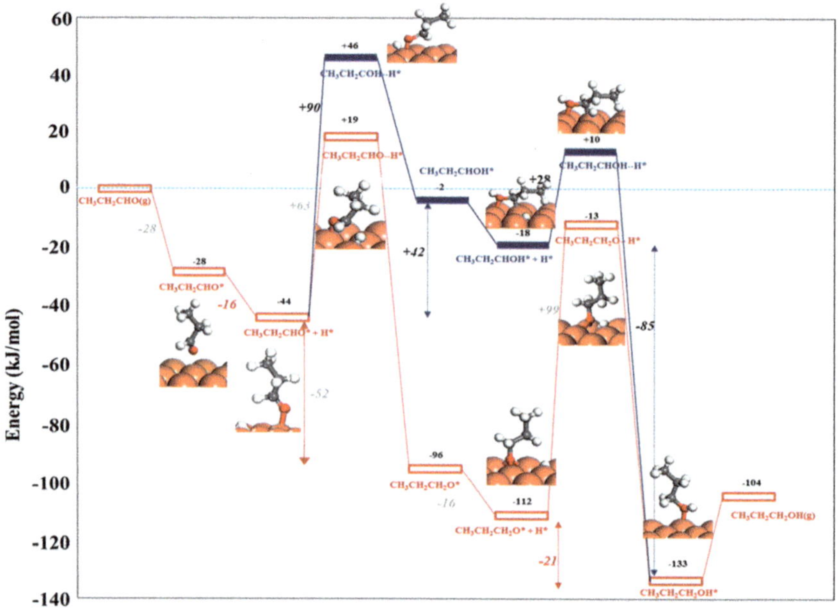

Fig. 1 DFT-calculated pathways for the hydrogenation of propanal to propanol and the reverse reaction involving the dehydrogenation of propanol to propanal and hydrogen. The propoxide path (unfilled rectangular symbols) involves the initial addition of hydrogen to the C_1 of the carbonyl to form the propoxide intermediate and the subsequent hydrogenation of the propoxide and the desorption of the propanol product. The hydroxypropyl path (filled rectangular symbols) proceeds instead by the initial addition of hydrogen to the O of the carbonyl to form the hydroxypropyl intermediate and the subsequent addition of hydrogen to the C of the hydroxypropyl to form propanol which desorbs.

calculated to be significantly higher at +90 kJ mol^{-1} taken with respect to the adsorbed propanal and H* as it requires the breaking of a strong Cu–O bond. The overall reaction energy to form the hydroxypropyl intermediates was calculated to be endothermic at +2 kJ mol^{-1}. The intrinsic activation barrier and overall reaction energy to subsequently hydrogenate the hydropropxyl intermediate to form adsorbed propanol were calculated to be 28 kJ mol^{-1} and −115 kJ mol^{-1} (taken with respect to the adsorbed hydropropxyl intermediate and H*), respectively. The results in Fig. 1 indicate that the rate for this path is likely limited by the initial hydrogen addition to the oxygen to form the hydroxyalkyl intermediate. A more detailed analysis of the energies for both the propoxide and the hydroxyalkyl paths shown in Fig. 1 suggests that the propoxide path is more favorable than the hydroxypropyl path, as the barrier to form the propoxide (+63) is 27 kJ mol^{-1} lower than that to form the hydroxypropyl (+90 kJ mol^{-1}). As such, the propoxide, regardless of the path, is readily formed and likely builds up on the surface and as a result is the kinetically dominant surface intermediate. This is consistent with previous analyses for the hydrogenation of different aldehydes and ketones over Ru.[31]

Propanol dehydrogenation, the microscopic reverse of propanal hydrogenation, is not likely to proceed over the Cu(111) surface as the intrinsic activation energies to activate the O–H or the C–H bond of propanol, taken with respect to adsorbed propanol, are 115 and 143 kJ mol^{-1}, respectively, as is shown in Fig. 1. This is consistent with ultrahigh vacuum studies which show that, in the absence of oxygen, methanol as well as other light alcohols desorb and do not dissociate over Cu(111), and with other previous theoretical studies of methanol dehydration on Cu(111).[27,28] The results by Sad *et al.*[1] carried out over Cu/SiO_2 at higher pressures with mixtures of propanal, hydrogen and propanol, however, show that mixtures readily equilibrate. The difference in the reported reactivity of the alcohol on Cu are likely the result of higher propanol coverages at relevant reaction conditions which can assist O–H activation or *via* the presence and reactivity of coordinatively unsaturated sites that exist at edges or corners on the Cu particles. The results reported in Fig. S1† show that the barrier to activate the O–H bond of propanol decreases to 70 kJ mol^{-1} with increasing propanol surface coverage, as higher coverages result in the formation of hydrogen bonding networks which stabilize the transition state and allow facile proton-transfer *via* a Grotthus-like mechanism similar to that reported for other alcohol decomposition reactions in the presence of water over different metals.[32–34] While the higher coverages help to promote the formation of surface propoxide intermediates, the subsequent C–H activation would still likely be prohibitive as the barrier is 115 kJ mol^{-1} and does not decrease at higher coverages.

Propanol dehydration may proceed more favorably at the edge and corner sites of Cu particles. Both the O–H and C–H activation barriers decrease significantly to 73 kJ mol^{-1} when carried out at the more coordinatively unsaturated edge site on Cu(110) (Fig. S2†) which is consistent with experimental results for the activation of methanol, ethanol and propanol to form methoxy, ethoxy and propoxy, respectively, over Cu(110) under ultrahigh vacuum conditions.[35–38] We find similar low O–H and C–H activation barriers (<80 kJ mol^{-1}) for the dehydrogenation of propanol to propanal at the edge and corner sites on a Cu_{201} cluster as presented in Fig. S3.† As such, propanol dehydrogenation to propanal and H_2 can occur at the edge and corner sites. The propanal and hydrogen that form can subsequently readsorb and react on the Cu(111) terrace sites to carry out condensation and esterification. This would still be consistent with the results from Sad *et al.*[1] which show that the rates increase with Cu particle size, thus indicating that the terrace sites are most active for the steps that control the rate.

3.2 Reactions of propanol and propanal

3.2.1 Base-catalyzed reactions on Cu. As discussed, esterification and aldol condensation reactions are thought to proceed over Cu *via* base-catalyzed reactions. The classic mechanism for a base catalyzed aldol condensation involves the base abstracting the weakly acidic proton from the C_α–H bond of an alkanal or alkanone to form an enolate intermediate that subsequently attacks the electrophilic carbon on the carbonyl group of a vicinal alkanal or alkanone to form a β-hydroxy alkanal (aldol) as shown in Scheme 3A. The base can then dehydrate the aldol to form an α,β-unsaturated alkanal. Esterification proceeds *via* the nucleophilic attack of an alkoxide at the electrophilic carbon of the carbonyl group on the alkanal which then deprotonates to form an ester product, as shown in Scheme 3B.

While base-catalyzed aldol condensation and esterification reactions are well-established, neither Cu or SiO_2 exhibit basic sites; yet Cu/SiO_2 catalyzes both

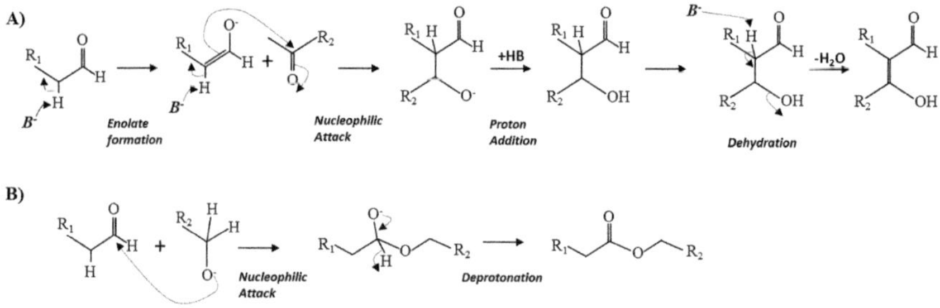

Scheme 3 Mechanism of base-catalyzed (A) aldol condensation and (B) esterification.

reactions.[1] The reaction instead has been proposed to proceed *via* the formation of a negatively-charged and basic propoxide intermediate on Cu, derived *in situ* from the interconversion of the equilibrated propanol–propanal–H_2 mixture. Alkoxides are electronically very similar to hydroxides which abstract electron density from group 11 metals to form weakly bound ($HO^{\delta-}*$) intermediates.[39–41] These weakly bound surface alkoxides can act as a base, thus enabling the activation of the weakly-acidic proton at the αC position of a vicinal aldehyde to initiate aldol condensation. In addition, the basic alkoxide intermediate can also nucleophilically attack the carbonyl of a vicinal alkanal to initiate esterification. Before describing the energetics of such reactions, we first present details on the electronic structure of these adsorbed alkoxide intermediates on transition metal surfaces.

3.2.2 Basic nature of RO* on Cu. Alkoxide anions (RO^-) in solution are bases that can readily abstract protons and carry out nucleophilic attack.[42] The properties and behavior of a bound alkoxide (RO*), on the other hand, is controlled by the electronic properties of the metal surface and the binding of the alkoxide to the metal, both of which are dictated by the direction and degree of charge transfer between the alkoxide and the metal. For noble metals such as Cu, Ag and Au, charge is transferred from the metal to the alkoxide. The charge transfer may be sufficient to allow for the formation of an anionic alkoxide intermediate, which would exhibit basicity comparable to solvated alkoxide anions. DFT-calculated electron densities for propoxide and propanal adsorbed at a 3-fold fcc site on Cu(111) are summarized in Table 1. Propoxide preferentially adsorbs to a three-fold fcc Cu site on the Cu(111) surface (Fig. 2a) with its C–O axis normal to the surface, resulting in a binding energy of -218 kJ mol^{-1}. The charge density

Table 1 Charge density differences on the oxygen of the alkoxide and alcohol before and after adsorption on the Cu(111) surface[a]

Species	Before adsorption (e^-)	After adsorption (e^-)
Cu_3[b]	0.13	0.71
O of $CH_3CH_2CH_2O*$	−0.33	−0.83
O of $CH_3CH_2CH_2OH*$	−0.62	−0.62

[a] The charge difference analysis is carried out using QUAMBO. [b] The total charge of the three Cu atoms of the 3-fold site.

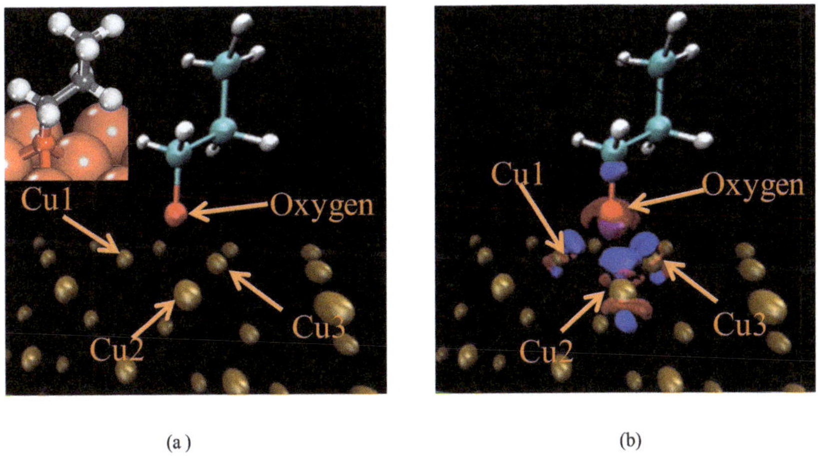

Fig. 2 (a) The adsorption structure for propoxide bound to a 3-fold fcc site on the Cu(111) surface; (b) charge density difference map between the adsorbed alkoxide and the alkoxide in the gas phase. The red isosurface depicts an increase of charge density whereas the blue isosurface reveals a decrease of charge density.

differences between the adsorbed and the separated propoxide/Cu system, depicted in Fig. 2b, show a significant shift in the electron density from the Cu to the oxygen upon adsorption. The results in Table 1 show a charge transfer of >0.5 e^- from the Cu surface to the O-atom in the bound propoxide thus resulting in a negative -0.83 e^- on the oxygen, which is consistent with anionic hydroxide and alkoxide species in solution, thus allowing it to behave as a base with properties similar to those of anionic alkoxides in solution. For comparison, we examined the charge transfer from the metal to the oxygen of molecularly adsorbed propanol. The results shown in Table 1 indicate that there is essentially no charge transfer from the metal to the propanol, as the charge on the O-atom following adsorption is less than 0.009 e^-.

The basicity of bound oxygen atoms and oxygen-containing intermediates on the group 11 (Cu, Ag, and Au) metals and other late transition metals including Pd, Pt, Rh and Ir has been discussed previously.[39–41,43–45] Hydroxide species (HO*) adsorbed to Au(111), Pt(111), or Pd(111) can readily abstract protons from water or from adsorbed alcohols during the dehydrogenation of alcohols to form aldehydes.[39–41] Adsorbed hydroxides (HO*) were also found to nucleophilically attack the carbonyl group at a vicinal alkanal to form the corresponding acid.[39–41] Herein, we examine the adsorption of the methoxide species as a simple probe alkoxide intermediate and compare the charges on the CH_3O* with those of OH*.

The charge transfer to the O of HO upon its adsorption to different 3d, 4d and 5d transition metals ranges from -0.77 e^- on the more oxophilic metals (Pd, Rh and Ru) which bind O strongly, to -1.04 to -1.07 e^- on the group 11 metals (Cu, Ag, Au) which bind O* weakly, as shown in Table S1 of the ESI.† Similarly, charge transfer from the metal to the O-atom in the adsorbed alkoxide (RO*) results in a charge on the oxygen that ranges from -0.54 on Rh and Ru to -0.83 e^- on Cu, Ag, and Au. The results indicate that both RO* and HO* species on group 11 metal (Cu, Ag and Au) surfaces act as bases as they extract electron density from the metal and become negatively charged (Table S1†). The charge transferred to the RO* and HO* species from other metals is significantly less than that from group 11 metals, likely due to the high electron density in the nearly-filled d-bands of these coinage metals (Cu: $3d^{10}$; Ag: $4d^{10}$; Au: $5d^{10}$).

3.2.3 C–H activation of propanal and propanol. Enolates, which are key intermediates in aldol condensation, form *via* the activation of the α-C–H bond of the aldehyde either by the metal surface or by an adsorbed alkoxide. $C_{2\alpha}$–H activation on Cu can proceed *via* an oxidative addition reaction where a Cu atom in the surface inserts into the C–H bond to form $CH_3CH^*CH{=}O$ and H* intermediates (eqn (5)).

$$CH_3CH_2CH_2OH^* \rightarrow CH_3CH^*CH_2OH + H^* \quad (5)$$

The weakly acidic $C_{2\alpha}$–H bond can also be activated by an adsorbed RO* intermediate which proceeds instead *via* proton abstraction by the alkoxide to form the corresponding alcohol and a surface enolate (eqn (6)).

$$CH_3CH_2CH_2OH^* + RO^* \rightarrow CH_3CH^*CH_2OH + ROH^* \quad (6)$$

The α and β carbons of the propanol and propanal refer to the carbon atoms at the C_2 and C_3 positions, respectively, as is shown in Scheme 2. The barrier to activate the C_1–H bond of propanal *via* Cu-insertion to form the adsorbed CH_3-$CH_2C^*O^*$ acyl and H* intermediates is 91 kJ mol^{-1}, which is 10 kJ mol^{-1} lower than the barrier for Cu to activate the C_2–H bond of propanal to form surface $CH_3CH^*CHO^*$ and H* (101 kJ mol^{-1}) intermediates. While the C_2–H bond is weaker than the C_1–H bond, there are greater steric limitations in activating the secondary C_2–H bond than the primary C_1–H bond over metal surface sites. The steric constraints compensate for the C–H bond energies, thus resulting in a lower barrier to activate the primary C_1–H bond.

The activation of the $C_{2\alpha}$–H can also proceed in a heterolytic manner, through a reaction in which a vicinal propoxide ($CH_3CH_2CH_2O^{(\delta-)}{}^*$) abstracts the weakly acidic α-H (with a charge of +0.166) to form propanol and an adsorbed enolate ($CH_3CH^{(-)}{}^*CHO^*$). The barrier for the heterolytic activation of the $C_{2\alpha}$–H bond is 80 kJ mol^{-1} which is 21 kJ mol^{-1} lower than that for the homolytic $C_{2\alpha}$–H activation and 52 kJ mol^{-1} lower than that for the heterolytic activation of the C_1–H, as the C_1–H hydrogen is non-acidic (with a charge of +0.041). This is consistent with previous theoretical gas phase calculations, which show that the heterolytic abstraction of the C_1–H from the propanal *via* a gas phase OH intermediate is 60 kJ mol^{-1} higher than the abstraction of the acidic $C_{2\alpha}$–H of the propanal. The activation of the terminal C_3–H bond of propanal was found to be unfavorable on the Cu surface as well as by the vicinal bound propoxide. These results are consistent with calculated gas phase deprotonation energies and the charges on the H and C atoms of propanal and propanol (reported in Table 2) which provide a measure and ranking of their acidity. This is also consistent with pK_a values reported in the literature that show pK_a values for the $C_{2\alpha}$–H of different aldehydes range from 16–20, while those for the C_3–H hydrogen range from 40–50.[46]

The preferential activation at the $C_{2\alpha}$ site *via* basic surface species is consistent with experimental results[1] that show only the formation of 2-methyl-3-pentanone,

Table 2 DFT-calculated reaction and activation energies for the activation of C–H bonds in propanal and propanol together with the gas phase deprotonation energies of the bond that is broken as well as the charges on the H and C atoms on the C–H bond that is activated

	Metal catalyzed		Alkoxide catalyzed				
C–H bond activation	ΔE_{rxn}, kJ mol^{-1}	ΔE_{act}, kJ mol^{-1}	ΔE_{rxn}, kJ mol^{-1}	ΔE_{act}, kJ mol^{-1}	DPE, kJ mol^{-1}	Charge on CH H	Charge on CH C
C_1–H of propanal	41	91	24	122	1668	0.09	+0.45
$C_{2\alpha}$–H of propanal	52	101	34	70	1556	0.23	−0.50
C_3–H of propanal	112	166	91	162	1687	0.20	−0.59
C_1–H of propanol	118	144	107	166	1756	0.15	−0.02
C_2–H of propanol	117	166	104	155	1656	0.20	−0.41
C_3–H of propanol	129	174	117	163	1760	0.19	−0.58

2-methyl-3-pentanal and 3-pentanone products which form only *via* the reactions between a $C_{2\alpha}$ enolate and a vicinal bound propanal. The detailed elementary steps to form these products are reported in the next section. The formation of the linear 1-hexanal and 3-hexanone products which would form *via* the attack of $C_{3\beta}$ enolate on a vicinally bound propanal were notably absent.

The activation of each type of C–H bond in 1-propanol was also examined. The C_1–H bond is the most reactive on Cu as the –OH group withdraws electron-density from the C–H bonds at the terminal C-atom resulting in the following inductive effect: $CH_3^{(\delta+)} \rightarrow CH_2^{(\delta++)} \rightarrow CH_2^{(\delta+++)} \rightarrow OH^{(\delta-)}$.[46] A detailed charge analysis shows that the C_1 has a partial charge of -0.02 e$^-$, while C_2 has a partial charge of -0.41 e$^-$ and the C_3 has a partial charge of -0.58 e$^-$. As such, the barriers for C–H activation over Cu(111) at these three carbons increase from 144, 166 and 174 kJ mol^{-1} for C_1–H, C_2–H and C_3–H, respectively.

The barriers for activating the C_1–H and C_2–H bonds of propanol by the adsorbed alkoxide were calculated to be significantly higher than those for activating the C_1–H and C_2–H bonds of bound propanal. This is the result of the conjugation on the carbonyl that stabilizes the negative charge that results in abstracting a proton at the C_1 and C_2 position. Propanol, however, lacks the conjugation of the C=O bond and thus offers little stabilization of the negative charge. As such the C_1–H and C_2–H bonds of the alcohol are non-acidic. The barriers to activate the C_3–H bond for both propanol and propanal are very similar to one another as the –CH=O and –CH_2OH groups are now far enough removed from the C_3–H bond to influence its properties and reactivity.

The alkoxide chain length was varied to determine its effect on the barrier to activate the αC_2–H bond in propanal. The results shown in Table 3 indicate that hydroxides are slightly more basic than methoxides on Cu surfaces and thus are more effective in the activation of the mildly acidic αC_2–H bond. This is likely due to the fact that the O-atom in OH* is more negative than the O in CH_3O* and thus more basic. For larger alkoxides, the activation barriers increase slightly with chain length, from 65 kJ mol^{-1} for methoxide to 72 and 70 kJ mol^{-1} for ethoxide and propoxide, respectively. These small increases are the result in steric hindrance between the alkoxide and aldehyde on the surface and will likely disappear with the inclusion of dispersion.

Table 3 Reaction and activation energies for the αC_2–H bond activation of propanal with different RO* species on Cu(111)

$C_{2\alpha}$–H bond activation	ΔE_{rxn}, kJ mol^{-1}	ΔE_{act}, kJ mol^{-1}
Metal catalyzed	52	101
HO* (R = H)	35	56
CH_3O^* (R = CH_3)	36	65
$CH_3CH_2O^*$ (R = CH_3CH_2)	34	72
$CH_3CH_2CH_2O^*$ (R = $CH_2CH_2CH_2$)	30	70

The theoretical analyses reported here show that C–H bond activation in propanal preferentially proceeds *via* proton-transfer to RO* bound to the Cu surface. The RO* species acts as a base and abstracts the most acidic hydrogen, which is located at the α-C-atom in propanal, to form enolate species that can subsequently react with a second adsorbed alkanal *via* aldol condensation to form a new C–C bond. The barriers to activate the C_1–H (99 kJ mol^{-1}) and C_3–H (166 kJ mol^{-1}) bonds in the aldehydes and the C_1–H (144 kJ mol^{-1}), C_2–H (166 kJ mol^{-1}) and C_3–H (174 kJ mol^{-1}) bonds in the alcohols with the bound propoxide were all significantly higher than the base-catalyzed activation of the $C_{2\alpha}$–H, as all of the other hydrogens are non-acidic and as such will not be activated by a base.

3.3 Aldol condensation paths on Cu

Base-catalyzed aldol condensation of two C_n alkanals typically proceeds through the activation of the first alkanal to form an enolate. The enolate subsequently attacks the carbonyl in the vicinal-bound alkanal to form a β-alkoxy-alkanal intermediate that rapidly protonates to form the C_{2n} β-hydroxy alkanal (aldol) as shown in the first step in Scheme 4. The aldol can readily dehydrate to form an α,β-unsaturated alkanal (a conjugated enone). In the propanal reactions examined here, the C_{2n} aldol (3-hydroxy-2-methyl-pentanal) dehydrates to form 2-methyl-2-pentenal (α,β-unsaturated alkanal), that can readily hydrogenate resulting in the observed 2-methyl-2-pentanal product, as shown in the upper

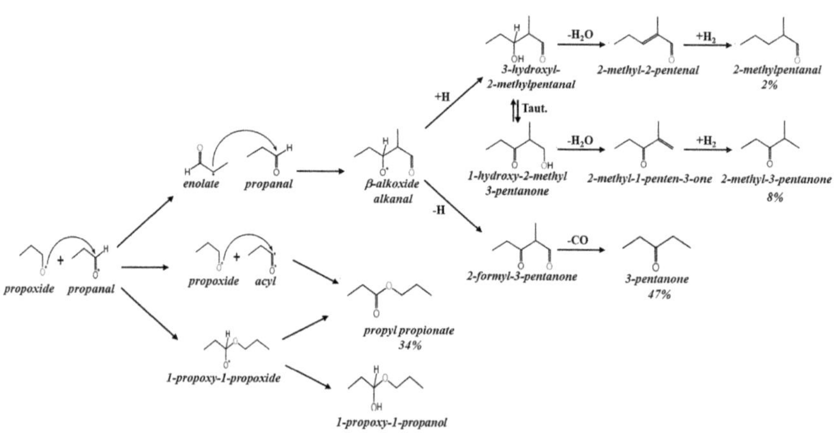

Scheme 4 Possible reaction pathways following aldol condensation.

branch in Scheme 4. The 2-methyl-2-pentanal product, however, only accounts for 2% of the products observed and neither the aldol (3-hydroxyl-2-methylpentanal) or the conjugated enone (2-methyl-2-pentenal) products were observed experimentally.[1] As such, this path is only a minor route for condensation. The second condensation product observed, 2-methyl-3-pentanone (8% selectivity) forms *via* tautomerization or intra-molecular hydrogen transfer reactions that convert the aldol (3-hydroxy-2-methylpentanal) to the β-hydroxy ketone (1-hydroxy-2-methyl-3-pentanone) which subsequently dehydrates and then hydrogenates to form 2-methyl 3-pentanone that desorbs from the surface.[1] This is the path in the center of Scheme 4.

The predominant condensation product is 3-pentanone which makes up ~82.5% of all of the condensation products formed and ~47% of the total products which include those from esterification as well as direct dehydration. The prevalence of the 3-pentanone product is consistent with previous studies that indicate that the most favorable coupling products are C_{2n-1} ketones.[47,48] This suggests that aldol condensation is followed by rapid decarbonylation or decarboxylation where oxygen is removed as either CO or CO_2. These decarbonylation/decarboxylation reactions can proceed *via* an intra or inter (*via* reactions with a vicinal alkanal) molecular hydrogen transfer to form the β-diketo intermediate (2-formyl-3-pentanone). The β-diketo can subsequently react *via* a retro-aldol reaction to eliminate CO and form 3-pentanone (shown as the lower condensation path in Scheme 4).

The first step in aldol condensation, which involves the propoxide-catalyzed activation of the weakly acidic C_α–H bond of propanal to form the enolate, was calculated to have an intrinsic activation barrier of 70 kJ mol^{-1} (taken with respect to the adsorbed propoxide and propanal) (Fig. 3A and Table 2). The enolate subsequently reacts with a vicinal bound alkanal to form the β-alkoxide alkanal surface intermediate with a barrier of only 19 kJ mol^{-1} and an overall reaction energy of −26 kJ mol^{-1} (Fig. 3B). The structures of the reactant, transition and product states for all of the initial aldol condensation steps are shown in Fig. 3.

3.3.1 Formation of 2-methyl-2-pentanal. The β-alkoxy-alkanal (β-alkoxy-2-methyl pentanal) intermediate formed *via* the C_3-enolate attack on the propanal can hydrogenate to give the 3-hydroxy-2-methyl-pentenal which can subsequently dehydrate and then hydrogenate to form the 2-methyl-2-pentanal product shown in the upper path in Scheme 4. This sequence proceeds by the protonation of the β-alkoxide alkanal as shown in Fig. 3C and has an intrinsic activation barrier of 69 kJ mol^{-1} calculated with respect to the adsorbed β-alkoxide and H*. While the subsequent dehydration and hydrogenation steps to form the minor 2-methyl-2-pentenal product readily proceed, they were not examined herein. The tautomerization and subsequent dehydration and hydrogenation of the 1-hydroxy-2-methyl 3-pentanone to form the 2-methyl-3-pentanone product (center condensation path in Scheme 4) were also not examined.

3.3.2 Formation of 3-pentanone. The predominant aldol product as discussed above is 3-pentanone, which proceeds *via* dehydrogenation of the β-alkoxide alkanal and its subsequent decarbonylation (bottom condensation path in Scheme 4). The C–H activation of the β-alkoxide alkanal over Cu (Fig. 3D) was calculated to be prohibitive with an intrinsic barrier of 105 kJ mol^{-1}.

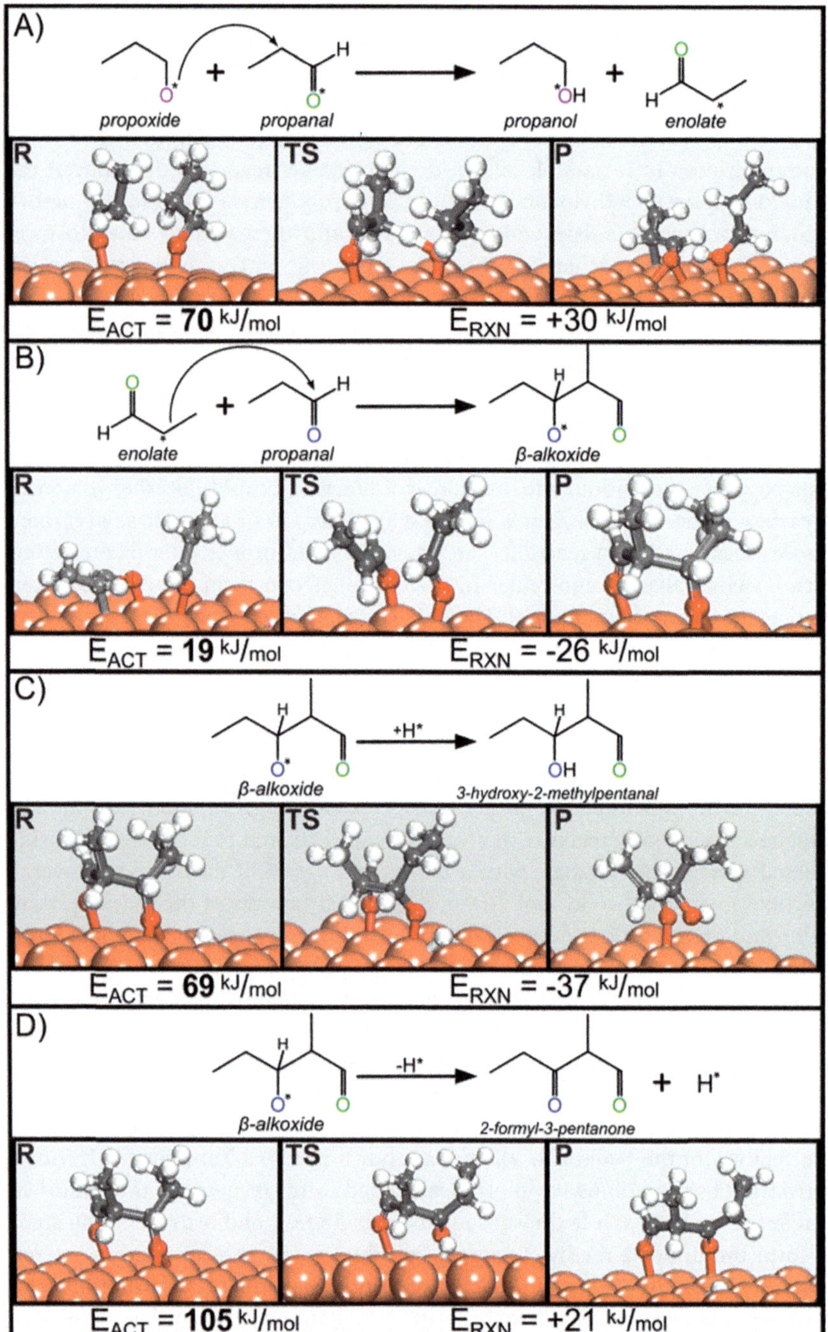

Fig. 3 DFT-calculated reactant, transition and product states along with the intrinsic activation barriers and the overall reaction energies for elementary steps in the aldol condensation of adsorbed propoxide and propanal over Cu(111) to form the 3-hydroxy-2-methylpentanal and 2-formyl-3-pentanone intermediates. These include: (A) activation of the weakly acidic $C_{2\alpha}$–H of propanal by a surface propoxide to form a surface enolate, (B) enolate attack on the C=O bond of a vicinal propanal to form β-alkoxide alkanal, (C) protonation of the β-alkoxide alkanal to form the aldol, and (D) C–H activation of β-alkoxide alkanal to form 2-formyl-3-pentanone.

The reaction instead appears to proceed *via* intermolecular hydrogen transfer. The C_1–H hydrogen of the alkoxide is hydridic in character, and as such, it can be readily abstracted by a vicinal-bound electrophile such as the bound propanal. The barrier to activate the C_1–H bond of the β-alkoxide alkanal by a coadsorbed propanal was calculated to be 74 kJ mol^{-1} ($\Delta E_{rxn} = 0$ kJ mol^{-1}), as shown in Fig. 4A. The intrinsic barrier for the propanal activation of the β-alkoxide alkanal to form 2-formyl-3-pentanone is 31 kJ mol^{-1} lower than that calculated for the Cu C_1–H activation (105 kJ mol^{-1}). The adsorbed Cu–O=CHR adduct behaves as a Lewis acid that assists H-transfer.

The surface enone (2-formyl-3-pentanone) can subsequently decarbonylate *via* a direct retro-aldol CO elimination from the C_1-alkoxide alkanal intermediate as shown in Scheme 4. The barrier for this path, however, was calculated to be >110 kJ mol^{-1}. Decarbonylation can instead proceed *via* the initial activation of the C_1–H bond of the 2-formyl-3-pentanone intermediate to form the surface 2-acyl-3-pentanone ($CH_3CH_2C(O)CH(CH_3)CO^*$) intermediate with an intrinsic activation barrier of +74 kJ mol^{-1} and an overall reaction energy of +42 kJ mol^{-1} (see Fig. 4B). This C_1–H activation subsequently weakens the C_1–C_2 bond thus allowing for direct C=C bond scission over Cu to form 3-pentene-2-olate ($CH_3CH_2C(O^*)CHCH_3$) and CO* products with an intrinsic activation energy of only +7 kJ mol^{-1} and an overall reaction energy of −40 kJ mol^{-1} (Fig. 4C). The resulting 3-pentene-2-olate ($CH_3CH_2C(O^*)CHCH_3$) intermediate subsequently hydrogenates to form 3-pentanone with a barrier of +68 kJ mol^{-1} and a reaction energy of −60 kJ mol^{-1} (Fig. 4D) which then desorbs from the surface.

The final aldol condensation path considered is rather different than the first two paths in that it proceeds *via* the initial C–H activation of the alkanal to form an acyl (RCO*) intermediate, that subsequently attacks a vicinal bound propanal to form the aldol product directly.[27] This path was found to be unlikely as the intrinsic activation barriers for the initial C–H activation and subsequent C–C bond formation steps were calculated to be very high at 133 and 153 kJ mol^{-1}, respectively.

The activation barriers and reaction energies for the elementary steps presented in Fig. 3 and 4 above were used to construct overall reaction energy diagrams for different pathways, determine rate expressions for each path and establish the lowest energy route. The results are presented in detail in ESI S.4.† In all paths, we assume that propanal and hydrogen are equilibrated with propanol (see Section 3.2). For simplicity, all of the paths are therefore referenced to the same adsorbed propanal and propoxide initial state, defined as $E = 0$.

The lowest energy paths which are shown in Fig. 5 proceed *via* the coupling of propoxide and propanal surface intermediates to form the 2-methyl-pentanal and 2-methyl-3-pentanone products. The condensation rates are ultimately dictated by the barrier to form the enolate (70 kJ mol^{-1}) *via* the abstraction of the weakly acidic $C_{2\alpha}$–H hydrogen by a vicinal propoxide, which is the highest point along the overall energy path taken with respect to adsorbed propanal and propoxide, the most abundant surface intermediates.

The enolate can readily attack the carbonyl of a vicinal propanal to form the β-hydroxide alkanal that subsequently reacts by: (1) H* addition to form the aldol that then reacts to form 2-methyl-pentanal and 2-methyl-3-pentanone products,

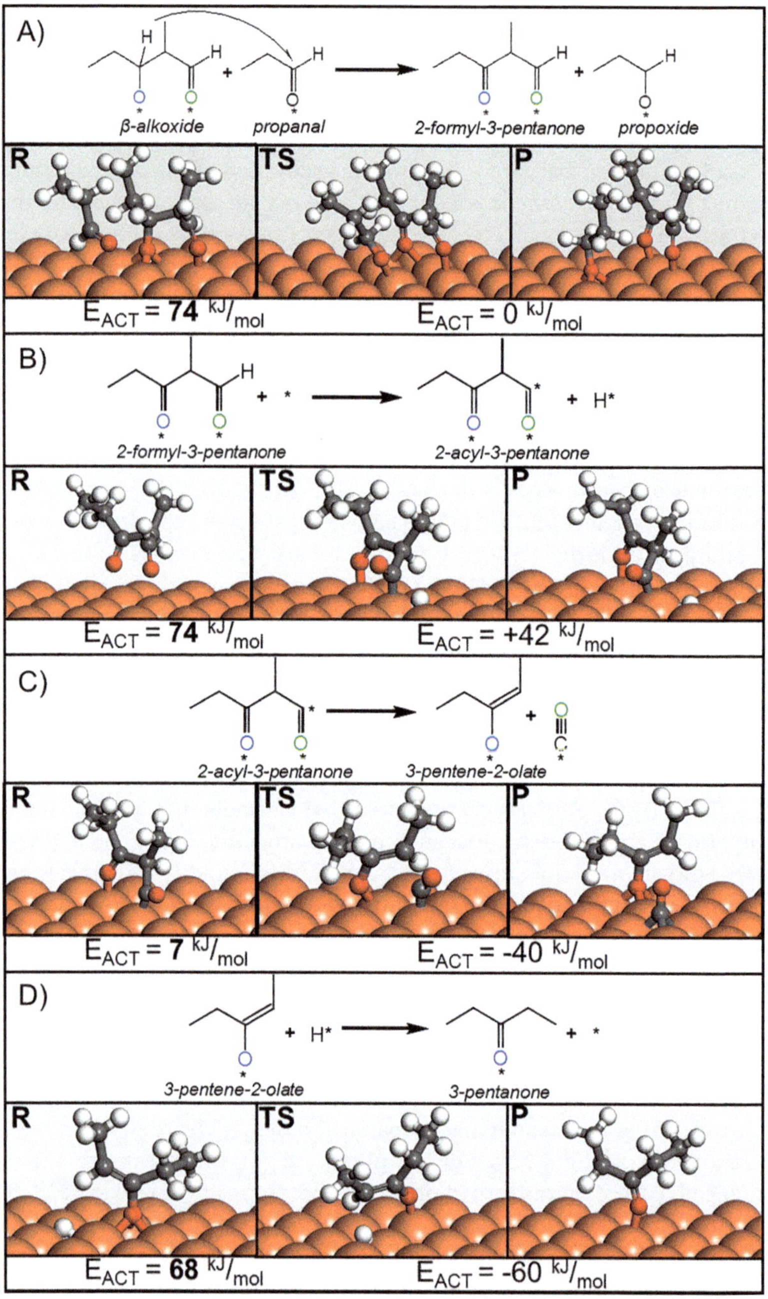

Fig. 4 DFT-calculated reactant, transition and product states along with the intrinsic activation barriers and the overall reaction energies for elementary steps in the conversion of the dehydrogenated aldol intermediate, β-alkoxy-2-methyl pentanal, to 3-pentanone over Cu(111). The steps include: (A) intermolecular hydride transfer from β-alkoxide alkanal to adsorbed propanal to form the 2-formyl-3-pentanone intermediate, (B) C_1–H scission of the 2-formyl-3-pentanone to form 2-acyl-3-pentanone, (C) decarbonylation of 2-acyl-3-pentanone to form 3-pentene-2-olate and CO and (D) hydrogenation of 3-pentene-2-olate to form 3-pentanone.

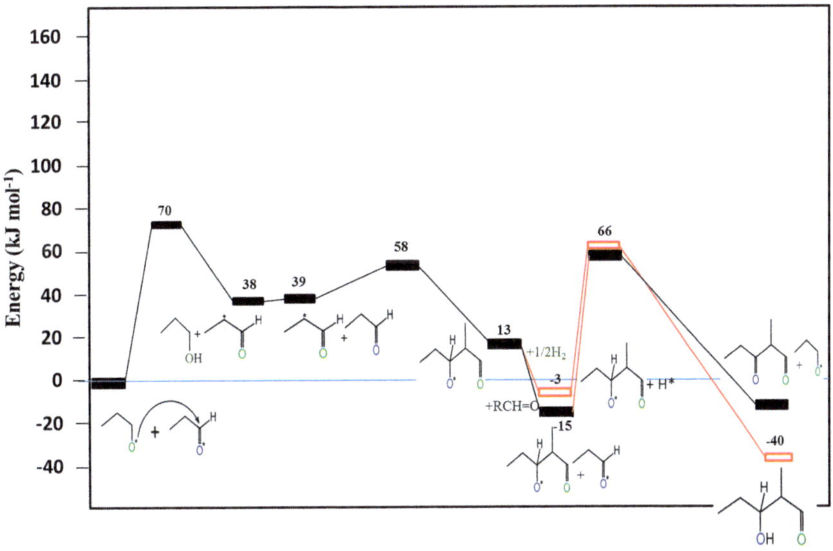

Fig. 5 The lowest energy DFT-calculated pathways for the aldol condensation reactions of propanol–propanal on Cu(111) proceed *via* the coupling of the enolate and propanal to form the β-alkoxide alkanal which can: (i) hydrogenate (shown with unfilled rectangular symbols) to form the aldol or (ii) dehydrogenate *via* hydride transfer from an adsorbed propanal to form 2-formyl-3-pentanone (shown with filled rectangular symbols).

or (2) hydride abstraction by a vicinal propanal to form a 2-formyl-3-pentanone intermediate which subsequently undergoes C–H activation and decarbonylates over Cu to form 3-pentanone (Fig. 6).

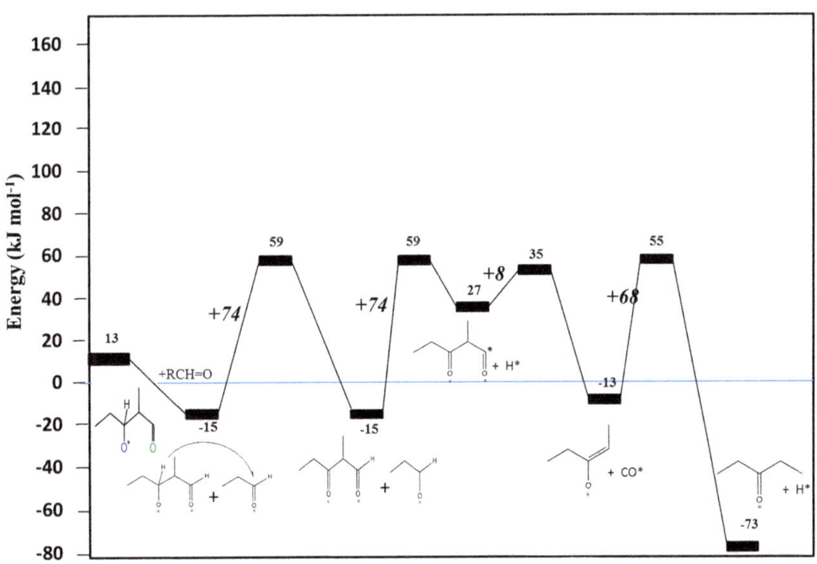

Fig. 6 DFT-computed pathways for the decarbonylation of the 1-propoxy-1-propoxide intermediate to form 3-pentanone on Cu(111). Activation barriers for each step are shown in larger font and in italic.

The rate for both propoxide and propanal coupling reactions is controlled by the initial activation of the weakly acidic $C_{2\alpha}$–H bond on the alkanal by a basic propoxide surface intermediate and can be written as:

$$\begin{aligned} r_{\text{cond}} &= k_{\text{cond}}[\text{propanal*}][\text{propoxide*}] \\ &= k_{\text{cond}} K_{\text{PAL}}{}^2 K_{\text{HYD}_1} K_{\text{H}_2}{}^{1/2} P_{\text{PAL}}{}^2 P_{\text{H}_2}{}^{1/2} / [1 + K_{\text{PAL}} P_{\text{PAL}} \\ &\quad + K_{\text{PAL}} K_{\text{HYD}} K_{\text{H}_2}{}^{1/2} P_{\text{PAL}} P_{\text{H}_2}{}^{1/2}]^2 \end{aligned} \tag{7}$$

where k_{cond} refers to the intrinsic rate constant for deprotonation of the weakly acidic $C_{2\alpha}$–H bond; K_{PAL}, K_{HYD_1}, and K_{H_2} are the equilibrium constants for the adsorption of propanal, the hydrogenation of propanal to the surface propoxide over Cu and the dissociative adsorption of hydrogen, respectively, and P_{PAL} and P_{H_2} refer to the partial pressure of propanal and hydrogen, respectively. The apparent activation barrier for this path calculated with respect to the adsorbed propanal and propoxide is simply the 70 kJ mol^{-1} barrier for the initial $C_{2\alpha}$–H activation by the propoxide to form the enolate (Fig. 3A). The theoretical results suggest that propanal and propoxide are the abundant surface intermediates. Experimentally-measured rates of reaction as a function of the pressures of propanal and hydrogen indicate that while both propanal and propoxide exist on the surface, the propanal is the most abundant surface intermediate.[1]

3.4 Esterification paths on Cu

Base-catalyzed esterification reactions typically proceed *via* Tischenko-type pathways involving the reaction between an adsorbed alcohol and aldehyde. Three potential paths by which the surface alkoxide and alkanal species can react over Cu are shown in Scheme 5. The first path (E_1 in Scheme 5) proceeds *via* nucleophilic addition of the surface alkoxide to the vicinal alkanal to form

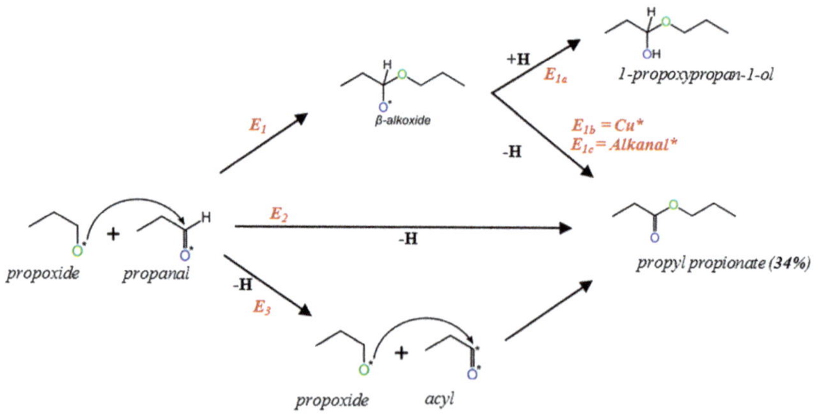

Scheme 5 Three possible reaction paths for the esterification paths involving C–O bond formation *via* the coupling of propanal and propoxide intermediates over Cu. E_1 paths proceed by sequential C–O formation followed by hydrogen addition (E_{1a}) or hydrogen elimination *via* Cu (E_{1b}) or *via* an adsorbed alkanal (E_{1c}). The E_2 path follows simultaneous C–O formation and H-elimination. E_3 proceeds *via* the initial activation of the propoxide to form an acyl intermediate.

a hemiacetalate (1-propoxy-1-propoxide) surface intermediate. The reaction proceeds by an initial shift of the alkoxide from its stable 3-fold adsorption site to a bridging site where it readily attacks the carbonyl C of the neighboring propanal intermediate to form the C–O bond, resulting in an intrinsic barrier of 60 kJ mol^{-1}. A significant portion of the activation barrier is associated with the incipient activation of the strong Cu–OC_3H_7 bond to form the more weakly bound and reactive propoxide. The Cu–OC_3H_7 weakening is compensated by the formation of a strong Cu–$OCH(OCH_2CH_2CH_3)(CH_2CH_3)$ bond in the resulting 1-propoxy-1-propoxide intermediate, as shown in Fig. 7A. The 1-propoxy-1-propoxide can subsequently dehydrogenate either by Cu-catalyzed C_1–H bond cleavage to form the propyl propionate product and a surface H* (path E_{1b} in Scheme 5) or *via* intermolecular hydride transfer from the C_1–H bond of the 1-propyl-1-propoxide to a vicinal propanal species (E_{1c}) thus resulting in the formation of propyl propionate and a surface propoxide. The Cu-catalyzed C_1–H activation of the hemiacetalate (1-propoxy-1-propoxide) was found to be rather difficult with a barrier of 90 kJ mol^{-1} (as shown in Fig. 7B) which is nearly identical to the barrier for propoxide C_1–H activation (91 kJ mol^{-1}) reported in Table 2. The propanal-catalyzed hydride transfer occurs *via* a Meerwein–Ponndorf–Verley (MPV) type mechanism as shown in Fig. 7C resulting in a barrier of only 31 kJ mol^{-1}, which is 59 kJ mol^{-1} lower than the barrier to activate the C_1–H bond over Cu. The lower barrier for propanal-catalyzed hydride transfer compared to that for Cu-catalyzed C–H activation is consistent with the results for the propanal- and Cu-catalyzed C_1–H activation of the β-alkoxide alkanal discussed in Section 3.3 for aldol condensation. The hydride-transfer barrier from the 1-propyl-propoxide in the esterification reaction, however, is 43 kJ mol^{-1} lower than that from the β-alkoxide alkanal in the condensation path, as the oxygen in the ester significantly increases the hydricity of the C_1–H bond by stabilizing the charge in the transition state.

In addition to dehydrogenation, the 1-propoxy-1-propoxide can also hydrogenate *via* protonation of the alkoxide to form the 1-propoxy-1-propanol hemiacetal (path E_{1a} in Scheme 5) resulting in a barrier of 104 kJ mol^{-1} (Fig. 7D). The 1-propoxy-1-propanol can desorb from the surface or dehydrogenate to form a propyl propionate product.

The surface propoxide and propanal can also react together in a second path (path E_2 shown in Scheme 5) that proceeds *via* a concerted S_N2 type reaction involving the nucleophilic attack of the surface propoxide on the carbonyl of the surface propanal (C–O bond formation) together with the simultaneous elimination of C_1 propoxyl hydrogen to the Cu surface to form propyl propionate directly. The transition state for this concerted path (Fig. 7E) involves the elongation of the C–O and C=O bonds in the ester (1.39 Å and 1.28 Å, respectively) relative to their product state (1.34 Å and 1.24 Å), which helps to assist the activation of the propoxyl C_1–H bond. The intrinsic barrier for this reaction taken with respect to the adsorbed propanal and propoxide was calculated to be 69 kJ mol^{-1} (Fig. 7E), which is slightly higher than that for the direct coupling of surface propanal and propoxide, thus making it a viable path for esterification and the direct formation of propyl propionate.

The third and final path to form propyl propionate is characteristically different than the first two paths in that it proceeds by initially activating the C_1–H bond of the alkanal to form an acyl intermediate. The acyl subsequently reacts

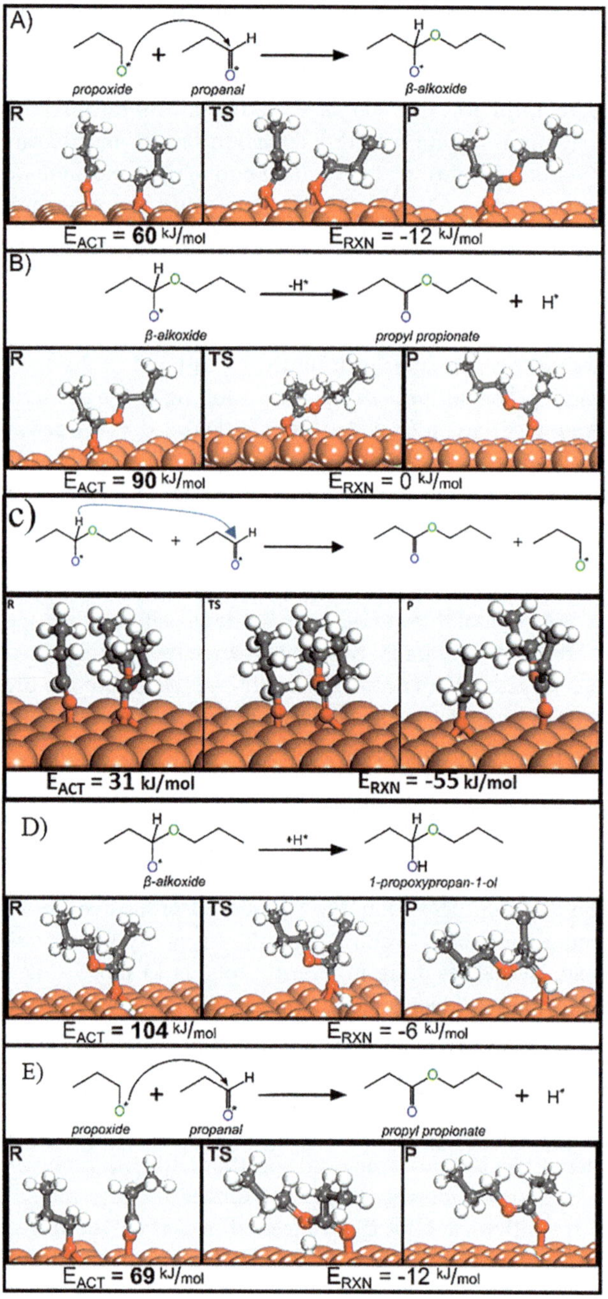

Fig. 7 DFT-calculated reactant, transition state and product state structures and the corresponding activation barriers and overall reaction energies for different elementary steps in the esterification of propanal and propanol over Cu(111) surface to form propyl propionate. The steps include: (A) the nucleophilic attack of the adsorbed propoxide on the adsorbed propanal (C–O bond formation), (B) C_1–H activation by a surface Cu site, (C) intermolecular hydride transfer from the C_1–H of the 1-propoxy-1-propoxide intermediate to a vicinal bound propanal, (D) hydrogenation of the 1-propoxy-1-propoxide to form the 1-propoxy-1-propanol hemiacetal, and (E) concerted S_N2-type nucleophilic attack of the adsorbed propoxide on the adsorbed propanal and C_1–H hydrogen elimination to Cu.

with a surface alkoxide to form the ester product directly (path E_3 in Scheme 5).[1] The intrinsic barrier to activate the alkanal to form the surface acyl intermediate is rather high at 101 kJ mol^{-1}. The acyl subsequently couples with a co-adsorbed propoxide with an intrinsic barrier of 81 kJ mol^{-1}, which is significantly higher than the coupling barrier between propanal and propoxide (53 kJ mol^{-1}). This does not appear to be a viable path to the esterification products that form.

The elementary steps and corresponding activation and reaction energies for the different esterification paths presented above are discussed in detail in ESI S.5.† The lowest energy routes in Table 1 were used to construct the two viable energy paths presented in Fig. 8 and the operative kinetic expressions for esterification. For simplicity, the paths presented in Fig. 8 are referenced to adsorbed propanal and propoxide as these species rapidly equilibrate, as was discussed in Section 3.2.

Both paths presented in Scheme 5 (and in Fig. 8) proceed *via* reactions involving surface bound propoxide and propanal intermediates, as shown in paths E_{1c} and E_2. These paths are similar in nature to those reported previously for methanol[49,50] and ethanol[51] esterification over Au. The results in Fig. 8 indicate that both paths proceed in a very similar manner *via* the coupling of the surface propoxide and propanal in the rate controlling step. The E_{1c} path involves sequential C–O formation and H-transfer steps whereas E_2 proceeds *via* simultaneous C–O formation and H-transfer to form the hemiacetal directly.

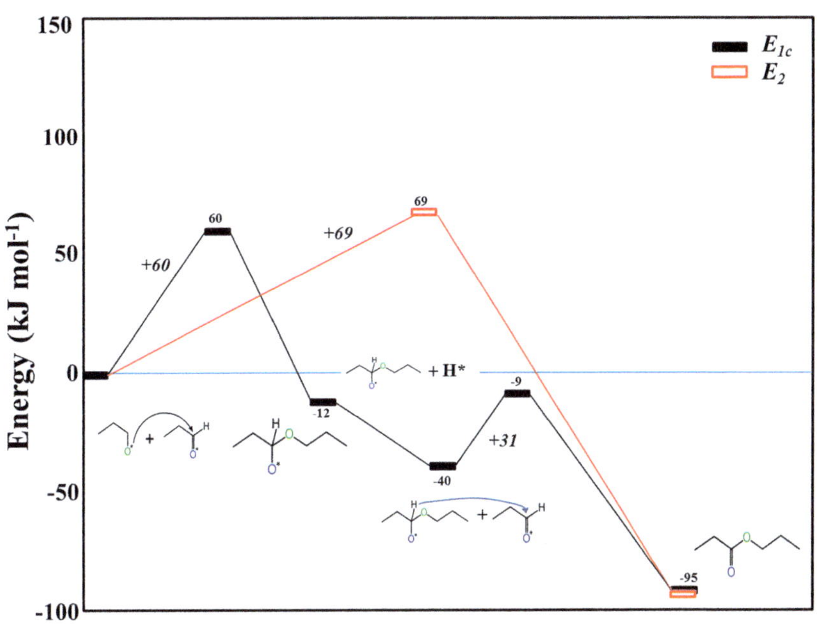

Fig. 8 DFT-calculated reaction paths for the esterification of propanol–propanal on Cu(111). The E_{1c} path (filled rectangular symbols) proceeds *via* the coupling of a surface alkoxide and surface alkanal to form the β-alkoxide hemiacetalate species which subsequently reacts with a second vicinal propanal to eliminate H to form the propyl propionate. The E_2 path (unfilled rectangular symbols) proceeds *via* a concerted coupling of the surface propyl and propanal intermediates together with the C_1–H activation by Cu.

The rate equations for these paths are identical and can be written as:

$$r_E = k_E[\text{propanal*}][\text{propoxide*}] = k_E K_{PAL}{}^2 K_{HYD_1} K_{H_2}{}^{1/2} P_{PAL}{}^2 P_{H_2}{}^{1/2} / [1 + K_{PAL} P_{PAL} + K_{PAL} K_{HYD_1} K_{H_2}{}^{1/2} P_{PAL} P_{H_2}{}^{1/2}]^2 \quad (8)$$

where k_E is the intrinsic rate constant for the coupling of surface propoxide and propanal for paths E_{1c} and E_2; K_{PAL}, K_{HYD_1} and K_{H_2} are the equilibrium constants for the adsorption of propanal, hydrogenation of adsorbed propanal to the adsorbed propoxide, and the dissociative adsorption of hydrogen, respectively, and P_{PAL} and P_{H_2} refer to the pressure of propanal and hydrogen, respectively.

Experimental results indicate that the propanal is the most abundant surface intermediate. As such the rate expression in eqn (8) would reduce to reduce to:

$$r_E = k_E K_{HYD_1} K_{H_2}{}^{1/2} P_{H_2}{}^{1/2} \quad (9)$$

The equilibrium constants for the surface and gas phase hydrogenation of propanal to propanal, $K_{H_2}{}^{1/2}$ and K_{HYD_1}, are directly included in the E_{1c} and E_2 barriers as they were measured with respect to the propanal and propoxide directly. The apparent activation barriers for the E_{1c}, and E_2 paths were therefore calculated to be 60 and 69 mol^{-1}, respectively and as such, both are viable paths for esterification.

The results reported here indicate that the ester is formed either by an E_{1c} path which involves a sequential mechanism where the rate-controlling C–O formation step, involving the coupling of a surface propoxide and propanal, precedes a rapid hydride transfer from the alkoxide C–H bond to the vicinal bound propanal or by an E_2 path which proceeds *via* the concerted coupling of a surface propoxide with coadsorbed propanal and direct hydrogen elimination to the surface. Both of these paths result in rate expressions and propanal and hydrogen dependencies that match those reported experimentally.[1]

3.5 Comparing free energies for condensation and esterification

In order to appropriately compare the rates of condensation and esterification we calculated the enthalpies, entropies and the free energies for the most favorable condensation and esterification paths. The activation energies reported earlier for condensation (70 kJ mol^{-1}) and esterification (60 kJ mol^{-1} for the E_{1c} and 69 kJ mol^{-1} for the E_2 paths) were calculated at 0 K. These values were subsequently corrected for zero point energies as well as changes that result in specific heats in moving from 0 to 298 K. The resulting activation enthalpies at 298 K for condensation (66 kJ mol^{-1}) and esterification (60 kJ mol^{-1}) were used together with the activation entropies reported in Table 4 to determine free energy barriers of 69 kJ mol^{-1} and 66 kJ mol^{-1} for the condensation and esterification reactions, respectively. The difference between the condensation and esterification free energy barriers is only 3 kJ mol^{-1} with the esterification being slightly favored.

Sad *et al.*[1] showed that rates of esterification and condensation follow identical dependencies on the partial pressures of propanal as well as propanol and hydrogen, as is shown in Fig. S5 in the ESI.† The ratio of the rate of condensation to the rate of esterification was found to be constant over a range of propanal

Table 4 Comparison of the activation enthalpies, entropies and free energies for the condensation and esterification of propanal and propoxide species on the Cu(111) surface

	Condensation	Esterification
$\Delta H^{\neq}$ (kJ mol^{-1})	66	60
$\Delta S^{\neq}$ (kJ (mol K)$^{-1}$)	−0.009	−0.013
$\Delta G^{\neq}$ (kJ mol^{-1})	69	66

pressures as well as propanol and hydrogen pressures as is shown in Fig. S5.† This suggests that rates of condensation and esterification obey similar rate equations but contain different rate constants,[1] indicating that their respective kinetically-relevant steps involve the same adsorbed propanal ($CH_3CH_2CHO^*$) and surface propoxide ($CH_3CH_2CH_2O^*$) intermediates. The two steps are limited by kinetically-relevant propoxide–propanal reactions mediated by bimolecular transition states, but lead to different products. This is fully consistent with the theoretical results discussed above and summarized in Fig. 9 that show that the rate-controlling step for condensation involves the abstraction of the weakly acidic H at the α-position of propanal by basic propoxide to form the enolate intermediate whereas that for esterification involves nucleophilic addition of the basic propoxide to bound propanal.

In condensation, the enolate species that forms in the rate-limiting step can readily attack a vicinal propanal to form a β-alkoxide alkanal which can undergo facile inter- or intra-molecular hydride transfer to form the aldol species that can subsequently dehydrogenate and decarbonylate to form 3-pentanone (shown in the upper path in Fig. 9). The 1-propoxy-1-propoxide intermediate that forms in the rate controlling step for esterification can undergo facile intermolecular hydride transfer to a surface propanal (path E_{1c}

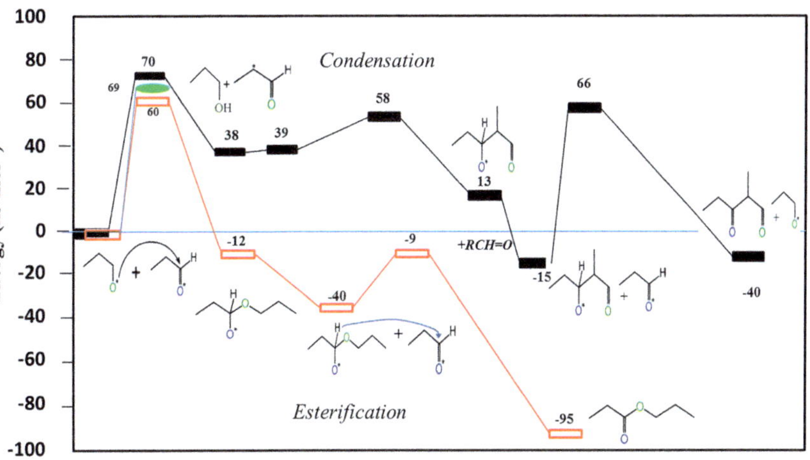

Fig. 9 A comparison between the energies for the elementary steps involved in aldol condensation and esterification reactions over Cu. The aldol condensation is shown in the top curve, denoted by dark filled rectangular symbols. Esterification can proceed either *via* the E_{1c} path involving sequential C–O formation and intermolecular hydride transfer which is shown in the unfilled rectangular symbols or *via* the E_2 concerted C–O formation H-elimination path shown in the green elliptical symbols.

which is the lower path in Fig. 9) or a concerted S_N2-like mechanism (path E_2 shown with filled oval symbol) involving simultaneous C–O formation *via* the coupling of the surface propoxide with coadsorbed propanal and direct H-elimination to form the propyl propionate product.

The ratio of the rates shown in Fig. S5† can be analyzed by writing out the ratio of the rate expressions for condensation (eqn (7)) to esterification (eqn (8)). This ratio simplifies to a ratio of rate constants for C–C bond formation (k_C) *versus* C–O bond formation (k_E):

$$\frac{r_C}{r_E} = \frac{k_C}{k_E} = \frac{A_C \exp(-E_{aC}/RT)}{A_E \exp(-E_{aE}/RT)} = \frac{A_C}{A_E} \exp(-(E_{aC} - E_{aE})/RT) \qquad (10)$$

where r_C, r_E, E_{aC}, E_{aE}, A_C, and A_E are the rates, activation energies and pre-exponential factors respectively and the subscripts C and E denote condensation and esterification routes.

The experimental rate data reported by Sad *et al.* (shown in Fig. S4†) indicate that condensation rates are about two-times larger than those for esterification at all pressures at 503 K.[43] This corresponds to an esterification barrier that is ~3 kJ mol^{-1} higher than that for condensation (*i.e.* $E_{aC} - E_{aE} = -3$ kJ mol^{-1}), provided that both reactions have similar pre-exponential factors. Theoretical treatments (Fig. 9) show that the barriers for aldol condensation and esterification are similar, with a condensation barrier that is 3 kJ mol^{-1} higher than that for esterification ($E_a^{Cond} - E_a^{Est} = 3$ kJ mol^{-1}). Although the sign is different between experimental and theoretical estimations, the difference between these two is well within the accuracy of density functional theory of ±5 kJ mol^{-1}.[52]

Previous experimental results show that Cu can uniquely catalyze condensation as well as esterification reactions. Such reactions are typically catalyzed by acids and bases for reactions that are carried out over metal oxides. The theoretical and experimental results presented here show that the exposed Cu metal surface sites act as Lewis acid sites that readily bind the oxygen of alkanal, alkanol and alkoxide intermediates and work cooperatively with basic sites formed *via* the *in situ* generation of propoxide intermediates to create acid–base site pairs that can rapidly equilibrate alkanal/hydrogen/alkanol mixtures and catalyze condensation and esterification reactions.

Alkoxy intermediates on Cu as well as other group 11 metals (Au and Ag) are more weakly bound to the metal than those on other transition metal surfaces as a result of significant Pauli repulsion with the nearly filled d-band of Cu and other group 11 metals. They act to abstract electron density from the metal, thus creating negatively charged alkoxides that behave as a base and work together with the Lewis acid sites to form acid–base site pairs that can readily activate acidic C–H and O–H bonds and carry out nucleophilic attack.

The nearly filled nature of the d-band of Cu, as well as other group 11 metals, limits their ability to readily activate C–H bonds alone. The Cu surface atoms, however, can act as Lewis acid sites and bind to the O of the alkanal thus creating an enol-like surface intermediate that can undergo electrophilic additions, readily accept hydridic H atoms and catalyze the activation of C–H bonds. These adsorbed alkanals are electron acceptors and thus behave as weak Lewis acids. They work closely with the basic alkoxide sites of the surface to provide acid–base site pairs, similar to those on metal oxides such as TiO_2,[53,54] and ZrO_2 (ref. 55–57) that can readily catalyze aldol condensation and esterification reactions. The cooperative influence of weak Lewis acid–base pairs for these reactions and others was pioneered by Tanabe.[58,59] Cu is unique in that it selectively carries out decarbonylation along with C–C bond formation. This is likely due to the fact that Cu does not form Brønsted acid sites upon the deprotonation of C–H or O–H bonds but instead delivers the proton to weakly held alkoxide intermediates to produce non-acidic alcohols. The Brønsted acid sites that form metal oxide catalysts, on the other hand, readily promote hemiacetal formation and in addition catalyze dehydration reactions which lead to the hemiacetal and deoxygenated alkanal and alcohol products.

4. Conclusions

First principles density functional theory calculations demonstrate that Cu can catalyze both aldol condensation and esterification reactions without the addition of basic oxide promoters or a basic oxide support. This chemistry is facilitated by the *in situ* production of adsorbed alkoxide species which act as a base as well as an active nucleophile, effectively replacing the role of the basic oxide support. In addition, the adsorbed alkanal species weakly bind to Cu to form enolic-type Lewis acid sites that catalyze the activation of hydridic C_1–H bonds for various different reactive intermediates. Propanal, propanol and H_2 readily react to form an equilibrated mixture that subsequently undergoes aldol condensation and esterification reactions. The primary product from aldol condensation is 3-pentanone which is formed *via* C–C coupling followed by decarbonylation. 2-methyl-pentanal and 2-methyl-3-pentanone, which are the predominant products over typical acid and base catalysts as a result of subsequent dehydration steps, are only produced in minor amounts (<10%), indicating that decarbonylation is facile on these Cu/SiO_2 catalysts. Cu was also found to catalyze the C–O bond formation for the reactions between propanol and surface propionate, resulting in the formation of propyl propionate as a second primary product.

The alkoxide that forms from the interconversion of propanal + H_2 and propanol mixtures withdraws electron density from the Cu substrate, thus resulting in the formation of an alkoxide anion that behaves as a base that catalyzes both aldol condensation and esterification. The alkoxide can directly abstract the acidic α-H of aldehyde to form the C_3 enolate during aldol condensation and can carry out the nucleophilic attack of the carbonyl of the adsorbed aldehyde to form C–O bonds during esterification. These two steps have the highest activation barriers and appear to be the kinetically-relevant steps for the condensation or esterification paths, respectively. Since the kinetically-relevant steps for both the condensation and esterification reactions proceed *via* identical reactants (*i.e.* propanal and propoxide), both reactions result in the same rate equations and have identical kinetic dependencies on the partial pressures of propanal and hydrogen which is consistent with the experimental results. The calculated difference in the activation barriers for rate-limiting steps for esterification (C–O bond formation) and condensation (enolate formation) is small at 3 kJ mol^{-1}. This is consistent with the small differences in the activation barriers found experimentally, but condensation is favored experimentally over esterification.

Acknowledgements

The authors gratefully acknowledge BP for the financial support of this work as part of the BP-XC2 program and the computational support and resources from Minnesota Supercomputing Institute and the Molecular Science Computing Facility (MSCF) in the William R. Wiley Environmental Molecular Sciences Laboratory, a national scientific user facility sponsored by the U.S. Department of Energy, Office of Biological and Environmental Research at the Pacific Northwest National Laboratory for computing resources. The authors also wish to thank Dr Maria Sad, Dr George Huff, Dr Glenn Sunley, Dr John Shabaker, Professor Jay Labinger, Professor John Bercaw and Professor Joe Topczewski for their helpful discussions.

References

1 M. E. Sad, M. Neurock and E. Iglesia, *J. Am. Chem. Soc.*, 2011, **133**, 20384.
2 C. A. Hamilton, S. D. Jackson and G. J. Kelly, *Appl. Catal., A*, 2004, **263**, 63.
3 J. N. Chheda and J. A. Dumesic, *Catal. Today*, 2007, **123**, 59.
4 R. M. West, Z. Y. Liu, M. Peter, C. A. Gartner and J. A. Dumesic, *J. Mol. Catal. A: Chem.*, 2008, **296**, 18.
5 T. Lowry and K. S. Richardson, *Mechanism and Theory in Organic Chemistry*, Harper and Row, NY, 1987.
6 H. Henschela, M. H. Prosencb and I. A. Nichollsa, *J. Mol. Catal. A: Chem.*, 2011, **351**, 76.
7 N. Kumagai, S. Matsunaga, N. Yoshikawa, T. Ohshima and M. Shibasaki, *Org. Lett.*, 2001, **3**, 10.
8 N. Kumagai, S. Matsunaga, T. Kinoshita, S. Harada, S. Okada, S. Sakamoto, K. Yamaguchi and M. Shibasaki, *J. Am. Chem. Soc.*, 2003, **125**, 2169.
9 R. Mahrwald and B. Ziemer, *Tetrahedron Lett.*, 2002, **43**, 4459.
10 H. Inoue, M. Kikuchi, J. Ito and H. Nishiyama, *Tetrahedron*, 2008, **64**, 493.
11 J. I. Di Cosimo, V. K. Diez and C. R. Apesteguia, *Appl. Catal., A*, 1996, **137**, 149.
12 G. Wang, Z. Zhang and Y. W. Dong, *Org. Process Res. Dev.*, 2003, **8**, 18.
13 Y. Shigemasa, K. Yokoyama, H. Sashiwa and H. Saimoto, *Tetrahedron Lett.*, 1994, **3**, 1263.
14 Z. Zhang, Y. W. Dong and G. W. Wang, *Chem. Lett.*, 2003, **32**, 966.
15 C. A. Hamilton, S. D. Jackson and G. J. Kelly, *Appl. Catal., A*, 2004, **263**, 63.
16 A. E. Palomares, G. Eder-Mirth and J. A. Lercher, *J. Catal.*, 1998, **180**, 56.
17 R. Zeng, X. Fu, C. Gong, Y. Sui, X. Ma and X. Yang, *J. Mol. Catal. A: Chem.*, 2005, **229**, 1.
18 B. M. Choudary, M. L. Kantam, P. Sreekanth, T. Bandopadhyay, F. Figueras and A. Tuel, *J. Mol. Catal. A: Chem.*, 1999, **142**, 361.
19 J. C. A. A. Roelofs, D. J. Lensveld, A. J. Dillen and K. P. de Jong, *J. Catal.*, 2001, **203**, 184.
20 M. J. Climent, A. Corma, S. Iborra and A. Velty, *J. Catal.*, 2004, **221**, 474.
21 M. J. L. Gines and E. Iglesia, *J. Catal.*, 1998, **176**, 155.
22 C. Gardner Swain, A. L. Powell, W. A. Sheppard and C. R. Morgan, *J. Am. Chem. Soc.*, 1979, **101**(13), 3576.
23 H. Berberich and P. W. Roesky, *Angew. Chem., Int. Ed.*, 1998, **37**(11), 1569.

24 (*a*) G. Kresse and J. Furthmuller, *Comput. Mater. Sci.*, 1996, **6**, 15; (*b*) J. Perdew, *Phys. Rev. B: Condens. Matter Mater. Phys.*, 1992, **46**, 6671; (*c*) D. Vanderbilt, *Phys. Rev. B: Condens. Matter Mater. Phys.*, 1990, **41**, 7892; (*d*) G. Mills, H. Jonsson and G. K. Schenter, *Surf. Sci.*, 1995, **324**, 305; (*e*) H. Jónsson, G. Mills and K. W. Jacobsen, in *Classical and quantum dynamics in condensed phase simulations*, World Scientific, Singapore, 1998; (*f*) G. Henkelman, B. P. Uberuaga and H. J. Jonsson, *J. Chem. Phys.*, 2000, **113**, 9901; (*g*) G. Henkelman and H. Jonsson, *J. Chem. Phys.*, 2000, **113**, 9978; (*h*) G. Henkelman and H. Jonsson, *J. Chem. Phys.*, 1999, **111**, 7010; (*i*) X. Qian, J. Li, L. Qi, C. Wang, T. Chan, Y. Yao, K. Ho and S. Yip, *Phys. Rev. B: Condens. Matter Mater. Phys.*, 2008, **78**, 245112.
25 S. M. Johnston, A. Mulligan, V. Dhanak and M. Kadodwala, *Surf. Sci.*, 2004, **548**, 5.
26 C. L. A. Lamont, W. Stenzel, H. Conrad and A. M. Bradshaw, *J. Electron Spectrosc. Relat. Phenom.*, 1993, **64**, 287.
27 Y. Ishikawa, M. Liao and R. C. Cabrera, *Surf. Sci.*, 2000, **463**, 66.
28 J. R. B. Gomes, J. Gomes and F. Illas, *J. Mol. Catal. A: Chem.*, 2001, **170**, 187.
29 S. K. Desai, M. Neurock and K. Kourtakis, *J. Phys. Chem. B*, 2002, **106**, 2559.
30 R. I. Masel, *Principles of Adsorption and Reaction on Solid Surfaces*, J. Wiley and Sons, New York, 1996.
31 N. K. Sinha and M. Neurock, *J. Catal.*, 2012, **295**, 31.
32 D. Cao, G. Q. Lu, A. Wieckowski, S. Wasileski and M. Neurock, *J. Phys. Chem. B*, 2005, **109**, 11622.
33 S. K. Desai and M. Neurock, *Phys. Rev. B: Condens. Matter Mater. Phys.*, 2003, **68**, 075420.
34 S. K. Desai, V. Pallassana and M. Neurock, *J. Phys. Chem. B*, 2001, **105**, 9171.
35 C. Ammon, A. Bayer, G. Held, B. Richer, T. Schmidt and H. P. Steinrück, *Surf. Sci.*, 2002, **507**, 845.
36 M. Bowker and R. J. Madix, *Surf. Sci.*, 1980, **95**, 190.
37 M. Bowker and R. J. Madix, *Surf. Sci.*, 1982, **116**, 549.
38 P. D. A. Pudney, S. A. Francis, R. W. Joyner and M. Bowker, *J. Catal.*, 1991, **131**, 104.
39 B. N. Zope, D. D. Hibbitts, M. Neurock and R. J. Davis, *Science*, 2010, **330**, 74.
40 D. Hibbitts and M. Neurock, *Surf. Sci.*, 2016, **650**, 210.
41 D. Hibbitts and M. Neurock, *J. Catal.*, 2013, **299**, 261.
42 W. B. Jensen, *The Lewis acid–base concepts: An overview*, J. Wiley, 1979.
43 F. Calaza, D. Stacchiola, M. Neurock and W. T. Tysoe, *J. Am. Chem. Soc.*, 2010, **132**(7), 2202.
44 D. Stacchiola, F. Calaza, L. Burkholder, A. W. Schwabacher, M. Neurock and W. T. Tysoe, *Angew. Chem., Int. Ed.*, 2005, **44**(29), 4572.
45 R. J. Madix and J. T. Roberts, in *Surface Reactions*, ed. R. J. Madix, Springer-Verlag, Berlin, 1994, p. 5.
46 Q. Xing, W. Pei and *et al.*, *Fundamental Organic Chemistry*, Higher Education Press, China, 2005.
47 T. Akashi, S. Sato and K. Inui, *Catal. Commun.*, 2002, **4**, 411.
48 D. J. Elliot and F. J. Pennella, *J. Catal.*, 1989, **119**, 359.
49 B. Xu, X. Liu, J. Haubrich, R. J. Madix and C. M. Friend, *Angew. Chem., Int. Ed.*, 2009, **48**, 4206.

50 B. Xu, J. Haubrich, T. A. Baker, E. Kaxiras and C. M. Friend, *J. Phys. Chem. C*, 2011, **115**, 3703.

51 X. Liu, B. Xu, J. Haubrich, R. J. Madix and C. M. Friend, *J. Am. Chem. Soc.*, 2009, **131**, 5757.

52 R. A. van Santen and M. Neurock, *Molecular Heterogeneous Catalysis: A Mechanistic and Computational Approach*, VCH-Wiley, Inc., 2006.

53 S. C. Luo and J. L. Falconer, *Catal. Lett.*, 1999, **57**, 89.

54 J. E. Rekoske and M. A. Barteau, *Ind. Eng. Chem. Res.*, 2011, **50**, 41.

55 V. V. Ordomsky, V. L. Sushkevich and I. Ivanova, *J. Mol. Catal. A: Chem.*, 2010, **333**, 85.

56 J. M. Sun, K. K. Zhu, F. Gao, C. M. Wang, J. Liu, C. H. F. Peden and Y. Wang, *J. Am. Chem. Soc.*, 2011, **133**, 11096.

57 K. Tanabe and T. Yamaguchi, *Catal. Today*, 1994, **20**, 185.

58 K. Tanabe and W. F. Holderich, *Appl. Catal., A*, 1999, **181**, 399.

59 K. Tanabe, *Appl. Catal., A*, 1994, **113**, 147.

Faraday Discussions

PAPER

Metal–support interaction on cobalt based FT catalysts – a DFT study of model inverse catalysts

Tracey van Heerden* and Eric van Steen

Received 3rd October 2016, Accepted 20th October 2016
DOI: 10.1039/c6fd00201c

It is challenging to isolate the effect of metal–support interactions on catalyst reaction performance. In order to overcome this problem, inverse catalysts can be prepared in the laboratory and characterized and tested at relevant conditions. Inverse catalysts are catalysts where the precursor to the catalytically active phase is bonded to a support-like ligand. We can then view the metal–support interaction as a ligand interaction with the support acting as a supra-molecular ligand. Importantly, laboratory studies have shown that these ligands are still present after reduction of the catalyst. By varying the quantity of these ligands present on the surface, insight into the positive effect SMSI have during a reaction is gained. Here, we present a theoretical study of mono-dentate alumina support based ligands, adsorbed on cobalt surfaces. We find that the presence of the ligand may significantly affect the morphology of a cobalt crystallite. With Fischer–Tropsch synthesis in mind, the CO dissociation is used as a probe reaction, with the ligand assisting the dissociation, making it feasible to dissociate CO on the dense fcc Co(111) surface. The nature of the interaction between the ligand and the probe molecule is characterized, showing that the support-like ligands' metal centre is directly interacting with the probe molecule.

1 Background

The term strong metal–support interactions (SMSI) was first used to describe the drastic reduction in the ability to chemisorb CO and H_2 on group VIII metals, when supported on reducible oxides.[1] The observed effect of reduced chemisorption ability could not be attributed to agglomeration, encapsulation, poisoning, or incomplete reduction.[1,2] The term SMSI has subsequently developed into an umbrella term in catalysis, used to describe many of the (generally negative) effects of close contact between the catalytic active phase and its support. However, there is not much real understanding on the nature of the interaction and its effects on the catalyst's characteristics and performance.

Centre for Catalysis Research, Chemical Engineering Department, University of Cape Town, Cape Town, South Africa. E-mail: tracey.vanheerden@uct.ac.za; Tel: +27 21 650 1621

SMSI are now frequently implied *via* differences in the reduction behaviour of catalysts due to the presence of the support, and indeed the TPR profile is often used as an indicator in terms of whether SMSI are present or not.[3–6] Other catalyst characteristics are also affected by the presence of a support, such as dispersion and crystallite size.[4,5] It is commonly believed that SMSI influence the activity, selectivity,[4,7–11] and stability behaviour[4,9,11–13] of catalysts. With this broad range of "symptoms" of SMSI, it should not be surprising that the exact cause (whilst under the umbrella term of SMSI) varies as well. It has been shown that in some cases the support can creep over the active metal (encapsulating the crystallite in severe cases),[14] and metal ions can migrate into the support structure yielding hard to reduce metal–support compounds (such as cobalt aluminates or cobalt silicates).[15–17] Electronic effects are potentially also present. Furthermore, the formation of special sites at the metal–support interface is possible.

When attempting to study SMSI, the most common approach is to alter some parameter in the preparation of the catalyst, in an effort to alter the level of interaction, and then to characterize and test the catalysts extensively. Unfortunately the results cannot be easily de-convoluted,[3,18] since any change to the preparation method may alter the interaction, then affecting other characteristics of the catalytically active material, *e.g.* dispersion or degree of reduction.

Mogorosi *et al.*[19,20] created 'inverse' catalysts for the iron/silica system to be able to isolate the interaction between metal and support. A support-like ligand was attached to the precursor of the catalytically active phase. In this viewpoint the metal–support interaction is viewed as a ligand interaction with the support acting as a supra-molecular ligand.

These type of ligands may indeed be present on real catalysts, as small quantities of support may dissolve during the wet chemical steps of the preparation, and may re-deposit onto the (precursor of the) catalytically active phase during the drying and calcination steps. Importantly, rather than altering the preparation conditions to study SMSI, Mogorosi *et al.* varied the quantity of silica precursor added during the preparation. As expected, the reducibility was negatively affected with the addition of the silica precursor. Additionally the level of sintering was decreased with increased silica concentration. They were able to show that the Fe–O–Si bonds (indicating the presence of the ligands) remained intact throughout the preparation and catalyst reduction steps. Catalyst testing revealed that increased concentrations of silica groups resulted in increased activity, methane selectivity, and a decrease in the olefin content. Changes in the carbiding process of the iron were also observed. Analysis of the data led to the conclusion that the silica groups increased the surface hydrogen availability, and a reduction in the CO adsorption strength.

Computational models provide a unique opportunity to gain additional insight into aspects of the catalytic process. Around SMSI specifically, very little modelling work has been published, as the inclusion of both active metal and support leads to computationally expensive systems, thereby reducing the value of computational studies. Mikhailov[21] studied six cobalt atoms on the alumina surface, and concluded that the presence of the support causes the cobalt cluster to partially oxidise. Valero *et al.*[22] investigated how the presence of alumina as a support affected the adsorption of carbon monoxide and ethylene on Pd_4 clusters. While the supported metal cluster is still very small in this system, the alumina surface is not a simple system, and the inclusion of adsorption resulted in numerous interesting observations. They found that upon adsorption of CO, the Pd cluster deforms significantly, allowing the CO to have a greater stability on the supported cluster than on the isolated Pd_4 cluster, or the Pd(111) extended surface. Their investigation using ethylene showed that the hydroxylation of alumina affected the adsorption mode of the ethylene on the supported Pd_4 cluster. This led to the conclusion that the hydroxylation can affect the reactivity of a catalyst, through differences in the strength of metal–support interactions. More recently, as computational resources continuously grow in capability, more complex investigations become possible. The same group has since expanded on their previous work to include Pd_{13} and Pt_{13} clusters.[23] They found that the order of stability of gas phase clusters did not translate to the order of stability when adsorbed on the alumina surface. The most stable adsorption site was also not predictable, and differed for different cluster morphologies. This confirms the idea that several metallic shapes will co-exist on a single support. Investigation into H and OH reverse spillover showed that only the migration of H is thermodynamically favorable. Comparison to the previous paper leads to the conclusion that smaller platinum clusters are more stable on the alumina surface.[22,23] There is also a strong indication that the hydroxylation of the alumina surface is a significant factor in the strength of the interaction. Matsutsu *et al.* conducted a study of Pt_{38} clusters supported on functionalised graphene, and was able to show how various functional groups alter the support's interaction with the platinum nano-particle.[24] To study SMSI of cobalt on supports such as alumina (or silica or titania) the model would need to be somewhat more complex, as the metal cluster would ideally need to be larger (a 1 nm cobalt crystallite consists of 47 atoms[25]) and the support cannot be described by a single layer (as done for graphene in the study by Matsutsu *et al.*, for example). With the inverse catalysts there is a great opportunity to take advantage of the simplicity of the system and be able to conduct meaningful DFT calculations that can be tied together with laboratory results on similar systems. With simplicity in mind, we looked specifically at the Al/Co system (as opposed to the more complex iron carbide). We studied the ligands on the cobalt surface, how these ligands may affect the morphology of the crystallite, and how the presence of these ligands may affect one of the key FT reaction steps, namely the dissociation of CO.

2 Methodology

2.1 General calculational setup

All calculations were conducted using VASP (Vienna *Ab initio* Simulation Package).[26–29] Bulk cobalt was modelled as a fcc structure. Cobalt catalyst nanoparticles primarily display the (111) and (100) facets. The fcc (111) and its equivalent hcp (0001) cobalt surface are the simplest and most commonly studied surfaces.[30] On an fcc cobalt crystallite, the (111) and (100) surfaces make up over 80% of the surface sites (although this is size dependent).[25] The RPBE functional was employed, together with PAW (Projector Augmented Wave) pseudopotentials. *k*-Point sampling was conducted using a Monkhorst–Pack scheme. For surface calculations 2nd order Methfessel and Paxton[31] smearing are utilized, with a width of 0.1 eV. For geometry optimization calculations, the conjugate gradient algorithm is used. Dipole corrections are applied perpendicular to the surface for all systems involving the cobalt surface. The electronic convergence criteria used was 10^{-4} eV and the force convergence was set to $|0.03|$ eV Å^{-1}.

Bulk fcc cobalt, the fcc cobalt(111) and (100) surface systems, and all required gas phase species were validated against both computational and experimental literature. A $p(2 \times 2)$ unit cell is used where possible, with larger $p(3 \times 3)$ unit cells used for the bulkier ligands and co-adsorbed systems. For adsorption on the fcc Co(111) surface a 4 layer slab with the bottom two layers fixed was used. A 6 layer slab with the bottom 2 layers fixed was used for the fcc Co(100) surface. Ligand feasibility was conducted by geometry optimization of a number of proposed ligands. The energies of each ligand in various positions were calculated, and vibrational analysis was used to confirm the stability of the lowest energy configurations.

Surface energies are calculated using:

$$\gamma = [E_{\text{slab}} - n_{\text{atoms}} E_{\text{bulk}}]/n_{\text{surface atoms}} \tag{1}$$

where E_{bulk} is the energy per atom in bulk fcc cobalt.

For systems including ligands, E_{slab} is replaced with $E_{\text{slab+ligand}}$.

Vibrational analysis is calculated in VASP. For these calculations the electronic convergence criteria is set to 10^{-5} eV. Finite differences are used to displace each (relaxed) ion in the positive and negative direction of each Cartesian co-ordinate, then determining the Hessian matrix from the forces, as well as all of the vibrational frequencies. The size of the displacement was used was 0.015 Å. All atoms of the adsorbates are relaxed.

Charge density difference plots for co-adsorbed systems are visualized using VESTA (Visualization for Electronic and STructural Analysis).[32] By subtracting the charge densities of various systems obtained from VASP, the redistribution of charge as a result of different species being absorbed near each other can be viewed. The method of subtracting charge is given by:

$$\rho_{\text{diff}}(x,y,z) = \rho_{\text{coadsorbed}} + \rho_{\text{slab}} - \rho_{\text{CO,ads}} - \rho_{\text{ligand,ads}} \tag{2}$$

Bader charge analysis is conducted on the valence charge, using the complete charge as a reference.[33–35] In order to calculate the Bader charge differences we compared the stable co-adsorbed system to the isolated adsorbates in their stable geometries (unlike the reference system for the charge density plots where adsorbates are frozen in the geometry of the co-adsorbed system). In other words, we used eqn (3) for each adsorbate atom to calculate the change in Bader charge.

$$\Delta q_{\text{i}} = q_{\text{i-coadsorbed}} - q_{\text{i-isolated}} \tag{3}$$

The nudged elastic band (NEB) method is used to search for the transition state (TSS).[36] Because the constraints of the NEB method are non-linear, we used the VASP recommended RMM-DIIS (residual minimization scheme, direct inversion in the iterative subspace) algorithm to ensure convergence of the calculation.[37] Vibrational analysis is used to confirm that the transition state found in the TSS is a true transition state.

3 Results and discussion

3.1 Ligand feasibility and selection on Co(111) surface

A set of different alumina based ligands were investigated. Ligands were limited to those containing single aluminium atoms, that are able to fit into a $p(3 \times 3)$ cell without interacting with the neighbouring cell, due to constraints on computational resources. Ligand feasibilities were assessed in terms of their energy, level of hydroxylation, and vibrational states. Based specifically on the use of vibrational analysis, only 4 of the numerous ligands investigated were found to have stable positions; these are displayed in Fig. 1. Two of the ligands (AM1 and AM1O) are monodentate ligands, with the anchoring oxygen sitting comfortably in the fcc hollow site. The anchor is then connected to the metal core, to which the oxygen and/or OH groups are attached. AM2 features two symmetric OH groups, while AM1O has a slightly tilted arrangement of an oxygen and an OH group. The bidentate AB2 ligand features two anchoring oxygen atoms sitting in the hollow sites, with two OH groups off the metal centre giving it a tetrahedral type shape. The other bidentate ligand, AB1, on the other hand only features 3 oxygen atoms, and is attached at a hollow site, and an atop site, with an OH group off the metal centre.

Relative stabilities are calculated at 0 K, by setting a consistent reference system that has an identical atomic composition to the ligand in question. Our reference system consists of a linear O–Al–O ligand in the atop position (an arbitrary choice). As many of our ligands have varied compositions, in terms of the number of hydrogen and oxygen atoms specifically, in order for the reference system to match the composition of the ligand in question, we included gaseous H_2O and H_2 as required as a reference. Relative stabilities are calculated such that a more negative value indicates greater stability. Every ligand was investigated in all of the adsorption sites available on the Co(111) surface (atop, bridge, fcc hollow, and hcp hollow sites), but we have only included data for ligands found to be stable, in their energetically most favoured position. It was found that for all of the monodentate ligands, the hollow sites were energetically the most stable. The 4 stable ligands described previously, and their relative stabilities are displayed in Fig. 1.

Ligands AB1, AB2, and AM2 may be considered favourable in the FT environment as they are fully hydroxylated. We are naturally most interested then in the stabilities of the M2 ligand. We take the feasibility study further by calculating the vibrational modes of the ligands on the surface (all cobalt atoms are fixed for these calculations). The vibrational analysis showed that the AM2 ligand is indeed

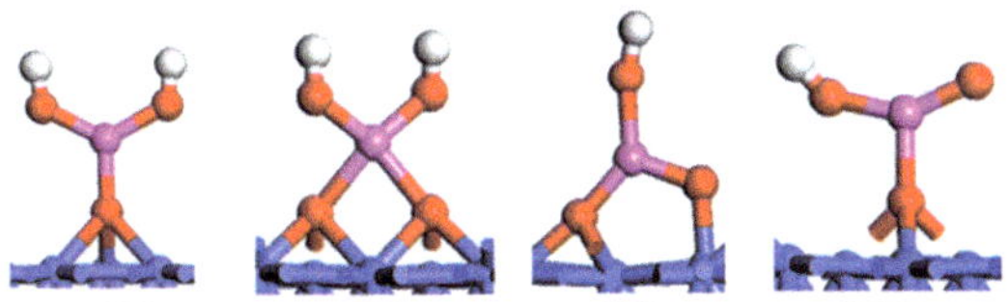

Fig. 1 Stable alumina based ligands on the Co(111) surface, displayed with their shorthand notation, as well as their stabilities relative to a linear O–Al–O ligand in the atop site.

a stable geometry. The equilibrium geometrical arrangement (bond lengths and angles) of the AM2 ligand described previously is included here in Table 1.

3.2 Ligand of interest characterization

AM2 is the most stable in H_2/H_2O containing atmospheres. With its relevance to FT, we conducted characterization calculations on the AM2 ligand. Further understanding in how the AM2 ligand attaches to the Co(111) surface is gained through the analysis of the charge distribution calculated using Bader charge analysis. The Bader charges of each atom in the ligand are tabulated in Table 2. If we sum up the valence charge of the ligand (which is concentrated at the oxygen atoms), we find that the cobalt surface donates 0.6 electrons worth of charge to the AM2 ligand ($OAl(OH)_2$).

While the focus is primarily on the Co(111) surface, we also found that on the Co(100) surface, the AM2 ligand adsorbs preferentially in the hollow site. Vibrational modes were calculated for the ligand on this system as well; all frequencies were real.

3.3 Ligand adsorption effect on cobalt surface energies

Using simple Wulff constructions together with the calculated surface energies, we can calculate a Co(100)/Co(111) surface area ratio (these being the two dominant surfaces for fcc cobalt crystallites). The (100)/(111) surface area ratio calculated for a clean crystallite is calculated to be 0.14, with a diagram of the construction shown in Fig. 2. Considering the simplicity of our calculations, this compares well to the paper by Van Helden *et al.* which implies a (100)/(111) surface area ratio of 0.27 for crystallites above 10 nm in size.[25] When the AM2 ligand is adsorbed on both surfaces, the surface energy is affected, and a new construction can be calculated. At a coverage of 0.25 ML of the AM2 ligand, the (100)/(111) surface area ratio increases significantly to 0.65. The AM2 ligand clearly has a much greater impact on the less dense Co(100) surface, stabilising it to some extent, and altering the morphology of the cobalt crystallites. As the Co(100) surface is known to be more reactive, this change in morphology should have significant benefits to the catalyst activity. It needs to be noted that this is a somewhat simplistic view, as there are some other surfaces present, as well as kinks and defects, however at 80% of the surface area consisting of the Co(111) and Co(100) surfaces, this result still has qualitative meaning, and points us to

Table 1 AM2 equilibrium bond lengths and bond angles

Labelled image	Bond	Length (Å)	Angle	Angle
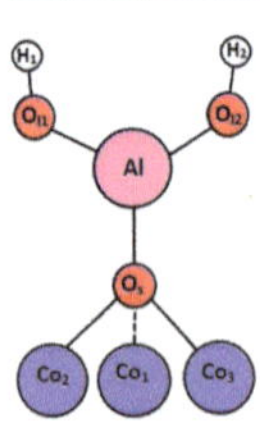	H_1–O_{11}	0.97	H_1–O_{11}–Al	124.0
	H_2–O_{11}	0.97	H_2–O_{11}–Al	123.5
	O_{11}–Al	1.71	O_{11}–Al–O_s	116.8
	O_{12}–Al	1.71	O_{12}–Al–O_s	117.0
	Al–O_s	1.72	O_{11}–Al–O_{12}	126.2
	O_s–Co_1	2.05	Al–O_s–Co_1	133.2
	O_s–Co_2	2.05	Al–O_s–Co_2	132.7
	O_s–Co_3	2.05	Al–O_s–Co_3	135.1

Table 2 AM2 ligand valence charge according to Bader charge analysis (given as q)

System	H1	H2	O1	O2	O3	Al
AM2	0.33	0.27	7.50	7.55	7.34	0.61

one of the mechanisms through which SMSI may affect catalyst characteristics and performance.

3.4 Co-adsorption of probe molecule (CO) and alumina based ligand (AM2) on fcc Co(111) surface

Our probe molecule CO was adsorbed in five different sites surrounding the AM2 ligand in a $p(3 \times 3)$ cell. Many other sites could be excluded due to symmetry or proximity to the ligand. Geometry optimization was performed on 5 different co-adsorbed systems. The obtained final geometries are shown in Fig. 3. We can compare the calculated energies and bond lengths of these co-adsorbed systems to a reference state. For the purpose of this section the reference state is chosen as the AM2 ligand adsorbed on the Co(111) surface, with carbon monoxide in the gas phase, and the results relative to this system are displayed in Table 3. When the CO is adsorbed 'far' from the ligand, we find that the system energy drops by 1.23 or 1.28 eV, which is comparable to the calculated heat of adsorption of CO on the clean cobalt surface. Once CO is adsorbed near enough to share a surface cobalt atom (system C), we see that the energy drops further, although the CO bond length remains unchanged. There is some adjustment to the ligand geometry. Once CO is less than 3 Å removed from the AM2 ligand, we see a strong interaction. The relative energy of the system drops to −2.22 eV, and the CO bond becomes elongated, with the oxygen in carbon monoxide forming a bond with aluminium in the ligand. We also see a very significant blue shift in the calculated CO stretch frequency, dropping from 1847 cm^{-1} on the clean surface, to 1715 cm^{-1} for co-adsorbed geometry A, and even further to 1483 cm^{-1} in co-adsorbed geometry E. This is likely due to the movement of electron density into the CO antibonding orbital.

The co-adsorbed systems were further characterized using Bader charge analysis and charge distribution calculations. Charge densities calculated in VASP can be subtracted such that we can visualize the interaction between the adsorbed

Fig. 2 Wulff shapes using calculated Co(100) and C(111) surface energies, for clean surfaces [left] and 0.25 ML adsorbed AM2 ligands [right].

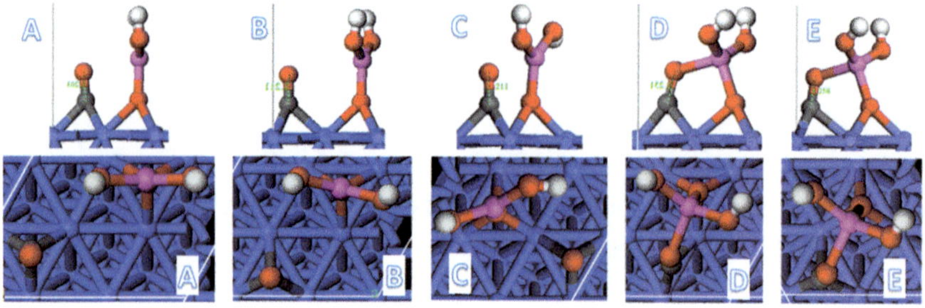

Fig. 3 Geometries resultant from CO and AM2 co-adsorbed on a Co(111) surface.

carbon monoxide and the AM2 ligand on the Co(111) surface. As expected from the results of the geometry optimization, the systems A, B, and C showed very little interaction. The interaction obtained from the subtraction for system A is shown in Fig. 4. Only a very subtle re-arrangement of charge on the oxygen of the carbon monoxide can be seen. A very similar visual was observed for systems B and C, with an increase in the interaction as the carbon monoxide nears the ligand. More interesting is the case of system E presented in Fig. 5. The interaction for the case where the species are close is of a much greater magnitude. For system E (and D which is almost identical), we see a shift in electrons from the C–O bond into the newly formed O–Al bond. There is also the loss of some charge from the bond between the aluminium atom and its anchoring oxygen. It also appears as if the ligand oxygens return some of the charge to the aluminium atom that they gained from the original ligand arrangement. There is no visible effect on the hydrogens, and the cobalt surface appears to have lost some electrons to the co-adsorbed adsorbates.

Bader charge analysis gives more useful quantitative information on the characteristics of the interaction between carbon monoxide and the ligand. A subtraction of the co-adsorbed system from the original stable geometry of the isolated adsorbates can be done for each atom. The results for all co-adsorbed systems are given in Table 4. Fig. 5 shows the charge density difference plot for

Table 3 Results of CO and AM2 ligand co-adsorption

Position	Energy (eV)	CO–ligand distance[a] (Å)	CO bond length (Å)	CO stretch vibrational frequency (cm^{-1})
Reference: AM2 + CO(g)	0	—	1.14	2149
CO(ads); no AM2	—	—	1.22	1847
A	−1.23	4.40	1.20	1715
B	−1.28	3.89	1.21	1673
C	−1.71	3.08	1.21	1670
D	−2.22	2.76	1.25	1527
E	−2.25	2.77	1.26	1483

[a] CO–ligand distance is measured as the distance between the C of the CO and the O that anchors the ligand.

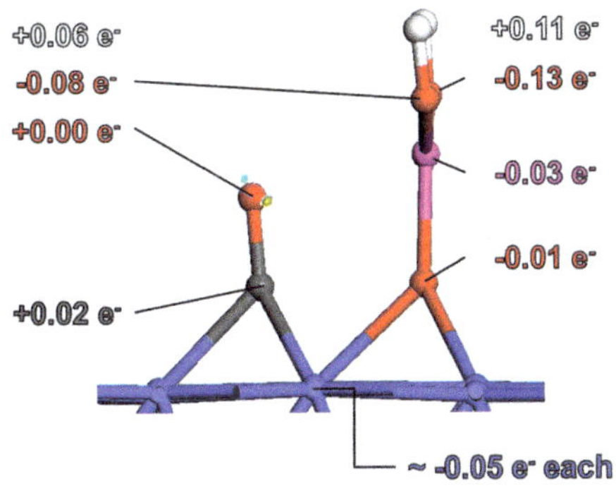

Fig. 4 Charge density difference plot for co-adsorbed system A. Blue indicates a loss of charge, yellow indicates a gain. Isosurface level = 0.0028. Atom labels indicate results of Bader charge analysis subtraction as calculated by eqn (3).

system E, with additional labels to clarify the naming of the ligand atoms in Table 4. There is a clear trend visible in the table, with the CO gaining as many as 0.36 electrons in the strongly interacting systems (D and E), but receiving less than 0.09 electrons for the non interacting systems. The anchoring oxygen on the ligand is not affected when sharing a surface cobalt with the carbon monoxide in system C where there is only a weak interaction, indicating that the interaction in systems D and E is likely not through the surface. While the non interacting systems (A, B, and C) show almost no effect on the CO, there is still some rearrangement of charge on the ligand. This is likely due to the highly rotational mobility of the OH groups, which are affected by the presence of the CO. The surface donated between 0.05 and 0.08 electrons per surface atom depending on the level of interaction. We are able to neatly tie the various results of calculation on the co-adsorbed systems together. Of the five co-adsorbed systems investigated, two are found to be highly interacting (D and E). This interaction appears to be directly related to the proximity of the carbon monoxide to the ligand, with a proximity below 3 Å (measured as the distance between the ligands anchoring oxygen and the carbon) leading to strong interactions. Despite the adsorbates

Table 4 Results of CO and AM2 ligand co-adsorption

Bader charge subtraction (nr of electrons)[a]									
System	H1	H2	C	O1	O2	O3	O4	Al	Co[b]
A	0.04	0.10	0.15	−0.05	−0.12	−0.11	0.21	−0.07	−0.08
B	0.07	0.11	0.11	−0.08	−0.13	−0.12	0.22	−0.07	−0.08
C	0.06	0.12	0.04	−0.09	−0.15	−0.02	0.04	−0.03	−0.05
D	0.05	0.12	0.05	−0.07	−0.14	−0.01	0.04	−0.03	−0.05
E	0.06	0.11	0.02	−0.08	−0.13	0.01	0.00	0.03	−0.05

[a] CO/Co used as reference system for C and O4, AM2/Co used as reference system for all other atoms. Bader subtraction = Bader charge of co-adsorbed system − reference Bader charge. Positive values indicate an increase in electron density around atom due to co-adsorption, and *vice versa*. [b] Average Co Bader charge (or Bader charge subtraction) for surface cobalt atoms.

sharing a surface cobalt, the configuration (C) where the CO–ligand distance is 3.08 Å shows a level of interaction similar to configurations that are even further away, and do not share any surface cobalt atoms (such as A – 4.4 Å). This presents a strong case against the interaction being through the surface of the cobalt. The geometry optimization results and the charge density difference plots both point to a weakening of the CO bond for systems D and E. For systems A, B, and C, the interaction primarily affects the geometry and charge distribution of the ligand, with the CO essentially unaffected in terms of both geometry and charge densities. Vibrational analysis showed that system E is stable, and we have thus used system E as the starting configuration for the dissociation studies that follow.

3.5 Diffusion of CO over Co(111) towards AM2 ligand

DFT work on the Co(111) surface shows that the diffusion barrier for CO moving across the surface is approximately 0.08 eV.

We used the diffusion of carbon monoxide from system A to system E as an equivalent of carbon monoxide diffusing towards the ligand. System A is deemed appropriate for this, as it had the greatest distance between the adsorbates, and its characterization displayed no significant interaction. Using system A as the starting image, and system E as the end point, a nudged elastic band calculation (using 3 intermediate images) was conducted. A transition state was found for the diffusion pathway. The geometries of the initial, transition, and final states of CO diffusing towards the ligand are displayed in Fig. 6. The energy of the transition state was calculated, and we found that the diffusion barrier is 0.06 eV. This confirms that the adsorbed carbon monoxide will easily find itself in an intimate position with the AM2 alumina ligand.

3.6 AM2 ligand assisted CO dissociation on fcc Co(111) surface

It was found that the dissociation of CO proceeds more easily when in the presence of the AM2 ligand, with the carbon subsequently easily removed from the system; however, the oxygen remains bonded to the ligand. The geometries of the initial, transition, and final states of CO dissociation in the presence of the AM2 ligand are displayed in Fig. 7. The energy of the transition state was calculated, and we found that the barrier is 1.7 eV. This is lower than the barrier of 1.8 eV

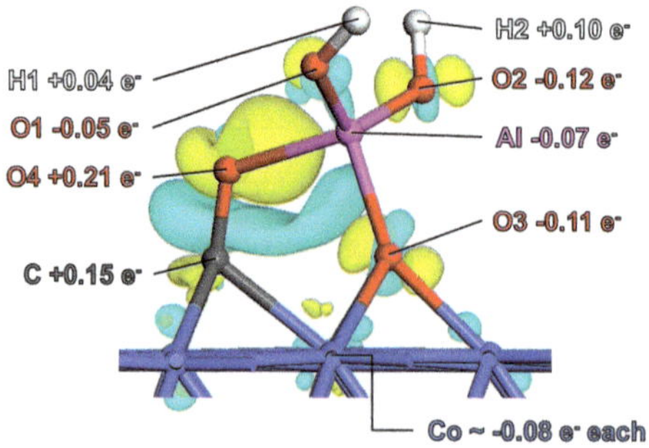

Fig. 5 Charge density difference plot for co-adsorbed system E. Blue indicates a loss of charge, yellow indicates a gain. Isosurface level = 0.0028. Atom labels indicate results of Bader charge analysis subtraction as calculated by eqn (3).

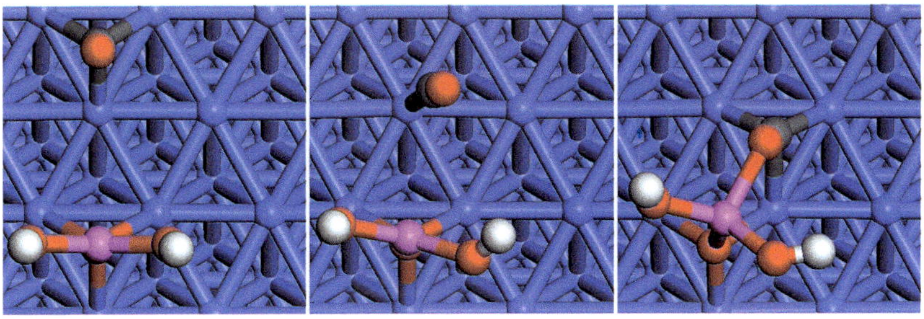

Fig. 6 Carbon monoxide diffusion from adsorption site (system A) to ligand (system E), *via* a single transition state, with a barrier of 0.06 eV.

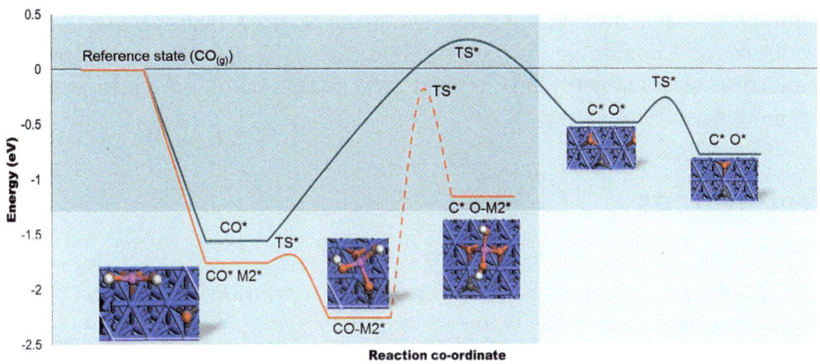

Fig. 7 Energy diagram of CO dissociation on Co(111), with and without the assistance of the AM2 ligand.

calculated for direct dissociation of CO on the Co(111) surface. The full energy diagram of CO dissociation on the Co(111) surface is shown in Fig. 7. It can be clearly seen that while the direct dissociation of CO is not feasible (as expected), in the presence of the ligand the dissociation becomes feasible despite the still somewhat large barrier. We must comment however on the fact that this does not yet represent a complete catalytic cycle, and we are conducting further calculations to determine the pathway for removal of the oxygen atom from the ligand after dissociating the CO.

4 Conclusions

DFT calculations were used to study metal–support interactions, specifically those relating to what has been termed the ligand effect.

In the context of a H_2/H_2O containing atmosphere, the bidentate AM2 ligand ($OAl(OH)_2$) is stable, and could feasibly be present on the synthesised model catalysts. If this ligand is present, it will alter the morphology of the crystallite, increasing the exposure of the less dense (and more reactive) Co(100) surface.

Further calculations on the Co(111) surface established that CO can easily diffuse into close proximity of this ligand, where they then interact strongly. Characterization of this co-adsorbed system showed that the interaction is not through the surface of the cobalt, but rather a direct effect of the metal atom at the centre of the ligand. DFT calculations confirm that the dissociation of CO without any assistance on a dense cobalt surface is not feasible, as the dissociation barrier is too high. Instead CO prefers to dissociate on steps, edges or defects, or through the assistance of surface hydrogen. This indicates that the assistance of the ligand in dissociating the CO can be considered similar to a promoter effect, rather than a metal–support interaction.

Acknowledgements

Financial assistance from the DST-NRF Centre of Excellence in Catalysis (c*change) is gratefully acknowledged. Computations were performed using facilities provided by the University of Cape Town's ICTS High Performance Computing team: http://hpc.uct.ac.za; and the South African Centre for High Performance Computing: http://www.chpc.ac.za. The valuable computational experience of Mr Molefi Matsutsu and Mr Thobani Gambu are also gratefully acknowledged.

References

1 S. Tauster, *J. Catal.*, 1978, **55**, 29–35.
2 S. J. Tauster, S. C. Fung and R. L. Garten, *J. Am. Chem. Soc.*, 1978, **100**, 170–175.
3 J. W. Bae, Y.-J. Lee, J.-Y. Park and K.-W. Jun, *Energy Fuels*, 2008, **22**, 2885–2891.
4 Z. Huang, F. Cui, H. Kang and J. Chen, *Chem. Mater.*, 2008, 5090–5099.
5 G. Jacobs, Y. Ji, B. H. Davis, D. Cronauer, a. J. Kropf and C. L. Marshall, *Appl. Catal., A*, 2007, **333**, 177–191.
6 W. Ma, E. L. Kugler and D. B. Dadyburjor, *Energy Fuels*, 2010, **24**, 4099–4110.
7 M. Comotti, W.-C. Li, B. Spliethoff and F. Schüth, *J. Am. Chem. Soc.*, 2006, **128**, 917–924.
8 C. Pliangos, I. V. Yentekakis, V. G. Papadakis, C. G. Vayenas and X. E. Verykios, *Appl. Catal., B*, 1997, **14**, 161–173.
9 S. Wang and G. Q. M. Lu, *Appl. Catal., B*, 1998, **16**, 269–277.
10 P. Weerachawanasak, O. Mekasuwandumrong, M. Arai, S. I. Fujita, P. Praserthdam and J. Panpranot, *J. Catal.*, 2009, **262**, 199–205.
11 J. Zhang, H. Wang and A. K. Dalai, *J. Catal.*, 2007, **249**, 300–310.
12 M. C. Bradford and M. A. Vannice, *Appl. Catal., A*, 1996, **142**, 73–96.
13 Z. Zhang, *J. Catal.*, 1996, **158**, 51–63.
14 V. A. de la Peña O'Shea, M. Consuelo Alvarez Galvan, A. E. Platero Prats, J. M. Campos-Martin and J. L. G. Fierro, *Chem. Commun.*, 2011, **47**, 7131–7133.
15 R. Riva, H. Miessner, R. Vitali and G. D. Piero, *Appl. Catal., A*, 2000, **196**, 111–123.
16 a. H. Kababji, B. Joseph and J. T. Wolan, *Catal. Lett.*, 2009, **130**, 72–78.
17 A. Sirijaruphan, A. Horváth, J. G. Goodwin and R. Oukaci, *Catal. Lett.*, 2003, **91**, 89–94.
18 L. Ji, J. Lin and H. C. Zeng, *J. Phys. Chem. B*, 2000, **104**, 1783–1790.
19 R. P. Mogorosi, N. Fischer, M. Claeys and E. Van Steen, *J. Catal.*, 2012, **289**, 140–150.
20 R. P. Mogorosi, M. Claeys and E. V. Steen, *Top. Catal.*, 2014, **57**, 572–581.

21 M. N. Mikhailov, G. M. Zhidomirov and a. Y. Krylova, *Russ. Chem. Bull.*, 2005, **54**, 2264–2269.
22 M. Corral-Valero, P. Raybaud and P. Sautet, *J. Catal.*, 2007, **247**, 339–355.
23 C. H. Hu, C. Chizallet, C. Mager-Maury, M. Corral-Valero, P. Sautet, H. Toulhoat and P. Raybaud, *J. Catal.*, 2010, **274**, 99–110.
24 M. Matsutsu, M. A. Petersen and E. V. Steen, *Phys. Chem. Chem. Phys.*, 2016, **18**, 25693–25704.
25 P. V. Helden, I. M. Ciobîc and R. L. J. Coetzer, *Catal. Today*, 2016, **261**, 48–59.
26 G. Kresse and J. Hafner, *Phys. Rev. B: Condens. Matter Mater. Phys.*, 1993, **47**, 558–561.
27 G. Kresse and J. Hafner, *Phys. Rev. B: Condens. Matter Mater. Phys.*, 1994, **49**, 14251–14269.
28 G. Kresse and J. Furthmuller, *Phys. Rev. B: Condens. Matter Mater. Phys.*, 1996, **54**, 11169–11186.
29 G. Kresse and J. Furthmuller, *Comput. Mater. Sci.*, 1996, **6**, 15–30.
30 S. J. Jenkins and D. A. King, *Surf. Sci.*, 2002, **504**, 138–144.
31 M. Methfessel and A. T. Paxton, *Phys. Rev. B: Condens. Matter Mater. Phys.*, 1989, **40**, 3616–3621.
32 K. Momma and F. Izumi, *J. Appl. Crystallogr.*, 2011, **44**, 1272–1276.
33 G. Henkelman, A. Arnaldsson and H. Jónsson, *Comput. Mater. Sci.*, 2006, **36**, 354–360.
34 E. Sanville, S. D. Kenny, R. Smith and G. Henkelman, *J. Comput. Chem.*, 2007, **28**, 899–908.
35 W. Tang, E. Sanville and G. Henkelman, *J. Phys.: Condens. Matter*, 2009, **21**, 084204.
36 G. Mills, H. Jonsson and G. K. Schenter, *Surf. Sci.*, 1995, **50**, 305–337.
37 P. Pulay, *Chem. Phys. Lett.*, 1980, **73**, 393–398.

Faraday Discussions

PAPER

Understanding FTS selectivity: the crucial role of surface hydrogen

C. J. Weststrate*[ab] and J. W. Niemantsverdriet[ac]

Received 15th September 2016, Accepted 24th October 2016
DOI: 10.1039/c6fd00191b

Monomeric forms of carbon play a central role in the synthesis of long chain hydrocarbons *via* the Fischer–Tropsch synthesis (FTS). We explored the chemistry of C_1H_{xad} species on the close-packed surface of cobalt. Our findings on this simple model catalyst highlight the important role of surface hydrogen and vacant sites for product selectivity. We furthermore find that CO_{ad} affects hydrogen in multiple ways. It limits the adsorption capacity for H_{ad}, lowers its adsorption energy and inhibits dissociative H_2 adsorption. We discuss how these findings, extrapolated to pressures and temperatures used in applied FTS, can provide insights into the correlation between partial pressure of reactants and product selectivity. By combining the C_1H_x stability differences found in the present work with literature reports of the reactivity of C_1H_x species measured by steady state isotope transient kinetic analysis, we aim to shed light on the nature of the atomic carbon reservoir found in these studies.

Introduction

In the Fischer–Tropsch synthesis (FTS) reaction, a mixture of hydrogen and carbon monoxide is converted into a complex mixture of hydrocarbon products. In the commonly assumed carbide mechanism,[1] C_1H_{xad} species play a central role: (i) new chains are initiated *via* $C_1 + C_1 \rightarrow C_2$ and (ii) chain growth proceeds *via* $C_1 + C_n \rightarrow C_{n+1}$, whereas (iii) methane is formed by complete hydrogenation of C_1H_{xad}. The selectivity of the reaction depends on the relative rates of reaction (i) and (ii) compared to the rate of methane formation, reaction (iii).

Steady state isotope transient kinetic analysis (SSITKA) provides information on the residence time of C_1H_x species under steady state reaction conditions. In studies where this technique was applied, C_1H_x residence times on cobalt catalysts are found to be in the order of 5–20 s, depending on the exact temperature

[a]*SynCat@DIFFER, Syngaschem BV, PO Box 6336, 5600 HH Eindhoven, The Netherlands. E-mail: c.j.weststrate@syngaschem.com*
[b]*Sasol Technology Netherlands B.V., Eindhoven University of Technology, P.O. Box 513, 5600 MB Eindhoven, The Netherlands*
[c]*SynCat@Beijing, Synfuels China Technology Co., Ltd., Beijing-Huairou, People's Republic of China*

and reaction conditions used.[2–5] Den Breejen *et al.*, for example, report a residence time of 10 s for a carbon nanofiber-supported (CNF) Co catalyst with ~8 nm large Co crystallites operated at 210 °C under low alpha conditions[2] (65% CH_4, $\alpha = 0.4$). Mims *et al.* work under conditions with a much higher chain growth probability of 0.84 and report a CH_x residence time of 20 s for a Co/SiO_2 catalyst.[6] The difference between the two studies is primarily the much higher CO partial pressure used in the study by Mims *et al.* as well as a lower operating temperature. The longer C_1H_x residence time can be attributed to the lower temperature in this case, 202 °C *vs.* 210 °C, rather than to a significant change of concentration of hydrogen on the surface. The question arises how such a selectivity shift upon a change of reaction conditions can be understood in terms of changes in the composition of the surface adsorbate layer.

To answer this question, information on the exact identity of the C_1H_x species and on the reactivity of the various C_1H_x species is needed. The value of θ_{CH_x} derived from SSITKA experiments is found to be in the order of 0.1–0.2 ML, even when methane is the dominant product.[2–5] As θ_{CH_x} represents the sum of all possible C_1H_x species adsorbed on a variety of adsorption sites exposed by a cobalt nano-crystallite, such detailed information cannot be obtained from *in situ* approaches. In contrast, by using a well-defined cobalt single crystal surface in controlled UHV conditions, the different elementary reaction steps of C_1H_{xad} species can be studied in isolation. Here we present results from this approach to obtain more information on the intrinsic reactivity of C_1H_{xad} species on the close-packed surface of cobalt, the most abundant surface exposed by fcc-Co catalysts.[7,8] As surface hydrogen takes part in most C_1H_{xad} surface reactions, we also explore how surface hydrogen is affected by the presence of co-adsorbed CO. In the discussion we then use these insights into the reactivity of C_1H_{xad} species as a function of hydrogen coverage, and in particular the potential influence of co-adsorbed CO, to explore the role of surface hydrogen in Fischer–Tropsch synthesis. Specific questions are: (i) how do reactant pressures affect the surface coverage of the reactants CO and H_{ad}, and how do these in turn affect the selectivity of the chain growth reaction? (ii) What is the origin of the long residence time of C_1H_x species under FTS-like conditions, even when methane is the dominant product?

Experimental methods

The two vacuum systems used for temperature programmed desorption/work function measurements and RAIRS measurements, have been described in detail elsewhere.[9] The (hcp) Co(0001) sample was clamped between two W wires in thermal contact with a liquid nitrogen reservoir, so that sample temperatures of ~90 K could be readily achieved. Sample temperature was measured using a K-type thermocouple, spotwelded to the backside of the sample. A typical cleaning cycle consisted of 20 min sputtering using 1 V Ar^+ at 630 K, followed by annealing under vacuum for 20 min. XPS measurements were performed at the MatLine beamline at ASTRID2 (Aarhus, Denmark), where a similar cleaning protocol was used. The C 1s spectra reported here were recorded using a photon energy of 380 eV.

The liquids, CH_3I and Cl_2CH_2, were cleaned by at least three freeze–pump–thaw cycles to remove dissolved gases. Due to the photosensitive nature of iodomethane in particular, care was taken to shield the liquid from light and additional freeze–pump–thaw cycles were regularly performed to ensure the purity of the CH_3I dosed onto the crystal. The hydrogen doses reported here were calculated using an ion gauge sensitivity factor of 0.46.

For the analysis of the isotopologues produced during the reaction of CH_{3ad} and D_{ad}, the m/z signals from $m/z = 20$ down to $m/z = 16$ were recorded. The $m/z = 20$ signal can be attributed solely to CD_4, and from the cracking pattern of methane the contribution of CD_4 to the $m/z = 18$ and $m/z = 16$ could be subtracted from these signals. Likewise, $m/z = 19$ can be solely attributed to D_3CH. Loss of one hydrogen yields 25% D_3C ($m/z = 18$) and 75% D_2CH ($m/z = 17$), and the loss of two hydrogens yields 25% D_2C ($m/z = 16$), 50% DCH ($m/z = 15$) and 25% CH_2 ($m/z = 14$). These numbers were used to correct the corresponding mass signals. The residual $m/z = 18$ signal was then attributed to D_2CH_2. From this the contribution of this species to $m/z = 17$ and $m/z = 16$ was calculated. The residual $m/z = 17$ signal was attributed to DCH_3, *etc.* In this way, we were able to eliminate the effect of overlapping masses and to report the 'real' isotopic composition of the produced methane with reasonable accuracy.

Results

C_1H_x chemistry on Co(0001)

Iodomethane is often used as a source of surface methyl groups, as it readily dissociates to I_{ad} and CH_{3ad} at low surface temperature.[10–12] We used iodomethane to produce methyl (CH_{3ad}) adsorbates at a low temperature[13] on the surface of Co(0001). In addition, dichloromethane was used to produce methylene adsorbates in a brief exploration of CH_{2ad} reactivity. Fig. 1(a) shows a series of H_2 TPD spectra for various CH_3I doses, producing a θ_{CH_3} of up to 0.14 ML, a coverage regime where H_2 is the only desorbing product. The H_2 desorption peak area after dosing H_2 to saturation (~500 L) at a low temperature corresponds to a H_{ad} coverage of 0.5 ML, and this information was used to derive the value of θ_{CH_3} reported in the figure.

Two separate H_2 desorption peaks, with a peak area ratio of 2 : 1, develop with increasing methyl coverage. The desorption-limited peak between 300–380 K is attributed to associative desorption of surface-bound hydrogen formed by methyl dehydrogenation at a low temperature, whereas the reaction-limited peak around 400 K is attributed to the dehydrogenation of the remaining C_nH_n adsorbate, which, according to the ratio of the desorption peak areas, must have a C/H ratio of 1. The H_2 desorption spectrum observed for a saturated Cl_2CH_{2ad} layer shows two hydrogen desorption peaks at similar temperatures, where 50% desorbs at <380 K and the other 50% at around 400 K.

Fig. 1(b) shows the temperature programmed work function measurements for CH_3I and CH_2Cl_2 adsorbed on Co(0001). Intact adsorption of iodomethane at 90 K causes a significant decrease of the work function. Upon heating, the work function changes by +290 meV at around 110 K, attributed to scission of the carbon–iodine bond,[12] which produces a θ_{CH_x} coverage of 0.14 ML. A second increase of the work function is found at around 230 K, whereas the third work function change at around 400 K coincides with the high temperature H_2 desorption peak. Intact CH_2Cl_2 adsorption at 90 K also decreases the work function. The C–Cl bond appears to be slightly more difficult to break than the C–I

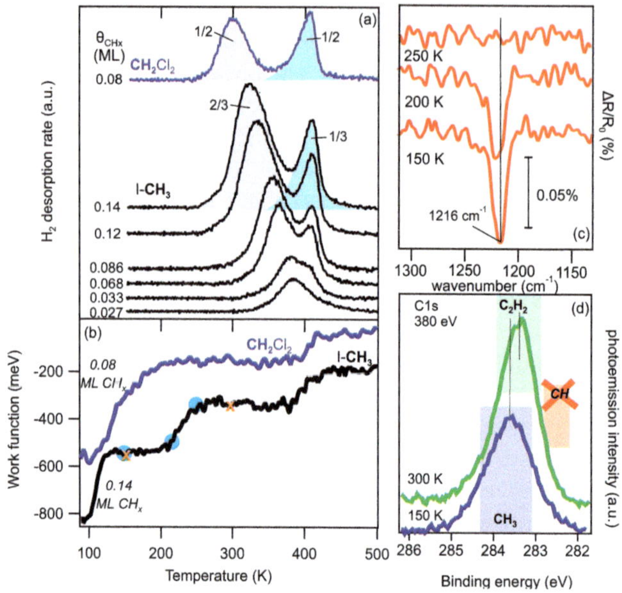

Fig. 1 (a) H_2 desorption spectra resulting from the decomposition of CH_3I-derived methyl and CH_2Cl_2-derived methylene adsorbates. (b) Work function measurements during heating of the indicated CH_x coverages both for methylene and methyl. (c) The CH bending region in the IR absorption spectrum of species derived from 0.14 ML CH_3I recorded at the indicated temperatures. (d) XPS spectra of the C 1s region after annealing 0.12 ML adsorbed CH_{3ad} to the indicated temperatures. A heating rate of 0.5 K s^{-1} was used for both the TPD and WF measurements.

bond, as the work function change assigned to this process is found at 150 K instead of at 110 K for the CH_3I case. In this case, a θ_{CH_x} of 0.08 ML is produced. The work function remains constant between 200–380 K, and increases further at around 400 K.

On Ni(111), adsorbed methyl groups, adsorbed in a threefold hollow site, give rise to a dipole active mode at 1220 cm^{-1}, attributed to their symmetric deformation mode, δ_s.[14] Fig. 1(c) shows the IR absorption spectrum after heating the CH_{3ad}-covered surface to the indicated temperatures. It can be seen that the CH_{3ad}-related peak persists up to 200 K, but disappears after heating to 250 K. Based on this spectroscopic observation we attribute the work function change at around 230 K to the decomposition of adsorbed methyl species. As no IR absorption bands were observed in the IR spectrum for higher temperatures, XPS measurements were used to further aid identification of the methyl decomposition products for $T > 200$ K. At 150 K we find a C 1s photoemission peak at 283.5 eV, close to the value found for methyl on the Ni(111) surface.[15] The XP spectrum recorded after heating to 300 K shows that the adsorbate formed at this temperature has a C 1s binding energy of 283.3 eV. On Ni(111) the CH_{ad} species appear at a binding energy (BE) of 282.5 eV,[15] around 1 eV lower than methyl species. As we instead find a ΔBE of only 0.2 eV, we attribute the C 1s spectrum at 300 K to acetylene instead. This is in line with the findings on Ni(111): CH, at 282.5 eV reacts at 250 K and forms acetylene, which has a binding energy of

283.3 eV. We confirmed this assignment by decomposing ethylene into acetylene,[16] which produced a peak at 283.3 eV.

These observations are in line with the hypothesis that the thermal behaviour of C_1H_x adsorbates on Co(0001) is practically identical to that on Ni(111), as described in ref. 14 and 15: on both surfaces adsorbed methyl groups decompose at around 230 K to produce CH_{ad} + $2H_{ad}$. This implies that CH_{2ad} is less stable than CH_{3ad} as it immediately reacts further after being formed from CH_{3ad}. The low stability of CH_{2ad} also explains why the work function measurements for CH_2Cl_2 show only one clear reaction step. CH_{2ad} being very unstable immediately dehydrogenates after its formation *via* C–Cl bond breaking, already at a temperature as low as 150 K. Like on Ni(111), the CH_{ad} species readily react further to acetylene, the most stable C_2H_x species on Co(0001) and significantly more stable than two separate CH species.[16,17] As CH_{ad} is only present in a narrow temperature window, we only identified the product of the $CH_{ad} + CH_{ad} \rightarrow C_2H_{2ad}$ reaction after heating to 300 K. It is however reasonable to assume that the coupling of two methylidyne adsorbates occurs at the same temperature as on Ni(111), that is, at around 250 K.[15] It follows from this reaction sequence that the H_2 desorption peak at 400 K is due to the dehydrogenation of acetylene. This is in line with studies of acetylene decomposition on Co(0001), which produces a (reaction-limited) H_2 desorption peak at 400 K as well.[18]

Methyl hydrogenation

In addition to H_2 formation due to CH_{3ad} decomposition, the formation of methane was also found when high CH_3I doses were used. Fig. 2(a) shows the temperature programmed work function measurements after saturating the surface with CH_3I (6 L at 100 K), whereas Fig. 2(b) shows the desorption spectra of H_2 as well as of CH_4 for increasingly high methyl coverages >0.14 ML. As multiple products desorb in this coverage regime, quantification of the initial CH_{3ad} concentration was not possible. Instead, we report the dose of CH_3I used in the experiment, as well as the quantity of CH_{3ad} that dehydrogenates as derived from the H_2 produced. The work function measurements show that a high initial methyl coverage (and an equally high quantity of I_{ad}) causes a ~45 K upward shift of the work function change attributed to CH_{3ad} decomposition. Due to the low vacancy concentration to accommodate hydrogen produced in the dehydrogenation step, only a small quantity of the methyl groups can decompose, whereas the majority remain intact up to 250 K. At this temperature they can react with surface hydrogen and leave the surface as methane. Methane formation terminates when desorption of surface hydrogen leaves the surface as H_2: at this point the vacancies needed for CH_{3ad} dehydrogenation become available and CH_{3ad} decomposition becomes the preferred reaction pathway.

By dosing 500 L H_2 at 100 K, the surface is covered by 0.5 ML H_{ad}.[19,20] Iodomethane can still adsorb and decompose onto this hydrogen-covered surface, as evident from the desorption of methane during a subsequent TPD experiment, shown in Fig. 2(c) and (d). The low quantity of vacancies for H_{ad} to adsorb suppresses CH_{3ad} decomposition and instead methane formation *via* $H_{ad} + CH_{3ad}$ occurs, with a peak maximum at 250 K. H_2 desorption during further heating leads to a selectivity shift from hydrogenation to dehydrogenation, as the surface sites required for dehydrogenation become available. The occurrence of CH_{3ad}

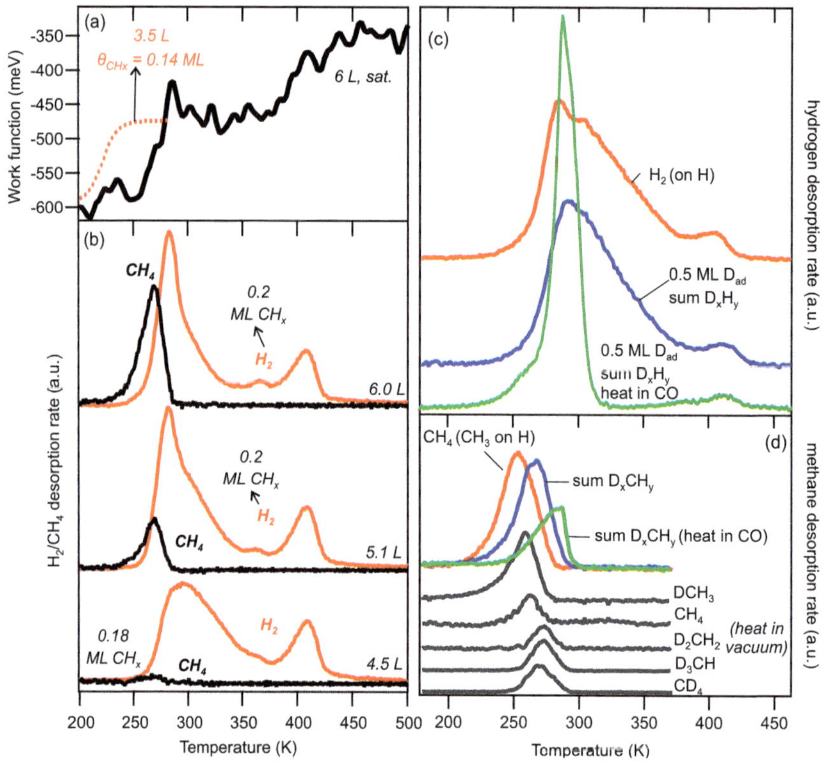

Fig. 2 (a) Work function measurements and (b) TPD spectra for high doses of CH_3I, showing methane formation alongside H_2 desorption due to decomposition. TPD spectra after dosing CH_3I onto a 0.5 ML H_{ad}(/D_{ad})-covered surface, showing desorption of hydrogen (c) and methane (d), respectively, both during heating in the absence and in the presence of 5×10^{-7} mbar CO. Grey lines in (d) show the isotopologues formed during heating under vacuum. A heating rate of 0.5 K s^{-1} was used in all experiments.

dehydrogenation is evident from the H_2 desorption peak at 400 K, attributed to acetylene dehydrogenation. Co(0001) and Ni(111) show a similar reactivity for $CH_{3ad} + H_{ad}$, which occurs at around 250 K in both cases, given that the hydrogen coverage is high enough so that methyl decomposition is suppressed.[21]

When the surface is pre-saturated using D_2 we find that both the onset and the peak maximum of the methane formation peak (D_xCH_y, the sum of all isotopologues) is shifted to higher temperatures, attributed to a kinetic isotope effect. The formation of all possible methane isotopologues, shown separately with grey lines in (d), is indicative of extensive exchange between surface-bound D_{ad} and CH_{3ad}. At the onset of the methane formation peak, at 220 K, DCH_3 is exclusively formed. At this point only D_{ad} and CH_{3ad} are present and isotope exchange between CH_{3ad} and D_{ad} cannot happen, as the concentration of vacancies needed to catalyse the dehydrogenation–rehydrogenation sequence to exchange hydrogen with deuterium are not available. As soon as both methyl and D_{ad} start to be consumed these vacancies start to appear, and isotope exchange becomes possible. As this generates H_{ad} it then becomes possible to form CH_4. With a further increase of temperature, more extensive isotope exchange can occur, and the high temperature end of the methane formation peak is due to the more D-rich isotopologues of methane. Like for the H_2 experiments, decomposition of CH_{3ad} is evident from the ~400 K desorption peak, which consist for around 70% of deuterium. The comparatively higher temperature with respect to the peak for the H_2 experiment is attributed to a kinetic isotope effect on the rate of C_nH_n dehydrogenation.

The influence of CO on the methane formation was studied in an experiment where CH_3I was dosed onto a 0.5 ML D_{ad}-covered surface. After heating to 160 K to break the C–I bond, a CO pressure of 5×10^{-7} mbar was introduced into the chamber and desorption of all hydrogen and methane isotopologues was measured during subsequent heating under a CO gas atmosphere. The presence of CO induces a 15 K upward shift of the onset of methane formation relative to the D_{ad}–CH_{3ad} co-adsorbed layer heated under vacuum. All possible isotopologues of methane were observed (not shown here), indicating that isotope exchange is not affected by the presence of CO. Methane formation is abruptly aborted by the desorption of deuterium, in a sharp desorption peak at around 290 K due to co-adsorbed CO, as discussed hereafter in more detail. Finally, the hydrogen desorption peak at around 400 K indicates that methyl dehydrogenation also occurs.

The influence of CO_{ad} on surface hydrogen

Surface hydrogen is an essential reactant in FTS surface chemistry and since CO_{ad} is abundantly present under FTS conditions,[4,22,23] we studied how adsorbed CO and H_{ad} mutually affect each other. To investigate how much CO and H_{ad} can co-adsorb on the Co(0001) surface, we first dosed hydrogen onto the surface held at 100 K, followed by a high CO dose to populate all remaining sites with CO_{ad}. A selection of TPD experiments for various initial hydrogen coverages is shown in Fig. 3(a), where the spectra in blue represent the pure H_{ad} (0.5 ML) and pure CO_{ad} (0.67 ML (ref. 9 and 24)) layers, respectively.

Hydrogen desorption is strongly affected by post-dosed CO. For a low (0.05 ML) hydrogen coverage a single H_2 desorption peak is found at around 280 K, around 100 K lower than for the same quantity of H_{ad} desorbing from the CO-free surface (see *e.g.* ref. 19 and 20). This peak gains intensity as a function of hydrogen pre-dose, but the typical downward shift of the desorption peak maximum expected for a 2nd order desorption process is not observed, and the peak maximum remains constant around 270–280 K. In this coverage regime the CO desorption spectrum shows only minor changes compared to the reference case without H_{ad}. Both desorption peaks <350 K decrease in intensity, and in particular the sharp desorption peak at 240 K shows a slight downward shift. For $\theta_H > 0.24$ ML a high temperature shoulder develops in the H_2 desorption spectrum. In this coverage regime the CO desorption spectrum is also more significantly affected.

Co-adsorbed CO and H_{ad} form segregated ad-islands on Co(0001) at a low temperature, similar to what has been reported for the close-packed surfaces of other metals.[25] Lewis *et al.*[26,27] show that adsorbed hydrogen is compressed into ad-islands, a (1×1) structure with a local coverage of 1 ML due to co-adsorbed CO. This means that the downward shift of the H_2 desorption peak maximum caused by CO_{ad} is due to H_2 desorption from a compressed (1×1) H_{ad} layer, which, according to DFT calculations, is a less stable structure than the one with

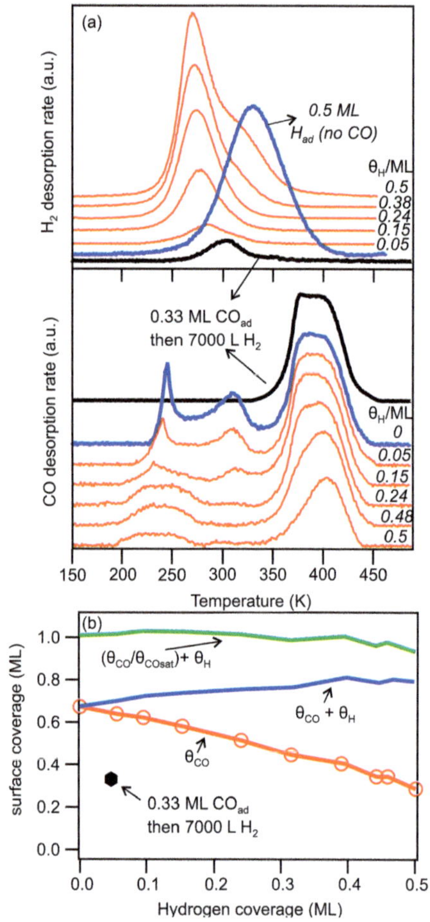

Fig. 3 (a) H_2 and CO TPD spectra (1 K s^{-1}) after pre-adsorbing increasing quantities of H_{ad}, followed by a saturation dose of CO. (b) Quantification of the $\theta_H-\theta_{CO}$ correlation from the spectra shown in (a). The TPD spectra in black in (a) and the data point (hexagonal marker) in (b) represent an experiment where H_2 was dosed onto a 0.33 ML CO_{ad}-covered surface.

a surface coverage of 0.5 ML.[19] The effect of CO becomes less pronounced for higher H_{ad}/CO_{ad} ratios, leading to the appearance of a high temperature shoulder in the H_2 desorption spectrum. However, a comparison with the H_2 desorption peak temperature in the absence of CO reveals a modest effect of CO_{ad} on this high temperature shoulder as well, as it is shifted by ~35 K to lower temperatures compared to the CO-free case.

Fig. 3(b) shows a plot of the quantity of CO adsorbed as a function of H_{ad} pre-coverage, where the quantity of hydrogen and CO were determined using TPD. Work function measurements corroborate the quantification of θ_{CO} derived from the TPD experiment.[9,24] The CO coverage decreases approximately linearly with increasing θ_H, but the sum of θ_{CO} and θ_H is not constant and in addition it is <1 ML. As the CO compresses the H_{ad} layer into (1 × 1) islands, the global value of θ_H corresponds to the surface fraction covered by H_{ad} islands. As the CO saturation coverage at 100 K in the pressure regime <10^{-4} mbar is 0.67 ML (ref. 9 and 24) at 100 K, the fraction of the surface covered by CO-saturated islands should be equivalent to $\theta_{CO}/0.67$. Indeed, as shown in Fig. 3(b) the sum of θ_H + ($\theta_{CO}/0.67$) is approximately constant at a value of 1.

The influence of CO_{ad} on the dissociative adsorption of H_2 was studied in an experiment where a 0.33 ML CO-covered surface was exposed to a high H_2 dose of 7000 L at 100 K, a dose that is one order of magnitude larger than the dose needed to saturate the surface in the absence of CO_{ad}. The H_2 desorption spectrum, shown in black in Fig. 3(a), reveals that hardly any hydrogen has adsorbed onto the surface. The quantity of H_{ad} found in this case, 0.05 ML, [black marker in Fig. 3(b)], is much lower than the 0.45 ML H_{ad} adsorbed alongside 0.33 ML CO when H_2 is dosed first. In other words, CO_{ad} lowers the dissociative sticking coefficient of hydrogen, *i.e.* it affects the kinetics of dissociative H_2 adsorption. Similar effects of CO have been reported on other metal surfaces. Morkel, Rupprechter and Freund,[28] for example, find that H_{ad} on Pd(111) suppresses CO adsorption at a low surface temperature, whereas hydrogen is replaced by CO when a higher surface temperature is used. They also find that CO_{ad} inhibits dissociative adsorption of H_2. Suppression of dissociative hydrogen adsorption by CO was also found in a temperature and pressure regime where both hydrogen and CO adsorption are in dynamic equilibrium with the gas phase. Johansson, Lytken and Chorkendorff[29] find that ppm levels of CO in a 99% H_2/1% D_2 mixture significantly suppresses the HD formation rate on evaporated metal films of various metals, an effect that can be traced back to a decreased dissociative sticking coefficient due to the presence of CO.

Discussion

The first set of experiments (Fig. 1 and 2) provides detailed information on the chemistry of C_1H_x surface species under model conditions and highlights that product selectivity depends on the concentration of surface hydrogen as well as on the concentration of vacancies where hydrogen can be accommodated. When free sites are present in abundance, methyl dehydrogenation to CH followed by coupling to C_2H_2 is the dominant path. However, when the sites available for hydrogen to adsorb are occupied to a large extent, hydrogenation of methyl adsorbates to methane is favoured instead.

The second set of experiments (Fig. 3) shows that surface hydrogen is in several ways affected by CO_{ad}. When hydrogen is adsorbed first at a temperature where desorption is impossible, we find that the adsorption capacity for CO decreases with increasing hydrogen coverage. As the coverages of the two adsorbates at saturation do not add up to unity, it is clear that a single site model is too simple to capture the complexity of CO-H_{ad} co-adsorption. For our simple model system, a reasonable quantitative description of the correlation between the values of θ_H and θ_{CO} could be obtained by taking adsorbate segregation into account. However, at the operation temperatures used for FTS, the entropy contribution due to mixing of adsorbates makes reactant segregation under these conditions unlikely. Moreover, under the dynamic conditions of an adsorption–desorption equilibrium, the site blocking effect most likely works the other way: CO adsorbs more strongly than H_{ad} so that the CO concentration dictates the quantity of hydrogen that can be adsorbed next to it. The high complexity of the co-adsorbed system makes reliable predictions of the surface hydrogen concentration under reaction conditions difficult, and instead the concentration of surface hydrogen under steady state reaction conditions can be more easily obtained from experiments. In the SSITKA approach adopted by Den Breejen *et al.*,[2] a hydrogen coverage of 0.1 ML was found to co-adsorb next to 0.43 ML CO on a cobalt catalyst operated at 210 °C with $p_{H_2} = 550$ mbar and $p_{CO} = 55$ mbar.

The critical role of H vacancies

Such experiments provide us with an estimate of the quantity of surface hydrogen available under reaction conditions, but the concentration of vacancies where hydrogen can adsorb is still unknown. Even simple model experiments show that for adsorbates of a different chemical nature, their concentrations at saturation do not add up to one, so we cannot assume that $\theta_* = 1 - \sum\theta_n$. Instead, we can estimate the vacancy concentration by assuming that site blocking by CO and all other strongly bound adsorbates such as CH_x and OH_x[2] leave some surface space open for hydrogen to adsorb, thereby setting the value of θ_{Hmax}. The hydrogen adsorption–desorption equilibrium is then established on the area available. To indicate that the free sites are specifically those available for hydrogen to adsorb we refer to them as θ_{*H} in the following discussion.

In a Langmuir model, the rate of adsorption is described by $r_{ads} = k_1 p_{H_2}(1 - \theta_H)^2$, and the rate of desorption by $r_{des} = k_{-1}\theta_H{}^2$, where k_1 and k_{-1} are the rate constants for adsorption/desorption, respectively, and θ_H is the fraction of occupied sites. The rate constant for adsorption can be written as $k_1 = \frac{S_0}{\sqrt{2\pi kTm}}$, where S_0 is the initial sticking coefficient, and a term to correlate molecular arrival rate per site per second with the gas pressure, which depends on the mass of the gas molecule (m) and the gas temperature (T). The rate constant for desorption, k_{-1}, depends on the barrier for H_2 desorption *via* $k_{-1} = A\exp(-E_a/RT)$. At FTS temperatures in the order of 200–250 °C and a hydrogen pressure of 1 bar, the collision frequency is in the order of $\sim 5 \times 10^8$ per Co surface atom per second (and $\sim 1.3 \times 10^8$ for CO due to its higher molar mass). With a flux of incoming molecules that is 10 orders of magnitude larger than the typical turn-over-frequency of the FTS reaction, in the order of $\sim 10^{-2}$ s^{-1} per surface site, it is rather likely that the hydrogen and free site coverage are determined by the adsorption–desorption equilibrium, and consumption *via* the reaction has a negligible effect. For such a quasi-equilibrium situation the following relations hold (see *e.g.* ref. 30, where $K = k_1/k_{-1}$):

$$\theta_H = \frac{\sqrt{Kp_{H_2}}}{1 + \sqrt{Kp_{H_2}}} \tag{1}$$

with $\theta_{*H} = 1 - \theta_H$ the free site concentration is then:

$$\theta_{*H} = \frac{1}{1 + \sqrt{Kp_{H_2}}} \tag{2}$$

and their ratio $\frac{\theta_H}{\theta_{*H}}$ becomes:

$$\frac{\theta_H}{\theta_{*H}} = \sqrt{Kp_{H_2}} \tag{3}$$

On the CO-free surface the barrier for hydrogen desorption is in the order of 90–100 kJ mol^{-1}, see ref. 19, but our experiments show a destabilizing effect of CO_{ad} on surface hydrogen and the effective barrier for H_2 desorption under FTS conditions is most likely lower. In addition to this, we find that CO suppresses dissociative adsorption of H_2, so it affects both the adsorption and desorption term in the equilibrium.

To qualitatively explore the potential influence of CO-induced effects on the adsorption–desorption equilibrium we used eqn (1) and (3) to calculate θ_H as well as θ_H/θ_{*H} as a function of the initial sticking coefficient, and for different values of the barrier for H_2 desorption, plotted in Fig. 4(a). It is clear that the hydrogen coverage changes very little with sticking coefficient, and, unless the desorption barrier for hydrogen becomes smaller than 80 kJ mol^{-1}, it is close to saturation. As $\theta_{*H} = 1 - \theta_H$ the free site concentration is typically much smaller than that of θ_H. The free site concentration as well as the ratio of θ_H/θ_{*H}, [plotted in Fig. 4(b)] is therefore much more strongly affected than θ_H by factors that influence either the adsorption or the desorption process, such as a change of the initial sticking coefficient and the desorption barrier of hydrogen.

Speciation of C_1H_x on close-packed terraces

The absolute hydrogen coverage determines the rate of irreversible hydrogenation reactions such as $CH_{3ad} + H_{ad} \rightarrow CH_4$, but the ratio between occupied and empty sites for hydrogen affects all reversible hydrogenation reactions, for example the interconversion of C_1H_x species explored here. Our data indicate that methylidyne (CH) is the most stable C_1H_x species on Co(0001), *i.e.* the rate constant for CH hydrogenation must be smaller than that of CH_{2ad} and CH_{3ad} dehydrogenation. However, with reaction rates being a function of both the reaction rate constant as well as reactant coverage a high value of θ_H/θ_{*H} counter-acts the stability difference and favours the less stable but more hydrogenated C_1H_x species. In kinetic terms, a low free site concentration diminishes the probability that a CH hydrogenation step is reversed. The interplay between θ_H/θ_{*H} and the relative stabilities within the C_1H_x pool can be explored most easily by assuming a quasi-

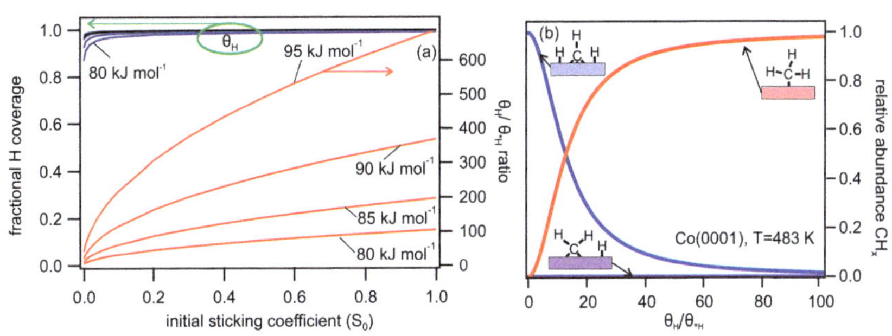

Fig. 4 (a) Fractional hydrogen coverage as well as the hydrogen/free site ratio as a function of the initial dissociative sticking coefficient for H_2, for a range of activation energies, assuming a pre-factor of 1×10^{13} and a temperature of 210 °C. $p_{H_2} = 1$ bar. (b) Relative abundance of the various C_1H_x species under equilibrium conditions as a function of θ_H/θ_{*H} at 210 °C and using ΔH values reported by Cheng *et al.*[31] to calculate equilibrium constants.

equilibrium between the different C_1H_{xad} species, a simplification which eliminates kinetic factors and only requires information on the relative stabilities to calculate the equilibrium constant. For the reaction $CH_{3ad} + {}^*H \leftrightarrow CH_{2ad} + H_{ad}$, the extensive isotope incorporation seen in the experiments shows that both reactions can occur, below room temperature even. The equilibrium constant K_{eq} is given by:

$$\frac{\theta_{CH_2}\theta_H}{\theta_{CH_3}\theta_{*H}} = e^{-\Delta G/RT} \tag{4}$$

Our experiments provide us with the following order of stability:

CH_{ad} (+2H_{ad}) > CH_{3ad} > CH_{2ad} (+H_{ad}), but numerical values for ΔG cannot be derived from our dataset. We can make use of the computed energy differences between CH_{3ad}, CH_{2ad} + H_{ad} and CH_{ad} + 2H_{ad} reported on Co(0001) by Cheng *et al.*[31] to explore trends in the interplay between C_1H_{xad} stability and θ_H/θ_{*H}.

Fig. 4(b) shows how the relative stabilities of the various C_1H_x adsorbates with $x = 1$–3 on Co(0001) translate to relative concentrations as a function of θ_H/θ_{*H}. It is interesting to note that the concentration of CH_{2ad}, the popular inserting monomer in the classical alkyl mechanism, is never present in significant concentrations, as it is less stable than either CH_{ad} or CH_{3ad}. The methyl group, CH_{3ad}, is less stable than CH_{ad} as well, so CH is dominant for low values of θ_H/θ_{*H}. However, a high ratio effectively counteracts the stability difference and CH_{3ad} becomes dominant, despite its comparatively lower stability. Hence, the C_1H_{xad} pool is either dominated by CH_{ad} for a low θ_H/θ_{*H} or by CH_{3ad} for high θ_H/θ_{*H}. In kinetic terms, conditions that favour CH_{3ad} mean that dehydrogenation reactions of CH_{2ad} and CH_{3ad} are slow, hydrogenation dominates, and methane selectivity is high. Instead, for conditions that favour CH_{ad}, the concentration of free sites to accommodate hydrogen is sufficiently large so that CH_{3ad} dehydrogenation to CH (a process with a high intrinsic rate constant, as we observed in our experiments at 220 K previously) becomes favoured over methyl hydrogenation. This is beneficial for selectivity: CH_{ad} is highly reactive for coupling reactions and we were therefore not able to observe it in our experiments as it readily forms acetylene below room temperature.[16,17] By extrapolation we can envisage a similar influence of θ_H/θ_{*H} on the speciation in the pool of C_{2+} species as well. The more hydrogenated surface intermediates, although less stable than hydrogen-lean species such as alkylidyne and alkynes,[16,17] are favoured at high θ_H/θ_{*H} values and consequently termination is fast and chains remain short. Conversely, a lower ratio favours alkyne and alkylidyne species. These are facile intermediates in the alkylidyne chain growth mechanism where chain growth proceeds *via* CH + CR; this reaction is equivalent to CH + CH which was found to occur below room temperature on the close-packed surfaces of cobalt and nickel.

Implications for selectivity

With these insights we can now rationalize the interplay between reaction selectivity and reactant partial pressures. Due to the high reactant pressures, which translate to collision frequencies in the order of 10^8 to 10^9 s^{-1} per surface atom, combined with a relatively low temperature and a high sticking coefficient, the CO coverage is expected to be high already at modest CO partial pressures.

The value of θ_{CO} therefore does not change very much when the CO partial pressure is increased.[9] As the available space for hydrogen to adsorb depends on the value of θ_{CO}, the value of θ_{Hmax} is expected to be only modestly lowered when p_{CO} increases. However, CO_{ad} also inhibits the dissociative adsorption of H_2. We can capture this effect by assuming that the adsorption probability of H_2 scales with the concentration of free sites left open by the CO adsorption–desorption equilibrium, θ_{*CO}. With $\theta_{CO} \gg \theta_{*CO}$ at FTS pressures and temperatures a change of CO partial pressure mainly affects the free site concentration, which has an equivalent impact on the dynamic H_2 adsorption–desorption equilibrium as a decrease of the initial sticking coefficient. Thus, a change of the CO partial pressure affects θ_H by site blocking, but the most pronounced effect is its impact on θ_H/θ_{*H}, with a corresponding impact on selectivity as discussed in the previous paragraph. When the hydrogen partial pressure is changed, the adsorption capacity for hydrogen is not affected, and it mainly affects the value of θ_H/θ_{*H} with corresponding influence on selectivity.

Surface residence times

A second issue concerns the long residence times of C_1H_x species. SSITKA experiments indicate a typical residence time of 10 seconds,[2,4,5] *i.e.* a rate constant of 0.1 s^{-1} at a temperature of 483 K. In our single crystal experiments we find methyl hydrogenation around 250 K, indicative of a barrier in the order of 60–70 kJ mol^{-1}. The latter translates to an elementary rate constant in the order of 10^5 s^{-1} at a reaction temperature of 483 K (210 °C). We find in our experiments that the presence of CO causes a modest (~15 K) upward shift of the methane formation temperature, but this is not sufficiently large to bridge the gap between long C_1H_x residence times and methyl hydrogenation occurring at or below room temperature already.

If C_1H_x is predominantly present as CH_{3ad}, adsorbed on terrace sites the residence time can be calculated from the elementary reaction barrier for CH_{3ad} hydrogenation *via* $\tau_{CH_3} = 1/(k\theta_H)$. Thus, a rate constant of 1×10^5 s^{-1} requires a hydrogen concentration of 10^{-6} ML to compensate for the high rate constant and produce a residence time of 10 s. This potential explanation that the long C_1H_x residence time is caused by a very low concentration of surface hydrogen under reaction conditions is at odds with experimental observations. Den Breejen *et al.*[2] report a surface hydrogen concentration in the order of 0.1 ML, which, combined with a rate constant of 1×10^5 s^{-1}, yields a methyl residence time of only 1×10^{-4} s.

Low values of θ_H/θ_{*H} favour CH_{ad} instead of CH_{3ad}, and this could be an alternative explanation for the long C_1H_x residence time. Assuming a pre-equilibrated C_1H_x pool, a rate constant of 1×10^5 s^{-1} for methyl hydrogenation combined with a θ_H of 0.1 ML would then require a CH concentration that is ~10^5 times higher than the CH_{3ad} concentration, a condition that, according to the evaluation shown in Fig. 4(b), is only fulfilled on terrace sites when θ_{*H} is around 20 times larger than θ_H. As this is physically impossible when θ_H is 0.1 ML, this explanation cannot account for the experimentally found long C_1H_x residence time of ~10 s. In addition, if C_1H_x is predominantly present in the form of CH_{ad} on terrace sites extensive chain growth would be expected, which is at odds with the experimental observation of a high (65%) methane selectivity and a chain growth probability of only 0.4.

Thus, the long residence times must be due to a C_1H_x species that is strongly bound and difficult to hydrogenate, and it must also be unreactive for coupling, to satisfy the experimental observation of a long residence time combined with high methane selectivity. These requirements exclude CH_{ad}, CH_{2ad} and CH_{3ad} adsorbed on terrace sites, since our experiments show that they are either very unstable (*e.g.* CH_2), readily couple (*e.g.* $CH_{ad} + CH_{ad} \rightarrow C_2H_{2ad}$), or react to form methane at a low temperature (CH_{3ad}).

DFT calculations show that atomic carbon is difficult to hydrogenate.[31] On a terrace site C_{ad} is less stable than CH_{ad}, but step edges stabilize atomic carbon comparatively more than the other C_1H_{xad} species, and consequently C adsorbed at a step edge is more stable than CH on terrace sites. Indeed, this is a likely source of atomic carbon when we take into consideration that carbon in FTS comes from CO. Step edge sites are known to be active for direct CO dissociation,[32] and, owing to its high adsorption energy, C_{ad} is rather immobile and hydrogenation to C_1H_{xad} species is required to provide enough to migrate to the adjacent terrace sites. As the CO concentration is high, the vacancy created at the step edge is quickly filled by a CO dissociation step, inhibiting re-attachment of the C_1H_{xad} species to the step site and trapping it to the terrace site. The hypothesis that the active sites for CO dissociation are filled to a large extent by strongly bound species such as C and O is in line with the high CH_x and OH_x concentrations in the order of 0.1–0.2 ML found by SSITKA. We therefore propose that hydrogenation of C at a step edge followed by CH_x migration to the terrace is the rate-limiting step that gives rise to a residence time of about 10 s. After this step, residence times of CH_x species on terrace sites are much smaller, <1 s, so that their contribution to the overall residence time is negligible. Now the value of θ_H/θ_{*H} comes into play: when this value is high, dehydrogenation reactions are suppressed and the lifetime of CH_{ad}, the monomer form required for coupling reactions, is small, *i.e.* there is a small chance for chain growth to occur before the monomer has become irreversibly hydrogenated to methane. Instead, for a low value of θ_H/θ_{*H} the time spent as CH_{ad} is larger, which increases the probability to meet with either another CH_{ad} or with a growing chain and makes chain growth more efficient.

Summary and conclusions

In this paper we discuss how information on the elementary reaction steps of C_1H_x surface species can be reconciled with observations on applied cobalt FTS catalysts under semi-realistic conditions, in particular regarding the selectivity of the catalyst as a function of the reactant pressures. An exploration of the chemistry of CH_{3ad} and CH_{2ad} adsorbates on a close-packed surface of Co reveals that surface hydrogen in combination with the availability of empty sites to accommodate hydrogen strongly affect product selectivity: starting with CH_{3ad}, a high number of empty sites leads to CH_{ad} formation, followed by $CH_{ad} + CH_{ad}$ coupling to produce adsorbed acetylene. This route is suppressed when free sites are scarce, either by a high initial coverage of CH_{3ad} and I_{ad} or by pre-adsorbing H_{ad}, and methane can instead be formed. Under FTS reaction conditions the value of θ_H/θ_{*H} is therefore held responsible for the selectivity of the reaction, where a high ratio favours the formation of methyl species and fast methane formation, whereas a low ratio promotes CH species with a high reactivity for chain growth reactions.

Experiments show that CO_{ad} strongly suppresses dissociative adsorption of H_2, preventing the experimental determination of the H_{ad} saturation point at a given CO pre-coverage using UHV-type pressures. By first adsorbing H_{ad}, we find that pre-adsorbed H_{ad} limits the available space for CO to adsorb. With CO being the most strongly adsorbed we propose that under the reaction conditions θ_{CO} is not affected by the presence of H_{ad}. Instead, the value of θ_{CO} defines θ_{Hmax}, the quantity of H_{ad} that can adsorb next to it. At the temperatures and pressures used under FTS conditions, the value of θ_{CO} does not change much with a change in the CO partial pressure, and θ_H is therefore also expected to vary only modestly with a change of CO partial pressure. Instead, the ratio of occupied over free sites, θ_H/θ_{*H}, depends more strongly on the CO pressure: CO suppresses dissociative adsorption of H_2 and the small number of unoccupied sites (θ_{*CO}) are essential as catalysts for H_2 dissociation. As their concentration varies significantly with p_{CO}, the CO pressure indirectly affects θ_H/θ_{*H} and thereby the selectivity.

Surface residence times of ~10 s for C_1H_{xad} species as indicated by SSITKA experiments, even under conditions that favour methane formation, cannot be attributed to C_1H_{xad} species adsorbed on terrace sites as they react on a much shorter time scale. We instead propose that the ~10 s residence time is due to the slow hydrogenation of step-bound carbon atoms formed by CO dissociation at these active sites. Their endothermic migration to the terrace is made irreversible due to subsequent CO dissociation blocking C_1H_x re-attachment to the step edge. Further reactions of terrace-bound C_1H_x species, be it coupling of CH species or methane formation, *via* CH_{3ad} + H_{ad} occurs on a much shorter time scale and changes in their relative rates do not contribute to the observed C_1H_x residence time in SSITKA experiments.

Acknowledgements

Sasol group Technology is acknowledged for allowing to use data collected during CJW's employment by Sasol. Several ideas in this paper arose during collaborations and discussions with colleagues from Sasol group Technology, Dr Ionel Ciobîcă, Pieter van Helden, Denzil Moodley, Abdool Saib and Jan van de Loosdrecht. We acknowledge ASTRID2, the synchrotron light source of the Department of Physics and Astronomy at the University of Aarhus for beam time at the MatLine end station. Gilbere Mannie, Zheshen Li and Yijia Li are acknowledged for their participation in the XPS measurement series. Hans Fredriksson is acknowledged for participating in the XPS measurement series and for providing critical comments on the manuscript. Johan den Breejen is acknowledged for providing additional information on the chain growth probability for the study reported in ref. 2. Syngaschem BV acknowledges funding from Synfuels China Technology Co. Ltd. Profs. Yong-Wang Li, Xiaodong Wen and Dr Jose Gracia and colleagues from SynCat@Beijing are acknowledged for valuable discussions.

References

1 R. A. van Santen, A. J. Markvoort, I. A. W. Filot, M. M. Ghouri and E. J. M. Hensen, *Phys. Chem. Chem. Phys.*, 2013, **15**, 17038.

2 J. P. Den Breejen, P. B. Radstake, G. L. Bezemer, J. H. Bitter, V. Frøseth, A. Holmen and K. P. de Jong, *J. Am. Chem. Soc.*, 2009, **131**, 7197–7203.

3 C. A. Mims, J. J. Krajewski, K. D. Rose and M. T. Melchior, *Catal. Lett.*, 1990, **7**, 119–125.
4 C. J. Bertole, G. Kiss and C. A. Mims, *J. Catal.*, 2004, **223**, 309–318.
5 V. Frøseth, S. Storsæter, Ø. Borg, E. A. Blekkan, M. Rønning and A. Holmen, *Appl. Catal., A*, 2005, **289**, 10–15.
6 C. A. Mims and L. E. McCandlish, *J. Phys. Chem.*, 1987, **91**, 929–937.
7 P. van Helden, I. M. Ciobîcă and R. L. J. Coetzer, *Catal. Today*, 2016, **261**, 48–59.
8 J. X. Liu, H. Y. Su, D. P. Sun, B. Y. Zhang and W. X. Li, *J. Am. Chem. Soc.*, 2013, **135**, 16284–16287.
9 C. J. Weststrate, J. van de Loosdrecht and J. W. Niemantsverdriet, *J. Catal.*, 2016, **342**, 1–16.
10 F. Zaera, *Chem. Rev.*, 1995, **95**, 2651–2693.
11 S. Tjandra, H. Guo and F. Zaera, *Top. Catal.*, 2011, **54**, 26–33.
12 S. Tjandra and F. Zaera, *J. Vac. Sci. Technol., A*, 1992, **10**, 404.
13 F. Steinbach, J. Kiss and R. Krall, *Surf. Sci.*, 1985, **157**, 401–412.
14 Q. Y. Yang, K. J. Maynard, A. D. Johnson and S. T. Ceyer, *J. Chem. Phys.*, 1995, **102**, 7734–7749.
15 R. Denecke, *Appl. Phys. A*, 2005, **80**, 977–986.
16 C. J. Weststrate, I. M. Ciobîcă, A. M. Saib, D. J. Moodley and J. W. Niemantsverdriet, *Catal. Today*, 2014, **228**, 106–112.
17 C. J. Weststrate, P. van Helden and J. W. Niemantsverdriet, *Catal. Today*, 2016, **275**, 100–110.
18 J. Vaari, J. Lahtinen and P. Hautojarvi, *Catal. Lett.*, 1997, **44**, 43–49.
19 P. van Helden, J.-A. van den Berg and C. J. Weststrate, *ACS Catal.*, 2012, **2**, 1097–1107.
20 K. M. E. Habermehl-Cwirzen, K. Kauraala and J. Lahtinen, *Phys. Scr.*, 2004, **108**, 28–32.
21 S. Tjandra and F. Zaera, *J. Catal.*, 1994, **147**, 598–600.
22 A. C. Kizilkaya, J. W. Niemantsverdriet and C. J. Weststrate, *J. Phys. Chem. C*, 2016, **120**, 4833–4842.
23 J. Yang, E. Z. Tveten, D. Chen and A. Holmen, *Langmuir*, 2010, **26**, 16558–18567.
24 J. Lahtinen, J. Vaari and K. Kauraala, *Surf. Sci.*, 1998, **418**, 502–510.
25 X. C. Guo and D. A. King, *Chem. Phys. Solid Surf.*, 1993, 114.
26 E. A. Lewis, D. Le, A. D. Jewell, C. J. Murphy, T. S. Rahman and E. C. H. Sykes, *Chem. Commun.*, 2014, **50**, 6537–6539.
27 E. A. Lewis, D. Le, A. D. Jewell, C. J. Murphy, T. S. Rahman and E. C. H. Sykes, *ACS Nano*, 2013, **7**, 4384–4392.
28 M. Morkel, G. Rupprechter and H. J. Freund, *J. Chem. Phys.*, 2003, **119**, 10853–10866.
29 M. Johansson, O. Lytken and I. Chorkendorff, *Surf. Sci.*, 2008, **602**, 1863–1870.
30 I. Chorkendorff and J. W. Niemantsverdriet, *Concepts of Modern Catalysis and Kinetics*, Wiley VCH, 2nd edn, 2007.
31 J. Cheng, X. Q. Gong, P. Hu, C. M. Lok, P. Ellis and S. French, *J. Catal.*, 2008, **254**, 285–295.
32 C. J. Weststrate, P. van Helden, J. van de Loosdrecht and J. W. Niemantsverdriet, *Surf. Sci.*, 2016, **648**, 60–66.

Faraday Discussions

PAPER

Computational investigation of the kinetics and mechanism of the initial steps of the Fischer–Tropsch synthesis on cobalt†

Pieter van Helden,*[a] Jan-Albert van den Berg,[a] Melissa A. Petersen,[a] Werner Janse van Rensburg,‡[a] Ionel M. Ciobîcă[b] and Jan van de Loosdrecht[a]

Received 20th September 2016, Accepted 26th September 2016
DOI: 10.1039/c6fd00197a

A multi-site microkinetic model for the Fischer–Tropsch synthesis (FTS) reaction up to C_2 products on a FCC cobalt catalyst surface is presented. This model utilizes a multi-faceted cobalt nanoparticle model for the catalyst, consisting of the two dominant cobalt surface facets Co(111) and Co(100), and a step site represented by the Co(211) surface. The kinetic parameters for the intermediates and transition states on these sites were obtained using plane-wave, periodic boundary condition density functional theory. Using direct DFT data as is, the microkinetic results disagree with the expected experimental results. Employing an exploratory approach, a small number of microkinetic model modifications were tested, which significantly improved correspondence to the expected experimental results. Using network flux and sensitivity analysis, an in-depth discussion is given on the relative reactivity of the various sites, CO activation mechanisms, the nature of the reactive chain growth monomer, the probable C_2 formation mechanism, the active site ensemble interplay and the very important role of CO* surface coverage. The findings from the model scenarios are discussed with the aim of guiding future work in understanding the FTS mechanism and subsequent controlling kinetic parameters.

1 Introduction

The Fischer–Tropsch synthesis (FTS) reaction converts a simple mixture of hydrogen and CO to a wide range of hydrocarbon products. Although this reaction is technologically well developed and extensively studied, the mechanism that proceeds on the active catalyst surface can be considered to be one of most

[a]*Sasol Group Technology, P.O. Box 1, Sasolburg 1947, South Africa. E-mail: pieter.vanhelden@sasol.com*
[b]*Sasol Technology Netherlands B.V., Vlierstraat 111, 7544 GG Enschede, The Netherlands*

† Electronic supplementary information (ESI) available: Additional descriptions of methodology and validation work, a summary of experimental $\acute{S}$ values, as well as additional tables and figures with micorkinetic model results can be seen here. See DOI: 10.1039/c6fd00197a

‡ Present address: Centre for High Performance Computing, 15 Lower Hope Road, Rosebank, Cape Town, South Africa, 7700.

complex.[1] In a basic process of FTS on a catalyst surface, the bond of the adsorbed CO molecule is somehow split and, together with dissociated H_2, the carbon atoms end up in a pool of hydrocarbon and oxygenated products. The O atoms end up as water, and to a lesser extent, CO_2. However, despite this apparent simplicity, the presence of these atoms and species on the surface allows for a very large variety of mechanisms to be proposed. In-depth reviews of all these possibilities can be found elsewhere[2–4] and only the main points of interest to this study are pointed out here. Some of the important points of the FTS mechanism that are currently debated include the mechanism of CO bond scission, the nature of the chain growth monomer, the mechanism of chain growth, and the nature of the rate-controlling reaction steps, intermediates and active site.

While empirical studies may provide some insight into these issues, as well as show how the activity and selectivity depend on process conditions, and surface science studies can isolate and study selected mechanistic steps on specific surface facets, the advent of computational chemistry has enabled the study of all the fundamental surface reactions steps at the atomic level. However, the vast amount of thermodynamic and kinetic parameters that can be obtained in this way makes the identification of key steps in the mechanism difficult, especially if considered in isolation. To interpret this data in a consistent framework, a microkinetic analysis[5] can be applied. The microkinetic analysis approach (also referred to as microkinetic modelling) has been applied over a wide range of catalytic systems in different types of applications. A number of studies utilizing these approaches have been attempted in the field of FTS. Some approaches[6,7] use estimation techniques such as the unity bond index-quadratic exponential potential (UBI-QEP) method to estimate the parameters in the model under a single active site assumption. Others build models with initial kinetic parameter estimates, where the activation enthalpies are refined *via* regression of experimental data to obtain a model.[6,8,9] The problem with this approach is that, while it gives a good description of the resulting kinetics, the capability to relate these effects back to the fundamental kinetic parameters is generally lost. Other sources of fundamental energy data are quantum chemical calculation approaches, such as density functional theory (DFT). DFT based microkinetic models have been attempted for FTS in the past. In one approach,[9] the DFT data on Co was collated from literature sources to obtain an initial model, which was once again refined *via* regression. A more consistent approach requires that all the DFT data are calculated using a single computational approach. An example of such a consistent model and thorough analysis has been shown for FTS on a Ru step site.[10,11] In the absence of consistent DFT model data, van Santen *et al.*[3] used an exploratory microkinetic model to do scenario modeling with the aim to discriminate between different mechanistic approaches. However, these models share the implicit assumption that one active site is responsible for all the FTS reaction steps. Multi-site microkinetic models are not readily seen, but one example of such a study is a microkinetic analysis of steam methane reforming on multifaceted Ni.[12]

Here we present a microkinetic model of the FTS reaction up to C_2 intermediates on a FCC cobalt nanocrystal catalyst surface. Supported cobalt catalysts are usually synthesized with average Co particle diameters typically between 1 nm and about 50 nm. These small crystallites are in general stable as FCC-Co (β).[13] The model presented here utilizes a multi-site FCC cobalt nanoparticle model[14] for the catalyst, consisting of the two dominant cobalt surface facets Co(111) and Co(100), and a step site represented by the Co(211) surface. The important role that defect or step sites can play in terms of the activity of FTS has been pointed out before[15–17] and one of the aims of the present study is to see whether this argument is indeed warranted based on the resulting reaction network.

In this model the kinetic and thermodynamic parameters have been calculated using quantum chemical calculations and the results have not been obtained by fitting to experimental FTS data.

2 Methods and models

2.1 Quantum chemical calculations

The kinetic parameters for these surfaces were calculated using plane-wave periodic boundary condition Density Functional Theory (DFT). The DFT calculations used in this paper were performed with the Vienna ab initio Simulation Package (VASP).[18–20] This code applies DFT to periodic systems, using plane waves and ultrasoft pseudopotentials (US-PP).[21,22] All energies were calculated using the generalized spin-polarized gradient approximation (GGA) with the Perdew and Wang exchange–correlation functional (GGA-PW91).[23] Electronic energy convergence was accelerated by describing the electron distribution at the Fermi level by Methfessel and Paxton[24] smearing with $\sigma = 0.2$ eV. The cut-off energy for the calculations was set to 400 eV and dipole corrections were applied.

Three surface facet models were used to obtain the kinetic parameters for the three considered sites: the (111), (100) and (211) surface facets of a FCC crystal of cobalt atoms at a bulk lattice parameter of 3.538 Å (experimental value of 3.550 Å (ref. 25)). In all cases a 10 Å vacuum layer was used to create the surfaces. The *k*-point sampling was generated by following the Monkhorst–Pack[26] procedure (see Table 1). The effective coverage for a single adsorbed intermediate in these unit cells represents 0.25 ML. Optimization of the considered systems was carried out and optimization parameters were converged to obtain adsorption energies that are converged to within 0.01 eV. The VASP implementation of the nudged elastic band (NEB) method[27] was used to locate transition states.

Partial Hessian vibrational analysis was performed on the calculated structures. From these calculations we obtained the vibrational frequencies (ν_i) within the harmonic approximation. These frequencies are used to confirm that the imaginary mode for each transition state corresponds to the desired reaction coordinate. From the list of frequencies the vibrational zero-point energy was calculated as $E_{\text{ZPVE}} = \frac{1}{2}\sum_i h\nu_i$, with h as the Planck constant. These frequencies

Table 1 Parameters of the surface unit cells used in the DFT calculations

Co facet	Layers	Unit cell	*k*-Point mesh
111	5	2 × 2	9 × 9 × 1
100	4	2 × 2	5 × 5 × 1
211	10	2 × 1	6 × 7 × 1

are also used to calculate the partition functions used in estimating the Gibbs free energies (see below).

2.2 Species thermochemistry

The enthalpy (H), entropy (S) and Gibbs free energy (G) are computed for each species as a function of temperature. To obtain these values, the partition functions (q) were calculated for all the species as outlined in the ESI.† These partition functions were applied to describe vibrational states, as well as the translation and rotation states as either free or frustrated (*e.g.* a freely translating and rotating gas phase *vs.* a two-dimensional gas with free translation in an adsorption transition state *vs.* a tightly bound adsorbate with a frustrated translation potential). The Gibbs free energy can in short be expressed as:

$$G_0(T) = H_0(T) - TS_0(T) = E_0 + E_{\mathrm{ZPVE}} - RT\ln(q(T))$$

where E_0 is the electronic energy, T is the temperature and R is the molar gas constant.

It must be noted that there is a known inherent problem in DFT in calculating the electronic state of CO.[28,29] As a result, the model can be thermodynamically inconsistent, adsorption energies incorrect and the results do not correlate with experimental values. If the calculated electronic energy of CO is used as is, there is a mismatch in the overall thermodynamics of all the gas phase reactions involving CO, as well as resulting in an overestimated adsorption energy. An in-depth discussion of the origins of this effect is outside the scope of the current work. However, it implies that we need to correct the energy level of the gas phase CO molecule. Other authors have applied such thermochemical corrections to the gas phase species before.[12,30] Following a similar approach, a thermochemical correction of −0.49 eV was applied to the CO gas phase electronic energy to improve the agreement with experimental thermodynamics.

Since adsorption and desorption rates, and the equilibrium determine the limits of the surface coverages of the basic reagents (which in turn determine surface reactivity through the other reaction rate equations), it is important to confirm whether the microkinetic model has a sufficiently accurate description of these processes. There are surface science literature and in-house work available on both H_2 and CO adsorption and desorption on the Co(111)/Co(0001) surface.[31–33] These were used in a validation of these processes as described in the ESI.†

CO* and H* have sufficiently large adsorption energies on the Co surfaces that can result in coverages in excess of the coverage used in the initial DFT energy calculations (equivalent to $\theta_i = 0.25$ ML). If the 0.25 ML DFT energies for CO* are used as is, the coverage of CO* is predicted to be greater than 0.95 ML. It is known that the CO adsorption process on the Co(111)/Co(0001) surface termination is coverage dependent due to adsorbate–adsorbate interactions.[32] The saturation coverage of CO on this surface is around 0.60 ML. This was also confirmed by in-house surface science investigations of the CO desorption process.[33] Consequently, the binding energies of CO* on the Co(111) surface facet are corrected for coverage-dependent effects with the energy function shown in Fig. 1. In a similar fashion the Co(100) surface also exhibits coverage dependent behaviour (see Fig. 1). This energy function has been calculated with DFT as outlined in the ESI.† It is assumed that, due to its significantly more open structure, coverage

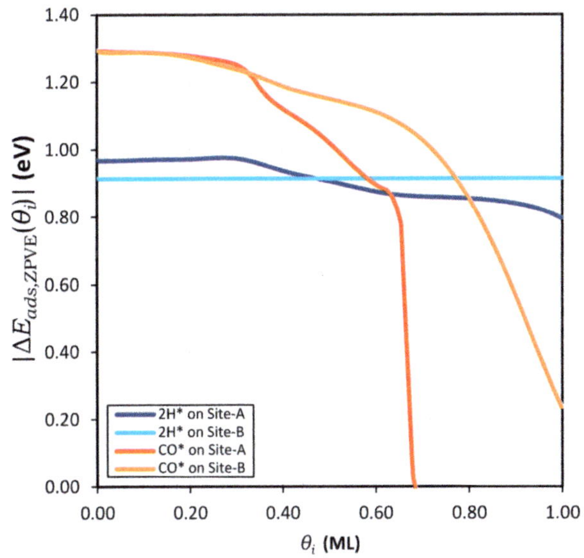

Fig. 1 Coverage dependent adsorption energies used in the model for i = H* and CO* on the sites Co(111) (A) and Co(100) (B).

dependent energy effects on the Co(211) are limited below 1.00 ML. The coverage dependent adsorption of H* on the Co surfaces has been reported earlier[31] and the energy function can be seen in Fig. 1. Although the partition functions of these species will also depend on its coverage, as a first approach the coverage dependent free energies of these species are expressed as:

$$G_{0,i}(T,\theta_i) = E_{0,\mathrm{ZPVE}}(\theta_i) - RT\ln(q(T))$$

Under the conditions studied below no other species were found with surface coverages exceeding 0.5 ML. Note that for these species no additional adsorbate–adsorbate interactions are included in the present model.

2.3 Microkinetic model

The aim here is to perform a microkinetic analysis[5] of the reactivity of Co at industrially relevant conditions. To this extent a mean-field microkinetic model is constructed by combining the thermochemical information of all considered species (intermediates and transition states) on all three considered surface sites. The subsequent analysis includes response, reaction network flux and sensitivity analysis to obtain an increased understanding of which sites, intermediates and reaction steps play a controlling role in determining the activity and the product distribution. All of the resulting differential equations described below were solved using an in-house developed code, *kinsolv*,[31] which solves these equations over time to give the resulting surface coverages, rates and reactant conversions of the model for a given set of initial reaction conditions. It also calculates the free energies of the species on-the-fly given the DFT input. It has also been developed to be able to efficiently deal with coverage dependent adsorption behavior.

2.3.1 Assumptions. To be able to employ this approach efficiently, a large number of assumptions and estimations are needed. The first assumption is the mean-field behaviour, resulting in a model where the gases, as well as the surface adsorbates behave as a mean-field. Each of these are thus described by their mean energy and concentration, no specific spatial information is included and no discreet atoms or molecular fragments are explicitly considered. This assumes that the adsorbates are ideally mixed and no spatially dependent effects (*e.g.* island formation) dominate the surface reactions. This is a large assumption since island formation of CO* and H* has been shown on metal surfaces, including on nanoparticles of Co.[34–36] There are a number of ideal aspects included in such a model, such as the use of ideal gases, ideal temperature control, idealized reactor models and ideal mass flow and mixing. Here, an ideal continuous-stirred tank reactor model (CSTR) containing the catalyst was used. This results in a pseudo-homogeneous catalyst mixing with the gas, and all surface sites are available to interact with the whole gas volume. Furthermore, it was assumed that gas adsorption/desorption transition states can generally be described by a two-dimensional gas-like state. Despite this model being a mean-field model, the diffusion of species between the reaction sites are treated as explicit reactions. This essentially links three separate microkinetic models together. Although this ensures thermodynamic consistency, it still has the assumption that there is one dominant pathway for a species to move between two sites. In the mean-field approach there exists the further assumption that these diffusions are sufficiently fast that there are no concentration gradients near the interface between sites. This implies that the sites are also homogeneously mixed. This differs from what is known about the site availability on Co nanoparticles.[14] This might overestimate the availability of species to undergo inter-site diffusion.

Two large assumptions remain: the sufficiency of the included reaction set and the assumption that the surface free energies are correctly represented. These two are quite large assumptions, since, if the included reactions are not sufficient, the rates would not correspond to those of the experiment. Therefore, this assumption should be critically reviewed based on the model results. The accuracy of the surface free energies is difficult to assess and depends directly on the accuracy of the DFT results. It is important to keep in mind that small variations in the energies of a critical step could have a significant effect on the model outcome. Since some of the reaction energies have been calculated by including various degrees of estimation of the free energy, it is important to review whether any of these specific steps ended up as a rate controlling step.

2.3.2 Rate expressions. The molar rate of change (r) for a species i on a surface site s, participating in various reactions (j) together with other species (l) can be expressed as:

$$r_{i,s} = \frac{\mathrm{d}C_{i,s}}{\mathrm{d}t} = \sum_{j \text{ on } s} \mu_{i,j} \left[k_{j,s} \prod_{l} C_{l,s}^{\mu_{l,j}} - k_{-j,s} \prod_{l} C_{l,s}^{\mu_{l,j}} \right]$$

where k is the rate constant of reaction j (or $-j$ for the reverse reaction) and μ are the stoichiometric coefficients, with C being the specific species concentration.

For gas phase species the rate of change is a sum of the gas flows into and out of the CSTR, and the adsorption and desorption reactions. The total molar rate of change of gas concentration is given by:

$$\frac{\mathrm{d}C_i}{\mathrm{d}t} = \sum_s r_{i,\mathrm{ads}-s} + C_{i,\mathrm{in}}Q_{\mathrm{in}} - C_{i,\mathrm{out}}Q_{\mathrm{out}}$$

where $r_{i,\mathrm{ads}-s}$ is the effective rate of adsorption and Q is the volumetric flow rate. The conversion of a specific reactant species is defined as:

$$\chi_i = \frac{C_{i,\mathrm{in}}Q_{\mathrm{in}} - C_{i,\mathrm{out}}Q_{\mathrm{out}}}{C_{i,\mathrm{in}}Q_{\mathrm{in}}}$$

To ensure that our reactor total pressure remains constant we employ the constraint:

$$\sum_{i=1}^{N_{\mathrm{gases}}} \frac{\mathrm{d}C_i}{\mathrm{d}t} = 0$$

The further constraint of the total number of active sites is applied:

$$\sum_s \sum_i \theta_{s,i} = 1$$

The reaction rate constants contain the kinetic parameters that define the model. We define these rate constants according to the transition state theory (TST):

$$k_{\pm j,s} = \frac{k_{\mathrm{B}}T}{h}\mathrm{e}^{-\Delta G_0^{\ddagger}/RT}$$

where h is the Planck constant, k_{B} the Boltzmann constant and $\Delta G_0^{\ddagger}$ the Gibbs free activation energy of the specific reaction step as calculated from the DFT energies and partition functions. Thus, the typically used pre-exponential factors are reformulated to be included in the Gibbs free activation energy. It has been noted that the reaction rate can be incorrectly estimated if the contribution of quantum tunneling of hydrogen atoms is not accounted for, since the small mass of the hydrogen atom causes its motions to be non-classical.[37] An ad hoc approach was implemented that accelerates both forward and reverse reactions by multiplying their rates with a tunneling factor (F_{tunnel}) according to the Wigner[38] correction:

$$F_{\mathrm{tunnel}} = 1 + \frac{1}{24}\left(\frac{\nu_{\mathrm{k}}h}{k_{\mathrm{B}}T}\right)^2$$

where ν_{k} the absolute value of the imaginary mode of the transition state. This tunneling correction is only applied to the rate constant of reactions involving hydrogen transfer processes.

2.4 Multi-site model

In order to incorporate different relative abundances of each surface site in a multi-site microkinetic model, the available fraction of each surface site is required. This can be obtained from the predicted site composition of Co, which is known to be a function of particle size. As a reasonable approximation of an average size of a Co nanocrystal in a supported catalyst, 8 nm can be used. In previous work,[14] we showed that for FCC Co nanocrystals there are five unique sites that need to be accounted for. Two of these sites ($2B_3$ and B_4) typically comprise terraces corresponding to the (111) and (100) facets of FCC Co. The other three sites can be considered to be "defect" sites, namely the two B_5 sites, that correspond to the (211) and (221) step sites, as well as the (321)-like B_6 site (kinked step site). As a first approach, and to limit computational effort, it was assumed that the B_5-A site (corresponding to the Co(211) surface) is a suitable proxy for all three step sites. In the rest of the paper we will refer to the sites according to the labels in Table 2.

As was mentioned above, species can be transferred between sites *via* inter-site diffusion. These diffusion reactions link the three respective sites into one reaction network. However, considering the typical structures of FCC nanocrystals[14] it is clear that sites B and C are not directly in contact, therefore only the following diffusion reaction types are included:

$$\text{A-}i^* + \text{B-}^* \rightleftharpoons \text{A-}^* + \text{B-}i^*$$

$$\text{A-}i^* + \text{C-}^* \rightleftharpoons \text{A-}^* + \text{C-}i^*$$

As a first attempt, and to limit the amount of estimation required, not all species have been included in the diffusion reactions. The full list of the included diffusion reactions can be seen in Table 5.

2.5 Reactions included on each site

A big component of building a microkinetic model is the choice of which reactions to include on the active site. The aim here was, as a first approach, to include a FTS reaction network that is sufficient to give all the required products, but that is at the same time as simple as possible. A schematic representation of the included reaction pathways can be seen in Fig. 2. Thus, the network remained close to the typical carbide mechanism pathways for chain growth (*i.e.* no CO insertion mechanisms). The two simplest possible H-assisted CO dissociation mechanisms have been included, but further hydrogenation of the COH* and HCO* intermediates have been excluded for now. However, these reactions do take place to some minor extent on a Co catalyst, producing a small amount of CH_3OH at the end.[39] The influence of these reactions on FTS, especially at high

Table 2 Labels and proportions used for active Co sites in the microkinetic model. Uncertainty ranges are indicated in square brackets

Site name	Facet	Label	$\theta_s^{8\text{ nm}}$ (ML) (ref. 14)	θ_s used (ML)
$2B_3$	Co(111)	A	0.65 [0.55; 0.77]	0.67
B_4	Co(100)	B	0.18 [0.16; 0.21]	0.18
B_5-A	Co(211)	C	0.05 [0.04; 0.07]	0.15

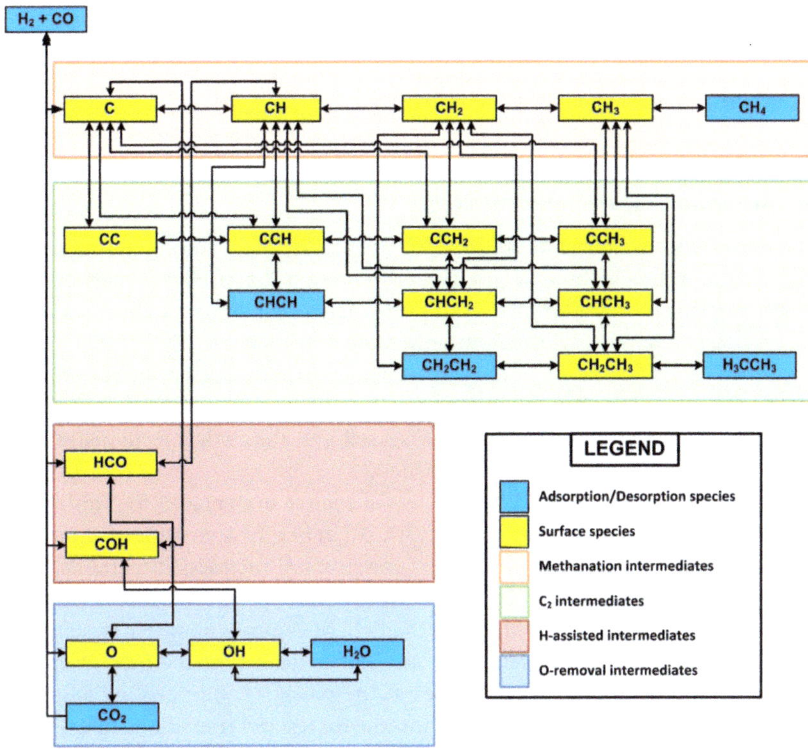

Fig. 2 Network of reactions considered on each surface site. Arrows indicate an included chemical reaction.

coverage, has been receiving recent attention.[40–42] Other bonding interactions between CH_x* ($x = 0$–3) and CO* have also been excluded for this first attempt. For O* removal we have included water formation pathways with both direct OH* hydrogenation and OH* disproportionation reactions. For CO_2 formation only direct CO* oxidation was included on each site. Although the FTS produces carbon products of high carbon numbers, here the reaction network was truncated at C_2 as a first attempt to study the early steps of FTS. All species and reactions that are included have been calculated using DFT or have been partially estimated. The DFT data for C_2 on an HCP Co terrace and step site have previously been reported,[43] and differ significantly from the results we have obtained. All the possible coupling reactions between CH_x* have been included except the coupling of two CH_3 groups, which was found to have excessively large reaction barriers in any case. The C_2 intermediates can be converted into other C_2 intermediates *via* one or several sequential hydrogenation or dehydrogenation reactions. However, internal isomerizations of the C_2 species have been excluded.

Due to the structure and location of site-C on Co nanoparticles[14] many of the reactions require an intermediate to be located in the site and another to approach from either the top or bottom terrace (site-A). To account for this we have described all site-C hydrogenation reactions in the following inter-site reaction form:

$$\text{A-H*} + \text{B-}i\text{*} \rightleftharpoons \text{A-*} + \text{B-}i\text{H*}$$

Carbon coupling and CO scission reactions on site-C have been described in a similar fashion.

2.6 Analysis of the microkinetic model

After obtaining a steady state solution for the model, the resulting output can be converted to a directed acyclic graph representation that assists in network flux analysis. In this representation each intermediate is considered to be a node and each reaction connecting these intermediates is a vertex. The direction and magnitude assigned to each of the vertices is obtained by calculating the effective rate of a reaction step by $r_{j,\text{eff}} = r_j - r_{-j}$. From this analysis the material flow through the reaction network can be assessed and the main mechanistic pathways through the network can be identified.

Model sensitivity analysis is a very useful tool to understand the underlying controlling features in the reaction network. An in-depth discussion on sensitivity analysis is given by Saltelli *et al.*[45] One of the simplest mentioned methods is the discrete local one-at-a-time (OAT) method. This approach is also known as Campbell's Degree of Rate Control[46] in the field of computational catalysis. Here, we employed a discrete OAT sensitivity analysis to determine the first order rate-controlling reactions and intermediates at the present conditions, by adjusting the Gibbs free energy of every species (intermediates and transition states) up by a factor of +0.01 eV after conducting a baseline run. The deviation in a model response from the baseline due to the perturbation is presented normalised to the absolute value of the largest deviation. A positive sensitivity result indicates an increase in the response relative to the baseline model and *vice versa* due to the positive energy perturbation.

2.7 Comparison to experimental results

Alfred Korzybski noted that "a map *is not* the territory."[47] By noting that a microkinetic model is an attempt to build a *map* of the FTS *territory* it should not be surprising if the comparison to experimental results can be quite challenging and that there will inevitably be discrepancies. Thus, when comparing a *map* to the *territory*, the *map* can be considered to be correct if it has similar "logical" characteristics and so reveals the most prominent features and relations of the territory.

When comparing the microkinetic model to experimental results, there are two important features to consider: rates of consumption or production of a species and the selectivities of the products. The most important rate to consider is the steady-state rate of CO consumption. Within this response there are three factors that should be compared with the experimental characteristics to assess whether this model is a sufficient description of the FTS kinetics.

The first is the absolute rate of CO consumption. Typically reaction rates are reported relative to the mass of the catalyst, while our model gives the turnover frequency as a result (TOF – number of molecules produced/consumed per active site per unit time). To convert between the two, an active metal surface area needs to be supplied. All rates will thus be expressed as TOFs. When rates are compared to the macrokinetic models (see below) the rates will be normalized to a single reference condition ($T = 503$ K; $P_{total} = 20$ bar; inlet $H_2 : CO = 2$; CO conversion $\chi_{CO} = 50\%$), unless noted otherwise. Steady-state isotopic transient kinetic analysis (SSITKA) experiments can give very useful information to compare microkinetic results to, including TOFs and surface coverages of selected species. Here we will also compare the rates and coverages we obtain to those presented in SSITKA experiments by Den Breejen *et al.*[48] ($T = 483$ K; $P_{total} = 1.85$ bar; inlet $H_2 : CO = 10$; low χ_{CO}). The Co in the catalysts presented by Den Breejen *et al.* are supported on carbon nanofibers (CNF) which are not known for very strong metal–support interactions, making it a good system for direct comparison to microkinetic data based on pure Co surfaces.

The second comparison for rates is the shape of the reactant partial pressure–rate response. To check the correspondence of the microkinetic model reaction network and rate controlling features to experiments, the partial pressure–rate response needs to be considered. A variety of macrokinetic models have been developed with the aim to describe these relations in a single closed mathematical form[4] within a defined window of reaction conditions. Instead of comparing the microkinetic model responses to a specific experimental set, we consider two macrokinetic models as representative proxies for, or summaries of, the experimental results, specifically related to the shape of the partial pressure–rate relations. These choices are not made with any reference to the underlying mechanistic assumptions, but due to the goodness of fit of these models. The first is an in-house model developed for a commercial Co catalyst[44] (M1) and the second is the recently proposed REGK model[4] (M2) developed to be fit for both Co and Fe FTS results. A summary of the two models can be seen in Table 3.

The third factor is the apparent activation energy. In this work we used both M1 and M2 to compensate for gas phase pressure changes in order to obtain k_{eff}. The slope of the plot of $\ln(k_{eff})$ *vs.* $1/T$ gives an indication of the apparent activation energy.

Comparing the resulting selectivities to experimental results is a formidable challenge, since FT catalysts make a large range of hydrocarbons (and oxygenates) with increasing carbon numbers, while the present model truncates the mechanism at C_2. This will affect the selectivity results due to the inherent reversibility of

Table 3 Considered summary models for pressure–rate relations, with their goodness of fit statistics and parameter ranges[4]

Model	Equation	MARR[a] (%)	R^2	$P_{25\%}$[b] (%)	k_b parameters Min[4]	Max[4]	Mean[4]	Used[c]
M1 (ref. 44)	$-r_{FT} = \dfrac{P_{H_2}^{0.75} \times P_{CO}^{0.5}}{(1 + k_b \times P_{CO}^{0.5})^2}$	9.2	0.982	95.8	0.144	2.15	1.01	1.58
M2 (ref. 4)	$-r_{FT} = \dfrac{P_{H_2}^{0.75} \times P_{CO}}{(1 + k_b \times P_{CO})^2}$	6.9	0.991	98.1	0.011	1.923	0.428	0.428

[a] MARR – mean absolute relative residual as previously defined.[4] [b] $P_{25\%}$ refers to the percentage of data that has absolute relative residual of less than 25% as calculated by Mousavi *et al.*[4] [c] For M1 the final value in the original paper was used and for M2 the mean value was used.

all reactions. However, a selectivity comparator value is used to assess whether the microkinetic model is producing a sufficiently meaningful amount of C_2 products compared to what is experimentally expected. To do this we use a C_2 to C_1 C atom-based ratio: $\acute{S} = \frac{2r_{C_2}}{r_{C_1}}$ where r_{C_2} is the sum total TOF of C_2 production and r_{C_1} is the TOF of methane production. The expected values of $\acute{S}$, as derived from a variety of experiments,[39,49–58] vary depending on the catalyst preparation and the reaction conditions, but they range typically between 0.1 and 0.3 with a mean around 0.15 (see ESI†). For CO_2 selectivities (S_{CO_2}) we use either the C% or O% values relative to the rate of CO consumption. For an industrially relevant Co catalyst system the expected values for S_{CO_2} are in the single digit percentage range or lower.

3 Results and discussion

The results for the microkinetic analysis of the multi-site Co microkinetic model will be presented in a number of subsections. Firstly we will discuss the intrinsic kinetic features of the DFT results on each site when considered in isolation, with reference to the calculated free energy surfaces. In this regard a summary of the calculated free energy barriers for all the reactions in the model is given in Tables 4 and 5. The second section will describe the results of the multi-site model where all three sites are integrated. Based on these results, specific modifications to model parameters are explored based on prior insight to their deficiencies. Through this exploration we will show the first attempt to establish which scenarios will give rise to sufficient chain growth and the correct pressure–rate relations.

3.1 Isolated reactive sites

The sections below give a basic overview of the main features of the three sites of the model when they are considered to be in isolation. This can be considered to be a "fingerprint" of each surface's activity and it gives an initial indication of what to expect in each case. Full microkinetic simulations were performed on each surface up to the steady state at the low conversion limit, considering only a single condition ($T = 503$ K, $P_{total} = 20$ bar, $H_2 : CO = 2$).

3.1.1 Site-A. A summary of the main rate, selectivity and coverage features of the isolated Site-A at steady state can be seen in Table 6. The free energy surface of methanation at 503 K of Site-A, obtained from the *kinsolv* output, can be seen in Fig. 3. Not much effort will be spent in interpreting the energy surfaces, but two observations can be made. Firstly, the slow overall rate on this surface is governed by the very high barrier of the CO scission process (around 1.75 eV). Secondly, the final step in water formation should be quite easy *via* the OH disproportionation process. The barrier of OH formation is around 1.2 eV on this surface, which should be low enough to produce water at a meaningful rate on this surface.

It is interesting to note that the overall CO TOF ($-r_{CO}$) is four orders of magnitude too slow on this surface (typical FT TOF is *ca.* 0.01 s^{-1}),[48] with no other significant surface species besides CO* and H*. Because of the slow rate of CO consumption, no meaningful amount of C_2 formation is observed. This is due to the low C_1 concentration on this surface due to the slow CO dissociation rate. If

Table 4 Effective steady state Gibbs free energy barriers for the reaction of each site in the unmodified multi-site Co microkinetic model at 503 K

#	Reaction	Site-A		Site-B		Site-C	
		$\Delta G^{\ddagger}_{0,f}$ (eV)	$\Delta G^{\ddagger}_{0,r}$ (eV)	$\Delta G^{\ddagger}_{0,f}$ (eV)	$\Delta G^{\ddagger}_{0,r}$ (eV)	$\Delta G^{\ddagger}_{0,f}$ (eV)	$\Delta G^{\ddagger}_{0,r}$ (eV)
1	$CO + * \rightleftharpoons CO*$	0.92	0.99	0.58	0.67	0.71	1.01
2	$H_2 + 2* \rightleftharpoons 2H*$	0.98	1.29	0.73	0.99	0.52	0.91
3	$CO_2 + * \rightleftharpoons CO_2*$	1.65	0.42	0.69	0.04	0.59	0.10
4	$H_2O + * \rightleftharpoons H_2O*$	0.60	0.15	0.60	0.12	0.57	0.25
5	$CH_4 + * \rightleftharpoons CH_4*$	1.37	0.70	1.16	0.35	1.36	0.66
6	$CHCH + * \rightleftharpoons CHCH*$	0.58	1.96	0.58	2.11	0.58	2.01
7	$CH_2CH_2 + * \rightleftharpoons CH_2CH_2*$	0.58	0.22	0.58	0.46	0.58	0.87
8	$CH_3CH_3 + * \rightleftharpoons CH_3CH_3*$	1.38	0.68	1.18	0.37	1.38	0.80
9	$CO* + * \rightleftharpoons C* + O*$	2.35	1.74	1.36	2.09	1.53	1.49
10	$H* + CO* \rightleftharpoons HCO* + *$	0.94	0.06	0.78	0.37	0.72	0.01
11	$CO* + H* \rightleftharpoons COH* + *$	1.45	0.58	1.42	0.87	1.35	0.43
12	$HCO* + * \rightleftharpoons CH* + O*$	0.92	1.55	1.11	2.07	0.60	1.13
13	$COH* + * \rightleftharpoons C* + OH*$	1.31	1.15	0.78	1.66	0.76	1.56
14	$C* + H* \rightleftharpoons CH* + *$	0.73	1.10	0.69	0.51	0.69	0.55
15	$CH* + H* \rightleftharpoons CH_2* + *$	0.63	0.22	0.74	0.09	0.75	0.25
16	$CH_2* + H* \rightleftharpoons CH_3* + *$	0.57	0.68	0.30	0.08	0.41	0.61
17	$CH_3* + H* \rightleftharpoons CH4* + *$	0.92	1.13	0.81	0.98	0.81	0.67
18	$O* + H* \rightleftharpoons OH* + *$	1.20	0.78	1.32	0.91	0.85	0.77
19	$OH* + H* \rightleftharpoons H_2O* + *$	1.29	0.90	1.45	0.81	1.38	0.49
20	$2OH* \rightleftharpoons H_2O* + O*$	0.32	0.34	0.66	0.42	1.07	0.26
21	$CO* + O* \rightleftharpoons CO_2* + *$	1.89	0.74	1.35	0.27	1.43	0.77
22	$2C* \rightleftharpoons CC* + *$	1.14	1.54	1.87	0.47	1.71	1.44
23	$C* + CH* \rightleftharpoons CCH* + *$	0.84	1.34	1.50	0.85	1.10	1.15
24	$C* + CH_2* \rightleftharpoons CCH_2* + *$	0.76	1.55	1.30	1.07	0.97	1.14
25	$C* + CH_3* \rightleftharpoons CCH_3* + *$	1.05	1.94	0.89	1.04	1.06	0.86
26	$2CH* \rightleftharpoons CHCH* + *$	0.62	0.98	1.31	0.76	1.19	1.01
27	$CH* + CH_2* \rightleftharpoons CHCH_2* + *$	0.83	0.91	1.35	0.77	1.04	0.96
28	$CH* + CH_3* \rightleftharpoons CHCH_3* + *$	1.26	1.11	1.23	0.81	1.36	0.79
29	$2CH_2* \rightleftharpoons CH_2CH_2* + *$	0.63	1.00	1.17	1.25	0.39	0.99
30	$CH_2* + CH_3* \rightleftharpoons CH_2CH_3* + *$	1.18	1.18	1.19	1.04	1.05	1.09
31	$CC* + H* \rightleftharpoons CCH* + *$	0.64	1.11	0.68	1.24	0.63	0.81
32	$CCH* + H* \rightleftharpoons CCH_2* + *$	0.90	0.77	0.53	0.32	0.73	0.36
33	$CCH* + H* \rightleftharpoons CHCH* + *$	0.67	0.89	0.75	0.67	0.61	0.23
34	$CCH_2* + H* \rightleftharpoons CCH_3* + *$	0.60	0.82	0.51	0.67	0.74	0.56
35	$CCH_2* + H* \rightleftharpoons CHCH_2* + *$	0.66	0.33	0.69	0.15	0.68	0.29
36	$CCH_3* + H* \rightleftharpoons CHCH_3* + *$	0.74	0.06	0.93	0.18	0.59	0.09
37	$CHCH* + H* \rightleftharpoons CHCH_2* + *$	1.23	0.54	0.88	0.20	0.82	0.44
38	$CHCH_2* + H* \rightleftharpoons CHCH_3* + *$	0.71	0.58	0.62	0.56	0.57	0.28
39	$CHCH_2* + H* \rightleftharpoons CH_2CH_2* + *$	0.68	0.56	0.49	0.51	0.73	0.91
40	$CHCH_3* + H* \rightleftharpoons CH_2CH_3* + *$	0.67	0.42	0.77	0.40	0.37	0.47
41	$CH_2CH_2* + H* \rightleftharpoons CH_2CH_3* + *$	0.57	0.31	0.70	0.25	0.64	0.28
42	$CH_2CH_3* + H* \rightleftharpoons CH_3CH_3* + *$	0.68	1.06	0.61	0.90	0.75	0.70

Table 5 Steady state Gibbs free energy barriers for the inter-site diffusion reactions in the unmodified multi-site Co microkinetic model at 503 K

		X = A; Y = B		X = C; Y = A	
#	Reaction	$\Delta G^{\ddagger}_{0,f}$ (eV)	$\Delta G^{\ddagger}_{0,r}$ (eV)	$\Delta G^{\ddagger}_{0,f}$ (eV)	$\Delta G^{\ddagger}_{0,r}$ (eV)
43	X-CO* + Y-* ⇌ X-* + Y-CO*	0.13[a]	0.14	0.36[a]	0.13
44	X-H* + Y-* ⇌ X-* + Y-H*	0.14[a]	0.11	0.23	0.18
48	X-C* + Y-* ⇌ X-* + Y-C*	0.39[a]	1.50	1.20	0.39
49	X-CH* + Y-* ⇌ X-* + Y-CH*	0.49[a]	1.01	0.73	0.44
50	X-CH_2* + Y-* ⇌ X-* + Y-CH_2*	0.31[a]	0.57	0.50	0.29
51	X-CH_3* + Y-* ⇌ X-* + Y-CH_3*	0.31[a]	0.22	0.57	0.27
54	X-O* + Y-* ⇌ X-* + Y-O*	0.46[a]	0.72	0.47[a]	0.17
55	X-OH* + Y-* ⇌ X-* + Y-OH*	0.17[a]	0.41	0.78[a]	0.15

[a] Estimated from terrace diffusion barriers.

the barriers of coupling are considered in Table 4, it is clear that this surface should be able to couple C_1 species effectively if the concentration was high enough, as there are five coupling reactions with barriers below 1.00 eV.

3.1.2 Site-B. The free energy surface at 503 K of Site-B can be seen in Fig. 3. The first observation here is that the overall rate on this surface is expected to be governed by the CO scission process, which is a combination of direct dissociation and both the HCO* and COH* mediated H-assisted CO dissociation processes. This can be seen in the energy surface where the overall barriers of these processes are all in the same range. Secondly, water formation should be relatively easy *via* the OH disproportionation process, but the formation of OH (barrier of OH formation is around 1.6 eV) will probably be a rate limiting step, leading to the O atoms to rather be removed by CO_2 formation than water formation (as is borne out in the selectivities).

A summary of the main rate, selectivity and coverage features at steady state can be seen in Table 7. The CO consumption rate for Site-B is significantly faster than that of Site-A. This rate is similar to the order expected for FT. Despite this rate correspondence, this surface has a large propensity to produce CO_2 (55 O% selectivity), due to the slow water formation rate. It is interesting to note that, in

Table 6 Outputs of the microkinetic analysis of isolated Site-A/Co(111) model at the low conversion limit (T = 503 K; P_{total} = 20 bar; inlet H_2 : CO = 2)

Response	Value	Unit
r_{CO}	-6.14×10^{-6}	s^{-1}
r_{CH_4}	5.90×10^{-6}	s^{-1}
r_{C_2}	$<1 \times 10^{-15}$	s^{-1}
$\acute{S}$	$<1 \times 10^{-9}$	
S_{CH_4}	98	C%
S_{CO_2}	2	C%
θ_{CO}	0.53	ML
θ_H	0.47	ML
θ_{A-*}	1.14×10^{-3}	ML
$\sum\theta_{rest}$	<0.01	ML

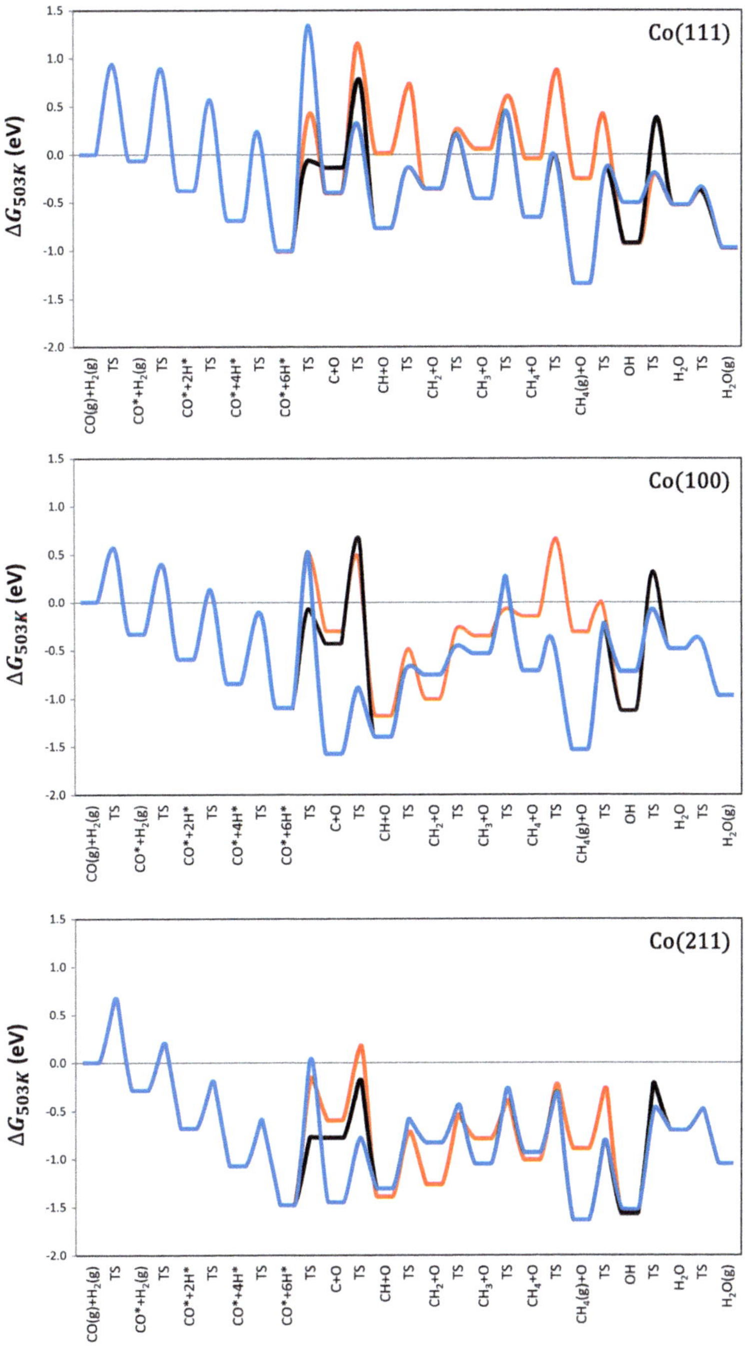

Fig. 3 Free energy surface diagrams for methanation at 503 K on Site-A/Co(111) (left), Site-B/Co(100) (middle) and Site-C/Co(211) (right). Curve colors: blue – methanation *via* direct CO dissociation and OH disproportionation; black – CO scission *via* the HCO intermediate and OH hydrogenation; red – pathway *via* the COH intermediate. Labels pertain only to the intermediates of the blue curve. Pathways to C_2H_x and CO_2 have been omitted.

Table 7 Outputs of the microkinetic analysis of isolated Site-B/Co(100) model at the low conversion limit ($T = 503$ K; $P_{total} = 20$ bar; inlet $H_2 : CO = 2$)

Response	Value	Unit
r_{CO}	-0.024	s^{-1}
r_{CH_4}	0.016	s^{-1}
r_{C_2}	1.07×10^{-5}	s^{-1}
$\acute{S}$	1.34×10^{-3}	
S_{CH_4}	65	C%
S_{CO_2}	55	O%
θ_{CO}	0.768	ML
θ_{H}	0.208	ML
θ_{HCO}	0.019	ML
θ_{C}	6.0×10^{-4}	ML
θ_{CH}	0.003	ML
θ_{B-*}	6.0×10^{-4}	ML

contrast to Site-A, intermediates other than CO* and H* have significant coverages (HCO* and CH*). Once again a low rate of C_2 production is observed, but at least the C_2 production rate on this surface is ten orders higher than on the isolated Site-A model, leading to an $\acute{S}$ value that is only an order of magnitude too low compared to the expected values. This increase compared to Site-A is due to the higher CH* coverage observed on this surface.

3.1.3 Site-C. To test the Site-C model in isolation, it was assumed that the whole surface is covered by only this type of site. The free energy surface of Site-C can be seen in Fig. 3. In this case the H-assisted CO dissociation pathway, proceeding *via* the HCO* intermediate, has a lower overall barrier than the other CO scission possibilities, with an overall free energy barrier of 1.31 eV at 503 K. Thus, a higher rate of CO consumption than on site-B is expected. Water formation on this surface should be quite easy *via* the OH disproportionation process. The formation of OH proceeds with a barrier of 0.85 eV and should be a relatively easy process. It is interesting to note how "flat" the energy profile of the methanation process is on this surface. This should result in a very high methane formation rate.

A summary of the main rate, selectivity and coverage features at steady state can be seen in Table 8. The CO consumption rate for Site-C is significantly faster than that of both Site-A and Site-B. In contrast to Site-B, this site makes almost no CO_2. Once again intermediates other than CO* and H* have significant coverages (especially O* and OH*). The low coverages of the CH_x* intermediates are due to the high rate of methane formation, as these intermediates do not have the time to build up significant concentrations. Once again no significant C_2 production is observed due to the low CH_x* coverages on this surface.

3.2 Multi-site Co microkinetic model

Considering these three sites in isolation shows that Site-A will not be active for CO dissociation, but Sites B and C are. However, on a catalyst Co nanoparticle, different sites are located next to each other and we have previously proposed[17] that certain species can be produced on one site and then be transferred to another site where they can react further. Thus, we included these three sites in

Table 8 Outputs of the microkinetic analysis of isolated Site-C/Co(211) model at the low conversion limit ($T = 503$ K; $P_{total} = 20$ bar; inlet $H_2 : CO = 2$)

Response	Value	Unit
r_{CO}	−1.877	s^{-1}
r_{CH_4}	1.877	s^{-1}
r_{C_2}	7.49×10^{-6}	s^{-1}
$\acute{S}$	7.98×10^{-6}	
S_{CH_4}	~100	C%
S_{CO_2}	1.00×10^{-5}	O%
θ_{CO}	0.647	ML
θ_H	0.161	ML
θ_{HCO}	6.20×10^{-4}	ML
θ_C	1.80×10^{-5}	ML
θ_{CH}	1.30×10^{-4}	ML
θ_O	1.80×10^{-5}	ML
θ_{OH}	0.008	ML
θ_{C-*}	0.183	ML

the first version of an integrated multi-site microkinetic model for Co by allowing diffusion of species between the surfaces (as described above). The results of the model using the DFT data as is will be discussed (also referred to as scenario S1).

3.2.1 Rates and coverages. As a first result the model is compared to the SSITKA results by Den Breejen *et al.*[48] ($T = 483$ K; $P_{total} = 1.85$ bar; inlet $H_2 : CO = 10$; low χ_{CO}) for Co sizes ranging from 8–16 nm. The results of the rate and the coverages of CO*, H*, OH_x* and CH_x* as compared to the experimental results can be seen in Fig. 4 (S1). The model results in Fig. 4 represent the rates and coverages at χ_{CO} between 0.1% and 30%. It is clear that the absolute CO consumption TOF is in good agreement with the experiment at these conditions. However, if we compare the coverages, the agreement is not as good. The value of θ_{CO} is higher by about 0.2 ML, while θ_H is significantly higher than what is obtained from SSITKA. The range for θ_{OH_x} is generally lower than expected, while the θ_{CH_x} value is not even close, with values below 1×10^{-5} ML over the whole conversion range. Thus, at first glance the model seems to produce the rate in the right order, but the coverages are not comparable. Although these conditions are typical for methane production, the SSITKA modeling shows that a C_1 selectivity of around 65 C% is expected. This implies that even at these conditions, Co catalysts will still have the ability to grow a meaningful amount hydrocarbon chains. However, the present model gives $\acute{S}$ values of less than 1×10^{-7} for the whole conversion range. This indicates that basically no chain growth is happening in the microkinetic model at these conditions.

Detailed results at the considered standard reference conditions ($T = 503$ K; $P_{total} = 20$ bar; inlet $H_2 : CO = 2$; $\chi_{CO} = 50\%$) can be seen in Table 9. From these results it can be seen that in terms of the CO consumption rate, the model is close to the correct order of magnitude. However, no C_2 is essentially produced. Another important deviation here is the fact that O* atoms are removed by CO_2 formation, rather than as water. Considering the coverages it can be seen that there are no significant C_1* or C_2* concentrations on any of the sites. The surface coverages are dominated by CO* and H*, together with B-HCO*, B-O* and C-OH*.

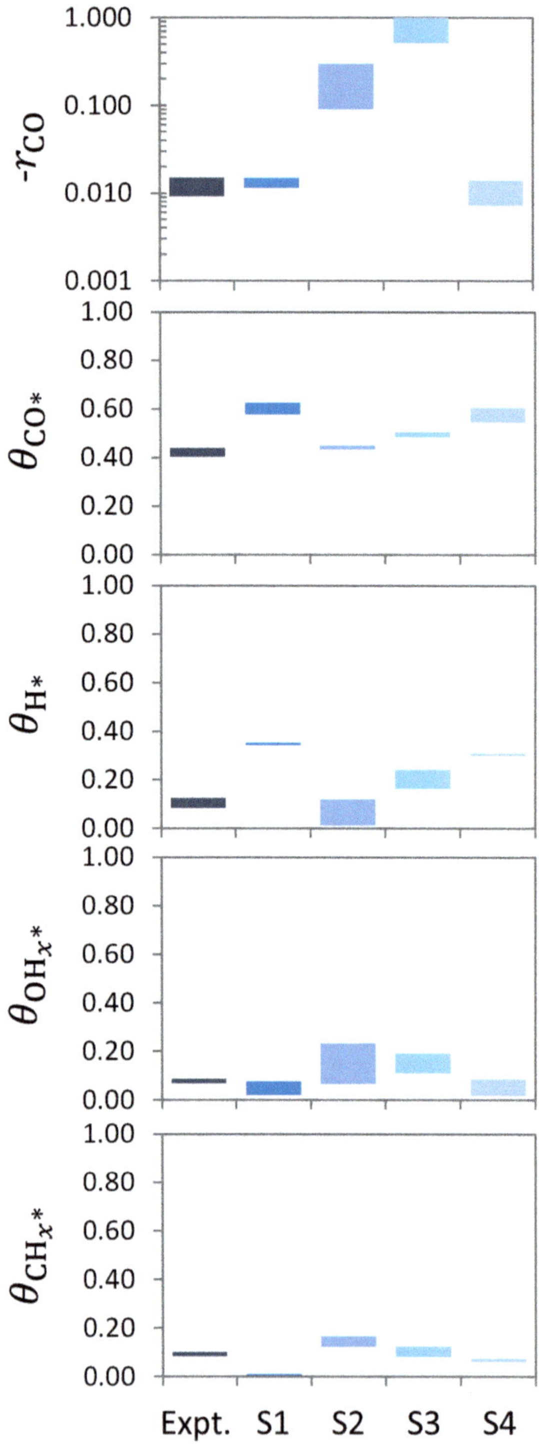

Fig. 4 Comparison of microkinetic calculated FTS rates and coverages for scenarios S1–S4 to the experimental results by Den Breejen *et al.*[48] ($T = 483$ K; $H_2 : CO = 10$). The microkinetic results reflect the rate and coverage results for a range of χ_{CO} of 0.1–30%. The coverages are the total for species *i* over all three sites.

Table 9 Outputs of the microkinetic analysis of the combined multi-site model at the reference condition ($T = 503$ K; $P_{total} = 20$ bar; inlet $H_2 : CO = 2$; $\chi_{CO} = 50\%$)

Response	Value		Unit
r_{CO}	−0.084		s^{-1}
r_{CH_4}	0.047		s^{-1}
r_{C_2}	2.63×10^{-9}		s^{-1}
$\acute{S}$	1.11×10^{-7}		
S_{CH_4}	56		C%
S_{CO_2}	44		C%
Coverages (ML)	Site-A	Site-B	Site-C
θ_s [a]	0.670	0.180	0.150
θ_{CO}	0.393	0.136	0.133
θ_H	0.276	0.034	0.002
θ_{HCO}	3.89×10^{-7}	0.002	6.18×10^{-6}
θ_{COH}	4.16×10^{-7}	1.55×10^{-5}	4.77×10^{-8}
θ_C	1.39×10^{-15}	1.03×10^{-5}	1.48×10^{-8}
θ_{CH}	1.86×10^{-9}	4.69×10^{-5}	3.25×10^{-7}
θ_{CH_2}	7.68×10^{-11}	4.86×10^{-9}	1.27×10^{-10}
θ_{CH_3}	4.44×10^{-7}	9.30×10^{-9}	9.56×10^{-7}
θ_O	1.25×10^{-4}	0.007	1.80×10^{-4}
θ_{OH}	4.67×10^{-6}	1.83×10^{-4}	0.016
θ_{C_2}	3.62×10^{-13}	4.25×10^{-9}	1.47×10^{-14}
θ_*	7.28×10^{-4}	1.16×10^{-4}	1.17×10^{-6}

[a] Initial amount of empty sites included in the model.

3.2.2 Response to reaction conditions. The response of the model to reaction conditions can be studied by considering the apparent activation energy and the pressure–rate relationship. A value of 1.21 eV is obtained for the apparent activation energy using both M1 and M2. This is somewhat higher than those proposed in literature, typically around 93–112 kJ mol^{-1} (0.96–1.16 eV).[59–62] The pressure–rate relationship parity plot compared to M1 can be seen in Fig. 5 (S1). The comparison to M2 is similar and is available in the ESI.† The R^2 values of

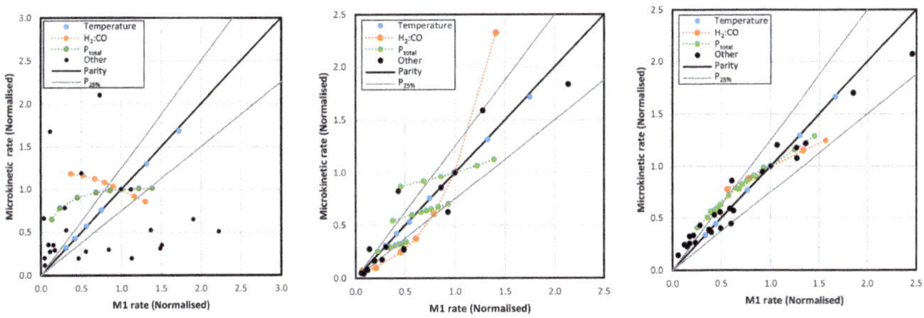

Fig. 5 Comparison between microkinetic calculated FTS rates and experimental summary macrokinetic model M1 for different scenarios: S1 (left); S2 (middle); S4 (right) All rates are normalized to a single condition ($T = 503$ K; $P_{total} = 20$ bar; inlet $H_2 : CO = 2$; $\chi_{CO} = 50\%$). Comparisons to M2 can be seen in the ESI.†

these parity plots are extremely low and even visually it can be seen that this model does not describe the expected pressure rate relationships at all. What is clear is that by varying the H_2 : CO ratio at 503 K the multi-site microkinetic model gives a response that is almost orthogonal to the expected parity line. Similarly the response to total pressure changes is also poor. The additional black points in this figure are obtained by considering a space-filling design of points over relevant ranges of P_{total}, H_2 : CO and T. It is clear that there is no good agreement between the summary of the experimental results and the micro-kinetic model. In all these calculations the $\acute{S}$ values were all lower than 9.7×10^{-7}. There are therefore no reaction conditions where this model produces meaningful chain growth.

3.2.3 Reaction network and sensitivity analysis. To understand why this model does not give a good correspondence to the experimental results we can consider the reaction network flux and sensitivity analysis results. The C atom flux reaction network (truncated at 1% of the overall CO consumption rate) at the standard reference conditions can be seen in Fig. 6. The bulk of the CO that adsorbs binds to site-B. Here 45% of the CO reacts with O* to produce the

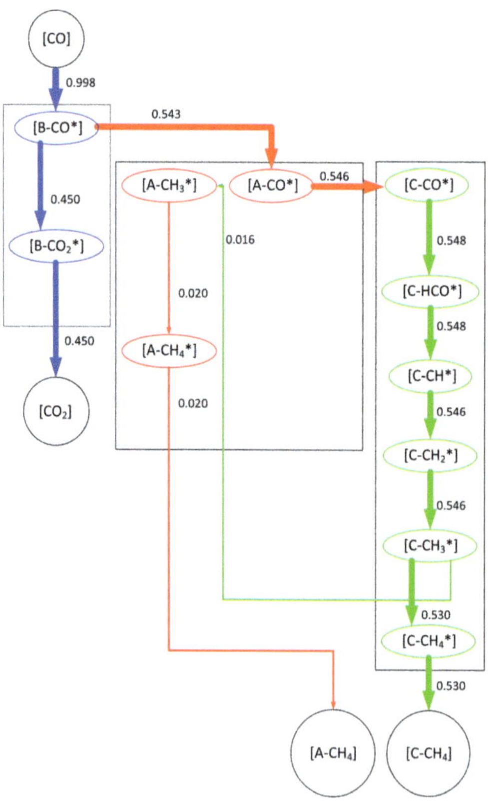

Fig. 6 The carbon atom reaction network for the multi-site Co microkinetic model at the standard reference condition (T = 503 K; P_{total} = 20 bar; inlet H_2 : CO = 2; χ_{CO} = 50%) using the DFT data as is (S1). The arrow thickness is an indication of the scale of the rate and the C mol% flow is indicated in the labels. The network was truncated at 1% of the total CO conversion rate.

observed CO_2. The remaining 54% of CO diffuses to site-A and subsequently to site-C. Here it undergoes H-assisted CO dissociation and the resulting CH* is mostly hydrogenated to CH_4, which desorbs from site-C (noted as C-CH_4 in the figure). A small amount of the CH_3* that was produced on site-C (~2% of C atoms) diffuses back to site-A where it also undergoes hydrogenation to CH_4 that desorbs from that site (A-CH_4 in the figure). To summarize, site-B acts as the CO adsorption site as well as a CO_2 producer. Site-A plays a minor transfer and hydrogenation role, while site-C gives the main CO dissociation path and acts mainly as a CH_4 producer. Not shown in this network is that the O* that results from the CO scission process diffuses *via* site-A to site-B where it partakes in the CO_2 formation reaction. Some of the O* is hydrogenated to OH* on this site, but eventually it also diffuses to other sites, where it is dehydrogenated to O* and moves to site-B.

The sensitivity analysis of these results (Fig. 7) indicates that the rate of CO consumption in this model is the most sensitive to the coverages of A-H*, C-CO*, and the empty sites on Site-A. The coverage of hydrogen on site A is expected since it appears in most of the hydrogenation reaction rate equations. The most prominent rate controlling barrier is that of H-assisted CO dissociation on site-C as proceeding *via* the HCO* intermediate. As there are no other barriers, except a little contribution by CO_2 formation on site-B, that controls the rate, this step can be considered to be the dominating feature of this reaction network. Since the scission of CO is the slow step here it determines the rate at which C_1* species are formed. As the hydrogenation steps do not influence the CO consumption rate, they will be significantly faster than the CO scission step. This implies that the moment C_1* species are formed, they are rapidly hydrogenated to methane and no significant pool of C_1* can build up on any surface site. Since the presence of C_1* on the surface sites is critical for chain growth, this would explain the fact that no chain growth is observed in this model. Considering the factors that control

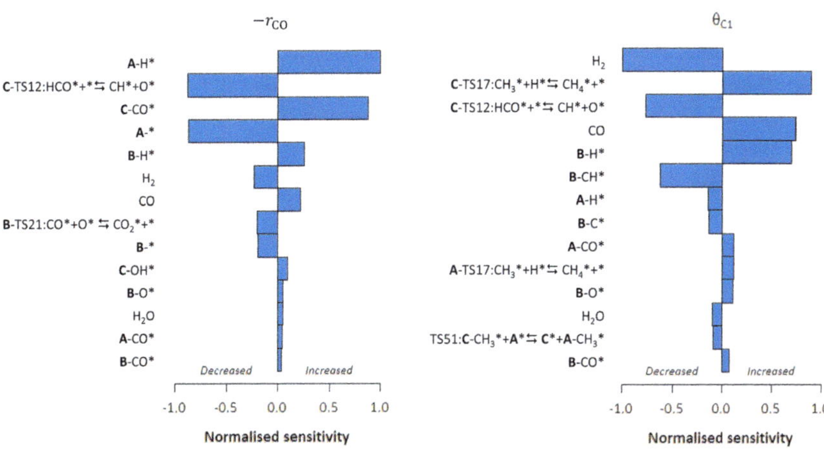

Fig. 7 Sensitivity analysis results of $-r_{CO}$ (left) and θ_{C_1} (right) for the multi-site Co microkinetic model at the standard reference condition ($T = 503$ K; $P_{total} = 20$ bar; inlet $H_2 : CO = 2$; $\chi_{CO} = 50\%$). The sensitivity factors have been normalized to the largest absolute value. Only the top few most sensitive factors are shown.

the C_1* concentration, it is clear that there is a correlation between this concentration and a number of other surface species. The main barriers controlling this concentration are the CO scission step and the hydrogenation of the methyl group on site-C. This sensitivity analysis already points towards two possible modifications of this model under which chain growth could be more likely. Either the CO scission process should be significantly faster so as to produce more C_1* or the hydrogenation process of the C_1* species should be more difficult. These two factors are explored in more depth in the following sections.

3.3 Exploratory modifications

Specific steps in the model were modified to assess how the model responds to these changes. These include lowering the barrier of the CO scission process and destabilizing the CH_x species energies (discussed further below). These modifications will result in multiple exploratory scenarios of the model. In the previous section it was noted that either the production of C_1* should be sped up, or the hydrogenation of C_1* should be slower, allowing C_1* species to build up and proper chain growth to occur. Although it is very easy to just modify any process in the microkinetic model, these exploratory modifications should be fundamentally justifiable. If not, the risk exists that the correspondence to experiment improves for reasons that are in conflict with the true underlying physical chemistry.

On one hand, the controlling factor for steady state C_1* production is the CO scission process. In this model the lowest overall CO scission process free energy barriers are 1.31 eV on site-C and 1.36 eV on site-B. It is known that there are some significant discrepancies with experiments when CO adsorption is studied with GGA-based DFT. The question arises whether these issues result in an overestimation of the CO dissociation process barriers as well. Another idea that has been explored is whether the high steady-state coverage of CO* can affect the barrier of CO dissociation. Studies on Ni surface sites[63] indicated that the presence of CO* increases the direct dissociation barrier on most sites, while H-assisted pathways could be inversely affected. In this model we use the Co(211)-based kinetic parameters as a proxy for all three defect sites on FCC Co. However, the other two sites will have different behavior towards the CO scission process. The Ni(321) kink site (which is structurally equivalent to Co(321)) has a geometric arrangement of atoms that facilitates the direct CO dissociation with a lower barrier than Ni(211).[63] In recent in-house work we showed that $\Delta E^{\ddagger}_{\text{COdiss}} = 1.20$ eV on Co(321).[64] Similarly, on the HCP Co(11$\bar{2}$4) a kink type site is exposed with a CO dissociation barrier of 1.20 eV.[65] This is even lower than the H-assisted route on site-C. The process barriers on the Co(221) surface are also generally equal to, or higher than those on Co(211). The possibility of a lower CO dissociation barrier than that of site-C would warrant further investigation. A further consideration should be that under FTS conditions other pathways that are not currently included in the model are at play.[41,42,66,67] Adding to these theoretical arguments are recent surface science experiments in which the energy barrier value for defect site facilitated direct CO dissociation was extracted.[33] These experiments showed that CO dissociation can already occur on defect sites at around 330 K, with an estimated barrier of between 90 and 104 kJ mol^{-1} (0.93–1.08 eV) without the assistance of hydrogen. These values are significantly lower than those in this model. It is thus important to assess if lowering the CO dissociation barrier to these levels will indeed produce meaningful chain growth.

On the other hand it was pointed out that the energy of the barrier of methyl hydrogenation is also a controlling factor for the C_1* concentration by controlling the rate of removal of these species from the surface. All of the DFT data used in this work (except the CO* and H* adsorption energies) have been calculated in a low coverage regime and no other lateral interaction have been included. However, CO* has a very high coverage under steady state conditions. The lateral interactions of CO* with other species such as the carbon containing fragments on the surface could play an important role. This has been proposed from surface science experiments[17] and initial DFT studies to this effect have been attempted.[66,68] To assess this, a probe DFT study was conducted on the Co(111) surface where adsorbed CH_x species were optimized in the presence of increasing CO* coverage. The result can be seen in Fig. 8. It is clear that by increasing θ_{CO} the CH_x species become destabilized relative to the most stable species and so move upwards in the energy landscape. The CH_3* intermediate at $\theta_{CO} = 0.50$ ML is 0.48 eV less stable compared to CH* than it was with no CO present. Although the DFT energies are not calculated here, applying the Brønstead–Evans–Polanyi[69] (BEP) relationship shows that the energy levels of the transition states of the hydrogenation reactions will move upwards as well. Similarly, in a study of the free energies of methanation on Ni surfaces it was found that the high CO* coverage destabilized the CH_x species relative to the most stable species.[63] Thus, there is good reason to believe that at least all the CH_x species in the model are too stable. This concept would apply to other intermediates and transition states as well and, if this indeed affects the microkinetic results, a more in-depth DFT analysis of these effects would be needed. To probe this destabilization, a single destabilization descriptor factor ($\Delta\acute{E}_{CH_x}$) was used. It is defined by the additional energy difference between the considered CH_x intermediate and the most stable CH_x intermediate on the specific surface. This factor is also multiplied by the number of additional H atoms in the considered CH_x intermediate. After the energies of the intermediates were rescaled the connecting barriers were adapted using the BEP relationship with $\alpha = 0.5$.

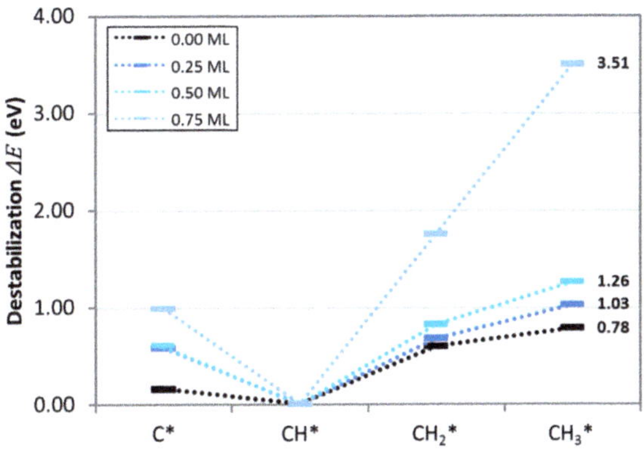

Fig. 8 Calculated electronic energies of 0.25 ML CH_x* on the Co(111) surface as destabilized relative to adsorbed CH* in the presence of various coverages of CO* (as indicated in the legend). Energies are balanced with separately adsorbed H*.

These two modification possibilities could also be present simultaneously. To assess this, the combined effects of lowering the CO scission barriers on site-C and of destabilizing the CH_x intermediates on all sites is added to the analysis. The resulting CO consumption rates and $\acute{S}$ values obtained by simultaneous modification of these two dimensions can be seen in Fig. 9. Considering the case where $\Delta\acute{E}_{CH_x} = 0$ eV, it can be seen that lowering the CO dissociation barrier by up to 0.3 eV has no effect. This is due to the fact that the H-assisted CO dissociation barrier is still the controlling factor in that range. However, lowering the barrier further shows the rate proceeding through a maximum and then dropping off to a slower rate at $\Delta G^{\ddagger}_{f,COdiss} < -0.63$ eV. At these barrier values the $\acute{S}$ values start to increase. More detailed results on this dimension can be found in the ESI.† If however we increase the value of $\Delta\acute{E}_{CH_x}$, while keeping the CO scission process unchanged, the rate decreases somewhat. The $\acute{S}$ values start to increase to the expected range from about $\Delta\acute{E}_{CH_x} = 0.3$ eV and proceed to values higher than the expected range when $\Delta\acute{E}_{CH_x}$ exceeds 0.4 eV. For the case of lowering the H-assisted CO dissociation barrier a similar result is achieved, however a smaller decrease in $\Delta G^{\ddagger}_{f,HCO\text{-}diss}$ is needed to affect a similar change. Using these contour plots of rate and $\acute{S}$ together, regimes that correspond broadly to the expected experimental results can be identified ($-r_{CO} < 0.05$ s^{-1}; $0.05 < \acute{S} < 0.35$). These regimes are highlighted in yellow in the overlay plots in Fig. 9. It is therefore clear that it is a case of either the CO scission process is very rapid, or the CH_x is significantly destabilized. The combined effect seems to overestimate the $\acute{S}$ values significantly. Given these results, a detailed analysis and comparison can be made for four scenarios of the microkinetic model:

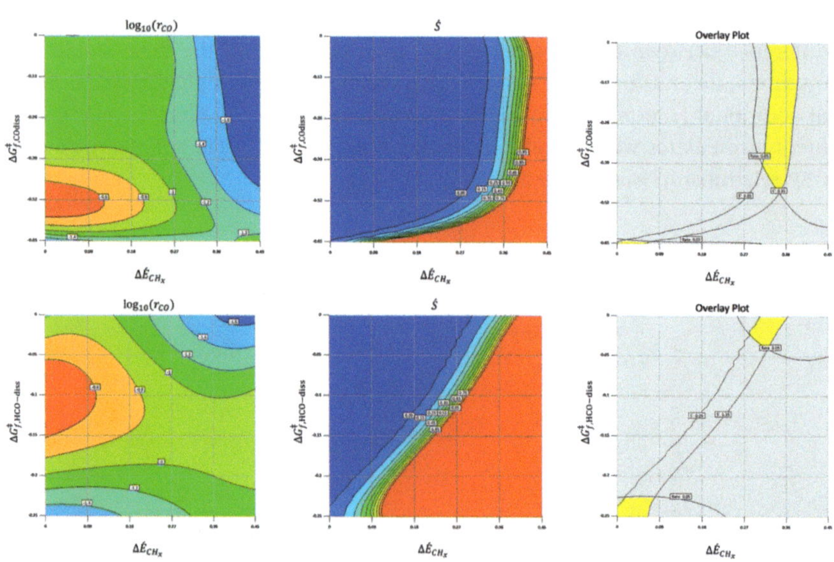

Fig. 9 Contour plots of the absolute CO consumption rate (represented by $\log_{10}(r_{CO})$) (left) and $\acute{S}$ (middle) as a function of the lowering of the of the CO scission barriers (direct CO dissociation $\Delta G^{\ddagger}_{f,COdiss}$ (top); HCO* route $\Delta G^{\ddagger}_{f,HCO-diss}$ (bottom)) and the ad hoc destabilization of the CH_x intermediates and transition states ($\Delta\acute{E}_{CH_x}$). Overlay plots (right) of the rate and $\acute{S}$ plots indicate the regions in yellow corresponding to regimes representing the expected experimental results.

S1 Using the DFT data as is (see above)
S2 Lowered CO dissociation barrier on site-C ($\Delta G^{\ddagger}_{f,\mathrm{COdiss}} = -0.7$ eV)
S3 Lowered H-assisted CO dissociation barrier on site-C ($\Delta G^{\ddagger}_{f,\mathrm{HCO-diss}} = -0.25$ eV)
S4 Destabilized CH_x on all sites ($\Delta \acute{E}_{CH_x} = 0.35$ eV)

3.3.1 Scenarios S2 & S3 – lowered CO scission barriers. A comparison of S2 and S3 to the SSITKA experiments can be seen in Fig. 4. From these results it is clear that in these two scenarios the correspondence of the coverages to the SSITKA results improved greatly, especially θ_{CH_x*}, while θ_{OH*} moved almost too high. The main discrepancy lies in the predicted CO consumption rates, which are an order of magnitude too high in these cases.

The full rate and coverage result of S2 and S3 at the standard reference conditions can be found in the ESI.† The main coverage feature that needs to be pointed out is that for both S2 and S3 the hydrogen coverage is quite low on all sites compared to S1. The whole coverage distribution is very similar between the two scenarios. The main other surface species, besides CO*, are O* and the C_2 pool on site-A, C*, CH*, O* and the C_2 pool on site-B and O* and OH* on site-C. Although the CO scission process is now much quicker, the apparent activation energy for S2 is around 1.23 eV, which is still somewhat higher than that proposed from experiments. The parity plot for the pressure–rate relations for S2 can be seen in Fig. 5 with R^2 values of 0.822 and 0.804 for M1 and M2, respectively. This is a significant improvement from the DFT data as is (S1), especially for response to changes in the H_2 : CO ratio. Still, the total pressure effects deviate quite far from the parity line. Scenario S3 responds similarly. Although this is an improvement, there are still some aspects of the reaction that seem to be incorrectly described under these scenarios.

The reaction networks for S2 and S3 can be seen in the ESI.† The main carbon flux reaction networks of these two scenarios have the exact same structural layout as that of S4 (see Fig. 11), except for S2 where the CH* is formed through direct dissociation of CO* instead of *via* the HCO* intermediate. The difference between these two networks and that of S4 (described below) is that the species flow through this network structure in a different way due to a difference in the rate controlling steps and intermediates. Exploring the main feature of this network it is clear that reactions on site-B are all generally quite slow, except for the formation of CO_2. This site is therefore the main CO_2 producing site. CO* adsorbs mainly on site-B, where after it diffuses to site-C *via* site-A. CO scission proceeds exclusively on the defect site, where after or during which CH* is formed. This CH* is mostly rapidly hydrogenated to CH_3* and subsequently a large fraction is hydrogenated to methane. Both CH* and CH_3* can diffuse back to site-A, where the CH_3* is mostly hydrogenated to methane and the CH* partakes in the formation C_2*. The details of the C_2 network will be discussed in scenario S4, but it is important to note that in all three scenarios ethylene is the primary gas phase product. However, ethylene binds very strongly to site-C and will very rapidly be readsorbed and hydrogenated to ethane on site-C. This step of the mechanism would explain the ethylene deviation§ in the typical FTS product distribution. The sensitivity analysis of S2 and S3 can be seen in Fig. 10. For both the CO

§ The ethylene deviation in FTS is where for the C_2 products the ethane production exceeds that of ethylene by around an order of magnitude, resulting in a low olefin-to-paraffin ratio,[39,57] while for C_3 to C_6 the olefin products are typically more or at a similar level.

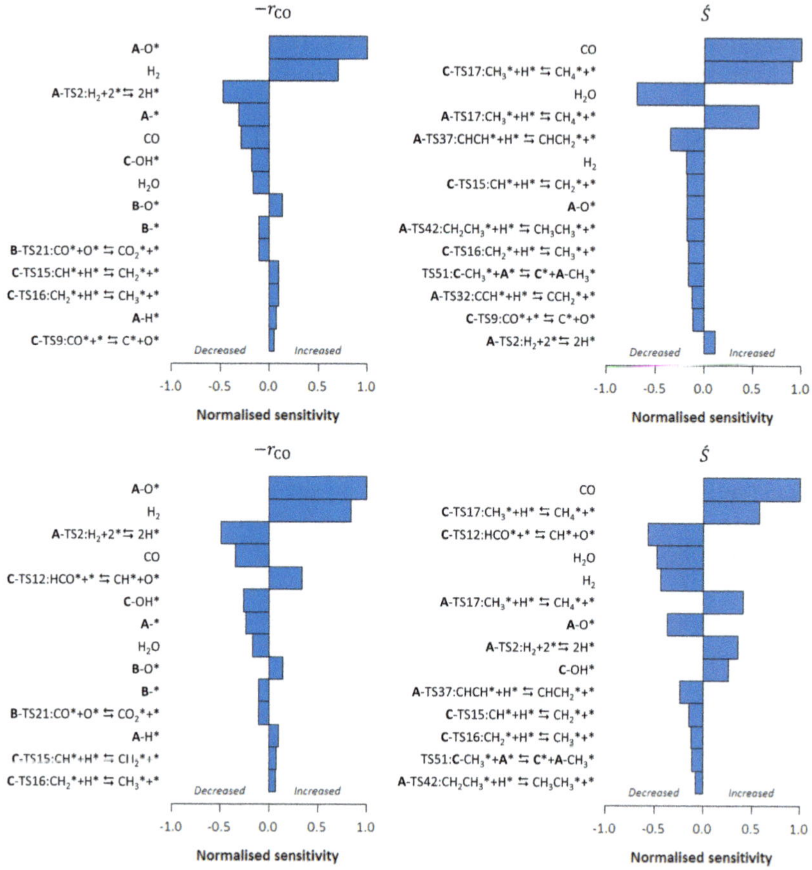

Fig. 10 Sensitivity analysis results of $-r_{CO}$ (left) and $\acute{S}$ (right) for the modified multi-site Co microkinetic model scenario S2 (top) and S3 (bottom) at the standard reference condition ($T = 503$ K; $P_{total} = 20$ bar; inlet $H_2 : CO = 2$; $\chi_{CO} = 50\%$). The sensitivity factors have been normalized to the largest absolute value.

consumption rate control and the control of selectivity ($\acute{S}$) these two scenarios have similar rate controlling species, albeit in a different order of importance. The main difference is that for S3 the HCO dissociation process barrier still has a partial controlling role. The main rate controlling factors in both of these scenarios are the coverage of O* on site-A, together with the coverage of OH* on site-C. This indicates that the stability and concentration of OH_x* (which needs to be removed as water) could be rate limiting. This has been proposed from surface science experiments.[33] Another interesting rate controlling step is the adsorption rate of H_2 on site-A. These steps would affect how this model responds to conversion and reaction conditions and probably underlies the discrepancies that are still observed in the rate–pressure relationship. In terms of the sensitivity of the C_2 growth it is clear that both the methyl hydrogenation steps on sites C and A play a big role. The main C_2 reaction network step in S2 that controls the selectivity directly is the hydrogenation of the acetylene intermediate on site-A. A number of other barriers are also involved in the rate control, but to a lesser extent. An interesting point worth observing is that these two scenarios are very similar and in both cases the CO scission process is not the main rate controlling feature. This would imply that if CO scission is not rate limiting, the exact mechanism of CO scission should not be of too much consequence. This should be considered in any further discussion in the debate on direct CO dissociation *vs.* the H-assisted route.

3.3.2 Scenario S4 – destabilized CH_x. The comparison of S4 to the SSITKA experiments can be seen in Fig. 4 (S4). From these results it is clear that in this scenario the correspondence of the coverages $\theta_{CH_x^*}$ and θ_{OH^*} to the SSITKA results is quite good although θ_{CO^*} and θ_{H^*} is somewhat overestimated. The predicted CO consumption rates are however in much better agreement than those of S2 and S3.

The full rate and coverage result of S4 at the standard reference condition can be found in the ESI.† The main coverage features include the fact that the hydrogen coverage is higher than expected. A similar result compared to SSITKA can be seen. The whole coverage distribution differs drastically from that of S2 and S3. The main other surface species besides CO* and H* are O*, the C_2 pool, and the empty surface sites on site-A. On site-B C*, CH*, O* and the C_2 pool dominate, while on site-C O* and OH* are the only other species with a significant concentration. Although the CO scission process has not been adapted, the actual apparent activation energy for S4 is lower than the other scenarios at 1.14 eV, which starts to enter the experimentally proposed range (0.96–1.16 eV). The parity plot for the pressure-rate relations for S4 can be seen in Fig. 5 with R^2 values of 0.949 and 0.911 for M1 and M2, respectively. This is a further improvement over the other scenarios. This would indicate that this model has most of the features to describe the FTS rate over a range of reactions conditions. Since this relationship depends on the rate controlling features of the reaction network, these should then be in better agreement with the true controlling steps of FTS. Although this is a significant agreement, there might still be some aspect of the reaction that is incorrectly described under these scenarios, especially since the destabilization that has been applied is not fundamental in nature. Thus, the destabilization of intermediates due to a high coverage of CO* on all sites needs to be investigated further with DFT.

The main carbon flux reaction network for this scenario can be seen in Fig. 11. Considering the sensitivity analysis that underlies this network (Fig. 12) it is clear that the CO consumption rate is controlled by three main features. The first and most important is the hydrogen adsorption process. In this case the coverage of H* on site-A and the empty site have opposing effects in terms of sensitivity. This points to the hydrogenation capacity of this site as can be expressed by the $\frac{\theta_H}{\theta^*}$ ratio. Since the destabilization of CH_x has been proposed to be modified by CO* coverage, we can also imagine that the CO* coverage could affect the energetics of H*, resulting in a change in the $\frac{\theta_H}{\theta^*}$ ratio. This all points to further work that needs to be done to elucidate the full set of FTS kinetic parameters under high CO* coverage conditions. The second rate controlling step is the methyl hydrogenation barrier on the step site. As this is one of the modified parameters, the exact DFT values based on high CO* coverage are required formally to pin down the extent of control by this feature of FTS. Thirdly other intermediates and barriers

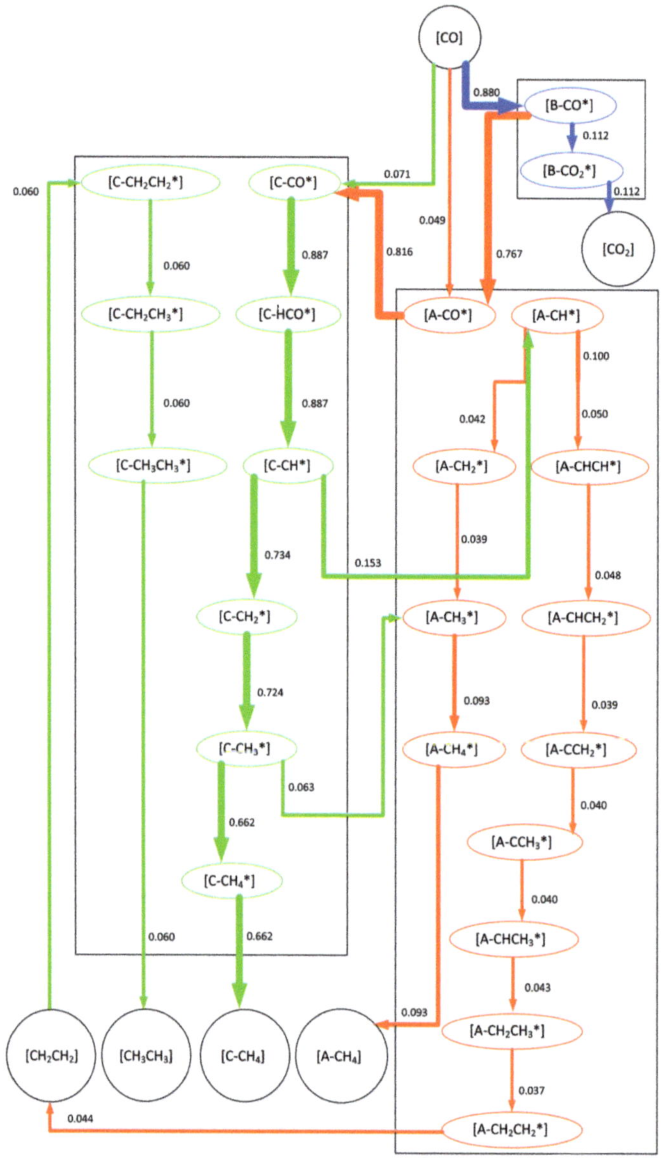

Fig. 11 The carbon atom reaction network for the multi-site Co microkinetic model under scenario S4 at the standard reference condition ($T = 503$ K; $P_{total} = 20$ bar; inlet $H_2 : CO = 2$; $\chi_{CO} = 50\%$). The arrow thickness is an indication of the scale of the rate and the C mol% flow is indicated in the labels. The network was truncated at 1% of the total CO conversion rate.

play a minor controlling role. These include OH* on site-C, which has a high coverage and would essentially block some of the CO scission capacity of this site. This also points towards the proposal that the water production process could be a rate controlling feature in FTS.[33] In terms of the chain growth features of this model, the $\acute{S}$ value is the most sensitive to the methyl hydrogenation step on site-

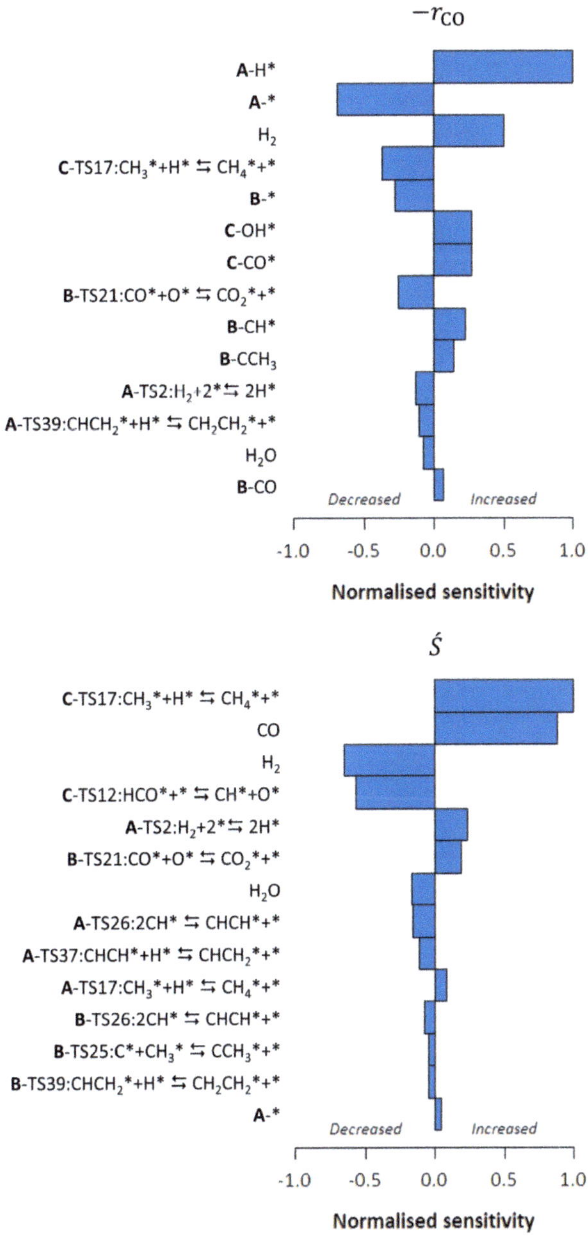

Fig. 12 Sensitivity analysis results of $-r_{CO}$ (top) and $\acute{S}$ (bottom) for the modified multi-site Co microkinetic model scenario S4 at the standard reference condition ($T = 503$ K; $P_{total} = 20$ bar; inlet $H_2 : CO = 2$; $\chi_{CO} = 50\%$). The sensitivity factors have been normalized to the largest absolute value.

C. If this step becomes slower, there are more CH_x species that can diffuse to the other sites where coupling can take place.

Based on the carbon flux analysis of the C_2 reaction network and the sensitivity analysis of this scenario, a simplified scheme of the initial coupling process of

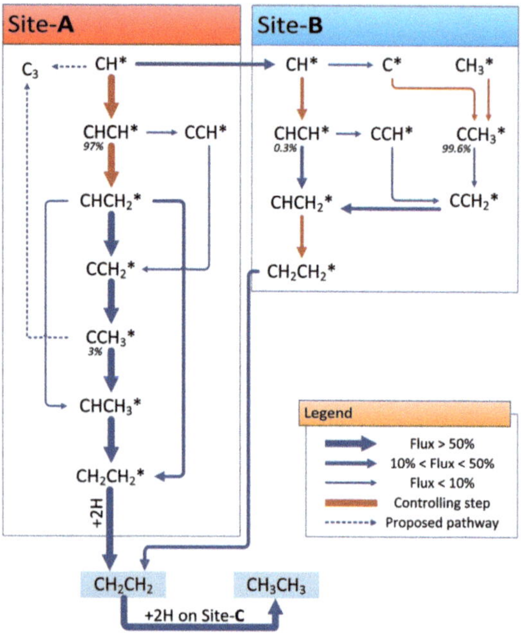

Fig. 13 Summary of the proposed C_2 reaction network for FTS starting from C_1 species based on S4. Percentages under species labels indicate typical fractional C_2 coverage. Arrow widths are proportional to the carbon flux in the network. Selectivity controlling steps are indicated.

FTS is envisaged in Fig. 13 under the condition that C_1* can be produced in a non-rate limiting fashion. From all the results thus far, coupling on site-C never featured. This would indicate that site-C has the main role of producing C_1*, methane and hydrogenating ethylene. After the C_1* diffused to a terrace like site, chain growth started to happen, with the close packed surface (site-A) dominating the reaction. This is in agreement with our previous mechanistic proposals.[17] The main coupling intermediate is CH* on both the terrace surfaces, with C* and CH_3* playing a minor role on the less active site-B. Thus, we propose that the main chain growth intermediate is CH* on site-A proceeding to couple with itself to form CHCH*. This is also the most stable C_1* intermediate on site-A at the considered DFT calculation coverage. The only other C_2 species with a meaningful coverage on this site is CCH_3*. If CH* would continue to couple to CHCH* if could eventually lead to a polymeric carbon species on the terrace, which has been proposed as a deactivation mechanism for FTS. The other option is for CH* to couple with CCH_3* to form a C_3* intermediate. According to surface science experiments,[17] CO* coverage can affect the relative stability of the CHCH* and CCH_3* species, leading to an increase in the relative stability of CCH_3. That would imply a larger relative coverage than the 3% of the C_2s noted here, which would lead to further chain growth. What is interesting to note is that the CCH_3* species has a significant coverage on site-B. Although this species is not very active on this site, this site could be viewed as a reservoir of these alkylidyne species. The stability under high CO* coverage and intersite diffusions of C_2* species between site-A and B should be investigated further with DFT. In terms of the steps that control the selectivity of C_2 formation it was found that these are the initial coupling step (due to the CH* concentration) as well as the first hydrogenation of CHCH* on site-A. However, these should be considered with caution, as this might change under a high coverage of CO*.

4 Implications for FTS and future mechanistic studies

Considering these results, a number of findings regarding the microkinetic description of FTS can be noted that can guide the focus of future FTS DFT studies and microkinetic model development:

(1) The single isolated surface sites of a Co nanocrystal in a low coverage regime cannot describe the kinetics of FTS correctly. Site-A cannot activate the CO*, site-B makes only CO_2 and CH_4 and site-C is a very efficient CH_4 producer and is inactive for chain growth.

(2) A multi-site model is required to give an appropriate microkinetic description of the FTS rates.

(3) The activity of FTS depends on the activity and availability of the defect sites (represented by site-C here).

(4) For chain growth to take place CO* scission cannot be the main controlling feature of the mechanism. This being the case, it would not be of significant consequence which CO scission mechanism (direct *vs.* H-assisted) is dominant.

(5) If the low coverage DFT data is used as is (S1), the FTS rate is in the right order, but the underlying mechanism is not correct, giving the incorrect pressure–rate relationships and selectivities.

(6) Lateral interaction of adsorbed species are of great importance at relevant FTS conditions. Here especially the role of CO* has to be considered. The high steady state coverage of CO* will at least induce destabilization of CH_x* species. This would apply to C_2* species as well. It could also affect the stability of H* as well as the hydrogen adsorption process, as was previously shown on an Fe surface.[70] For a consistent microkinetic description of FTS, future mechanistic studies should focus on obtaining and utilizing the kinetic parameters of these species and processes at high CO* coverage.

(7) In the modified scenarios the OH_x* species played prominent roles. Some of these species were present at relatively high coverages. The kinetic parameters of these species at high CO* coverage should be further investigated. In this regard, initial studies towards understanding the importance of OH_x* have recently been undertaken.[42]

(8) For chain growth to take place, a meaningful amount of C_1* species should be present on the Co surfaces. This could be accomplished by having a very rapid CO* scission process (like S2 and S3), or a slower rate of CH_4 formation (S4). In the former case, the pressure–rate relationships still have significant discrepancies, while in the latter case (*via* the destabilization of CH_x*) this relationship is better described. However, in this last case the species coverage distribution still has a too high H* concentration.

(9) From the underlying reaction network structures of C_2 growth it is clear that CH* on site-A is the main chain initiating and growth intermediate. Processes

controlling this concentration will be critical in terms of the FTS selectivity profile.

(10) Although not included in the present work, CO insertion mechanisms for chain growth do not seem to be necessary, as the coupling of CH* and the subsequent hydrogenations seem to be sufficient to produce chain growth under the condition that enough CH* can build up on the surface.

(11) Currently the inter-site diffusion of C_2* species was omitted from the model. The mobility of these species (and other longer chain growth intermediates) between sites could be an important feature, since some of these species can build up meaningful concentrations on other less active sites (such as site-B).

(12) The FTS mechanism can be split up into sub-processes that generally proceed on different sites. The defect sites are responsible for the C_1* production through CO* scission, the main source of CH_4 production and the hydrogenation of olefin products. The terraces are responsible for the main chain growth activity. The main CO_2 production is related to site-B. Thus, the selectivity profile of FTS will primarily depend on the structure of the supported Co catalysts nanocrystals, especially in the ratio of steps to terraces. Besides the possibility of direct chemical promotional effects, support interactions (and also gas phase conditions) can modify the nanocrystal morphology, which would result in differences in the site composition and therefore the selectivities. Understanding the different site compositions under actual reaction conditions will therefore be a key to understanding the changes in the selectivity profile.

(13) From the reaction networks it is clear that ethylene is the primary C_2 gas phase product. This would probably also be the case for the rest of the FTS chain growth products. The ethylene anomaly is explained by the strong adsorption of ethylene onto the defect site (C) where it is rapidly hydrogenated to ethane. The implication for the rest of the FTS product distribution is that the olefin : paraffin ratio will depend on the effective adsorption energy of the olefin product on the defect sites of the catalyst.

Given the good correspondence of S4 with experimental results it is our opinion that the study of FTS reaction steps at high CO* coverage should be the first focus in improving the microkinetic description of FTS. This would confirm the current proposals of the relevance of the CO scission mechanisms, the chain growth mechanism and monomer, and the controlling factors for activity and selectivity. Initial efforts in this regard are already underway.[40,66,68]

5 Conclusions

A multi-site CSTR microkinetic model for Co FTS has been presented based on a consistent DFT dataset. The results indicated that, using the low coverage DFT data set as is, the correspondence to the typical range of experimental results is quite poor. From a reaction network analysis it is clear that not only one site is responsible for the total reaction network, but that intermediates can migrate from one site to another and undergo different reaction steps. Using this model as a base-case, a number of fundamentally justifiable modifications to the model are considered to explore, under which regimes a meaningful amount of C_2 products, together with a more appropriate pressure–rate relationship can be found. From these exploratory results the relative reactivity of the various sites, CO activation mechanisms, the nature of the reactive chain growth monomer, the probable C_2

formation mechanism, the active site ensemble interplay, and the role of CO surface coverage have been discussed. Future focus areas for the DFT studies of FTS and microkinetic model development have been highlighted, especially the role that the high coverage of CO* has in determining the appropriate kinetic parameters under relevant steady state conditions.

Acknowledgements

The authors would like thank following people: Drs Gideon Botes, Matthys Janse van Vuuren and Denzil Moodley for their assistance with experimental results; Profs Rutger van Santen, Hans Niemantsverdriet and Eric van Steen, as well as Drs Kees-Jan Weststrate, Gregory Govender, Ashriti Govender and Braam van Dyk for all the discussions on mechanisms and kinetic modelling; Ivan Bester for supporting the high performance computing efforts.

References

1 H. Schultz, *Appl. Catal., A*, 1999, **186**, 3.
2 P. M. Maitlis and V. Zanotti, *Chem. Commun.*, 2009, **13**, 1619–1634.
3 R. A. van Santen, A. J. Markvoort, I. A. W. Filot, M. M. Ghouri and E. J. M. Hensen, *Phys. Chem. Chem. Phys.*, 2013, **15**, 17038.
4 S. Mousavi, A. Zamaniyan, M. Irani and M. Rashidzadeh, *Appl. Catal., A*, 2015, **506**, 57–66.
5 J. A. Dumesic, D. F. Rudd, L. M. Aparicio, J. E. Rekoske and A. A. Teviño, *The Microkinetics of Heterogeneous Catalysis*, The American Chemical Society, Washington D.C., 1993.
6 M. Moqadam, M. Rahmani, Z. Karimi and A. Naderifar, *Procedia Eng.*, 2012, **42**, 34–44.
7 S. Storsæter, Ph.D. thesis, Norwegian University of Science and Technology, 2005.
8 G. Lozano-Blanco, J. Thybaut, K. Surla, P. Galtier and G. Marin, *Oil Gas Sci. Technol.*, 2006, **61**, 489–496.
9 P. Azadi, G. Brownbridge, I. Kemp, S. Mosbach, J. S. Dennis and M. Kraft, *ChemCatChem*, 2015, **7**, 137–143.
10 I. A. W. Filot, Ph.D. thesis, Technische Universiteit Eindhoven, 2015.
11 I. A. W. Filot, R. A. van Santen and E. J. M. Hensen, *Angew. Chem., Int. Ed.*, 2014, **53**, 12746–12750.
12 D. W. Blaylock, Y.-A. Zhu and W. H. Green, *Top. Catal.*, 2011, **54**, 828–844.
13 O. Kitakami, H. Sato, Y. Shimada, F. Sato and M. Tanaka, *Phys. Rev. B: Condens. Matter Mater. Phys.*, 1997, **56**, 13849.
14 P. van Helden, I. M. Ciobîcă and R. L. J. Coetzer, *Catal. Today*, 2016, **261**, 48.
15 P. van Helden, J.-A. van den Berg and I. M. Ciobîcă, *Catal. Sci. Technol.*, 2012, **2**, 491–494.
16 C. Chen, Q. Wang, G. Wang, B. Hou, L. Jia and D. Li, *J. Phys. Chem. C*, 2016, **120**, 9132–9147.
17 C. Weststrate, P. van Helden and J. Niemantsverdriet, *Catal. Today*, 2016, **275**, 100–110.
18 G. Kresse and J. Hafner, *Phys. Rev. B: Condens. Matter Mater. Phys.*, 1993, **47**, 558–561.

19 G. Kresse and J. Furthmüller, *Phys. Rev. B: Condens. Matter Mater. Phys.*, 1996, **54**, 11169–11186.
20 G. Kresse and J. Furthmüller, *Comput. Mater. Sci.*, 1996, **6**, 15–50.
21 D. Vanderbilt, *Phys. Rev. B: Condens. Matter Mater. Phys.*, 1990, **41**, 7892–7895.
22 G. Kresse and J. Hafner, *J. Phys.: Condens. Matter*, 1994, **6**, 8245.
23 J. P. Perdew and Y. Wang, *Phys. Rev. B: Condens. Matter Mater. Phys.*, 1992, **45**, 13244.
24 M. Methfessel and A. T. Paxton, *Phys. Rev. B: Condens. Matter Mater. Phys.*, 1989, **40**, 3616.
25 N. W. Aschroft and N. D. Mermin, *Solid State Physics*, Holt, Rinehart and Winston, New York, 1976.
26 H. J. Monkhorst and J. D. Pack, *Phys. Rev. B: Solid State*, 1976, **13**, 5188.
27 H. Jónsson, G. Mills and K. W. Jacobsen, in *Classical and Quantum Dynamics in Condensed Phase Simulations*, ed. B. J. Berne, G. Ciccotti and D. F. Coker, World Scientific, Singapore, 1998, p. 385.
28 S. E. Mason, I. Grinberg and A. M. Rappe, *Phys. Rev. B: Condens. Matter Mater. Phys.*, 2004, **69**, 161401.
29 F. Abild-Pedersen and M. P. Andersson, *Surf. Sci.*, 2007, **601**, 1747.
30 L. C. Grabow and M. Mavrikakis, *ACS Catal.*, 2011, **1**, 365.
31 P. van Helden, J.-A. van den Berg and C. J. Weststrate, *ACS Catal.*, 2012, **2**, 1097–1107.
32 J. Lahtinen, J. Vaari and K. Kauraala, *Surf. Sci.*, 1998, **418**, 502.
33 C. J. Weststrate, P. van Helden, J. van de Loosdrecht and J. W. Niemantsverdriet, *Surf. Sci.*, 2016, **648**, 60–66.
34 P. van Helden and E. van Steen, *J. Phys. Chem. C*, 2008, **112**, 16505–16513.
35 X.-C. Guo and D. A. King, in *The Chemical Physics of Solid Surfaces*, ed. D. A. King and D. P. Woodruff, Elsevier, 1993, vol. 6, ch. 4, pp. 113–155.
36 E. Lewis, D. Le, A. Jewell, C. Murphy, T. Rahman and E. Sykes, *ACS Nano*, 2013, 7, 4384–4392.
37 R. A. van Santen and J. W. Niemantsverdriet, *Chemical Kinetics and Catalysis*, Plenum Press, New York, 1995, pp. 233–238.
38 E. Wigner, *Z. Phys. Chem., Abt. B*, 1932, **19**, 203.
39 F. Bertoncini, M. Marion, N. Brodusch and S. Esnault, *Oil Gas Sci. Technol.*, 2009, **64**, 79–90.
40 Y. Qi, J. Yang, D. Chen and A. Holmen, *Catal. Lett.*, 2015, **145**, 145–161.
41 D. D. Hibbitts, B. T. Loveless, M. Neurock and E. Iglesia, *Angew. Chem., Int. Ed.*, 2013, **52**, 12273–12278.
42 G. T. K. K. Gunasooriya, A. P. van Bavel, H. P. C. E. Kuipers and M. Saeys, *ACS Catal.*, 2016, **6**, 3660–3664.
43 J. Cheng, X.-Q. Gong, P. Hu, C. M. Lok, P. Ellis and S. French, *J. Catal.*, 2008, **254**, 285–295.
44 F. Botes, B. van Dyk and C. McGregor, *Ind. Eng. Chem. Res.*, 2009, **48**, 10439–10447.
45 A. Saltelli, M. Ratto, S. Tarantola and F. Campolongo, *Chem. Rev.*, 2005, **105**, 2811–2828.
46 C. Stegelmann, A. Andreasen and C. T. Campbell, *J. Am. Chem. Soc.*, 2009, **131**, 8077–8082.
47 A. Korzybski, *Science and Sanity*, Institute of General Semantics, 1994.

48 J. P. Den Breejen, P. B. Radstake, G. L. Bezemer, J. H. Bitter, V. Fröseth, A. Holmen and K. P. De Jong, *J. Am. Chem. Soc.*, 2009, **131**, 7197.
49 C. N. Satterfield, *Indirect Liquefaction Contractor's Review Meeting Proceedings*, 1989, pp. 125–155.
50 E. Iglesia, S. Reyes, R. Madon and S. Soled, *Adv. Catal.*, 1993, **39**, 221.
51 P. J. van Berge, Ph.D. thesis, Potchefstroom University for Christian Higher Education, 1994.
52 M. Claeys, Ph.D. thesis, University of Karlsruhe, 1997.
53 E. Iglesia, *Appl. Catal., A*, 1997, **161**, 59–78.
54 S. Zwane, M.Sc. thesis, University of Cape Town, 2003.
55 M. Dry, in *Fischer-Tropsch Technology*, ed. A. Steynberg and M. E. Dry, Elsevier, 2004, vol. 152, ch. 3.
56 C. G. Visconti, E. Tronconi, L. Lietti, P. Forzatti, S. Rossini and R. Zennaro, *Top. Catal.*, 2011, **54**, 786–800.
57 B. Todic, W. Ma, G. Jacobs, B. H. Davis and D. B. Bukur, *J. Catal.*, 2014, **311**, 325–338.
58 A. Kasht, R. Hussain, M. Ghouri, J. Blank and N. O. Elbashir, *Am. J. Anal. Chem.*, 2015, **6**, 659–676.
59 I. C. Yates and C. N. Satterfield, *Energy Fuels*, 1991, **5**, 168–173.
60 C. G. Cooper, T.-H. Nguyen, Y.-J. Lee, K. M. Hardiman, T. Safinski, F. P. Lucien and A. A. Adesina, *Catal. Today*, 2008, **131**, 255–261.
61 Z. Yan, Z. Wang, D. B. Bukur and D. W. Goodman, *J. Catal.*, 2009, **268**, 196–200.
62 M. Mansouri, H. Atashi, A. A. Mirzaei and M. Karimi, *J. Thermodyn. Catal.*, 2012, **3**, 113.
63 M. Andersson, F. Abild-Pedersen, I. Remediakis, T. Bligaard, G. Jones, J. Engbæk, O. Lytken, S. Horch, J. Nielsen, J. Sehested, J. Rostrup-Nielsen, J. Nørskov and I. Chorkendorff, *J. Catal.*, 2008, **255**, 6–19.
64 M. A. Petersen, J.-A. van den Berg, I. M. Ciobîcă and P. van Helden, *ACS Catal.*, 2017, DOI: 10.1021/acscatal.6b02843.
65 Q. Ge and M. Neurock, *J. Phys. Chem. B*, 2006, **110**, 15368–15380.
66 Y. Qi, J. Yang, X. Duan, Y.-A. Zhu, D. Chen and A. Holmen, *Catal. Sci. Technol.*, 2014, **4**, 3534–3543.
67 I. M. Ciobîcă, P. van Helden and R. van Santen, *Surf. Sci.*, 2016, **653**, 82–87.
68 M. Zhuo, A. Borgna and M. Saeys, *J. Catal.*, 2013, **297**, 217–226.
69 R. A. van Santen, M. Neurock and S. G. Shetty, *Chem. Rev.*, 2010, **110**, 2005–2048.
70 E. van Steen and P. van Helden, *J. Phys. Chem. C*, 2010, **114**, 5932–5940.

Faraday Discussions

PAPER

Kinetic aspects of chain growth in Fischer–Tropsch synthesis†

Ivo A. W. Filot,[ab] Bart Zijlstra,[a] Robin J. P. Broos,[ab] Wei Chen,[a] Robert Pestman[a] and Emiel J. M. Hensen*[ab]

Received 4th October 2016, Accepted 3rd November 2016
DOI: 10.1039/c6fd00205f

Microkinetics simulations are used to investigate the elementary reaction steps that control chain growth in the Fischer–Tropsch reaction. Chain growth in the FT reaction on stepped Ru surfaces proceeds *via* coupling of CH and CR surface intermediates. Essential to the growth mechanism are C–H dehydrogenation and C hydrogenation steps, whose kinetic consequences have been examined by formulating two novel kinetic concepts, the degree of chain-growth probability control and the thermodynamic degree of chain-growth probability control. For Ru the CO conversion rate is controlled by the removal of O atoms from the catalytic surface. The temperature of maximum CO conversion rate is higher than the temperature to obtain maximum chain-growth probability. Both maxima are determined by Sabatier behavior, but the steps that control chain-growth probability are different from those that control the overall rate. Below the optimum for obtaining long hydrocarbon chains, the reaction is limited by the high total surface coverage: in the absence of sufficient vacancies the CHCHR → CCHR + H reaction is slowed down. Beyond the optimum in chain-growth probability, CHCR + H → CHCHR and OH + H → H_2O limit the chain-growth process. The thermodynamic degree of chain-growth probability control emphasizes the critical role of the H and free-site coverage and shows that at high temperature, chain depolymerization contributes to the decreased chain-growth probability. That is to say, during the FT reaction chain growth is much faster than chain depolymerization, which ensures high chain-growth probability. The chain-growth rate is also fast compared to chain-growth termination and the steps that control the overall CO conversion rate, which are O removal steps for Ru.

Introduction

The widespread availability of cheap natural gas resources in the past few decades in combination with the dwindling supplies of readily available crude oil has led

[a]*Laboratory of Inorganic Materials Chemistry, Schuit Institute of Catalysis, Department of Chemical Engineering and Chemistry, Eindhoven University of Technology, P. O. Box 513, 5600 MB, Eindhoven, The Netherlands. E-mail: e.j.m.hensen@tue.nl*

[b]*Netherlands Center for Multiscale Catalytic Energy Conversion, Universiteitsweg 99, 3585 CG, Utrecht, The Netherlands*

† Electronic supplementary information (ESI) available. See DOI: 10.1039/c6fd00205f

to commercialization of gas-to-liquids (GTL) processes in the fuels and chemicals industry. The Fischer–Tropsch (FT) reaction plays a central role in these GTL efforts: it converts synthesis gas into clean transportation fuels and chemicals. Despite significant research efforts in the last decade, many aspects of the underlying mechanism need to be better understood. Such insight will facilitate the guided design of improved catalysts. Modern approaches such as computational advances and development of model systems are tools that may provide such insights. In this work, we will discuss our recent findings regarding the dominant factors governing the chain growth, one of the essential reaction steps in the FT reaction, for Ru model catalysts.

The FT reaction is a polymerization reaction of C_1 monomer species, which are *in situ* generated on the catalytic surface by CO dissociation.[1,2] Chain growth proceeds through association reactions of C_1 monomers with growing hydrocarbon chains. Products leave the surface as olefins (β-H elimination), paraffins (hydrogenation) or aldehydes (CO insertion).[1,3] Termination of C_1 monomers by hydrogenation yields methane, which is an undesired reaction product as it is the source of synthesis gas. In commercial practice, the aim is usually to minimize methane yield and maximize the formation of liquid hydrocarbons that can be converted to transportation fuels in downstream hydrocracking processes.[4] This can be achieved by increasing the chain-growth probability (α), defined as the rate of propagation (r_p) over the sum of the rates of propagation and termination (r_t).

$$\alpha = \frac{r_p}{r_p + r_t} \tag{1}$$

As termination is slow compared to propagation – a necessary condition to obtain long hydrocarbon chains in the overall FT mechanism, the chain-growth probability can be raised by lowering the reaction temperature. Consequently, the FT reaction is usually operated below the CO conversion rate optimum.

Recently, we have explored the mechanism of the FT reaction on Ru surfaces using microkinetics simulations of CO hydrogenation, involving kinetic parameters of all relevant elementary reaction steps over Ru surfaces determined by density functional theory calculations.[5] The main findings of this work are as follows: CO hydrogenation on a planar Ru surface is slow and the main hydrocarbon product is methane. Step-edge sites are required for the FT reaction. The low barrier of CO dissociation ensures sufficient rapid supply of C_1 monomers to maintain a high rate of chain propagation. On stepped Ru surfaces, propagation involves CH surface intermediates so that the initial chain is formed by CH + CH coupling and further chain growth involves the reaction of CH with CR (R = alkyl) fragments. This implies that chain growth involves hydrogenation and dehydrogenation steps to transform the direct CHCR coupling product into the CCH_2R intermediate that can further couple to CH. Further salient details worth mentioning of the mechanism are the reversibility of chain growth, the predominance of olefins as the main hydrocarbon reaction products and the removal of oxygen in the form of water. A mechanism involving CO insertion into growing chains followed by C–O bond scission in RCO fragments contributes little to the overall FT reaction.

By using the degree of rate control approach[6,7] and applying linear scaling laws and Brønsted–Evans–Polanyi relationships, characteristic kinetic regimes for the FT reaction were identified as a function of metal–carbon and metal–oxygen binding energies. For Ru this leads to three kinetic regimes.[5] Too weak binding with carbon and oxygen results in CO dissociation being rate controlling. Too strong binding of oxygen makes water removal the most difficult step in the FT mechanism, while too strong binding of carbon shifts the rate controlling step to chain growth. The latter will limit the chain-growth probability. Whilst Ru operates in the water removal limited regime, the FT reaction on Co is limited by CO dissociation and water removal.

Currently, there is a lack of computational studies that describe in detail the parameters that influence the chain-growth probability. Understanding the selectivity in the FT reaction is of fundamental importance to improve the current generation of FT catalysts, especially with respect to the question of how activity and selectivity are interrelated. Van Santen and co-workers have discussed several aspects about FT selectivity by constructing theoretical models based on the steady-state solutions of coupled rate expressions, Monte Carlo simulations and density functional theory calculations.[8–13] Understanding chain-growth probability is also important for novel processes such as the conversion of synthesis gas to olefins. The group of De Jong developed an FT process using a modified Fe catalyst that produces selectively lower olefins.[14] They speculated that such a selectivity is possible by the unique bonding characteristics of an iron-carbide phase compared to that of metals such as Co or Ru.

In this work, we contribute to the current understanding of the selectivity issues in the FT reaction by exploring the origin of the hydrocarbon selectivity with a focus on chain-growth probability. In particular, we elucidate the elementary reaction steps that control the chain-growth probability. For this purpose, we introduce a mathematical tool, which is akin to the degree of selectivity concept introduced by Campbell.[15,16] By analysis of the surface composition, the kinetic network and the individual rates of the elementary reaction steps contributing towards the overall reaction, we can relate the predicted product selectivity, *viz.* chain-growth probability, to underlying factors that govern these processes.

Method

A first-principles microkinetic model to simulate the FT reaction over stepped Ru surfaces was constructed using activation barriers and corresponding vibrational frequencies derived from density functional theory calculations as previously published.[5,17] From these barriers and frequencies, the forward and backward rate constants for the elementary reaction steps were constructed using the Eyring equation:

$$k = \frac{k_b T}{h}\frac{Q^{\ddagger}}{Q}e^{-\frac{E_a}{k_b T}} \tag{2}$$

where k is the rate constant, k_b is the Boltzmann constant, T is the temperature in K, h is Planck's constant, $Q^{\ddagger}$ is the partition function of the transition state complex, Q is the partition function of the complex in the pre-activated (initial) state, and E_a is the electronic activation energy. The partition functions for the activated and pre-activated complexes were taken as the products of the translational, rotational, and vibrational partition functions corresponding to the configurational degrees of freedom of the surface complexes.

For non-activated molecular adsorption, it was assumed that the complex loses one translational degree of freedom with respect to the gas phase in the transition state. The changes in the rotational degrees of freedom were neglected. Accordingly, the following expression was employed for the rate of molecular adsorption:

$$k = \frac{PA}{\sqrt{2\pi m k_b T}} S \tag{3}$$

where P is the partial pressure of the adsorbate in the gas phase, A is the surface area of the adsorption site, m is the mass of the adsorbate, and S is the sticking coefficient. The surface area A was set to the area of a threefold site. The sticking coefficients were set to unity in the current simulations.

For desorption, it was assumed that the complex has three rotational degrees of freedom and two translational degrees of freedom in the activated state, while it has only vibrational degrees of freedom in the adsorbed state. Accordingly, the rate of desorption is given by:

$$k = \frac{k_b T^3}{h^3} \frac{A(2\pi m k_b)}{\sigma \theta_{rot}} e^{-\frac{E_{des}}{k_b T}} \tag{4}$$

where σ is the symmetry number, θ is the characteristic temperature for rotation, and E_{des} is the desorption energy.

The approach for the microkinetics simulations has been extensively described before and is briefly discussed here for clarity.[18] Differential equations for all surface reaction intermediates were constructed using the rate constants of the elementary steps. Given a system of N elementary reaction steps, $2N$ rate expressions (*i.e.*, both forward and backward reactions) were obtained with the form:

$$r_j = k_j \prod_i c_i^{\nu_i{}^j} \tag{5}$$

where c_i is the concentration of species i in the elementary reaction step j on the surface, and ν_i is the stoichiometric coefficient of species i in elementary reaction step j. These rate expressions were used to derive an ordinary differential equation for each component on the surface with the form:

$$\frac{dc_i}{dt} = \sum_j \nu_{i,j} r_j \tag{6}$$

where c_i is the concentration of species i on the surface, and $\nu_{i,j}$ is the stoichiometric coefficient of species i in elementary reaction step j.

The in-house developed C++ program MKMCXX[19] was employed to determine the steady-state coverages by integrating this set of ordinary differential equations with respect to time using the backward differentiation formula method. The steady-state surface coverage values were used to compute the rates of the individual elementary reaction steps and the overall rate per active site.

Result and discussion

Microkinetics simulations were performed to investigate the influence of the elementary reaction steps on the chain-growth probability α. The underlying microkinetic model contains kinetic parameters for elementary reaction steps determined by quantum-chemical DFT calculations for the stepped Ru(11$\bar{2}$1) surface.[5,17] The dataset is available in the ESI.† The total pressure in the current microkinetics simulations was set to 20 atm and a H_2/CO ratio of 2 was used. These conditions are close to the conditions used in commercial practice. First, we will present rates, the chain-growth probability and the product distribution as a function of temperature. Next, we explore the influence of the rates of the elementary reaction steps on the chain-growth probability by modifying the reaction barrier of the elementary reaction steps in the system. This approach is derived from the concept of degree of rate control as introduced by Campbell and co-workers.[15,16] We then rationalize these results by relating the variations in chain-growth probability to the rates of the elementary reaction steps and the surface coverages.

We first determined how temperature affects the overall CO consumption rate (Fig. 1a) and the chain-growth probability (Fig. 1b). From Fig. 1a, it can be seen that the optimum temperature for CO conversion lies around 650 K, while the optimum reaction temperature for obtaining higher hydrocarbons is below the Sabatier maximum for CO conversion. This is in line with experimental results.[1,20–22] Fig. 1b further underpins this interpretation by showing the optimum temperature for chain-growth probability to be around 500 K. This optimum also finds its origin in Sabatier's principle. As we will demonstrate below, the chain growth is limited by hydrogenation, one of the essential steps in the growth mechanism, at low temperature. At high temperature, the reversibility of the chain-growth process is at the origin of the decreased chain-growth probability in addition to faster hydrogenation of chains and C_1 monomers. These two effects also lead to increased methane selectivity. An analysis of the degree of rate control with respect to the CO conversion rate shows that the overall conversion of CO is limited by water removal from the catalytic surface. While O hydrogenation and water desorption are controlling the rate at temperatures below the optimum, at high temperature the rate-controlling step shifts to the reaction of two OH groups to produce adsorbed water.

Fig. 2 shows the product distribution in more detail as a function of temperature. At the chain-growth probability optimum, the amount of C_{40+} hydrocarbons is highest. In commercial applications, high chain-growth probability and low

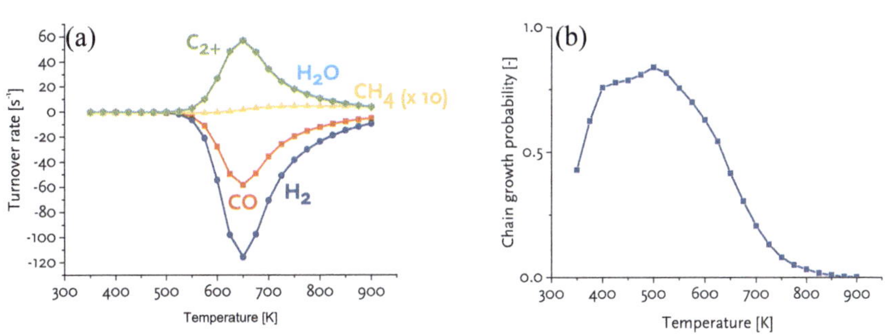

Fig. 1 Microkinetics simulations of the FT reaction on stepped Ru(11$\bar{2}$1) surface at p = 20 atm and a H_2/CO ratio of 2: (a) CO and H_2 consumption rates and CH_4, H_2O and C_{2+} production rates as a function of temperature. Note that the C_{2+} trace overlaps the H_2O trace. (b) The chain-growth probability as a function of temperature, displaying a maximum at ~500 K.

methane production is targeted. Therefore, we will focus our more detailed analysis on the influence of the elementary reaction steps on the chain-growth in the temperature regime between 400 K and 600 K. In order to do so, we first define a degree of chain-growth control (DCGC) as follows:

$$\chi_{i,\text{DCGC}} = \left(\frac{\partial \alpha}{\partial \ln k_i}\right)_{i \neq j, K_i} \tag{7}$$

in which α is the chain-growth probability, k_i is the rate constant of elementary reaction step i, and K_i is the equilibrium constant of elementary reaction step i. In essence, DCGC probes the absolute change in chain-growth probability relative to a relative change in the rate constant. We do this by changing the height of the reaction barrier such that the equilibrium constant of the elementary reaction step is not changed. This is done to satisfy the microscopic reversibility condition. We find that the elementary steps that control the overall CO consumption rate are not necessarily the same as those that control the chain-growth probability.

Fig. 3a shows the DCGC for all relevant elementary reaction steps within the chosen temperature range. A positive DCGC value implies that increasing the rate constant of a particular elementary reaction step increases the chain-growth probability and *vice versa*. In the former case, we can say that the elementary reaction step is controlling the chain growth, and in the latter case it is inhibiting chain growth. As several elementary reaction steps share similar features, we changed the barriers for all of these steps together in assessing the influence of chain growth. For instance, for evaluation of the influence of monomer insertion into the growing chain on the chain-growth process, we changed the barriers for all C-insertion reactions, instead of changing only one, such as that of $C + CCH_2CH_3$ coupling. The same holds for other CH_x insertion reactions and chain hydrogenation reactions. The discussion of the results in Fig. 3a is supported by the reaction network showing the rates per elementary reaction step (Fig. 3b). We first discuss (sets of) elementary reaction steps that lead to an increase of the chain-growth probability. Thereafter, we will discuss the reactions that inhibit the chain-growth probability.

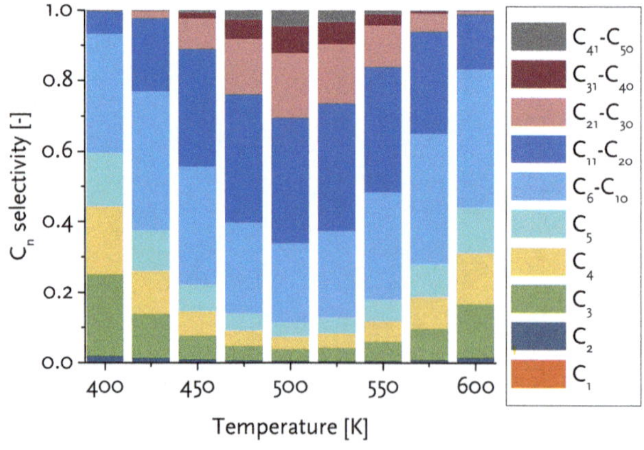

Fig. 2 Hydrocarbon product selectivity for C_1–C_{50} as a function of temperature. The selectivity towards C_{40+} hydrocarbons is highest at 500 K, consistent with the highest chain-growth probability determined at this temperature. Data shown for the FT reaction on stepped Ru at $p = 20$ atm and a H_2/CO ratio of 2.

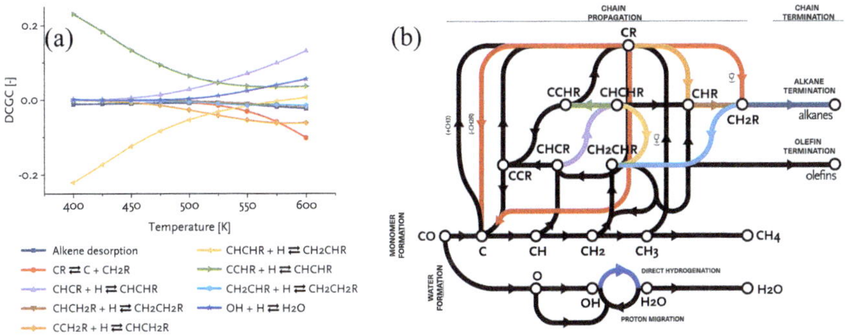

Fig. 3 (a) Degree of chain-growth control (DCGC) as a function of temperature for relevant elementary reaction steps, whose DCGC value is larger than 0.01. Positive values imply that decreasing the barrier for an elementary reaction step increases the chain-propagation probability. (b) Reaction network diagram of the relevant elementary reaction steps and the direction of chemical conversion. Data shown for the FT reaction on stepped Ru at $p = 20$ atm and a H_2/CO ratio of 2.

At low temperature, the chain-growth probability is mainly influenced by the CCHR + H → CHCHR elementary reaction step. From Fig. 3b, we infer that this elementary reaction step proceeds in the reverse direction, or in other words, CHCHR is dehydrogenated to CCHR (CHCHR → CCHR + H). This is in line with the earlier discussed reaction sequence following CH coupling to CR surface intermediates as the dominant coupling reaction.[5] For this reaction to propagate after the CH insertion reaction, the α-carbon atom in CHCR needs to be dehydrogenated followed by hydrogenation of the β-carbon to CH_2. Thus, dehydrogenation of the α-carbon is kinetically important to chain growth. The strong influence of this elementary reaction step below 500 K is next discussed in terms of the surface coverages (Fig. 4). The α-dehydrogenation of CHCHR requires a vacant site on the surface. At low temperature, there are almost no free sites, which limits chain growth. The increase in surface vacancies with temperature results in faster α-carbon dehydrogenation. Beyond the optimum in chain-growth probability, two other reactions limit the chain growth, namely CHCR + H → CHCHR and OH + H → H_2O. The latter reaction also contributes to the control of the overall CO consumption rate next to OH hydrogenation and the proton transfer between two OH groups to form adsorbed water. The barrier for the CHCR + H → CHCHR reaction is the highest of all kinetically relevant chain-propagation reactions. The OH + H reaction is also important as it removes both O and H atoms from the surface. Faster OH hydrogenation results in a lower H coverage. As chain growth is limited by only one hydrogenation step in this regime, whereas it is inhibited by two hydrogenation steps (*vide infra*), a lower H coverage results in increased chain-growth probability.

Three elementary reaction steps strongly inhibit chain growth. At low temperature, this is the CHCHR + H → CH_2CHR reaction. This step converts growing hydrocarbon chains that are a part of the dominant chain propagation cycle (CR → CHCR → CHCHR → CCHR → CR) to the CH_2CHR surface intermediate. This intermediate can readily desorb from the surface as a 1-olefin; in other words, it leads to chain-growth termination. With increasing temperature, CCH_2R + H → $CHCH_2R$ and CR → CH_2R + C reactions become rate-inhibiting.

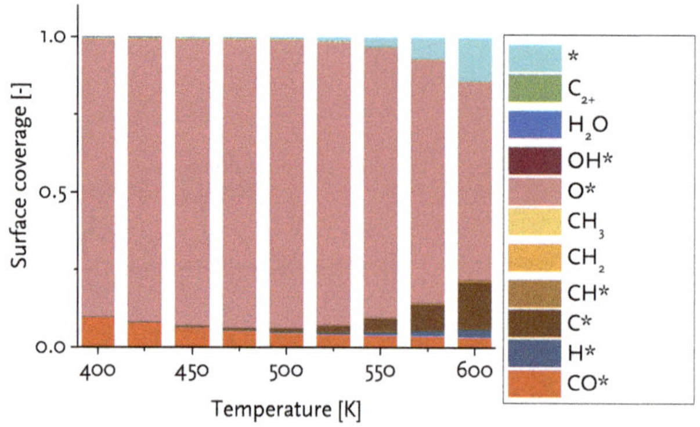

Fig. 4 Surface coverage as a function of temperature for the FT reaction on stepped Ru at $p = 20$ atm and a H_2/CO ratio of 2. The total coverage of growing chains on the surface is around 3%. The surface is predominantly covered with O.

Similarly to the CHCHR + H → CH_2CHR reaction, these two elementary reactions convert intermediates that are part of the chain-growth cycle to intermediates that can desorb from the surface as an olefin, either directly or *via* two additional hydrogenation steps. At low temperature, these two reactions are not rate-inhibiting for the following reasons. The CCH_2R + H → CHR reaction requires at least two additional hydrogenation steps for chain-growth termination. As the surface coverage of H is relatively low, this chain-growth termination pathway is slow. The CR → CH_2R + C is not rate-inhibiting at low temperature, because this reaction requires a surface vacancy, which are scarce.

Our analysis of the chain-growth controlling elementary reaction steps in conjunction with the composition of the adsorbed layer reveals the delicate interplay between the coverage of H and the kinetic network of the growing chains. To further analyze this interplay, we have performed a thermodynamic degree of chain-growth control analysis, which we define as follows:

$$\chi_{i,\mathrm{TDCGC}} = \left(\frac{\partial \alpha}{\partial\left(-\frac{\Delta G}{RT}\right)}\right)_{G^{\mathrm{TS}}_{i\neq j},G^{0}_{m}} \tag{8}$$

where ΔG is the Gibbs free energy of a specific surface species.[15] In this case, the surface stability of one particular surface adsorbate is changed, while keeping the height of the activation barrier and the stability of all other species the same. The effect of this change on the chain-growth probability is investigated. A positive value of $\chi_{i,\mathrm{TDCGC}}$ indicates that stabilizing the particular surface species increases chain-growth probability. The results of this analysis are depicted in Fig. 5. For the FT process to occur, it is often assumed that a small amount of free sites should be available for CO dissociation, the first step in the overall FT process. We already showed that the FT reaction has several kinetic regimes and the Ru-catalyzed FT reaction operates in a kinetic regime where CO dissociation is not controlling the overall rate.[5] This is supported by the results in Fig. 5, where it can be seen that the surface stability of CO does not influence the chain-growth probability. The interesting consequence is that when CO dissociation is not rate-controlling, an increase in the number of free surface sites does not necessarily increase the overall CO consumption rate and the chain-growth probability. On the contrary, an increased number of free sites on the surface has a negative influence on the chain-growth probability. This is primarily the result of an increased rate of chain depolymerization. The relevant elementary reaction step is CR → CH_2R + C, which requires free sites. Furthermore, at low temperature chain growth is inhibited by one hydrogenation step, *i.e.* CHCHR + H → CH_2CHR, and limited by one dehydrogenation step, *i.e.* CHCHR → CCHR + H. Increasing the surface stability of H atoms, therefore, lowers the rate of hydrogenation and increases the rate of dehydrogenation reactions. This provides a rationale for the positive TDCGC for H at low temperature (Fig. 5). At high temperature, chain growth is limited by two hydrogenation reactions, *i.e.*, OH + H → H_2O and CHCR + H → CHCHR with DCGC values of 0.06 and 0.13, respectively, and one dehydrogenation reaction (CHCHR → CCHR + H; DCGC = 0.04) and inhibited by one hydrogenation reaction (CR + H → CHR, DCGC = −0.06). This would suggest that destabilizing H atoms at the surface results in faster hydrogenation reactions and higher chain-growth probability. However, the opposite is seen in Fig. 5 with the TDCGC value being strongly negative for H. This is because at high temperature, destabilization of H results in a faster rate of H_2 desorption, which results in a higher number of vacant sites. These vacant sites have a negative impact on chain growth, as they facilitate depolymerization. This is confirmed by the DCGC value of the CR → CH_2R + C reaction (−0.10). These results would suggest that increasing the H_2 pressure would result in an increase of the chain-growth probability. Typically, the reaction order of the FT reaction with respect to H_2 is positive. In our microkinetic model, the H_2 reaction order is also positive, because more H atoms on the surface remove O from the surface, allowing higher coverage with CH intermediates. This can in principle increase the amount of methane and higher hydrocarbons produced. From our kinetic analysis, we predict that the chain-growth probability will be increased in such case, as the increased H coverage will suppress the rate of chain depolymerization. All these results emphasize the critical role of the coverage of H and free sites on the chain-growth

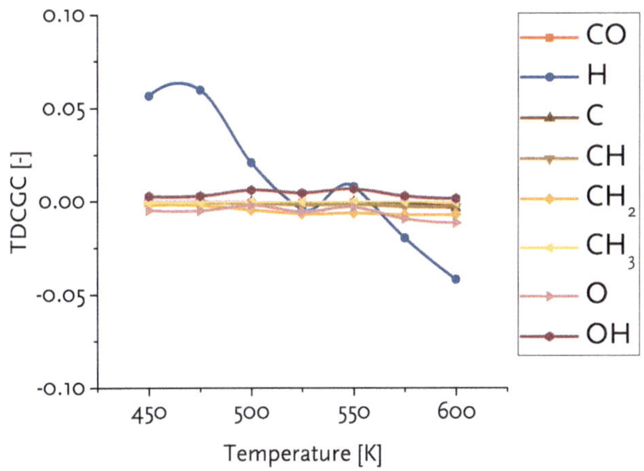

Fig. 5 Thermodynamic degree of chain growth control (TDCGC) as a function of temperature. A positive value indicates that stabilizing this species on the surface results in a higher chain-propagation probability and *vice versa*. Data shown for the FT reaction on stepped Ru at p = 20 atm and a H_2/CO ratio of 2.

mechanism during the FT reaction. The optimum in chain-probability finds its origin in Sabatier behavior. At low temperature, chain growth is limited by a lack of surface vacancies limiting the dehydrogenation step of a reaction intermediate, while at high temperature chain depolymerization occurs due to the increased number of surface sites. Overall, we infer that a necessary condition for the production of long hydrocarbon chains in the FT reaction is high surface coverage to suppress chain depolymerization. The H surface coverage also plays an important role in chain growth as hydrogenation and dehydrogenation steps are key in the interconversion of the growing chains for the next coupling step.

Conclusions

Microkinetics simulations of the complex FT reaction provide detailed insight into the elementary reaction steps that control the CO conversion rate and the chain-growth probability, a key selectivity parameter in commercial FT technology. Chain growth in the FT reaction on stepped Ru surfaces proceeds *via* coupling of CH and CR surface intermediates. Essential to the growth mechanism are C–H dehydrogenation and C hydrogenation steps, whose kinetic consequences have been examined by formulating two novel kinetic concepts, the degree of chain-growth probability control and the thermodynamic degree of chain-growth probability control. These concepts are based on Campbell's degree of rate control. For Ru the CO conversion rate is controlled by the removal of O atoms from the catalytic surface. The Sabatier maximum of the CO conversion rate is predominantly determined by the surface coverage. Chain-growth probability shows a maximum at a lower temperature than the CO conversion maximum. The steps that control chain-growth probability are different from those that control the overall rate. These are necessarily decoupled as chain growth cannot be rate-controlling in the overall FT process. Below the optimum for obtaining long hydrocarbon chains, the reaction is limited by the high total surface coverage: in the absence of sufficient vacancies the CHCHR → CCHR + H reaction is slowed down. Beyond the optimum chain-growth probability, CHCR + H → CHCHR and OH + H → H_2O limit the chain-growth process. The thermodynamic degree of chain-growth probability control emphasizes the critical role of the H and free-site coverage and shows that at high temperature chain depolymerization contributes to the decreased chain-growth probability. That is to say, during the FT reaction chain growth is much faster than chain depolymerization, which ensures high chain-growth probability. The chain-growth rate is also fast compared to chain-growth termination and the steps that control the overall CO conversion rate, which are O removal steps for Ru.

Acknowledgements

The authors thank the Netherlands Organization for Scientific Research for access to the national high-performance computing facilities.

References

1 G. P. Van Der Laan and A. A. C. M. Beenackers, Kinetics and Selectivity of the Fischer–Tropsch Synthesis: A Literature Review, *Catal. Rev.*, 1999, **41**, 255–318.

2 A. A. Adesina, Hydrocarbon Synthesis *Via* Fischer–Tropsch Reaction: Travails and Triumphs, *Appl. Catal., A*, 1996, **138**, 345–367.
3 E. van Steen and H. Schulz, Polymerisation Kinetics of the Fischer–Tropsch Co Hydrogenation Using Iron and Cobalt Based Catalysts, *Appl. Catal., A*, 1999, **186**, 309–320.
4 A. P. Steynberg, Chapter 1-Introduction to Fischer–Tropsch Technology, in *Studies in Surface Science and Catalysis*, ed. S. André and D. Mark, Elsevier, 2004, vol. 152, pp. 1–63.
5 I. A. W. Filot, R. A. van Santen and E. J. M. Hensen, The Optimally Performing Fischer–Tropsch Catalyst, *Angew. Chem., Int. Ed.*, 2014, **53**, 12746–12750.
6 S. Kozuch and S. Shaik, A Combined Kinetic-Quantum Mechanical Model for Assessment of Catalytic Cycles: Application to Cross-Coupling and Heck Reactions, *J. Am. Chem. Soc.*, 2006, **128**, 3355–3365.
7 S. Kozuch and S. Shaik, Kinetic-Quantum Chemical Model for Catalytic Cycles: The Haber–Bosch Process and the Effect of Reagent Concentration, *J. Phys. Chem. A*, 2008, **112**, 6032–6041.
8 R. A. van Santen, I. M. Ciobîcă, E. van Steen and M. M. Ghouri, Chapter 3-Mechanistic Issues in Fischer–Tropsch Catalysis, in *Advances in Catalysis*, ed. C. G. Bruce and K. Helmut, Academic Press, 2011, vol. 54, pp. 127–187.
9 R. A. van Santen, M. Ghouri and E. J. M. Hensen, Microkinetics of Oxygenate Formation in the Fischer–Tropsch Reaction, *Phys. Chem. Chem. Phys.*, 2014, **16**, 10041–10058.
10 R. A. van Santen, M. M. Ghouri, S. Shetty and E. J. M. Hensen, Structure Sensitivity of the Fischer–Tropsch Reaction; Molecular Kinetics Simulations, *Catal. Sci. Technol.*, 2011, **1**, 891–911.
11 R. A. Van Santen, A. J. Markvoort, I. A. W. Filot, M. M. Ghouri and E. J. M. Hensen, Mechanism and Microkinetics of the Fischer–Tropsch Reaction, *Phys. Chem. Chem. Phys.*, 2013, **15**, 17038–17063.
12 R. A. van Santen, A. J. Markvoort, M. M. Ghouri, P. A. J. Hilbers and E. J. M. Hensen, Monomer Formation Model *Versus* Chain Growth Model of the Fischer–Tropsch Reaction, *J. Phys. Chem. C*, 2013, **117**, 4488–4504.
13 R. A. van Santen and A. J. Markvoort, Chain Growth by Co Insertion in the Fischer–Tropsch Reaction, *ChemCatChem*, 2013, **5**, 3384–3397.
14 H. M. Torres Galvis, J. H. Bitter, T. Davidian, M. Ruitenbeek, A. I. Dugulan and K. P. de Jong, Iron Particle Size Effects for Direct Production of Lower Olefins from Synthesis Gas, *J. Am. Chem. Soc.*, 2012, **134**, 16207–16215.
15 C. Stegelmann, A. Andreasen and C. T. Campbell, Degree of Rate Control: How Much the Energies of Intermediates and Transition States Control Rates, *J. Am. Chem. Soc.*, 2009, **131**, 8077–8082.
16 C. Stegelmann, N. C. Schiødt, C. T. Campbell and P. Stoltze, Microkinetic Modeling of Ethylene Oxidation over Silver, *J. Catal.*, 2004, **221**, 630–649.
17 I. A. W. Filot, R. A. van Santen and E. J. M. Hensen, Quantum Chemistry of the Fischer–Tropsch Reaction Catalysed by a Stepped Ruthenium Surface, *Catal. Sci. Technol.*, 2014, **4**, 3129–3140.
18 I. A. W. Filot, R. J. P. Broos, J. P. M. van Rijn, G. J. H. A. van Heugten, R. A. van Santen and E. J. M. Hensen, First-Principles-Based Microkinetics Simulations of Synthesis Gas Conversion on a Stepped Rhodium Surface, *ACS Catal.*, 2015, **5**, 5453–5467.
19 I. A. W. Filot, B. Zijlstra and E. J. M. Hensen, http://www.mkmcxx.nl.

20 B. W. Wojciechowski, The Kinetics of the Fischer–Tropsch Synthesis, *Catal. Rev.*, 1988, **30**, 629–702.
21 R. Zennaro, M. Tagliabue and C. H. Bartholomew, Kinetics of Fischer–Tropsch Synthesis on Titania-Supported Cobalt, *Catal. Today*, 2000, **58**, 309–319.
22 H. Schulz and M. Claeys, Kinetic Modelling of Fischer–Tropsch Product Distributions, *Appl. Catal., A*, 1999, **186**, 91–107.

Faraday Discussions

DISCUSSIONS

Catalysis for Fuels: general discussion

Hans Niemantsverdriet, Pieter van Helden, Emiel Hensen, David Lennon, Katherine Holt, Graham Hutchings, Michael Bowker, Richard Catlow, Mzamo Shozi, Linda Jewell, Michael Claeys, James Hayward, Neil Coville, Nico Fischer, Alberto Roldan, Evjeniy Redekop, Thobani Gambu, Letisha Deeplal, Thabiso Perfect Oscar Mkhwanazi, Kees-Jan Weststrate, Detlef Bahnemann, Matthew Neurock, Hans Schulz, Ding Ma, Simon Kondrat, Paul Collier, Abhishek Kumar Gupta, Avelino Corma, Paul Akomeah, Enrique Iglesia, Eric van Steen, Nora de Leeuw, Moritz Wolf and Tracey van Heerden

DOI: 10.1039/c7fd90010d

Nico Fischer opened a general discussion of the paper by Nora de Leeuw: You have chosen Cu as a potential catalytic material and your calculations show that Cu is not the ideal surface to decompose hydrazine to H_2. Which materials would you be looking into in the future? Have you found any design parameters that should be followed?

Nora de Leeuw replied: We have made in-depth studies of three different mechanisms for N_2H_4 decomposition and these could be followed up using other potential catalysts, for example including Ir and Cu alloys.

Katherine Holt asked: In your model you neglect lateral adsorbate-adsorbate interactions. How can these be incorporated into the model and how would this influence the results?

Nora de Leeuw answered: We could incorporate lateral interactions by analysing the coverage effect. However, we have already reported previously that the interaction between species is insignificant at less than 0.01 eV, except for N_2H_4 + CH_2, which is explicitly included in the microkinetic simulations.

Michael Bowker remarked: In Table 1 (DOI: 10.1039/C6FD00186F) there appear to be no net barriers in the adsorption-desorption processes from the gas phase. For example, dissociative adsorption of hydrogen (step R7) and desorption (step R6) have exactly the same energetics. It is known that this reaction is highly net activated from the gas phase, with very low sticking probabilities,[1] and hence the forward and reverse barriers are very different. This will have a significant effect on H populations on the surface and on H_2 production rates.

1 C. T. Rettner, D. J. Auerbach, and H. A. Michelsen, *Phys. Rev. Lett.*, 1992, **68**, 1164.

Alberto Roldan responded: We have considered the reaction barriers for H_2 dissociative adsorption and associative desorption in the microkinetics simulation, although they are missing in the pre-printed version of the paper. We have now updated the final paper to include these barriers.

Nora de Leeuw replied: The adsorption and desorption energies have the same absolute value and opposite sign. The activation energies of these processes are the ones that will show different values and they have been considered in the simulation process.

Emiel Hensen noted: The use of microkinetics modeling to predict the performance of different Cu surfaces in the decomposition of hydrazine is certainly a worthwhile approach. Within this approach, it is possible to interrogate the kinetics for the step(s) that determine the overall reaction rate as well as the steps that determine the selectivity. The latter may help to understand which aspect of the surface precludes formation of hydrogen.

An example of the use of the degree of rate control can be found in our work on CO hydrogenation on a Ru surface.[1] An example of the use of the degree of selectivity control and its derivation can be found in work on CO hydrogenation on Rh surfaces.[2]

1 I. A. W. Filot, R. A. van Santen and E. J. M. Hensen, *Angew. Chem., Int. Ed.*, 2014, **53**, 12746–12750.
2 I. A. W. Filot, R. J. P. Broos, J. P. M. van Rijn, G. J. H. A. van Heugten, R. A. van Santen and E. J. M. Hensen, *ACS Catal.*, 2015, **5**, 5453–5467.

Nora de Leeuw responded: Certainly, the microkinetic model employed here was derived entirely from computational results without any external parameters. This approach has been shown to be efficient and reliable.[1,2] However, on a reaction network like the one presented here, it is not easy to find the rate limiting step and to analyze each degree of rate control will be extremely useful to find model descriptors.

1 A. Roldan, G. Novell, J. M. Ricart and F. Illas, *J. Phys. Chem. C*, 2010, **114**, 5101–5106.
2 N. Y. Dzade, A. Roldan and N. H. de Leeuw, *J. Chem. Phys.*, 2013, **139**, 124708–124708.

Thobani Gambu questioned: The simulated TPR (Fig. 2, DOI: 10.1039/C6FD00186F) seems to indicate that NH_3 coverage is zero (or trace compared to other species). Moreover, NH_3 production/desorption onset temperature coincides well with the decrease in N_2H_4 coverage. This may simply mean that NH_3 does not adsorb strongly enough on the Cu(111) surface and as a result it desorbs almost instantaneously upon formation from N_2H_4.

Therefore, could we potentially use the adsorption energy of NH_3 as a descriptor in screening for catalysts with improved H_2 yield/selectivity from N_2H_4? A similar approach has been presented in the literature.[1]

1 J. K. Nørskov, F. Abild-Pedersen, F. Studt and T. Bligaard, *Proc. Natl. Acad. Sci. U.S.A.*, 2011, **108**, 937–943.

Nora de Leeuw answered: We would like to analyse the degree of rate control to confirm that the adsorption/desorption is the limiting step in the production of H_2. We recognise that both the adsorption of NH_3 and H_2 are important rates in the evolution of hydrogen, but more analysis is still required.

Hans Niemantsverdriet enquired: From the point of view of making hydrogen, copper does not appear as a good choice, as the main product is ammonia. I wonder if it would make sense to look at metals known for decomposing ammonia, for which *e.g.* ruthenium is the best choice. On the other hand, this may not work either, because iridium is known as the best system for hydrazine activation. With that in mind, wouldn't it be useful to understand better why iridium is so special for hydrazine utilisation? Would you have any insights on this?

Nora de Leeuw responded: We initiated this project with the aims of explaining the use of hydrazine as a reducing agent in the synthesis of Cu nanoparticles. Certainly, exploring the reaction mechanism on Ir would be the next step to compare directly with the Cu and perhaps optimise the limiting rates using metal alloys to develop a more reactive catalyst.

Pieter van Helden said: Ir is an interesting metal, since it has C and O bond strengths in the same range as those of Co, Ni and Ru, while having a much lower H bond strength.[1] This would imply that it would quite easily liberate hydrogen gas, while maintaining the bonds to the other adsorbates at the operating temperature. This could be one of the features that makes it a good catalyst for the liberation of hydrogen gas from hydrazine.

1 J. Greeley and M. Mavrikakis, *J. Phys. Chem. B*, 2005, **109**, 3460–3471.

Nora de Leeuw answered: That is right, the hydrogen adsorption cannot be strong if we aim to evolve H_2. However, other species adsorbing much more strongly than H can poison the catalyst or decrease its performance. We should seek for the right balance between the different adsorption energies.

Matthew Neurock said: This is an interesting reaction system. I have a comment and a question. The theoretical calculations that were carried out are consistent with the results we presented in our paper (DOI: 10.1039/C6FD00226A) in that the coordinatively-saturated Cu sites do not readily activate N–H bonds. The presence of NH_2* species on Cu that form as a result of the activation of hydrazine, however, can act as a base and readily catalyze N–H bond activation *via* hydrogen transfer to NH_2*. This, however, will more selectively lead to ammonia rather than H_2 production, as you found in your simulations. Iridium, which is used experimentally, however, will readily activate N–N bonds of the partially dehydrogenated N_2H_x species as well as the N–H bonds, thus allowing for the activation of hydrazine and the formation of hydrogen as a product. More active catalytic materials to form H_2 would likely require metals or metal alloys that have higher N* binding energies and can activate N–N as well as N–H bonds and allow for facile H recombination. My question concerns a comparison of the experimental results and those from the simulations carried out at higher temperatures.

The N_2 production from the simulation appears to be about 10% lower than that reported experimentally. Can you speculate what the differences may be due to? Is this possibly due to the role of coordinatively unsaturated Cu sites that can more readily activate N–H bonds but are not currently included in the simulation?

Nora de Leeuw replied: That is what we believe. The surface in our model is ideal, without vacancies, ad-atoms, steps and any other defects that would modify the coordination and the electronic structure of the substrate. These low-coordinated atoms accumulate electron density and could bind N* stronger than high-coordinated ones (at the terrace), favouring the deprotonation process. The 10% disparity with the experiment may be due to the methodology and approximations taken into account: exchange–correlation functional and transition state theory among others.

Richard Catlow commented: The methodology you have developed is clearly powerful and predictive. Could you now apply it to other metals and alloys to identify which might be most effective. I am not suggesting a large scale screening exercise, but rather a focussed study of a number of systems which we expect to be promising.

Nora de Leeuw responded: That is correct; the next step would be identification of the degree of rate control, focussing only on the parameters modifying the steps that largely control the kinetics and selectivity of the process. In this way, we would significantly reduce the computational cost.

Paul Collier remarked: As a comment, this paper makes an interesting comparison between two modes of operation for hydrazine decomposition: batch and flow reactor. There is often great interest in which type of operation is best in other processes and perhaps DFT can illuminate this debate. Additional question: in reality, copper would be used as a supported catalyst in this process (such as Cu/Al_2O_3 or Cu/SiO_2). How would you expect this to change the modelling results, if at all?

Nora de Leeuw answered: Thanks for the comment. I would like to add that different sections of the flow reactor can be explored considering different amounts of products, *e.g.* an early section of the reactor will not contain products in the inlet flow, while late sections of the reactor will have small quantities of reactants. The validity of our model in a real supported catalyst depends on the extension of non-defective (111) terraces on the Cu particles. Small particles would have vertexes and other defects with reactivity not included in our model.

Thabiso Perfect Oscar Mkhwanazi enquired: Have you investigated how the change of supports influences the metal charge transfer of copper and the effect of this charge density or transfer on selectivity? Can one take advantage of this charge transfer to select a specific product?

Nora de Leeuw replied: We have focused on Cu only but we agree that the charge transfer from the metal to the adsorbed species will influence their interaction and, therefore, the conversion and selectivity.

Ding Ma opened a general discussion of the paper by Matthew Neurock: As the processes share similar intermediates; is there a way to control the selectivity to selectively get one of the products?

Matthew Neurock replied: Controlling selectivity is somewhat of a challenge as aldol condensation and esterification both proceed through the direct reactions of propanal and propoxide intermediates in the rate controlling step. One could examine other metals, which would alter the binding energies or binding configuration of the propoxide or propanal, in an effort to increase the selectivity. I suspect that it would be easier to alter the selectivity by changing the binding energies of propanal rather than those of the propoxide, as C_α–H activation and nucleophilic attack on the propanal occur at different sites on the molecule. A second and perhaps more sensitive approach would be to change the nature of the support, which would allow you to carry these reactions out at different sites.

Avelino Corma asked: In proton abstraction, the type of metal and support can be selected to maximise the reaction. Can you look at the nature of the interations and the efficiency?

Matthew Neurock replied: Indeed, the activation of the weakly acidic Cα–H bond is critical for the condensation reaction and this can be influenced by the metal as well as the support. We can carry out calculations on both the metal as well as the oxide separately to explore the binding energies as well as the activation barriers and overall reaction energies for the proton abstraction step. We can also examine the metals anchored to the support and explore the activation at the interfacial sites. In addition to the adsorption, reaction energies, and activation barriers at these sites, we can also examine how charge is transferred in the reaction to provide more mechanistic insight into how to drive this reaction.

Graham Hutchings enquired: Alcohols are being used as models for biomass, but biomass has many O atoms present; what happens if you use a diol rather than an alcohol? Would dehydration occur preferentially with a 1,2-diol?

Matthew Neurock responded: Previous experiments with 1,3-propanediol over $Cu/ZnO/Al_2O_3$[1] show that the diol undergoes both dehydration and dehyrogenation to form equilibrated mixtures of propene,[2] propen-1-ol, acrolein, as well as C_5–C_6 type oxygenated products presented here. The ZnO/Al_2O_3 support used in this work, however, can also catalyze similar reactions.

For 1,2-propanediol, I would suspect dehydration to predominantly occur at the secondary C center to form 1-propanol, as the formation of a secondary carbenium ion intermediate would be more stable than the primary intermediate. The dehydration reactions, however, may be occurring over sites on the ZnO/Al_2O_3 support and not on Cu/SiO_2. As such, the nature of the oxide support can play an important role in controlling the selectivity for specific products as it can actively carry out different reactions. In terms of dehydrogenation, I would suspect that the terminal CH_2OH groups are selectively dehydrogenated, as there is less steric hindrance for the metal to attach to the C–H bonds. This is consistent with previous experimental results for base (OH–) catalyzed oxidation of glycerol over Au, which show the selective dehydrogenation at the terminal hydroxyl sites and the absence of dehydration.[1] This is also in line with other previous theoretical and experimental work that shows the metal catalyzed dehydrogenation preferentially occuring at the terminal CH_2OH sites. This, of course, is a speculation for the Cu system examined here.

1 W. C. Ketchie, M. Murayama and R. J. Davis, *Top. Catal.*, 2007, **44**, 307–317.
2 M. Ide, B. Hao, M. Neurock, and R. J. Davis, *ACS Catal.*, 2012, **2**, 671–683.

Michael Bowker asked: Propanal is weakly bound at ~ 28 kJ mol^{-1}, so it is physisorbed and difficult to activate. Do you think, as water is being produced in this reaction, it can have an impact on the binding of such species and the selectivities, and can you calculate these? Maybe the OH present can act as a base-promoter in these reactions? What is the binding energy of the propanol?

Matthew Neurock responded: Little water is produced in this reaction as esterification proceeds without the loss of oxygen and the predominant aldol condensation path proceeds *via* decarbonylation/decarboxylation with loss of oxygen as CO or CO_2. The results show that about 10% of the products would lead to the formation of water.[1] Water is thought to be activated on Cu (similar to propanol) and forms a basic surface hydroxyl that can carry out the same nucleophilic attack or H abstraction as the alkoxide on Cu, which is consistent with your statements. We reported in the manuscript (DOI: 10.1039/C6FD00226A) that the OH* groups are somewhat more basic than the alkoxide interediates and show that the barrier to activate the weakly acidic C–H bond on the bound propanal is ~14 kJ mol^{-1} more favorable than that of the alkoxide. We calculated the binding energy for propanol to be 29 kJ mol^{-1}.

1 M. E. Sad, M. Neurock and E. Iglesia, *J. Am. Chem. Soc.*, 2011, **133**, 20384–20398.

Enrique Iglesia added: The reactions were carried out in the presence of H_2 and we observed no detectable deactivation. The kinetic response is consistent with a limiting step mediated by a bimolecular transition state for both condensation and esterification.[1] We observed no effects of water, suggesting that neither O* nor OH* are involved as nucleophiles.

1 M. E. Sad, M. Neurock and E. Iglesia, *J. Am. Chem. Soc.*, 2011, **133**, 20384–20398.

Abhishek Kumar Gupta asked: Can Cu nanoparticles be used as a catalyst for *trans*-esterification of vegetable oil or waste oil? Can we synthesise fatty acid methyl esters or biodiesel from vegetable oil or waste oil?

Matthew Neurock answered: In theory, yes, as the base-catalyzed *trans*-esterification in the homogeneous phase proceeds *via* the formation of an alkoxide of the carbonyl carbon on the ester, thereby, replacing the existing –OR group on the ester. This is very similar to the esterification mechanism observed on Cu, where the alkoxide ion attacks the carbonyl carbon, thus forming a C–O bond. There are various studies of Cu supported on basic oxides such as Cu/ZnO[1] and Cu/Mg/Al,[2] which carry out this chemistry. The Lewis base and acid sites on the support likely contribute significantly to the reactivity.

1 G. Baskar and A. Ravi, *Bioresour. Technol.*, 2015, **188**, 124–127.
2 X. Deng, Z. Fang, Y. Liu and C. L. Yu, *Energy*, 2011, **36**, 777–784.

Evjeniy Redekop queried: Aldol condensation (*e.g.* adsorbed propoxide and propanal) involves an elementary step with two adjacent oxygen-bound intermediates. How will the mobility of such intermediates affect the probability of their co-location on the Cu surface under the reaction conditions?

Matthew Neurock responded: Propanal and propoxide are thought to be present at high coverages on the surface. In addition, propanal and propanol are shown to be rapidly equilibrated on the surface. As such propanal and propoxide would be equilibrated and there should not be any issues concerning mobility.

Paul Collier remarked: (a) In the industrial process of acetylene hydrogenation, oligomerisation of reactive species leads to fouling of the catalyst by green oil formation. Are there any parallels between this and catalytic esterification/condensation?

(b) The results of your study are impressive; how might this learning be incorporated into a catalyst design led approach to develop new catalysts?

Matthew Neurock responded: (a) Aldol condensation and esterification reactions are carried out in the presence of hydrogen. The hydrogen tends to prevent or suppress the formation of more unsaturated intermediates, C–C bond formation and hydrocarbon chain growth. For reactions carried out over Cu/SiO_2, the hydrogen appears to prevent the continued chain growth as the reaction appears to stop at the coupled C_5–C_6 products. (b) Our results indicate that we want to use group 11 metals (Cu, Ag, Au) as well as large particles with highly coordinatively saturated metal sites, as this weakens the binding energy of the alkoxide, thus increasing its basicity and its ability to activate the weakly acidic C_α–H bonds and the nucleophilic attack on the carbonyl group. One could also look at moving to metal alloys to not only improve the activity but also enable increases in the selectivity. One could look at the binding energy of a probe alkoxide as well as perhaps the charge on the alkoxide to provide descriptors for the reactivity. Weaker alkoxides and higher degrees of electron density on the alkoxide would suggest more basic and reactive alkoxides. One could then readily screen a wide range of metals, metal alloys and different potential surface structures. The differences in selectivity are more challenging, as both condensation and esterification proceed *via* the same intermediates. One could use instead the overall reaction energies of the Ca–H activation and nucleophilic attack reactions to discern the most active metals and sites. One can and should analyze the influence of the oxide support. This would require more detailed calculations, however, to probe the individual elementary C_α–H and C–O formation steps.

Graham Hutchings remarked: Following on from the discussion point raised by Paul Collier on catalyst design, you mentioned that the small nanoparticles are less effective than the larger nanoparticles. This implies that the peripheral sites are not effective; is there a reason for this? Also it means we need to design catalysts where we have a raft of atoms as a large nanoparticle does not use the metal atoms very effectively.

Matthew Neurock replied: Experimental results show that the turnover rates for propanal condensation and esterification rates carried over Cu/SiO_2 increase linearly as the Cu cluster sizes increase from 5 to 15 nm and level out as sizes > 20 nm.[1] Both aldol condensation and esterification reaction rates are controlled by the reactions between a surface alkanal and a surface alkoxide. The rate controlling step for esterification involves C_α–H activation of the propanal *via* the surface propoxide, while esterification proceeds *via* the nucleophilic attack of the alkoxide on the bound alkanal. The weakly bound alkoxide acts as a base that can abstract the weakly acidic C_α–H hydrogen of propanal in the rate controlling step for the condensation reaction, or carry out a nuclophillic attack on the carbonyl of the bound alkanal in the rate controlling step for esterification. The basicity and reactivity of the alkoxide are controlled by the binding energy of the alkoxide to the metal surface, where the more weakly bound alkoxides are more basic and reactive. The binding energies are controlled by the coordination number of the metal to which the alkoxide binds. Increasing the particle size increases the number of the more basic and more reactive coordinatively saturated low index terrace sites over the less basic and less reactive corner and edge sites. This is analogous to the high reactivity of large Au particles for oxidation when carried out in the presence of base. In terms of design, this would require increasing the number of coordinatively saturated sites. In order to save on the use of an expensive metal, one could think of the rafts of metal atoms as you propose, or core-shell particles where the active metal is deposited on a high surface area support or core.

1 M. E. Sad, M. Neurock and E. Iglesia, *J. Am. Chem. Soc.*, 2011, **133**, 20384–20398.

Mzamo Shozi communicated: Since TiO_2 exhibits both basic and acidic sites, could it be a better support than SiO_2 for the aldol condensation and esterification reactions?

Matthew Neurock communicated in reply: TiO_2 is known to readily carry out the aldol condensation but deactivates as result of carbon buildup. Cu supported on TiO_2 is indeed a very good catalyst as Cu aids in the equilibration of alkanals and alkanols, dehydrogenation of unstable hemiacetal intermediates and hydrogenation of the unsaturated intermediates that lead to carbon build up that otherwise deactivate the TiO_2.[1]

1 S. Wang, K. Goulas and E. Iglesia, *J. Catal.*, 2016, **340**, 302–320.

Letisha Deeplal communicated: To what extent would the support contribute towards aldol condensation? For example, γ-Alumina is known to be amphoteric, which can promote acid and base catalyzed aldol condensation.

Matthew Neurock communicated in response: In the reactions discussed here, SiO_2 was used as a support, thus allowing us to rule out the reactions that might come from the support. Indeed, many of the oxide supports with acid and base sites can also contribute to the activity. In more recent efforts, Cu/TiO_2 was used to take advantage of moderate Lewis acid–base sites on TiO_2 to carry out C–O and C–C bond formation, whereas Cu carried out the facile hydrogenation and dehydrogenation steps to equilibrate the alkanal + H_2 and alkanol mixtures, and to avoid deactivation by the formation of higher hydrocarbons. There are a range of other oxides that can also be used to take advantage of the sites on the metal and the oxide.

Letisha Deeplal communicated: How applicable would this study be towards the selective hydrogenation of α,β-unsaturated aldehydes,whereby the adsorption and activation of either the C=C or C=O bond can be controlled?

Matthew Neurock communicated in reply: A few of the mechanistic concepts in this work are similar to those involved in selective hydrogenation. On most transition metal surfaces the atomic hydrogen that forms extracts electron density from the surface and becomes hydridic. The hydride can act as a base and carry out nucleophilic attack on the carbon of the C=O (or C=C) to form an alkoxide (or alkyl) species similar to the rate controlling attack of the alkoxide on the C=O in the esterification reaction. The most active metals are those that bind hydrogen strong enough to allow for H_2 dissociation but weak enough to allow hydrogen to add to the carbon or the oxygen. This is consistent with the surfaces necessary for esterification and condensation, which require metals that can activate H_2 but are weak enough to allow the alkoxide to behave as a base. The subsequent addition to the oxygen, however, would be different as it would involve the addition of a positively charged hydrogen, together with an electron transfer.

Hans Niemantsverdriet opened a general discussion of the paper by Tracey van Heerden: Your work is very useful to get a feel for what a support in combination with a metal might do. It looks like the O–Al–$(OH)_2$ moiety acts a bit like a potassium promoter. It sets up a dipole on the surface, which somewhat weakens the C–O internal bond, and as a result you find a stabilisation of the CO adsorption plus a slight weakening of its internal bond. The activation energy for dissociation is a bit lowered, although probably not enough to break it, as the entropic advantage of desorption is hard to beat. Nevertheless, the result provides valuable insight. Do you have any insight into what would happen if instead of this molecular Al species, you had a more extended island of Al_2O_3 (perhaps OH terminated) on the cobalt surface? I noticed that the Al in your cluster is not quite in the 3+ state, while it would be in Al_2O_3. Would we still see similar effects for CO bonded adjacent to such Al_2O_3 islands?

Tracey van Heerden responded: I can only speculate with this question. I expect that with a more extended island of alumina we would encounter a reduced effect. Partially due to the different environment the Al finds itself in, partially due to the Al being in the 3+ state in the island like you mentioned. I would also wonder though if the CO would still have as easy access to the metal centre of the Al on an extended island than what it does on the ligand, in order for it to be affected in the first place.

Nico Fischer remarked: In your paper you discuss the different types of ligands you have investigated as models for metal support interaction based effects on activity and selectivity. In a realistic supported cobalt catalyst, where would these ligands be found? Do you think they – or similar species – might exist at the metal support interface, or are there other formation mechanisms possible? Also, how mobile would these species be under realistic FTS conditions, *i.e.* could they possibly migrate to the edge and kink sites and possibly block certain active centers there?

Tracey van Heerden replied: We know from laboratory work that on an inverse catalyst, the alumina based ligands are highly dispersed and do not form segregated amorphous alumina. Some calculations on edges and kinks are still needed to truly determine the location of these ligands, as it is feasible that they may occupy active centres at these sites. It is worth noting though that experimentally the inverse catalysts show increased activity when the loading of the support-based ligand is increased. On a real catalyst, one of the formation mechanisms being proposed is the dissolution of the support during the wet chemistry steps, which then deposits back onto the support or the metal oxide during drying, with it having been demonstrated that the reduction of the metal oxide does not destroy these ligands.

Pieter van Helden noted: This is a very interesting approach that has been presented here with a lot of future potential in studying promoter/support interface effects. The presented charge density difference plots remind me somewhat of early computational work on how potassium atoms bind to oxygen atoms on a Rh surface.[1] The key there was also the direct bonding interaction and charge transfer. Typically for promoters, changes in the electronic structure of the metal are invoked, but here there seems to be a direct bond-forming interaction. Looking at a typical Co catalyst, the Co is present as nanoparticles (NPs), with a significant percentage of exposed defect/step sites. Here you show Alumina ligands on a single Co surface. However, if these ligands were to bind to some of the defect/step sites on a realistic Co NP, would it still have the same effect on CO? Would it not bind stronger to these sites and poison the sites?

1 Z.-P. Liu and P. Hu, *J. Am. Chem. Soc.*, 2001, **123**, 12596–12604.

Tracey van Heerden answered: It is possible that the ligands may prefer to bind to defect/step sites, however, inverse catalysts that were produced in the lab show a net positive influence of these ligands on the CO dissociation, so it is evidently not a simple matter of poisoning when working on realistic NPs.

Michael Bowker said: I am impressed by this work and approach. Sometimes we think of inverse catalysts as 'model' catalysts, but of course they are not. Ammonia synthesis Fe catalysts (and Fe FT catalysts) have minor components of Al and K (and others), which preferentially segregate to the surface. They are unusual in having very high levels (~95%) of the active metal component in the catalyst. Although Al is often called a 'structural promoter' it must act in a very different way from a traditional support and have more than one role. It may be that, since it is intimately connected with the surface, it has a specific chemical role. Thus, the work examining the influence of oxidised Al species on reactivity is quite relevant to 'real' systems.

Simon Kondrat remarked: Fischer–Tropsch synthesis involves aggressive reaction conditions that will likely change the metal–support interaction. Do you feel your model reflects the possibility of an alumina overlayer under reaction conditions? Or does the non-reducible nature of alumina make this unlikely? An example of this effect is the Cu–ZnO methanol synthesis catalyst system, where there is evidence that under reducing conditions Zn or ZnO_x forms at the Cu particle interface.[1]

1 M. Behrens, F. Studt, I. Kasatkin, S. Kühl, M. Hävecker, F. Abild-Pedersen, S. Zander, F. Girgsdies, P. Kurr, B. Kniep, M. Tovar, R. W. Fischer, J. K. Nørskov and R. Schlögl, *Science*, 2012, **336**, 893–897.

Tracey van Heerden responded: Yes, it is unlikely due to the nature of the support. It is much more likely that a support like titania would form overlayers on a Co FT catalyst, whereas alumina and silica would not.

Emiel Hensen continued: The presented work describes a novel approach to tune CO dissociation on a cobalt surface. It links to other approaches where for instance a reducible oxide is used to increase the rate of CO dissociation. In this particular case, the coordinative unsaturation of the AM2 ligand can explain the lowered CO dissociation barrier. It would be worthwhile to compute the regeneration of the active site by removing the O atom of the Lewis acidic Al site. Was this route explored? What are the typical barriers?

Another approach following the suggestion of Prof. Niemantsverdriet would be to explore patches of aluminium oxide which contain Al atoms at their edges with a lower degree of coordinative unsaturation.

Tracey van Heerden replied: This route is currently being explored, so I cannot yet comment on typical barriers. The suggestion of Hans Niemantsverdriet is I believe computationally too expensive to be properly explored at this point.

Alberto Roldan remarked: As you showed in the pictures, the presence of Al activates the CO bond (the C–O distance increases). Is it an effect of the Al low coordination? Do you think CO will also be affected at the interface where the particle-support mismatch may affect the Co–Co distance and electronic structure?

Tracey van Heerden answered: I do believe CO would be similarly affected at the interface, even with possible effects on the Co structure. The interaction is very direct between the ligand and the CO and may well be somewhat due to the Al coordination.

Graham Hutchings addressed Tracey van Heerden and Paul Collier: What happens if you start with a preformed Co/SiO_2 or Co/TiO_2 catalyst, perhaps you would expect a positive effect by adding the aluminium complexes to these catalysts? I would expect this to be an easy experiment to try and would provide support for the effect you are proposing.

Paul Collier responded: Decorating the cobalt surface with different well-defined aluminium complexes is an interesting idea. If these could be tailored to sit on specific cobalt sites it might be possible to influence the process positively as described.

Tracey van Heerden replied: Yes, I would expect a positive effect.

Matthew Neurock commented: Your calculations show that Al may play a role in activating the CO bond. The model that you have used, however, involves a 3-coordinate Al site that binds to the O of CO. I would expect the Al–O bond here to be rather strong as the Al site is coordinatively-unsaturated. While this would aid in the initial activation of the CO bond, these sites would be very difficult to regenerate the 3-coordinate Al, as the Al–O bond is much stronger than the 4- or 5-coordinate Al–O bond. As such, these sites would be difficult regenerate. In addition, your theoretical results show that while the CO is more strongly held to these sites, the barrier to activate the CO* is a little higher than that on the Co surface alone. Since CO is covering the surface, the activation barrier would be measured from the lowest adsorbed CO–M2 state (~−2.25 eV) to the TST (~0.25 eV), which is about 2 eV. This looks to be higher than the barrier to dissociate CO* on Co (solid cure), which is about 1.7 eV.

Tracey van Heerden responded: I found that the 3-coordinate M2 ligand was more stable than the 4- and 5-coordinate ligands, indicating that regeneration is possible.

Simon Kondrat asked: In the paper you discuss the effect of the aluminium ligands affecting the structure of the Co nanoparticle, and so changing the exposure of (111) and (100) planes. In your opinion, how significant is this effect?

Tracey van Heerden replied: The very simple calculations I presented show a significant effect on the structure of a Co nanoparticle. With a more realistic inclusion of particularly the edge sites, and the inclusion of a high CO coverage, this effect may be less severe.

Mzamo Shozi communicated: What difference would you see, if any, on your ligand valence charge according to Bader charge analysis if you used a silica based ligand?

Tracey van Heerden communicated in reply: I do not yet have results on the Bader analysis of the silica system, however I can say that the stable geometries differ to those of the alumina system, and so I do expect there to be differences in the characterization as well.

Michael Claeys communicated regarding the paper by Enrique Iglesia: You describe a rate equation which is applicable to cobalt and ruthenium based FTS. For iron based FTS we normally need to include a water term in the denominator to account for the inhibiting effect of water with these catalysts. In your opinion, is the mechanism on iron based catalysts fundamentally different compared to that on cobalt or ruthenium, or is the above merely a reflection of the dynamic interplay between magnetite and carbide, which is determined by local water to syngas ratios? This was also noted by B. Jager *et al.*,[1] where he stated that: "It may occur during macrokinetic studies, that the wide experimental range – statistically necessary – of the concentration of reactants used, leads to surface changes in the catalyst. In that case the selection of the 'best' rate of reaction based upon the least squares criterium becomes merely a statistical exercise."

1 B. Jager and R. Espinoza, *Catal. Today*, 1995, **23**, 17–28.

Enrique Iglesia communicated in reply: Indeed, the effects of water, both as an inhibitor and as an occasional promoter of Fischer–Tropsch synthesis rates, differ among Co, Ru, and Fe. The rate equations to which I refer have been typically reported based on data at differential CO conversion and thus low prevalent water pressures relative to those of CO and H_2. These inhibition effects are stronger in Fe than in Co and even weaker on Ru, the expected trend from their oxygen binding energies and their noble/base character. In the case of Fe, the complexities of such surfaces during catalysis and their exchange of C and O with gaseous species during turnovers makes their surface composition and possibly even their surface and bulk structures sensitive to the O and C chemical potentials imposed by the elementary steps required for turnover and their reversibility. As such Langmuirian treatments become often an exercise in statistical analysis. This is what I showed in my lecture (DOI: 10.1039/C7FD00018A) by the inconsistency between the ability of such rate equations to accurately describe measured rates at FTS conditions, but with adsorption constants for CO that represent mere statistical fits and are unrelated to the CO adsorption properties of such surfaces as a result of the dense adlayers present during catalysis.

Neil Coville opened the general discussion: The use of hydrazine as a ligand has been explored *e.g.* in organometallic chemistry over many years. If I remember, they can even form dimers. Do your studies allow for these different complexes in your mechanism? Are they ruled out? The early studies may also provide information on $M–N_2H_2$ interactions?

Nora de Leeuw responded: We have previously investigated the assembly of up to 9 hydrazines on Cu surfaces with multiple arrangements.[1] The interaction between them is very small (<0.01 eV) and it will not influence in the microkinetic model. However, comparison with organometallic clusters would indeed be interesting.

1 S. S. Tafreshi, A. Roldan and N. H. de Leeuw, *Surf. Sci.*, 2015, **637-638**,140–148.

Nico Fischer continued: From DFT based studies preferred reaction mechanisms can often be postulated. From these results, can active site requirements be deducted, extending beyond the ideal surface? If so, could these aid experimentalists in the actual design of catalysts?

Nora de Leeuw replied: From reaction mechanism and electronic structure calculations we can find predictors that direct the limiting rates. The extrapolation of these predictors to defects on pure and alloy systems should be possible.

Matthew Neurock added: The insights on the nature of the active sites can be used to help aid in future design efforts. While I did not discuss the results, we have shown in the manuscript (DOI: 10.1039/C6FD00226A) that the binding of the alkoxide is over 26 and 40 kJ mol^{-1} stronger at the coordinatively unsaturated step and corner sites. The stronger binding energy leads to much less basic surface intermediates and significantly higher barriers for carrying out the rate determining C–H activation and nucleophilic attack. This suggests that these sites are not active in carrying out the catalysis. This is consistent with the experimental results that show that small particles are much less active than the larger particles and that the rates increase with increasing particle size, as the larger particles increase the number of coordinatively saturated terrace sites.[1] The results indicate that the most active sites are those that are coordinatively saturated and weakly bind the alkoxide, thus allowing the alkoxide to behave as a base. One could potentially increase catalytic activity by moving to Ag or Au, alloying Cu, Ag or Au with other metals, or producing core-shell metal alloys, which decrease the binding energy of the alkoxide and increase its basicity and reactivity. In addition, one could also improve activity and selectivity by altering the support to create ideal bifunctional sites.

1 M. E. Sad, M. Neurock and E. Iglesia, *J. Am. Chem. Soc.*, 2011, **133**, 20384–20398.

Neil Coville remarked: The use of the alkoxy group on the surface is proposed to act like an alkoxy group in a homogeneous base-catalysed organic reaction; how do organic chemists perceive this use? The idea is that staying bound to the surface impacts on the alkoxy group but that it still acts like a base. It seems then that there is a spectrum in using OR– (*i.e.* unattached) to surface bound OR. One could imagine that ion-pairing, for instance, would also modify this interaction and hence lead to variations in activity. Can this variation be exploited in terms of activity/selectivity?

Matthew Neurock answered: I don't know how the organic chemists would perceive this, but I suspect favorably as the mechanisms proposed are consistent with classic base-catalyzed organic reactions in solution as well as in homogeneous catalysis. In the organic systems, the reactive alkoxide intermediates are stabilized by their interactions with the solvent as well as the cation. In the gas phase work carried over Cu discussed herein (DOI: 10.1039/C6FD00226A), the electron-rich metal surface acts like the solvent and the cation as it allows for charge transfer to and stabilization of the alkoxide intermediate.

You raise a very interesting point in that one way in which to begin to tune the reactivity would be *via* the cationic promoters. For the gas phase reactions discussed herein, metal cations in the form of MO_x clusters deposited on the surface or on the support may be used to promote the reaction or participate themselves in carrying out the reaction. This is similar to what happens for Cu supported on basic metal oxide supports such at TiO_2, ZrO_2 and MgO, where these supports themselves can readily carry out the reaction.[1] The reactions can also be carried out in the liquid phase, where one can indeed generate alkoxide ions that may participate in solution reactions. We have discussed such solution phase and surface reactions for the oxidation of alcohols and carbonyl compounds over supported Au particles in the presence of base and aqueous media.[2] This would open up a number of opportunities in which one could engineer the active sites to control activity and selectivity.

1 M. E. Sad, M. Neurock and E. Iglesia, *J. Am. Chem. Soc.*, 2011, **133**, 20384–20398.
2 B. N. Zope, D. D. Hibbitts, M. Neurock and R. J. Davis, *Science*, 2010, **330**, 74–78.

Enrique Iglesia enquired: Is your microkinetic model accurate enough to explore the following possibility: explore the best catalysts for the reverse reactions – the synthesis of hydrazine from N_2 and H_2 – then combine the most optimal catalysts for these reactions with those that you have discarded as useful for hydrazine decomposition because they form ammonia? This would provide for a bifunctional system to form ammonia from N_2 and H_2, possibly at temperatures much lower (and also at much lower pressures) than in current practice. The two functions would not have to co-exist in the same catalyst particle because hydrazine would act as a molecular shuttle between functions residing within diffusion distances.

Nora de Leeuw answered: This is a very interesting point worth exploring. Indeed, using principal component analysis and limiting rate control techniques may give us the properties of these catalysts.

Graham Hutchings addressed Nora de Leeuw and Enrique Iglesia: Following on from the point raised by Enrique Iglesia concerning the possibility of finding a low pressure, low temperature route to ammonia, from Fig. 7 of Nora de Leeuw's paper (DOI: 10.1039/C6FD00186F) it is clear there is a facile route from hydrazine to ammonia. Taking into account the comment by Hans Niemantsverdriet that Ir is a much better catalyst than Cu for the reaction, perhaps an alloy of Ir woud be a good starting point. However, the second component would need to activate nitrogen and hydrogen.

Nora de Leeuw responded: This is an interesting point worth considering. We could move to the formation of NH_3, which also has commercial interest, or the formation of H_2 by alloying different metals, preferably cheap ones, whose properties fit the selectivity for each process.

Ding Ma opened a general discussion of the paper by Kees-Jan Weststrate: What will happen if you begin with CH_3I or CH_3 + CO? Will CO dominate the surface?

Kees-Jan Weststrate responded: We have not yet performed experiments for the CH_3/CO system in the absence of extra surface hydrogen. This informative experiment would provide interesting information on how CO affects dehydrogenation of CH_3 as well as the coupling reaction.

David Lennon asked: In Fig. 1(c) (DOI:10.1039/C6FD00191B) you present the RAIRS spectrum for CH_3I on Co(0001) which corresponds to the symmetric deformation mode of adsorbed methyl groups.

(i) Did you see any evidence for other CH_x species in these RAIRS measurements?

(ii) Did you examine CH_2Cl_2 adsorption on Co(0001) by RAIRS? If so, did you see any evidence for adsorbed methylene species?

Kees-Jan Weststrate responded: (i) The RAIRS experiment only showed the presence of methyl species and we did not resolve any other intermediates such as CH or acetylene. This could be due to the relatively poor signal-to-noise ratio in the CH_x stretching region. We did not, for example, get a clear signal for the CH_3-related stretch vibration. Both CH and acetylene would only have vibrational bands in this region.

(ii) We did not perform IR experiments using CH_2Cl_2. It is however expected that CH_2 groups are highly reactive and would be difficult to observe in significant concentrations.

Paul Akomeah remarked: To what extent has research on Fischer–Tropsch synthesis been carried out by replacing H_2 with D_2 and what were the findings, if any?

In case the reaction of D_2 with CO turned out well, what would be the fate of the resultant fuel obtained? Have such fuels been used in machines before and what was their effect on the systems they were used in?

Kees-Jan Weststrate responded: When hydrogen is replaced by deuterium, the CO conversion activity typically increases and the chain growth probability increases as well, see *e.g.* ref. 1. I am not aware of studies where deuterated FTS products have been produced in high enough quantities to be used as fuels, as such studies are typically done to obtain fundamental insights. From the perspective of surface reactions, replacing hydrogen by deuterium has an impact in various ways. Due to the higher mass of D_2 compared to H_2, the collision frequency at a given pressure will be 1.4 times smaller. Thermal desorption experiments from a Co(0001) surface, Fig. 1, show that deuterium desorption *via* 2

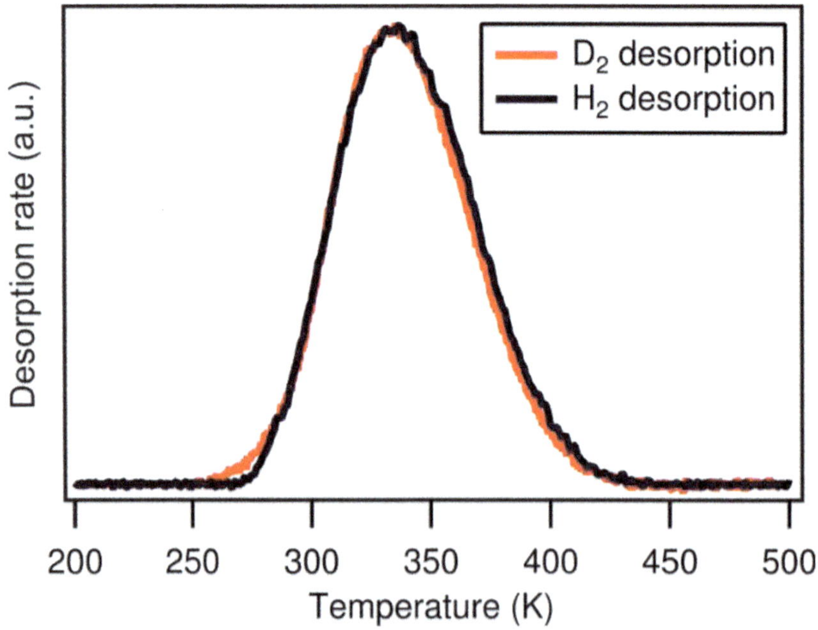

Fig. 1 H_2 and D_2 desorption from H_{ad} or D_{ad}-saturated Co(0001) showing the identical desorption behaviour.

$D_{ad} \rightarrow D_2$ occurs at the same temperature as the desorption of H_2, indicating that the desorption rate constant is not significantly affected by a kinetic isotope effect. The combination of a decreased collision frequency and a constant desorption rate constant translates to a higher free site concentration. In addition, both primary and secondary kinetic isotope effects will alter reaction rate constants of reactions that involve hydrogen. An example of the complexity that is introduced by replacing hydrogen by deuterium can be found in Fig. 2d in the article (DOI: 10.1039/C6FD00191B). The formation of CH_4 *via* CH_3 + H occurs at a significantly lower temperature than the formation of CH_3D *via* CH_3 + D. The formation of the more deuterated isotopologues, such as CD_4, occur at an even higher temperature.

1 M. Ojeda, A. Li, R. Nabar, A. U. Nilekar, M. Mavrikakis and E. Iglesia, *J. Phys. Chem. C*, 2010, **114**, 19761–19770.

Michael Bowker addressed Eric van Steen and Kees-Jan Weststrate: In your case, halide is present with the adsorbates. Do you know if anyone has run experiments to look at the effect of halide in FT? I presume halide will be a poison for the reaction.

Eric van Steen responded: The effect of halides on the performance of FT-catalysts has been explored in particular by the Davis group due to the interest of converting sygas originating from biomasss.

Over iron-based catalysts, Ma *et al.*[1] showed that the addition of up to 100 ppm of NaCl or KCl to the feed of the Fischer–Tropsch synthesis did not affect the activity or selectivity. The addition of hydro halide acids showed strong deactivation if the concentration increased to beyond 20 ppm (Ma *et al.*[2]), which was ascribed to competitive adsorption. In experiments in our lab, we showed that the addition of 50 ppm of chlorobutane resulted in complete, irreversible deactivation, which was ascribed to a phase transformation to FeOCl.

Over cobalt based catalysts, it has been observed that the addition of hydro halide acids increases the rate of deactivation (Gnanamani *et al.*[3]) proportionally to the amount present in the feed (in the range of 0.5–1 ppm), although the data is less clear than the data at higher levels for Fe-based catalysts. This has been ascribed to site blockage.

1 W. Ma, G. Jacobs, J. Kang, D. E. Sparks, M. K. Gnanamani, V. R. R. Pendyala, W. D. Shafer, R. A. Keogh, U. M. Graham, G. A. Thomas, B. H. Davis, *Catal. Today*, 2013, **215**, 73–79.
2 W. Ma, G. Jacobs, G. A. Thomas, W. D. Shafer, D. E. Sparks, H. H. Hamdeh and B. H. Davis, *ACS Catal.*, 2015, **5**, 3124–3136.
3 M. K. Gnanamani, V. R. R. Pendyala, G. Jacobs, D. E. Sparks, W. D. Shafer and B. H. Davis, *Catal. Lett.*, 2014, **144**, 1127–1133.

Kees-Jan Weststrate responded: The influence of halogens has been studied on cobalt in the context of biomass derived synthesis gas by *e.g.* Rytter and coworkers[1] and on both Fe and Co catalysts by Davis and co-workers.[2,3] In both cases, chlorine concentations in the ppm range were used. Reported effects of halide addition range from very little effect[1] to mild poisoning.[3,4] In the context of the paper where halogen-containing C_1 species serve as precursors for C_1H_x adsorbates, the main question is whether the halide product affects the C_1H_x chemistry. Various studies use the approach of halogenated precursor compounds, and comparison with results from more complex but halogen-free precursors shows only small differences induced by the presence of halogen adsorbates alongside the C_1H_x moieties.[5,6] In the discussion on this question, Prof. E. van Steen referred to the interesting study by van Barneveld and Ponec, where halogenated C_1 species were used in a fundamental study of the FTS mechanism on supported Co catalysts.[7] They find that $CH_3Cl + H_2$ over Co catalysts only produced methane, whereas chain growth was found when CH_2Cl_2 and $CHCl_3$ were used instead.

1 Ø. Borg, N. Hammer, B. C. Enger, R. Myrstad, O. A. Lindvåg, S. Eri, T. H. Skagseth, E. Rytter, *J. Catal.*, 2011, **279**, 163–173.
2 W. Ma, G. Jacobs, J. Kang, D. E. Sparks, M. K. Gnanamani, V. R. R. Pendyala, W. D. Shafer, R. A. Keogh, U. M. Graham, G. A. Thomas, B. H. Davis, *Catal. Today*, 2013, **215**, 73–79.
3 M. K. Gnanamani, V. R. R. Pendyala, G. Jacobs, D. E. Sparks, W. D. Shafer and B. H. Davis, *Catal. Lett.*, 2014, **144**, 1127–1133.
4 A. Paredes-Nunez, D. Lorito, Y. Schuurman, N. Guilhaume, F.C. Meunier, *J. Catal.*, 2015, **329**, 229–236.
5 B. E. Bent, *Chem. Rev.*, 1996, **96**, 1361-1390.
6 C. Chao-Ming and B. E. Bent, *Surf. Sci.*, 1992, **279**, 79–88.
7 W. A. A. van Barneveld and V. Ponec, *J. Catal.*, 1984, **88**, 382–387.

Paul Collier queried: In the Fischer–Tropsch process there is a lot of water present. This isn't treated in your model, would you expect this to change the conclusions?

Kees-Jan Weststrate replied: Water is indeed an important product in FTS and several studies show an impact on the catalyst activity and selectivity. A recent surface science study shows that water adsorbs weakly onto the cobalt surface and defects can induce dissociation.[1] Molecular water therefore most likely plays only a minor role due to its weak adsorption, and the main effect should be attributed to O and OH groups on the surface, in particular at step edge sites. As these sites play an important role in important steps in the FTS mechanism, such as CO dissociation[2] as well as dissociative adsorption of H_2,[3] an increased concentration of O and/or OH will have a strong impact on FTS activity and selectivity.

1 L. Xu, Y. Ma, Y. Zhang, B. Chen, Z. Wu, Z. Jiang, W. Huang, *J. Phys. Chem. C*, 2010, **114**, 17023–17029.
2 C. J. Weststrate, P. van Helden, J. van de Loosdrecht and J. W. Niemantsverdriet, *Surf. Sci.*, 2016, **648**, 60–66.
3 P. van Helden, J. A. van den Berg and C. J. Weststrate , *ACS Catal.*, 2012, **2**, 1097–1107.

James Hayward said: You show the monomers quickly dehydrogenating and forming acetylene on your surface. Do you see any evidence of the formation of any other C_2 products, or further chain growth products?

Kees-Jan Weststrate responded: In the spectroscopic studies that we did, we do not find any indication of products other than acetylene. If one looks at the stability diagram of C_2H_x species reported for example in ref. 1 and 2, one can see that acetylene is the most stable species and it is therefore the dominant product. Acetylene can react further by dehydrogenation to atomic carbon or by cyclo-polymerization to form benzene or graphene.[1,2]

Further growth would, according to the alkylidyne growth mechanism,[3,4] require ethylidyne species and CH species to co-exist on the surface. Formation of acetylene consumes all CH present on the surface, and ethylidyne species do not form under the conditions used since they are less stable than acetylene for low surface coverages. When the surface coverage becomes high, other minor products might form as well. In ref. 4 and 5 we have recently shown that high surface coverage or a high coverage of CO co-adsorbates can lead to stabilization of ethylidyne species. The TPD experiment for the highly covered surface, shown in Fig. 2(b) in the article (DOI: 10.1039/C6FD00191B), shows a small hint for the formation of a minor C_2H_x product other than acetylene, characterized by a H_2 formation peak at around 360 K. It is tentatively attributed to the formation of a minor quantity of ethylidyne species, but further experiments would be needed to test this hypothesis.

1 Q. Y. Yang, A. D. Johnson, K. J. Maynard and S. T. Ceyer, *J. Am. Chem. Soc.*, 1989, **111**, 8748–8749.
2 C. J. Weststrate, A. C. Kizilkaya, E. T. R. Rossen, M. W. G. M. Verhoeven, I. M. Ciobîcă, A. M. Saib and J. W. Niemantsverdriet, *J. Phys. Chem. C*, 2012, **116**, 11575–11583.
3 C. J. Weststrate, I. M. Ciobîcă, A. M. Saib, D. J. Moodley and J. W. Niemantsverdriet, *Catal. Today*, 2014, **228**, 106–112.
4 C. J. Weststrate, P. van Helden and J. W. Niemantsverdriet, *Catal. Today*, 2016, **275**, 100–110.
5 C. J. Weststrate, I. M. Ciobîcă, J. van de Loosdrecht and J. W. Niemantsverdriet, *J. Phys. Chem. C*, 2016, **120**, 29210–29224.

Emiel Hensen opened a general discussion of the paper by Pieter van Helden: It is very interesting to learn about microkinetics simulations starting from a multi-site model. The model also involves migration of species between different sites, for which an activation barrier is considered. The implicit assumption is that diffusion of the species and mixing over the sites is much faster than the exchange itself. Is this assumption reasonable? Did you check diffusion barriers and estimate their effect on the multisite microkinetic model?

Pieter van Helden answered: This is an important observation about the inter-site diffusion reactions. There are two types of diffusion in our model. The first is that on a surface site. These are described using the typical mean field assumptions as outlined in the paper, together with the inclusion of the constrained diffusion potential in the species entropy description (See the ESI of our paper for more details, (DOI: 10.1039/C6FD00197A). We still need to check whether the on-site diffusion of species is significantly constrained in the presence of a high CO coverage. With a very few exceptions, surface diffusion processes typically have barriers significantly lower than those of the rate controlling surface processes at the relevant FT temperatures. Thus, as a first study of the FT multi-site mechanism, we accept the assumption as reasonable. The second type of diffusion is the inter-site diffusion, which have all been expressed as explicit reactions in the reaction network. Most of these barriers were calculated using DFT, while a small number were estimated based on the reaction energy and the terrace diffusion barriers. Since this second type of diffusion is an explicit reaction in the model, it was also included in the sensitivity analysis. At present, none of the inter-site diffusion reactions showed significant rate or selectivity control in any of the four considered scenarios. However, since we conclude that the high CO* coverage is essential for the correct FT kinetic regime to exist, the inter-site diffusion of species in the presence of a high coverage of CO* still needs to be assessed with DFT calculations.

Pieter van Helden commented: In our paper (DOI: 10.1039/C6FD00197A) we have employed sensitivity analysis to assess which steps and intermediates are rate and selectivity controlling. Although it does not play the biggest controlling role, the stability of the OH* species (especially on the step site) plays a smaller role in controlling the overall CO consumption rate. The direct steps of OH_x hydrogenation are therefore not rate limiting in the best fitting scenario (S4), but the thermodynamic stability of OH* seems to play a role. Thus the coverage of OH* on the step site is important for controlling the rate of CO consumption.

Moritz Wolf added: Our experimental results (DOI: 10.1039/C6FD00200E) lead to similar conclusions. Co-feeding water to synthesis gas (220 °C, $p_{H_2} = 0.15$ bar, $p_{CO} = 0.07$ bar, $p_{Ar} = 0.78$ bar, $p_{H_2O} = 0.02$–1.45 bar) increased the magnetisation of a cobalt based catalyst when compared to the magnetisation during exposure to dry synthesis gas. This increase indicates the presence of OH* species on the cobalt surface, which are known to increase the magnetisation of surface cobalt atoms.[1] The conversion of CO decreased, while the amount of co-fed water was increased. Before oxidation of cobalt was observed, the conversion of CO was already less than 50% of the initial conversion at dry synthesis gas conditions, possibly due to blockage of active sites by adsorbed, water originated and thermodynamically stable OH* species. Hence, the presence of OH* on the cobalt surface does affect the CO consumption.

1 M. Claeys, M. E. Dry, E. van Steen, P. J. van Berge, S. Booyens, R. Crous, P. van Helden, J. Labuschagne, D. J. Moodley, and A. M. Saib, *ACS Catal.*, 2015, **5**, 841–852.

Matthew Neurock asked: The model that you have developed involves 3 different sites that must all cooperate. This becomes difficult to understand. This may, in part, be due to the DFT results, which require 3 different environments. If I have this correct, you are assuming that CO dissociates *via* H-assistance at the step sites. The CH* species that form must then diffuse away from the step and onto the terrace sites. The CH* species would then diffuse across the CO-covered terraces until they find another CH* species that they can react with. This appears to be rather complicated. You indicate that C–O activation occurs more readily at the step sites. While this is true at low coverages, I don't think this would be the case at higher coverages as the more coordinatively-unsaturated Co sites here would bind the CO* or CH_x* and OH* intermediates much more strongly. As such, the higher coverages at these sites would significantly increase the barrier required for CO* to dissociate. In previously reported work on Ru,[1,2] we showed that the CO binding energies and CO activation barriers significantly increase at the more coordinatively unsaturated step sites as well as edge and corner sites. I believe the DFT results that used in your simulations are those taken from low coverage DFT calculations. They may not be reflective of the barriers or the coverages present under actual FT conditions.

1 B. T. Loveless, C. Buda, M. Neurock and E. Iglesia, *J. Am. Chem. Soc.*, 2013, **135**, 6107–6121.
2 D. Hibbitts, E. Dybeck, T. Lawlor, M. Neurock and E. Iglesia, *J. Catal.*, 2016, **337**, 91–101.

Pieter van Helden replied: Our starting point in this work was the nanocrystal structure of FCC Co at a relevant size for the Fischer–Tropsch (FT) synthesis.[1] FCC Co nancrystals at a representative size of 8 nm expose meaningful amounts of at least five unique sites, of which two are terraces and the other three are defects/step sites. These step sites are generally associated with the close-packed terrace. We approached the microkinetic model by not assuming upfront that any one of these sites favour any specific reactions. Furthermore, we allowed these sites to exchange surface intermediates, but did not assume upfront that they must cooperate. Thus, the interchange and cooperation we discuss in the paper emerge from the included kinetics at the considered conditions. As you note, the CH* species is formed *via* a H-assisted CO dissociation on the step site in the final scenario of our model (although it does not need to be this specific mechanism). The results indicate that the CH* species can indeed diffuse to the terrace surface. This happens with a free energy barrier of only 0.73 eV at 500 K. This is significantly lower than the overall FT process barrier. We agree that this process could indeed be affected by the coverage of other adsorbates at the step or terrace. Our results indicate that there is a high steady state coverage of both CO* and OH* at the step. This has to be explored further. As we note in our final conclusion, we demonstrated that the lateral interactions of reactive species with CO* are of importance, especially in the role that CO* could play in determining the kinetic parameters that need to be considered. As you rightly point out, CO dissociation process barriers have been shown to become higher in the presence of CO* (although there are inverse cases for H-assisted pathways on Ni). From our models it is clear that if breaking the CO bond is rate limiting, chain growth will be severely limited due to the hydrogenation of monomers being more rapid than the production thereof. Thus, if the barrier of the dissociation process of CO* at the step becomes too high, other possible pathways for splitting the CO bond could indeed be followed (maybe even similar to what was shown in Prof. Iglesia's lecture). However, such pathways should still not be rate limiting, and as such, are not the main concern in determining the controlling features of the FT reaction rate. The rate constant of the dissociation process could still be a controlling factor in the chain growth selectivity, by controlling the monomer concentration. As we conclude our discussion in the paper, we noted that the study of these steps under an appropriately high coverage of CO* should be the focus of continued work in this regard.

It is also important to take note of the fact that this model does not include all three of the possible defect sites available on the FCC Co particle. Thus, other CO dissociation pathways on these sites should not be disregarded, but explored further.

Lastly, the DFT data for CO dissociation barriers we used in the model are obtained at a 0.25 ML coverage equivalent. However, if we compare these barriers to the most recent experimental work on direct CO dissociation on defect sites, these barriers seem to still be overestimated even by the low coverage DFT calculations. I refer you to section 3.3 of our paper (DOI: 10.1039/C6FD00197A) for a more extensive discussion.

1 P. van Helden, I. M. Ciobîcă and R. L. J Coetzer, *Catal. Today*, 2016, **261**, 48–59.

Detlef Bahnemann remarked: Which are the active hydrogen species actually attacking carbon monoxide? Are these just hydrogen atoms or are reactive intermediates, such as hydride, also involved? Does the initial attack take place at the carbon atom or at the oxygen atom of the CO molecules?

Pieter van Helden responded: To my knowledge, there are no significant charge transfers happening in the dissociative hydrogen adsorption process on Co. Thus, the surface hydrogen that participates in the hydrogenation reactions are adsorbed atomic hydrogen. In the mechanisms of hydorgen assisted CO dissociation included in our model we considered both the attachment of H* on the C and the O atoms of adsorbed CO. On the two sites that can dissociate CO with any meaningful rate, the hydrogen assisted pathways seem to proceed preferably *via* attack on the C atom of the adsorbed CO, forming HCO*, which can subsequently dissociate into CH* + O*.

Enrique Iglesia asked: Let's address the experimental facts. Surfaces are saturated with chemisorbed CO at pressures between 5 kPa to 2 MPa as shown from infrared spectra; kinetic isotope effects and rate equations are consistent with H-assisted CO dissociation; DFT calculations indicate that direct CO dissociation is not feasible on low-index planes present on large particles that give highest turnover rates. Now you state that CO dissociation is fast and equilibrated based on isotopic exchange, even though the thermodynamics of such reactions would then lead to surfaces covered with C* and O* instead of CO*; this seems improbable. Have you considered that (i) such exchange may not require dissociation and that (ii) since you support the presence of two or more sites that one of those sites may be able to carry out the fast exchange without undergoing catalytic turnovers itself?

Pieter van Helden replied: The dissociation of the CO bond in FT is an important step, whether it is rate controlling or not. In our work on Co we claim that, given the outcomes of our microkinetic models, the CO scission process (*via* direct dissociation, HCO mediated or otherwise) should have a rate constant that could be considered to be "fast". This essentially should be understood as a fundamental rate constant higher than that of the rate limiting processes. That would mean that the CO dissociation step and its mechanisms is not such a key in understanding FT rates (although it might still play a role in the control of the selectivity profile). With this in mind "fast" CO dissociation does not in all cases lead to a surface covered with C* and O*. There are more factors at play in the steady state FT reaction network, such as availability of empty sites, coverages of other species, reversibility, rate of hydrogenation *etc.* For non-steady state results in the absence of hydrogen (not shown in the paper) C* is rapidly deposited on the surface, with an initial burst of CO_2 that forms, but this is also self-limiting as the C* quickly occupies most Co sites. This is not the scenario that is valid under the FT conditions studied here.

As is shown in Prof. Hensen's paper, the site he considered for Ru has a very high coverage of O* under steady state FT, due to the high rate constant of CO dissociation and the "slow" removal of O* to water. However, on Co the O* is rapidly hydrogenated to OH* on the step sites. These OH* species are quite stable on the B5 step sites. This stability ensures a moderately high coverage of OH* on the step sites, but still in line with the experimental SSITKA results (see our paper for more details). So in the case of Co FT, the surface is not rapidly covered with C* and O*, even if the CO dissociation is "fast". On Co nanoparticles there are a number of different exposed sites. Indeed the possibility exists that exchange could take place on one of the sites, without being part of the dominant FT reaction network. However, we do not any obvious indication from our microkinetics that would indicate this. In our model the most active CO scission process is directly linked to the rest of the FT reaction network while adsorbed CO* can easily migrate between the surface sites.

Matthew Neurock remarked: In the model you presented, you assume that the CH_x* species that form diffuse from the step edges and then across the terraces, where they can then find and react with another CH_x species. The diffusion of CH_x species through a dense adlayer of CO, however, would be difficult as it would occur *via* a random walk requiring the CH_x* species to diffuse through the formation of neighboring vacancies. These vacancies are random and as such, one would need to carry out a number of steps before the CH_x species would ever encounter one another. Even if the CH_x species are nearby, this would still require a number of diffusion steps for them to encounter one another. The mean-field microkinetic model assumes that diffusion is fast and would not address the stochastic nature of such processes.

Pieter van Helden replied: The mean field model we employed in this work does assume that diffusion on a terrace site is significantly faster that the controlling steps and thus is described by the mean-field limit. The stochastic nature of such processes in the presence of species which modify the diffusion potentials through lateral interactions will attenuate the rate given by the mean field model. In our model the diffusion of a species between dissimilar sites are explicitly included and will also be affected to an extent by the presence of a high coverage of CO*. For two CH* species to find one another and start chain growth on a terrace surface, they have to randomly traverse the terrace sites until they find each other, before they are hydrogenated to CH_4. The probability of this would depend on a number of factors: The effective diffusion barrier in the presence of CO*, the size of the terrace facet on the nanocrystal, the rate of CH* production and delivery to the surface facet. The effect of CO* on the barrier of diffusion of species is a valid concern and the detail thereof needs to be assessed with DFT calculations. However, it is important for all active species in the FT reaction in any case under high CO* coverage. The criterion for only two species to find each other on a terrace, without being the rate limiting step, would be a diffusion barrier that is more than 30 kJ mol^{-1} lower than the overall barrier for the FT process (CO consumption rate). This would ensure that each CH* can undergo over a 1000 random hops in the time one CO is converted to a monomer. Furthermore, in the opening paper of the conference (DOI: 10.1039/C7FD00018A), the flexibility of a high coverage CO* layer towards accommodating other intermediates was demonstrated. This would ensure that the CH* will not necessarily be fully blocked by the CO* overlayer. Typical supported Co FT catalysts have a distribution of Co nanocrystals in the nm range. An 8 nm FCC Co particle (as we used in our model) will expose a number of Co(111) terraces with an exposed diameter of about 16 to 20 sites (if no further overlayers of Co atoms are present). This would give a terrace area of about 400 sites that needs to be traversed by the CH* species. This is quite small and a random walk of over 100 steps would easily traverse all these sites. Furthermore, the rate constants for the process of producing CH* needs to be high enough that sufficient CH* is formed to promote chain growth over termination. Co SSITKA work[1] indicated that under steady state (even in methanation mode) the CH_x concentration is as high as 0.1 ML. Even if we assume a lower coverage of 1%, this would still ensure that there are around four monomers on the typical 400 site terrace at steady state participating in the random walk. It would be quite feasible that they find each other. The overall rate of CO dissociation and production of monomers on the surface would control this concentration. In our model this would imply that the ratio of defect to terrace sites would be important.

If indeed the CO* would impede the diffusion of CH* over the terrace, this would mean that a diffusion gradient would develop from the defect site. One can imagine that the local higher concentration of monomers near the defect site would enhance the probability of coupling. Despite these arguments, this is a clear area for further fundamental study.

1 J. P. Den Breejen, P. B. Radstake, G. L. Bezemer, J. H. Bitter, V. Frøseth, A. Holmen and K. P. de Jong, *J. Am. Chem. Soc.*, 2009, **131**, 7197–7203.

Michael Claeys opened the discussion of the paper by Emiel Hensen: One of the major take home messages of your paper is that, based on your theoretical work, the chain growth probability with ruthenium goes through a maximum. Typically, in polymerization reactions, including the Fischer–Tropsch synthesis, a decrease of the chain growth probability is observed experimentally with an increase of reaction temperature. I am therefore surprised that you predict a decrease of alpha with decreasing temperature. This is particularly surprising seeing that ruthenium is known to produce a high molecular weight product ('polymethylene', see early work by Pichler[1]) at low temperature conditions. Do you have, or are you aware of, any experimental evidence that supports your prediction that chain growth probabilities can indeed go through a maximum?

1 H. Pichler, B. Firnhaber, D. Kioussis and A. Dawallu, *Macromol. Chem. Phys., 1964*, **70**, 12–22.

Emiel Hensen answered: Indeed, we highlighted the occurrence of two Sabatier maxima, one with respect to CO consumption rate and one with respect to chain-growth probability. This is the outcome of the microkinetic model and the parameters obtained by DFT calculations. The latter values suffer of course from accuracy issues, which at least limit the absolute predictions of rates and also of the indicated maxima. So, the predicted maximum for chain-growth probability occurs at temperatures close to the FT conditions. We are not aware of an experimental gas-phase study showing that by lowering temperature the chain-growth probability goes down. However, we have published a paper on aqueous phase FTS,[1] in which we report that below 200 °C we produce both long-chain hydrocarbons and long-chain oxygenates. Their alpha values are very different and for the long-chain oxygenates alpha goes through a maximum. Note the very low temperatures used, not sure if gas-phase studies have explored such low temperatures. We have also discussed the kinetics behind this behaviour. Of course, two types of products with different alpha values suggests two different sites – we have speculated about this in the indicated paper, but based on our current insights about what these two different sites might be, we may have to rethink the origin and then also the relation to the maximum predicted by the microkinetics simulations.

1 X.-Y. Quek, Y. Guan, R. A. van Santen and E. J. M. Hensen, *ChemCatChem*, 2011, **3**, 1735–1738.

Matthew Neurock asked: You indicate that the barrier for CO to dissociate at the step site is low and as such, CO dissociation does not limit the rates. The references that were cited, however, were for calculations that were carried out on bare Ru surfaces at low coverage and have a barrier of ~65 kJ mol^{-1}. As such, there are few repulsive interactions in the adlayer and the barriers to dissociate CO are low. The surfaces under working FT conditions, however, are covered in CO that may range from 0.67 ML to 1 ML of CO. You also note that the active surfaces are highly covered under reaction conditions. As such the barriers for CO dissociation would be higher. We have carried out calculations on large Ru particles and examined the barriers to dissociate CO on the terrace, B5-step sites and corner and edge sites. The barrier to dissociate CO on the step sites that you considered here at high CO coverages have an intrinsic barrier of at least 150 kJ mol^{-1}. The apparent barriers for CO dissociation at the step sites are significantly higher at 269 kJ mol^{-1} as they require the desorption of CO to create a vacancy to dissociate the CO.[1] If one considers the C* and O* that form at these sites, the barriers would be even higher. Even if one assumes that C* and O* are removed from these sites, they would be filled by CO* as CO is equilibrated. How do the simulation results change if you change the barriers that would be more reflective of the higher CO* coverages?

1 B. T. Loveless, C. Buda, M. Neurock and E. Iglesia, *J. Am. Chem. Soc.*, 2013, **135**, 6107–6121.

Emiel Hensen answered: The barrier energies as calculated by Neurock and co-workers are quite different from ours. The reason is that our model is a real B5 site as first coined by van Hardeveld *et al.*,[1] while Neurock[2] used a stepped site model (to be sure, and which is not a B5 site) on which CO dissociation is unfavorable. As already shown by us,[3] the barrier at low CO coverage is 174 kJ mol^{-1} for such sites.

In your question, you correctly indicate that we have used a CO dissociation barrier at low coverage. The CO coverage is about $\frac{1}{12}$ ML. We have performed additional calculations up to $\frac{5}{12}$ ML coverage of CO. The CO dissociation barrier does not increase substantially; for instance, we find a CO dissociation barrier of 82 kJ mol^{-1} at $\frac{5}{12}$ ML. The underlying reason for this is that the topology of a step-edge site is such that significant lateral interactions are avoided up to a reasonable CO coverage (in our view lower than under real FT conditions). Furthermore, in your question you state that CO coverages range from 0.67 ML to 1 ML. We have not seen solid experimental evidence to support such high coverages. Our own SSITKA data show that under typical FT reaction conditions, the CO coverage is below 0.5 ML. This is in line with surface science experiments[4] and coverages calculated by us and others.[5,6] We have carried out some microkinetics simulations based on barriers reported by Neurock and Iglesia *et al.*,[2] including hydrogen-assisted elementary reaction steps. In our hands, this leads to very slow CO dissociation and consequently methane as the only hydrocarbon product. The fact that this should be the case can already be appreciated from the activation barriers (CO dissociation barrier is very high compared to other steps, which is inconsistent with conditions for chain growth). The apparent activation energy in these simulations is very high (> 150 kJ mol^{-1}), which is inconsistent with the experiment. Our microkinetic simulations, on the other hand, produce activation energies below 100 kJ mol^{-1}, consistent with experimental data being in the 80–120 kJ mol^{-1} range.

1 R. van Hardeveld and A. van Montfoort, *Surf. Science*, 1966, **4**, 396–430.
2 B. T. Loveless, C. Buda, M. Neurock and E. Iglesia, *J. Am. Chem. Soc.*, 2013, **135**, 6107–6121.
3 X.-Y. Quek, I. A. W. Filot, R. Pestman, R. A. van Santel, V. Petkov and E. J. M. Hensen, *Chem. Commun.*, 2014, **50**, 6005–6008.
4 G. A. Beitel, A. Laskov, H. Oosterbeek and E. W. Kuipers, *J. Phys. Chem.*, 1996, **100**, 12494–12502.
5 G. T. K. K. Gunasooriya, A. P. van Bavel, H. P. C. E. Kuipers and M. Saeys, *Surf. Sci.*, 2015, **642**, L6– L10.
6 M. Zhuo, A. Borgna and M. Saeys, *J. Catal.*, 2013, **297**, 217–226.

Kees-Jan Weststrate enquired: One of the main themes in the various presentations during the conference is that the CO concentration under reaction conditions is expected to be high, and that this should be taken into account when considering the surface chemistry of FTS. It is for example an essential ingredient of the studies by the groups of Iglesia *et al.*[1] and Saeys *et al.*[2] on Ru and Co catalysts, respectively. In Fig. 4 of your article (DOI: 10.1039/C6FD00205F) you show that the CO coverage at 600 K is only 0.03 ML, whereas the vacancy concentration is 0.14 ML and the remaining sites are covered by O, C and H. With an impingement rate of CO molecules in the order of 5×10^8 (=6.7 bar at 600 K) the barrier for desorption (with an assumed pre-factor of 1×10^{15}) then has to be around 60 kJ mol^{-1} to arrive at a $\theta CO/\theta^*$ of ~0.22. This seems like a very low adsorption energy for CO on Ru. In your presentation you mentioned that the CO concentration at step sites is low because of the high reactivity of these sites. But the reaction that consumes CO can only significantly impact the value of $\theta CO/\theta^*$ produced by the adsorption–desorption equilibrium when its rate is in the same order of magnitude as the adsorption and desorption processes, that is, 10^8 s^{-1}.

The question is twofold: 1(a) why are there so many free sites at 600 K (Fig. 4, DOI: 10.1039/C6FD00205F) despite the presence of 6.7 bar of CO? 1(b) Does the statement that the CO concentration at the step edges is low because of the high reactivity of step sites imply that the CO consumption reaction there is in the order of 108 or does it mean something else?

2 In the article you mention that the rates of the elementary reaction steps that convert the alkyne coupling product (HC–CR → C–CH_2R), a process which requires two hydrogenation steps and one dehydrogenation step, strongly affect the selectivity of the FTS reaction. In our experimental study we find that that the conversion of acetylene to ethylidyne (C_2H_2 + H → CCH_3) occurs with surprising ease on Co(0001) when CO is present on the surface as well.[3] Due to the abundance of CO, the free site concentration is expected to be low, yet the reaction proceeds at around 250 K already. With such facile reaction kinetics, how can the kinetics of these steps involved still be so important at typical reaction temperatures of 500 K?

1 D. Hibbitts, E. Dybeck, T. Lawlor, M. Neurock and E. Iglesia, *J. Catal.*, 2016, **337**, 91–101.
2 M. Zhuo, A. Borgna and M. Saeys, *J. Catal.*, 2013, **297**, 217–226.
3 C. J. Weststrate, P. van Helden and J. W. Niemantsverdriet, *Catal. Today*, 2016, **275**, 100–110.

Emiel Hensen replied: Let's start off by discussing your assumptions. We calculate, including lateral interactions, an adsorption energy of around 80 kJ mol^{-1} at 600 K. Combined with an activation energy of 65 kJ mol^{-1} for CO dissociation, this results in a situation where the CO desorption rate matches the CO dissociation rate. Hence, one can no longer apply the quasi-equilibrium assumption.

Answer to question 1: in your question you refer to high CO coverage as suggested by several papers. Let me first comment on these references.

As to ref. 1 in your question, Iglesia *et al.* use a very high coverage of CO in their discussion, higher than one ML. In the mentioned paper, a value of 1.04 ML is used, which is attributed to full coverage of planar surface atoms with an unlikely 1 CO per surface metal atom assumption augmented by coverage of steps and corners with dicarbonyls. There is no solid experimental evidence presented for such high coverage. As to the references mentioned in the Iglesia paper to support this high coverage, it can only be concluded that CO is the most common adsorbate.

Ref. 4 of this paper[1] discusses SSITKA data, showing that the surface is mainly covered by CO, some C_1 species and a very small amount of growing chains. The absolute coverage of CO is not calculated as the dispersion is unknown.

1 C. A. Mims and L. E. McCandish, *J. Phys Chem.*, 1987, **91**, 929–937.

Ref. 5–7 of this paper[2–4] use just the fitting of L–H rate equations that could be interpreted as CO being the most abundant surface species. Of course, kinetics do not prove a mechanism.

2 R. S. Dixit and L. L. Tavlarides, *Ind. Eng. Chem. Process Des. Dev.*, 1983, **22**, 1–9.
3 M. Ojeda, R. Nabar, A. U. Nikelar, A. Ishikawa, M. Mavrikakis and E. Iglesia, *J. Catal.*, 2010, **272**, 287–297.
4 I. C. Yates and C. N. Satterfield, *Energy Fuels*, 1991, **5**, 168–173.

Ref. 8 of this paper[5] presents IR data that confirms the presence of carbonyls, but without quantification. Quantification is only done using theoretical calculations. As outlined in our answers to Prof. Neurock, we also contend that the di- and tri-carbonyl bands reported for supported Ru nanoparticles upon CO adsorption are in fact positively charged Ru species due to partial oxidation. As to ref. 2, Saeys and co-workers found that at Fischer–Tropsch conditions (at 6 bar) the maximum coverage is $\frac{1}{3}$ ML. Higher coverages might theoretically be possible (up to a maximum of 0.58 ML) but are unstable. So there will always be abundant free sites. We conclude that there is no convincing evidence for a very high CO coverage during FTS. Of course, when such an assumption is made, there is a need for unusual mechanisms, as explored by the group of Prof. Neurock, that should explain Fischer–Tropsch synthesis on a densely CO-covered surface.

5 B. T. Loveless, C. Buda, M. Neurock and E. Iglesia, *J. Am. Chem. Soc.*, 2013, **135**, 6107–6121.

Our own SSITKA data show that under typical FT reaction conditions, the CO coverage is below 0.5 ML. This is in line with surface science experiments and our own and others' DFT calculations (coverage).[6,7] Also, we have never found indications for dicarbonyls present when Co nanoparticles are exposed to CO (nor did the spectra presented by Prof. Iglesia in his lecture provide evidence for such CO species on a Co nanoparticle catalyst, in which he also mentioned that the surface is already saturated at relatively low CO pressure). We argue that all this is not at odds with our microkinetic simulations as SSITKA will measure all reversibly adsorbed CO species and reversibly dissociated CO for that matter. Our micro-kinetic model pertains to step-edge sites, which on a real nanoparticle are present as a minority site together with many more low reactive sites (different terraces, *etc.*), where steady-state coverage will be much higher. To take this properly into account, a multi-site model is needed, which brings its own challenges but seems doable (see the work of Dr van Helden, DOI: 10.1039/C6FD00197A).

Further transient data emphasize that the rate of desorption is of the same order as the rate of CO dissociation. Accordingly, we expect that the rate of adsorption is also of the same order.

Question 2: Firstly, we would like to point out that we looked at a Ru step and not at a flat Co surface. Furthermore, we did not study the equilibrium of the reaction but the kinetics and thus selectivity. The experiments[8] you are referring to emphasize the easy conversion of acetylene to ethylidene. In the same paper, experiments are cited that show that CH to CH coupling also runs at 250 K. In the paper these two are added together to describe the chain growth. It does, however, not explain why a temperature 500 K is needed for the Fischer–Tropsch reaction. We calculate that on Ru(11–21):

The pathway HC–CH → HC–CH_2 → C–CH_2 → C–CH_3 has an overall barrier of 83 kJ mol^{-1} (21 kJ mol^{-1} from HC–CH_2) and results in a chain fragment.

The pathway HC–CH → HC–CH_2 → H_2C–CH_2 has the same overall barrier of 83 kJ mol^{-1} (45 kJ mol^{-1} from HC–CH_2) and results in products that can desorb.

As shown in the paper, these energies explain well that a temperature of 500 K is needed for the Fischer–Tropsch reaction to occur.

6 G. T. K. K. Gunasooriya, A. P. van Bavel, H. P. C. E. Kuipers and M. Saeys, *Surf. Sci.*, 2015, **642**, L6– L10.
7 M. Zhuo, A. Borgna and M. Saeys, *J. Catal.*, 2013, **297**, 217–226.
8 C. J. Weststrate, P. van Helden and J. W. Niemantsverdriet, *Catal. Today*, 2016, **275**, 100–110.

Michael Bowker commented: In your CO scrambling experiments why doesn't the reaction poison itself with carbon *via* the Boudouard reaction? I find it surprising that C isn't formed in this way because I would imagine the CO oxidation reaction is relatively easy once some CO is dissociated.

Emiel Hensen replied: During the CO scrambling reaction, CO_2 is produced, but this reaction stops relatively quickly. When the surface is exposed to $^{12}C^{16}O$ only, then obviously CO_2 is also observed and its production stops soon after CO exposure. It means that during CO exposure, C atoms remain at the surface. Titration of these C atoms by H_2 is possible and shows that with increasing exposure time to CO, the C/CO(surface) increases but never becomes very high. Despite this laydown of C atoms, the surface is still able to reversibly dissociate CO. Moreover, Fischer–Tropsch rates before and after carbon laydown are very similar, even with a significant part of the carbon remaining there (as it can only be removed by hydrogenation above 300 °C).

Our explanation is as follows. CO dissociates reversibly on a minority site (we presume these are step-edges). Part of the C and O atoms migrate to the terraces, which are more ubiquitous and are covered with spectator CO. CO reacts with O to form CO_2, but the C atoms, in the absence of H, remain and start blocking the terraces such that CO_2 production ceases. The nature of the C overlayer is unclear, although, as I said, part of it requires reduction above 300 °C to be removed, suggestive of a graphene-like layer being formed.

Pieter van Helden asked: It seems that oxygen species are quite stable on the sites of both Co and Ru. In our Co microkinetic model, oxygen is most stable as OH* on the step site. In the scenario where we find the best fit with experiments (S4), the OH* covers about half the step sites. In your Ru model, on the single step site the O* coverage is quite dominant under FT conditions. In our Co case, hydrogenation of O* to OH* is quite rapid and therefore you would not expect a high probability of forming CO_2 *via* the CO* + O* reaction on that site. This is indeed what you see experimentally in Fischer–Tropsch, that CO_2 selectivities are in the single percentage points under relevant conditions. However, a significant amount of CO_2 is still produced in our Co model, predominantly at the Site-B/Co(100).

Emiel Hensen answered: The kinetic data show that O hydrogenation is more difficult on Ru than on Co. This explains the difference in our data. We have used a single-site model so we do not include a CO_2 formation pathway on a low-reactive surface. We would argue that on a low-reactive surface the CO coverage is high so that any O migrating there will be removed as CO_2. An important aspect/challenge in simulating multiple sites with migration might be to correctly describe diffusion/mixing rates.

Enrique Iglesia enquired: Given the very high rates of reactions of CO* with O* and your proposal that CO dissociation is fast and equilibrated, how does one avoid stranding C* behind and causing rates to decrease precipitously with time? Even a very infrequent O*–CO* event would cause almost instantaneous high coverages of C* in light of your very high turnover rates for exchange both with and without H_2 present. How do you explain the absence of deactivation and the CO_2 formed *via* Boudouard reactions as the predominant product?

Emiel Hensen answered: Recent kinetic experiments involving isotopic CO scrambling in the presence of H_2 show that CO dissociation is reversible, but not totally equilibrated. Changing the scrambling measurements from pulsing mode to continuous flow mode indeed leads to build up of C species and deactivation. A key point is that most of the activity already disappears by formation of about 10–20% of the surface sites, suggesting that CO dissociation only occurs at a minority site. This C lay-down is suppressed in the presence of hydrogen, leading to formation of CH_4 and other hydrocarbons and H_2O (more easy to form H_2O than CO_2).

Linda Jewell communicated: One of the things that you said is that chain growth is reversible under FT conditions. This suggests that equilibrium can be established between alkanes of different lengths, but in general the distribution obtained is not an equilibrium distribution.

If chain growth is reversible, why do you think alkanes are not at equilibrium under FT conditions?

Emiel Hensen communicated in response: The fact that chain growth is reversible does not mean it is equilibrated. The chain length distribution is determined by the kinetics in this case, not by equilibrium. At low temperatures, there are few vacancies so that the forward reaction that propagates the chain is much faster than the reverse decoupling. Hence, long chains are formed. At high temperatures, there are vacancies so that the rate of decoupling is much higher than the rate of chain growth and mostly methane is obtained.

Hans Niemantsverdriet opened the general discussion and addressed Emiel Hensen, Pieter van Helden and Enrique Iglesia: Comparing the lecture of Prof. Iglesia with that of Prof. Hensen and Dr van Helden, we interestingly see the difference in approaches between the conventional 'one-site does it all' Langmuir–Hinshelwood approach and that of micro-kinetic modelling, where you allow for the cooperation of step sites with terrace sites. Prof. Iglesia cited the first order hydrogen dependence in the context of his LH rate expression over a range of different conditions as compelling evidence that hydrogen must be involved in the rate determining step, namely CO dissociation. Dr van Helden and Prof. Hensen, in agreement with our own work,[1] do not consider CO dissociation as rate determining. In the latter view, the first order dependence in $p(H_2)$ could well be due to the removal of O-atoms from the surface, or from other hydrogenation steps in the reaction sequence. However, the fact that many steps in the hydrocarbon surface chemistry can occur at room temperature or below,[1,2] points to O-removal, which indeed has a high barrier.[3] What is your opinion?

Secondly, is it possible to derive an overall rate equation from your microkinetic modelling equivalent to the LH expression, so that a direct comparison might be possible?

1 C. J. Weststrate, P. van Helden and J. W. Niemantsverdriet, *Catal. Today*, 2016, **275**, 100–110.
2 C. J. Weststrate, I. M. Ciobîcă, A. M. Saib, D. J. Moodley, J. W. Niemantsverdriet, *Catal. Today*, 2014, **228**, 106–112.
3 C. J. Weststrate, J. van de Loosdrecht and J. W. Niemantsverdriet, *J. Catal.*, 2016, **342**, 1–16.

Emiel Hensen responded: Let us answer the second question of Prof. Niemantsverdriet first – with microkinetic modeling it is very easy to compute CO conversion rates as a function of temperature and partial pressure and then fit them with LH equations, just like one would do with experimental data. We have not done so for our predicted Ru data, but power rate law fits of these data produce reaction orders with respect to CO and H_2 of −0.2 and ~1 in good agreement with the LH-type equation shown by Prof. Iglesia. Then, a degree of rate control analysis for the Ru-catalyzed FT reaction shows that CO consumption rate is completely controlled by O removal. Note that this pertains to Ru (see our Angewandte Chemie paper).[1] Thus, the observed kinetic parameters do not prove CO dissociation is rate limiting. Prof. Iglesia assumes that CO dissociation is rate limiting (probably he has some indications for this, different from the kinetics). If the reaction order in H_2 is 1 then indeed this implies that CO dissociation – the first step in the kinetic scheme – should be H-assisted and one would have to introduce a HCOH or H_2CO species, for which the C–O bond cleavage is then rate limiting.

In the Ru simulations, we find that O removal is rate limiting. This requires one or two H atoms, which can explain the reaction order in H_2 being 1. In the Ru case, the rate control of O removal is high so that one could say there is one rate limiting step.

Based on our microkinetic modeling of Co, we predict that both O removal and CO dissociation are rate controlling – this makes the use of LH rate equations cumbersome as there is not one rate limiting step. Note that these predictions are in line with the works mentioned by Prof. Niemantsverdriet. Although the CH_x (de)hydrogenation steps are indeed facile (*e.g.* accessible at room temperature or below) under FT and methanation conditions, they are not fast as coverages are high or the surface gets depleted in H, respectively. Prof. Iglesia raised a solid point that if O removal is rate limiting, the surface should be full with O – experimentally the surface is observed to be full with CO. Here is the basis of our very different view: to explain this we require at least two sites, and these are in our opinion step-edges for low-barrier CO dissociation, where nearly all the FT chemistry takes place, and terraces that contain spectator CO (and produce CO_2 with O atoms spilling over from the step-edges to the terraces).

1 I. A. W. Filot, R. A. van Santen and E. J. M. Hensen, *Angew. Chem., Int. Ed.*, 2014, **53**, 12746–12750.

Pieter van Helden added: I am of the opinion that it is very difficult to directly relate the reaction orders of a multi-step surface reaction only to one single rate limiting step. From our sensitivity analysis, it is clear that there are a number of rate controlling features at typical Fischer–Tropsch conditions. However, we showed in our contribution that both the concentration of H* and the final hydrogenation of the CH_3* species on the step site plays a important rate controlling role. This would indeed point to a positive dependence of the rate on the pressure of hydrogen. In our paper (DOI: 10.1039/C6FD00197A) we argue that the CO scission process cannot be rate limiting and it is therefore my opinion that the first order in hydrogen relates to a hydrogenation step. It is also worth noting that both the two published macrokinetic models, which describe the experimental pressure-rate relationship, the best overall (referred to in our paper, DOI: 10.1039/C6FD00197A) has the pressure of hydrogen expressed to the power 0.75. It is interesting to note that in the assumed mechanisms underlying the derivation of these models, one employed direct CO dissociation and the other used a hydrogen assisted CO dissociation. In both cases they resulted in the same order in hydrogen (although the CO orders differ somewhat). Regarding the second part of the question: I believe it is possible to derive an overall rate equation from our microkinetic modelling work. One way is to take the proposed reaction network, rate-limiting steps and dominant surface species, and derive a mechanistically based macrokinetic model. In my experience this could become a difficult task if you use two active sites explicitly in your model, resulting in having to solve a high order polynomial to be able to ensure you have a closed form equation. Thus, although it is surely possible, you would need some smart mathematics to keep it simple. Thus far, we have not put significant effort into this in our team.

Enrique Iglesia replied: The kinetic relevance of the oxygen removal step would not be consistent with the well-demonstrated presence of CO as the predominant chemisorbed species during catalysis, while quasi-equilibrated CO dissociation would lead to the predominance of C* and O* as the adsorbed species (instead of CO* in the molecular form). Of course, at steady-state, carbon removal must occur at the same rate as oxygen removal, thus the kinetic relevance of one is not possible without that of the other one, returning us once again to the earlier proposals of H-addition to C* and the subsequent step as the sole kinetically relevant ones. A first-order dependence on H* would require equilibration between O* and H* and a kinetically relevant addition of the second H* to OH*; this is inconsistent with the very fact of isotopic exchange between H_2O and D_2O (and between both and H_2 or D_2) during the practical conditions of Fischer–Tropsch catalysis. Such proposals would also fail to account for the measured kinetic isotope effects and for theoretical assessments that demonstrate the high barriers for CO dissociation on low-index planes, on corners and edges, and on all except possibly unique step sites, the prevalence of which would be difficult to envision to be the same on large particles of varying size, which show similar turnover rates at the conditions of the Fischer–Tropsch synthesis.

Pieter van Helden addressed Emiel Hensen, Kees-Jan Weststrate and Matthew Neurock: Regarding the coverage of CO at steady state Fischer–Tropsch (FT) conditions, it is important to note the typical rate of desorption of CO at these temperatures. The pre-exponential factor of desorption and the desorption heats are such that at FT conditions the temperature is above the desorption temperature of CO. Thus, there will be a rapid exchange of CO on the surface with that of the gas phase. This rapid exchange will create transient, short-lived empty sites on the surface. The adsorbed CO layer on the surface can therefore not be considered to be statically blocking the sites. It has been shown through in-house computational work and surface science experiments, as well as previous well-known surface science work (referred to in our paper, DOI: 10.1039/C6FD00197A) that the CO surface coverage on the close packed Co surface saturates to a value lower than a monolayer of Co. This value is noted to be around 0.67 ML. This value is not as high as was noted by Prof. Iglesia in his paper (DOI: 10.1039/C7FD00018A). At such a saturation coverage there will be some sites still available for the smaller adsorbates. Once the surface reaches this saturation coverage, it becomes difficult to push more CO onto the surface due to the strong repulsive lateral interactions, which negate the effect of CO binding to the surface. It is these lateral interactions of CO with other adsorbates that pushes up the energies of these adsorbates and transition states into a different kinetic regime.

Emiel Hensen replied: Thank you for this comment. Our calculations and those of Mark Saeys[1] emphasize that the (0001) surface saturates at 0.67 ML. Also, in response to an earlier question from Dr Weststrate, I would like to comment on fast adsorption/desorption. Indeed, using the usual considerations we find that CO adsorption/desorption rates must be fast compared to CO consumption. However, a simple transient experiment in the spirit of Prof. Kruse[2] tells us something different. If we switch the CO/H_2 feed for Co/SiO_2 to a H_2 only feed, all the CO on the surface will be consumed and form hydrocarbons – CO does not desorb. We determined CO coverages (determined by SSITKA, range 0.1–0.4) as a function of the partial pressure of CO and then carried out the transient experiment (switched CO/H_2 to H_2). We counted the carbon atoms coming out in the form of hydrocarbons and they matched the total amount of CO on the surface. Thus, the rate of desorption is certainly not fast compared to the rate of CO dissociation. It is an important experimental observation, which is inconsistent with the simple adsorption/desorption kinetics we assume in our modeling and in our considerations.

1 M. Zhuo, A. Borgna and M. Saeys, *J. Catal.*, 2013, **297**, 217–226.
2 J. Schweicher, A. Bundhoo and N. Kruse, *J. Am. Chem. Soc.*, 2012, **134**, 16135–16138.

Kees-Jan Weststrate added: In a recent review we have summarized experimental findings about CO adsorption on cobalt surfaces, in particular the close-packed surface.[1] We find that at 90 K a coverage of 0.83 ML ($\frac{6}{7}$) can be reached when using CO pressures of 1×10^{-3} mbar. At lower pressures a coverage of 0.67 ML ($\frac{2}{3}$) can be reached. When we extrapolate our data measured in pressures up to 10^{-3} mbar CO to FTS pressures and temperatures, we find that those two states are not populated, and instead a coverage of around 0.5 ML is found, associated with an adsorption energy of around 90 kJ mol^{-1}.[2] This prediction matches the SSITKA results reported by Den Breejen *et al.*; they use 55 mbar CO in 550 mbar CO at 483 K and report a CO coverage of 0.43 on the larger particles which exhibit good FTS performance.[3] By exploring the adsorption–desorption equilibrium we can get some insight into how adsorption energies translate to coverage, and by looking at the rate constants we can get an idea about the dynamics of the system.[4] When assuming a pre-factor for desorption of 1×10^{15} and $T = 500$ K, an adsorption energy of 67 kJ mol^{-1} would produce a desorption rate constant equal to the influx of CO molecules at 1 bar, approx. 1×10^{8} per surface atom per second. Vice-versa, for the surface state corresponding to an adsorption energy of 90 kJ mol^{-1}, 99.6% would be occupied. Yet, the desorption rate constant at this point would still be 4×10^{5} with an average residence time per CO molecule of 2.5×10^{-6} s. A state with an adsorption energy of 100 kJ mol^{-1} would be 99.97% occupied, yet the rate constant would still be 3×10^{4}, with a residence time of 3.3×10^{-5}. At 120 kJ mol^{-1} we obtain a residence time of 3.3×10^{-3} s. To summarize, CO coverage on cobalt is expected to be around 0.5 ML, a notion that is supported by SSITKA data. The situation is dynamic, with random appearance and disappearance of vacant sites on the millisecond timescale or faster.

1 C. J. Weststrate, J. van de Loosdrecht and J. W. Niemantsverdriet, *J. Catal.*, 2016, **342**, 1–16.
2 H. Papp, *Surf. Sci.*, 1983, **129**, 205–218.
3 J. P. Den Breejen, P. B. Radstake, G. L. Bezemer, J. H. Bitter, V. Frøseth, A. Holmen and K. P. de Jong, *J. Am. Chem. Soc.*, 2009, **131**, 7197–7203.
4 C. J. Weststrate, P. van Helden and J. W. Niemantsverdriet, *Catal. Today*, 2016, **275**, 100–110.

Matthew Neurock answered: High pressure CO adsorption surface science studies carried out on Ru(0001) show that the CO coverage monotonically increases from 0.67 ML at typical UHV conditions (10^{-8} torr) to ~0.9 ML of CO at 0.1 torr.[1] This study, which was carried out at CO pressures of up to 0.1 torr is still 3–4 orders of magnitude lower than the 1–10 atm CO pressures used for FTS. At the high pressure conditions and FTS temperatures, CO is quasi-equilibrated on the surface.

CO readily desorbs at these temperatures but is rapidly replaced as a result of the high gas phase pressures. The rate of CO adsorption is significantly higher than the rates of reaction. The high saturation coverages on CO are fully consistent with reported *in situ* infrared analyses,[2,3] isotopic labeling studies[4] and kinetic studies[5] that indicate that surfaces are fully covered at the conditions of Fischer–Tropsch. The CO saturation coverage on metal particles is also significantly higher than that on single crystal surfaces, as the metal particles expand radially to relieve the stress induced by the repulsive interactions between the oxygen atoms on neighboring CO molecules, thus creating more curvature along the surface and allowing the CO molecules to point outward and away from one-another. Single crystal surfaces cannot accommodate the high stress that would result from the repulsive interactions. As such, single crystal surfaces either saturate at lower coverages (0.67–0.9 ML) or reconstruct. Theoretical and experimental results for Ru particles show CO coverages > 1 ML.[2,6]

1 D.E. Starr and H. Bluhm, *Surf. Science*, 2013, **608**, 241–248.
2 B. T. Loveless, C. Buda, M. Neurock and E. Iglesia, *J. Am. Chem. Soc.*, 2013, **135**, 6107–6121.
3 P. Winslow and A.T. Bell, *J. Catal.*, 1984, **86**, 158–172.
4 C.A. Mims and L.E. McCandlish, *J. Phys. Chem.*, 1987, **91**, 929–937.
5 M. Ojeda, R. Nabar, A.U. Nilekar, A. Ishikawa, M. Mavrikakis and E. Iglesia, *J. Catal.*, 2010, **272**, 287–297.
6 D. Hibbitts, E. Dybeck, T. Lawlor, M. Neurock and E. Iglesia, *J. Catal.*, 2016, **337**, 91–101.

Matthew Neurock commented: Many appear to agree that under FT conditions the metal Ru and Co surfaces have high CO coverages. There is some debate, however, as to what the actual coverages are. Single crystal surface science studies indicate that the coverages on Co(0001) and Ru(0001) surfaces are ~0.67 ML. We calculate very similar saturation coverages to those from experiments on Co(0001) and Ru(0001) surfaces (~0.67 ML).[1] On particles, however, the surface coverages are significantly higher as the surfaces of the metal particles expand. This is consistent with EXAFS data for CO on particles. In addition, the CO molecules can fan away from one another to relieve some of the steric repulsion from the CO groups. Single crystal surfaces, however, are constrained laterally and as such cannot relieve themselves from the strain that results from the repulsion from the repulsive CO* interactions. At high enough coverages, however, they release the strain by reconstruction. Our previous calculations indicate that one can go up to 1.05 ML on 201 atom[1] and 586 atom Ru particles.[2] The Ru particles expand radially to allow for higher packing. In addition, the coordinatively unsaturated edge and corner Ru atoms will bind 2 and 3 CO molecules, respectively, to complete their coordination sphere. This is consistent with experimental, kinetic and *in situ* IR spectroscopic studies, which indicate coverages > 1 ML and the presence of di- and tri- carbonyls.[1]

1 B. T. Loveless, C. Buda, M. Neurock and E. Iglesia, *J. Am. Chem. Soc.*, 2013, **135**, 6107–6121.
2 D. Hibbitts, E. Dybeck, T. Lawlor, M. Neurock and E. Iglesia, *J. Catal.*, 2016, **337**, 91–101.

Kees-Jan Weststrate answered: Before discussing CO on ruthenium nanoparticles, it is important to realize that cobalt and ruthenium are different in various ways. Fig. 1 shows a comparison of the TPD spectra for CO from Co(0001) and Ru(0001), respectively (obtained in the same vacuum system, with the same heating rate, *etc.*) It is clear that, although in both cases the total coverage is 0.67 ML, the desorption temperature found on Ru is much higher. This higher adsorption energy on Ru would translate to a higher saturation coverage at reaction conditions for Ru compared to Co.[1] On the close-packed Ru, a CO coverage of up to 0.9 ML was reported at room temperature[2] for pCO = 1 Torr, but it is not obvious that this state can be reached at 500 K, even when the pressure is 4 orders of magnitude higher. Secondly, because of the 10% smaller diameter of cobalt atoms, the situation regarding lateral interactions is different. This can be seen in infrared experiments: on Ru(0001) only the top sites are occupied up to a coverage of 0.67 ML, the low pressure saturation point.[3] Instead, on Co(0001) we find top-only up to 0.33 ML, top-hollow between 0.33–0.5 ML, bridge-top between 0.5–0.67 ML and hollow-top between 0.67–0.83 ML, *i.e.* much more complex.[1] Finally, Co nanoparticles in the relevant regime for FTS often have the bulk fcc structure[4] and thus expose predominantly (111) surfaces, with (100) facets being second most abundant.[5] Instead, Ru nanoparticles adopt a hcp structure. The lower symmetry of the hcp unit cell implies that spherical particles can only expose the close-packed surface on two opposing faces, and the other surfaces must have a different, more open structure. As a consequence, the close-packed surface accounts for only a small part of the surface, and a large quantity of more open surfaces, such as (10–10) and (11–20) are exposed.[6–8] It is therefore not obvious at all that the small (2.5 nm), cubo-octahedral particles used in ref. 9,10, which are fully terminated by close-packed surfaces, are representative of 7 nm hcp-Ru catalyst particles. The presence of significant quantities of surface structures, such as (10–10) and (11–20), has important consequences: CO coverages up to 1.25 ML have been reported for the (10–10) surface,[11] which would translate to a global coverage >0.67 ML in a Ru nanoparticle. In addition, surfaces such as (11–20) will be exposed, which have been found to be very active for CO dissociation.[12] To add to the complexity, some of the ordered adsorbate structures found for high coverages on extended flat single crystal surfaces might not form on the small terraces exposed by nanoparticles, see *e.g.* the discussion in ref. 13. Thus, surface science reveals the complexity of CO adsorption and the pressure–coverage relation even on well-defined cobalt and ruthenium surfaces. On small nanoparticles the convex surface and lattice expansion can serve to relax repulsive adsorbate interactions, thereby stabilizing higher surface coverages. Calculations in ref. 9,10 suggest that such relaxation indeed takes place for small clusters, but whether this extends to 7 nm particles is not yet clear.

1 C. J. Weststrate, J. van de Loosdrecht and J. W. Niemantsverdriet, *J. Catal.*, 2016, **342**, 1–16.
2 D.E. Starr and H. Bluhm, *Surf. Science*, 2013, **608**, 241–248.
3 H. Pfnur, D. Menzel, F. M. Hoffmann, A. Ortega and A. M. Bradshaw, *Surf. Sci.*, 1980, **93**, 431–452.
4 O. Kitakami, H. Sato, Y. Shimada, F. Sato and M. Tanaka, *Phys. Rev. B*, 1997, **56**, 13849.
5 P. van Helden, I. M. Ciobîcă and R. L. J Coetzer, *Catal. Today*, 2016, **261**, 48–59.
6 R. M. Nielsen, S. Murphy, C. Strebel, M. Johansson, I. Chorkendorff and J. H. Nielsen, *J Nanopart. Res.*, 2010, **12**, 1249–1262.
7 J. Gavnholt and J. Schiøtz, *Phys. Rev. B*, 2008, **77**, 035404.
8 X.-Y. Quek, I. A. W. Filot, R. Pestman, R. A. van Santel, V. Petkov and E. J. M. Hensen, *Chem. Commun.*, 2014, **50**, 6005–6008.
9 B. T. Loveless, C. Buda, M. Neurock and E. Iglesia, *J. Am. Chem. Soc.*, 2013, **135**, 6107–6121.
10 D. Hibbitts, E. Dybeck, T. Lawlor, M. Neurock and E. Iglesia, *J. Catal.*, 2016, **337**, 91–101.
11 G. Lauth, T. Solomun, W. Hirschwald and K. Christmann, *Surf. Sci.*, 1989, **210**, 201–224.
12 J. Wang, Y. Wang and K. Jakobi, *Surf. Sci.*, 2001, **488**, 83–89.
13 S. Murphy, C. Strebel, S. B. Vendelbo, C. Conradsen, Y. Tison, K. Nielsen, L. Bech, R. M. Nielsen, M. Johansson, I. Chorkendorff and J. H. Nielsen, *Phys. Chem. Chem. Phys.*, 2011, **13**, 10333–10341.

Pieter van Helden replied: Without further dwelling on the differences between Co and Ru, we agree that there are some structural changes that can be expected under exposure to CO gas. The extent of these still needs to be assessed for Co. In our specific model the different sites have different saturation coverages. For the Co(111) type sites the experimental saturation coverage of 0.67ML was implemented, while on the defect sites the only constraint was applied (allowing for up to 1 ML coverage). There is a strong possibility that CO* might saturate at higher values on the defect sites. Thus, the overall particle coverage maximum that can be attained in our model is around 0.75 ML in total. In the simulations, these are never attained due to the build up of C_2* and O* and OH* species, which compete with CO* for both terrace and defect sites. In this way it also limits the amount of CO* under steady state FT conditions.

The example of a 201 atom Ru particle in the question relates to the expansion of extremely small particles. For particles of this size, 30% or more of the atoms are located at the surface. This results in a very high total energy of the particle (due to the high surface energy) which can be significantly reduced due to binding with CO (or other adsorbates). With such high surface energies, these particles will expand and restructure in the presence of CO to lower the total energy. At larger particle sizes (as the 8 nm particle we considered in our work), less than 15% of the atoms will be located at the surface, giving a much lower total energy of the particle. These particles will not have such a large driving force to lower their surface energies as those in the range lower than 3 nm, thereby not needing excessive amounts of adsorbed CO. We are of the opinion that sites on such particles will start to approach the regime where the single crystal saturation coverages become relevant on the terrace sites. However, we concede that this still needs to be confirmed. Another indication of Co CO coverage under steady state FT is given by SSITKA work such as that shown in ref. 1. Here the steady state total coverage of CO is between 0.4 and 0.5 ML for Co size larger than 6 nm. All of the scenarios in our model give results of total Co coverage between 0.4 and 0.6 ML. In ref. 1 they show that there is an increase in the total CO coverage as the Co size decreases below 6nm (accompanied by a decrease in activity). This could point to very small Co particles accommodating higher CO coverages, in line with the examples in the question. Alternative experimental results summarized in ref. 2 showed that di-carbonyls are formed on Co defect sites in the presence of CO pressure at low temperatures. However, it was also noted that with the addition of H_2 and increased temperature, this feature decreased significantly.

1 J. P. Den Breejen, P. B. Radstake, G. L. Bezemer, J. H. Bitter, V. Frøseth, A. Holmen and K. P. de Jong, *J. Am. Chem. Soc.*, 2009, **131**, 7197–7203.
2 H. Oosterbeek, *Phys. Chem. Chem. Phys.*, 2007, **9**, 3570–3576.

Emiel Hensen replied: Our point of view regarding the high surface coverage is expressed in the answer to an earlier question from Kees-Jan Weststrate. In brief, high coverages are highly unlikely based on both theoretical and experimental observations. There is also no solid experimental evidence for high (> 1 ML, or even > 0.7 ML) CO coverage on Co and Ru nanoparticles. The IR bands in the work of Loveless at higher wavenumbers derive from CO adsorbed to positively charged Co (see ref. 1 and references therein). The samples in Loveless *et al.*[2] were passivated and then re-reduced, which appears reasonable, but then cooled in He flow, meaning some Ru will be oxidized by trace O_2 in this way. We also refer to the IR data presented by Prof. Iglesia which clearly show that there are no di- and tricarbonyls at all and also that the CO coverage saturates at low CO pressure (much less than 1 bar), consistent with the view of Weststrate.

1 X.-Y. Quek, Y. Guan, R. A. van Santen and E. J. M. Hensen, *ChemCatChem*, 2011, **3**, 1735–1738.
2 B. T. Loveless, C. Buda, M. Neurock and E. Iglesia, *J. Am. Chem. Soc.*, 2013, **135**, 6107–6121.

Enrique Iglesia asked: You have proposed that surfaces are not at full monolayer at typical Fischer–Tropsch synthesis conditions. In my Introductory Lecture (DOI: 10.1039/C7FD00018A), I presented infrared spectra that show saturation coverage temperatures even higher than for Fischer–Tropsch synthesis and pressures significantly lower (5–100 kPa). Volumetric uptakes of CO are consistent with CO/M surface stoichiometries of unity for >5 nm diameter Ru and Co particles and even greater than unity for smaller particles (as evident from *gem*-dicarbonyl bands in the infrared spectra). Saturation coverages are indeed about 0.6 ML on flat extended surfaces because of their inability to relax laterally, but what is the evidence that small or large clusters of metals, and of Co and Ru in particular, cannot chemisorb stoichiometric amounts of CO?

Kees-Jan Weststrate responded: As mentioned in the reply to a related question from Prof. Neurock, there are significant differences between Ru and Co regarding which comes to the interaction with CO. As there is more data available for cobalt-based catalysts in FTS, we limit the discussion here to experimental findings on cobalt. In ref. 1 we have previously shown that on the close-packed terrace of Co the CO coverage can be pushed to 0.89 ML using a combination of low temperature and relatively high pressure. However, when we extrapolate our data to FTS temperatures and pressures, a coverage of around 0.5 is realistic. Step edge sites can indeed accommodate higher CO coverages. This is illustrated by the work of Falo, Cano and Salmeron reported in ref. 2. For small [fcc-(100)] cobalt islands on top of a Cu(100) surface, CO quantities up to 3 per cobalt surface atom were reported, but this value converged to 0.65 ML with increasing Co coverage. Likewise, on the Co(10–10) surface which exposes rows of atoms similar to step edge sites, Toomes and King report CO coverages slightly higher than 1 ML,[3] confirming the notion that a local high coverage can exist at step edges. Due to this, even without invoking lattice relaxation, global CO coverages higher than 0.5 can be accommodated on cobalt nanoparticles. SSITKA measurements provide *in situ* information about the CO coverage under working conditions. Den Breejen *et al.* report a CO coverage of 0.43 ML on ~8 nm Co particles under methanation conditions, with $p_{CO} = 55$ mbar at 210 C.[4] In ref. 5, Mims *et al.* report similar data, but measured at 0.4 bar CO, and 202 °C. Their Co/SiO_2 catalyst has a cobalt metal loading of 23% with a dispersion of 2.8% (~35 nm). This translates to a cobalt surface atom concentration of 109 000 nmol g^{-1} catalyst. The reported concentration of reversibly bound CO is 60000 nmol per g catalyst, yielding a 0.55 ML CO coverage in this case. Thus, *in situ* quantification of the CO coverage provides evidence for a CO coverage lower than 1 ML under FTS conditions. On flat Co single crystal surfaces a CO concentration of 1 ML is highly unfavourable due to adsorbate interactions. Yet several authors report FTS activity on such surfaces,[6,7] and indicate that long chains can grow.[8] This illustrates that a CO concentration close to unity is not essential for the growth of long chains.

1 C. J. Weststrate, J. van de Loosdrecht and J. W. Niemantsverdriet, *J. Catal.*, 2016, **342**, 1–16.
2 F. Falo, I. Cano and M. Salmerón, *Surf. Sci.*, 1984, **143**, 303–313.
3 R.L. Toomes and D.A. King, *Surf. Sci.*, 1996, **349**, 1–18.
4 J. P. Den Breejen, P. B. Radstake, G. L. Bezemer, J. H. Bitter, V. Frøseth, A. Holmen and K. P. de Jong, *J. Am. Chem. Soc.*, 2009, **131**, 7197–7203.
5 C.A. Mims, L.E. McCandlish, *J. Phys. Chem.*, 1987, **91**, 929–937
6 J.J.C. Geerlings, M.C. Zonnevylle and C.P.M. de Groot, *Surf. Sci.*, 1991, **241**, 315–324.
7 M. Ehrensperger and J. Wintterlin, *J. Catal.*, 2014, **319**, 274–282.
8 V. Navarro, M.A. van Spronsen, J.W.M. Frenken, *Nat. Chem.*, 2016, **8**, 929–934.

Michael Bowker remarked: We appear to have a disagreement between the single crystal work and the catalysts work with regard to coverages by CO, however, it seems to me that there is not so much of a problem. 'Curved' surfaces, such as on small nanoparticles, enable better separation of the O ends of adjacent CO molecules, which then reduces O–O repulsion and results in a higher strength of adsorption, all other effects ignored, at the same coverage in the high coverage regime. Furthermore, it means that the same repulsion energy (which reduces adsorption energy) only occurs at a higher coverage on curved nanoparticles. In a similar vein, steps act like curves surfaces, at least in the sense that they provide the facility for the staggering of CO-surface bond angles to minimise repulsion.

Matthew Neurock replied: The different coverages reported on particles and single crystal surfaces are the result of: 1) the different pressures used under UHV studies on single crystal surfaces and catalytic studies on supported particles, and 2) the differences in the surface structures of the single crystals and metal particles. Most of the surface science studies carried at UHV conditions (10^{-7} -10^{-8} torr) report coverages on Co(0001) and Ru(0001) surfaces of ~0.67 ML CO. A recent UHV study of CO adsorption on Ru(0001) showed that increasing the CO pressure from 10^{-8} torr to 0.1 torr increased the CO coverage from 0.67 ML to ~0.9 ML of CO.[1] It is important to note that Fischer–Tropsch is carried out at significantly higher pressures (> 1 atm). In addition to the changes in pressure, the repulsive interactions between neighboring COs on the particles is different than that on the surface. Surface curvature and edge sites present on the particles allow the CO to 'fan' outwards away from the particle and neighboring COs to reduce repulsive O–O interactions as you point out. A second factor that leads to higher coverages on the particles over those found on single crystal surfaces is the ability of the particle to expand radially to reduce the repulsion between the coadsorbed CO. The outer metal atoms relax outward and result in more 'curved' surfaces, even for the larger metal particles. This enables the COs to 'fan' away from one another and significantly reduce the repulsive O–O interactions. This is clearly seen in EXAFS experiments carried out at higher pressures of CO on different metals reported in the literature. Single crystal surfaces can only expand perpendicular to the surface at high CO coverages, which does not allow for a significant reduction in repulsion from O–O interactions. As such, single crystal surfaces reconstruct to relieve the strain that results from repulsive interactions that occur at higher coverage or limit surface coverages to 0.6–0.7 ML CO. The coverage and interactions of CO on edges and corner sites may be rather different as the coordinatively unsaturated metal sites can take up more than 1 CO to form di- and tri-carbonyls analogous to what is seen on single metal atom organometallic systems. Infrared spectral data taken at high CO pressures on supported Ru particles show clear evidence of di-carbonyl bands. This is supported by theoretical calculations, which report the presence of di-carbonyls and tri-carbonyls at edge and corner sites on Ru_{201} clusters.[2]

1 D.E. Starr and H. Bluhm, *Surf. Science*, 2013, **608**, 241–248.
2 B. T. Loveless, C. Buda, M. Neurock and E. Iglesia, *J. Am. Chem. Soc.*, 2013, **135**, 6107–6121.

Avelino Corma enquired: How many active sites do you need to continue the reaction and how many types of sites? From a kinetic point of view there are fewer species on the surfaces – they need to find themselves to react. Frequency factor of the reaction goes down a lot.

Emiel Hensen answered: Remembering our live discussion, the issue raised by Prof. Corma is interesting. Does a chain grow using CH_x species from one CO dissociation site or are more sites involved? This is of course very difficult to probe experimentally. It would require titrating the step-edge sites and then establishing the relation between chain-growth probability and step-edge density. In our previous work on aqueous phase FTS we have found indications that more step-edge sites does not lead to a different chain-growth probability so that it seems that one CO dissociation site is linked to one chain growth site. Of course, we should admit that we understand even less about aqueous phase FTS compared with conventional FTS.[1]

Microkinetics simulations cannot answer such question as we assume mean field. We expect that diffusion on the crowded surface will be rather slow so that it is reasonable to assume that CO dissociation and chain growth are closely linked at the surface. CO dissociation needs to be sufficiently fast, and chain growth needs to be fast in comparison with chain growth termination. Computationally, kinetic Monte Carlo could help to simulate the role of diffusion, which is of course exceedingly challenging for this particular reaction.

1 X.-Y. Quek, R. Pestman, R. A. van Santen and E. J. M. Hensen, *Catal. Sci. Technol.*, 2014, **4**, 3510–3523.

Pieter van Helden also replied: In our microkinetic work we did not assume a single site *a priori*, but tried to simulate the actual FCC Co particle. What we do see in our model is that most of the surface sites have a high coverage of CO, thus the effective number of active sites are much lower than the total number of a specific site. Allowing for interaction between sites, we see, arising from the simulation results, that at least two types of sites seem to be necessary for FT on FCC Co: one to facilitate the CO scission process and one where chain growth occurs. On closer inspection of our results, we also see another site being active for chain growth, but at an order of magnitude slower than the overall reaction rate. This site also becomes a store of spectator species. This gives an indication of the mechanistic complexity that underlies the FT reaction on actual nanoparticles.

Hans Schulz commented: Molecular modeling in Fischer–Tropsch synthesis is challenging because of extreme complexity. Advanced models are being developed, however further pertinent questions prevail. With cobalt as catalyst the steady state of reaction is only attained during a certain time, with changing rate and selectivity, indicating self-organization of the FT-regime (see Fig. 2, left-side).

(1) Increasing CO-conversion from 10 to 30%; (2) Chain growth probability (range C_3 – C_7), increasing from 60 to 90%, (3) Olefin content (in C_3), increasing from 30 to 50%, (4) Olefin-1 among n-olefins in C_4, increasing from 20 to 50%, (5) Chain branching probability (pbr4), decreasing from 10 to 4%.

As shown in Fig. 2, right-side, the typical pattern of chain branching probability over the carbon number (chain length) at steady state (8620 min on stream) exhibits an exponential decline with an increasing carbon number, as is thought to reflect increasing spatial constraints with increasing size of chains. The first value (at C_3) is exceptionally low, as assumed to be caused by a particular steric constraint to desorption, applying to the involved tertiary C-atom (for desorption of isobutene, respectively isobutane). Initially, *e.g.* at 100 min reaction time, branching probability is much higher than at steady state, indicating less spatial constraints on growth sites. Also, a different pattern is observed, indicating a further branching reaction, which increases with carbon number (from C_5 to higher C-numbers). This opposite C-number-dependence is thought to result from the carbon number increasing and olefin re-adsorption on growth sites being now possible with the carbon atom 2 because of lower spatial constraints prevailing.

As a further controlling principle, specific barriers against migration of intermediates between catalyst sites can be imagined.

1 H. Schulz, *Catal. Today*, 2014, **228**, 113–122.

Kees-Jan Weststrate replied: Model experiments and theory calculations on well-defined surfaces focus on the elementary surface reaction steps and micro-kinetics are used to explore the interplay between various elementary reaction steps to highlight and provide insights into the factors that determine the activity and selectivity. With respect to olefin selectivity, some studies exist where the effect of chain length was studied by theoretical means, see *e.g.* ref 1. Also, in our recent publication we report distinct reactivity differences between ethene and propene on a Co surface, which can contribute to the typical deviation of C_2 products from the typical selectivity pattern. I am unaware of theoretical studies

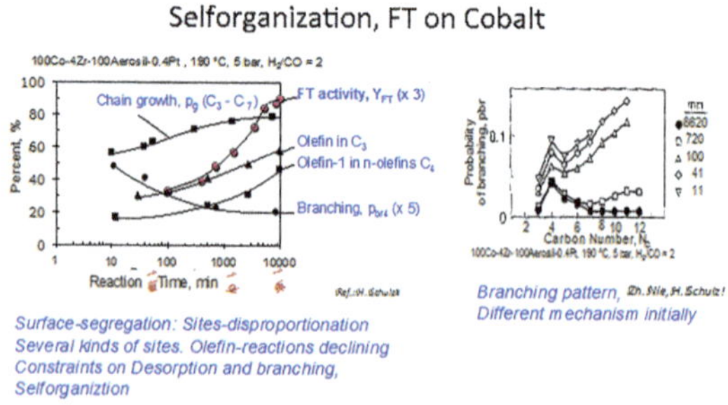

Fig. 2 Selforganization FT on cobalt.

that explore the mechanism of the formation of branched products on cobalt catalysts. In the initial phases, the catalyst is modified in several ways. One notable effect is the build-up of wax inside the pores, and on the surface of the catalyst particles as well[3], on a time scale of hours. Further relatively slow changes that occur on the time scale of days include carbon deposition and sintering, which would both affect the catalytic surface. The impact of these factors is typically not taken into account in the model studies at present. The high coverage of the surface under reaction conditions was identified as an important factor during the discussion. The influence of how lateral interactions, both steric and electronic, affect the stability of surface intermediates which translates to selectivity differences appears to be an important direction for future research.

1 J. Cheng, T. Song, P. Hu, C.M. Lok, P. Ellis and S. French, *J. Catal.*, 2008, **255**, 20–28.
2 C. J. Weststrate, I. M. Ciobîcă, J. van de Loosdrecht and J. W. Niemantsverdriet, *J. Phys. Chem. C*, 2016, **120**, 29210–29224.
3 V. Navarro, M. A. van Spronsen and J. W. M. Frenken, *Nat. Chem.*, 2016, **8**, 929–934.

Faraday Discussions

PAPER

Carbon induced selective regulation of cobalt-based Fischer–Tropsch catalysts by ethylene treatment†

Peng Zhai,‡[a] Pei-Pei Chen,‡[bc] Jinglin Xie,[a] Jin-Xun Liu,[b] Huabo Zhao,[a] Lili Lin,[a] Bo Zhao,[a] Hai-Yan Su,[b] Qingjun Zhu,[d] Wei-Xue Li*[bc] and Ding Ma*[a]

Received 19th September 2016, Accepted 10th October 2016
DOI: 10.1039/c6fd00194g

Various carbonaceous species were controllably deposited on Co/Al_2O_3 catalysts using ethylene as carbon source during the activation process for Fischer–Tropsch synthesis (FTS). Atomic, polymeric and graphitic carbon were distinguished by Raman spectroscopy, thermoanalysis and temperature programmed hydrogenation. Significant changes occurred in both the catalytic activity and selectivity toward hydrocarbon products after ethylene treatment. The activity decreased along with an increase in CH_4 selectivity, at the expense of a remarkable decrease of heavy hydrocarbon production, resulting in enhanced selectivity for the gasoline fraction. *In situ* XPS experiments show the possible electron transfer from cobalt to carbon and the blockage of metallic cobalt sites, which is responsible for the deactivation of the catalyst. DFT calculations reveal that the activation barrier (E_a) of methane formation decreases by 0.61 eV on the carbon-absorbed Co(111) surface, whereas the E_a of the CH + CH coupling reaction changes unnoticeably. Hydrogenation of CH_x to methane becomes the preferable route among the elementary reactions on the Co(111) surface, leading to dramatic changes in the product distribution. Detailed coke-induced deactivation mechanisms of Co-based catalysts during FTS are discussed.

[a]*Beijing National Laboratory for Molecular Sciences, College of Chemistry and Molecular Engineering, Peking University, Beijing 100871, China. E-mail: dma@pku.edu.cn*
[b]*State Key Laboratory of Catalysis, Dalian Institute of Chemical Physics, Chinese Academic of Sciences, Dalian 116023, China*
[c]*College of Chemistry and Material Sciences, iChEM (Collaborative Innovation Center of Chemistry for Energy Materials), CAS Center for Excellence in Nanoscience, Hefei National Laboratory for Physical Sciences at the Micro-Scale, University of Science and Technology of China, Hefei, China. E-mail: wxli70@ustc.edu.cn*
[d]*National Institute of Clean-And-Low-Carbon Energy, Future Science & Technology City, Changping District, Beijing 102209, China*

† Electronic supplementary information (ESI) available. See DOI: 10.1039/c6fd00194g

‡ These authors contributed equally to this work.

Introduction

The catalytic performance of a catalyst is determined by the structure of the catalyst. Therefore, efforts have been made in the last few decades to investigate the control and modulation of size, composition, phase, morphology, as well as the exposed facets of the active catalytic material, and therefore to tune the activity and the selectivity of the catalyst.[1] The construction of a stable catalyst with a designated surface structure, or so-called catalytic-active surface ensembles, is the ultimate goal in catalyst design. Indeed, different approaches have been used for the control of the surface structure. For example, by using a 2nd metal, it is possible to modulate the surface pattern of metal or metal oxide catalysts. The added 2nd metal can form an intermetallic compound,[2] an alloy,[3] a core–shell structure,[4] or an island-/atomically decorated structure[5] with the main metal components, which can alter the surface structure as well as the electronic structure of the catalyst, leading to the demonstration of superior catalytic performance in specific reactions. Besides the use of a 2nd metal, the surface of metal or metal oxide catalysts could be modified by relatively soft materials including surfactants,[6] protection agents[7] as well as carbonaceous species[8] on the surface of the catalyst. The carbonaceous species could be added during the preparation process of the catalyst, or be generated during the reaction process spontaneously. Normally, those carbonaceous species on the metal surface were treated as inactive species that block the active sites of the metal catalysts. However, it has been realized recently that they may have a major impact on the catalytic performance as well.[9] In this paper, we will use the carbon on a Co catalyst as an example to demonstrate the effect of this previously neglected species in heterogeneous catalysis.

Fischer–Tropsch synthesis is a catalytic process that transforms syngas, derived from coal, natural gas or biomass, to a wide spectrum of hydrocarbons involving diesel, gasoline and lower olefins. The design of efficient catalysts to produce desired products is one of the most important research topics. In contrast to the bulk iron-based catalysts used in industry, supported cobalt catalysts remain competitive because of their higher activity and selectivity towards long-chain hydrocarbons at low temperature.[10] Particle size,[11,12] crystallographic structure[13] and the interactions between cobalt and supports or promoters[14,15] are directly related to the catalytic properties in FTS.[16] It is reported that when the cobalt particle size is smaller than 6–8 nm, the activity as well as the selectivity for methane increase with the increase in cobalt particle size.[17] Moreover, hexagonal close packed (hcp) Co shows higher intrinsic activity than face centred cubic (fcc) Co in most cases,[18] and it was revealed by a first-principles kinetic study that the former prefers a direct CO dissociation route whereas H-assisted CO dissociation is a more plausible route on the fcc Co surface.[19] Since the metallic cobalt is identified as the active site,[20,21] one effective strategy to design better Co catalysts is to vary the pretreatment conditions to change the initial physico-chemical state of cobalt species before the reaction to regulate the selectivity of FTS.[22]

Although cobalt has excellent properties in liquid fuel production at low temperature, the high cost of cobalt and noble metal promoters is of considerable burden to industry. Therefore, improving the stability of the catalyst is crucial to ensure that the process is economically feasible. Various factors are recognized to cause a decrease in activity including particle sintering, the re-oxidation and carburization of cobalt, poisoning and carbon formation.[23,24] Carbon deposition occurs slowly in hydrogen-rich syngas, but it is a negligible problem under a hydrogen deficient atmosphere.[25] Both carbidic and polymeric carbon species were observed on Co/Al_2O_3 after extended runs; their accumulation blocked the active sites, resulting in a decrease in activity.[26] A long reaction time is required to observe a considerable amount of deposited carbon,[27] and the used catalysts are always contaminated by a large amount of heavy hydrocarbon products, which are difficult to be eliminated.[28] In order to investigate the role of carbon, some catalyst models have been established to simulate the process.[29,30] Weststrate *et al.* successfully constructed carbonaceous species on Co(0001), and found that the existence of atomic carbon weakens the adsorption of CO and H_2 on the cobalt surface and graphene completely blocks the active sites.[31] It has also been conjectured that the carbide reconstructed or graphene-covered Co(111) surface is the stable form during the reaction by density functional theory (DFT) calculations.[32] In contrast, carbon atoms are also suggested to induce the reconstruction of Co(0001), leading to the formation of step edge sites with low CO dissociation activation energy.[33] The correlations between carbonaceous species and catalytic performance in FTS are still under debate.

Ethylene is a favorable carbon source for graphene growth because of its low decomposition temperature.[34–36] It is interesting to note that cobalt is a feasible substrate for layered graphene deposition.[37] Carbon filament formation was observed on a reduced Co surface after being treated in a C_2H_4 and H_2 (1 : 1) atmosphere.[38] Therefore, ethylene was chosen to build up a deactivated Co/Al_2O_3 catalyst model in our research. In this report, we systematically investigated the detailed role of ethylene in the pre-treatment process, offering a new perspective for facilitating the reduction of cobalt. More importantly, coupled with various characterizations and DFT calculations, we attempt to reveal the deactivation mechanism of the cobalt catalyst in FTS.

Experimental

Preparation of Co/Al_2O_3 catalysts

The Co/Al_2O_3 catalysts were prepared by an impregnation method. Typically, 8.5 mmol of $Co(NO_3)_2\cdot 6H_2O$ (AR) was dissolved in 20 mL ethanol, then the solution was added dropwise into 5 g γ-Al_2O_3 powder (high surface area, Alfa Aesar). After being stirred and dried at 60 °C, the mixture was calcined at 400 °C for 5 hours in a muffle furnace with a heating ramp of 2 °C min^{-1}. The obtained black powder was denoted as Co/Al. Prior to FTS, the catalysts were reduced by hydrogen or pretreated in 10% C_2H_4/H_2 at 450 °C for 3 hours. A number of Co/Al_2O_3 catalysts were prepared using various conditions (Table 1) to investigate the ethylene treatment effect on cobalt catalysts. Samples for *ex situ* characterizations were passivated by 1% O_2/N_2 at room temperature.

Catalytic tests

The Fischer–Tropsch reaction was carried out in a stainless continuous flow fixed-bed reactor with a 12 mm inner diameter quartz tube inside. 0.2 g Co/Al_2O_3

Table 1 Detailed synthesis conditions of the Co/Al_2O_3 catalysts

Sample	Reduction condition	Treatment
Co–H	H_2	
Co–H–C1	H_2	C_2H_4 250 °C/1 h
Co–H–C3	H_2	C_2H_4 250 °C/3 h
Co–M	10% C_2H_4/H_2	
Co–M–H	10% C_2H_4/H_2	H_2 450 °C/3 h

catalyst sieved into 40–60 mesh was diluted with 1 g inert SiC. Prior to the catalytic test, the catalyst was *in situ* pretreated at atmospheric pressure with a gas flow rate of 20 mL min^{-1}. The pretreated conditions were identical to those used in the catalyst preparation. After cooling down below 100 °C, the system was switched to 3.0 MPa syngas with the composition of $H_2/CO/Ar = 64/32/4$ (molar ratio). The reaction was operated at 230 °C with a weight hourly space velocity (WHSV) of 6000 mL $g_{cat.}^{-1}$ h^{-1}.

The gas products were analyzed on-line by an Agilent 7890A GC. CO, H_2, CH_4, CO_2 and C_2H_6 were analyzed using a thermal conductivity detector (TCD) and the other gas products were analyzed through an alumina capillary column with a hydrogen flame ionization detector (FID). Liquid products including oxygenates collected from a cold trap were analyzed off-line by an Agilent 7820A GC. The selectivity of solid wax products was determined by weight. All the selectivity was determined on a carbon basis.

Characterization and computation method

Inductively coupled plasma atomic emission spectrometry (ICP) was performed to determine the cobalt content. X-ray diffraction (XRD) and transmission electron microscopy (TEM) were used to observe the morphology of cobalt. Thermogravimetry and differential scanning calorimetry (TG/DSC) and Raman spectroscopy were employed to analyze the carbonaceous species. Detailed information about the characterization is referred to in the ESI.†

The X-ray absorption spectroscopy spectra (XAFS) were acquired on a beamline BL14W1-XAFS at Shanghai Synchrotron Radiation Facility (SSRF). The storage ring was operated at 3.5 GeV and 300 mA using a Si(111) double-crystal monochromator. XAFS signals were measured in transmission mode at the Co absorption K-edge. Co_3O_4 powder grounded with hexagonal boron nitride was used as the reference material.

Temperature programmed hydrogenation (TPH) experiments were carried out in a quartz tube fixed bed micro-reactor. 50 mg of sample was hydrogenated in 5% H_2/N_2 (50 mL min^{-1}) ranging from 50 °C to 800 °C with a ramp rate of 10 °C min^{-1}. A Pfeiffer GSD320 mass spectrometer was used to monitor the CH_4 fragments as a function of temperature.

The X-ray photoelectron spectroscopy (XPS) measurements were conducted on an Axis Ultra Imaging Photoelectron Spectrometer equipped with an Al Kα (1486.7 eV) quartz monochrometer source. The sample wafers were pre-treated inside a treatment chamber with the reaction gas (H_2, C_2H_4 or 10% C_2H_4/H_2). The treatment conditions were identical to the conditions for the catalyst preparation. The chamber was connected to the vacuum system; thus the samples were transferred to the measurement chamber by a slide without exposure to air.

Periodic spin-polarized density functional theory (DFT) calculations were conducted to study the CO activation, methanation and C_{2+} formation on pure and carbon-decorated Co(111) surfaces (denoted as Co and C/Co surfaces). The top two Co layers and adsorbates were allowed to fully relax and the bottom two metal layers were fixed. The calculated lattice constant of bulk Co with a face-centered cubic structure is $a = b = c = 3.52$ Å, which is consistent with the experimental value of 3.55 Å.[39] Detailed information about the calculations is referred to in the ESI.†

Results and discussion

Structure characterization of cobalt

The XRD patterns of calcined Co/Al_2O_3 catalysts are displayed in Fig. 1. All of the diffraction peaks of Co/Al are indexed to the Co_3O_4 phase except for the γ-Al_2O_3 peaks. The crystallite size of Co_3O_4 is estimated by the Scherrer equation ($D = K\lambda/B\cos\theta$) to be 11.8 nm. Nevertheless, Co–H exhibits peaks at 2θ values of 44.22° and 51.52°, which correspond to fcc Co (JCPDS # 15-0806), after the reduction in pure H_2. In addition, a peak at a 2θ value of 42.40° assigned to CoO (JCPDS # 43-1004) is also observed, implying that Co_3O_4 is partially reduced under these conditions. Co–H–C1 and Co–H–C3 which are further treated with ethylene after reduction show similar patterns to Co–H. However, the characteristic peaks of metallic Co become more significant with the increase in ethylene treatment time, indicating that ethylene facilitates the reduction of cobalt oxide. Compared with Co–H, there is an apparent decrease of the metallic Co peak intensity on Co–M. Since no increase in CoO diffraction peak intensity is detected, the difference is attributed to the weakened degree of crystallinity of the cobalt particles.

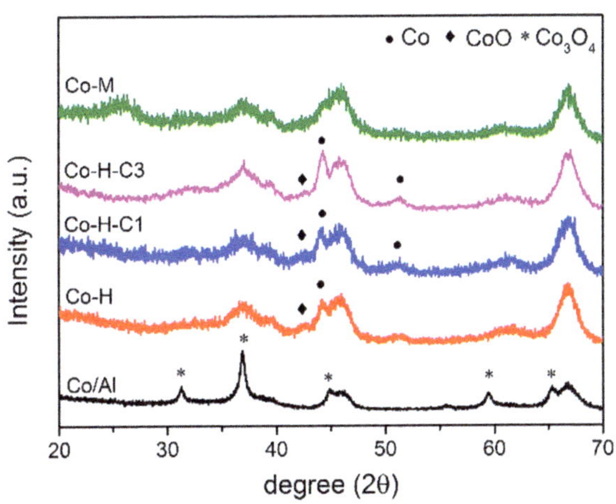

Fig. 1 XRD patterns of Co/Al_2O_3 catalysts.

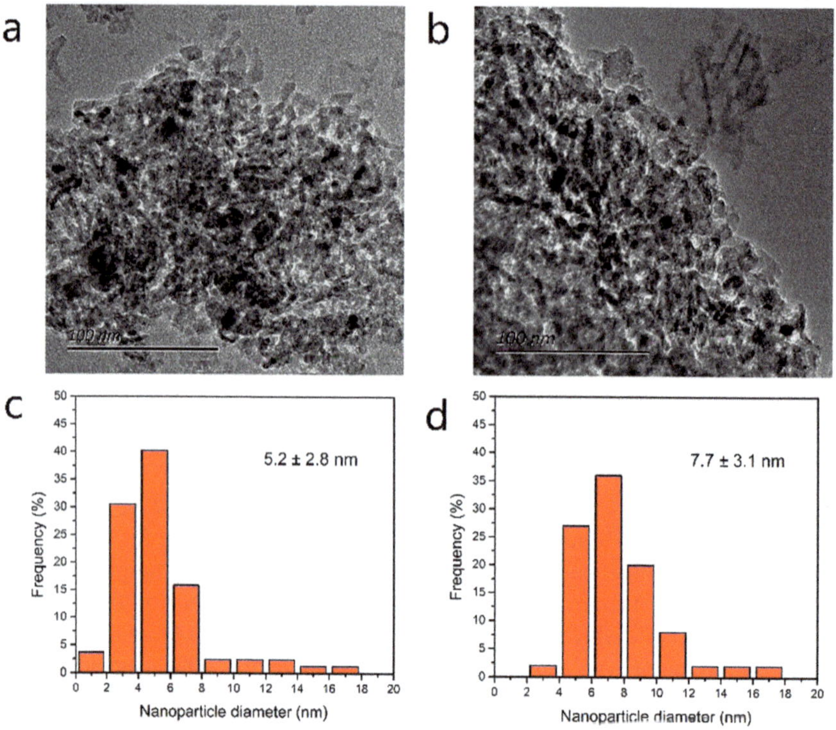

Fig. 2 TEM images and particle size distribution of Co/Al_2O_3 catalysts: (a and c) Co–H; (b and d) Co–M.

Fig. 2 presents TEM images of the Co/Al_2O_3 catalysts. It was found that Co/SiO_2 dried at 100 °C showed a smaller particle size and better aggregation-resisting properties compared to the catalyst dried at 175 °C,[40] and it was reported that ethoxyl groups that formed on the silica surface resulted in strong interactions between the cobalt and support.[41] In this work, the cobalt nitrate ethanol solution, instead of common aqueous solution, not only lowers the removal temperature of the solvent, but also hinders the sintering process of the precursor salt decomposition. Small and homogeneously dispersed cobalt nanoparticles are obtained consequently. In the case of Co–H, 85% of cobalt particles are in the range of 2–8 nm with an average diameter of 5.2 nm. Most of the particles are irregular spheres although several larger ones above 10 nm are also observed. When 10% C_2H_4/H_2 is used instead of pure H_2 during the pre-treatment process, the average particle size increases to 7.7 nm diameter. Moreover, the particle distribution becomes narrower and several pieces of graphitic carbon are identified in the micrographs.

The oxidation state of the cobalt species is investigated by XANES at the Co K-absorption edge (Fig. 3). In the spectra of the Co/Al_2O_3 catalysts, the entire peak shapes resemble the patterns between CoO and Co foils, indicating that the oxidation states of the three Co/Al_2O_3 catalysts are between Co(0) and Co(II), which is in accordance with the XRD patterns. The spectrum of Co–H is similar to the CoO reference, implying that this sample possesses the lowest reducibility of

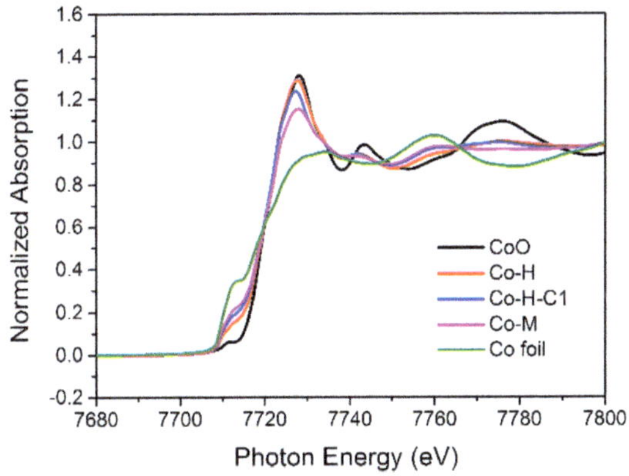

Fig. 3 Co K-edge XANES spectra of Co/Al_2O_3 catalysts.

cobalt out of all the catalysts. The degree of cobalt reduction, derived from the intensity of white lines at 7725 eV, shows a tendency as follows: Co–H < Co–H–C1 < Co–M. Table S1 and Fig. S1† exhibit the fitting results of the Fourier transform of the EXAFS data for the Co/Al_2O_3 catalysts. According to the coordination number of Co–Co which is a statistical average value for cobalt nanoparticles, Co–M has the largest cobalt particle size among the catalysts. The average diameter for Co–H is much smaller than for the samples treated by 10% C_2H_4/H_2, which is confirmed by TEM. Since the nucleation and growth of particles mainly occurs during calcination and reduction, the cobalt size of Co–H–C is similar to Co–H, which is verified by their similar Co–Co coordination numbers.

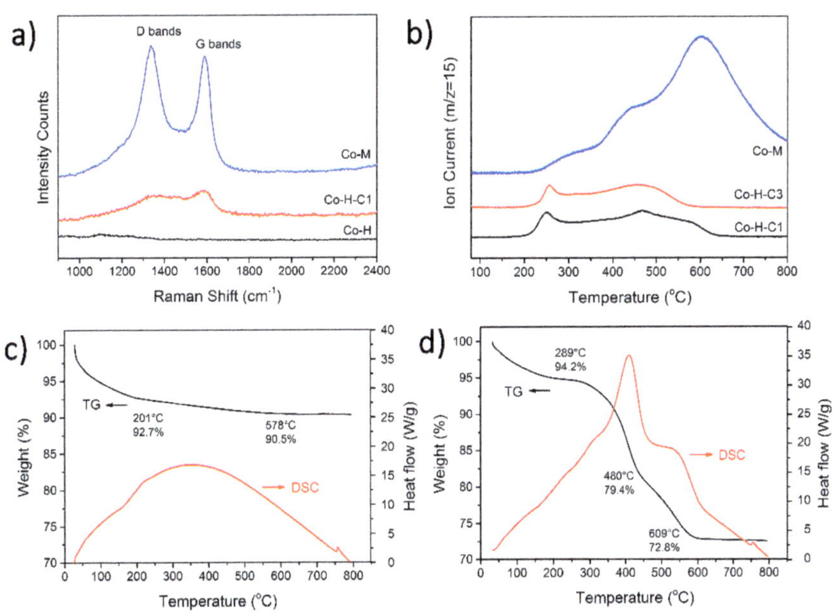

Fig. 4 Characterization of carbon on Co/Al_2O_3 catalysts: (a) Raman spectra, (b) TPH profiles, (c) TG and DSC behavior of Co–H–C1, (d) TG and DSC behavior of Co–M.

Characterization of carbon on catalysts

The Raman spectra of Co/Al_2O_3 catalysts are exhibited in Fig. 4a. No remarkable bands are visible for Co–H. Two relatively broad bands at about 1344 and 1591 cm^{-1}, which correspond to D (disordered) and G (graphitic) bands, are observed on Co–H–C1, pointing to the occurrence of carbon growth with poor crystallinity during the ethylene treatment. Moreover, the intensity of both the G-band and the D-band significantly increased on Co–M, indicating that the carbons are relatively well graphitized. TPH profiles are used to distinguish the different carbonaceous species because of their different reactivities toward H_2 (Fig. 4b).[26,42] The peak at 250 °C is assigned to the hydrogenation of atomic carbon species, indicating that carbide species may exist on the 250 °C ethylene carburized samples. Additionally, two broad peaks at around 450 °C and 600 °C on Co–M are ascribed to the hydrogenation of polymeric carbon and graphitic carbon species, respectively. It is generally recognized that the different positions of C 1s XPS peaks correspond to different forms of carbon species (Fig. S2†).[43,44] A slightly increased amount of carbon is observed on Co–H–C1, while the spectrum of Co–M shows typical characteristics of graphitic carbon.

The thermal behavior of the Co/Al_2O_3 catalysts was studied by TG-DSC measurements in an air atmosphere to elucidate the content and species of carbon (Fig. 4c and d). It is concluded that *ca.* 2.5% carbon is deposited (>200 °C) on Co–H–C1 during the carburization process. In comparison, the TG diagram presents two stages between 280 °C and 610 °C for Co–M, resulting in a 21.4% weight loss, which correspond to two forms of carbon. The first degradation stage with a 14% weight loss is assigned to the oxidation of amorphous carbon in the presence of cobalt, as evidenced by the heat release peak displayed on the DSC curve.[22] Pyrolysis of graphitic carbon seems to be a mild process in air, resulting in a second weight loss procedure above 450 °C.

Catalytic performance in Fischer–Tropsch synthesis

The Co/Al_2O_3 catalysts were investigated in FTS at 230 °C and 3.0 MPa (Fig. 5 and Table S2†); 28% CO conversion was achieved on Co–H at steady state. After being treated by C_2H_4, an obvious decrease in activity was observed on Co–H–C1 and

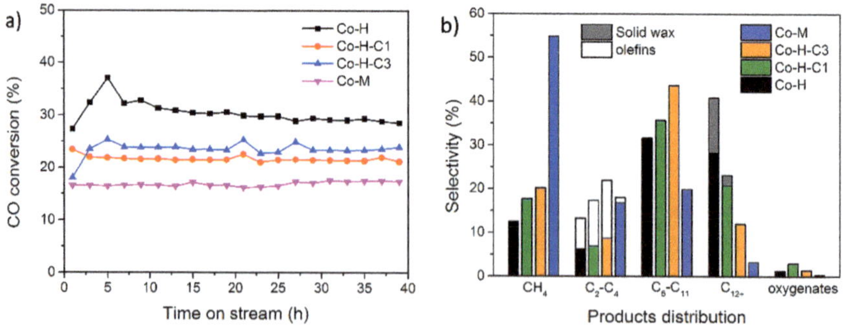

Fig. 5 The catalytic performance of Co/Al_2O_3 catalysts in FTS at 230 °C, 3.0 MPa, and a H_2/CO ratio of 2. (a) CO conversion as a function of time, (b) product distribution. The oxygenate component contains C1–C5 alcohols.

Co–H–C3. The lowest CO conversion (18% at steady state) was detected on Co–M, implying that there was a poisoning effect caused by the ethylene pre-reduction.

One of the notable characteristics that distinguish the catalysts is their different selectivities toward hydrocarbon products. As is shown in Fig. 5b, remarkable changes were observed after Co–H was treated with ethylene at 250 °C. The catalyst favored the formation of light fractions, and the tendency to form large molecular weight hydrocarbons was suppressed. The methane selectivity increased from 13% to 18% over the Co–H–C1 catalyst. Correspondingly, the selectivity in terms of C_{12+} products declined considerably from 41% to 23%. When the treatment time was prolonged to 3 hours, the production of hydrocarbons with a carbon number higher than 12 continuously decreased, the solid wax completely disappeared, and only 12% of diesel components remained. Interestingly, the gasoline fraction was enhanced from 31.6% to 43.7% on Co–H–C3 compared to Co–H, so the quality of the F–T products was improved by the ethylene treatment to some extent.

On the other hand, dramatic changes were observed on the 10% C_2H_4/H_2 treated sample. Co–M exhibited high selectivity toward methane (54.9%) and an extremely low C_{12+} production (3.3%). Furthermore, the production of olefins was almost completely inhibited which resulted in a significantly low olefin to paraffin (o/p) ratio (from 1.1 to 0.07). The enhanced CO_2 selectivity was related to the occurrence of the water–gas shift reaction (WGS, $CO + H_2O \rightarrow CO_2 + H_2$). It seems that the ethylene reduction atmosphere leads to a catalyst that prefers CH_4 formation, which is completely different from the diesel favoring catalyst. The continuously declining chain growing probability (α, from 0.82 to 0.71) clearly demonstrated that the chain growth ability was weakened by the ethylene treatment.

Discussion

We have investigated the influence of ethylene pre-treatment on the structure of the Co/Al_2O_3 catalysts, as well as the corresponding catalytic performance in FTS. The pure hydrogen reduced sample showed low methane selectivity and a strong capability of producing long-chain hydrocarbon products including solid wax. CO conversion decreased significantly and the CH_4 selectivity increased when ethylene participated in the activation process. The remarkable decrease of C_{12+} production indicates that the carbon chain propagation process is inhibited. A high o/p ratio is obtained on C_{5+} preferring catalysts, whereas a low olefin selectivity appears on the methane favoring catalyst. In general, the FTS performance of a cobalt-based catalyst depends on several factors: (1) the morphology and size of cobalt nanoparticles; (2) the phase of cobalt; (3) the degree of reduction; and (4) the surface state of the catalyst. Since CO hydrogenation is a structure-sensitive reaction, FTS performance is affected by the morphology and particle size of cobalt. It was reported that the FTS activity decreased with a decrease in the diameter of cobalt nanoparticles on carbon nanofibers, and CH_4 was the main product on particles smaller than 6 nm.[17,45] Our TEM images show that most of the cobalt particles are approximately irregular spheres between 5–10 nm in diameter. Although the particle diameter remained unchanged after being treated in ethylene at 250 °C, the CO conversion significantly declined. The small particles showed high selectivity toward C_{12+} products while the larger ones produced a large amount of CH_4. It seems that the size effect is not the key factor in this system.

On the other hand, it has been confirmed that the crystal phase of cobalt plays a crucial role in the activity of the catalyst. Hcp Co reduced from cobalt carbide showed an enhanced FTS activity compared with fcc Co which is usually obtained in a reduced cobalt catalyst.[46] The carburization of cobalt oxide was always induced by CO at low temperature, and light hydrocarbons including ethylene were also believed to be effective carburizers.[47] Cobalt carbide is metastable under reducing atmospheres even at 220 °C,[46] and DFT calculations suggested it should have a higher methane selectivity than metallic Co.[48] However, no experimental evidence demonstrates that hcp Co is beneficial for CH_4 formation. Since it is difficult to discriminate between cubic and hexagonal phases of cobalt in XAFS,[49] and no clear diffraction fringe assigned to hcp Co domains is observed in HR-TEM and XRD, the change in catalytic selectivity is tentatively ascribed to other factors.

It is well known that it is difficult for Co_3O_4 to be completely reduced to the metallic phase without any other auxiliary metals such as Pt, Ru.[50,51] Because it is generally believed that metallic Co is responsible for the activity of the CO hydrogenation reaction, the Co metal concentration of cobalt species should be directly related to the catalytic activity. Both XRD and XAFS data offer information about the bulk structure of the particles; therefore, it is necessary to investigate the Co metal concentration on the surface where the FTS reaction occurs. Co 2p XPS spectra were collected *via* an *in situ* technique without exposure to air (Fig. 6). The noticeably low ratio of the surface to bulk cobalt suggests that the cobalt particles are preferentially located in alumina pores (Table S3†). It is also inferred that the majority of carbon is deposited on the catalyst surface, as is evidenced by the remarkably high ratio of surface to bulk carbon.

For Co–H, the intense peak at 777.6 eV is generally identified to be characteristic of metallic cobalt, and the broad shoulder at higher energy (~5 eV) suggests the existence of unreduced CoO.[52] A considerable amount of metallic Co exists on the surface, which is different from that in the bulk. Co–H–C1 shows a dramatic decrease in the intensity of the metallic Co peak after the carburation process. Combined with the 17% percentage increase of carbon on the C 1s spectra, carbon deposition is the most plausible reason for this phenomenon.

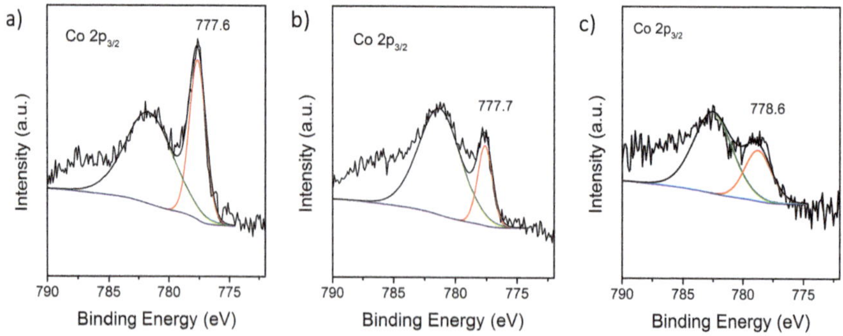

Fig. 6 Co $2p_{3/2}$ XPS spectra of Co/Al_2O_3 catalysts: (a) Co–H (b) Co–H–C1 (c) Co–M.

However, the relative peak of metallic Co in the Co–M spectrum shifts to 778.6 eV, indicating a change in the chemical environment of cobalt. The possible electron transfer from cobalt to graphitic carbon contributes to the apparent binding energy shift to high field.[53] In our previous work, we found that the electron-rich iron carbide prefers carbon chain growth and olefin production.[54] Correspondingly, the electron-deficient cobalt seems to facilitate the formation of low paraffin, and especially methane production in this work. Moreover, the remarkably weakened signal of Co–M spectrum demonstrates that the sample suffers severely from carbon deposition. As confirmed by Fig. 7, some cobalt nanoparticles in Co–M are surrounded by amorphous carbon with a low degree of graphitization, which decreases the cobalt concentration in the subsurface area. It is reported that carbon prefers to nucleate on step and defect sites of cobalt and then migrates to form large particles.[55] The growth of atomic carbon and carbon islands block partial active sites on Co–H–C, leading to a slight decrease in activity. When 10% C_2H_4/H_2 is used instead of H_2, the thicker graphitized carbon layers inhibit the reactant gas from coming into contact with the active sites, and thus a significantly decreased level of CO conversion is observed on Co–M.

To elucidate the influence of carbon on the selectivity of cobalt, we present a DFT study of methane formation and C–C coupling with or without pre-adsorbed carbon on fcc Co(111), which is observed in TEM and is suggested to be the most abundant facet exposed in many studies.[56–58] Since polymeric and graphitic carbons mainly cover and deactivate active sites,[59] atomically adsorbed carbons can be more relevant to tune the electronic structure of the substrate, which in turn change the relative activity of competing reaction pathways. Therefore, two models of pure and atomic carbon-decorated fcc Co(111)-(2 × 3) surfaces (denoted as Co and C/Co, respectively) are constructed. To simulate the carbon-decorated C/Co(111)-(2 × 3) surface, two atomic carbons (1/3 monolayer) were pre-adsorbed on the surfaces, as shown in Fig. 8.

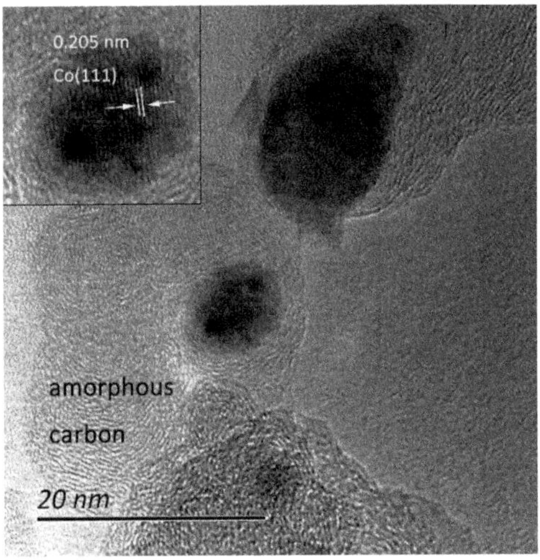

Fig. 7 HR-TEM picture of Co–M catalyst.

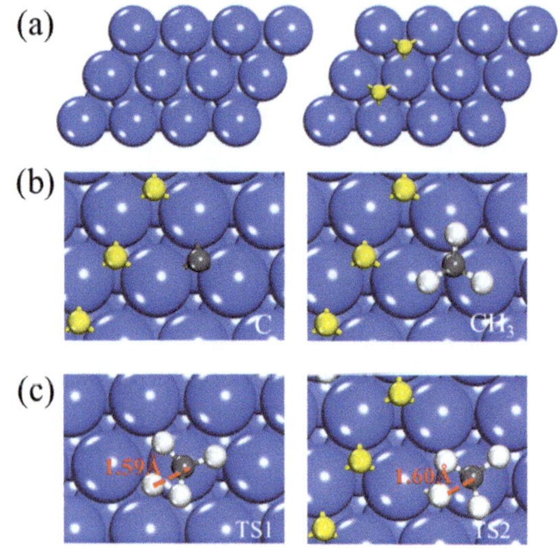

Fig. 8 (a) The schematic structures (top view) of Co(111) surfaces without (a) and with (b) atomic carbon in the $p(2 \times 3)$ unit cell. (b) Optimized structures of C and CH_3 species on C/Co surface (O has the same structure as C). (c) Transition states of CH_3 hydrogenation in Fig. 9. The blue and yellow balls represent Co and pre-adsorbed C atoms, respectively.

We first calculate the adsorption energies ΔE_{ads} of the reactive intermediates for methanation and C–C coupling on Co and C/Co, and the results are listed in Table S4.† In general, the C/Co surface binds the intermediates less strongly than the Co surface. C, O and CH_3 show the largest reduction in ΔE_{ads} by 0.46 eV, 0.59 eV and 0.52 eV, respectively, and the corresponding structures are shown in Fig. 8b. It can be seen that the intermediates share metal atoms with the pre-adsorbed C, leading to substrate-mediated repulsive interactions between adsorbates. In addition, CH_3 is tilted away from the pre-adsorbed C upon adsorption, which suggests that steric hindrance results in additional repulsion and a further decrease in ΔE_{ads}. The weakened intermediate binding not only facilitates the thermochemistry of methanation, as can be seen from the large reduction in the reaction heat ΔH by 1.03 eV for carbon hydrogenation, but also lowers the apparent activation energy by 0.61 eV, as seen in Fig. 9. These results indicate that the existence of carbon greatly accelerates methane formation on the cobalt surface, which is consistent with our experimental findings.

Various C–C couplings between CH_i ($i = 0, 1, 2,$ and 3) were considered to study the effect of the pre-adsorbed C, and corresponding formation energies of intermediates E_i with respect to the most favorable CH were calculated. The formation energy ($E_i + E_j$) of intermediates between CH_i and CH_j and the corresponding activation barrier E_a for subsequent CH_i–CH_j couplings are listed in Table 2. The effective barriers (E_{eff}) are then achieved by the summation of E_a and $E_i + E_j$, which allows a direct comparison of the relative activity of the C–C coupling steps.[60] The E_a for the C–C coupling steps are generally lower on C/Co than on Co, which indicates that the weaker binding of intermediates with the pre-adsorbed C leads to more facile bond-forming reactions. When it comes to the effective barrier (E_{eff}), C*-involved couplings have a relatively high E_{eff} on both

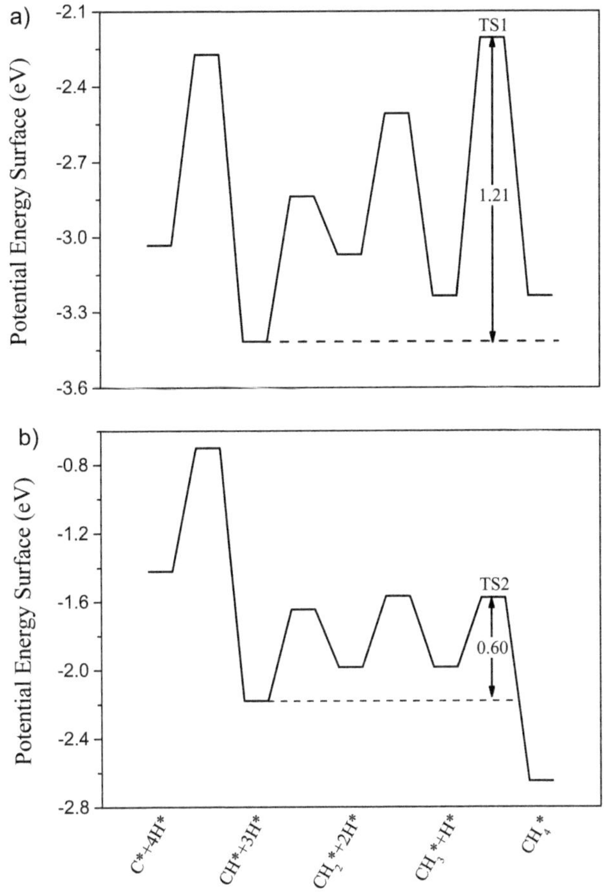

Fig. 9 The potential energy surfaces for CH_4 formation on Co (a) and C/Co (b), respectively. The apparent activation energies are denoted.

Table 2 The activation energies E_a, formation energies $E_i + E_j$ and E_{eff} for CH_i–CH_j ($i, j = 0, 1, 2, 3$) coupling on Co and C/Co

	Co			C/Co		
Reaction	E_a	$E_i + E_j$	E_{eff}	E_a	$E_i + E_j$	E_{eff}
C–C	1.09	0.76	1.85	0.87	1.52	2.40
C–CH	0.84	0.38	1.22	0.65	0.76	1.41
C–CH_2	0.73	0.73	1.46	0.47	0.96	1.43
C–CH_3	1.00	0.56	1.56	0.48	0.95	1.43
CH–CH	0.72	0.00	0.72	0.74	0.00	0.74
CH–CH_2	0.66	0.35	1.01	0.67	0.20	0.87
CH–CH_3	1.11	0.18	1.29	0.89	0.20	1.09
CH_2–CH_2	0.41	0.70	1.11	0.41	0.39	0.81
CH_2–CH_3	1.01	0.53	1.54	0.58	0.39	0.97

cobalt catalysts owing to their high formation energies, and thus this is less relevant for C–C coupling processes. However, for the remaining C–C coupling steps (Fig. S3†), the E_a and $E_i + E_j$ are basically lowered on C/Co, which dramatically decreases E_{eff}. Based on the E_{eff} calculations, the CH–CH coupling stands out with the lowest E_{eff} values of 0.72 and 0.74 eV on Co and C/Co, respectively. We have chosen this pathway as the most probable pathway for the C–C coupling process.

The comparison of the apparent barriers for methane formation and C–C coupling is shown in Fig. 10. Despite the similar apparent barriers of the C–C coupling process (0.72 eV *vs.* 0.74 eV), methane formation is more favorable on C/Co than on Co (0.60 *vs.* 1.21 eV), leading to greatly improved methane selectivity, which agrees well with our experimental studies.

The presence of carbonaceous species significantly changes the catalytic performance in FTS, leading to the deactivation of cobalt catalysts. To further investigate the regeneration properties of the cobalt catalyst under relevant conditions, Co–M was treated in H_2 at 450 °C for 3 hours to obtain Co–M–H. CO conversion on Co–M–H increased to 32% which was slightly higher than that on Co–H (Fig. 11). More importantly, the CH_4 selectivity decreased to 18% and the C_{12+} selectivity increased to 34%. A relative enhancement of the o/p ratio and inhibited WGS reactivity were also observed. It seems that the catalyst almost recovers the ability for chain growth and olefin desorption, which is very similar to Co–H. Although both the C 1s XPS spectrum and TG analysis demonstrated that only a small amount of carbon (11%) was eliminated during the hydrogenation treatment, metallic cobalt was exposed again on the surface of the catalyst, as revealed in the Co 2p XPS spectrum (Fig. S4†). Unlike the graphitic carbon deposited on Al_2O_3 which is resistant to hydrogen,[61] the carbon located on the cobalt is mostly eliminated, leading to recovered C–C coupling ability over Co–M–H. The carbon on Al_2O_3 is proposed to mainly act as a spectator during the FTS process (Fig. S6†).

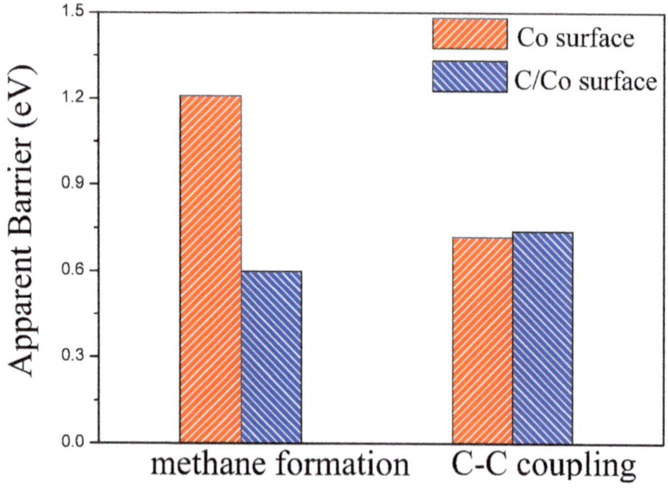

Fig. 10 Comparison of apparent barriers for methanation and C–C coupling. Red and blue columns represent Co and C/Co systems, respectively.

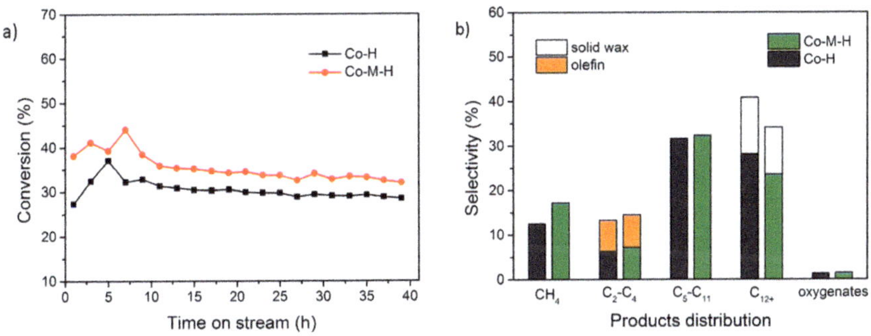

Fig. 11 Comparison of Co–M–H and CO–H catalytic performance in FTS at 230 °C, 3.0 MPa, and H_2/CO ratio of 2. (a) CO conversion as a function of time, (b) product distribution.

Conclusions

This report demonstrates that ethylene treatment has a remarkable influence on the structure of Co/Al_2O_3 catalysts through multiple aspects. Firstly, compared to the pure H_2 activated sample, the particle size tends to be larger in the 10% C_2H_4/H_2 treated sample. Secondly, although bulk cobalt carbide hardly exists under a hydrogen-rich environment, the possible carburization process contributes to enhance the reduction degree of cobalt. Finally, the different pretreatment conditions produce distinct carbon species on the surface of cobalt. Dispersed atomic carbon or polymeric carbons are generated at 250 °C in C_2H_4, and graphitic carbons are the predominant species on Co–M that are activated in 10% C_2H_4/H_2. Accordingly, the deposited carbons on the Co catalysts regulate the FTS selectivity by enhancing the low paraffin selectivity at the expense of decreasing C_{12+} production, especially the solid wax. The blockage of active sites either by atomic or graphitized carbon is responsible for the decreased catalyst activity. Nevertheless, the catalytic performance can be almost recovered by the hydrogenation of carbon atoms on cobalt sites at 450 °C. The treatment of ethylene is an efficient method to regulate the structure of cobalt nanoparticles for their catalytic activities in FTS and the coke deposition plays a dramatic role in terms of the product selectivity. Further DFT calculations suggest that carbon atoms on Co(111) not only increase the CO dissociation barrier but also change the activation energy of elementary reactions on the cobalt surface. Methane formation is preferred over C–C coupling on the carbon-adsorbed Co(111) surface, and the electron-deficient cobalt facilitates the hydrogenation of intermediates to produce low carbon number paraffins.

Statement of contribution

Ding Ma and Qingjun Zhu designed the study. Peng Zhai synthesized and evaluated the catalysts and participated in most of the characterization experiments and data analysis for the catalysts. Jinglin Xie did the in-situ XPS experiments. Huabo Zhao carried out the XAFS experiment. Lili Lin carried out the TEM observation and Bo Zhao did the TG analysis. Pei-Pei Chen did most of the DFT calculations and Jin-Xun Liu constructed the model and discussed the results. Hai-Yan Su and Wei-Xue Li carried out analysis and discussed the DFT results.

Acknowledgements

This work was supported by the Natural Science Foundation of China (21473003, 21222306, 21473229, 91545121, 21273224, 21321002, 21225315), and the 973 Project of China (2013CB933100, 2013CB834603). We thank the SSRF for providing the beamline and Siyu Yao for data discussion.

Notes and references

1 K. Zhou and Y. Li, *Angew. Chem., Int. Ed.*, 2012, **51**, 602–613.
2 B. Y. Xia, H. B. Wu, X. Wang and X. W. Lou, *J. Am. Chem. Soc.*, 2012, **134**, 13934–13937.
3 F. Besenbacher, I. Chorkendorff, B. S. Clausen, B. Hammer, A. M. Molenbroek, J. K. Nørskov and I. Stensgaard, *Science*, 1998, **279**, 1913–1915.
4 M. B. Gawande, A. Goswami, T. Asefa, H. Guo, A. V. Biradar, D. L. Peng, R. Zboril and R. S. Varma, *Chem. Soc. Rev.*, 2015, **44**, 7540–7590.
5 B. Qiao, A. Wang, X. Yang, L. F. Allard, Z. Jiang, Y. Cui, J. Liu, J. Li and T. Zhang, *Nat. Chem.*, 2011, **3**, 634–641.
6 N. R. Jana, Z. L. Wang and T. Pal, *Langmuir*, 2000, **16**, 2457–2463.
7 M. I. Redondo and C. B. Breslin, *Corros. Sci.*, 2007, **49**, 1765–1776.
8 S. S. Chen, L. Brown, M. Levendorf, W. W. Cai, S. Y. Ju, J. Edgeworth, X. S. Li, C. W. Magnuson, A. Velamakanni, R. D. Piner, J. Y. Kang, J. Park and R. S. Ruoff, *ACS Nano*, 2011, **5**, 1321–1327.
9 C. Wang, P. Zhai, Z. C. Zhang, Y. Zhou, J. K. Zhang, H. Zhang, Z. J. Shi, R. P. S. Han, F. Q. Huang and D. Ma, *J. Catal.*, 2016, **334**, 42–51.
10 A. Y. Khodakov, W. Chu and P. Fongarland, *Chem. Rev.*, 2007, **107**, 1692–1744.
11 G. Prieto, A. Martinez, P. Concepcion and R. Moreno-Tost, *J. Catal.*, 2009, **266**, 129–144.
12 A. Tuxen, S. Carenco, M. Chintapali, C. H. Chuang, C. Escudero, E. Pach, P. Jiang, F. Borondics, B. Beberwyck, A. P. Alivisatos, G. Thornton, W. F. Pong, J. H. Guo, R. Perez, F. Besenbacher and M. Salmeron, *J. Am. Chem. Soc.*, 2013, **135**, 2273–2278.
13 D. I. Enache, B. Rebours, M. Roy-Auberger and R. Revel, *J. Catal.*, 2002, **205**, 346–353.
14 O. Borg, P. D. C. Dietzel, A. I. Spjelkavik, E. Z. Tveten, J. C. Walmsley, S. Diplas, S. Eri, A. Holmen and E. Ryttera, *J. Catal.*, 2008, **259**, 161–164.
15 H. Wang, W. Zhou, J. X. Liu, R. Si, G. Sun, M. Q. Zhong, H. Y. Su, H. B. Zhao, J. A. Rodriguez, S. J. Pennycook, J. C. Idrobo, W. X. Li, Y. Kou and D. Ma, *J. Am. Chem. Soc.*, 2013, **135**, 4149–4158.
16 P. Zhai, G. Sun, Q. J. Zhu and D. Ma, *Nanotechnol. Rev.*, 2013, **2**, 547–576.
17 G. L. Bezemer, J. H. Bitter, H. P. C. E. Kuipers, H. Oosterbeek, J. E. Holewijn, X. D. Xu, F. Kapteijn, A. J. van Dillen and K. P. de Jong, *J. Am. Chem. Soc.*, 2006, **128**, 3956–3964.
18 M. K. Gnanamani, G. Jacobs, W. D. Shafer and B. H. Davis, *Catal. Today*, 2013, **215**, 13–17.

19 J. X. Liu, H. Y. Su, D. P. Sun, B. Y. Zhang and W. X. Li, *J. Am. Chem. Soc.*, 2013, **135**, 16284–16287.

20 K. H. Cats, I. D. Gonzalez-Jimenez, Y. Liu, J. Nelson, D. van Campen, F. Meirer, A. M. van der Eerden, F. M. de Groot, J. C. Andrews and B. M. Weckhuysen, *Chem. Commun.*, 2013, **49**, 4622–4624.

21 N. E. Tsakoumis, R. Dehghan, R. E. Johnsen, A. Voronov, W. van Beek, J. C. Walmsley, Ø. Borg, E. Rytter, D. Chen, M. Rønning and A. Holmen, *Catal. Today*, 2013, **205**, 86–93.

22 L. Shi, Y. Z. Jin, C. Xing, C. Y. Zeng, T. Kawabata, K. Imai, K. Matsuda, Y. S. Tan and N. Tsubaki, *Appl. Catal., A*, 2012, **435**, 217–224.

23 N. E. Tsakoumis, M. Ronning, O. Borg, E. Rytter and A. Holmen, *Catal. Today*, 2010, **154**, 162–182.

24 A. M. Saib, D. J. Moodley, I. M. Ciobica, M. M. Hauman, B. H. Sigwebela, C. J. Weststrate, J. W. Niemantsverdriet and J. van de Loosdrecht, *Catal. Today*, 2010, **154**, 271–282.

25 P. G. Menon, *J. Mol. Catal.*, 1990, **59**, 207–220.

26 K. F. Tan, J. Xu, J. Chang, A. Borgna and M. Saeys, *J. Catal.*, 2010, **274**, 121–129.

27 D. J. Moodley, J. van de Loosdrecht, A. M. Saib, M. J. Overett, A. K. Datye and J. W. Niemantsverdriet, *Appl. Catal., A*, 2009, **354**, 102–110.

28 D. Pena, A. Griboval-Constant, V. Lecocq, F. Diehl and A. Y. Khodakov, *Catal. Today*, 2013, **215**, 43–51.

29 M. C. Zonnevylle, J. J. C. Geerlings and R. A. Vansanten, *Surf. Sci.*, 1990, **240**, 253–262.

30 J. C. W. Swart, E. van Steen, I. M. Ciobica and R. A. van Santenc, *Phys. Chem. Chem. Phys.*, 2009, **11**, 803–807.

31 C. J. Weststrate, A. C. Kizilkaya, E. T. R. Rossen, M. W. G. M. Verhoeven, I. M. Ciobica, A. M. Saib and J. W. Niemantsverdriet, *J. Phys. Chem. C*, 2012, **116**, 11575–11583.

32 M. C. Valero and P. Raybaud, *J. Phys. Chem. C*, 2014, **118**, 22479–22490.

33 X. Q. Zhang, R. A. van Santen and E. J. M. Hensen, *ACS Catal.*, 2015, **5**, 596–601.

34 H. Zhang, Q. Fu, Y. Cui, D. L. Tan and X. H. Bao, *J. Phys. Chem. C*, 2009, **113**, 8296–8301.

35 J. Lahiri, T. Miller, L. Adamska, I. I. Oleynik and M. Batzill, *Nano Lett.*, 2011, **11**, 518–522.

36 S. Gunther, S. Danhardt, M. Ehrensperger, P. Zeller, S. Schmitt and J. Wintterlin, *ACS Nano*, 2013, **7**, 154–164.

37 H. Ago, Y. Ito, N. Mizuta, K. Yoshida, B. Hu, C. M. Orofeo, M. Tsuji, K. Ikeda and S. Mizuno, *ACS Nano*, 2010, **4**, 7407–7414.

38 M. S. Kim, N. M. Rodriguez and R. T. K. Baker, *J. Catal.*, 1993, **143**, 449–463.

39 N. W. Ashcroft and N. D. Mermin, *Solid State Physics*, Holt, Rinehart and Winston, New York, 1976, p. 70.

40 P. Munnik, P. E. de Jongh and K. P. de Jong, *J. Am. Chem. Soc.*, 2014, **136**, 7333–7340.

41 S. W. Ho and Y. S. Su, *J. Catal.*, 1997, **168**, 51–59.

42 J. Xu and C. R. Bartholomew, *J. Phys. Chem. B*, 2005, **109**, 2392–2403.

43 J. M. Xiong, Y. J. Ding, T. Wang, L. Yan, W. M. Chen, H. J. Zhu and Y. Lu, *Catal. Lett.*, 2005, **102**, 265–269.

44 Y. J. Gao, D. Ma, G. Hu, P. Zhai, X. H. Bao, B. Zhu, B. S. Zhang and D. S. Su, *Angew. Chem., Int. Ed.*, 2011, **50**, 10236–10240.

45 J. P. den Breejen, P. B. Radstake, G. L. Bezemer, J. H. Bitter, V. Froseth, A. Holmen and K. P. de Jong, *J. Am. Chem. Soc.*, 2009, **131**, 7197–7203.

46 G. Kwak, M. H. Woo, S. C. Kang, H. G. Park, Y. J. Lee, K. W. Jun and K. S. Ha, *J. Catal.*, 2013, **307**, 27–36.

47 H. Karaca, O. V. Safonova, S. Chambrey, P. Fongarland, P. Roussel, A. Griboval-Constant, M. Lacroix and A. Y. Khodakov, *J. Catal.*, 2011, **277**, 14–26.

48 J. Cheng, P. Hu, P. Ellis, S. French, G. Kelly and C. M. Lok, *J. Phys. Chem. C*, 2010, **114**, 1085–1093.

49 J. C. Mohandas, M. K. Gnanamani, G. Jacobs, W. P. Ma, Y. Y. Ji, S. Khalid and B. H. Davis, *ACS Catal.*, 2011, **1**, 1581–1588.

50 N. Kumar, M. L. Smith and J. J. Spivey, *J. Catal.*, 2012, **289**, 218–226.

51 J. Y. Park, Y. J. Lee, P. R. Karandikar, K. W. Jun, J. W. Bae and K. S. Ha, *J. Mol. Catal. A: Chem.*, 2011, **344**, 153–160.

52 C. R. Brundle, T. J. Chuang and D. W. Rice, *Surf. Sci.*, 1976, **60**, 286–300.

53 J. Deng, P. J. Ren, D. H. Deng, L. Yu, F. Yang and X. H. Bao, *Energy Environ. Sci.*, 2014, **7**, 1919–1923.

54 P. Zhai, C. Xu, R. Gao, X. Liu, M. Li, W. Li, X. Fu, C. Jia, J. Xie, M. Zhao, X. Wang, Y. W. Li, Q. Zhang, X. D. Wen and D. Ma, *Angew. Chem., Int. Ed.*, 2016, **55**, 9902–9907.

55 H. S. Bengaard, J. K. Norskov, J. Sehested, B. S. Clausen, L. P. Nielsen, A. M. Molenbroek and J. R. Rostrup-Nielsen, *J. Catal.*, 2002, **209**, 365–384.

56 J. C. W. Swart, L. M. Ciobica, R. A. van Santen and E. van Steen, *J. Phys. Chem. C*, 2008, **112**, 12899–12904.

57 C. B. Chen, Q. Wang, G. R. Wang, B. Hou, L. T. Jia and D. B. Li, *J. Phys. Chem. C*, 2016, **120**, 9132–9147.

58 B. H. Li, Q. J. Zhang, L. A. Chen, P. Cui and X. Q. Pan, *Phys. Chem. Chem. Phys.*, 2010, **12**, 7848–7855.

59 C. Klink, I. Stensgaard, F. Besenbacher and E. Lagsgaard, *Surf. Sci.*, 1995, **342**, 250–260.

60 J. Cheng, P. Hu, P. Ellis, S. French, G. Kelly and C. M. Lok, *J. Phys. Chem. C*, 2009, **113**, 8858–8863.

61 I. Czekaj, F. Loviat, F. Raimondi, J. Wambach, S. Biollaz and A. Wokaun, *Appl. Catal., A*, 2007, **329**, 68–78.

Faraday Discussions

PAPER

Structural and elemental influence from various MOFs on the performance of Fe@C catalysts for Fischer–Tropsch synthesis†

Tim A. Wezendonk,[a] Quirinus S. E. Warringa,[a] Vera P. Santos,[b] Adam Chojecki,[b] Matthijs Ruitenbeek,[c] Garry Meima,[c] Michiel Makkee,[a] Freek Kapteijn*[a] and Jorge Gascon*[a]

Received 28th September 2016, Accepted 17th October 2016
DOI: 10.1039/c6fd00198j

The structure and elementary composition of various commercial Fe-based MOFs used as precursors for Fischer–Tropsch synthesis (FTS) catalysts have a large influence on the high-temperature FTS activity and selectivity of the resulting Fe on carbon composites. The selected Fe-MOF topologies (MIL-68, MIL-88A, MIL-100, MIL-101, MIL-127, and Fe-BTC) differ from each other in terms of porosity, surface area, Fe and heteroatom content, crystal density and thermal stability. They are re-engineered towards FTS catalysts by means of simple pyrolysis at 500 °C under a N_2 atmosphere and afterwards characterized in terms of porosity, crystallite phase, bulk and surface Fe content, Fe nanoparticle size and oxidation state. We discovered that the Fe loading (36–46 wt%) and nanoparticle size (3.6–6.8 nm) of the obtained catalysts are directly related to the elementary composition and porosity of the initial MOFs. Furthermore, the carbonization leads to similar surface areas for the C matrix (S_{BET} between 570 and 670 $m^2\ g^{-1}$), whereas the pore width distribution is completely different for the various MOFs. The high catalytic performance (FTY in the range of $1.9–4.6 \times 10^{-4}$ $mol_{CO}\ g_{Fe}^{-1}\ s^{-1}$) of the resulting materials could be correlated to the Fe particle size and corresponding surface area, and only minor deactivation was found for the N-containing catalysts. Elemental analysis of the catalysts containing deliberately added promoters and inherent impurities from the commercial MOFs revealed the subtle interplay between Fe particle size and complex catalyst composition in order to obtain high activity and stability next to a low CH_4 selectivity.

Introduction

Fischer–Tropsch synthesis (FTS) is a process that catalytically converts syngas, a mixture of CO and H_2, towards longer chain hydrocarbon products such as

[a]Catalysis Engineering, ChemE, Delft University of Technology, Van der Maasweg 9, 2629 HZ Delft, The Netherlands. E-mail: f.kapteijn@tudelft.nl; j.gascon@tudelft.nl

[b]Core R&D, Dow Benelux B.V., P.O. Box 48, 4530 AA, Terneuzen, The Netherlands

[c]Hydrocarbons R&D, Dow Benelux B.V., P.O. Box 48, 4530 AA, Terneuzen, The Netherlands

† Electronic supplementary information (ESI) available: Additional MOF identification, Fe@C characterization, FTS results and spent catalyst results. See DOI: 10.1039/c6fd00198j

gasoline, diesel and plastics precursors. The FTS process was designed nearly a century ago to cope with the limitations in petroleum feedstock, allowing the production of fuels and chemicals from coal, natural gas and biomass.[1,2] Currently, several large plants worldwide operate the FTS process through syngas obtained from steam methane reforming or coal gasification.[3–6] Fast-growing worldwide energy consumption and conventional oil reserve challenges provide a continuous drive to develop improved catalysts for the FTS process.

In the last few years, metal–organic frameworks (MOFs) have emerged from the deep waters of scientific research and have been increasingly proposed as heterogeneous catalysts for many applications, such as Knoevenagel condensation, "click" chemistry, selective hydrogenations, Friedländer coupling, electrocatalysis and photocatalysis.[7] However, we have shown recently that MOFs are not only potent catalysts in their final form, but they also serve a purpose as excellent precursors for high-temperature FTS (HTFTS) catalysts.[8] The concept of the synthesis technique is a simple pyrolysis, heat treatment in an inert atmosphere, which carbonizes the MOF towards embedded metal nanoparticles in a porous C matrix (M@C). The resulting nanoparticle–carbon composites possess outstanding properties for applications in electrodes, electrocatalysts and H_2/CO_2 adsorbents;[9–13] these materials among many others have been comprehensively reviewed.[14] Recent investigations into MOF-derived catalysts for FTS by Wang's group confirmed the high potential of the Fe@C system using the MIL-88B topology as catalyst precursor.[15] Recently, we reported an in-depth study of the nature of the Fe species during pyrolysis and the subsequent FTS reaction, elucidating the carbonization mechanism of the commercial Fe-BTC MOF with the MIL-100 topology.[16] *In situ* DRIFTS-MS and XAFS applied during the pyrolysis process revealed the decarboxylation of the framework and auto-reduction of the Fe^{3+} species towards active Fe carbide. *In situ* Mössbauer emission spectroscopy and HR-TEM allowed for the correlation of the pyrolysis temperature and the Fe nanoparticle size and showed that the evolution of the Hägg carbide phase was dependent on the particle size as the FTS reaction progressed.

Herein, we determine how the MOF topology is related to activity and selectivity in HT-FTS in order to optimize the Fe@C catalyst. By systematically varying the structure and elemental composition prior to pyrolysis, we gain understanding in the transformation of the framework towards an active Fe@C catalyst. The commercially available Fe-MOFs selected in this study (MIL-68, MIL-88A, MIL-100, MIL-127, MIL-101-NH_2 and Fe-BTC) comprise of carboxylate linkers that

Table 1 Fe-MOF properties taken from reported literature studies,[21–35] comprising BET area (S_{BET}), Fe loading (w_{Fe}), calculated crystal density (ρ_{crys}), pore diameter (d_{pore}) and decomposition temperature in air (T_d)

Fe-based MOF	S_{BET}, $m^2\ g^{-1}$	w_{Fe}, %	ρ_{crys}, g cm^{-3}	d_{pore}, Å	T_d, °C
MIL-68	355	23.6	0.87	6–16	290
MIL-88A	40	26.2	1.55	6	195
MIL-100	2150	25.8	0.69	25–29	300
MIL-101-NH_2	4100	22.5	0.62	29–34	320
MIL-127	1400	21.3	0.97	6–10	360
F300 Fe-BTC	1040	25	na	22	295

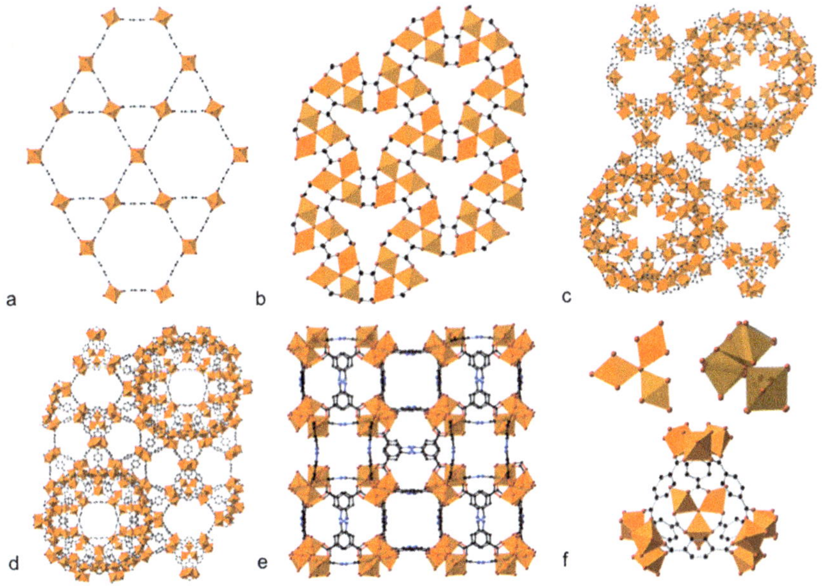

Fig. 1 Selected Fe-MOF topologies: (a) MIL-68, (b) MIL-88A, (c) MIL-100, (d) MIL-101, (e) MIL-127 and (f) the configuration of Fe trimers and super tetrahedra present in the different structures.

differ both in size and connectivity, in addition to heteroatoms such as N. Therefore, the MOFs show large differences in porosity, thermal stability and heteroatom content, which is summarized in Table 1 and Fig. 1. The size of the biggest voids within these structures ranges from 6 to 34 Å with corresponding BET areas between 40 and 4100 $m^2 g^{-1}$. The thermal stability in air, depending on the strength of the metal–ligand interaction, varies between 200–360 °C. We deliberately assess the use of commercial MOFs as a route towards industrial catalyst manufacturing. Several MOF producers exist nowadays (*e.g.* MOF Technologies, NuMat Technologies, theMOFcompany, all spin-off companies from academia[17–20]), each using different manufacturing technologies, bearing an aspect that is to be investigated. Prices of MOFs are expected to come down and more focus on industrial scale production increases the chances of the industrial implementation of MOFs.

Results

Characterization of MOF precursors

N_2 physisorption at 77 K, TGA and XRD confirmed the presence of MIL-68,[21–23] MIL-88A,[24–26] MIL-100,[27–29] MIL-101,[30–32] and MIL-127 (ref. 33–35) (see Table S1, Fig. S1 and S2†).‡ XRD of the F300 Fe-BTC MOF revealed a very low degree of crystallinity, but the MIL-100 topology was confirmed since the isotherm contained an additional step in the micropore regime corresponding to the large

‡ Limited data was available for MIL-101-NH_2 due to practical issues between characterization, pyrolysis and catalytic testing.

cages in the MTN-type framework.[27,28] This behaviour was more pronounced in the MIL-100 sample, and values for porosity exceeded the F300 as well. As reported, MIL-88A heavily contracts upon degassing and thus, a near-zero N_2 uptake was found.[23,24] In general, no significant deviation from literature values was found.

Characterization of Fe@C catalysts

Texture of C matrix and crystalline Fe phases. N_2 physisorption at 77 K of the Fe@C materials showed highly microporous catalysts with significant contributions of mesopores and external surface area (Table S2† and Fig. 2a). To estimate the porosity of the C matrix, these values were normalized by the C content correcting for the Fe mass assuming a Fe_2O_3 phase (Table 2). The resulting BET areas are closely aligned, with the lowest value of 570 m^2 g^{-1} for the low-crystalline starting MOF Fe-BTC. It is clear that the coordinated solvent molecules located in

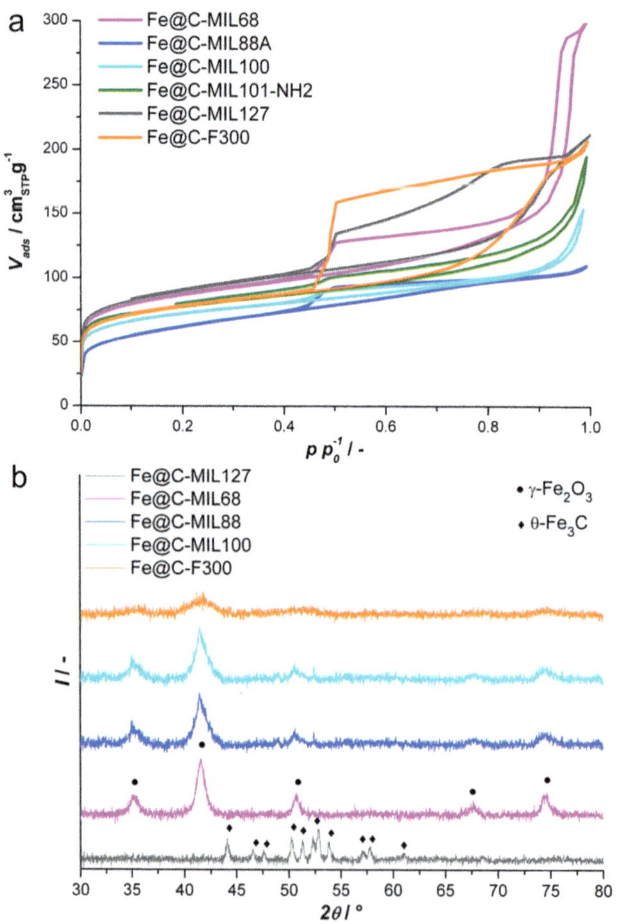

Fig. 2 (a) Nitrogen uptake at 77 K per gram of Fe@C catalyst and (b) XRD of the Fe@C catalysts, indicating the presence of nanosized γ-Fe_2O_3 and θ-Fe_3C phases present after passivation.

the intra-framework sites (MIL-68, MIL-88A) are released during the carbonization, liberating the porosity of the MOF. In the specific case of MIL-88A, which is not porous prior to pyrolysis, the pore volume is increased to 0.47 $cm^3 g^{-1}$ of the C matrix with a corresponding BET area of 655 $m^2 g^{-1}$. This is an impressive result, especially keeping in mind that these values are one third of carbonized ZIF-8 at 800 °C.[10] Although values derived from the BJH pore size distribution have to be taken as merely qualitative, clear trends can be obtained from this analysis (Fig. S3†). The MIL-68 derived material, Fe@C-MIL-68, contains most of its pore volume in large meso-macropores of around 50 nm in size, whereas for Fe@C-MIL-100 and -127, well-defined mesopores exist in a narrow distribution between 10–20 nm in addition to the micropores. XRD results demonstrated that nanocrystalline Fe oxides (γ-Fe_2O_3) were formed for Fe@C-MIL-68, -88, -100 and -F300 after passivation of the material (Fig. 2b). However, the diffraction pattern of Fe@C-MIL-127 revealed the presence of a single Fe carbide phase (cementite), without Fe oxide phases present. MIL-127 possesses the lowest amount of Fe, although the C/Fe ratio is similar to that of MIL-68 and MIL-101-NH_2, suggesting that either the N functionality or the high decomposition temperature play a role in the carburization of Fe during the pyrolysis.

Bulk and surface Fe loading. TGA and ICP analyses were performed to determine the bulk Fe loading of the pyrolyzed MOFs, establishing values between 36–46 wt% of Fe (Table 2 and Fig. S5†). Both techniques suggest a nonlinear relation between the Fe loading of Fe@C and initial Fe-MOF loading. Only Fe@C-MIL-100 shows an unexpectedly large deviation between the TGA and ICP results, indicating that non-C impurities can be present. Elemental analysis of Fe and C allowed for the calculation of the remainder phase, which was assumed as O (Fig. 3a).

By accounting for the Fe loading and associated O content as Fe_2O_3, we calculated the C/O ratio of the matrix in order to estimate the degree of carbonization (Table 2). Remarkably, for all the different MOFs, except for MIL-127, the mass balance of the leaving groups is $CO_2 + (H_2)O$, indicating a very similar degree of carbonization. On the other hand, the decomposition mass balance for MIL-127 suggests the release of only $2(H_2)O$ and no CO_2 upon carbonization, thus confirming the higher C content in this particular pyrolyzed MOF. Furthermore, the C/O/N ratio suggests that either the N=N bond is maintained, or scission followed by incorporation into the C has taken place. XPS analysis showed that the encapsulation of Fe by the C matrix varied greatly between 45–96% (Fig. 3b)

Table 2 Fe@C properties determined by ICP (Fe loading and O-functionalization of C, w_{Fe} and C/O), XPS (Fe surface/bulk ratio Fe/Fe S/B), N_2-physisorption (BET area, S_{BET}) and HR-TEM (average particle diameter, d_{Fe})

Catalyst	S_{BET}, $m^2 g_C^{-1}$	w_{Fe}, %	C/O (bulk), —	Fe/Fe (S/B), %	d_{Fe}, nm
Fe@C-MIL-68	675	37.3	6.9	31	6.8
Fe@C-MIL-88A	655	46.1	3.3	55	5.9
Fe@C-MIL-100	610	38.1	2.8	53	4.2
Fe@C-MIL-101-NH_2	635	39.8	na	na	4.2
Fe@C-MIL-127	585	34.5	3.9–0.5 N	4	4.3
Fe@C-F300	570	35.7	4.2	50	3.6

and that the Fe content at the exposed Fe@C surface is a linear function of the Fe loading of the MOF. Interestingly, the surface of Fe@C-MIL-127 consists almost exclusively of N-functionalized C, with only 1.3 wt% Fe on the surface. Fe@C-MIL-100 and -F300 contain nearly the same fraction of Fe on the surface, in addition to a very similar bulk Fe loading value.

Fe oxidation state and C matrix functionalization. Fe core-level XPS spectra illustrated that the Fe surface consists mainly of Fe_2O_3, as witnessed by the binding energy of 710.9 eV for the $2p_{3/2}$ peak and multiplet splitting with satellite features due to the high spin Fe^{3+} phase.[36] These findings indicate that the θ-Fe_3C phase of Fe@C-MIL-127 is covered by a protective oxide layer, most likely formed during the passivation of the sample, implying that the Fe surface is at least

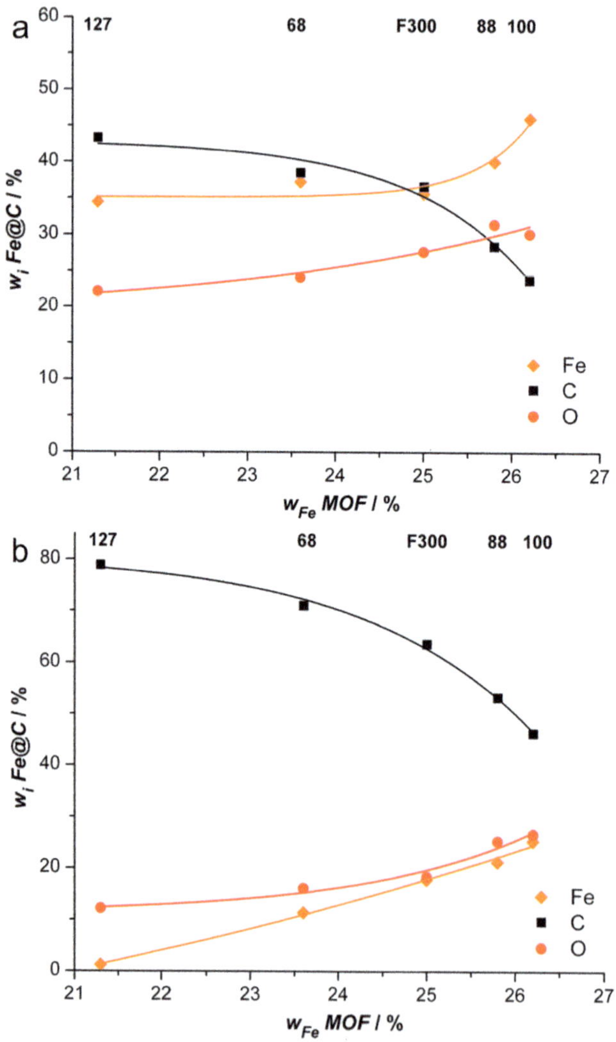

Fig. 3 (a) Bulk and (b) surface Fe@C catalyst composition as function of Fe loading of parent MOF calculated from ICP and XPS, respectively.

partially accessible. The XPS survey spectra were used to calculate the degree of O-functionalization of the C surface by accounting for the Fe loading and accompanying O content (Fig. 3b). We can divide the catalysts into high and low degrees of O-functionalized carbons, with Fe@C-MIL-88 and -100 containing between 23 and 26 O atoms per 100 C atoms, and Fe@C-MIL-68, -127 and -F300 only 11 to 13 O atoms per 100 C atoms. Additionally, the surface of the C matrix of Fe@C-MIL-127 was found to contain 9 N atoms per 100 C atoms, and 3 contributions could be distinguished from the N core-level spectra. The binding energies of the peaks are located at 398.7, 399.4 and 400.4 eV, demonstrating the presence of pyrrolic, pyridinic and graphitic N species.[37,38] The surface N content was even higher in comparison to the bulk value, 7.1 wt% *versus* 5.6 wt%, respectively. Furthermore, the survey spectra exposed the presence of impurities in addition to the expected Fe, C, O and N contributions. Additional N signals were obtained from the spectra of Fe@C-MIL-68 and -100, but were below 0.5 wt%. Both Fe@C-MIL-88 and -127 contained significant amounts of Cl (1.5 and 1.0 wt%, respectively), possibly originating from the $FeCl_3$ reactant in the synthesis process. Fe@C-MIL-100 contained a large amount of Cr and Mo surface impurities (both 2.8 wt%), which explains the large deviation between the Fe loading obtained from ICP and that calculated from TGA analysis (~7 wt%). The higher bulk content of these metal impurities suggests that they are not encapsulated by C, affirming the origin of the impurities in the MOF synthesis.

Morphology and Fe particle size. SEM analysis illustrates the conservation of the MOF morphology after pyrolysis, showing large differences in the

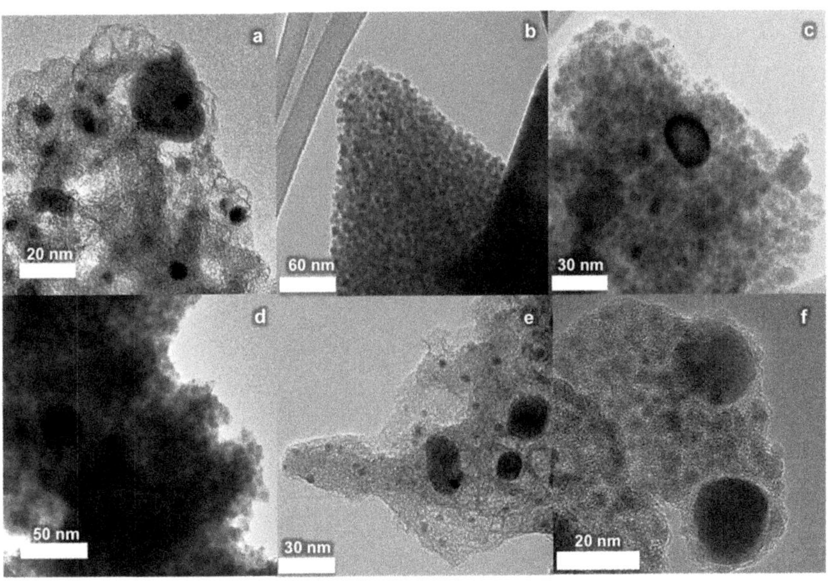

Fig. 4 TEM images of (a) Fe@C-MIL-68, (b) Fe@C-MIL-88, (c) Fe@C-MIL-100, (d) Fe@C-MIL-101-NH_2, (e) Fe@C-MIL-127 and (f) Fe@C-F300. Small Fe nanoparticles (medium contrast spheres) can be distinguished from the carbonaceous matrix (higher transmission), and are often accompanied by a number of agglomerates (large spheres). The high dispersion is still visible in larger fragments where several layers of particles overlap.

Table 3 Catalytic performance of unpromoted Fe@C catalysts

Catalyst	X_{CO}, %	FTY, 10^{-4} mol_{CO} g_{Fe}^{-1} s^{-1}	S_{CH_4}, %	S_{CO_2}, %	α, —
Fe@C-MIL-88	66	2.59	11.6	46	0.54
Fe@C-MIL-101-NH_2	65	2.90	17.2	47	0.43
Fe@C-F300	76	3.73	14.0	47	0.48

crystallinity and grain size (Fig. S6†). However, the maximum size of the crystallites did not exceed a few micrometers, and thus we expect no diffusion limitations in the HT-FTS performance of the various Fe@C catalysts. TEM images (Fig. 4 and S8†) were analysed to calculate the average Fe nanoparticle size (Table 2) and to generate particle size distributions (PSD) of the Fe@C catalysts (Fig. S7†). The average particle size, excluding the agglomerate fraction, was estimated to be between 3.6–6.8 nm. The PSDs show that the majority of the particles are below 10 nm, implying a homogeneous carbonization of the framework towards the embedded Fe on C system. No dependence of the Fe loading on the Fe particle size exists, contrary to conventional catalyst synthesis routes.

K-Promotion of Fe@C catalysts. Elemental analysis confirmed that the targeted promoter loading of 0.6 wt% K was obtained for the Fe@C catalysts with values varying from 0.53 to 0.59 wt%. EDX mapping confirmed the high dispersion of the promoter throughout the bulk phase of the catalysts. K 2p core-level contributions were confirmed in the tail of the C 1s core-level XPS spectra.

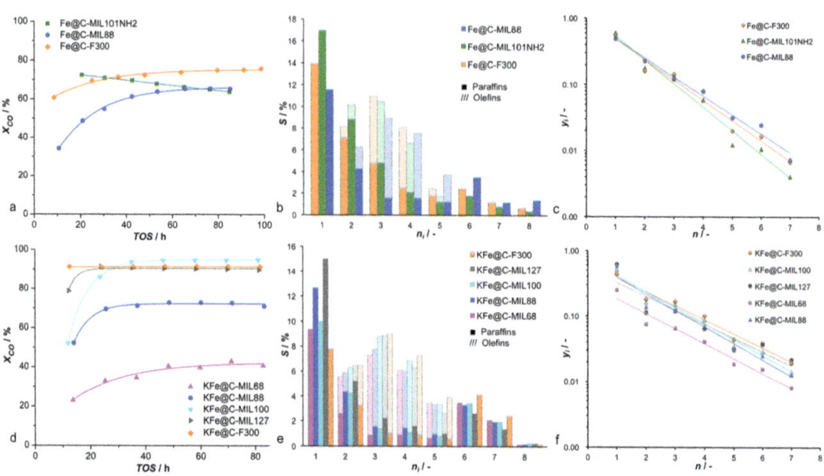

Fig. 5 HT-FTS performance of (a–c) unpromoted and (d–f) K-promoted Fe@C catalysts, displaying CO conversion levels *versus* time on stream, C-based selectivity with product distributions and Anderson–Schulz–Flory plots to assess the chain growth probability. Reactions were carried out at 340 °C, 20 bar, $H_2/CO = 1$ and employing a space velocity of 15 cm_{syngas}^3 g_{cat}^{-1} s^{-1}. Product analysis of the unpromoted and promoted catalysts was carried out at steady state conversion after 80 h time on stream (TOS). Paraffin and olefin selectivity is lumped from C_6 onward.

Catalytic testing

Unpromoted Fe@C catalyst performance. Previous results had shown improved activity and product selectivity for K-promoted Fe@C catalysts,[8] however, 3 unpromoted samples were tested in HT-FTS as well; Fe@C-MIL-88, -101-NH_2 and -F300 (Fig. 5 and Table 3). Although Fe@C-MIL-101-NH_2 showed the highest activity after 20 h (X_{CO} = 72.6%), a linear deactivation rate resulted in the lowest CO conversion after 80 h. Fe@C-MIL-88 and -F300 displayed an activation period over 50 h after which the CO conversion levels stabilized. Fe@C-MIL-88 exhibited lower CH_4 selectivity levels than Fe@C-F300, whereas Fe@C-MIL-101-NH_2 reached a value over 17%. This catalyst additionally produced the lowest O/P ratios over the entire product slate, combined with the lowest value for the chain growth probability (α = 0.43). These results are indicative of increased hydrogenation, possibly by enhanced readsorption of unsaturated hydrocarbons by the N-functionalized C. The enhanced hydrogenation might also explain the significant deactivation, as CO methanation is far more exothermic than Fischer–Tropsch ASF-type hydrocarbon production ($\Delta H_f^0 = -247$ kJ mol_{CO}^{-1} *vs.* -177 kJ mol_{CO}^{-1} for $\alpha = 0.5$, respectively[39,40]). Fe@C-MIL-88 displayed the highest α-value of 0.54, while Fe@C-F300 displayed the highest activity (FTY = 3.73×10^{-4} $mol_{CO}\ g_{Fe}^{-1}\ s^{-1}$).

K-Promoted Fe@C catalyst performance. In general, and in good agreement with the literature,[41,42] the potassium promotion of the catalysts resulted in a significant increase in CO conversion, decrease in methane selectivity, increased O/P ratios and higher values for α in the HT-FTS experiments (Fig. 5). However, large differences in the CO conversion levels for the different KFe@C catalysts are observed, as KFe@C-MIL-68 and -88 convert significantly less CO and consequently, their FTY values are the lowest. KFe@C-MIL-100 and -F300 display very similar CO conversion and FTY values, though KFe@C-MIL-100 exhibits slightly higher methane selectivity. The activation periods of KFe@C vary between 0 to 50 h, with a trend toward longer activation for the catalysts with lower activity. Again, the catalyst with N-functionalization exhibits deactivation, however, to a much smaller extent compared to the unpromoted Fe@C-MIL-101-NH_2 (1.9% *vs.* 8.9%). Remarkably, the KFe@C-MIL-127 catalyst shows the highest methane selectivity combined with the highest value for alpha, implying some deviation from ideal ASF behaviour and suggesting the presence of Fe sites with enhanced hydrogenation activity. KFe@C-F300 combines the highest FTY with the lowest CH_4 selectivity while operating at very high conversion levels, indicating that its performance does not suffer from higher local hydrogen concentrations due to limiting CO reactant depletion or diffusion limitations (Table 4).

Table 4 Catalytic performance of K-promoted Fe@C-catalysts

Catalyst	X_{CO}, %	FTY, 10^{-4} $mol_{CO}\ g_{Fe}^{-1}\ s^{-1}$	S_{CH_4}, %	S_{CO_2}, %	α, —
KFe@C-MIL-68	41.1	1.86	9.4	55	0.60
KFe@C-MIL-88A	71.3	2.62	12.7	49	0.58
KFe@C-MIL-100	94.9	4.23	10.0	47	0.56
KFe@C-MIL-127	89.5	4.46	15.0	49	0.65
KFe@C-F300	91.7	4.59	7.8	47	0.59

Discussion

Characterization of Fe@C catalysts

Key in this work was examining the influence of the MOFs on the Fe@C formation in terms of the C matrix properties and Fe nanoparticle size, as it would allow for the formulation of structure–activity relations to Fischer–Tropsch synthesis. The porosity and surface area are very similar when transforming the values to per gram of C matrix. These results indicate comparable pyrolysis behaviour, implying that the framework in itself does not have an effect on the resulting porosity, and that other factors like temperature and the duration of pyrolysis influence the properties of the C matrix. These claims are supported by the limited porosity and surface area observed by pyrolysis of MIL-88A for just 1 h, obtaining a BET area of 115 $m^2\ g^{-1}$, only half of the value reported in this work.[43] The similar pyrolysis behaviour for the Fe@C catalysts was confirmed by analysing the degree of carbonization, calculated from C/O ratios of the C matrix, indicating that one molecule of CO_2 is removed from the framework next to an additional O molecule, assumed to be in the form of a water molecule. In previous work, we showed that the structure collapse by initial decarboxylation is followed by carboxylate side reactions, supposedly forming anhydride species as intermediates towards more graphite-like structures.[16] Additionally, higher C/O ratios were found on the surface than in the bulk, suggesting that surface C is more prone to releasing O from its structure. The presence of graphitic shells on the surface of the Fe nanoparticles, witnessed in HR-TEM, suggest that Fe plays an important role in the carbonization process by scavenging O from the C matrix. Such C morphology control and graphitization have been illustrated previously in Fe-MOF pyrolysis.[44] These generalizations seem valid for Fe carboxylates, but extrapolation to non-carboxylate MOFs with different metal–ligand interactions is speculative. The average Fe nanoparticle size obtained from the Fe-MOF pyrolysis is not a function of the Fe loading, but seems to be related to the structure and porosity of the original MOF precursor. A trend towards smaller Fe nanoparticles is visible with increasing surface area and pore volume (Fig. S9†). Shorter nearest-neighbour distances of Fe clusters may very well play a role in the formation of larger Fe nanoparticles, as is clearly observed for the highly contracted and non-porous MIL-88A, having an average nanoparticle size of 5.9 nm. An additional factor at play is the decomposition temperature, being the lowest for MIL-68 and -88A, forming the largest Fe nanoparticles.

Catalytic testing

A linear dependence of the activity of the Fe@C catalysts on the size of the Fe nanoparticles was displayed (Fig. 6). For the unpromoted samples, the particle size effect on the activity was less pronounced than for the promoted samples, in line with previously reported FTY values for a similar range of particle sizes produced by the pyrolysis of F300 Fe-BTC at different temperatures.[16] These findings indicate that particle size is the key factor in Fe@C catalyst design, and this effect is even more pronounced in the K-promoted samples. The surface specific activity, or areal time yield (ATY), provides clearer insight into the relative catalytic activity of the different samples. All unpromoted Fe@C show similar performance, whereas higher ATY values are obtained for the promoted catalysts

(Fig. 6). The exceptions to this trend are KFe@C-MIL-68 and -88, which display ATY values similar to those of the unpromoted catalysts. From these results, together with the identical O/P ratios (C_3–C_4 = 0.8 and C_5 = 0.7 for both) and even slightly higher methane selectivity than that observed for the unpromoted Fe@C-MIL-88, we conclude that K promotion in these samples is not as effective as in the other cases. Furthermore, we suspect that the lower Fe surface area of KFe@C-MIL-68 results in K overpromotion, giving rise to some CO disproportionation (S_{CO_2} = 55%).[45] In general, the alkali promotion resulted in decreased hydrogenation ability of KFe@C catalysts, showing a linear correlation toward far higher O/P ratios, and additionally, providing much higher activity (Fig. 7). To put into perspective, the productivity of lower olefins as chemical precursors is maximized

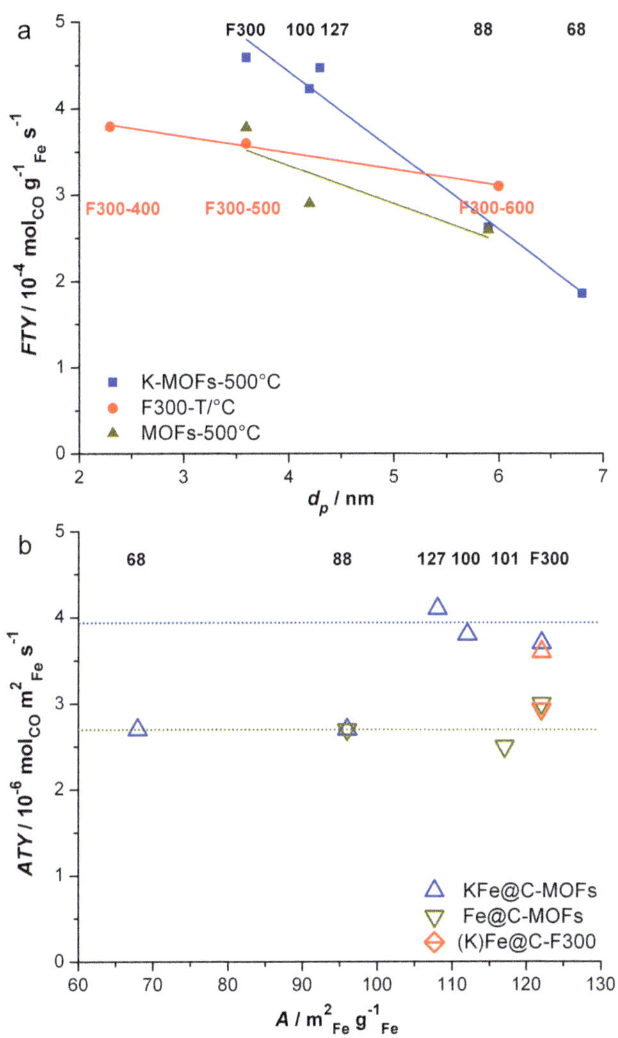

Fig. 6 (a) Correlation between Fe nanoparticle size and catalytic activity normalized for the Fe mass (FTY), including data from Fe@C-F300 synthesized at different temperatures[16] and (b) catalytic activity normalized for the Fe surface area (ATY) of several Fe@C catalysts.[8]

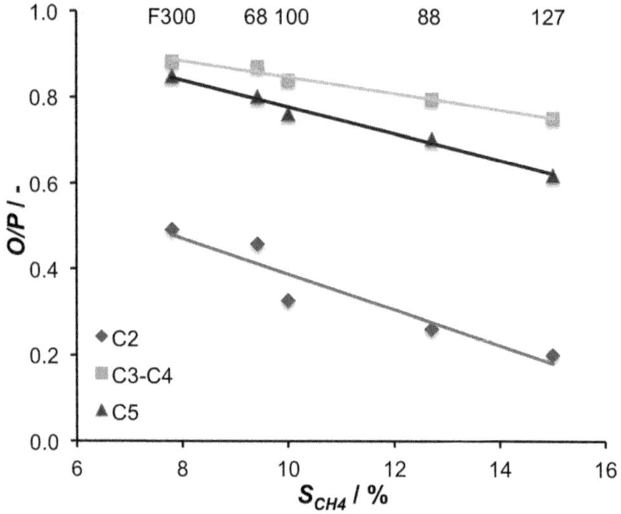

Fig. 7 Correlation of the hydrogenation ability of KFe@C catalysts, displaying olefin to paraffin ratio (O/P) and methane selectivity (S_{CH_4}).

with the best KFe@C catalysts, obtaining 50–90% of olefins in the C_2–C_5 range with CO_2-free selectivity over 53% without showing deactivation after 80 h. As discussed, deactivation was only observed for N-functionalized Fe@C catalysts, although this apparent relationship seems somewhat counterintuitive. For example, in studies on Fe/CNTs for use as FTS catalysts, N-functionalization is often used to enhance the anchoring of Fe particles in the CNTs, thus increasing their activity and stability.[46,47] Furthermore, the increased addition of N groups to the support often decreases the CH_4 selectivity in FTS over these N-doped carbons.[48] Therefore, EDX mapping of large clusters of catalysts was performed on the samples that contained impurities in the XPS spectra (Table 5, Fig. S10 and S11†). EDX results confirmed the K loading obtained by ICP analysis and showed the high dispersion of the alkali, and additionally confirmed bulk impurities for KFe@C-MIL-88, -100, -127 and -F300. The EDX analysis of KFe@C-MIL-88 and -127 contained 0.5 and 1.0 wt% of highly dispersed Cl. The EDX mapping for KFe@C-MIL-100 demonstrated that large amounts (4.9 wt% total) of Cr and Mo

Table 5 Results of EDX mapping studies on KFe@C catalysts

X/Fe, mol%	KFe@C-MIL-68	KFe@C-MIL-88A	KFe@C-MIL-100	KFe@C-MIL-127	KFe@C-F300
K/Fe	2.10	2.18	1.57	1.73	1.97
Cl/Fe		2.34		4.79	
Cr/Fe			5.58		
Mo/Fe			2.40		
N/Fe				73.2	
S/Fe					0.85
Cu/Fe					0.85
Mn/Fe					1.55

were located in agglomerates, most likely in the form of bulk metal oxides. Interestingly, the KFe@C-F300 catalyst with the highest activity and lowest CH_4 selectivity contained Cu, Mn and S in amounts known as the optimum for promoting FTS activity and selectivity.[41,42,49–52] Minute amounts of S are reported to titrate the active sites for hydrogenation, increasing the O/P ratio and activity, but not affecting the alpha value.[49,50] Additionally, increased activity from S promotion was shown by DFT calculations to originate from a decrease in the Fe–CO bond strength.[52] Cu is reported to increase the reduction of Fe_2O_3 by a combination of textural and H_2 spillover effects, and promotes the reaction rates in FTS and WGS.[41,53] Apparently, it is the exact combination of the above-mentioned promoters that distinguishes KFe@C-F300 as the best catalyst tested in this study.

Conclusions

In the preparation of FTS catalysts through the MOF mediated synthesis approach, the structure and elementary composition of the MOF play key roles in the FTS activity and selectivity of the resulting Fe on C composites. On one hand, the exact composition of the linker seems to play a small role in the surface area of the resulting C matrix (S_{BET} between 570 and 670 $m^2\ g^{-1}$). On the other hand, the pore size distribution and average Fe particle size of the final catalysts are highly dependent on the original topology. All Fe@C samples prepared following this approach displayed outstanding activities in HT-FTS, especially after promotion with K (FTY in the range of 1.9–4.6 $\times\ 10^{-4}\ mol_{CO}\ g_{Fe}^{-1}\ s^{-1}$). The obtained activities correlate well with the surface of the Fe particles, though the ATY of KFe@C-MIL-68 and -88 revealed that K promotion did not improve their activity. When it comes to selectivity towards short chain olefins and catalyst deactivation, our results highlight the importance of impurities and the presence of other heteroatoms in the original MOF precursor. The use of linkers containing N resulted in less stable catalysts, while an optimum final composition is obtained when the commercial Fe-BTC MOF is used as precursor. In addition to the added K, the presence of traces of S, Mn and Cu, most likely originating from the synthesis conditions, result in lower hydrogenation activity and a decreased Fe–CO bond strength. The combined effects result in decreased methane selectivity, an increased O/P ratio and increased activity. Altogether, our results further demonstrate the complexity of FTS and the potential of commercial MOFs for the preparation of a new generation of catalysts able to fulfil most of the requirements of this challenging reaction.

Experimental

Synthesis

The Fe-based MOFs used in this research were commercially acquired from BASF (Basolite F300®) and from MOFTECH (MIL-68, MIL-88A, MIL-100, MIL-101-NH_2 and MIL-127). They were used without further purification in both the pyrolysis and characterization. Pyrolysis of the MOFs was performed in a quartz tubular reactor (L = 1.0 m × ID = 5.0 cm) located horizontally in a ceramic fiber oven (Carbolite, Sheffield) where between 50–100 mg of the MOF was placed in a ceramic crucible and subjected to heat treatment in a nitrogen atmosphere

($150\ cm^3\ min_{STP}^{-1}$, gas hourly space velocity $4.5\ h^{-1}$). The pyrolysis temperature was fixed at 500 °C, and the ramp rate and dwell time were set to 2 °C min^{-1} and 8 h, respectively. Before exposure to ambient conditions, each material was passivated at room temperature using a stream of 5 vol% O_2 in N_2 for 2 h. The synthesized catalysts are denoted herein as "Fe@C-X", with X representing the MOF used as the catalyst precursor. Incipient wetness impregnation (IWI) was carried out with a 50 : 50 vol% methanol/water solution in which K_2CO_3 was dissolved. The amount dissolved is determined by the iron loading of the sample with molar ratio K/Fe = 0.0226, aiming at 0.6 wt% of K loading. The IWI volume used is 1.0 $cm^3\ g^{-1}$ and is pipetted onto the sample. After careful mixing, the sample is put in an ultrasonic bath for 90 min and dried in a N_2 atmosphere for 4 h at 250 °C. K-Promoted samples are denoted as KFe@C-X, where K is added to distinguish the promoted samples.

Characterization

Thermogravimetric analysis (TGA). Thermogravimetric analysis (TGA) was performed on a Mettler Toledo TGA/SDTA1 with sample robot (TSO 801RO) and gas control (TSO 800GC1). The temperature was linearly increased from 25 to 1000 °C at a heating rate of 5 °C min^{-1} under air flow. The Fe loading of the carbon-matrix catalysts was calculated using the Fe_2O_3 phase residual mass.

Powder X-ray diffraction (PXRD). Powder X-ray diffraction (PXRD) was performed with a Bruker D8 Advance diffractometer in Bragg–Brentano geometry with Co Kα radiation and a Lynxeye position sensitive detector. The slit used was a Divergence slit V12. The scatter screen height was 5 mm with 45 kV at 40 mA. Data was analysed by Bruker software Diffrac. EVA version 4.0.

N_2-Physisorption. N_2-Physisorption was performed on a Micromeritics Tristar II apparatus at 77 K to determine the textural properties of compounds. All samples were evacuated at 175 °C for 16 h prior to measurement. Data analysis was performed with Microactive software V3.00. The BET area was determined by the direct fitting of points in the $P/P_0 = 0.05–0.15$ domain, constraining the upper boundary of the relative pressure window using the two-point BET method.[54] Total pore volume was measured at $P/P_0 = 0.95$ where no inter-particle condensation occurred, the external surface area and microporous volume were determined by the *t*-method, obtaining a linear fit in the De Boer thickness of at least 10 data points.

Elemental analysis. Elemental analysis was carried out with inductively coupled plasma optical emission spectroscopy (ICP-OES) instrumentation. Firstly, the samples underwent destruction in a closed flask under high temperature and pressure. C was measured with a CHNOS-Analyzer from Elementar; Fe was measured after acidic destruction with an Analyst 200 AAS from Perkin Elmer. About 5 mg of sample was used per element measured.

X-ray photoelectron spectroscopy (XPS). X-ray photoelectron spectroscopy (XPS) measurements were performed on a K-alpha Thermo Fisher Scientific spectrometer using a monochromatic Al Kα X-ray source. The measurements were carried out using point analysis with auto-height signal optimization, with each point having a spot size of 300 microns at ambient temperature and a chamber pressure of about 10^{-8} mbar. A flood gun was used for charge compensation. All the spectra measured were corrected by setting the reference binding energy of C

1s at 284.8 ± 0.025 eV. The electron energy analyzer was operated with a pass energy of 200 eV and 0.25 eV energy spacing for the survey spectrum, and a pass energy of 50 eV and 0.1 eV energy spacing for the high-resolution spectrum was used. Each spectrum reported is the statistical average of 10 measured scans. The spectra were analyzed and processed using Thermo Avantage v5.903 software (Thermo Fisher Scientific). Smart background subtraction, derived from the Shirley background, was used over the peak width. By applying full width integration over the core-level signals and using tabulated atomic sensitivity factors (ASF), relative atomic contributions on the surface were calculated.

Scanning electron microscope (SEM). Scanning electron microscope (SEM) images and energy dispersive X-ray (EDX) spectra of the samples were recorded by a JEOL JSM-7500F field emission scanning electron microscope equipped with a Noran System Six spectral imaging system and a 30 mm^2 Novar detector. For SEM, an acceleration of 5 kV was used with a spot size of 20–50 nm. For EDX, the acceleration was increased to 20 kV and scanning times of 5 min are used to determine chemical composition. Elemental mapping was performed with similar settings employing a mapping time of approximately 15 min.

High-resolution transmission electron micrographs (HRTEM). High-resolution transmission electron micrographs (HRTEM) were collected on a JEOL model JEM-2010 working at 200 kV with a LaB_6 filament. It reaches a resolution between layers of 0.14 nm and between points of 0.25 nm. It was equipped with an Orius 831 camera from Gatan. Ethanol suspensions containing the samples were deposited onto a Au grid covered with lacey carbon and left to dry in air prior to measuring.

Catalyst testing

The catalyst performance tests were conducted in the high-pressure reactor assembly module-I (HPRAM-I) system at the Dow Chemical Company. It consists of 48 reactors that can be operated simultaneously at various process conditions, gas feed systems, exit modules with flow splitters, and gas chromatograph analyzers (Siemens Maxum-II). For activity testing, ~10 mg (20 mm^3) of the fresh catalyst with the particle size of 177–420 μm was diluted with 100 mm^3 SiC particles of similar size range. First, samples were activated *in situ* in a 90/10 vol% H_2/He mixture at 425 °C for 3 h at 3 bar followed by cooling to 340 °C under N_2 flow at the same pressure. After increasing the pressure to the process set point of 20 bar, a 10 cm^3 min_{STP}^{-1} flow consisting of CO 45 vol%, H_2 45 vol% and He 10 vol % was introduced. These operating conditions are hereby referred to as high-temperature Fischer–Tropsch (HT-FTS) conditions. Catalytic activity is expressed as Fe time yield (FTY, defined as the number of moles of CO converted to hydrocarbons per gram of Fe per second), and as areal time yield (ATY, moles of CO converted to hydrocarbons per surface area of Fe particles per second). CO conversion and carbon selectivity are defined by eqn (S1) and (S2), respectively, where X_{CO} stands for CO conversion and S stands for the carbon selectivity towards a product with n carbon atoms.

$$X_{CO} = \left(1 - \frac{C_{He,d,R} \times C_{CO,d,R}}{C_{He,d,blk} \times C_{CO,d,blk}}\right) \times 100\% \tag{S1}$$

where $C_{\text{He,d,blk}}$, $C_{\text{He,d,R}}$, $C_{\text{CO,d,blk}}$, $C_{\text{CO,d,R}}$ are the concentrations determined by GC analysis of He in the blank, He in the diluted reactor effluent, CO in the blank, and CO in the diluted reactor effluent, respectively.

$$S_{\text{C}n} = \frac{n \times \left(\dfrac{C_{\text{C}n\text{,d,R}}}{C_{\text{He,d,R}}}\right)}{\left(\dfrac{C_{\text{CO,d,blk}}}{C_{\text{He,d,blk}}} - \dfrac{C_{\text{CO,d,R}}}{C_{\text{He,d,R}}}\right)} \times 100\% \tag{S2}$$

where $C_{\text{C}n\text{,d,R}}$ is the concentration in the diluted reactor effluent determined by GC analysis of a product with n carbon atoms.

Acknowledgements

We would like to thank Harrie Jansma, Bart van der Linden, Kevin Mouthaan, Willy Rook, Bart Boshuizen, Ben Norder and Duco Bosma for the technical support during this project.

Notes and references

1 H. Schulz, *Appl. Catal., A*, 1999, **186**(1–2), 3–12.
2 A. N. Stranges, A history of the fischer-tropsch synthesis in Germany 1926–45, in *Studies in Surface Science and Catalysis*, B. H. Davis and M. L. Occelli, Elsevier, Amsterdam, 2007, vol. 163, pp. 1–27.
3 M. E. Dry, *Catal. Today*, 2002, **71**(3–4), 227–241.
4 J. R. Rostrup-Nielsen, *Catal. Today*, 1993, **18**(4), 305–324.
5 J. R. Rostrup-Nielsen, *Catal. Today*, 1994, **21**(2–3), 257–267.
6 A. P. Steynberg, R. L. Espinoza, B. Jager and A. C. Vosloo, *Appl. Catal., A*, 1999, **186**(1–2), 41–54.
7 J. Gascon, A. Corma, F. Kapteijn and F. X. Llabrés i Xamena, *ACS Catal.*, 2013, **4**(2), 361–378.
8 V. P. Santos, T. A. Wezendonk, J. J. D. Jaén, A. I. Dugulan, M. A. Nasalevich, H.-U. Islam, A. Chojecki, S. Sartipi, X. Sun, A. A. Hakeem, A. C. J. Koeken, M. Ruitenbeek, T. Davidian, G. R. Meima, G. Sankar, F. Kapteijn, M. Makkee and J. Gascon, *Nat. Commun.*, 2015, **6**, 1–8.
9 Y.-Z. Chen, C. Wang, Z.-Y. Wu, Y. Xiong, Q. Xu, S.-H. Yu and H.-L. Jiang, *Adv. Mater.*, 2015, **27**(34), 5010–5016.
10 H.-L. Jiang, B. Liu, Y.-Q. Lan, K. Kuratani, T. Akita, H. Shioyama, F. Zong and Q. Xu, *J. Am. Chem. Soc.*, 2011, **133**(31), 11854–11857.
11 T. K. Kim, K. J. Lee, J. Y. Cheon, J. H. Lee, S. H. Joo and H. R. Moon, *J. Am. Chem. Soc.*, 2013, **135**(24), 8940–8946.
12 J. H. Lee, B. Moon, T. K. Kim, S. Jeoung and H. R. Moon, *Dalton Trans.*, 2015, **44**(34), 15130–15134.
13 K. J. Lee, T.-H. Kim, T. K. Kim, J. H. Lee, H.-K. Song and H. R. Moon, *J. Mater. Chem. A*, 2014, **2**(35), 14393–14400.
14 K. Shen, X. Chen, J. Chen and Y. Li, *ACS Catal.*, 2016, **6**(9), 5887–5903.
15 B. An, K. Cheng, C. Wang, Y. Wang and W. Lin, *ACS Catal.*, 2016, **6**(6), 3610–3618.
16 T. A. Wezendonk, V. P. Santos, M. A. Nasalevich, Q. S. E. Warringa, A. I. Dugulan, A. Chojecki, A. C. J. Koeken, M. Ruitenbeek, G. Meima,

H.-U. Islam, G. Sankar, M. Makkee, F. Kapteijn and J. Gascon, *ACS Catal.*, 2016, **6**(5), 3236–3247.
17 D. Crawford, J. Casaban, R. Haydon, N. Giri, T. McNally and S. L. James, *Chem. Sci.*, 2015, **6**(3), 1645–1649.
18 D. E. Crawford and J. Casaban, *Adv. Mater.*, 2016, **28**(27), 5747–5754.
19 M. Taddei, D. A. Steitz, J. A. van Bokhoven and M. Ranocchiari, *Chem.–Eur.J.*, 2016, **22**(10), 3245–3249.
20 C. E. Wilmer, M. Leaf, C. Y. Lee, O. K. Farha, B. G. Hauser, J. T. Hupp and R. Q. Snurr, *Nat. Chem.*, 2012, **4**(2), 83–89.
21 A. Fateeva, P. Horcajada, T. Devic, C. Serre, J. Marrot, J.-M. Grenèche, M. Morcrette, J.-M. Tarascon, G. Maurin and G. Férey, *Eur. J. Inorg. Chem.*, 2010, **2010**(24), 3789–3794.
22 C. Volkringer, M. Meddouri, T. Loiseau, N. Guillou, J. Marrot, G. Férey, M. Haouas, F. Taulelle, N. Audebrand and M. Latroche, *Inorg. Chem.*, 2008, **47**(24), 11892–11901.
23 C. Serre, C. Mellot-Draznieks, S. Surblé, N. Audebrand, Y. Filinchuk and G. Férey, *Science*, 2007, **315**(5820), 1828.
24 C. Mellot-Draznieks, C. Serre, S. Surblé, N. Audebrand and G. Férey, *J. Am. Chem. Soc.*, 2005, **127**(46), 16273–16278.
25 C. Serre, F. Millange, S. Surblé and G. Férey, *Angew. Chem., Int. Ed.*, 2004, **43**(46), 6285–6289.
26 S. Surble, C. Serre, C. Mellot-Draznieks, F. Millange and G. Ferey, *Chem. Commun.*, 2006, **3**, 284–286.
27 G. Férey, C. Serre, C. Mellot-Draznieks, F. Millange, S. Surblé, J. Dutour and I. Margiolaki, *Angew. Chem., Int. Ed.*, 2004, **43**(46), 6296–6301.
28 P. Horcajada, S. Surble, C. Serre, D.-Y. Hong, Y.-K. Seo, J.-S. Chang, J.-M. Greneche, I. Margiolaki and G. Ferey, *Chem. Commun.*, 2007, **27**, 2820–2822.
29 C. Volkringer, D. Popov, T. Loiseau, G. Férey, M. Burghammer, C. Riekel, M. Haouas and F. Taulelle, *Chem. Mater.*, 2009, **21**(24), 5695–5697.
30 S. Bauer, C. Serre, T. Devic, P. Horcajada, J. Marrot, G. Férey and N. Stock, *Inorg. Chem.*, 2008, **47**(17), 7568–7576.
31 G. Férey, C. Mellot-Draznieks, C. Serre, F. Millange, J. Dutour, S. Surblé and I. Margiolaki, *Science*, 2005, **309**(5743), 2040.
32 K. M. L. Taylor-Pashow, J. D. Rocca, Z. Xie, S. Tran and W. Lin, *J. Am. Chem. Soc.*, 2009, **131**(40), 14261–14263.
33 H. Chevreau, A. Permyakova, F. Nouar, P. Fabry, C. Livage, F. Ragon, A. Garcia-Marquez, T. Devic, N. Steunou, C. Serre and P. Horcajada, *CrystEngComm*, 2016, **18**(22), 4094–4101.
34 D. Cunha, M. Ben Yahia, S. Hall, S. R. Miller, H. Chevreau, E. Elkaïm, G. Maurin, P. Horcajada and C. Serre, *Chem. Mater.*, 2013, **25**(14), 2767–2776.
35 Y. Liu, J. F. Eubank, A. J. Cairns, J. Eckert, V. C. Kravtsov, R. Luebke and M. Eddaoudi, *Angew. Chem., Int. Ed.*, 2007, **46**(18), 3278–3283.
36 M. C. Biesinger, B. P. Payne, A. P. Grosvenor, L. W. M. Lau, A. R. Gerson and R. S. C. Smart, *Appl. Surf. Sci.*, 2011, **257**(7), 2717–2730.
37 F. Kapteijn, J. A. Moulijn, S. Matzner and H. P. Boehm, *Carbon*, 1999, **37**(7), 1143–1150.
38 J. R. Pels, F. Kapteijn, J. A. Moulijn, Q. Zhu and K. M. Thomas, *Carbon*, 1995, **33**(11), 1641–1653.

39 P. Chaumette, C. Verdon and P. Boucot, *Top. Catal.*, 1995, **2**(1), 301–311.
40 G. A. Mills and F. W. Steffgen, *Catal. Rev.*, 1974, **8**(1), 159–210.
41 D. B. Bukur, D. Mukesh and S. A. Patel, *Ind. Eng. Chem. Res.*, 1990, **29**(2), 194–204.
42 S. Li, A. Li, S. Krishnamoorthy and E. Iglesia, *Catal. Lett.*, 2001, **77**(4), 197–205.
43 Y. Li, Y.-X. Zhou, X. Ma and H.-L. Jiang, *Chem. Commun.*, 2016, **52**(22), 4199–4202.
44 Q. Li, P. Xu, W. Gao, S. Ma, G. Zhang, R. Cao, J. Cho, H.-L. Wang and G. Wu, *Adv. Mater.*, 2014, **26**(9), 1378–1386.
45 F. Kapteijn and J. A. Moulijn, *J. Chem. Soc., Chem. Commun.*, 1984, **5**, 278–279.
46 J. Lu, L. Yang, B. Xu, Q. Wu, D. Zhang, S. Yuan, Y. Zhai, X. Wang, Y. Fan and Z. Hu, *ACS Catal.*, 2014, **4**(2), 613–621.
47 H. J. Schulte, B. Graf, W. Xia and M. Muhler, *ChemCatChem*, 2012, **4**(3), 350–355.
48 H. Xiong, M. Moyo, M. A. Motchelaho, Z. N. Tetana, S. M. A. Dube, L. L. Jewell and N. J. Coville, *J. Catal.*, 2014, **311**, 80–87.
49 T. C. Bromfield and N. J. Coville, *Appl. Catal., A*, 1999, **186**(1–2), 297–307.
50 H. M. Torres Galvis, A. C. J. Koeken, J. H. Bitter, T. Davidian, M. Ruitenbeek, A. I. Dugulan and K. P. de Jong, *J. Catal.*, 2013, **303**, 22–30.
51 W. L. van Dijk, J. W. Niemantsverdriet, A. M. van dar Kraan and H. S. van der Baan, *Appl. Catal.*, 1982, **2**(4), 273–288.
52 J. Cheng, P. Hu, P. Ellis, S. French, G. Kelly and C. M. Lok, *J. Phys. Chem. C*, 2010, **114**(2), 1085–1093.
53 E. de Smit, F. M. F. de Groot, R. Blume, M. Havecker, A. Knop-Gericke and B. M. Weckhuysen, *Phys. Chem. Chem. Phys.*, 2010, **12**(3), 667–680.
54 M. F. De Lange, T. J. H. Vlugt, J. Gascon and F. Kapteijn, *Microporous Mesoporous Mater.*, 2014, **200**, 199–215.

Faraday Discussions

PAPER

Size dependent stability of cobalt nanoparticles on silica under high conversion Fischer–Tropsch environment†

Moritz Wolf, Hendrik Kotzé, Nico Fischer and Michael Claeys*

Received 3rd October 2016, Accepted 18th October 2016
DOI: 10.1039/c6fd00200e

Highly monodisperse cobalt crystallites, supported on Stöber silica spheres, as model catalysts for the Fischer–Tropsch synthesis were exposed to simulated high conversion environments in the presence and absence of CO utilising an in house developed *in situ* magnetometer. The catalyst comprising the smallest crystallites in the metallic state (average diameter of 3.2 nm) experienced pronounced oxidation whilst the ratio of H_2O to H_2 was increased stepwise to simulate CO conversions from 26% up to complete conversion. Direct exposure of this freshly reduced catalyst to a high conversion Fischer–Tropsch environment resulted in almost spontaneous oxidation of 40% of the metallic cobalt. In contrast, a model catalyst with cobalt crystallites of 5.3 nm only oxidised to a small extent even when exposed to a simulated conversion of over 99%. The largest cobalt crystallites were rather stable and only experienced measurable oxidation when subjected to H_2O in the absence of H_2. This size dependency of the stability is in qualitative accordance with reported thermodynamic calculations. However, the cobalt crystallites showed an unexpected low susceptibility to oxidation, *i.e.* only relatively high ratios of H_2O to H_2 partial pressure caused oxidation. Similar experiments in the presence of CO revealed the significance of the actual Fischer–Tropsch synthesis on the metallic surface as the dissociation of CO, an elementary step in the Fischer–Tropsch mechanism, was shown to be a prerequisite for oxidation. Direct oxidation of cobalt to CoO by H_2O seems to be kinetically hindered. Thus, H_2O may only be capable of indirect oxidation, *i.e.* high concentrations prevent the removal of adsorbed oxygen species on the cobalt surface leading to oxidation. However, a spontaneous direct oxidation of cobalt at the interface between the support and the crystallites by H_2O forming presumably cobalt silicate type species was observed in the presence and absence of CO. The formation of these metal–support compounds is in accordance with conducted thermodynamic predictions. None of the extreme Fischer–Tropsch conditions initiated hydrothermal sintering. Seemingly, the formation of metal–support

*Catalysis Institute and c*change (DST-NRF Centre of Excellence in Catalysis), Department of Chemical Engineering, University of Cape Town, Private Bag X3, Rondebosch 7701, South Africa. E-mail: michael.claeys@uct.ac.za*

† Electronic supplementary information (ESI) available. See DOI: 10.1039/c6fd00200e

compounds stabilised the metallic crystallites and/or higher partial pressures of CO are required to increase the concentration of mobile, cobalt oxide-type species on the metallic surface.

Introduction

Catalyst deactivation usually manifests itself through a loss in activity and/or selectivity over time on stream (TOS). If the degree of deactivation is too high, catalyst replacement or regeneration is required to maintain an industrially viable process. In order to successfully design and develop catalysts with a higher resistance to deactivation, the understanding of the relevant mechanisms is of utmost importance.[1] Timescales of the deactivation of catalysts vary widely and are strongly dependent on the process conditions and the type of catalyst. They range from up to 10 years for nickel contacts in the methanation process[2] to only a fraction of seconds in the case of zeolite based catalysts in fluid catalytic cracking.[3] For short timescales, the catalysts have to be replaced frequently or re-activated.[2] The dominant processes contributing to the deactivation of supported cobalt Fischer–Tropsch (FT) catalysts in the industrial application are reported to be site blockage through the formation of coke, oxidation of the active phase, loss of surface area *via* sintering, and operational deactivation such as poisoning by H_2, sulphur containing compounds, As, or NH_3.[1,4] Furthermore, aromatic coke precursors and high molecular mass waxes in the usually porous structures of the support material inhibit adsorption and slow down diffusion rates.[5] Besides coke, there are several other possible processes of deactivation by the formation of carbon in different forms: atomic subsurface carbon, carbides, or the deposition of elemental carbon as a product of the Boudouard reaction. However, the relative degree of contribution of these deactivation mechanisms in the FT synthesis is under debate.[5–7]

The FT product H_2O is reported to have a severe effect on the deactivation of cobalt based FT catalysts.[5,8–10] Three mechanisms are proposed in the literature: oxidation of metallic cobalt to FT inactive cobalt oxides, hydrothermal sintering of cobalt particles leading to a loss of specific surface area, and the formation of FT inactive cobalt–support compounds.[5] In particular, the occurrence and extent of oxidation is a controversial issue in the literature.[6,11–13] A widely accepted concept of the FT mechanism on cobalt based catalysts includes the dissociation of the carbon–oxygen bond in CO on the metal surface. Hydrocarbons are formed *via* successive hydrogenation steps of the carbon atom (eqn (1)). The adsorbed oxygen atoms are highly electronegative and associated with partially positive charged cobalt atoms. This can locally be described as a cobalt oxide type interaction.[5] If the FT inactive cobalt oxide plus H_2, compared to metallic cobalt and gaseous H_2O, are energetically favoured under FT conditions, the cobalt site is not regenerated *via* removal of surface oxygen (eqn (2)) and can therefore not partake in any subsequent reaction – it is deactivated (eqn (3)). If the removal of the adsorbed oxygen atom is suppressed in the presence of a high partial pressure of H_2O or the reverse reaction (dissociation of H_2O on the cobalt surface) is kinetically allowed, an indirect and/or direct oxidation of cobalt by gaseous H_2O is expected.[5,14] Thus, the ratio of the partial pressures of the FT product H_2O to the reactant H_2 (p_{H_2O}/p_{H_2}) is expected to be an essential factor affecting the stability of cobalt.[13,14] As this is a ratio between product and reactant, it consequentially shows a strong dependency on the conversion in the FT synthesis, especially at high conversion levels (Fig. S1†).

$$CO + H_2 + Co + * \rightarrow Co + O* + CH_2 \quad (1)$$

$$Co + O* + H_2 \rightarrow Co + * + H_2O \quad (2)$$

$$Co + O* \rightarrow CoO + * \quad (3)$$

Theoretical studies show that an oxidation of bulk metallic cobalt to cobalt oxide or cobalt hydroxide is not feasible under realistic FT synthesis conditions.[13–16] A ratio of the partial pressures of H_2O to H_2 of 128 is thermodynamically required for the oxidation of bulk α-cobalt (hcp) to CoO at 220 °C,[17] while a typical value for the low temperature FT synthesis is $p_{H_2O}/p_{H_2} = 0.5–2$ (Fig. 1). According to van Steen *et al.* and although unfeasible in bulk, the oxidation of cobalt crystallites in the nanometre size regime may become feasible due to the higher relative contribution of the surface energies compared to that of the bulk, and due to a reduction of said energy in CoO compared to metallic cobalt.[14] In addition, the degree of coordination of the exposed surface cobalt atoms decreases with smaller cobalt crystallites and potentially enhances the formation of an oxide shell or even bulk oxidation.[5] The theoretical calculations show a strong size dependent oxidation behaviour of cobalt, significantly dependent on the reaction temperature (Fig. 2).[14] Smaller metallic cobalt crystallites are expected to be less stable under realistic FT conditions. The calculations resulted in a critical diameter of 4.4 nm for the oxidation of metallic β-cobalt (fcc), which has been reported to be the stable allotrope in the nanometre size regime,[18] to CoO at 220 °C and $p_{H_2O}/p_{H_2} = 1.5$. This partial pressure ratio corresponds to a synthesis gas conversion of 75%. Crystallites below this diameter are expected to undergo oxidation, which is in qualitative agreement with the estimation of 5–6 nm by Iglesia *et al.*[19] Of the two main cobalt allotropes, α-cobalt has a higher surface

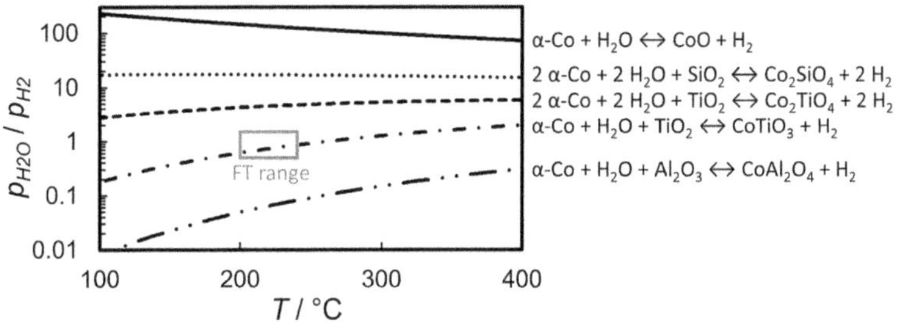

Fig. 1 Thermodynamic equilibria for selected oxidation reactions of bulk α-cobalt (hcp) by gaseous H_2O calculated with thermodynamic data from Knacke *et al.*:[21] oxidation to CoO (black, solid), formation of cobalt silicate (dotted), formation of cobalt titanates (dashed and dash-dotted), and formation of cobalt aluminate (dash-dot-dotted). The direct oxidation of α-cobalt by H_2O to Co_3O_4 is not shown as the required p_{H_2O}/p_{H_2} ratio under typical conditions for the cobalt based low temperature FT synthesis is two orders of magnitude higher than for the formation of CoO. The inset (grey, solid) shows typical operation conditions for the cobalt-based low temperature Fischer–Tropsch synthesis.

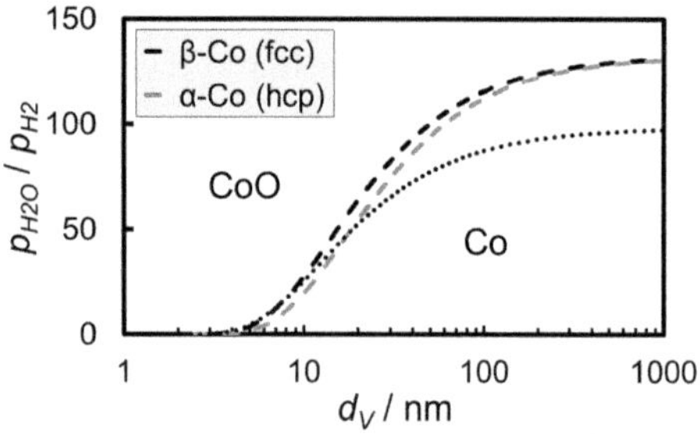

Fig. 2 Size dependency of the thermodynamic equilibrium of the oxidation of cobalt particles in the nanometre size regime to CoO by H_2O based on crystallite size originated changes in surface energy. The equilibria are calculated for α-cobalt (hcp; grey, dashed) and β-cobalt (fcc) for 220 °C (black, dashed), as well as for β-cobalt for 300 °C (black, dotted) according to van Steen *et al.*[14]

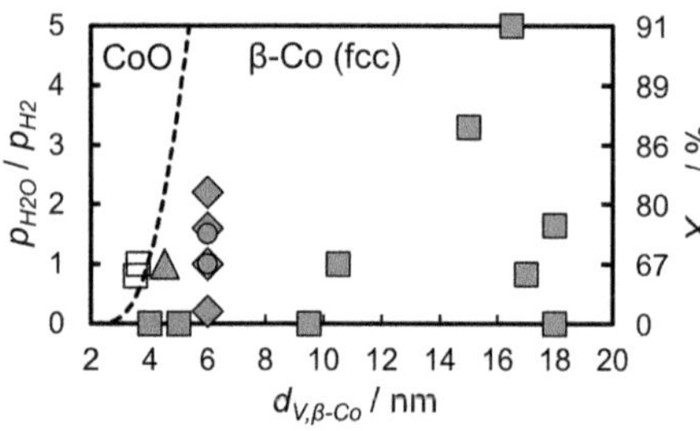

Fig. 3 Summary of literature data (squares,[29] triangle,[31] diamonds,[32] circles[10,26]) obtained *via in situ* characterisation techniques reporting oxidation (open markers) or no oxidation (grey markers) of supported metallic cobalt nanoparticles in the presence of H_2O and H_2 at the particular CO conversion levels of the FT synthesis. The data are superimposed on the size dependent thermodynamic equilibrium of the oxidation of β-cobalt (fcc) crystallites in the nanometre size regime[14] (dashed line) calculated for 220 °C.

energy[20] and is expected to be less stable than β-cobalt resulting in a larger critical diameter for its oxidation (Fig. 2).[14]

Various experimental studies report seemingly contradictory observations regarding the oxidation of cobalt in supported catalysts under FT conditions. Some studies report on the deactivation by oxidation of cobalt,[13,22–24] while no oxidation was observed by others.[10,25–27] This discrepancy in the literature is most likely due to a lack of direct characterisation of the cobalt oxide phase and/or quantitative measurement of loss in metallic cobalt under reaction conditions.[10,28] A complete differentiation between CoO, Co_3O_4, cobalt hydroxides, and cobalt–support compounds was not always possible.[7,24] In addition, catalysts with a variety of supports, different cobalt crystallite sizes and in some cases broad size distributions were compared under different conditions and especially varying p_{H_2O}/p_{H_2} ratios. Regarding these deficits, van de Loosdrecht *et al.* compared only data of directly observed oxidation with the thermodynamic calculations from van Steen *et al.*[10,14] A good agreement of several reports from the literature with thermodynamic predictions based on the proposed dependency of the oxidation behaviour on the crystallite size and the p_{H_2O}/p_{H_2} ratio was obtained. These results were confirmed for alumina-supported, monodisperse β-cobalt crystallites in the nanometre size regime in a more recent study utilising adequate *in situ* characterisation techniques for the identification of cobalt oxidation (Fig. 3). *In situ* magnetic measurements, validated by *in situ* X-ray diffraction (XRD), provided for the first time direct experimental evidence for the size dependency of the oxidation of cobalt crystallites. Further, this phase transformation was more pronounced at higher concentrations of H_2O.[29] Besides the reported oxidation of a silica-supported cobalt catalyst with a volume mean crystallite size of 12 nm at a p_{H_2O}/p_{H_2} ratio of 113 corresponding to 99.6% conversion,[30] other *in situ* characterisation based studies only confirm the absence of oxidation for crystallite sizes larger than 4.5 nm at p_{H_2O}/p_{H_2} ratios below 2.2 (81.5% conversion; Fig. 3).[10,26,31,32]

Though a good agreement is obtained between thermodynamic predictions and results of *in situ* characterisation techniques, the strong influence of the support on the physical and chemical nature, as well as the catalytic behaviour of the nanoparticles is not taken into account. The literature data (Fig. 3) were obtained with oxidic supports, namely alumina and silica, and the formation of cobalt–support compounds cannot be excluded. Thermodynamically, the formation of cobalt aluminate, cobalt silicate, and cobalt titanate is favoured over the oxidation to CoO for bulk α-cobalt (Fig. 1). According to these thermodynamic predictions, in particular cobalt aluminates are expected to form spontaneously and at much lower p_{H_2O}/p_{H_2} ratios,[13,16] but are reported to be kinetically restricted under commercial FT synthesis conditions.[13] However, cobalt aluminates might form at metal–support interfaces leading to a pronounced stabilisation of the cobalt nanoparticles.

Three general mechanisms of sintering of supported metal nanoparticles are reported: atomic migration *via* inter-particle transport of atoms on the support surface, crystallite migration on the support surface combined with collision and coalescence, and vapour transport.[3] These mechanisms are generally accepted to be accelerated at increased temperatures.[1] An index for the thermal stability of bulk and surface atoms are the Tammann ($0.5T_{melt}$ in K; Co: 877 °C)[33] and the Hüttig temperature ($0.3T_{melt}$ in K; Co: 526 °C),[33] respectively, above which the particular atoms exhibit mobility.[34] A structural melting point suppression was firstly reported by Thomson *et al.*, describing a size dependency of the melting point of solid particles, which becomes very significant for particle sizes below

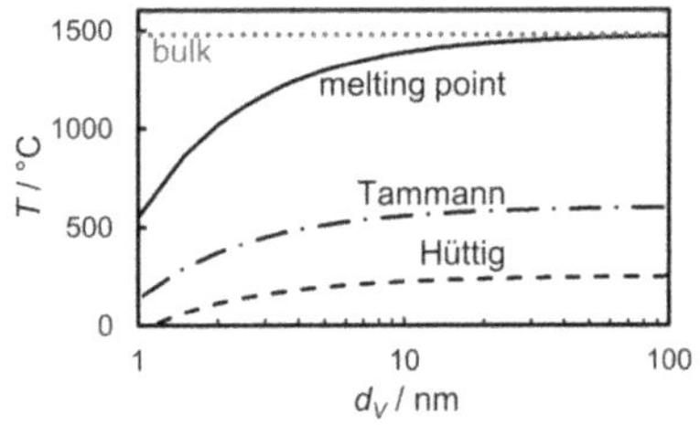

Fig. 4 Structural melting point suppression (black, solid), size dependent Tammann (black, dash-dotted) and Hüttig temperature (black, dashed) of cobalt spheres in the nanometre size regime according to the Gibbs–Thomson equation and compared to the melting point of bulk cobalt (grey, dotted). Physical properties of bulk cobalt used for the calculation: melting point of bulk cobalt, 1480 °C; specific gravity, 8900 kg m^{-3}; specific latent heat of melting, 259 844.91 J kg^{-1}; interfacial tension between the solid and liquid interface, 0.306 N m^{-1}.[33,89]

100 nm (Fig. 4) and can be predicted with the Gibbs–Thomson equation.[35,36] Application of this thermodynamic model on the calculation of the Tammann and Hüttig temperature shows that the critical temperature for sintering of cobalt particles decreases drastically with their size, especially below 10 nm. At 220 °C, a temperature significantly lower than the melting point of bulk cobalt, 1480 °C,[33] bulk cobalt atoms and surface atoms in particles of 1.2 nm and 8.5 nm can be expected to become mobile, respectively. Couchman *et al.* proposed a more complex thermodynamic approach which leads to similar predictions which are in agreement with experimental results for several metals.[37]

Cobalt nanoparticles in the range of 5–22 nm correspond to a Hüttig temperature range within typical conditions for the low temperature FT synthesis (200–240 °C).[38] Surface atoms of particles in this size range and smaller are presumed to be thermodynamically unstable by experiencing a high mobility. A higher susceptibility to sintering of smaller nanoparticles in the FT synthesis was also shown experimentally[7,27] and can be further explained by higher diffusion rates and higher relative surface energies of smaller nanoparticles.[39] Deactivation of cobalt based FT catalysts by sintering of the active cobalt phase is widely accepted and established.[40–42] However, it has to be noted that sintering does not necessarily lead to a loss in activity, especially with crystallites in the nanometre size regime with a reported size dependent optimum of the mass specific activity above the initial size in the cobalt based FT synthesis.[43,44] Besides the temperature, the atmosphere, the support, the metal dispersion, and promoters have a strong influence on the stability of particles with respect to sintering.[1,9,45,46] In the cobalt based FT synthesis, the surface oxidation/reduction cycle of CO cleavage and H_2O formation (eqn (1) and (2)) is hypothesised to be accompanied by a continuous structural re-arrangement leading to higher mobility of the cobalt atoms and resulting in crystallite growth.[24] This so called hydrothermal sintering leads to a loss of active surface area due to increased crystallite diameters.[5] The correlation between the presence of H_2O and cobalt crystallite growth during FT synthesis has been reported in several studies[24,25,39,42,47–49] and was recently summarised by Claeys *et al.*[9] Sadeqzadeh *et al.* proposed a H_2O assisted sintering mechanism *via* a highly mobile, surface oxidised intermediate facilitating particle migration, collision, and coalescence.[39] In contrast to this, Claeys *et al.* observed sintering to only occur at a combination of high H_2O and high CO partial pressures when studying an alumina-supported cobalt catalyst *in situ*.[9] From their results they concluded that an Ostwald ripening sintering mechanism may have been at play where cobalt sub-carbonyl species form under high CO partial pressure conditions and these would migrate on the hydroxyl rich alumina surface formed under the influence of H_2O. They also pointed out that this mechanism would be strictly dependent on the support material used in the catalyst and the applied reaction conditions.

In the present study, parameters affecting the stability of cobalt nanoparticles were minimised. Model catalysts with well-defined sizes were reduced and exposed to different ratios of partial pressures of H_2O to H_2 in the presence and absence of CO to clarify the feasibility of direct and indirect oxidation of cobalt crystallites in the nanometre range by H_2O. Stöber silica spheres were utilised as a support as, thermodynamically, the formation of metal–support compounds requires relatively high p_{H_2O}/p_{H_2} ratios if compared to other common oxidic supports such as alumina or titania (Fig. 1). In the absence of CO, the only parameters expected to affect the oxidation and sintering of monodisperse cobalt crystallites are the crystallite size, the temperature, the p_{H_2O}/p_{H_2} ratio, and the physicochemical properties of the support.

Experimental

Materials

Acetone (≥99.3% purity), aqueous ammonium hydroxide solution (min 25 wt%), diethyl ether (≥99.0% purity), ethanol (≥99.9% purity), ethanol (70% in H_2O), and iso-propanol (≥99.98% purity) were purchased from Kimix (South Africa); benzyl alcohol (ACS reagent, ≥99.0% purity), cobalt(II) acetate tetrahydrate (reagent grade), and tetraethyl orthosilicate (≥98% purity) were purchased from Sigma-Aldrich; bulk Co_3O_4 (type 2544) was purchased from Merck. All chemicals were used as received.

Synthesis of Stöber silica spheres

Stöber silica spheres[50] were prepared at room temperature according to a procedure published by Wang *et al.*[51] In short, 300 mL iso-propanol and 22.5 mL H_2O were mixed and magnetically stirred at 1000 rpm in an Erlenmeyer flask. 3.63 g of a 25 wt% aqueous ammonium hydroxide solution was added after 10 min. After stirring the mixture for another 10 min, 41.88 mg of tetraethyl orthosilicate was added over 1 min, which is much quicker than the dropwise addition in the reported procedure[51] to yield spheres of smaller diameter. The solution started turning opaque after 30 min and was stirred for a total of 5 h. Stöber spheres were collected *via* centrifugation, washed several times with 70% ethanol in H_2O, and dried in an oven at 120 °C.

Synthesis of Co_3O_4 nanoparticles and the parent catalysts

Monodisperse Co_3O_4 crystallites were synthesised *via* a surfactant-free, nonaqueous heat treatment of dissolved cobalt acetate in benzyl alcohol in the presence of ammonium hydroxide.[52] The amount of added ammonium hydroxide, the precursor concentration, and the reaction volume were varied to control the size (Table 1).[53] Cobalt acetate tetrahydrate was dissolved in benzyl alcohol under magnetic stirring (500 rpm) in a round bottom flask. After 2 h of stirring, a 25 wt% aqueous ammonium hydroxide solution was added dropwise to the pink to purple solution, forming a brown emulsion. Once the addition of ammonium hydroxide was completed, the flask was transferred to a preheated oil bath at 165 °C in a rotary evaporator for 3 h. The pressure was set to 900 mbar at 180 rpm and air was bubbled through the solution to provide adequate mixing.[53] A brown to black

Table 1 Varied parameters for the size controlled synthesis of Co_3O_4 nanoparticles

Parameter	CAT A	CAT B	CAT C
Benzyl alcohol/mL	140	70	140
Cobalt acetate tetrahydrate/mg	800	1600	3200
NH_4OH_{aq} (25 wt%)/mL	35	70	140

suspension was obtained and the volume was tripled with diethyl ether. The mixture was centrifuged for 1 hour at 7000 rpm. The centrifugate was re-dispersed in ethanol and washed several times with acetone.

Co_3O_4 crystallites were dispersed in ethanol in an ultrasonic bath until all particles were in dispersion. Stöber silica spheres for a target loading of 0.5 wt% metallic cobalt were sonicated for 4 h in ethanol after which the dispersion of Co_3O_4 crystallites in ethanol was added dropwise. After sonicating for another 4 h, the dispersion was transferred to a rotary evaporator and further mixed for 1 h at 240 rpm at 80 °C. Subsequently, ethanol was evaporated from the parent catalyst at 462 mbar. The absence of a calcination step during the supporting of the crystallites minimises the chances of the formation of cobalt silicates in the prepared model catalysts.

Ex situ characterisation

X-ray diffraction was conducted at 35 kV and 40 mA in a Bruker D8 Advance apparatus, equipped with a cobalt source ($\lambda_{K\alpha 1} = 0.178897$ nm) and a position sensitive detector (LYNXEYE XE, Bruker AXS) from 20–120° at a step size of 0.025° with an exposure time of 1 s per step. The diffraction patterns were compared to reference patterns of the Powder Diffraction File of the International Centre for Diffraction Data (PDF-2 Release 2008)[54] and volume mean crystallite sizes were determined by Rietveld refinement of the XRD patterns (TOPAS 4.1, Bruker AXS),[55] as well as through a peak broadening analysis of the 311 peak *via* the Scherrer equation.[56] A shape factor of 0.9 and an instrumental line broadening for correction of the obtained full width at half maximum were used.[57]

Analysis of Co_3O_4 nanoparticles *via* Raman spectroscopy was conducted in a Renishaw inVia Raman microscope ($\lambda = 532$ nm) fitted with a custom sapphire probe. Spectra were taken at a laser power of 40 mW and an exposure time of 60 s.

Unsupported samples and the synthesised Stöber silica spheres were dispersed in ethanol *via* ultrasonication for 30 min and deposited onto carbon-coated copper grids for analysis *via* transmission electron microscopy (TEM). To prevent separation of the deposited nanoparticles from the support in the parent catalysts, these samples were sonicated in acetone for 3 min and deposited onto identical grids. TEM micrographs of the Stöber silica spheres and supported samples were obtained using a Philips Tecnai F20 microscope equipped with a field emission gun and operated at 200 kV. Imaging was done with a Gatan US4000 4k × 4k CCD camera. The synthesised Co_3O_4 crystallites were analysed in a FEI Tecnai 20 equipped with a LaB6 emitter and operated at 200 kV. The TEM is fitted with a Gatan Tridiem GIF with a 2k × 2k CCD camera. Number based size distributions were determined by measuring the size of over 300 crystallites or Stöber silica spheres using the open-source ImageJ 1.51a software package.[58] The distributions were converted to volume based and/or surface area based size distributions allowing the calculation of volume mean and area mean sizes, respectively.

The BET surface area of the support was obtained in a Micromeritics TriStar II 3020 with N_2 as analysis adsorptive and a degassing temperature of 200 °C.

The parent catalysts were microwave digested for 24 h in order to conduct an elemental analysis *via* inductively coupled plasma optical emission spectrometry (ICP-OES) in a Varian ICP-OES 730 to obtain the cobalt loading of the samples.

In situ magnetic measurements

An in-house developed *in situ* magnetometer, which is based on the Weiss extraction method, was utilised for the *in situ* characterisation of the supported cobalt catalysts.[29,59,60] In this set-up, a field-controlled electromagnet provides field strengths of up to ±2 T (20 kOe). Infra-red heaters are capable of reaching temperatures up to 800 °C inside a 1/2″ stainless steel fixed bed reactor and operation at pressures in excess of 50 bar is possible. A resistance temperature detector (Pt100) placed inside the catalyst bed measures and controls the temperature of the reactor.

Methodology of the magnetometer. In ferromagnetic samples such as metallic cobalt, the parallel arrangement of electron spins leads to a strong magnetic moment. The samples show magnetisation in an external field and a remnant magnetisation upon removal of this field for temperatures below the Curie temperature (T_C). Above this characteristic temperature these materials lose their ferromagnetic features and the associated strong magnetic susceptibility. Oxidic cobalt species are anti-ferromagnetic (Table 2) and behave similar to ferromagnetic material, but have an antiparallel arrangement of the electron spins resulting in no magnetic net moment. Antiferromagnetic materials can also be magnetised in an external field and keep remnant magnetisation upon removal of this field for temperatures below the material specific Néel temperature (T_N).[61] The low Néel temperatures of oxidic cobalt species and the high Curie temperature of metallic cobalt leads to the latter being the only species present under FT conditions showing significant magnetisation susceptibility (mass specific saturation magnetisation (σ_S) of cobalt: 163.9 emu g^{-1})[62] and a remnant magnetisation after the removal of an external field (Table 2). Cobalt carbides are only weakly ferromagnetic, Co_2C in particular, which is the phase that has been detected in working cobalt catalysts[59] (Co_2C: 13 emu g^{-1}; Co_3C: 55 emu g^{-1}).[63] Furthermore, the Curie temperature of 237 °C is within the typical temperature range of the cobalt based low temperature FT synthesis leading to even lower magnetisations.[63] Thus, the measured magnetisation is directly proportional to the amount of metallic cobalt in the reactor.

Below a certain material specific size, ferromagnetic materials become superparamagnetic. The exact value for metallic cobalt is still under debate (approximately 15–20 nm for cobalt at room temperature).[71–73] Superparamagnetism occurs in nanoparticles which comprise a single-domain and the magnetic moments align in an external field, but cannot retain this alignment

Table 2 Magnetic properties of selected bulk cobalt species

Species	Magnetism	T_C or T_N/°C	Ref.
Co	Ferromagnetic	1131	64
CoO	Anti-ferromagnetic	18	61 and 65
Co_3O_4	Anti-ferromagnetic	−241	66
$Co(OH)_2$	Anti-ferromagnetic	−261	67
Co_2C	Ferromagnetic	237	68
Co_3C	Ferromagnetic	237	68
$CoSiO_3$	Anti-ferromagnetic	−223	69
Co_2SiO_4	Anti-ferromagnetic	−224	70

upon removal of this field due to the counteracting thermal agitation. Thus, superparamagnetic samples do not show a remnant magnetisation upon removal of an external field.[74] The mass fraction of material "large crystallites" displaying remnant magnetisation (γ) can be calculated from the saturation magnetisation (M_S) and the remnant magnetisation (M_{rem}; eqn (4)).[71] The saturation magnetisation is calculated *via* extrapolation of the measured magnetisations at high external field strengths to $1/H = 0$ and the remnant magnetisation is obtained upon removal of the external magnetic field.[71] The γ value describes the relative mass of ferromagnetic domains larger than the critical diameter. An increase in γ can be interpreted as an increased mass mean crystallite size and/or a decreased overall fraction of smaller crystallites in the metallic phase. The saturation magnetisation can be compared to the saturation magnetisation of 100 mg metallic bulk cobalt of a calibration curve ($M_{S,cal}$) to obtain a degree of reduction (DOR; eqn (5)). In the case of a single-phase superparamagnetic sample ($\gamma = 0$ wt%), crystallite size distributions can be deducted by measuring the sample magnetisation as a function of the external field strength, yielding a sigmoid type curve without hysteresis behaviour, and applying the Langevin formula (eqn (6)):‡[,71,75,76]

$$\gamma = \frac{2M_{\text{rem}}(T)}{M_S(T)} \times 100\% \tag{4}$$

$$\text{DOR} = \frac{M_S(T)}{M_{S,\text{cal}}(T)} \frac{100 \text{ mg}}{\text{catalyst loaded in mg} \times \text{cobalt loading}} \times 100\% \tag{5}$$

$$\frac{M}{M_S} = \coth\left(\frac{\rho\sigma_S\pi d^2 H}{6kT}\right) - \frac{6\,kT}{\rho\sigma_S\pi d^2 H} \times 100\% \tag{6}$$

where the ratio M/M_S is the magnetisation relative to the saturation magnetisation, ρ is the density of the magnetic material, d is the diameter of the magnetic crystallites, H is the external field strength, and σ_S is the mass specific saturation magnetisation of the magnetic material.

Experimental procedure. An amount of 2003.6 mg of the oxidic parent catalyst was placed in the reactor and reduced for 24 h in a 50% H_2 in an Ar atmosphere at a gas hourly space velocity (GHSV) of 6000 mL ($g_{catalyst}$ h)$^{-1}$ and a heating rate of 1 °C min^{-1}. The reduction temperature was increased stepwise (CAT B, CAT C: between 300 °C and 360 °C; CAT A: between 320 °C and 400 °C; Fig. S4†) to minimise the thermal strain during reduction with possible accompanying sintering. After cooling to 220 °C, the flow rates of H_2 and Ar were set to a GHSV of 369 mL_{H_2} ($g_{catalyst}$ h)$^{-1}$ and 2126 mL_{Ar} ($g_{catalyst}$ h)$^{-1}$, respectively, and kept constant throughout all steps, while H_2O was co-fed *via* a vaporiser to simulate FT conversion. Subsequently, the partial pressure of H_2O was incrementally

‡ Application of the Langevin equation(s) on sigmoid type curves as obtained *via* the measurement of the sample magnetisation as a function of external field strength comprises the superimposition of several theoretical magnetisation curves of a specific size range,[71] *e.g.* 1–15 nm with a bin size of 1 nm. Data processing provides the best mathematical fit, but defining an assumed log-normal distribution as a boundary condition[76] (or two superimposed log-normal distributions, *e.g.* in the case of sintered samples) provides more reasonable results while the goodness of fit hardly changes. Even though the application of the Langevin equation is only feasible for sigmoid type curves, the associated error for hysteresis type curves is reasonably small for samples with γ values of up to 10 wt%.

Table 3 Applied conditions for the incremental increase of the simulated Fischer–Tropsch conversion levels during *in situ* magnetic measurements upon reduction

H_2O level	p_{H_2O}/bar	p_{H_2}/bar	$p_{Ar}{}^{a}$/bar	p_{H_2O}/p_{H_2}	$X_{CO,sim}$/%	$d_{crit,\beta\text{-}Co}$/nm (ref. 14)	p/bar
1	0.02	0.15	0.85	0.15	25.93	3.09	1.02
2	0.22	0.15	0.85	1.50	77.78	4.20	1.22
3	0.73	0.15	0.85	5.00	92.11	5.36	1.73
4	1.45	0.15	0.85	10.00	95.89	6.51	2.45
5	2.90	0.15	0.85	20.00	97.90	8.48	3.90
6	4.35	0.15	0.85	30.00	98.59	10.46	5.35
7	5.80	0.15	0.85	40.00	98.94	12.68	6.80
8	7.25	0.15	0.85	50.00	99.15	15.29	8.25
9	7.25	—	0.85	∞	100.00	—	8.10
—	—	0.15	0.85	0.00	—	—	1.00

^{a} 0.07 bar of Ar was replaced by CO when synthesis gas was fed.

increased by regulating the amount of co-fed H_2O and the absolute pressure until pure H_2O in Ar was co-fed in the absence of H_2 (Table 3). All levels were monitored for at least two hours before applying the following conditions. To elucidate a possible effect of CO, Ar was partially replaced by CO to obtain a H_2/CO ratio of 2.11 in an additional run for CAT B, which introduced the exposure to synthesis gas as an additional first step. For comparison, a second sample of CAT A was exposed to H_2O level 7 immediately after reduction and subsequently CO was introduced. The reversibility of possible oxidation was tested in a final step without co-feeding H_2O. To distinguish between reducible oxidic cobalt and possibly formed cobalt silicates, the samples were re-reduced while heating at 1 °C min^{-1} to the respective maximum reduction temperature. After cool-down to room temperature, the catalysts were passivated in 1% O_2 in N_2 at a GHSV of 1500 mL ($g_{catalyst}$ h)$^{-1}$ for several hours[77] and collected for additional *ex situ* analysis.

When synthesis gas was fed, product gas was sampled *via* the ampoule method[78] to monitor the actual conversion of CO. The ampoules were analysed in an off-line Varian CP 3900 gas chromatograph equipped with a flame ionisation detector (GC-FID). Cyclohexane (0.17 vol% in N_2) was used as an internal standard and partially replaced Ar. The low CO conversion, a result of the high GHSV in combination with the little amount of loaded metallic cobalt in the reactor, was calculated from the formation rates of C_1–C_6 hydrocarbons only.

Results and discussion

Characterisation of the model catalysts

The XRD patterns of the as prepared Co_3O_4 nanoparticles show distinct peaks indicating a high crystallinity (Fig. 5 left) as reported for the applied preparation method.[52,53] The size dependent diffraction line[79] is clearly visible and leads to an overlap of adjacent peaks for smaller crystallite sizes. Rietveld refinement of the obtained XRD patterns leads to crystallite sizes of 2.8 nm, 5.5 nm, and 7.1 nm, which are in accordance with the particular sizes calculated *via* the Scherrer equation[56] (Table 3). The Raman spectra are normalised to the constant intensity

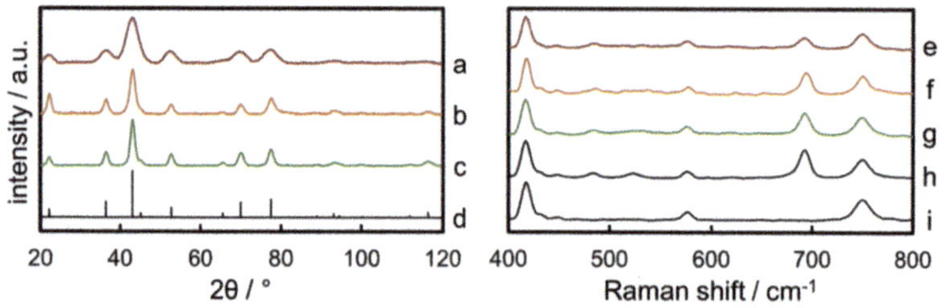

Fig. 5 XRD patterns ($\lambda_{K\alpha1}$ = 0.178897 nm; left) and Raman spectra (right) of the Co_3O_4 nanoparticles for the parent catalysts CAT A (a and e), CAT B (b and f), and CAT C (c and g), as well as the reference XRD pattern for Co_3O_4 (d), the Raman spectrum of a commercial bulk Co_3O_4 sample (h), and of the sapphire probe (i).

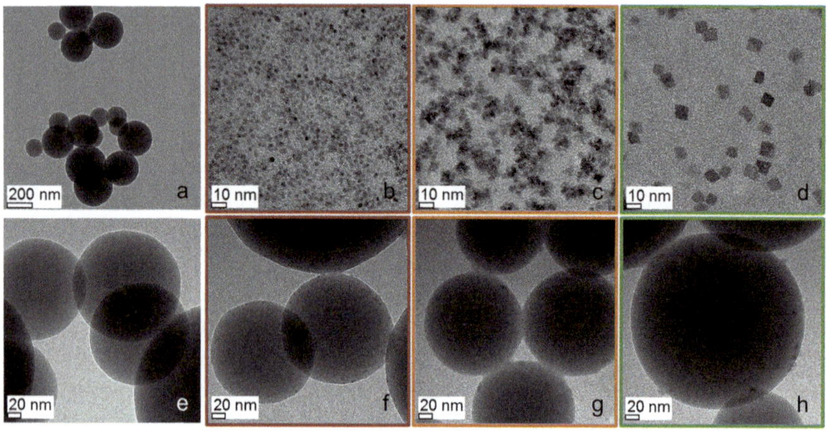

Fig. 6 TEM micrographs of the synthesised Stöber silica spheres (a and e) and as prepared Co_3O_4 nanoparticles before (b–d) and after deposition on the support (f–h) for the parent catalysts CAT A, CAT B, and CAT C.

of the broadening of the sapphire probe at 419 cm^{-1} and match with a commercial bulk Co_3O_4 sample (Fig. 5 right). The different sizes lead to varying degrees of band broadening as reported in the literature for other metal oxides.[80] This can be easily seen for the broadening at 695 cm^{-1} and by comparing the normalised intensity to the constant intensity of the main broadenings related to the sapphire probe (419 cm^{-1} and 753 cm^{-1}).

Analysis of the prepared Co_3O_4 nanoparticles *via* TEM confirms the successful preparation of differently sized crystallites (Fig. 6 and Table 4). Larger crystallites are of cubic shape, while smaller ones appear spherical. The obtained size distributions are narrow with relative size deviations below 17% (Fig. 7). The prepared silica spheres have an area mean diameter of 237.8 ± 70.8 nm, which corresponds to a specific surface area of 11.5 $m^2 g^{-1}$ using a density of 2200 kg m^{-3} for amorphous silica[33] and assuming ideal spheres. The latter assumption neglects the real uneven surface of the Stöber silica spheres. Thus, the BET surface area of 19.3 ± 0.1 $m^2 g^{-1}$ matches the approximation *via* TEM analysis. The small diameter, the amorphous structure, and its low density lead to a high

Table 4 Volume mean crystallite sizes of the as prepared Co_3O_4 nanoparticles for the parent catalysts

Method of size determination	$d_{V,CAT\ A}$/nm	$d_{V,CAT\ B}$/nm	$d_{V,CAT\ C}$/nm
Rietveld refinement	2.8	5.5	7.1
Scherrer equation (311 peak)	3.0	5.0	6.5
TEM	3.2 ± 0.5	3.8 ± 0.7	7.6 ± 1.1

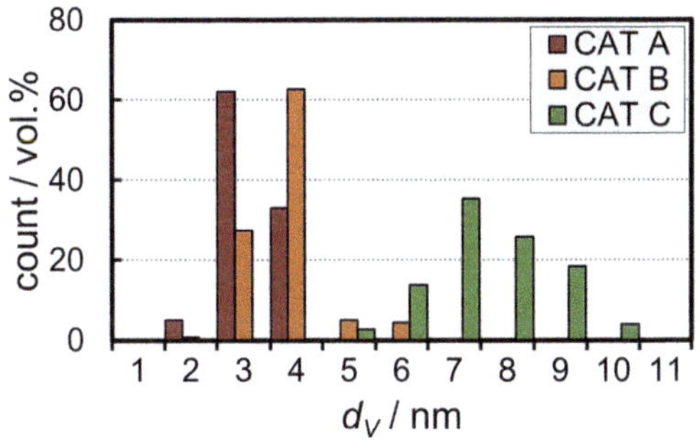

Fig. 7 Volume based size distributions of the Co_3O_4 nanoparticles of the parent catalysts as obtained *via* TEM.

contrast between cobalt or oxidic cobalt crystallites and the prepared silica support. Furthermore, the electron beam of the TEM is able to transmit the silica spheres, rendering most of the deposited nanoparticles visible (Fig. 6). The Co_3O_4 nanoparticles are uniformly distributed over the support and only a small fraction is present in agglomerates.

Magnetic measurements

The high GHSV and stepwise increase of the temperature during the reduction prevented major sintering of the catalysts. The smaller oxidic crystallites in CAT A were harder to reduce than the larger crystallites in CAT B and CAT C (Fig. S4† and Table 5). The necessity of increased temperatures for the reduction of smaller oxidic cobalt crystallites has been reported for a similar model catalyst.[81] However, an expected pronounced presence of silicates in the catalysts containing the smallest crystallites and varying preparation methods yielding different cobalt oxides may have affected the results by Saib *et al.*[81] The volume mean crystallite sizes, as calculated using the Langevin equation on the measurement of the sample magnetisation as a function of external magnetic field strength after reduction (Fig. 8), are comparable to the as prepared oxidic crystallite sizes§ (Table 4). The small γ values obtained for CAT A and CAT B further support this result. With 24 wt% of the crystallites in CAT C being larger than the critical diameter for superparamagnetism, which might be due to agglomeration and sintering or due to the larger initial size with particles being already larger than

§ The theoretical reduction of the diameter due to the higher density of metallic cobalt is negligible and within the error of the calculation and measurement.

Table 5 Physical properties of the model catalysts after reduction as obtained *via* ICP-OES of the parent catalyst and from the measurement of the sample magnetisation as a function of external field strength

Catalyst	Cobalt loading/wt%	DOR/%	γ/wt%	$d_{V,Co}$/nm
CAT A	0.42	43.8	0.2	3.2 ± 0.7
CAT B	0.43	56.4	4.3	5.3 ± 1.2
CAT C	0.41	82.7	23.7	n/a

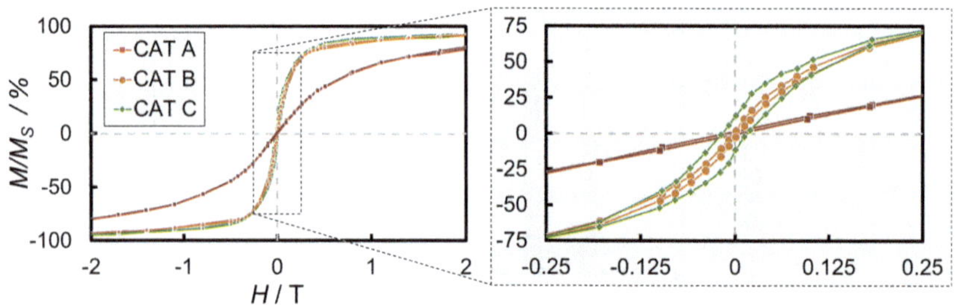

Fig. 8 Magnetic sigmoid type curves of the sample magnetisation as a function of external field strength with (CAT B and CAT C) and without (CAT A) hysteresis type behaviour (left) and an inset showing the magnetisation at low external field strength (right) after reduction. The data points for an external field strength below 0.1 T (absolute) were omitted for CAT A as the low loading of magnetic material in the reactor led to a low signal-to-noise ratio at low external field strengths.

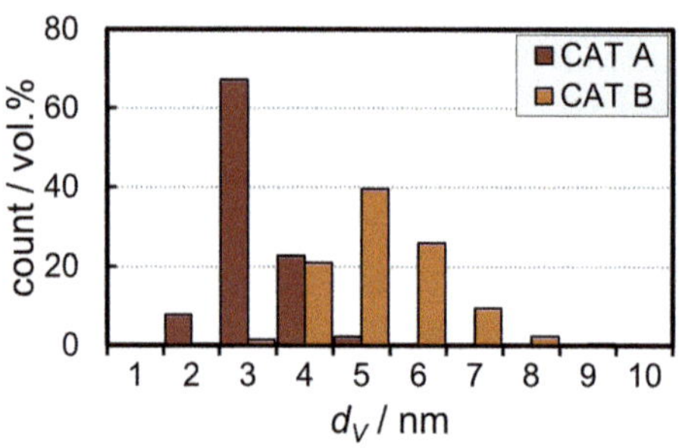

Fig. 9 Calculated size distributions of the metallic cobalt crystallites after reduction of CAT A and CAT B obtained *via* the application of the Langevin equation.

the vague critical diameter from the literature, the calculation of the size distribution for this catalyst is not possible and the average size is actually larger than 6.5 nm as obtained from crudely applying the Langevin equation. The size distributions of reduced CAT A and CAT B remained narrow, allowing the investigation of possible size dependencies (Fig. 9).

Exposure to high conversion levels in the absence of CO. The magnetisation of CAT C slightly increased during exposure to H_2O levels 1–4 (Fig. 10 left). This is most probably due to adsorbed hydroxyl species originating from co-fed H_2O, which are reported to increase the magnetisation of surface cobalt atoms due to

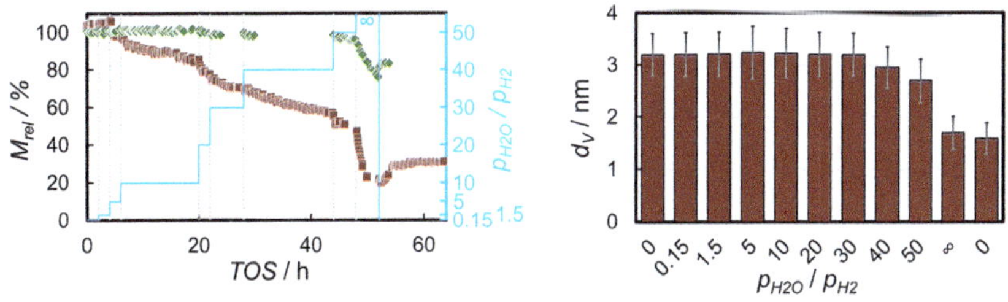

Fig. 10 Magnetisation at maximal field strength relative to the magnetisation after reduction of CAT A (red squares) and CAT C (green diamonds) as a function of time on stream with a stepwise increase of the ratio of partial pressures of H_2O to H_2 (blue, solid; left), and volume mean sizes with standard deviations of the cobalt crystallites in CAT A at the end of exposure to particular ratios of partial pressures of H_2O to H_2 obtained *via* the application of the Langevin equation (right).

changes in their electronic properties.[9] No oxidation was observed for these simulated conversion levels. Increasing the partial pressure of H_2O, or the H_2O to H_2 ratio respectively, resulted in a slow oxidation of the crystallites. A more pronounced increase of the relative magnetisation was observed for CAT A during H_2O levels 1–2, which can be explained by the higher specific surface area of the smaller crystallites allowing more hydroxyl species to adsorb. Oxidation was only observed after increasing the ratio of partial pressures p_{H_2O}/p_{H_2} to 5, still before oxidation of the larger crystallites in CAT C. Stepwise increasing the amount of co-fed H_2O revealed the expected size dependency of the re-oxidation of cobalt crystallites under simulated high FT conversion environment. While almost 50 wt% of the smaller metallic cobalt crystallites in CAT A were oxidised after H_2O level 8 ($p_{H_2O}/p_{H_2} = 50$), only 4 wt% were oxidised in CAT C containing larger crystallites. This decrease for CAT C is less than expected for the associated drop in magnetisation for oxidation of a monolayer of cobalt atoms if compared to theoretical calculations according to a shell model (Fig. S2†).¶ Thus, only the outer atoms of some, probably smaller crystallites oxidised. Exposing the catalysts to H_2O in the absence of H_2 (H_2O level 9) drastically decreased the relative magnetisation, showing the highly oxidising nature of H_2O. Therefore, even very low levels of H_2 (H_2O level 8) seem to inhibit oxidation. This is in accordance to results obtained by Claeys *et al.*[9] using the *in situ* magnetometer and Bezemer *et al.*[25] using *in situ* Mössbauer spectroscopy. Only a fraction of the loss in magnetisation was recovered during the final exposure to H_2 in the complete absence of H_2O, *i.e.* the oxidation was partially reversible at 220 °C. Some of the reducible oxidic cobalt re-reduced within the first hour, while afterwards only a very little fraction of metallic cobalt was recovered over time.

Hydrothermal sintering was not observed for both catalysts. CAT C maintained a constant γ value of 22–25 wt% throughout all steps indicating no change in size. The volume mean crystallite size of CAT A, as obtained using the Langevin method, was constant and eventually decreased most likely due to the formation

¶ The shell model comprises the oxidation of an outer layer of an ideal sphere. The volume of this layer relative to the original volume of the sphere defines an expected loss for particular layer thicknesses. The difference between the oxide and metal density is neglected.

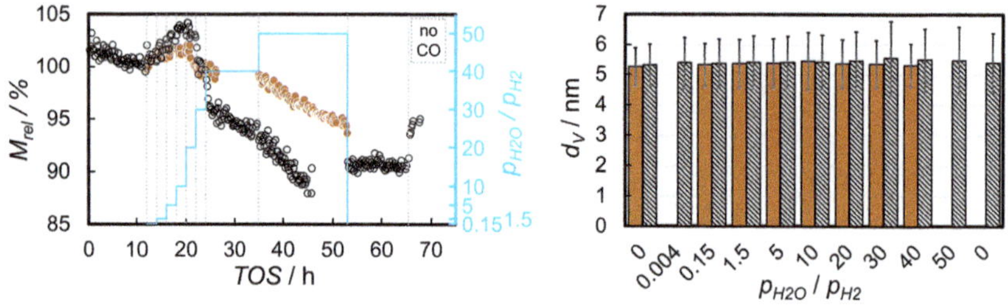

Fig. 11 Magnetisation at maximal field strength relative to the magnetisation before co-feeding H_2O as a function of time on stream with a stepwise increase of the ratio of partial pressures of H_2O to H_2 (blue, solid; left), and volume mean sizes with standard deviations of the cobalt crystallites at the end of exposure to particular ratios of partial pressures of H_2O to H_2 obtained *via* the application of the Langevin equation (right) of CAT B in the absence of CO (filled circles and bars) and presence of CO (open circles and patterned bars).

of an oxide shell (Fig. 10 right). Further oxidation was observed while increasing the concentration of H_2O until the drop in relative magnetisation corresponds to an oxide layer of 0.25 nm at H_2O level 7, 0.33 nm at H_2O level 8, and 0.64 nm at H_2O level 9. These theoretical estimations from a shell model (Fig. S2†) are confirmed by the decrease of the volume mean crystallite size from 3.2 nm to 2.7 nm to 1.7 nm, respectively. These theoretical layer thicknesses correspond to several layers of oxidised surface cobalt atoms. Upon removal of H_2O after level 9, the calculated volume mean crystallite size is further reduced. However, the overall change is within the error of calculation and measurement. A decrease might be due to a Kirkendall type re-dispersion into smaller crystallites upon re-reduction.[82]

The initial increase in magnetisation of CAT B is, once again, possibly due to adsorbed hydroxyl species (Fig. 13 left). The change is less pronounced than for the smaller crystallites in CAT A, but more pronounced than for the larger crystallites in CAT C. These results show that the initial change upon exposure to H_2O in all catalysts is due to adsorption as the trend correlates to the different sizes leading to different specific surface areas of cobalt. Similarly as for CAT C, oxidation of CAT B was commencing at H_2O level 5 and ongoing during the subsequent simulated conversions (Fig. 11 left). The reduction of magnetisation during H_2O level 8 was still ongoing when the run was stopped indicating slow, continuous oxidation of cobalt. Therefore, a potential for a more severe oxidation surpassing the observed drop in magnetisation of 6 wt% can be assumed, confirming the lower stability of smaller cobalt crystallites.

Exposure to high conversion levels in the presence of CO. To elucidate the effect of CO, CAT B was also exposed to H_2O levels 1–8 in the presence of synthesis gas with a H_2/CO ratio of 2.11 (Fig. 11 left). Under dry conditions, *i.e.* in the absence of co-fed H_2O, a slow decrease in magnetisation was observed. This initial change is possibly related to surface reconstruction upon first exposure to synthesis gas, namely re-oxidation, carbidisation, and/or CO adsorption on the metallic cobalt crystallites.[9,10,12,29,59] Adsorbed hydroxyl species increased the magnetisation and/or initiated the reversal of this process in the following H_2O levels 1–4. As for CAT B and CAT C in the absence of CO, the first decrease in

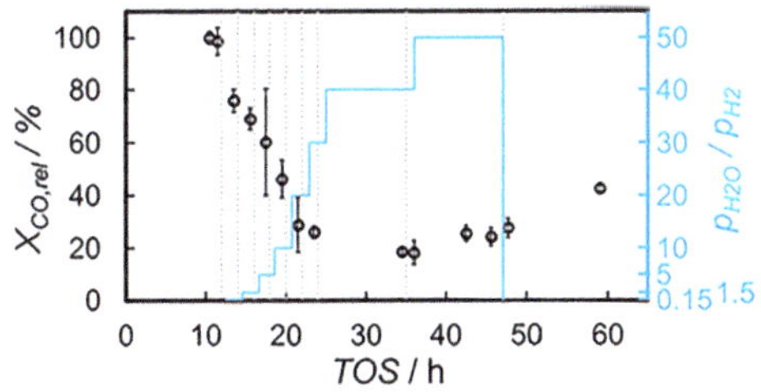

Fig. 12 Actual conversion of CO relative to the steady-state conversion before co-feeding H_2O on CAT B in the presence of CO (black circles) as a function of time on stream with a stepwise increase of the ratio of partial pressures of H_2O to H_2 (blue, solid).

magnetisation was observed at H_2O level 5, possibly due to re-oxidation of cobalt. However, as CO is present in this run, the decrease in magnetisation might also be a result of carbidisation, *i.e.* the formation of cobalt carbides with a lower magnetic susceptibility than metallic cobalt. No hydrothermal sintering was observed (Fig. 11 right). The decrease in magnetisation was more pronounced, especially at subsequent p_{H_2O}/p_{H_2} ratios, leading to a higher overall loss for CAT B in the presence of CO than in the absence of CO. The dependency on the concentration of H_2O indicates re-oxidation of cobalt as a reason for the loss in magnetisation rather than possibly ongoing carbide formation. The drop in magnetisation theoretically corresponds to the oxidation of a full monolayer of cobalt when applying the shell model. The drop in magnetisation in CAT B in the presence of CO is accompanied by a significant decrease in CO conversion (Fig. 12). However, the conversion dropped continuously from 0.71% under dry FT conditions (corresponds to a p_{H_2O}/p_{H_2} ratio of 0.004) before H_2O level 1 to below 0.15% at H_2O level 7 and 8. Deactivation occurred even before oxidation and/or carbidisation is detected, which is presumably due to competitive adsorption of H_2O originated hydroxyl species blocking the cobalt surface. Note that the H_2O levels 3–8 simulate CO conversions exceeding 93%, significantly higher than achieved during industrial FT synthesis. With such high p_{H_2O}/p_{H_2} ratios, the regeneration of the cobalt surface by H_2 (eqn (2)) is decelerated and eventually cobalt oxide is formed (eqn (3)). As the oxidation was only partially reversible, the CO conversion only recovered to a fraction of the initial conversion level of the catalyst as well. Sintering as a deactivation mechanism can be excluded as the volume mean crystallite size was constant throughout exposure to the different H_2O levels. Removing CO from the feed gas in a final step spontaneously increased the magnetisation indicating carbide decomposition. FT inactive carbides have been reported to decompose quickly in H_2 at 150–200 °C.[59] However, desorption of CO and re-reduction of possibly formed cobalt oxide might also contribute to the increase in magnetisation.

To highlight the impact of the highly oxidising extreme FT conversion environment, a freshly reduced CAT A was directly exposed to H_2O level 7, initially in the absence of CO. Rapid oxidation was observed within the first two measurements (<10 min TOS) showing the almost spontaneous oxidation of cobalt by H_2O (Fig. 13). In contrast, the stepwise increase of the partial pressure of H_2O in the previous runs functioned as a gentle stepwise passivation-like treatment of the cobalt crystallites with a slowly forming oxidic layer. The cobalt crystallites being

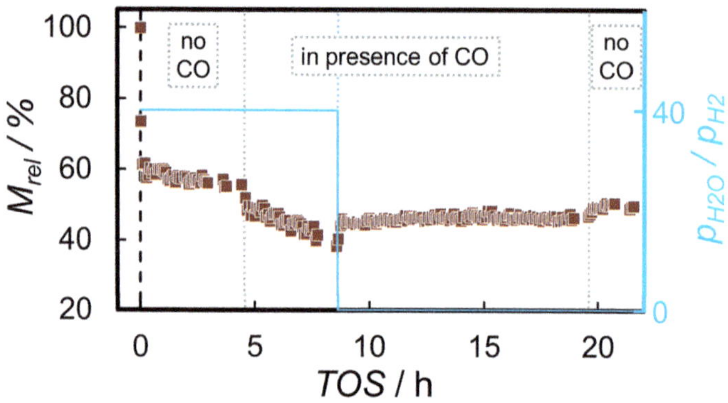

Fig. 13 Magnetisation at maximal field strength relative to the magnetisation after reduction (red squares) as a function of time on stream of reduced CAT A being exposed to different atmospheres with a high initial ratio of partial pressures of H_2O to H_2 (blue, solid).

directly exposed to a high partial pressure of H_2O oxidised to a higher extent. Afterwards, the rate of oxidation slowed down and another, less pronounced drop in magnetisation was observed when CO was co-fed. The rather quick loss in magnetisation during the subsequent 3 h on stream shows the exacerbating effect of CO, which was discussed in the comparison of CAT B in the presence and absence of CO. Exposure to synthesis gas in the absence of H_2O caused a spontaneous increase in magnetisation. As the partial pressure of CO did not change, this increase rather originates from the reversibility of unstable oxidic cobalt under the H_2 atmosphere than from the spontaneous decomposition of potentially formed cobalt carbides or the desorption of CO. However, the overall pressure was decreased, possibly resulting in the decomposition of cobalt carbides which may have formed during FT synthesis.[59] The further re-reduction of formed oxidic cobalt is very slow and after 5 h on stream no more metallic cobalt was formed. Exposing the catalyst to pure H_2 led to a further, non-spontaneous re-reduction, again excluding the decomposition of carbides[59] and therefore supporting the above conclusion that the loss of magnetisation is indeed due to oxidation.

Re-reduction of the catalysts. Heating the catalysts to higher temperatures during a re-reduction further recovered metallic cobalt (Fig. 14). However, full recovery was achieved for none of the samples indicating the formation of irreducible/hard-to-reduce metal support compounds such as cobalt silicates. A higher percentage of oxidised cobalt was re-reducible for the larger crystallites in CAT C in comparison to CAT A. This trend drives the hypothesis that smaller crystallites are forming cobalt silicates more readily based on their larger mass specific surface area interfacing with the support. CAT A directly exposed to H_2O level 7 with subsequent co-feeding of CO showed a higher re-reducibility than CAT A, which oxidised in the absence of CO. Furthermore, CAT B exposed to several H_2O levels in the presence of CO showed the highest recovery, despite consisting of smaller crystallites than CAT C. Note that the formation of carbides, which are expected to decompose at conditions preceding the re-reduction,[59] cannot be excluded for this run, but are included in this calculation. Nevertheless, one can conclude that a larger fraction of the formed oxidic cobalt was reducible when CO

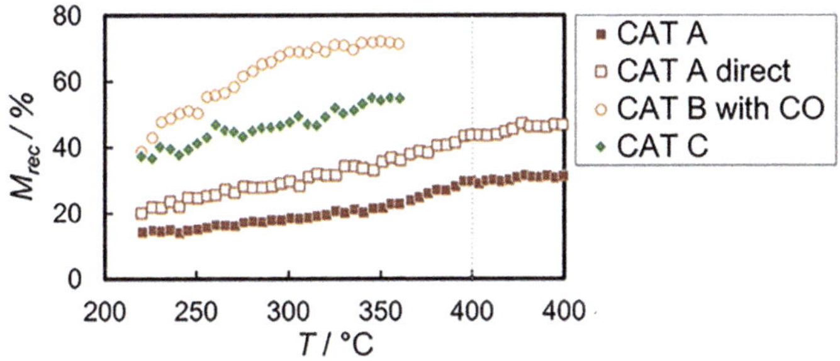

Fig. 14 Recovered magnetisation at maximal field strength during re-reduction relative to the overall loss in magnetisation between reduction and the highest H_2O level as a function of re-reduction temperature.

was present during the oxidation process. This is in accordance with the previously discussed exacerbating effect of CO, which seems to preferentially/exclusively support the formation of reducible cobalt oxide species, possibly CoO, *versus* hard-to-reduce cobalt silicate species. The re-reduction of CoO in the presence of a metallic cobalt core has been reported to occur at low temperatures in a study on passivation conditions.[77] In contrast, several temperature programmed reduction studies on silica-supported cobalt catalysts revealed a reduction of cobalt silicate species at temperatures significantly exceeding 400 °C.[81,83,84]

Experimental results *versus* thermodynamic predictions and literature. The relative change in magnetisation after exposing the model catalysts to the particular H_2O levels, which is mostly corresponding to the phase transition from metallic to oxidic cobalt, highlights the size dependent oxidation even though oxidation was still ongoing in some instances (Fig. 15). CAT A lost close to 80 wt% of its magnetic phase during exposure to H_2O levels 1–9, while CAT B only oxidised to a small extent and CAT C was rather stable, with the exception of its exposure to H_2 free gaseous H_2O. Saib *et al.* reported similar results for the oxidation of comparable model catalysts by H_2O in a fundamental *in situ* X-ray absorption near-edge spectroscopy (XANES) study.[81] The catalyst containing the largest cobalt crystallites of 28 nm was stable when exposed to H_2O in He

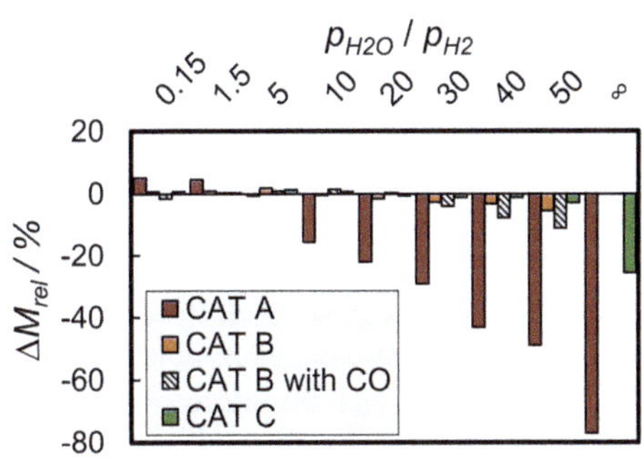

Fig. 15 Loss in magnetisation at maximum external field strength at the end of H_2O levels 1–9 with a stepwise increase of the partial pressure of H_2O relative to the initial magnetisation before H_2O level 1. Note that CAT B was not exposed to H_2O in the absence of H_2 ($p_{H_2O}/p_{H_2} = \infty$).

Table 6 Loss and recovery of magnetisation for both experiments with CAT A relative to the overall loss in magnetisation

Sample	H_2O levels	Slow loss in magnetisation/%	Recovered magnetisation/%
CAT A	1–9	41.5	47
CAT A[a]	7	34	31

[a] CAT A was directly exposed to H_2O level 7 after reduction.

(equivalent to H_2O level 9 at significantly lower pressures and temperatures up to 300 °C), while the catalyst with 13 nm crystallites was found to undergo significant oxidation. Surprisingly, the smallest crystallites of 4 nm did not show any oxidation at temperatures up to 400 °C. They concluded that the reduction at elevated temperatures possibly resulted in migration of the silica, which eventually formed encapsulated cobalt crystallites.[81] In addition, the introduction of a calcination step during the preparation of the catalyst may have resulted in similar effects. The lower stability against oxidation of the smaller crystallite sizes in the present study is in accordance with thermodynamic predictions.[14] However, a high fraction of the lost magnetisation of CAT A is presumably due to the formation of cobalt silicates at the interface between the support and metallic crystallites, as only about one third of the formed oxidic cobalt, most probably CoO, was recovered during re-reduction. The preferential formation of cobalt silicates is also supported by thermodynamic considerations (Fig. 1). Defining two distinct oxidation rates, spontaneous and slow oxidation, for the experiments with CAT A‖ potentially allows to differentiate between the formation of cobalt silicates and CoO. The loss in magnetisation due to slow oxidation was compared to the recovered magnetisation in the reversibility steps (Table 6), *i.e.* recovery during exposure to atmospheres in the absence of H_2O at 220 °C and the re-reduction. As the loss due to slow oxidation matches the recovered amount, CoO is presumably formed slowly if compared to the oxidation to cobalt silicate species. The discrepancy for CAT A exposed to H_2O levels 1–9 presumably originates in the simultaneous formation of cobalt silicates and CoO at H_2O level 9.

CAT A oxidised to a high extent during exposure of the catalyst to the H_2O levels, but the formation of CoO for such a small crystallite size becomes feasible before oxidation was observed (Table 3 and Fig. 16). CAT B and CAT C were stable during exposure to H_2O level 1–4 in the absence of CO and only oxidised at higher concentrations of H_2O. This is in accordance with thermodynamic calculations[14] for CAT C as the critical diameter for oxidation at this level is 6.5 nm. In contrast, CAT B was expected to oxidise at H_2O level 4. In addition, the observed loss in magnetisation of both catalysts was less than expected, *i.e.* the samples should have experienced full surface oxidation according to thermodynamic predictions.[14] Therefore, it can be concluded that direct oxidation of metallic cobalt by H_2O is kinetically hindered leading to extenuated oxidation rates, explaining the observed slow oxidation rate to CoO (Table 6). Adsorbed atomic oxygen on the metallic cobalt surface (Co + O*) is an intermediate of the oxidation to CoO and present under real

‖ The loss in magnetisation in the experiments with CAT B and CAT C was too small, excluding reasonable differentiation between spontaneous and slow oxidation.

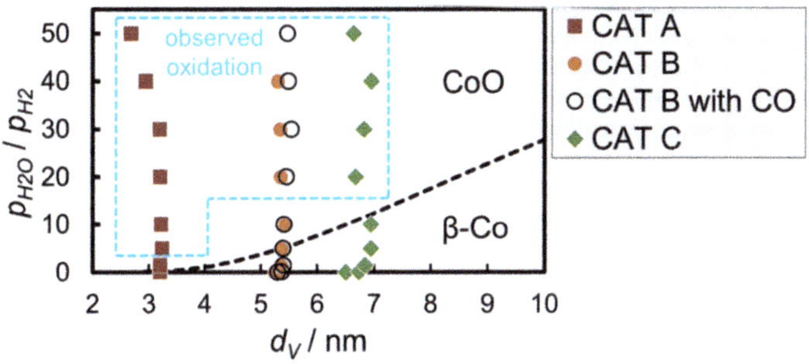

Fig. 16 Summary of experimental results with volume mean crystallite sizes of cobalt as obtained from the Langevin method after exposure to H_2O level 1–8 superimposed on the size dependent thermodynamic equilibrium of the oxidation of β-cobalt (fcc) crystallites in the nanometre size regime[14] (black, dashed line) calculated for 220 °C. The volume mean crystallite sizes for CAT C are obtained from crudely applying the Langevin equation. This procedure results in presumably smaller crystallite sizes than are actually present in the catalysts. The data within the marked area (blue, dashed) show a loss in magnetisation which can mostly be assigned to oxidation of cobalt.

FT synthesis conditions (eqn (1) and (3)). The magnetisation increased at low H_2O levels indicating the formation of hydroxyl species on the cobalt surface, but only little oxidation to CoO was observed as these hydroxyl species were either removed by adsorbed hydrogen forming water before dissociation and formation of the Co + O* intermediate, or the hydroxyl species are very stable on the catalytic surface. The latter is proposed in the literature based on mechanistic studies of the FT synthesis.[85,86] These conclusions from the exposure of the model catalysts to exclusively H_2O and H_2 are supported by the experiments in the presence of CO, which showed a pronounced oxidation with a higher extent of reversibility. As CO is proposed to play a crucial role in the oxidation to CoO, the partial pressure, as well as the absolute pressure, is a fundamental parameter and higher partial pressures of CO are expected to initiate more and/or accelerated oxidation to CoO. At the low partial pressures of CO present during the exposure of CAT B to a simulated high conversion FT environment, competitive adsorption of H_2O/hydroxyl species and CO, as shown *via* kinetic measurements, prevented a higher extent of oxidation. This hypothesis may also explain the oxidation behaviour of an alumina-supported cobalt catalyst during exposure to simulated high conversions reported by Fischer *et al.*[29] Despite a low p_{H_2O}/p_{H_2} ratio of 1, over 65 wt% of the small cobalt crystallites (volume mean diameter of 4.1 nm) oxidised almost spontaneously, which was probably due to the significantly higher CO partial pressure. Furthermore, cobalt aluminates are expected to form to a higher extent than cobalt silicates (Fig. 1). Claeys *et al.* exposed an alumina-supported cobalt catalyst to high concentrations of H_2O in the absence and presence of CO.[9] In contrast to the present study, the catalyst was more prone to oxidation in the absence of CO even though the partial pressure of CO was significantly higher than in the present study. However, the observations are not conflictive as the alumina support allowed an Ostwald ripening sintering mechanism with CO playing a crucial role. This sintering led to larger crystallite sizes which show a higher stability to oxidation. Furthermore, the relative loss in magnetisation due to oxidation in the absence of CO was low and possibly also due to the formation of cobalt aluminates, which might support the hypothesis that CO dissociation is a prerequisite for significant oxidation to CoO.

The high resistance of the cobalt crystallites in the synthesised model catalysts to hydrothermal sintering was confirmed by TEM analyses after passivation of the spent samples (Fig. S11 and S12†) and may be explained with the drawn conclusions: (1) competitive adsorption of H_2O/hydroxyl species and CO not only prevented a higher extent of oxidation, but also sintering. The Co + O* intermediate, which is reported to have an increased mobility leading to a pronounced sintering,[39] is absent or only present at very low concentrations. In addition, the CO partial pressures may have been too low to allow for sub-carbonyl type sintering as proposed for alumina-supported cobalt crystallites.[9] (2) The possibly much quicker and almost spontaneous oxidation of cobalt to metal–support compounds at the interface between silica and cobalt stabilises the crystallites suppressing their mobility. (3) The cobalt crystallites present in each model catalyst are highly monodisperse, which is reported to increase the sintering resistance of other metals.[87,88] (4) The low cobalt loading of below 0.5 wt% allows a high dispersion of the crystallites on the support, lowering their susceptibility to sintering.[45]

Conclusions

The mechanism of H_2O induced oxidation of metallic cobalt to CoO appears to be an indirect one with H_2O hindering the removal of adsorbed oxygen on the surface of metallic cobalt. Significant oxidation to CoO requires actual Fischer–Tropsch synthesis on the surface of the cobalt crystallites as the dissociation of CO is the fundamental step in supplying adsorbed oxygen species for oxidation. Direct oxidation by H_2O seems to be kinetically hindered, but occurs spontaneously at the interface between the silica support and cobalt crystallites *via* the formation of cobalt silicate species leading to a pronounced loss in metallic cobalt, especially for smaller crystallites. Larger cobalt crystallites have higher stability. The presence of metal–support compounds drastically increases the stability of the crystallites, which, at least for low partial pressures of CO, overcomes the drive to minimise the surface energy *via* a crystallite migration sintering mechanism. This absence of sintering was crucial for the present study as it allowed the isolated examination of the oxidation process.

Acknowledgements

The authors would like to express their gratitude to the organising committee for giving the opportunity to present their work at 'Faraday Discussion – Catalysis for Fuels' in Cape Town, South Africa, from 24th–26th of January 2017. Financial support from the University of Cape Town, the DST-NRF Centre of Excellence in Catalysis (c*change), the German Academic Exchange Service (DAAD), and the Catalysis Society of South Africa (CATSA) is gratefully acknowledged. Special thanks goes to Stephanie La Grange and Zulfa Le Riche of the Analytical Laboratory for conducting the physisorption experiment and the elemental analyses.

References

1 C. H. Bartholomew, *Appl. Catal., A*, 2001, **212**, 17–60.

2 M. S. Spencer, in *Catalyst Handbook*, ed. M. V. Twigg, Wolfe, Frome, England, 2nd edn, 1989, pp. 17–84.
3 J. A. Moulijn, A. E. van Diepen and F. Kapteijn, in *Handbook of Heterogeneous Catalysis*, ed. G. Ertl, H. Knözinger, F. Schüth and J. Weitkamp, Wiley-VCH, Weinheim, Germany, 2nd edn, 2008, pp. 1829–1847.
4 D. J. Moodley, J. van de Loosdrecht, A. M. Saib and H. J. W. Niemantsverdriet, in *Advances in Fischer–Tropsch Synthesis, Catalysts, and Catalysis*, ed. B. H. Davis and M. L. Occelli, CRC Press, Boca Raton, USA, 2012, pp. 49–81.
5 M. E. Dry, in *Fischer–Tropsch Technology*, ed. A. P. Steynberg and M. E. Dry, Elsevier, Amsterdam, 2004, vol. 152, pp. 533–600.
6 N. E. Tsakoumis, M. Rønning, Ø. Borg, E. Rytter and A. Holmen, *Catal. Today*, 2010, **154**, 162–182.
7 G. Jacobs, P. M. Patterson, Y. Zhang, T. Das, J. Li and B. H. Davis, *Appl. Catal., A*, 2002, **233**, 215–226.
8 A. Tavasoli, A. Nakhaeipour and K. Sadaghiani, *Fuel Process. Technol.*, 2007, **88**, 461–469.
9 M. Claeys, M. E. Dry, E. van Steen, P. J. van Berge, S. Booyens, R. Crous, P. van Helden, J. Labuschagne, D. J. Moodley and A. M. Saib, *ACS Catal.*, 2015, **5**, 841–852.
10 J. van de Loosdrecht, B. Balzhinimaev, J.-A. Dalmon, J. W. Niemantsverdriet, S. V. Tsybulya, A. M. Saib, P. J. van Berge and J. L. Visagie, *Catal. Today*, 2007, **123**, 293–302.
11 A. K. Dalai and B. H. Davis, *Appl. Catal., A*, 2008, **348**, 1–15.
12 I. M. Ciobîcă, R. A. van Santen, P. J. van Berge and J. van de Loosdrecht, *Surf. Sci.*, 2008, **602**, 17–27.
13 P. J. van Berge, J. van de Loosdrecht, S. Barradas and A. M. van der Kraan, *Catal. Today*, 2000, **58**, 321–334.
14 E. van Steen, M. Claeys, M. E. Dry, J. van de Loosdrecht, E. L. Viljoen and J. L. Visagie, *J. Phys. Chem. B*, 2005, **109**, 3575–3577.
15 R. B. Anderson, in *Hydrocarbon synthesis, hydrogenation and cyclization*, ed. P. H. Emmett and J. K. Dixon, Reinhold, New York, 1956, vol. 4, pp. 29–255.
16 C. Hou, G. Xia, X. Sun, Y. Wu, C. Jin, Z. Yan, M. Li, Z. Hu, H. Nie and D. Li, *Catal. Today*, 2016, **264**, 91–97.
17 *CRC handbook of thermophysical and thermochemical data*, ed. D. R. Lide and H. V. Kehiaian, CRC Press, Boca Raton, USA, 1994.
18 O. Kitakami, H. Sato, Y. Shimada, F. Sato and M. Tanaka, *Phys. Rev. B: Condens. Matter Mater. Phys.*, 1997, **56**, 13849–13854.
19 E. Iglesia, *Appl. Catal., A*, 1997, **161**, 59–78.
20 D. I. Enache, B. Rebours, M. Roy-Auberger and R. Revel, *J. Catal.*, 2002, **205**, 346–353.
21 *Thermochemical properties of inorganic substances*, ed. O. Knacke, O. Kubaschewski and K. Hesselmann, Springer, Berlin, 2nd edn, 1991.
22 D. Schanke, A. M. Hilmen, E. Bergene, K. Kinnari, E. Rytter, E. Ådanes and A. Holmen, *Catal. Lett.*, 1995, **34**, 269–284.
23 M. Rothaemel, K. F. Hanssen, E. A. Blekkan, D. Schanke and A. Holmen, *Catal. Today*, 1997, **38**, 79–84.
24 A. Tavasoli, R. M. Malek Abbaslou and A. K. Dalai, *Appl. Catal., A*, 2008, **346**, 58–64.

25 G. L. Bezemer, T. J. Remans, A. P. van Bavel and A. I. Dugulan, *J. Am. Chem. Soc.*, 2010, **132**, 8540–8541.
26 A. M. Saib, A. Borgna, J. van de Loosdrecht, P. J. van Berge and J. W. Niemantsverdriet, *Appl. Catal., A*, 2006, **312**, 12–19.
27 G.-Z. Bian, N. Fujishita, T. Mochizuki, W.-S. Ning and M. Yamada, *Appl. Catal., A*, 2003, **252**, 251–260.
28 H. Karaca, O. V. Safonova, S. Chambrey, P. Fongarland, P. Roussel, A. Griboval-Constant, M. Lacroix and A. Y. Khodakov, *J. Catal.*, 2011, **277**, 14–26.
29 N. Fischer, B. Clapham, T. Feltes, E. van Steen and M. Claeys, *Angew. Chem., Int. Ed.*, 2014, **53**, 1342–1345.
30 J. Scalbert, C. Legens, I. Clémençon, A.-L. Taleb, L. Sorbier and F. Diehl, *Chem. Commun.*, 2014, **50**, 7866–7869.
31 A. M. Saib, A. Borgna, J. van de Loosdrecht, P. J. van Berge and J. W. Niemantsverdriet, *J. Phys. Chem. B*, 2006, **110**, 8657–8664.
32 D. J. Moodley, A. M. Saib, J. van de Loosdrecht, C. A. Welker-Nieuwoudt, B. H. Sigwebela and J. W. Niemantsverdriet, *Catal. Today*, 2011, **171**, 192–200.
33 *Perry's Chemical Engineers' Handbook*, ed. D. W. Green and R. H. Perry, McGraw-Hill, New York, 8th edn, 2008.
34 J. A. Moulijn, A. E. van Diepen and F. Kapteijn, *Appl. Catal., A*, 2001, **212**, 3–16.
35 W. Thomson, *Philos. Mag.*, 1871, **42**, 448–452.
36 E. Roduner, *Chem. Soc. Rev.*, 2006, **35**, 583–592.
37 P. R. Couchman and W. A. Jesser, *Nature*, 1977, **269**, 481–483.
38 M. E. Dry, *Catal. Today*, 2002, **71**, 227–241.
39 M. Sadeqzadeh, J. Hong, P. Fongarland, D. Curulla-Ferré, F. Luck, J. Bousquet, D. Schweich and A. Y. Khodakov, *Ind. Eng. Chem. Res.*, 2012, **51**, 11955–11964.
40 T. K. Das, G. Jacobs, P. M. Patterson, W. A. Conner, J. Li and B. H. Davis, *Fuel*, 2003, **82**, 805–815.
41 A. M. Saib, D. J. Moodley, I. M. Ciobîcă, M. M. Hauman, B. H. Sigwebela, C. J. Weststrate, J. W. Niemantsverdriet and J. van de Loosdrecht, *Catal. Today*, 2010, **154**, 271–282.
42 M. Sadeqzadeh, S. Chambrey, J. Hong, P. Fongarland, F. Luck, D. Curulla-Ferré, D. Schweich, J. Bousquet and A. Y. Khodakov, *Ind. Eng. Chem. Res.*, 2014, **53**, 6913–6922.
43 G. L. Bezemer, J. H. Bitter, H. P. C. E. Kuipers, H. Oosterbeek, J. E. Holewijn, X. Xu, F. Kapteijn, A. J. van Dillen and K. P. de Jong, *J. Am. Chem. Soc.*, 2006, **128**, 3956–3964.
44 N. Fischer, E. van Steen and M. Claeys, *J. Catal.*, 2013, **299**, 67–80.
45 G. Prieto, J. Zečević, H. Friedrich, K. P. de Jong and P. E. de Jongh, *Nat. Mater.*, 2012, **12**, 34–39.
46 J. Sun, D. Ma, H. Zhang, X. Liu, X. Han, X. Bao, G. Weinberg, N. Pfänder and D. Su, *J. Am. Chem. Soc.*, 2006, **128**, 15756–15764.
47 G. Kiss, C. E. Kliewer, G. J. DeMartin, C. C. Culross and J. E. Baumgartner, *J. Catal.*, 2003, **217**, 127–140.
48 H. Karaca, J. Hong, P. Fongarland, P. Roussel, A. Griboval-Constant, M. Lacroix, K. Hortmann, O. V. Safonova and A. Y. Khodakov, *Chem. Commun.*, 2010, **46**, 788–790.
49 M. Sadeqzadeh, S. Chambrey, S. Piché, P. Fongarland, F. Luck, D. Curulla-Ferré, D. Schweich, J. Bousquet and A. Y. Khodakov, *Catal. Today*, 2013, **215**, 52–59.

50 W. Stöber, *J. Colloid Interface Sci.*, 1968, **26**, 62–69.
51 X.-D. Wang, Z.-X. Shen, T. Sang, X.-B. Cheng, M.-F. Li, L.-Y. Chen and Z.-S. Wang, *J. Colloid Interface Sci.*, 2010, **341**, 23–29.
52 N. Shi, W. Cheng, H. Zhou, T. Fan and M. Niederberger, *Chem. Commun.*, 2015, **51**, 1338–1340.
53 M. Wolf, N. Fischer and M. Claeys, 2016, in preparation.
54 *ICDD, PDF-2 (Database)*, International Centre for Diffraction Data, Newtown Square, USA, 2008.
55 A. A. Coelho, *J. Appl. Crystallogr.*, 2003, **36**, 86–95.
56 P. Scherrer, *Nachr. Ges. Wiss. Goettingen, Math.-Phys. Kl.*, 1918, 98–100.
57 G. Bergeret and P. Gallezot, in *Handbook of Heterogeneous Catalysis*, ed. G. Ertl, H. Knözinger, F. Schüth and J. Weitkamp, Wiley-VCH, Weinheim, Germany, 2nd edn, 2008, pp. 738–765.
58 W. Rasband, *ImageJ 1.51a*, National Institute of Mental Health, Bethesda, USA, 2016.
59 M. Claeys, M. E. Dry, E. van Steen, E. du Plessis, P. J. van Berge, A. M. Saib and D. J. Moodley, *J. Catal.*, 2014, **318**, 193–202.
60 M. Claeys, E. van Steen, J. Visagie and J. van de Loosdrecht, *US Pat.*, 8,773,118 B2, 2014.
61 C. Kittel, *Introduction to solid state Physics*, John Wiley Sons, Inc, Hoboken, USA, 8th edn, 2005.
62 L. F. Bates, *Modern Magnetism*, The Syndics of the Cambridge University Press, Cambridge, 4th edn, 1963.
63 K. J. Carroll, Z. J. Huba, S. R. Spurgeon, M. Qian, S. N. Khanna, D. M. Hudgins, M. L. Taheri and E. E. Carpenter, *Appl. Phys. Lett.*, 2012, **101**, 12409.
64 H. P. Myers and W. Sucksmith, *Proc. R. Soc.London, Ser. A*, 1951, **207**, 427–446.
65 C. G. Shull, W. A. Strauser and E. O. Wollan, *Phys. Rev.*, 1951, **83**, 333–345.
66 W. L. Roth, *J. Phys.*, 1964, **25**, 507–515.
67 H. Miyamoto, *Bull. Inst. Chem. Res., Kyoto Univ.*, 1966, **44**, 420–429.
68 V. G. Harris, Y. Chen, A. Yang, S. Yoon, Z. Chen, A. L. Geiler, J. Gao, C. N. Chinnasamy, L. H. Lewis, C. Vittoria, E. E. Carpenter, K. J. Carroll, R. Goswami, M. A. Willard, L. Kurihara, M. Gjoka and O. Kalogirou, *J. Phys. D: Appl. Phys.*, 2010, **43**, 165003.
69 A. Sawaoka, S. Miyahara and S.-I. Akimito, *J. Phys. Soc. Jpn.*, 1968, **25**, 1253–1257.
70 S. Nomura, R. Santoro, J. Fang and R. Newham, *J. Phys. Chem. Solids*, 1964, **25**, 901–905.
71 J.-A. Dalmon, in *Catalyst Characterization: Physical Techniques for Solid Materials*, ed. B. Imelik and J. C. Vedrine, Plenum Press, New York, 1994, pp. 585–609.
72 A. Barbier, A. Hanif, J.-A. Dalmon and G. A. Martin, *Appl. Catal., A*, 1998, **168**, 333–343.
73 A. Barbier, G.-A. Martin, P. Ramirez de la Piscina and N. Homs, *Appl. Catal., A*, 2001, **210**, 75–81.
74 V. Sechovsky, in *Encyclopedia of Materials: Science and Technology*, ed. K. H. J. Buschow, R. W. Cahn, M. C. Flemings, B. Ilschner, E. J. Kramer and S. Mahajan, Elsevier, Amsterdam, 2001, vol. 5, pp. 5018–5032.
75 A. Hernando, P. Crespo, P. Marìn and A. González, in *Encyclopedia of Materials: Science and Technology*, ed. K. H. J. Buschow, R. W. Cahn, M. C. Flemings, B.

Ilschner, E. J. Kramer and S. Mahajan, Elsevier, Amsterdam, 2001, vol. 5, pp. 4780–4787.
76 P. A. Chernavskii, J.-A. Dalmon, N. S. Perov and A. Y. Khodakov, *Oil Gas Sci. Technol.*, 2009, **64**, 25–48.
77 M. Wolf, N. Fischer and M. Claeys, *Catal. Today*, 2016, **275**, 135–140.
78 H. Schulz, W. Boehringer, C. Kohl, N. Rahman and A. Will, in *DGMK Forschungsbericht 320*, DGMK, Hamburg, 1984, pp. 320–329.
79 V. K. Pecharsky and P. Y. Zavalij, *Fundamentals of powder diffraction and structural characterization of materials*, Springer, New York, 2005.
80 J. Zuo, C. Xu, Y. Liu and Y. Quian, *Nanostruct. Mater.*, 1998, **10**, 1331–1335.
81 A. M. Saib, A. Borgna, J. van de Loosdrecht, P. J. van Berge, J. W. Geus and J. W. Niemantsverdriet, *J. Catal.*, 2006, **239**, 326–339.
82 C. J. Weststrate, M. M. Hauman, D. J. Moodley, A. M. Saib, E. van Steen and J. W. Niemantsverdriet, *Top. Catal.*, 2011, **54**, 811–816.
83 E. van Steen, G. S. Sewell, R. A. Makhothe, C. Micklethwaite, H. Manstein, M. de Lange and C. T. O'Connor, *J. Catal.*, 1996, **162**, 220–229.
84 G. Bagnasco, C. Cammarano, M. Turco, S. Esposito, A. Aronne and P. Pernice, *Thermochim. Acta*, 2008, **471**, 51–54.
85 I. A. W. Filot, R. A. van Santen and E. J. M. Hensen, *Angew. Chem., Int. Ed.*, 2014, **53**, 12746–12750.
86 A. C. Kizilkaya, J. W. Niemantsverdriet and C. J. Weststrate, *J. Phys. Chem. C*, 2016, **120**, 4833–4842.
87 K. Wettergren, F. F. Schweinberger, D. Deiana, C. J. Ridge, A. S. Crampton, M. D. Rötzer, T. W. Hansen, V. P. Zhdanov, U. Heiz and C. Langhammer, *Nano Lett.*, 2014, **14**, 5803–5809.
88 D. A. Chen, M. C. Bartelt, S. M. Seutter and K. F. McCarty, *Surf. Sci.*, 2000, **464**, L708–L714.
89 Q. Jiang and H. M. Lu, *Surf. Sci. Rep.*, 2008, **63**, 427–464.

Faraday Discussions

PAPER

Effect of crystallite size on the performance and phase transformation of Co_3O_4/Al_2O_3 catalysts during CO-PrOx – an *in situ* study†

Thulani M. Nyathi,[a] Nico Fischer,[a] Andy P. E. York[b] and Michael Claeys*[a]

Received 11th October 2016, Accepted 17th October 2016
DOI: 10.1039/c6fd00217j

The preferential oxidation of carbon monoxide has been identified as an effective route to remove trace amounts of CO (approx. 0.5–1.0 vol%) in the H_2-rich reformate gas stream after the low-temperature water–gas shift. Instead of noble metal-based catalysts, Co_3O_4-based catalysts were investigated in this study as cheaper and more readily available alternatives. This study aimed at investigating the effect of crystallite size on the mass- and surface area-specific CO oxidation activity as well as on the reduction behaviour of Co_3O_4. Model Co_3O_4 catalysts with average crystallite sizes between 3 and 15 nm were synthesised using the reverse micelle technique. Results from the catalytic tests revealed that decreasing the size of the Co_3O_4 crystallites increased the mass-specific CO oxidation activity in the 50–200 °C temperature range. On the other hand, the surface area-specific CO oxidation activity displayed a volcano-type behaviour where crystallites with an average size of 8.5 nm were the most active within the same temperature range. *In situ* characterisation in the magnetometer revealed that the Co_3O_4 crystallites are partially reduced to metallic Co above 225 °C with crystallites larger than 7.5 nm showing higher degrees of reduction under the H_2-rich environment of CO-PrOx. *In situ* PXRD experiments further showed the presence of CoO concurrently with metallic fcc Co in all the catalysts during the CO-PrOx runs. In all experiments, the formation of fcc Co coincided with the formation of CH_4. Upon decreasing the reaction temperature below 250 °C under the reaction gas, both *in situ* techniques revealed that the fcc Co previously formed is partially re-oxidised to CoO.

Introduction

With the present demand for clean alternative energy sources, the production of high purity hydrogen (H_2) for fuel cells has been gaining increasing attention over

[a]*Catalysis Institute and c*change (DST-NRF Centre of Excellence in Catalysis), Department of Chemical Engineering, University of Cape Town, Rondebosch 7701, South Africa. E-mail: Michael.Claeys@uct.ac.za*
[b]*Johnson Matthey Technology Centre, Sonning Common, Reading RG4 9NH, UK*
† Electronic supplementary information (ESI) available. See DOI: 10.1039/c6fd00217j

the past decades.[1–5] As one of the means of manufacturing H_2, the steam reforming of natural gas followed by a high- and low-temperature water–gas shift (WGS) reaction affords a reformate gas stream with H_2 in the 40–75 vol% range.[5] Water (H_2O), carbon dioxide (CO_2) and carbon monoxide (CO) are also produced with CO, in particular, in the concentration range of 0.5–1.0 vol%.[4,5] This amount of CO in the H_2-rich reformate gas stream has been reported to be detrimental to the platinum-based anode catalyst of proton-exchange membrane fuel cells (PEMFCs) and therefore there is a great need for finding a process which will further reduce the concentration of CO to below 10 ppmv.[1–5]

Processes including pressure swing adsorption, membrane separation, selective methanation and the preferential oxidation of CO (CO-PrOx) have been proposed and explored for the above-mentioned purpose.[1,2,4,6,7] Among these, CO-PrOx is regarded as the most efficient and cost-effective.[1,2,4,6,7] CO-PrOx makes use of co-fed oxygen (O_2) in the form of air to preferentially oxidise the CO present in the H_2-rich reformate gas to CO_2 with a minimal loss of H_2 *via* oxidation to H_2O.[4,8–10] Therefore, it is of utmost importance to find a cheap, active, stable and selective catalyst for this process.[4,8–10] Noble metal-based catalysts (*e.g.*, Pt, Rh, Pd and Ru) are being widely investigated for this process and have been shown to have very good activity and selectivity.[1–4,8–10] However, these catalysts as well as their precursors are relatively expensive. Therefore, cheaper alternatives such as base metal oxides (*e.g.*, CuO and Co_3O_4) are of great research interest currently.[11–13]

In this study, Co_3O_4 crystallites supported on Al_2O_3 were investigated as catalysts for the CO-PrOx reaction. Co_3O_4 has been shown to have very good CO oxidation activity in the absence of H_2.[14–16] Furthermore, the mass- and surface area-specific CO oxidation rates of Co_3O_4 are affected by crystallite size in the absence of H_2.[14–16] To the best of our knowledge, studies investigating the effect of crystallite size on the catalytic activity of Co_3O_4 during CO-PrOx have not been reported.

As a deactivation mechanism under the H_2-rich environment of CO-PrOx, evidence based on post-run *ex situ* PXRD analysis pointing towards the reduction of Co_3O_4 to metallic Co has been reported.[13] Others have observed the formation of methane (CH_4) during CO-PrOx at temperatures above 225 °C at atmospheric pressure, which is also indicative of the reduction of Co_3O_4 to metallic Co.[13,17,18] More recently, Lukashuk *et al.*,[19] reported their findings after performing operando X-ray absorption spectroscopy (XAS) and near atmospheric pressure X-ray photoelectron spectroscopy (NAS-XPS) experiments to monitor the bulk and surface changes, respectively, of Co_3O_4 during CO-PrOx. They were able to confirm that indeed Co_3O_4 reduces to CoO and ultimately to metallic Co during CO-PrOx due to the presence of H_2. They also showed that these phase changes negatively affect the CO_2 yield and selectivity because upon metallic Co formation, more CO is hydrogenated to CH_4 and less is oxidised to CO_2. However, from our review of the literature, there has not been a report on how crystallite size affects the bulk reduction of Co_3O_4 during CO-PrOx, leading to the undesired formation of CH_4.

Therefore, the study presented here aimed at investigating the effect of the size of Co_3O_4 crystallites on the mass- and surface area-specific CO oxidation activity as well as on the reduction behaviour of Co_3O_4 during CO-PrOx. The reduction of Co_3O_4 was monitored *in situ* using two complementary in-house developed techniques, *viz.* a magnetometer[20–23] and a PXRD capillary cell[22–26] while capturing kinetic data.

Experimental

Catalyst preparation

Five catalysts with varying Co_3O_4 crystallite sizes were synthesised using the reverse micelle technique as previously described.[27,28] The exact composition of the reverse micelle solutions prepared is shown in Table 1. Briefly, to obtain a stable and homogeneous reverse micelle solution at room temperature and atmospheric pressure, an aqueous solution of $Co(NO_3)_2 \cdot 6H_2O$ (Sigma Aldrich, reagent grade 98% purity) was added to a stirring mixture of hexane (AR grade, Kimix) and the non-ionic surfactant pentaethylene glycol dodecylether (PEGDE) (Akzo Nobel). Thereafter, aqueous NH_3 (25 wt%, Kimix) at a 1 : 4 Co^{2+} : NH_3 molar ratio was added to the prepared reverse micelle solution to initiate the precipitation process, followed by the dropwise addition of acetone (AR grade, Kimix) to destabilise the reverse micelles and liberate the precipitated particles. Acetone was further used to wash the precipitated particles and rid them of excess surfactant. The particles were dried and calcined at 120 °C and 200 °C respectively. The obtained Co_3O_4 powder was re-dispersed in distilled water under ultrasonication at room temperature and atmospheric pressure for 60 minutes and thereafter the suspension was transferred to a pre-weighed dry powder of Al_2O_3 (Puralox, SCCa 5-150, Sasol Germany: $S_{BET} = 162$ m^2 g^{-1}, $V_{pore} = 0.47$ cm^3 g^{-1}, $\bar{d}_{pore} = 11.5$ nm, $\bar{d}_{particle} = 150$–200 μm) in order to achieve a loading of 10 wt % Co_3O_4. The water was evaporated under reduced pressure in a rotary evaporator. The oil-to-surfactant ratio, water-to-surfactant ratio and amount of the cobalt precursor were carefully varied in order to achieve different sizes of Co_3O_4 crystallites between 3 and 15 nm (see Table 1).

Table 1 Composition of the reverse micelle solutions prepared at room temperature and atmospheric pressure for obtaining different crystallite sizes of Co_3O_4

Sample name	Targeted size (nm)	Hexane (g)	PEGDE (g)	Water (g)	o/s[a] (mol/mol)	w/s[b] (ω, mol/mol)	$Co(NO_3)_2 \cdot 6H_2O$ (g)
TN_01	13.0–15.0	250.0	22.0	15.0	53.6	15.4	4.4
TN_02	10.0–12.0	250.0	25.0	15.0	47.2	13.7	3.3
TN_03	7.0–9.0	500.0	68.3	26.4	34.5	8.7	2.6
TN_04	4.0–6.0	500.0	68.4	18.2	34.5	6.0	1.6
TN_05	3.0–5.0	875.0	161.0	12.0	25.6	1.7	1.3

[a] Oil-to-surfactant molar ratio. [b] Water-to-surfactant molar ratio (ω).

Ex situ catalyst characterisation

Powder X-ray diffraction (PXRD) was performed using a Bruker D8 Advance X-ray diffractometer equipped with a cobalt source ($\lambda_{K\alpha1} = 1.78897$ Å) and a position sensitive detector (Bruker Vantec). For all samples, the optics were set to parallel beam geometry so as to minimize possible peak shifts due to sample height difference (sample displacement). All recorded diffraction patterns were compared to known diffraction patterns from the International Centre for Diffraction Data PDF-2 database to determine the species present. To further identify phases and to obtain the average crystallite size, Rietveld refinement utilising the software package Topas 4.2 was carried out. As the alumina support material used in this study is known to be present as a mixed phase of γ-Al_2O_3 and δ-Al_2O_3, and as the crystal structure of δ-Al_2O_3 is not known at this stage, an approach for partial or not known crystal structures (PONKCS), first described by Scarlett and Madsen,[29] was used.

Transmission electron microscopy was performed for each fresh and spent sample using a Tecnai F20 transmission electron microscope operated at 200 kV with a field emission gun. The obtained micrographs were analysed using the freeware ImageJ in order to ultimately obtain average crystallite sizes and size distributions.

The Co_3O_4 loading was determined by energy-dispersive X-ray (EDX) spectroscopy using a LEO 1450 SEM/EDX instrument.

The reducibility of the individual catalysts was assessed by hydrogen temperature-programmed reduction (H_2-TPR) performed in a Micromeritics AutoChem 2910 instrument equipped with a thermal conductivity detector (TCD). 0.1 g freshly prepared supported catalyst was loaded in a U-tube quartz reactor and heated from 60–920 °C using a 10 °C min^{-1} ramp rate in a flow of 5 vol% H_2 in Ar (50 mL (NTP) per min).

In situ catalyst characterisation and testing

Two novel in-house developed *in situ* instruments were used for the catalyst testing experiments, *viz.* a low-frequency vibrating sample magnetometer with a maximum field strength of 2 T (or 20 kOe)[20–23] as well as a PXRD capillary cell mounted on the Bruker D8 Advance Laboratory X-ray diffractometer[22–26] already mentioned above. The magnetometer is based on the Weiss extraction method and was specifically designed for the detection of ferromagnetic and superparamagnetic materials. In the present study, this is only metallic Co since Co_3O_4 and CoO are antiferromagnetic[30–32] (see Section A in the ESI† for the definitions of ferromagnetism, superparamagnetism and antiferromagnetism). On the other hand, the PXRD capillary cell instrument can detect the crystalline phases Co_3O_4, CoO and Co up to the intrinsic instrument limitations regarding crystallite size and concentration.

In the magnetometer, the temperature was held for 60 minutes at every 25 °C between 50 and 350 °C upon heating and cooling, while magnetisation measurements were taken at field strengths of 2 T and 0 T (*i.e.*, 20 kOe and 0 kOe). This and data from a previously performed calibration of a 0.1 g freshly reduced metallic cobalt sample enabled the calculation of the degree of reduction, defined as the amount of metallic cobalt formed relative to the amount of cobalt present as Co_3O_4 in the starting material (see Fig. S.1 as well as eqn (S.1) and (S.2) in the ESI†). Similarly, PXRD patterns were also recorded during the 60 minute holding time every 10 minutes. A 2θ range of 40–90° was chosen for each scan in *in situ* mode.

Both instruments are based on a single fixed bed reactor and for the experiments performed in each instrument, the gas composition and the weight-hourly space velocity (WHSV) were kept constant. The gas was composed of 0.9 vol% CO, 0.9 vol% O_2 and 52.1 vol% H_2 in N_2. The gases CO_2 and H_2O were not co-fed as this study aimed at investigating the sole effect of H_2 on the catalytic performance and reduction of Co_3O_4. The presence of CO_2 and H_2O may mask this effect. However, the effect of both H_2O and CO_2 will be a topic of a future study. The amount of the supported catalyst loaded into the magnetometer reactor was varied between 1.0 and 1.2 g to compensate for the slight differences in the Co_3O_4 loadings (see the EDX results in

Table 2) and the gas flow rate was kept at 100 mL (NTP) per min. The mass of catalyst loaded into the PXRD capillary cell reactor was 0.015 g with the gas flow rate varied between 1.3 and 1.6 mL (NTP) per min. As mentioned above, the temperature was varied between 50 and 350 °C under the reacting gas using a 1 °C per min heating and cooling rate. The products were analysed using an Agilent Technologies 490 micro-GC fitted with three analysis modules equipped with thermal conductivity detectors for detecting CO, O_2, H_2, CO_2, CH_4 and N_2. As columns to separate the gases, two Molecular Sieve 5 Å PLOT columns of 10 and 20 m lengths were employed. In the shorter column, Ar was chosen as the carrier gas to allow for the detection of H_2 while H_2 was chosen as the carrier gas in the 20 m column separating O_2, N_2, CH_4 and CO. The third column was a 10 m PoraPlot Q column with H_2 as the carrier gas and was mainly used for the detection of CO_2. It is worth noting that the conversion of H_2 was too low to be accurately measured during the experiments, however, the H_2O selectivity based on the conversion of O_2 ($S^{O_2}_{H_2O}$) can be expected to be $100 - S^{O_2}_{CO_2}$ where $S^{O_2}_{CO_2}$ is the CO_2 selectivity based on the conversion of O_2.

Results and discussion

Ex situ catalyst characterisation

The five catalysts were characterised by means of PXRD, TEM, EDX and H_2-TPR. The PXRD patterns of the five supported catalysts are shown in Fig. 1. The obtained patterns clearly show that the catalysts consist only of Co_3O_4 and the Al_2O_3 support as expected. The differences in the intensity and width of the Co_3O_4 peaks between the individual diffraction patterns indicate that the catalysts have Co_3O_4 crystallites of different average sizes.

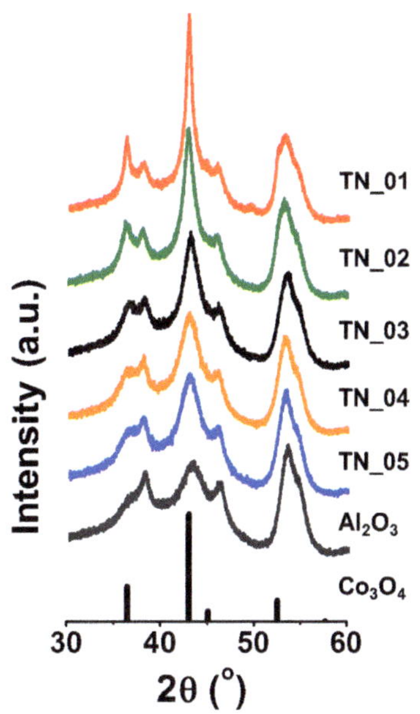

Fig. 1 PXRD patterns of the supported fresh samples of TN_01–TN_05 as well as the recorded diffraction pattern of the Al_2O_3 support and the reference diffraction lines of Co_3O_4 (ICDD card no. 01-073-1701).

Selected TEM micrographs of TN_01 and TN_05 in the unsupported state and the number-based size distributions for all five catalysts *i.e.*, TN_01–TN_05 are shown in Fig. 2 and 3 respectively. Although the crystallites appear agglomerated in the TEM micrographs in Fig. 2, they are less agglomerated on the Al_2O_3 support (see Fig. S.2 in the ESI†) as a result of the 60 minute ultrasonication performed prior to their deposition onto the support (see Experimental section). However, due to the poor contrast between the crystallites of Co_3O_4 and those of the Al_2O_3 support in the micrographs of the supported catalysts, obtaining a size distribution and average size was challenging (see Fig. S.2 in the ESI†). Therefore, it was seen best to extract size information from the micrographs of the unsupported material. It should be noted that the average size of the unsupported and supported crystallites is not expected to be different due to the mild conditions used in the supporting procedure followed in this work (see Experimental section). The TEM size distributions of the unsupported Co_3O_4 crystallites appear fairly narrow and support the claim that the reverse micelle technique can tune and control the size of the crystallites, in this case, to be between 3 and 15 nm.

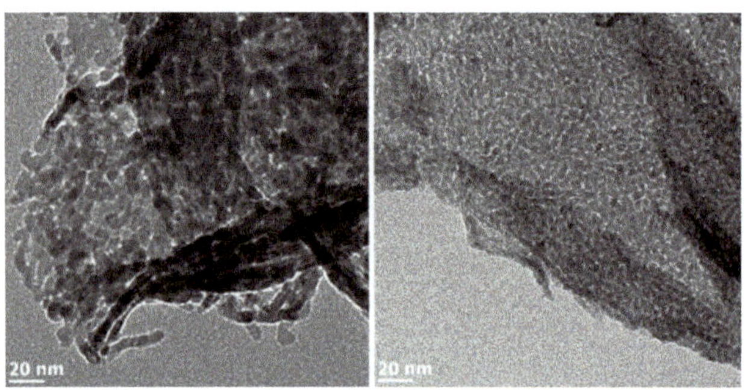

Fig. 2 Selected TEM micrographs of the unsupported fresh Co_3O_4 samples of (left) TN_01 and (right) TN_05.

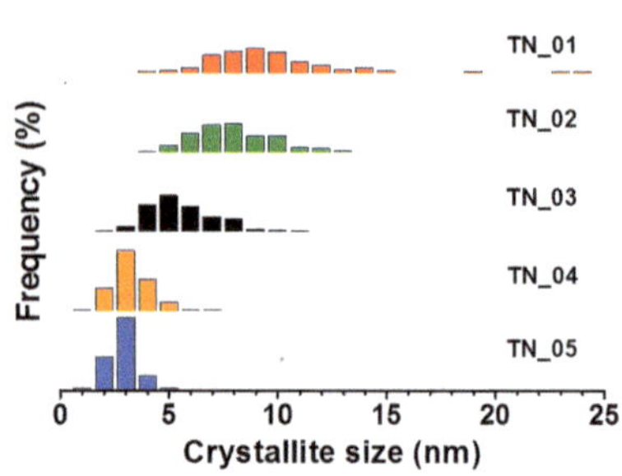

Fig. 3 Number-based size distributions of the unsupported fresh samples of TN_01–TN_05 obtained by means of TEM analysis.

Table 2 PXRD and TEM volume-based average crystallite sizes for TN_01–TN_05 as well as the EDX-derived Co_3O_4 loadings

Sample name	Targeted size (nm)	PXRD size[a] (nm)	TEM size[b] (nm)	TEM size[c] (nm)	Targeted Co_3O_4 loading (wt%)	EDX Co_3O_4 loading (wt%)
TN_01	13.0–15.0	15.3	10.1 ± 2.8	13.8 ± 3.2	10.0	10.9
TN_02	10.0–12.0	10.2	8.5 ± 1.7	9.5 ± 1.9	10.0	8.8
TN_03	7.0–9.0	7.5	6.1 ± 1.5	7.2 ± 1.6	10.0	9.2
TN_04	4.0–6.0	5.1	3.8 ± 0.8	4.5 ± 1.0	10.0	8.5
TN_05	3.0–5.0	4.5	3.3 ± 0.6	3.7 ± 0.7	10.0	9.4

[a] Volume-based average crystallite size obtained by Rietveld refinement with Topas 4.1. [b] TEM number-based average crystallite sizes obtained by: $d_{c,n} = \sum_{i=1}^{N} n_i d_i / N$. [c] TEM volume-based average crystallite size calculated by: $d_{c,v} = \sum_{i=1}^{N} n_i d_i^4 \Big/ \sum_{i=1}^{N} n_i d_i^3$.

PXRD- and TEM-derived Co_3O_4 volume-based average sizes are compared in Table 2. The sizes derived from these two techniques were in good agreement and fall in the desired size range of 3–15 nm. The obtained results also show how varying the different parameters of the reverse micelle solutions (*i.e.*, the o/s and w/s ratios as well as the amount of the cobalt precursor used (see also Table 1)) affects the final size of the crystallites. Also shown in Table 2 are the EDX-derived Co_3O_4 loadings, which are all close to the targeted 10 wt%.

To assess the reducibility of each catalyst, potentially giving some indication on the reduction behaviour which can be expected during the catalyst testing, H_2-TPR was conducted with all prepared catalysts (see Fig. 4). The profiles show that the catalysts with smaller crystallites (*i.e.*, TN_03–TN_05) start reducing at 180 °C while TN_01 and TN_02 start reducing at a slightly higher temperature of 200 °C. However, TN_01 and TN_02 are reduced over a narrower temperature range of 200–350 °C whereas TN_03–TN_05 are reduced over a much wider range of 180–480 °C. This implies that larger crystallites are much easier to reduce to metallic Co than the smaller crystallites, which contradicts previous work done on a similar system.[22]

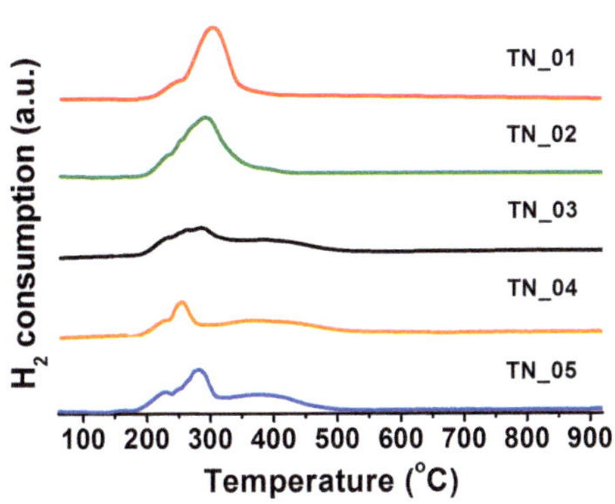

Fig. 4 H_2-TPR profiles of the supported fresh samples of TN_01–TN_05.

This observation could be explained by differences in the strength of the metal–support interactions in each catalyst with the smaller crystallites in TN_03–TN_05 interacting more strongly with the support. Crystallites smaller than the average pore diameter of the Al_2O_3 support used ($\bar{d}_{pore}$ = 11.5 nm) can enter these pores, thus enhancing the strength of the metal–support interaction.[27]

In situ catalyst characterisation and testing

CO-PrOx tests were performed in a fixed bed reactor in each of the two *in situ* instruments. The normalised outlet flow rates of CO, CO_2 and CH_4 are plotted as a function of temperature for the catalysts TN_01 and TN_05 in Fig. 5 and 6 respectively (see also Fig. S.3–S.5 in the ESI† for the kinetic data of TN_02–TN_04 respectively). These results originate from the tests performed in the magnetometer and can be interpreted as being specific to the mass of catalyst loaded in the reactor as Co_3O_4, which was kept constant at 0.1 g in each case. Due to higher absolute reactant flows and higher catalyst loading, the kinetic data obtained from the magnetometer displays a higher signal to noise ratio than the data obtained from the *in situ* PXRD experiments. Under the H_2-rich environment of CO-PrOx, Co_3O_4 is susceptible to reduction to form CoO and ultimately metallic Co at elevated reaction temperatures.[13,17–19] The two *in situ* instruments were used to capture these phase changes as a function of temperature in all five catalysts.

It can be seen from Fig. 5 and 6 that CO oxidation is the only CO-consuming pathway observed within the 50–225 °C temperature range over both TN_01 and TN_05. Also in this range, Co_3O_4 is the only crystalline phase present in each catalyst according to *in situ* PXRD. Summarised in Table 3 is (i) the temperature at which 50% CO conversion to CO_2 (T_{50}) is achieved before any catalyst reduction events, (ii) the temperature at which the highest CO_2 yield ($T_{CO_2\ max.}$) is achieved upon heating, and (iii) the onset temperature for CH_4 formation (T_{CH_4}) over each of the five tested catalysts. A decrease in the starting crystallite size of Co_3O_4 resulted in an increased mass-specific CO oxidation activity; *i.e.*, TN_05 showed the highest CO_2 yields at temperatures below 225 °C while TN_01 showed the lowest CO_2 yields in the same temperature range. To reduce the CO concentration in the H_2-rich reformate for PEMFCs to below 10 ppmv, a conversion of CO to CO_2 of at least 99.999% is required. TN_05 shows a maximum CO conversion to CO_2 of 98% at 175 °C which implies that there is still 200 ppmv of CO unreacted at this temperature.

In situ PXRD analysis was particularly useful in revealing the transformation of Co_3O_4 to CoO in each catalyst since the magnetometer cannot capture this transformation. As shown in Fig. 5 and 6 for TN_01 and TN_05 respectively, the formation of CoO is first observed at 225 °C. This transformation almost coincides with the decrease in the CO_2 yield over all catalysts after having reached a maximum (see also Fig. S.3–S.5†). Both the decrease in the CO_2 yield and onset of CoO formation may be explained by the complete consumption of O_2 in the feed caused by the H_2. Provided in Table 3 are the temperatures at which 100% O_2 conversion is first observed (*i.e.*, $T_{X_{O_2}=100\%}$) during each experiment. With the exception of TN_05, this temperature coincides with the temperature at which the maximum amount of CO_2 is formed upon heating ($T_{CO_2\ max.}$). A temperature increase of 25 °C from $T_{X_{O_2}=100\%}$ significantly increases the H_2 conversion which not only causes a decrease in the CO_2 yield but also induces the phase change from Co_3O_4 to CoO as there is not enough feed O_2 to keep the catalyst oxidised at this point.[19,33–37] Initial fcc Co formation is observed in the *in situ* PXRD at 250 °C for TN_01 and is confirmed by the magnetometer (see Fig. 5 and Table 3) which enabled the calculation of the degree of reduction from Co_3O_4 to fcc Co. Changes in the crystallite size and mass fraction of each observed cobalt phase as a function of temperature could not be accurately obtained from the *in situ* PXRD patterns using Rietveld refinement and PONKCS. This is attributed to the relatively low concentration of each cobalt phase formed and the interfering diffraction pattern of the Al_2O_3 support.

The onset for fcc Co formation in TN_05 is detected at 275 °C during *in situ* PXRD while in the magnetometer, this phase is detected slightly earlier *i.e.*, at 250 °C. It should be noted that the magnetometer is inherently more sensitive to low concentrations of ferromagnetic material (in this case, metallic Co) than the PXRD. Furthermore, the signal strength induced by ferromagnetic materials in the magnetometer is less dependent on crystallite size than in PXRD.[22,23,25] The formation of fcc Co in each catalyst results in the formation of CH_4. This observation does not only underline the necessity of metallic Co species for the methanation reaction[38,39] but also shows the importance of the oxidic species for the CO oxidation reaction.[33–35] As the temperature is increased stepwise from 250 to 350 °C, the catalysts with larger starting Co_3O_4 crystallites, *i.e.* TN_01 and TN_02, reach higher degrees of reduction compared to the other three catalysts, *i.e.* TN_03–TN_05 (see Table 3). During *in situ* PXRD, a comparison of the intensity of the CoO and fcc Co main diffraction peaks at 49.7° and 51.9°, respectively, also suggests that the degree of reduction is higher in TN_01 while TN_05 retains a significant percentage of CoO. This behaviour is similar to what was observed

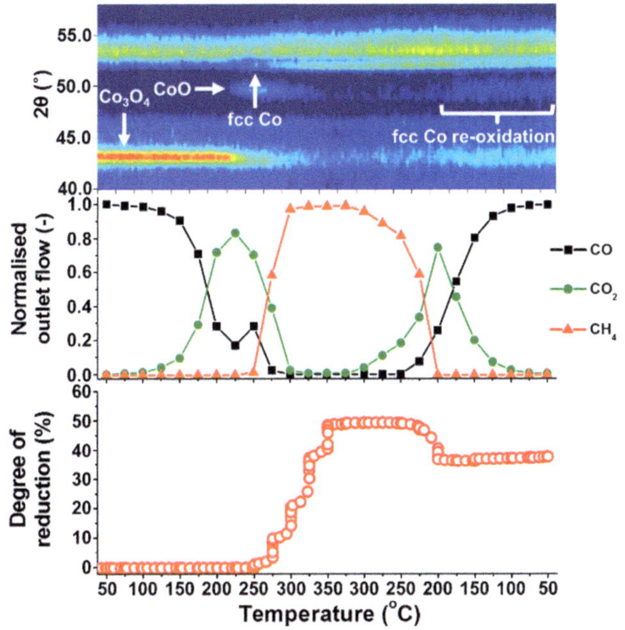

Fig. 5 (top) On-top view of the PXRD patterns recorded for TN_01, (middle) the normalised outlet flow rates of CO, CO_2 and CH_4, (bottom) changes in the degree of reduction calculated from the magnetometer-derived data.

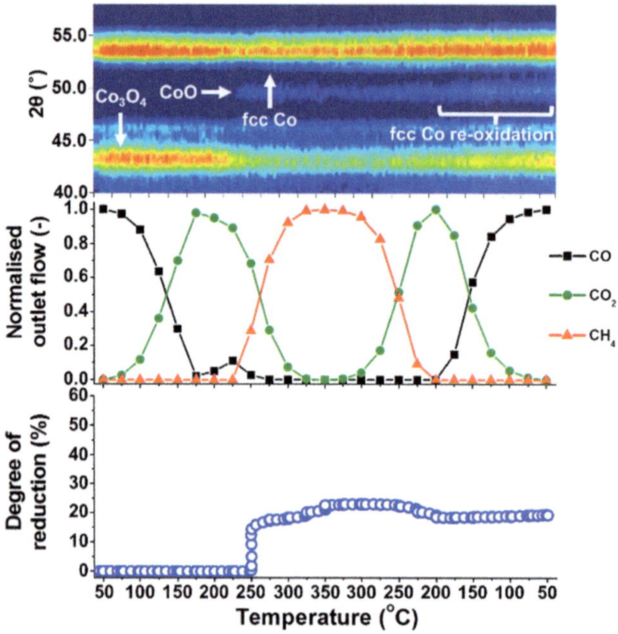

Fig. 6 (top) On-top view of the PXRD patterns recorded for TN_05, (middle) the normalised outlet flow rates of CO, CO_2 and CH_4, (bottom) changes in the degree of reduction calculated from the magnetometer-derived data.

Table 3 Summary of the kinetic and *in situ* characterisation data for all five tested catalysts

Sample name	T_{50}[a] (°C)	$T_{CO_2\,max}$[b] (°C)	$T_{X_{O_2}=100\%}$[c] (°C)	T_{CoO}[d] (°C)	T_{CH_4}[e] (°C)	T_{cobalt}[f] (°C)	DOR[g] (%) at 350 °C	Post-reaction DOR[g] (%) at 50 °C	Post-reaction m_{cobalt}[h] (wt%) at 50 °C
TN_01	187	225 (83.2%)	225	225	250	250	45.7	37.6	3.0
TN_02	149	200 (96.0%)	200	225	225	225	45.6	38.1	2.5
TN_03	154	200 (95.7%)	200	225	250	250	16.9	13.6	0.9
TN_04	148	200 (93.2%)	200	225	250	250	17.1	12.1	0.8
TN_05	135	175 (98.0%)	200	225	250	250	22.0	19.2	1.3

[a] Temperature at which 50% CO is converted to CO_2 while the catalyst is in the Co_3O_4 state. [b] Temperature at which maximum amount of CO_2 is formed upon heating. In parentheses are the maximum CO_2 yields obtained. [c] Temperature at which 100% O_2 conversion is first observed. [d] Onset temperature for CoO formation as observed during *in situ* PXRD. [e] Onset temperature for CH_4 formation. [f] Onset temperature for fcc Co formation as observed in the magnetometer. [g] Degree of reduction (DOR) at 350 °C and 50 °C (post-reaction) calculated from the magnetometer-derived data. [h] Mass fraction of fcc Co calculated from the DOR at 50 °C (post-reaction) assuming that the only Co-containing species in each sample are fcc Co and CoO.

during the H_2-TPR experiments where TN_01 and TN_02 reduced to metallic Co within a narrower temperature window, suggesting weaker metal–support interactions compared to the small crystallites of Co_3O_4.[27]

Upon decreasing the reaction temperature below 350 °C, CH_4 formation begins to decrease while the formation of CO_2 gradually increases until reaching a maximum at 200 °C for all five catalysts (see Fig. 5 and 6 as well as S.3–S.5†). The magnetometer shows a decrease in the degree of reduction at temperatures from 250 to 200 °C. The observed decrease in the degree of reduction is not an effect of the temperature dependent magnetic susceptibility of metallic Co (see also Section A in the ESI†),[40] but suggests that a fraction of the previously formed metallic Co is being re-oxidised to CoO. *In situ* PXRD also confirms the re-oxidation process *via* an increase in peak intensity of CoO at temperatures below 200 °C (see Fig. 5 and 6 as well as S.3–S.5†). In parallel, CO_2 formation decreases below 200 °C for all catalysts, most likely caused by the decrease in reaction temperature. The overall CO oxidation activity of the catalysts upon cooling is lower compared to that observed along the heating stages, which shows that the catalysts have undergone irreversible chemical changes as confirmed by the magnetometer and PXRD. Nonetheless, a relationship between the CO oxidation activity and the starting Co_3O_4 crystallite size is still observed *i.e.*, the catalysts with smaller starting Co_3O_4 crystallite sizes still showed superior mass-specific activity upon cooling.

The incomplete reduction observed at 350 °C could imply the presence of core–shell structured crystallites with a CoO core and a metallic Co shell. This would explain the loss of CO oxidation activity at high temperatures as the required surface active sites have been lost as a result of the reduction.[13,17–19,33–35] As for the partial re-oxidation of the crystallites upon cooling the reactor below 250 °C, this may imply the presence of double shell-structured crystallites with a CoO outer-shell, a metallic Co inner-shell and a CoO core. Note that the magnetometer and PXRD detect bulk phase changes and therefore, it is also possible that the re-oxidation of metallic Co may have begun above 250 °C (upon cooling) but was only limited to the surface of the crystallites. This would explain the gradual decrease in the CH_4 formation and the concurrent increase in the CO_2 formation between 350 and 250 °C upon cooling. However, it remains possible that the decrease in CH_4 formation and increase in CO_2 formation is a result of the decreasing reaction temperature which favours the CO oxidation pathway. Also, the stability of the proposed structured crystallites (especially those with a double-shell structure) remains questionable.

A different explanation for the observed partial phase transformations is that a certain fraction of the crystallites in each sample (probably the more weakly-bound relatively large crystallites) undergo reduction to metallic Co upon heating. During cooling below 250 °C, the previously formed metallic crystallites are then passivated *i.e.*, a thin layer of CoO is formed around the metallic crystallites due to the presence of the low concentration of O_2.[41–43] Wolf *et al.*[43] recently reported that such a CoO layer is formed around freshly reduced Co crystallites upon their exposure to 1 vol% O_2 in N_2. Note that the relative concentration of O_2 in the study by Wolf *et al.* is similar to that used in this study (see Experimental section). The formation of the CoO layer was confirmed by a rapid initial decrease in the measured magnetic signal of the freshly reduced Co sample within 15 minutes of their exposure to the passivation gas mixture. The presence of a relatively small amount of CoO was detected together with metallic Co by conventional PXRD analysis after the passivation procedure. It is worth mentioning that the formation of this CoO layer, in contrast to the present study, took place in the absence of H_2 and CO which are both reducing agents.

Changes in the CO_2 selectivity based on the conversion of O_2 as a function of temperature are shown in Fig. 7 for TN_01 and TN_05. Increasing the reaction temperature decreases the CO_2 selectivity as at higher temperatures, H_2 oxidation dominates over CO oxidation.[4,8–10,13,17–19] It should be noted that a 1 : 1 CO : O_2 volume ratio was used in all the experiments conducted which means that an excess of O_2 was fed into the system. To completely oxidise CO, an amount of O_2 equivalent to 50% of the CO is required (*i.e.*, 2 : 1 CO : O_2). This leaves the remaining 50% of the O_2 in these experiments, even at 100% CO_2 yield, available for H_2 oxidation. Feeding excess amounts of O_2 has been shown to negatively affect the CO_2 selectivity during CO-PrOx despite the reported positive effect this has on the CO conversion to CO_2.[18,44] In addition to the effect of H_2 oxidation, the methanation reaction (as a result of metallic Co formation at temperatures above 225 °C) also affects the CO_2 yield and selectivity.[13,17–19] The CO_2 selectivity did not show any obvious dependency on the starting crystallite size of Co_3O_4 (see Fig. 7 and S.6†).

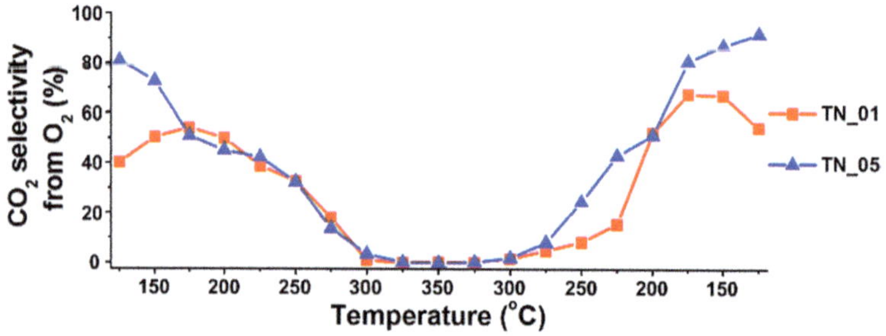

Fig. 7 CO_2 selectivity based on the conversion of O_2 as a function of temperature for the catalysts TN_01 and TN_05.

As mentioned earlier, decreasing the crystallite size of Co_3O_4 increased the mass-specific CO oxidation activity between 50 and 225 °C upon heating. In addition, TN_05 showed the lowest T_{50} (135 °C) and $T_{CO_2\ max.}$ (175 °C) within this temperature range. The stability of this catalyst was then monitored at $T_{CO_2\ max.}$ *i.e.*, at 175 °C, by performing a time-on-stream experiment over a 36 hour period using the same feed and gas flow as in the temperature sweep experiments already discussed above. Shown in Fig. 8 is the CO_2 yield, O_2 conversion, CO_2 selectivity from O_2 and degree of reduction calculated over the 36 hour period. The catalyst reaches full CO conversion to CO_2 after two hours and maintains a stable CO_2 yield of 100%, O_2 conversion of 91% and CO_2 selectivity of 53% throughout the whole experiment. There was no CH_4 formed during the course of the experiment as there was no metallic Co formation detected in the magnetometer (*i.e.*, DOR = 0%).

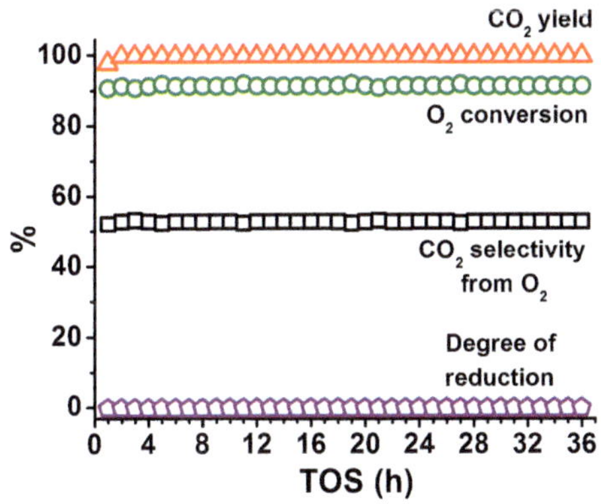

Fig. 8 CO_2 yield, O_2 conversion, CO_2 selectivity from O_2 and degree of reduction as a function of time at 175 °C for TN_05.

Characterisation of spent catalysts

TEM analysis was able to give insight on the sizes of the crystallites in each spent catalyst sample. Although distinguishing between the cobalt-containing crystallites and those of the support material in the micrographs of the spent samples was challenging (see Fig. S.7 in ESI†), extraction of size information was still attempted (see Fig. 9). No significant changes in the number-based average crystallite size and the size distribution for each sample compared to its fresh counterpart are observed, ruling out significant sintering events during catalytic testing (see Table 4).

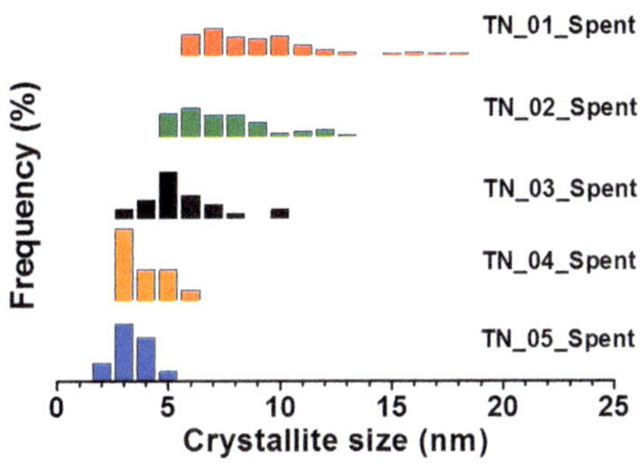

Fig. 9 Number-based size distributions of the spent samples of TN_01–TN_05 obtained by TEM analysis.

Table 4 Number-based crystallite sizes of the fresh and spent samples of TN_01–TN_05 obtained by TEM analysis

Sample name	Fresh (nm)	Spent (nm)
TN_01	10.1 ± 2.8	9.5 ± 3.1
TN_02	8.5 ± 1.7	7.9 ± 2.0
TN_03	6.1 ± 1.5	6.0 ± 1.7
TN_04	3.8 ± 0.8	4.4 ± 1.1
TN_05	3.3 ± 0.6	3.8 ± 0.7

Discussion on the effect of crystallite size on CO-PrOx activity

Crystallite size dependency refers to the effect that the crystallite size has on the activity of an active site on the surface of the crystallite. The origin of these effects can arise from the surface crystalline anisotropy of different-sized crystallites or from single crystals exposing different facets (possibly due to their differences in shape).[36,45,46] Therefore, it is common practice to normalise the activity by the exposed mass-specific surface area or number of exposed surface atoms to obtain a surface area-specific rate or turnover frequency (TOF).[36,45,46] In this study, the CO_2 yields between 75 and 200 °C were normalised with the available mass-specific surface area calculated from the TEM-derived number-based average Co_3O_4 crystallite sizes (assuming no significant sintering and also a spherical shape for the crystallites) and the EDX-based loadings, and thereafter plotted against the average Co_3O_4 crystallite size (Fig. 10). In the temperature range chosen, CO oxidation was the only CO-consuming reaction that took place and no Co_3O_4 reduction had taken place affecting bulk composition/oxidation and crystallite size (see Fig. 5 and 6 as well as S.3–S.5†).

The surface area-specific CO_2 formation rate increased with temperature and starting Co_3O_4 crystallite size, reaching an apparent maximum at 8.5 nm. At a crystallite size of 10.1 nm, the formation rate of CO_2 decreased. The decrease of the CO_2 formation rate over crystallites with an average size above and below 8.5 nm has been explained to some extent by Iablokov *et al.*[15] where, in contrast to the present study, the size-dependency of CO oxidation over Co_3O_4 in the absence of H_2 was investigated. An increase in the concentration of Co^{3+} and O^{2-} ions on the surface with an increase in crystallite size was reported with crystallite sizes ranging from 4 to 9 nm based on the analysis of the Co 2p and O 1s core level XPS spectra. Above 9 nm, the concentration of the respective species is reported to decrease. This size-dependent surface concentration of Co^{3+} and O^{2-} remains elusive to the authors. Nonetheless, this trend agrees reasonably well with the size-dependent CO oxidation activity trend they reported. The previous claim that Co^{3+} ions represent the "active site" for CO oxidation[33–35] is therefore further substantiated by Iablokov *et al.* The concentration of O^{2-} was particularly reported since CO oxidation over Co_3O_4 is generally accepted to follow a Mars–van Krevelen-type mechanism where lattice oxygen is believed to oxidise the CO to CO_2 and thereafter the partially reduced catalyst surface is re-oxidized by adsorbed and dissociated O^{2-} originating from the gas phase.[33–37] In conclusion, we

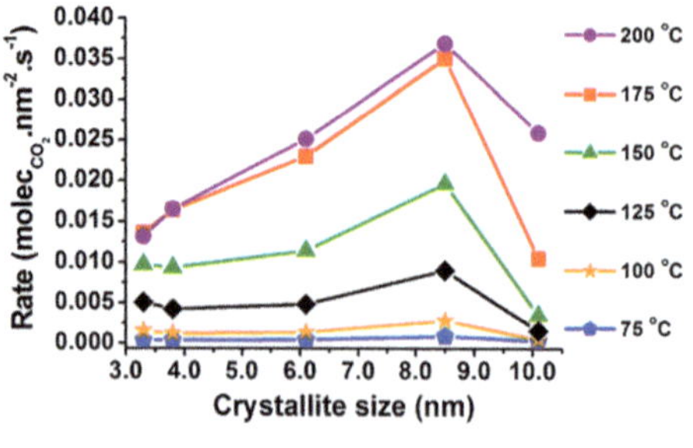

Fig. 10 Surface area-specific formation rates of CO_2 as a function of Co_3O_4 crystallite size and temperature.

propose that the size-dependent CO oxidation activity trends we observed during CO-PrOx conditions may also be explained by the differences in the surface concentration of the Co^{3+} and O^{2-} ions as observed by Iablokov *et al.*

Conclusions

The present study was aimed at giving insight into the effect of crystallite size on the overall CO-PrOx activity of Co_3O_4 catalysts. The mass-specific CO oxidation activity was shown to increase with a decrease in the starting crystallite size of Co_3O_4. However, the surface area-specific CO_2 formation rate displayed a volcano-type behaviour *i.e.*, it increased with an increase in the starting crystallite size of Co_3O_4 up to 8.5 nm and thereafter slightly decreased. The reduction behaviour of the different-sized Co_3O_4 crystallites as a function of temperature under CO-PrOx conditions was also investigated. The reduction of Co_3O_4 to CoO between 200–225 °C was evidenced with *in situ* PXRD for all catalysts, while the formation of fcc Co was observed above 225 °C in both the magnetometer and PXRD. The catalysts with larger crystallites showed higher degrees of reduction which was possibly due to the presence of slightly stronger metal–support interactions between the small crystallites and the support as previously proposed for a similar catalytic system.[27] Also upon decreasing the reaction temperature below 250 °C, metallic Co was partially re-oxidised to CoO but not to Co_3O_4. Furthermore, the CO oxidation activity observed upon decreasing the reaction temperature had diminished compared to that observed upon heating which could be attributed to the irreversible chemical changes that each catalyst underwent.

Acknowledgements

The authors would like to express their gratitude to the organising committee for giving the opportunity to present their work at ‘Faraday Discussion – Catalysis for Fuels’ in Cape Town, South Africa, from 24th–26th of January 2017. They would like to acknowledge the financial support from Johnson Matthey and the DST-NRF Centre of Excellence in Catalysis research (c*change). Finally, they would like to also thank the team from the Centre of Imaging and Analysis based at the University of Cape Town (UCT) for their assistance with TEM and EDX and the team in the Analytical Laboratories of the Chemical Engineering Department at UCT for their assistance with H_2-TPR.

References

1 S. Golunski, *Platinum Met. Rev.*, 1998, **42**, 2–7.
2 N. Edwards, S. R. Ellis, J. C. Frost, S. E. Golunski, A. N. J. van Keulen, N. G. Lindewald and J. G. Reinkingh, *J. Power Sources*, 1998, **71**, 123–128.
3 P. G. Gray and M. I. Petch, *Platinum Met. Rev.*, 2000, **44**, 108–111.
4 S. H. Oh and R. M. Sinkevitch, *J. Catal.*, 1993, **142**, 254–262.
5 A. F. Ghenciu, *Curr. Opin. Solid State Mater. Sci.*, 2002, **6**, 389–399.
6 E. D. Park, D. Lee and H. C. Lee, *Catal. Today*, 2009, **139**, 280–290.
7 A. Mishra and R. Prasad, *Bull. Chem. React. Eng. Catal.*, 2011, **6**, 1–14.
8 M. L. Brown Jr, A. W. Green, G. Cohn and H. C. Andersen, *Ind. Eng. Chem.*, 1960, **52**, 841–844.

9 M. L. Brown Jr and A. W. Green, *US Pat.* No. 3,088,919, 1963.
10 J. G. E. Cohn, *US Pat.* No. 3,216,783, 1965.
11 D. Gamarra, C. Belver, M. Fernández-García and A. Martínez-Arias, *J. Am. Chem. Soc.*, 2007, **129**, 12064–12065.
12 A. Hornés, A. B. Hungria, P. Bera, A. L. López Cámara, M. Fernández-García, A. Martínez-Arias, L. Barrio, M. Estrella, G. Zhou, J. J. Fonseca, J. C. Hanson and J. A. Rodriguez, *J. Am. Chem. Soc.*, 2010, **132**, 34–35.
13 Y. Teng, H. Sakurai, A. Ueda and T. Kobayashi, *Int. J. Hydrogen Energy*, 1999, **24**, 355–358.
14 C.-B. Wang, C.-W. Tang, S.-J. Gau and S.-H. Chien, *Catal. Lett.*, 2005, **101**, 59–63.
15 V. Iablokov, R. Barbosa, G. Pollefeyt, I. van Driessche, S. Chenakin and N. Kruse, *ACS Catal.*, 2015, **5**, 5714–5718.
16 C. Liu, Q. Liu, L. Bai, A. Dong, G. Liu and S. Wen, *J. Mol. Catal. A: Chem.*, 2013, **370**, 1–6.
17 M. M. Yung, Z. Zhao, M. P. Woods and U. S. Ozkan, *J. Mol. Catal. A: Chem.*, 2008, **279**, 1–9.
18 M. P. Woods, P. Gawade, B. Tan and U. S. Ozkan, *Appl. Catal., B*, 2010, **97**, 28–35.
19 L. Lukashuk, K. Föttinger, E. Kolar, C. Rameshan, D. Teschner, M. Hävecker, A. Knop-Gericke, N. Yigit, H. Li, E. McDermott, M. Stöger-Pollach and G. J. Rupprechter, *J. Catal.*, 2016, **344**, 1–15.
20 M. Claeys, E. van Steen, J. L. Visagie and J. van de Loosdrecht, *US Pat.* 8,773,118 B2, 2014.
21 M. Claeys, M. E. Dry, E. van Steen, P. J. van Berge, S. Booyens, R. Crous, P. van Helden, J. Labuschagne, D. J. Moodley and A. M. Saib, *ACS Catal.*, 2015, **5**, 841–852.
22 N. Fischer, B. Clapham, T. Feltes, E. van Steen and M. Claeys, *Angew. Chem., Int. Ed.*, 2014, **53**, 1342–1345.
23 M. Claeys, M. E. Dry, E. van Steen, E. du Plessis, P. J. van Berge, A. M. Saib and D. J. Moodley, *J. Catal.*, 2014, **318**, 193–202.
24 M. Claeys and N. Fischer, *US Pat.* 8,597,598, 2013.
25 N. Fischer, B. Clapham, T. Feltes and M. Claeys, *ACS Catal.*, 2015, **5**, 113–121.
26 N. Fischer and M. Claeys, *Catal. Today*, 2016, **275**, 149–154.
27 N. Fischer, E. van Steen and M. Claeys, *Catal. Today*, 2011, **171**, 174–179.
28 N. Fischer, T. Feltes and M. Claeys, Micelles: Reverse Micelles—Designer Nanoparticles for Investigative Catalysis, in *CRC Concise Encyclopedia of Nanotechnology*, ed. B. I. Kharisov, O. V. Kharissova and U. Ortiz-Mendez, CRC Press, Boca Raton, 2015, p. 547.
29 N. V. Y. Scarlett and I. C. Madsen, *Powder Diffr.*, 2006, **21**, 278–284.
30 K. H. J. Buschow, *Encyclopedia of Materials: Science and Technology*, Elsevier, Amsterdam, Netherlands, 2001, vol. 5, p. 5021.
31 T. Mousavand, T. Naka, K. Sato, S. Ohara, M. Umetsu, S. Takami, T. Nakane, A. Matsushita and T. Adschiri, *Phys. Rev. B: Condens. Matter Mater. Phys.*, 2009, **79**, 144411.
32 C. Kittel, *Introduction to Solid State Physics*, John Wiley & Sons, Inc., London, 2009.
33 P. Broqvist, I. Panas and H. J. Persson, *J. Catal.*, 2002, **210**, 198–206.

34 J. Jansson, A. E. C. Palmqvist, E. Fridell, M. Skoglundh, L. Osterlund, P. Thormahlen and V. Langer, *J. Catal.*, 2002, **211**, 387–397.

35 K. Omata, T. Takada, S. Kasahara and M. Yamada, *Appl. Catal., A*, 1996, **46**, 255–267.

36 B. R. Cuenya and F. Behafarid, *Surf. Sci. Rep.*, 2015, **70**, 135–187.

37 P. Mars and D. W. van Krevelen, *Chem. Eng. Sci.*, 1954, **3**(1), 41–59.

38 H. Pichler, *Adv. Catal.*, 1952, **4**, 271–341.

39 M. E. Dry, *Appl. Catal., A*, 2004, **276**, 1–3.

40 R. Serway and J. Jewett Jr, *Physics for Scientists and Engineers with Modern Physics*, Brooks/Cole, Boston, 2014.

41 J. Haber, J. H. Block and B. Delmon, *Pure Appl. Chem.*, 1995, **67**, 1257–1306.

42 F. Pinna, *Catal. Today*, 1998, **41**, 129–137.

43 M. Wolf, N. Fischer and M. Claeys, *Catal. Today*, 2016, **275**, 135–140.

44 J. Yan, J. Ma, P. Cao and P. Li, *Catal. Lett.*, 2004, **93**, 55–60.

45 M. Boudart, A. Aldag, J. E. Benson, N. A. Dougharty and C. G. Harkins, *J. Catal.*, 1966, **6**, 92–99.

46 M. Boudart, *Chem. Rev.*, 1995, **95**, 661–666.

Faraday Discussions

PAPER

The effect of sodium species on methanol synthesis and water–gas shift Cu/ZnO catalysts: utilising high purity zincian georgeite

Simon A. Kondrat,* Paul J. Smith, James H. Carter, James S. Hayward, Geoffrey J. Pudge, Greg Shaw, Michael S. Spencer, Jonathan K. Bartley, Stuart H. Taylor and Graham J. Hutchings

Received 3rd October 2016, Accepted 24th October 2016
DOI: 10.1039/c6fd00202a

The effect of sodium species on the physical and catalytic properties of Cu/ZnO catalysts derived from zincian georgeite has been investigated. Catalysts prepared with <100 ppm to 2.1 wt% Na^+, using a supercritical CO_2 antisolvent technique, were characterised and tested for the low temperature water–gas shift reaction and also CO_2 hydrogenation to methanol. It was found that zincian georgeite catalyst precursor stability was dependent on the Na^+ concentration, with the 2.1 wt% Na^+-containing sample uncontrollably ageing to malachite and sodium zinc carbonate. Samples with lower Na^+ contents (<100–2500 ppm) remained as the amorphous zincian georgeite phase, which on calcination and reduction resulted in similar CuO/Cu particle sizes and Cu surface areas. The aged 2.1 wt% Na^+ containing sample, after calcination and reduction, was found to comprise of larger CuO crystallites and a lower Cu surface area. However, calcination of the high Na^+ sample immediately after precipitation (before ageing) resulted in a comparable CuO/Cu particle size to the lower (<100–2500 ppm) Na^+ containing samples, but with a lower Cu surface area, which indicates that Na^+ species block Cu sites. Activity of the catalysts for the water–gas shift reaction and methanol yields in the methanol synthesis reaction correlated with Na^+ content, suggesting that Na^+ directly poisons the catalyst. *In situ* XRD analysis showed that the ZnO crystallite size and consequently Cu crystallite size increased dramatically in the presence of water in a syn-gas reaction mixture, showing that stabilisation of nanocrystalline ZnO is required. Sodium species have a moderate effect on ZnO and Cu crystallite growth rate, with lower Na^+ content resulting in slightly reduced rates of growth under reaction conditions.

Introduction

$Cu/ZnO/Al_2O_3$ is an important catalyst used for the synthesis of methanol from syn-gas (CO, CO_2 and H_2) and for the low-temperature water–gas shift reaction

Cardiff Catalysis Institute, Cardiff University, Main Building, Park Place, Cardiff, CF10 3AT, UK. E-mail: KondratSA@Cardiff.ac.uk

(LTS).[1,2] Both reactions are industrially important, with methanol having a worldwide demand in excess of 50 Mtons per year and the LTS reaction being used upstream of many industrial processes for H_2 production and CO removal.[2] In addition, both reactions have found new applications in green technology, for methanol synthesis from captured CO_2 and renewably sourced H_2,[3,4] and the use of LTS to remove trace catalytically poisonous CO from fuel cell gas streams.[5] These contemporary applications' requirements and their by-products limit the effectiveness of the conventional $Cu/ZnO/Al_2O_3$ catalyst, resulting in significant research into other metal/support combinations.[6–9] However, it is also hypothesised that alteration of a $Cu/ZnO/Al_2O_3$ catalyst, by modifying the Cu–ZnO interaction, introducing dopants and also replacing Al_2O_3, could result in enhanced effectiveness for CO_2 hydrogenation.[10,11]

Historically the practical optimisation of the $Cu/ZnO/Al_2O_3$ catalyst has advanced further than the fundamental understanding of the reaction mechanism and the active site. Studies have started to retrospectively understand the fundamental origin of catalytic activity, often with a focus on the simpler binary Cu–ZnO system. Such catalysts are conventionally prepared by the co-precipitation of metal nitrate salts with sodium carbonate to produce a precursor phase, which is then calcined to form CuO and ZnO, before being reduced to Cu/ZnO. It has been established that the preparation of specific copper and zinc hydroxycarbonate precursor phases, analogous to minerals such as zincian malachite, aurichalcite or rosasite, is a requirement for high catalytic activity.[12,13] It has been shown that, with suitable calcination and reduction conditions, the final catalyst properties are defined by the precursor structure.[12] Specifically, these desired properties are high Cu surface area (imparted by small particle size and meso-porosity),[12,14] good Cu–ZnO[15,16] or even surface Cu–Zn interaction,[17] and minimal contamination from alkali metals introduced during the co-precipitation process.

During the co-precipitation process an initial precipitate is often observed prior to the ageing step, which rapidly ages into crystalline zincian malachite. The study of this initial precipitate has often been neglected due to its amorphous nature and its instability, which leads to rapid uncontrolled ageing to form malachite. In one of the few studies related to this phase, Spencer and co-workers hypothesised that this initial precipitate was analogous to the highly rare mineral georgeite, but were unable to stabilise this phase long enough to perform detailed characterisation and catalytic testing.[18] Recently we succeeded in preparing stable zincian georgeite using supercritical antisolvent (SAS) precipitation, allowing its detailed characterisation to be undertaken and most importantly its use as a precursor for methanol synthesis and LTS catalysts to be evaluated.[19] It was found that the catalyst derived from zincian georgeite had distinct microstructural properties compared to conventionally prepared catalysts, with small 5 nm Cu crystallites embedded in a highly disordered sub 2 nm matrix of ZnO crystallites. In addition, the zincian georgeite derived catalyst was found to have a low Na^+ content, as the process requires no addition of base, unlike co-precipitation. Importantly, catalysts derived from zincian georgeite were found to be highly active for methanol synthesis and the LTS reaction, with exceptional stability for the latter reaction.

As previously mentioned, significant Cu–ZnO interaction has been reported to be important for high catalytic activity[15,17] and the observed microstructure of zincian georgeite catalysts complements these observations.[19] However, there have been far fewer studies on the effect of trace alkali metal content, specifically sodium species, on methanol synthesis and LTS catalyst activity and stability. The previous studies that have investigated the effect of Na^+ have used co-precipitation, with Na_2CO_3, to form zincian malachite and varied the number and/or duration of washing steps to remove residual Na^+.[20,21] The key findings from these studies showed that the presence of sodium species in the catalysts resulted in increased CuO and ZnO crystallite size during calcination and subsequently resulted in lower copper surface areas. The presence of sodium nitrate within these zincian malachite precursors resulted in the evolution of NO_x species during the calcination step, which caused CuO and ZnO crystal growth.[20,22] De Jongh and co-workers also showed that even when a Na^+-free preparation method (using ammonium carbonate instead of sodium carbonate) was used, residual species resulted in CuO crystallite growth during calcination (unless a specific calcination environment is used).[22] Their observation of similar CuO crystallisation with sodium nitrate or ammonium nitrate residue indicates that the effect of Na^+ on catalyst performance was being masked by residual nitrate species decomposition.

In this study we investigate the effect of Na^+ on the characteristics and catalytic performance of SAS precipitated zincian georgeite. The advantage of using SAS precipitation is that it requires no base (*i.e.* no sodium carbonate or ammonium carbonate) resulting in high purity catalyst precursors. Therefore, the content of Na^+ can be controlled by doping into the catalyst precursor, as opposed to washing it out after precipitation, which gives greater control over the composition. In addition to studying the importance of Na^+ content in zincian georgeite derived catalysts, we further studied the effectiveness of these novel catalyst structures in the LTS reaction and the synthesis of methanol from H_2 and CO_2 in the absence of CO.

Experimental

Materials

Copper(II) acetate monohydrate (99.99% trace metals basis and puriss. p. a., ≥99.0% grades), zinc(II) acetate dihydrate (99.999% trace metals basis and puriss. p. a., ≥99.0% grades) and sodium acetate trihydrate (reagent grade, ≥99.0%) were all purchased from Sigma-Aldrich. Ethanol (absolute 99.8%, Certified AR) was purchased from Fischer Scientific and CO_2 (CP grade) was provided by BOC. All purchased materials were used as received. Deionised water was provided in-house (determined Na^+ concentration < 1 ppm).

Catalyst preparation

The SAS precipitated materials were prepared as follows; a series of mixed $Cu(OAc)_2 \cdot H_2O$ (4 mg ml^{-1}) and $Zn(OAc)_2 \cdot 2H_2O$ (2.13 mg ml^{-1}) solutions, giving a nominal Cu : Zn molar ratio of 2 : 1, were prepared using a 10 vol% H_2O/ethanol mixture (1000 ml). High purity metal acetate salts (≥99.99% trace metals basis) were used to prepare Na-free zincian georgeite whilst lower purity grades (puriss. p. a., ≥99.0%) were used to make standard zincian georgeite. These samples will be referred to as Na-free-zincian georgeite and Std-Na-zincian georgeite respectively. Low, medium and high Na dosed zincian georgeite materials were also prepared by the addition of 0.0036 g, 0.0364 g and 0.3637 g $Na(OAc)_2 \cdot 3H_2O$ respectively. Target Na^+ concentrations in the calcined samples are shown in Table 1. These samples will be referred to as low-Na-zincian georgeite, med-Na-zincian georgeite and high-Na-zincian georgeite.

SAS precipitation experiments were performed using apparatus manufactured by Separex. Liquefied CO_2 was pumped to give a flow rate of 6.5 kg h^{-1} and the whole system was pressurised up to 110 bar and held at 40 °C. Initially pure solvent (10 vol% H_2O/ethanol) was pumped through the fine capillary into the precipitation vessel, with a flow rate of 6.5 ml min^{-1} for 15 minutes, in co-current mode with supercritical carbon dioxide ($scCO_2$) in order to obtain steady state conditions inside the precipitation vessel. After this initial period the flow of liquid solvent was stopped and the mixed acetate solution was delivered at a flow rate of 6.5 ml min^{-1}. This gave a $scCO_2$/metal solution molar ratio of 22 : 1. After the solution had been processed a drying step was carried out, which was achieved by pumping pure ethanol at 6.5 ml min^{-1} co-currently with $scCO_2$ for 30 min, before leaving with just $scCO_2$ to pump for a further 60 min. This step was required to remove residual solvent from the precipitation process. Furthermore, it was essential to completely dry the georgeite phases to prevent subsequent ageing to form malachite. When the drying step was completed the $scCO_2$ flow was stopped, the vessel was depressurised to atmospheric pressure and the precipitate was collected. Experiments were conducted for approximately 3.5 hours, which resulted in the synthesis of *ca.* 2.2–2.6 g of solid. Recovered materials were then calcined at 300 °C in static air (ramp rate 1 °C min^{-1} for 4 h).

Catalyst characterisation

Powder X-ray diffraction measurements were performed using a PANalytical X'pert Pro diffractometer with a Ni filtered CuK_α radiation source operating at 40 kV and 40 mA. Patterns were recorded over a 2θ range of 10–80° using a step

Table 1 Elemental composition of SAS prepared zincian georgeite derived catalysts

Sample name	Target Cu : Zn molar ratio	Precipitated Cu : Zn molar ratio	Target Na^+ concentration[a] (ppm)	Precipitated Na^+ concentration[a] (ppm)
Na-free-CuO/ZnO	2 : 1	1.9 : 1	<15	<100[b]
Std-Na–CuO/ZnO		2 : 1	150	320
Low-Na–CuO/ZnO		1.7 : 1	250	580
Med-Na–CuO/ZnO		1.9 : 1	2500	2520
High-Na–CuO/ZnO		1.9 : 1	25 000	21 200

[a] Values calculated from calcined samples. [b] Values calculated from blank solution of nitric acid solution gave *ca.* 100 ppm Na.

size of 0.016°. All patterns were matched using the ICDD database. An *in situ* Anton Parr XRK900 cell (internal volume of ~0.5 L) was used to monitor the formation of metallic Cu and ZnO during the reduction of the CuO/ZnO materials. A flow of 2% H_2/N_2 (50 ml min^{-1}) was passed through the sample bed while the cell was heated to 225 °C (ambient to 125 °C ramp rate = 10 °C min^{-1}, 125–225 °C ramp rate = 1 °C min^{-1}). After 1 h, under these conditions, a 20–80° 2θ scan was performed. The 2% H_2/N_2 gas environment was then switched to syn-gas (6% CO/9.2% CO_2/66% H_2 and 17.8% N_2) and held at 225 °C for 3 h, while repeat 20–80° 2θ scans were performed. To investigate the effect of water content on the catalysts, water was then introduced into the gas stream, *via* a bubbler (set at 85 °C to give a moisture content of 15%) for a further 3 h, while repeat 20–80° 2θ scans were performed.

Infrared spectroscopy was performed on the SAS precipitates using a Bruker Vertex 70 equipped with a single reflection diamond attenuated total reflectance accessory and a mercury cadmium telluride (MCT) detector. Cu and Zn content within the calcined materials was determined using microwave plasma atomic emission spectroscopy (MP-AES) using a 4100 MP-AES manufactured by Agilent Technologies. Na content was determined using inductively coupled plasma-mass spectrometry (ICP-MS). Solid samples and effluents were digested in 20 vol% HNO_3 (Sigma Aldrich-TraceSELECT purity)/H_2O solutions and compositions quantified against calibration standards.

Thermal gravimetric analysis (TGA) and differential thermal analysis (DTA) were performed using a Setaram Labsys 1600 instrument. Samples (20–50 mg) were loaded into alumina crucibles and heated to 600 °C (at 1 or 5 °C min^{-1}) in a flow of synthetic air (50 ml min^{-1}). TGA and evolved gas analysis (EGA) were also run on a Pyris 1 TGA linked to a Perkin Elmer Frontier, followed by a Clarus 580 GC-MS, using a TL-9000 interface. TGA was performed under helium, from 30–600 °C (at 5 °C min^{-1}) using *ca.* 25 mg of sample. For all specified TGA runs, blank runs were subtracted from the relevant data to remove buoyancy effects.

Cu surface area analysis was carried out on a Quantachrome ChemBET chemisorption analyser equipped with a thermal-conductivity detector (TCD). Calcined samples (100 mg) were reduced to catalysts using 10% H_2/Ar (30 ml min^{-1}) with heating to 140 °C at 10 °C min^{-1}, and then to 225 °C at 1 °C min^{-1}. For Cu surface area analysis, catalysts were cooled to 65 °C under He for N_2O pulsing. 12 N_2O pulses (113 µl each) were followed with 3 N_2 pulses for calibration. The amount of N_2 emitted was assumed to amount to half a monolayer coverage of oxygen and it was assumed that the surface density of Cu is 1.47×10^{19} atoms per m^2.

Catalyst testing

Water–gas shift testing was carried out on a custom made fixed-bed flow reactor. The catalyst (0.1 g) was suspended between two pieces of glass wool in a stainless steel reactor tube fitted with a stainless steel mesh to hold the catalyst in place within the bed. The catalyst was reduced under a feed of 2% H_2/N_2 at 225 °C for 1 h. Each catalyst was heated to 130 °C at 5 °C min^{-1} followed by 1 °C min^{-1} up to 225 °C before reduction. The gas feed consisted of 1.1% CO, 4.3% CO_2, 24.0% H_2O, 13.8% H_2 with N_2 as balance. The gases were introduced to the catalyst bed using mass flow controllers (Bronkhorst). The water was passed through a liquid-flow controller (Bronkhorst) into a controlled evaporator mixer heated to 140 °C where it was mixed with the carrier gas, N_2. The total flow rate was 50 ml min^{-1} with the reaction at ambient pressure. Products were quantified using an on-line Gasmet Dx4000 Fourier transform infrared spectrometer (FTIR). A spectrum was recorded every minute.

The catalytic performance of the catalysts for CO_2 hydrogenation was determined in a fixed-bed continuous-flow reactor. The catalyst (0.1 g, 425–600 μm) was placed in a stainless steel tube reactor with an internal diameter of 4.57 mm. Prior to the reaction, the catalysts were pre-reduced in a flow of H_2 (5 ml min^{-1}) for 1 h at 225 °C under atmospheric pressure. The reactor was then allowed to cool to room temperature before the gas flow was switched to the reactant mixture ($CO_2 : H_2 : N_2$ 20 : 60 : 20 molar%). The pressure was increased to 20 bar using a back-pressure regulator before the flow was set to 6.25 ml min^{-1} to give a GHSV of 1000 h^{-1}. The reactions were conducted at temperatures of 225 and 250 °C. All post-reactor lines and valves were heated at 110 °C to avoid product condensation. The gas products were analysed *via* online gas chromatography using an Agilent 7890 system with a flame ionisation detector (FID) and TCD. Nitrogen was used as an internal standard. Samples were taken every 7 minutes, with 20 samples being taken at each temperature.

Results

As can be observed in Table 1, the SAS precipitated materials for all Na^+ doping levels had Cu : Zn molar ratios close to the target of 2 : 1. The Na^+ content within the samples (determined from calcined products) varied slightly from the target compositions, with lower loadings having higher discrepancies from the target loadings. The reason for this is that the lowest target loadings will be more significantly affected by multiple contamination sources (both during preparation and also during elemental analysis). Despite this discrepancy, the set of precipitated samples successfully had a range of Na^+ between 2.12 wt% and less than <100 ppm (lowest Na^+ value obtained similar to experimental blank).

Precursor characterisation

The first notable effect of Na^+ content was that it influenced the nature of the SAS precipitated species. While all samples were initially blue, the high-Na-georgeite sample changed colour to green within a 24 h period. Infrared spectroscopy (Fig. 1a) showed that all samples with the exception of high-Na-georgeite had bands (OH band at 3400–3300 cm^{-1} and CO_3^{2-} bands at 1463, 1395 and 830 cm^{-1}) associated with zincian georgeite.[23] As to be expected from the green colour, the high-Na-georgeite sample had alternative IR bands more closely associated with malachite (two distinct OH bands at 3404 and 3275 cm^{-1} along with CO_3^{2-} bands at 1647, 1503, 1375, 1100, 1049, 842, 828, 817 and 740 cm^{-1}).[24] In all samples no bands associated with sodium acetate or sodium carbonate were observed, due to sodium species being a minor component of the samples. To gain an understanding of what the sodium species could be, the SAS precipitation of sodium acetate under the same conditions as the zincian georgeite preparation was performed. The resulting precipitate was found to have IR bands associated with sodium hydrogen carbonate (Fig. 1b). The presence of the sodium hydrogen

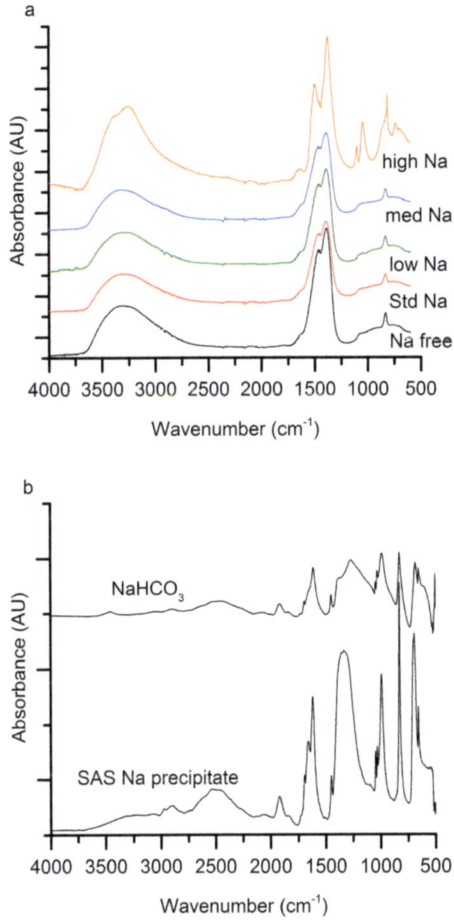

Fig. 1 (a) Infrared spectra of SAS precipitated materials with various Na^+ doping (lowest to highest Na content; <100 ppm, 320 ppm, 580 ppm, 2520 ppm to 21 200 ppm). (b) Infrared spectra of SAS precipitated sodium acetate and reference sodium hydrogen carbonate.

carbonate shows that the acetate cation was replaced, removing the potential exothermic nature of its decomposition during zincian georgeite calcination.[25] It is worth highlighting the importance of the removal of exothermically decomposing salts from the catalyst precursor, as previous studies have been significantly affected by the exothermic decomposition of sodium nitrate.[20,21]

XRD analysis of the precursor phases (Fig. 2) confirmed the observations seen by IR, with all SAS precipitated materials, with the exception of high-Na-georgeite, being amorphous, as expected for georgeite.[19,23] The crystalline phases present in the high-Na-georgeite precipitate were malachite[26] (unlabelled reflections in Fig. 2) and surprisingly sodium zinc carbonate ($Na_2Zn_3(CO_3)_4 \cdot 3H_2O$)[27] (labelled with arrows). It is generally acknowledged that zincian malachite (where *ca.* 27% of the Cu can be substituted for Zn) produces highly active methanol synthesis catalysts, due to the intimate mixing of Cu and Zn and the porous microstructure imparted from its needle like morphology.[12] It has been demonstrated that the

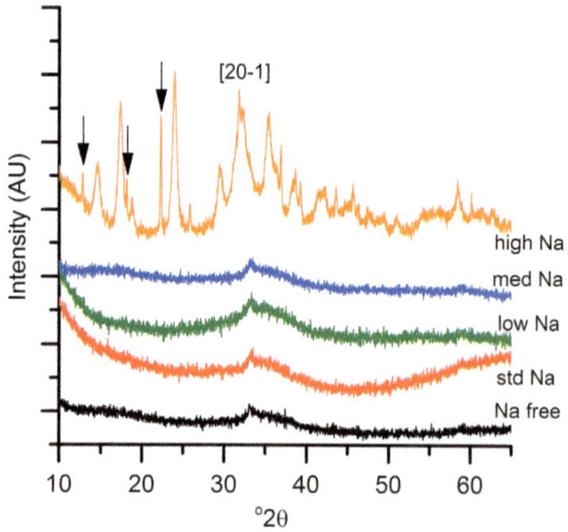

Fig. 2 Powder X-ray diffraction patterns of SAS precipitated materials with various Na^+ doping (lowest to highest Na content; <100 ppm, 320 ppm, 580 ppm, 2520 ppm to 21 200 ppm). The high Na sample contains zincian malachite (unlabelled reflections, with the exception of the [20−1] reflection used to calculate Zn content) and sodium zinc carbonate labelled with arrows.

degree of Zn incorporation into the malachite unit cell (maximum 27 at% Zn) can be determined by reduction in the [20−1] *d* spacing, associated with Zn replacing Jahn–Teller distorted Cu.[12] The extent of Zn incorporation has been directly correlated with methanol synthesis activity, with greater incorporation resulting in higher activity.[12] The [20−1] *d* spacing in malachite observed in the high-Na-georgeite material was found to be 2.81 Å relative to 2.86 Å in Cu only malachite, which indicates an approximate 10 at% Zn substitution. The extent of Zn incorporation has been found to be dependent on pH control during co-precipitation methods, with constant pH 6–7 processes (*i.e.* co-addition of base and metal precursor solution) having much greater Zn incorporation than varying pH methods (*i.e.* addition of base to metal precursor solution or *vice versa*).[28] It is therefore unsurprising that during the uncontrolled aging of zincian georgeite, Zn incorporation into the malachite phase was poor. The low activity of previously reported unaged catalysts,[15] which likely initially contained zincian georgeite and Na^+ impurities, can be explained by similar uncontrolled ageing to low Zn containing zincian malachite. As mentioned previously, the presence of sodium zinc carbonate is also interesting as it has been hypothesised as a Zn reservoir required for the formation of zincian malachite during co-precipitation ageing steps.[27] It is unclear whether sodium zinc carbonate is present from the point of precipitation, or forms during the ageing process. However, its presence will result in the segregation of Zn and Cu in the precursor material.

Another method of differentiating zincian georgeite and malachite is from TGA, with zincian malachite having a single mass loss (occasionally with a high temperature shoulder) and zincian georgeite having three distinct mass loss events.[19] The results (Fig. 3a) corroborate the observations of XRD and IR analysis, with all samples showing the distinct mass loss profile for zincian georgeite, except for high-Na-georgeite with the single mass loss associated with malachite. An interesting effect of Na^+ content on the zincian georgeite samples

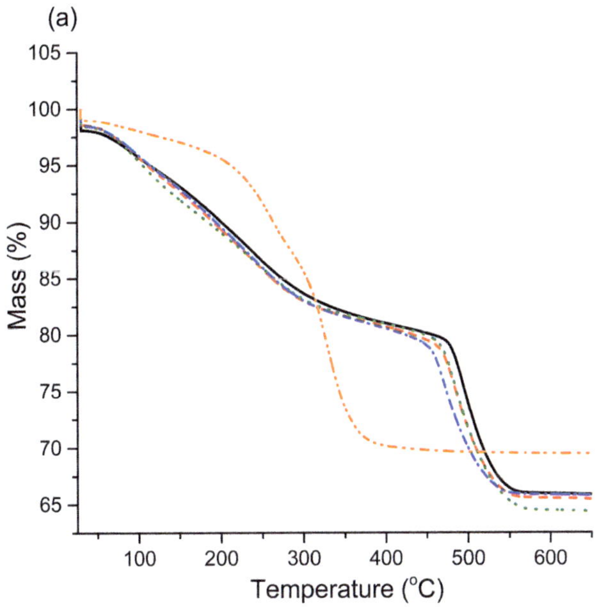

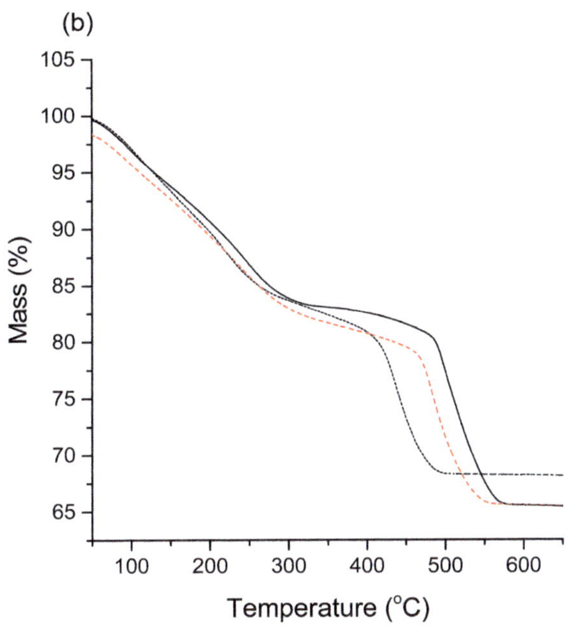

Fig. 3 Thermogravimetric analysis of SAS prepared zincian georgeite precursors. (a) Mass loss profiles of various Na doped samples. Solid black line, Na-free; red dashed, Na-Std; green dotted, Na-low; blue dashed dotted, Na-med; orange dashed dotted, high-Na (lowest to highest Na content; <100 ppm, 320 ppm, 580 ppm, 2520 ppm to 21 200 ppm). (b) Mass loss profiles of different Cu:Zn Std-Na zincian. Solid black line, 1 : 1 Cu : Zn; red dashed, 2 : 1 Cu : Zn; black dotted, Cu only.

(med-Na-georgeite to Na-free-georgeite) was that the final mass loss temperature subtly decreased with increasing Na^+ content. This final mass loss has previously been attributed to occluded carbonate between CuO and ZnO crystallites and has been associated with high Cu surface areas and high catalytic activity.[26] As the position of the high temperature decomposition is dependent on Zn content in zincian georgeite (Fig. 3b) it can be rationalised that the mixing of Cu and Zn subtly decreases with increasing Na^+ content.

In summary, the SAS precipitation of zincian georgeite with a significant Na^+ content of *ca.* 2 wt% results in a precipitate that rapidly ages to malachite, with Zn being partially incorporated into malachite and the remaining Zn being as a separate sodium zinc carbonate. The separation of Cu and Zn can possibly be seen in the other Na^+ containing samples from the reduction in temperature of the carbonate mass loss in TGA. Potentially trace sodium zinc carbonate is present in these samples, which subtly reduces Cu and Zn mixing but not to an extent that results in ageing into malachite.

Calcination and Cu surface area analysis

The chosen calcination temperature of the SAS precipitated materials was 300 °C, which retains the high temperature carbonate phase, preventing the formation of large CuO and ZnO crystallites. XRD of the calcined materials (Fig. 4) confirms this, with all zincian georgeite derived materials only displaying ill defined and broad reflections associated with CuO. If our hypothesis of increased Cu/Zn segregation in all samples due to sodium zinc carbonate is correct, a small increase in CuO particle size with Na^+ content could be expected. This would be due to CuO growth being less efficiently inhibited by dispersed ZnO particles. Also the presence of residual sodium acetate species could result in a local exotherm during its decomposition and result in CuO growth (similar to nitrate decomposition in previous studies). However, the crystallite sizes (calculated by the Scherrer equation) of the CuO were all found to be *ca.* 2 nm. It is important to note that crystallites of this size are on the detection limit of XRD and so subtle changes may not be detected. Clearly the effect of Na^+ at these levels is low, but to further investigate the hypothesis, techniques capable of looking at disordered or nanocrystalline materials such as pair distribution function analysis or X-ray absorbance spectroscopy are required. Nevertheless, the current XRD analysis shows that sintering of CuO and ZnO crystallites through exothermic decomposition of residual metal salt precursors (in this case acetates) has been mitigated by our approach.

To reiterate, the effect of Na^+ on CuO crystallite sizes for Na-free-CuO/ZnO through to med-Na–CuO/ZnO samples is not detectable by XRD. By contrast, the highest Na^+ containing sample, which unintentionally aged into malachite was found to comprise significantly larger (6 nm) CuO crystallites, which is attributed to the crystallinity of the precursor and the phase separated nature of the Cu and Zn in the sample. In an additional experiment, the high-Na-georgeite material was calcined immediately after SAS precipitation to limit the effect of uncontrolled ageing. As shown in Fig. 4b, this material was found to have small *ca.* 2 nm particles of similar size to the other zincian georgeite derived materials, which suggests that the previously observed increased crystallinity was primarily not caused by sodium species directly, but by uncontrolled ageing to malachite. From

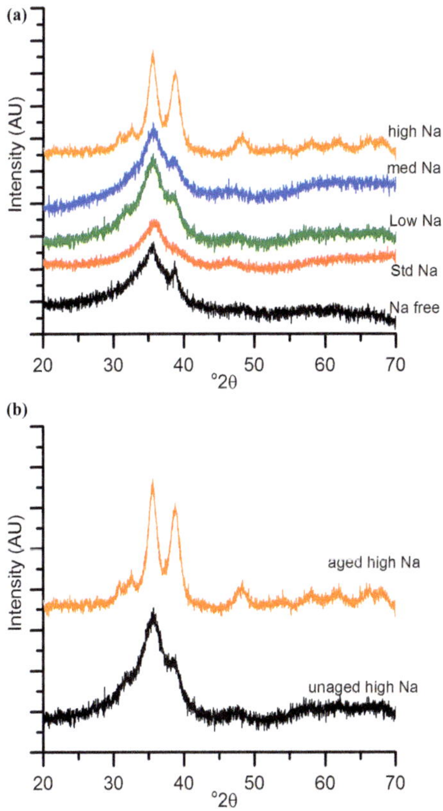

Fig. 4 (a) Powder X-ray diffraction patterns of calcined (300 °C, 4 h, static air) SAS precursors with various Na doping (lowest to highest Na content; <100 ppm, 320 ppm, 580 ppm, 2520 ppm to 21 200 ppm). (b) Powder X-ray diffraction patterns of calcined (300 °C, 4 h, static air) high-Na aged and unaged.

herein we refer to these samples as high-Na–CuO/ZnO and aged-high-Na–CuO/ZnO for the unaged and aged samples respectively. Previous studies into the effect of Na^+ in CuO/ZnO catalysts have highlighted that increased Na^+ content resulted in more crystalline oxide phases on calcination.[20] While we observe subtle indications of Na^+ causing Cu and Zn phase separation from TGA analysis, we did not detect (by XRD) changes in CuO crystallite size, except when Na^+ facilitated phase transformation of the precursor from georgeite to malachite.

Copper surface area analysis of the reduced SAS prepared catalysts is shown in Fig. 5. The small *ca.* 2 nm CuO crystallites seen for the Na-free-CuO/ZnO to medium-Na–CuO/ZnO samples corresponded to comparable Cu surface areas for all of these samples, at *ca.* 20 $m^2 g^{-1}$. As expected, the 6 nm CuO seen in the uncontrolled aged-high-Na–CuO/ZnO resulted in a dramatically lower Cu surface area of 9 $m^2 g^{-1}$. Interestingly, the Cu surface area of the high-Na–CuO/ZnO sample, which was calcined immediately after precipitation, was also significantly lower at 12 $m^2 g^{-1}$. This might be considered unexpected given that the CuO crystallite size determined by XRD for this sample was similar (*ca.* 2 nm) to those observed with the Na-free-CuO/ZnO to medium-Na–CuO/ZnO samples.

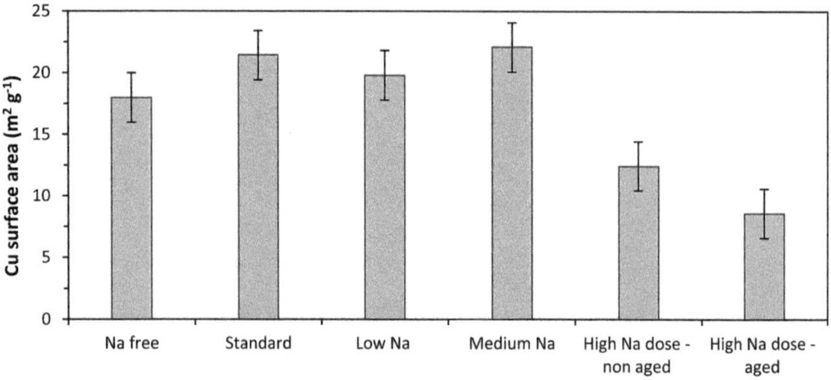

Fig. 5 Copper surface area analysis using N_2O titration of catalysts prepared from various Na doped catalyst precursors. Catalysts reduced under 10% H_2.

One possible explanation is that the Na^+ content in the high-Na–CuO/ZnO reduces the Cu–ZnO interaction, resulting in greater Cu sintering during reduction. However, *in situ* XRD analysis of Cu crystallite sizes of unaged high-Na–CuO/ZnO and Na-free-CuO/ZnO after reduction at 225 °C showed that this was not the case, with the Cu crystallites being almost identical at 6.9 and 6.8 nm respectively. As a point of reference the aged-high-Na–CuO/ZnO Cu crystallite size was notably bigger at 12.7 nm. Alternatively, several studies state that N_2O titration oxidises (in addition to Cu) reduced Zn or ZnO_x species at the periphery of Cu nanoparticles.[29] A reduced Cu–ZnO interaction would also reduce the number of these ZnO_x sites and therefore result in lower apparent Cu surface area values. Another possibility is that sodium species actively block Cu surface sites and reduce the Cu surface area. A rough calculation of the surface area of monolayer sodium, based on the 2.12 wt% present in the catalyst and assuming a sodium footprint of 1.2×10^{-19} m^2 g^{-1},[30] gives a value of *ca.* 70 m^2 g^{-1} in the reduced high-Na–CuO/ZnO catalyst. Whilst this calculation makes the improbable assumption of atomic layer metallic sodium on the catalyst, it illustrates that 2.12 wt% sodium species could cover a significant proportion of the surface of the 80 m^2 g^{-1} CuO/ZnO catalyst (general surface area determined by BET analysis).

Water–gas shift reaction

The catalysts were tested for LTS activity and stability (Fig. 6) at steady state conditions. The conversions of the Na-free-CuO/ZnO, Std-Na–CuO/ZnO, low-Na–CuO/ZnO and med-Na–CuO/ZnO steadily decreased with Na^+ content, but were all above 90% conversion after 100 min time on-line, while the high-Na–CuO/ZnO gave a dramatically lower CO conversion of 15%. CO conversion with the aged high-Na–CuO/ZnO was also approximately 15% at 100 min time on-line but dropped to 5% after 500 min time on-line. A number of previous studies have noted that LTS catalytic activity with Cu/ZnO catalysts was found to correlate with Cu surface area,[31,32] while others reported no correlation[33] and finally Hadden *et al.* suggested correlation with Cu surface area, but only within families of samples prepared by the same methodology.[34] No strong correlation with initial Cu surface area values and CO conversion was observed between the zincian

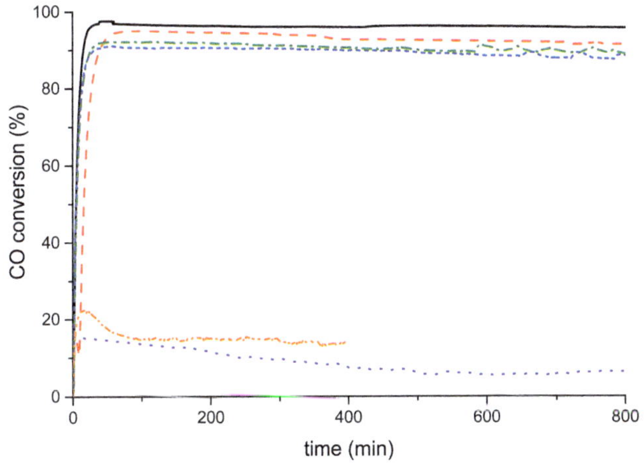

Fig. 6 Low temperature water–gas shift catalytic testing of Cu/ZnO catalysts derived from SAS zincian georgeite doped with various Na loadings. Solid black line, Na-free; red dashed, Na-Std; green dotted, Na-low; blue short dash, Na-med; orange dashed dotted, unaged-high-Na; purple dotted, aged-high-Na (lowest to highest Na content, <100 ppm, 320 ppm, 580 ppm, 2520 ppm to 21 200 ppm).

georgeite derived catalysts in this study, with discernibly different conversions being seen for catalysts (Na-free-CuO/ZnO, Std-Na-CuO/ZnO, low-Na–CuO/ZnO and med-Na–CuO/ZnO) despite them having similar Cu surface areas. The exceptionally low activity of high-Na–CuO/ZnO could be correlated with the reduced Cu surface area, but a simple proportional relationship between Cu surface area and CO conversion is not present. Specifically, the CO conversion of high-Na–CuO/ZnO was 15% that of Na-free-CuO/ZnO, while the Cu surface area was 66% that of Na-free-CuO/ZnO. It can be rationalised that Cu surface area is not a suitable indicator of activity under the LTS reaction conditions employed in this study, due to the aggressive reaction conditions (particularly the high water content) causing significant sintering before steady state is reached. Previously we showed that catalysts under similar LTS reaction conditions demonstrated a substantial reduction of Cu surface area after 40 h time on-line.[19]

While no correlation between initial copper surface area and activity was observable, there was a correlation with regards to Na^+ content. One potential explanation is that while the Cu surface area decreases during the reaction, the sodium species remain dispersed on the surface and therefore higher Na^+ concentrations more effectively block active Cu sites in the sintered catalyst. Alternatively, sodium species act as an intrinsic catalyst poison by modifying the active site of the catalyst, possibly by enhancing the number of basic sites and increasing the stability of adsorbed CO_2. Finally, a third consideration is that sodium species affect the Cu–ZnO interaction by causing phase separation of Cu and Zn within the catalyst precursor. In the case of the aged high-Na–CuO/ZnO catalyst this phase segregation was evident from the precursor phase itself. Interestingly this phase separated catalyst clearly showed significant deactivation with time on-line. Although the exact cause of catalytic deactivation is not clear, the effect of Na^+ content on the LTS activity of zincian georgeite derived catalysts is evident, with even trace content affecting performance. Conventional co-precipitation methods, using basic metal salts as precipitation agents, cannot provide the degree of purity (with respect to Na^+ content) required to highlight this phenomenon as materials prepared by the SAS precipitation process can.

While the effect of Na^+ poisoning has been shown to be important, previous work has shown that other structural properties in Cu/ZnO catalysts are also important, specifically the Cu surface area (although this appears to significantly change under reaction conditions) and the Cu–ZnO interaction.[16] Previously we reported approximately 10% greater activity for a zincian georgeite derived catalyst compared to a zincian malachite derived catalyst and attributed the improvement partly to enhanced Cu–ZnO interaction, through smaller and more disordered ZnO crystallites.[19] Importantly, this difference in activity was greater than the 5% difference between zincian georgeite derived Na-standard-CuO/ZnO and Na-low-CuO/ZnO, which has a Na^+ content comparable to the co-precipitated zincian malachite derived catalyst. Improved activity between catalysts derived from different precursor phases does involve more than simply Na^+ content, but its effect is clearly significant.

Methanol synthesis from CO_2

The activity and selectivity of the different Na^+ doped catalysts for CO_2 hydrogenation was tested at 20 bar pressure and 225–250 °C, with data shown in Table 2. At 225 °C all zincian georgeite derived catalysts had CO_2 conversions of between 2 and 7%. No clear trend with Na^+ content was observed, although the reaction with the unaged-Na-high-CuO/ZnO had the lowest conversion, attributable to the low Cu surface area of this catalyst. Analysis of gaseous products showed that in all reactions methanol and carbon monoxide were produced, with no detectable methane or higher alcohol formation. The methanol selectivities for the different zincian georgeite catalysts were broadly similar with values between 30–39%, with the clear exception being for the Na-free-CuO/ZnO catalyst, which had a significantly higher methanol selectivity of 67%. All catalysts were found to be stable over the 2 h testing

Table 2 CO_2 hydrogenation with reaction zincian georgeite derived Cu/ZnO catalysts, prepared by SAS process with various levels of sodium doping[a]

Sample name	Reaction temperature (°C)	CO_2 conversion (%)	Methanol selectivity (%)	Methanol yield (%)
Na-free-CuO/	225	4.4	67	2.9
ZnO	250	8.6	45	3.9
Std-Na–CuO/	225	6.2	39	2.4
ZnO	250	10.3	27	2.8
Low-Na–CuO/	225	4.4	39	1.7
ZnO	250	9.8	24	2.4
Med-Na–CuO/	225	7.0	21	1.5
ZnO	250	9.7	12	1.2
High-Na–CuO/	225	2.5	32	0.8
ZnO	250	4.4	18	0.8

[a] Reaction conditions: 0.1 g catalyst, 6.25 ml min^{-1} flow of 3 : 1 H_2 : CO_2 gas, 20 bar. Catalytic activity measured after 2 h reaction time.

time with regard to CO_2 conversion, although far longer reaction times are required to make a meaningful comment on catalyst deactivation.

Increasing the reaction temperature to 250 °C resulted in the expected increase in CO_2 conversion coupled with a reduction in selectivity towards methanol. Again, no strong trend between CO_2 conversion and Na^+ content was observed with conversions varying between 8 and 10% for all catalysts, except for the unaged-Na-high-CuO/ZnO catalyst which gave 4% conversion. The general reduction in selectivity seen for all catalysts is attributable to the shift towards the endothermic reverse water–gas shift reaction resulting in higher carbon monoxide production. Although the methanol selectivity of the Na-free-CuO/ZnO catalyst drops on increasing temperature it was still significantly higher at 45% compared to the other catalysts, which all had selectivities below 30%.

At both 225 and 250 °C methanol yields were found to correlate with Na^+ content. The lowest methanol yield (0.8%) was observed for the unaged-Na-high-CuO/ZnO catalyst and the highest yield (3.9%) was seen with the Na-free-CuO/ZnO catalyst. Previously Lee and co-workers showed a similar trend for conventionally prepared $Cu/ZnO/Al_2O_3$ with various levels of sodium nitrate contamination.[20] They attributed the reduced activities and selectivities primarily to an increase in CuO and ZnO crystallite size, although they did note that at very high Na^+ concentrations (4 wt%) the basicity of the catalyst also increased. In the current study we observed that initial CuO and reduced Cu crystallite sizes were comparable across the entire range of Na^+ doped samples (with the exception of the aged high-Na-georgeite sample). Cu surface areas were similar for all samples, except the high-Na–CuO/ZnO sample, where Na^+ potentially blocked Cu sites. The exceptionally poor CO_2 conversion and methanol selectivity for this catalyst can be explained by a reduced number of Cu active sites. However, this does not sufficiently explain the correlation between methanol yield and Na^+ content for the samples with moderate levels of Na^+ doping, as all evidence suggests that the number of Cu active sites is broadly the same. As noted in the discussion on LTS activity, Na^+ appears to be an intrinsic catalyst poison, possibly by increasing catalyst basicity or by subtly modifying Cu–ZnO interactions.

From a practical perspective comparison of the best zincian georgeite derived catalyst with the literature is difficult, due to the strong influence of reaction pressure on methanol yields, with high pressures thermodynamically favouring methanol synthesis over the reverse water–gas shift reaction.[35] However, methanol yields of 6% and 10%, at the same temperature, pressure and $CO_2 : H_2$ ratio as the current study (250 °C, 20 bar and 3 : 1 $CO_2 : H_2$) have been reported for Pd/ZnO[6] and $Cu/Ga_2O_3/ZrO_2$ (ref. 11) catalysts respectively. These results show that the zincian georgeite derived catalysts in this study are reasonable CO_2 hydrogenation catalysts (best yield of 3.9%) but are currently far from exceptional. However, optimisation of this catalyst has not yet been performed, with alterations in reduction procedures and also catalyst composition potentially being able to improve performance.

In situ XRD analysis

For both the methanol synthesis and LTS reaction several studies have shown that reaction conditions and gas environments influence catalyst microstructure. Specifically, there is strong evidence of metal–support interaction between Cu and

ZnO, with enhanced wetting of the ZnO by Cu under strong reducing conditions.[36] Also shown by studies with model impregnated Cu on crystalline ZnO is that Cu wetting dramatically reduces in the presence of water.[37] The presence of water is clearly important for the LTS reaction (with 24% water in the gas stream) and, maybe less obviously, in the synthesis of methanol from CO_2 as one mole of water is produced per mole of methanol (see eqn (1))

$$CO_2 + 3H_2 \Rightarrow CH_3OH + H_2O \quad (1)$$

To investigate the effect of Na^+ on the microstructure of Cu/ZnO catalysts under environments similar to reaction conditions, Na-free-CuO/ZnO and unaged high-Na–CuO/ZnO catalysts were subjected to syn-gas for 3 h followed by wet syn-gas (15% H_2O) at 225 °C, while being monitored by *in situ* XRD (Fig. 7). It is recognised that the water content within these experiments is below that used for the LTS reactions, but is sufficient to gain a basic understanding of its effect on catalyst structure. Both the Na-free-CuO/ZnO and unaged high-Na–CuO/ZnO catalysts broadly followed the same evolution of structure during the catalyst treatment, with no new phases being observed and crystallite growth at a rate dependent on the environment. Specifically, Cu crystallite size was noted to immediately increase on the addition of the syn-gas with no noticeable change in ZnO crystallite size. Then during the 3 h hold under syn-gas, a slow but noticeable increase in Cu crystallite size is observed, while the ZnO crystallite size remains constant. This growth of Cu particles can be attributed to simple sintering at the reaction temperature. Interestingly, the subsequent addition of the 15% water then resulted in an immediate increase in Cu and ZnO crystallite size growth rate.

While previous *in situ* studies have noted the sintering of Cu particles under methanol synthesis conditions,[36] the direct effect of water on ZnO and Cu crystallites has not been reported. Most studies concerning the deactivation of methanol synthesis catalysts emphasise Cu sintering or poisoning (from chlorine or sulphur) and do not consider ZnO crystallite size, or changes in its interaction with Cu.[38] In light of recent studies that state that partially reduced ZnO, or even metallic Zn, on the surface or at the interface of Cu particles aid activity for methanol synthesis, the evolution of ZnO particles under reaction conditions is important. Even if one assumes that ZnO is not involved in the catalytic mechanism, its role of dispersing Cu crystallites means that the observed impact of water on its agglomeration is important in preventing Cu sintering. Indeed, the concurrent increase in the rate of Cu particle sintering can be attributed to agglomeration of the inter-dispersing ZnO particles.

Given that the water content in LTS testing was greater than in the *in situ* XRD experiment, it can be assumed that the extent of ZnO crystal growth was far more significant. The lack of deactivation observed in the time on-line data suggests that sintering had occurred at the very start of the reaction, before reaching steady state conditions. While no specific growth in any particular plane was observed in these experiments, extensive growth under LTS reaction conditions could lead to changes in particle morphology and potentially alteration of the ratio of polar and non-polar ZnO planes, which has previously been reported to influence catalytic activity.[39] Further *in situ* studies with higher water partial pressures would provide greater detail.

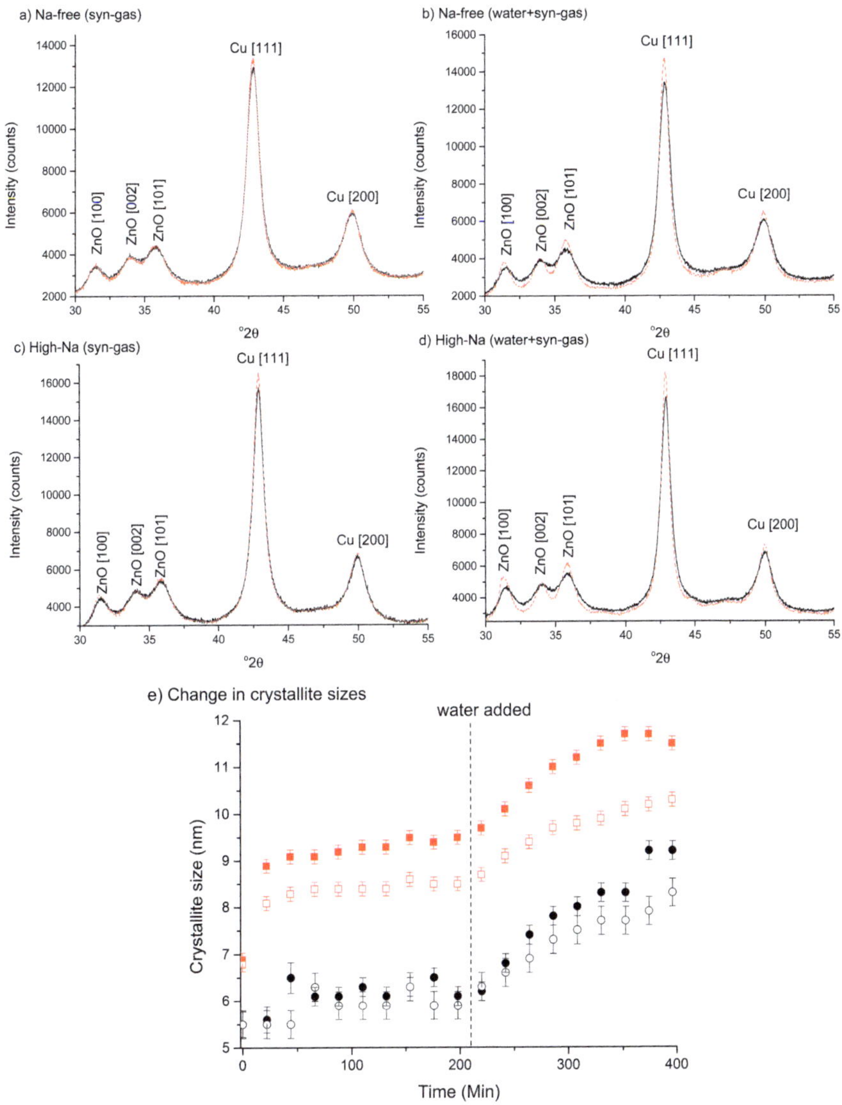

Fig. 7 *In situ* XRD analysis of high-Na–CuO/ZnO and Na-free-CuO/ZnO catalysts under syn-gas and syn-gas + water. Catalysts reduced *in situ* using 2% H_2 at a 1 °C min^{-1} ramp rate. Reaction conditions of 225 °C, 50 ml min^{-1} gas flow of syn-gas 6% CO/9.2% CO_2/66% H_2 and 17.8% N_2 followed by syn-gas + water (15% water). (a–d) XRD patterns at start (black line) and end (red dashed) of 3 h gas cycle. (a) Na-free catalyst under syn-gas, (b) Na-free catalyst under syn-gas + water, (c) high-Na catalyst under syn-gas and (d) high-Na catalyst under syn-gas + water. (e) ZnO (●, high-Na; ○, Na-free) and Cu (■, high-Na; □, Na-free) calculated crystallite sizes (using ZnO [100] and Cu [111]) with reaction time on-line. Dashed line indicates introduction of water to the gas mixture.

Although the water content was below that present in LTS reactions, its concentration is comparable to amounts produced from CO_2 hydrogenation to form methanol and also the reverse water–gas shift reaction. The presence of water in this reaction is already known to act as a poison, and we now show that it accelerates ZnO sintering and consequently Cu sintering; the implications of which are firstly a reduction in Cu surface area, and secondly a reduction in Cu–ZnO_x interaction that has been shown to be dependent on ZnO crystallite size.[17] Both of these micro-structural properties are important for high catalytic activity and the loss of these properties will be detrimental to activity. It is hypothesised that higher activity can be achieved by stabilising small and defective ZnO crystallites, possibly by the addition of Ga^{3+} or Al^{3+}.[11,40]

With regard to the effect of Na^+ on the changes in Cu/ZnO microstructure, the first observable difference between catalysts was that the increase in Cu particle size, with the addition of syn-gas, was slightly more significant for the unaged high-Na–CuO/ZnO catalyst. The rate of Cu crystallite growth under syn-gas was also marginally greater for the unaged high-Na–CuO/ZnO catalyst. During this time period the calculated crystallite sizes for ZnO were identical, with no observable growth for either catalyst. The effect of water addition was also slightly more pronounced in the unaged high-Na–CuO/ZnO catalyst with respect to both ZnO and Cu crystallite growth rates. Tentatively, these observations can be interpreted as evidence of Na^+ catalyst poisoning being associated with phase separation and particle growth of Cu and ZnO under reaction conditions. However, it must be considered that the observed differences were small, especially when it is considered that the *in situ* study compared the catalysts with the most significant difference in Na^+ content.

Conclusions

A range of Na^+ doped zincian georgeite materials were prepared, with content being varied from effectively <100 ppm to 21 200 ppm (2.12 wt%). For the high Na^+ content (2.12 wt%) precipitated material the uncontrolled ageing of zincian georgeite to phase separated sodium zinc carbonate and a low Zn containing zincian malachite was apparent. This sample was found to form large CuO and Cu particles, after calcination and reduction, making it a poor catalyst precursor. The materials containing less Na^+ remained zincian georgeite after precipitation. Small changes in the decomposition temperature of high temperature carbonate in these lower Na^+ zincian georgeite materials suggested that trace Na^+ might slightly reduce Cu/Zn mixing. Calcination of the 2.12 wt% doped sample immediately after precipitation limited the uncontrolled ageing to malachite.

Both the LTS and methanol synthesis reactions were found to be sensitive to Na^+ content. LTS activity was found to correlate directly with Na^+ content, with the Na^+ free sample having the highest activity and the high Na^+ containing catalysts the lowest. Trends with methanol synthesis activity were more complicated, with significant contribution from the reverse water–gas shift reaction. It was clear that catalysts with very low Na^+ content had far lower reverse water–gas shift activity and therefore superior methanol selectivity (highest methanol yield of 3.9%), while the high Na^+ content (2.12 wt%) almost completely deactivated the catalyst for CO_2 activation (lowest methanol yield of 0.8%). Methanol yields were found to correlate with Na^+ content, with intermediate Na^+ content catalysts giving moderate yields.

Previous studies into the effect of sodium species removal (*via* washing procedures) on methanol synthesis catalysts stated that Na^+ increased CuO crystallite size and reduced Cu surface area, which was attributed to be its primary effect on activity. This was due to the exothermic decomposition of residual nitrate species. As the counter ion for Na^+ in this study was carbonate (which decomposes endothermically), we found that CuO and reduced Cu crystallite sizes were unaffected by Na^+ content, except when high Na^+ content facilitated uncontrolled precursor ageing that consequently resulted in large CuO and Cu crystallite sizes. Also no variation in Cu surface area was observed for Na doping between <100 and 2500 ppm. When the aging of the 2.12 wt% Na^+ loaded sample was limited, reduced Cu surface area values were observed and were attributed to the blocking of surface sites. The discrepancy between this study and others is that we do not have residual sodium nitrate species, which are thought to be responsible for CuO crystallisation during calcination.

Given that discernable changes in activities and selectivities were observed for the different Na^+ doped catalysts with comparable Cu surface areas, it was concluded that Na^+ acts directly as a poison. Potentially this is through increasing surface basicity, blocking active sites or by inducing phase separation between Cu and ZnO. *In situ* XRD analysis showed that the addition of water to a syn-gas environment resulted in a marked growth in ZnO and Cu crystallite size, which indicates that under significant water partial pressures ZnO and Cu are becoming more phase separated. It was noted that this phase separation was slightly greater for the high Na^+ doped sample, indicating that sodium species may disrupt the Cu metal–support interaction.

Finally the results demonstrate that advanced material synthesis procedures, such as SAS precipitation, can be used to prepare catalysts with exceptionally low levels of impurity and allow for the fundamental study of these trace elements in catalytic performance.

Acknowledgements

We would like to thank the EPSRC and UK Catalysis Hub for funding (grants EP/K014714/1, EP/K014714/1, EP/K014668/1, EP/K014706/1, EP/H000925/1, EP/I019693/1 and EPSRC Grant EP/L027240/1).

References

1 G. C. Chinchen, P. J. Denny, J. R. Jennings, M. S. Spencer and K. C. Waugh, *Appl. Catal.*, 1988, **36**, 1–65.
2 C. Rhodes, G. J. Hutchings and A. M. Ward, *Catal. Today*, 1995, **23**, 43–58.
3 M. Behrens, *Angew. Chem., Int. Ed.*, 2014, **53**, 12022–12024.
4 M. Behrens, *Recycl. Catal.*, 2013, **2**, 78–86.
5 R. Burch, *Phys. Chem. Chem. Phys.*, 2006, **8**, 5483–5500.
6 H. Bahruji, M. Bowker, G. Hutchings, N. Dimitratos, P. Wells, E. Gibson, W. Jones, C. Brookes, D. Morgan and G. Lalev, *J. Catal.*, 2016, **343**, 133–146.
7 F. Studt, I. Sharafutdinov, F. Abild-Pedersen, C. F. Elkjaer, J. S. Hummelshoj, S. Dahl, I. Chorkendorff and J. K. Norskov, *Nat. Chem.*, 2014, **6**, 320–324.
8 E. M. Fiordaliso, I. Sharafutdinov, H. W. P. Carvalho, J.-D. Grunwaldt, T. W. Hansen, I. Chorkendorff, J. B. Wagner and C. D. Damsgaard, *ACS Catal.*, 2015, **5**, 5827–5836.

9 J. H. Carter, S. Althahban, E. Nowicka, S. J. Freakley, D. J. Morgan, P. M. Shah, S. Golunski, C. J. Kiely and G. J. Hutchings, *ACS Catal.*, 2016, **6**, 6623–6633.
10 X.-M. Liu, G. Q. Lu and Z.-F. Yan, *Appl. Catal., A*, 2005, **279**, 241–245.
11 M. M.-J. Li, Z. Zeng, F. Liao, X. Hong and S. C. E. Tsang, *J. Catal.*, 2016, **343**, 157–167.
12 M. Behrens, *J. Catal.*, 2009, **267**, 24–29.
13 T. Fujitani and J. Nakamura, *Catal. Lett.*, 1998, **56**, 119–124.
14 G. C. Chinchen, K. C. Waugh and D. A. Whan, *Appl. Catal.*, 1986, **25**, 101–107.
15 M. Behrens, F. Studt, I. Kasatkin, S. Kuehl, M. Haevecker, F. Abild-Pedersen, S. Zander, F. Girgsdies, P. Kurr, B.-L. Kniep, M. Tovar, R. W. Fischer, J. K. Norskov and R. Schloegl, *Science*, 2012, **336**, 893–897.
16 F. Studt, M. Behrens and F. Abild-Pedersen, *Catal. Lett.*, 2014, **144**, 1973–1977.
17 S. Kuld, M. Thorhauge, H. Falsig, C. F. Elkjaer, S. Helveg, I. Chorkendorff and J. Sehested, *Science*, 2016, **352**, 969–974.
18 A. M. Pollard, M. S. Spencer, R. G. Thomas, P. A. Williams, J. Holt and J. R. Jennings, *Appl. Catal., A*, 1992, **85**, 1–11.
19 S. A. Kondrat, P. J. Smith, P. P. Wells, P. A. Chater, J. H. Carter, D. J. Morgan, E. M. Fiordaliso, J. B. Wagner, T. E. Davies, L. Lu, J. K. Bartley, S. H. Taylor, M. S. Spencer, C. J. Kiely, G. J. Kelly, C. W. Park, M. J. Rosseinsky and G. J. Hutchings, *Nature*, 2016, **531**, 83–87.
20 K.-W. Jun, W.-J. Shen, K. S. Rama Rao and K.-W. Lee, *Appl. Catal., A*, 1998, **174**, 231–238.
21 J. Wu, S. Luo, J. Toyir, M. Saito, M. Takeuchi and T. Watanabe, *Catal. Today*, 1998, **45**, 215–220.
22 G. Prieto, K. P. de Jong and P. E. de Jongh, *Catal. Today*, 2013, **215**, 142–151.
23 P. J. Bridge, J. Just and M. H. Hey, *Mineral. Mag.*, 1979, **43**, 97–98.
24 J. A. Goldsmith and S. D. Ross, *Spectrochim. Acta, Part A*, 1968, **24**, 2131–2137.
25 Z. R. Tang, S. A. Kondrat, C. Dickinson, J. K. Bartley, A. F. Carley, S. H. Taylor, T. E. Davies, M. Allix, M. J. Rosseinsky, J. B. Claridge, Z. Xu, S. Romani, M. J. Crudace and G. J. Hutchings, *Catal. Sci. Technol.*, 2011, **1**, 740–746.
26 M. Behrens, F. Girgsdies, A. Trunschke and R. Schlögl, *Eur. J. Inorg. Chem.*, 2009, **10**, 1347–1357.
27 S. Zander, B. Seidlhofer and M. Behrens, *Dalton Trans.*, 2012, **41**, 13413–13422.
28 B. Bems, M. Schur, A. Dassenoy, H. Junkes, D. Herein and R. Schlögl, *Chem. – Eur. J.*, 2003, **9**, 2039–2052.
29 S. Kuld, C. Conradsen, P. G. Moses, I. Chorkendorff and J. Sehested, *Angew. Chem., Int. Ed.*, 2014, **53**, 5941–5945.
30 F. Yang and J. H. Hamilton, *Modern Atomic and Nuclear Physics*, World Scientific Publishing Company, 2010, p. 78.
31 M. J. L. Ginés, N. Amadeo, M. Laborde and C. R. Apesteguía, *Appl. Catal., A*, 1995, **131**, 283–296.
32 T. Shishido, M. Yamamoto, D. Li, Y. Tian, H. Morioka, M. Honda, T. Sano and K. Takehira, *Appl. Catal., A*, 2006, **303**, 62–71.
33 M. S. Spencer, *Top. Catal.*, 1999, **8**, 259–266.
34 R. A. Hadden, P. J. Lambert and C. Ranson, *Appl. Catal., A*, 1995, 122.
35 A. Bansode and A. Urakawa, *J. Catal.*, 2014, **309**, 66–70.
36 B. S. Clausen and H. Topsøe, *Catal. Today*, 1991, **9**, 189–196.
37 P. L. Hansen, J. B. Wagner, S. Helveg, J. R. Rostrup-Nielsen, B. S. Clausen and H. Topsøe, *Science*, 2002, **295**, 2053–2055.

38 M. V. Twigg and M. S. Spencer, *Top. Catal.*, 2003, **22**, 191–203.
39 C. Mateos-Pedrero, H. Silva, D. A. Pacheco Tanaka, S. Liguori, A. Iulianelli, A. Basile and A. Mendes, *Appl. Catal., B*, 2015, **174–175**, 67–76.
40 M. Behrens, S. Zander, P. Kurr, N. Jacobsen, J. Senker, G. Koch, T. Ressler, R. W. Fischer and R. Schlögl, *J. Am. Chem. Soc.*, 2013, **135**, 6061–6068.

Faraday Discussions

PAPER

PdZn catalysts for CO_2 hydrogenation to methanol using chemical vapour impregnation (CVI)†

H. Bahruji,[a] M. Bowker,*[ab] W. Jones,[ab] J. Hayward,[a] J. Ruiz Esquius,[a] D. J. Morgan[a] and G. J. Hutchings[a]

Received 14th September 2016, Accepted 12th October 2016
DOI: 10.1039/c6fd00189k

The formation of PdZn bimetallic alloys on ZnO, TiO_2 and Al_2O_3 supports was investigated, together with the effect of alloy formation on the CO_2 hydrogenation reaction. The chemical vapour impregnation (CVI) method produced PdZn nanoparticles with diameters of 3–6 nm. X-ray photoelectron spectroscopy and X-ray diffraction revealed the changes in the structure of the PdZn alloy that help stabilise formate intermediates during methanol synthesis. PdZn supported on TiO_2 exhibits high methanol productivity of 1730 mmol kg_{cat}^{-1} h^{-1} that is associated with the high dispersion of the supported PdZn alloy.

Introduction

In recent years the effects of global warming have attracted efforts to tackle the rising level of CO_2 in the atmosphere by capture, storage and utilisation of this abundant carbon source. One such effort is the transformation of CO_2 into methanol. Olah *et al.* introduced the anthropogenic carbon cycle concept for utilising carbon dioxide for synthetic hydrocarbon products.[1] In this concept methanol is used as a green starting chemical for the synthesis of fuels and various synthetic hydrocarbon products, and therefore CO_2 conversion to methanol is a crucial initial step.[1] However, the thermodynamic stability of CO_2 means that significant amounts of energy are required for activation and conversion into value-added products. The choice of catalysts is important to provide active sites for C–O bond dissociation and H-insertion in order to ensure high yields of methanol. For the methanol synthesis to be considered green, the captured CO_2 has to combine with H_2 produced from a sustainable source.

[a]*Cardiff Catalysis Institute, School of Chemistry, Cardiff University, Main Building, Park Place, CF10 3AT Cardiff, UK. E-mail: bowkerm@cardiff.ac.uk*
[b]*The UK Catalysis Hub, Research Complex at Harwell, Harwell, Oxon OX11 0FA, UK*

† Electronic supplementary information (ESI) available. See DOI: 10.1039/c6fd00189k

The first commercial methanol synthesis has been carried out since 1920s using a ZnO/Cr_2O_3 catalyst.[2] Cr_2O_3 was added to ZnO to improve the catalytic activity and prolong the life of the catalyst. Currently, a $Cu/ZnO/Al_2O_3$ catalyst developed by ICI has been used for methanol synthesis using a mixture of CO_2, CO and H_2 as the feedstock.[2,3] The reaction was carried out at 50–100 bar and 200–300 °C. Studies on Cu catalysts concluded that the methanol productivity was strongly correlated with the specific surface area of Cu,[4] with Cu oxidation states[5] and the Cu–Zn interface[6,7] also being significant factors. Cu–Zn species possess a bifunctional catalytic property which is created through interaction between Cu and ZnO in close proximity.[8] Deactivation over time is observed mainly due to the sintering of Cu metal.[9] Among the catalysts that have the potential for CO_2 hydrogenation, Pd has attracted interest due to the malleability property derived from the presence of a nearly full d-band. Pd/ZnO has higher stability than Cu catalysts for the methanol steam reforming reaction.[10,11] We previously found that Pd/ZnO catalysts can be active for CO_2 hydrogenation to methanol with the choice of preparation method and Pd precursor affecting the selectivity of the products.[12] The catalytic activity of Pd is strongly associated with the type of metal oxide support including ZnO,[13–15] CeO_2 (ref. 16) and Ga_2O_3.[17–19] Pd has a strong inclination to form intermetallic species with ZnO when exposed at a high temperature under a reducing environment.[20] Pd/ZnO transformed into PdZn alloy under a reducing environment, which subsequently altered the course of the catalytic reaction to favour methanol production over the reverse water gas shift reaction.[12] For the CO_2 hydrogenation reactions, the proximity between the metal and substrate stabilized the intermediate formate species, which is reported as the intermediate on Cu–Zn[21] and PdZn alloy.[12] In Cu–Zn systems, the formate intermediate can only be effectively converted to methanol on the Cu–Zn surface but not on Cu alone.[22] The hydrogenation of formate occurs at temperatures below its decomposition, which subsequently allows for complete hydrogenation to methanol.[23]

As in many heterogeneous catalytic systems, the dispersion of metal nanoparticles is important for enhancing the number of active sites. The prevention of highly dispersed catalysts from area loss by sintering is a challenging task, and researchers are conducting investigations into new catalytic preparation methods that improve nanoparticle dispersion and stability. The chemical vapor impregnation (CVI) method has been previously studied for nanoparticle deposition and can produce high metal dispersion on support materials.[24–26] In this method an organometallic precursor is heated under reduced pressure to induce sublimation, but not at sufficient temperatures to cause decomposition. In this manner, the organometallic is introduced from the gas phase, is adsorbed on the metal oxide support and tends to form highly isolated metal nuclei. This method also has the potential to enable the tailoring of the composition and morphology of supported nanoparticles in an easily accessible and scalable manner.[26] In the CVI method, the precursor sublimation step and deposition onto the support are not spatially separated since the precursor and support are in intimate contact during the sublimation–deposition step. Studies previously reported on Pd and Pt nanoparticles deposited *via* the CVI method onto TiO_2 reveal narrow particle size distributions of 1.5–2.8 nm.[24,26]

Previous investigations into the reactivity of Pd/ZnO catalysts for CO_2 hydrogenation reaction found that the activity of the catalyst is strongly influenced by the method of catalyst preparation, the type of palladium precursor, and the annealing treatment of the catalyst.[12] In that work we compared the behavior of materials produced by incipient wetness impregnation and sol immobilization methods. Here we report the activity of Pd catalysts on ZnO, TiO_2 and Al_2O_3 supports to clarify the importance of the formation of the Pd–Zn alloy and the catalysts were synthesized using the CVI method with non-chloride palladium precursors. This is important to eliminate the chlorine poisoning which can occur.[4] Chloride poisoning has been shown to promote particle agglomeration and shift the products towards a reverse water gas shift reaction.[12]

Experimental

Chemical vapour impregnation (CVI) preparation

Catalysts were prepared using a CVI method based on previous work by Forde *et al.*[26] This method was chosen in part to avoid the use of solvents and chloride precursors, and in part for the highly dispersed metal species it produces. For 5% Pd catalysts, ZnO, TiO_2 or Al_2O_3 and palladium acetylacetonate ($Pd(acac)_2$) were placed into a glass vial and shaken vigorously for 30 s to 1 min. This mixture was then transferred to a 50 mL Schlenk flask with the tube sealed. The tube was connected to a Schlenk line and was evacuated at room temperature ($\sim 10^{-3}$ mbar). The mixture was heated at 145 °C for 1 hour. For PdZn catalysts, zinc acetylacetonate ($Zn(acac)_2$) was added to the initial precursor mixture at Pd : Zn atomic ratios of 1 : 1 to 1 : 10, followed by a similar procedure. The sample was calcined in air at 500 °C for 16 hours to decompose the acetylacetonate precursors.

Catalyst characterisation

The characterisation of the samples was performed using a range of techniques. Powder X-ray diffraction (PXRD) patterns were obtained at room temperature with an Enraf Nonus FR590 diffractometer fitted with a hemispherical analyzer, using Cu Kα radiation ($\lambda = 1.54$ Å). X-ray photoelectron spectra (XPS) were recorded on a Kratos Axis Ultra-DLD XPS spectrometer with a monochromatic Al Kα source (75–150 W) and analyser pass energies of 160 eV (for survey scans) or 40 eV (for detailed scans). Samples were mounted using a double-sided adhesive tape and binding energies referenced to the C (1s) binding energy of adventitious carbon contamination which was taken to be 284.7 eV. Data were analyzed using Casa XPS software.[27]

To provide detailed morphological and compositional information at micro and nano-scales, high-resolution transmission electron microscope (HRTEM) system JEOL 2100 (LaB6) is employed. For the analysis, the material ground and mixed with water was placed on the TEM grid and dried. The instrument is equipped with a high-resolution Gatan digital camera (2k × 2k) with a maximum resolution of 0.2 Å, giving us the ability to see crystal lattices, to obtain diffraction patterns and giving accurate measurement of the lattice *d* spacing utilising Digital Micrograph software. In the scanning transmission electron microscopy (STEM) mode, a dark field HAADF/Z-contrast detector was used to provide excellent compositional contrast. An energy dispersive X-ray spectrometer (EDS) system (Oxford Instruments) was equipped with a large-area 80 mm^2 SDD (Silicon Drift

Detector). X-MaxN 80 T was employed for the elemental analysis in line scans, and elemental mapping modes.

Catalytic testing

The catalytic performance of the catalysts for CO_2 hydrogenation was determined in a fixed-bed continuous-flow reactor. A catalyst (0.5 g, 425–500 μm) was placed in a stainless tube reactor with i.d. of 0.5 cm and length 50 cm with the catalysts packed in the middle of the tube, occupying a length of ~10 cm. Prior to the reaction, catalysts were pre-reduced in a flow of 5% H_2/Ar gas (30 mL min^{-1}) for 1 h at 400 °C under atmospheric pressure. The reactor was cooled to 250 °C and a mixture of CO_2 and H_2 gases, ($1CO_2 : 3H_2$ molar ratios) was introduced with a flow rate of 30 mL min^{-1}. The pressure was increased to 20 bar using a back-pressure valve. All the post-reactor lines and valves were heated at 130 °C to avoid product condensation. The gas products were analysed *via* online gas chromatography using an Agilent 7890 system with an FID and TCD. Nitrogen was used as an internal standard. Reported values are given after 3 h of reaction under steady state unless otherwise stated.

Results

Catalyst characterization

XRD analysis. Catalysts were primarily characterized using X-ray diffraction (XRD) analysis to gain information about PdZn formation. Previous studies on Pd/ZnO catalysts revealed that PdZn alloy was formed under H_2 at 200 °C but achieved high methanol productivity when pre-reduced at 400 °C.[12] Fig. S1–S3 shows (ESI†) the XRD patterns of 5% Pd/ZnO, 5% Pd : Zn (1 : 10)/TiO_2 and 5% Pd : Zn (1 : 10)/Al_2O_3. Maud analysis software[28] has been used to resolve the XRD peak into different components. Detailed characterisation of the catalysts using XRD reveals that the tetragonal β-PdZn phase alloy was formed on ZnO and TiO_2 as seen by the peaks appearing at 41° and 44°, and only β-PdZn (111) was clearly seen on Al_2O_3. The intensity of the peak varied on the different supports. The peaks corresponding to ZnO and the PdZn alloy were observed on 5% Pd : Zn (1 : 10)/TiO_2 and 5% Pd : Zn (1 : 10)/Al_2O_3 catalysts, implying that some of the Zn precursor was reduced to form the alloy meanwhile some of the Zn was oxidised to ZnO during calcination. Based on the intensity of the ZnO peaks at 31.7, 34.4, 36.2, 47.5, 66, 68 and 69°, 5% Pd : Zn (1 : 10)/TiO_2 shows less crystalline ZnO in the structure in comparison to 5% Pd : Zn (1 : 10)/Al_2O_3. Fig. 1 shows the XRD patterns for the PdZn catalysts on TiO_2 with varied Pd : Zn molar ratios (a–d) and 5% Pd : Zn (1 : 10)/Al_2O_3 (e) at diffraction angles of 30–46°. For all of the catalysts, the PdZn peak at 41° (111) was clearly visible, but the peak at 44° (200) was only apparent for PdZn on the TiO_2 support, and was not so clear on Al_2O_3. At low Pd : Zn molar ratios (1 : 1 and 1 : 2), no peaks corresponding to ZnO were observed, indicating that the ZnO precursor was reduced to form a PdZn alloy. For high Pd : Zn ratios (1 : 5 and 1 : 10), the ZnO peaks were observed at diffraction angles of 32° and 34°.

XPS and XAES. XPS and XAES analysis were used to investigate the formation of the PdZn alloy and the level of carbon impurities on the catalyst surface. In what follows, it's important to note that the samples have been exposed to air

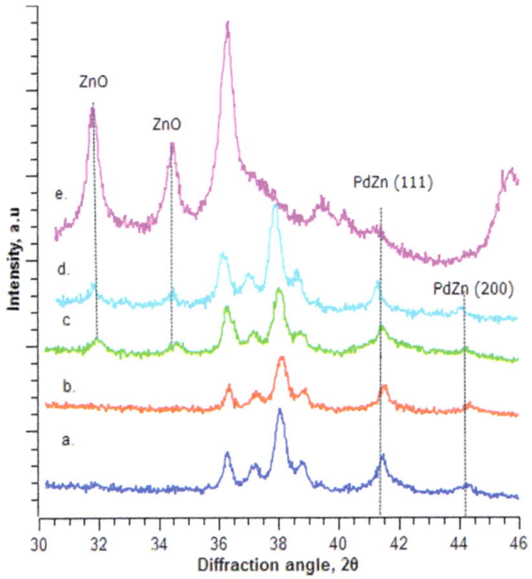

Fig. 1 X-ray diffraction pattern for (a) 5% Pd : Zn (1 : 1)/TiO_2, (b) 5% Pd : Zn (1 : 2)/TiO_2; (c) 5% Pd : Zn (1 : 5)/TiO_2; (d) 5% Pd : Zn (1 : 10)/TiO_2 and (e) 5% Pd : Zn (1 : 10)/Al_2O_3. All the catalysts were calcined in air at 500 °C for 16 h, followed by reduction in 5% H_2/Ar at 400 °C.

prior to XPS analysis. As is well known, this can lead to surface contamination by carbonaceous species, and may lead to oxidation of reduced species, and we will consider these effects below.

Annealing the as-synthesised catalyst in air or hydrogen significantly affects the catalytic activity, as reported below. The 5% Pd/ZnO catalyst was separately treated in air and in hydrogen, and was characterised using XPS analysis. The carbon 1s spectra were used to determine the level of carbon contamination from

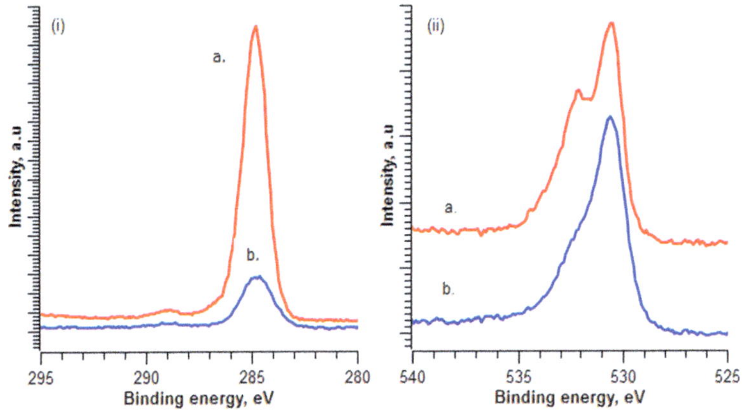

Fig. 2 (i) XPS C 1s spectra and (ii) O 1s spectra of (a) 5% Pd/ZnO that was only treated under 5% H_2/argon at 400 °C for 3 hours; and (b) 5% Pd/ZnO that was calcined in air at 500 °C for 16 h, followed by reduction at 400 °C for 1 h.

acetylacetonate decomposition on the surface. Fig. 2i reveals the effect of two different treatments on the level of the surface carbon in the 5% Pd/ZnO catalyst. A significant decrease in the intensity of the C 1s signal was observed when the catalyst was first subjected to air calcination at 500 °C, followed by reduction when compared to treatment in hydrogen alone. The observation was also supported by the carbon content from wide survey XPS analysis (ESI Fig. S4†). The quantification of the C 1s peak shows 73% of carbon on the catalysts that was only subjected to reduction at 400 °C, compared to only 37% of carbon on the catalysts that were calcined at 500 °C. Whilst carbon contamination is inevitable after exposure of such samples to air, it is clear that calcination in static air managed to remove a significant amount of the carbon. This has an effect on the activity of 5% Pd/ZnO catalysts as discussed below. Fig. 2(ii) also shows the XPS O 1s spectra of the catalysts. The O 1s signal shows peaks at 531 eV and 532.5 eV that corresponded to ZnO and $Zn(OH)_2$, respectively. However, for 5% Pd/ZnO that has only been subjected to reduction in H_2 at 400 °C, it shows a significant broadening of the $Zn(OH)_2$ peak at ~532.5 to high binding energy. This is probably due to the formation of carbonate species on the surface during acetylacetonate decomposition that has not been completely removed during the annealing treatment.

The influence of Pd : Zn molar ratios on the formation of the PdZn alloy on TiO_2 support was investigated. Fig. 3 shows the XPS Pd 3d spectra of these materials. These catalysts were air calcined first, followed by reduction at 400 °C, and revealed a peak at 335.0 eV binding energy, corresponding to Pd^0 with

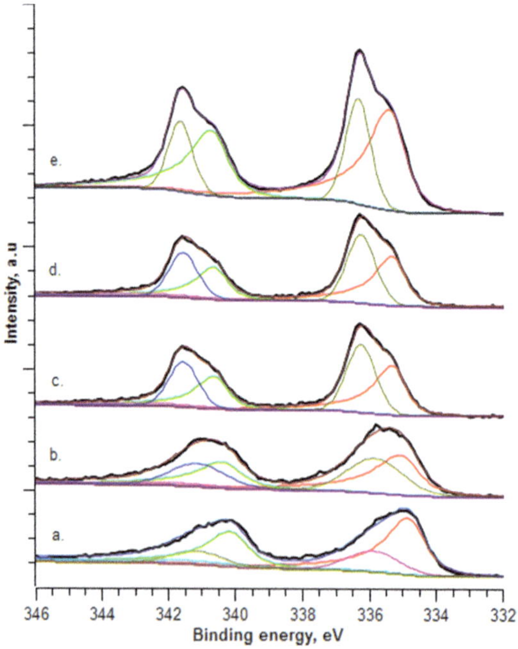

Fig. 3 XPS Pd 3d spectra of (a) 5% Pd : Zn (1 : 1)/TiO_2, (b) 5% Pd : Zn (1 : 2)/TiO_2; (c) 5% Pd : Zn (1 : 5)/TiO_2; (d) 5% Pd : Zn (1 : 10)/TiO_2 and (e) 5% Pd : Zn (1 : 10)/Al_2O_3. All the catalysts were calcined in air at 500 °C for 16 h, followed by reduction in 5% H_2/Ar at 400 °C. Tabulated Pd 3d XPS data are given in Table S1.†

a shoulder peak at ~336.0 eV.[20] The peak at 336.0 eV corresponds to the formation of the PdZn alloy on the catalyst following reductive transfer of Zn from ZnO,[29] although the Pd peak is still a dominant species for the 5% Pd : Zn (1 : 1)/TiO_2 sample. These shifts are probably caused by Pd (4d) → Zn (4p) charge transfer and Pd (4d) → Pd (5s, 5p) rehybridization.[30] Further increasing the Pd : Zn ratios to 1 : 2, the intensity of the PdZn peak at ~336 eV is enhanced and becomes dominant at Pd : Zn 1 : 5. Although the XRD analysis confirmed the presence of the β-PdZn alloy in PdZn deposited on TiO_2 catalysts, the alloy is formed when the Zn mole fraction is within the range of 0.4–0.6.[31] These results suggest that increasing the ratio of Pd : Zn in the preparation increases the amount of Zn that is incorporated into the Pd lattice and so may simultaneously enhance the amount of PdZn alloy on the surface.

In order to try to learn more about the alloying process, we examined the Zn Auger spectra, since these may show changes in level structure better than the core levels of Zn. We have also measured the spectra of Zn metal, partially oxidised Zn metal in air at 80 °C, and ZnO powder as the reference (Fig. S5†). It is possible that the surface of the samples may become oxidised when exposed to air, especially when the surface has highly oxidisable species such as Zn^0. The Zn $L_3M_{24}M_{24}$ Auger electron spectra of these samples are shown in Fig. S5.† For Zn metal it was reported that there are peak positions of 992 (major) and 996 (minor),[32] while for the oxide they are 988 and 991, respectively.[33] Zn metal shows major peaks at kinetic energy of 996 eV and 992.5 eV and minor peaks at 989.3 eV and 986 eV (Fig. S5a†). Apart from the peaks at 996 eV and 992.5 eV that correspond to Zn metallic state, the presence of peaks at 998.9 eV and 986 eV implies that a thin layer of oxidised Zn occurs on the metal surface. On partially oxidised Zn (Fig. S5b†), the metal peaks are still visible at 996 eV and 992.5 eV, meanwhile the oxidised peak at 998.5 becomes broader. Analysis on ZnO powder reveals a peak at 988 eV and a shoulder at 991 eV. The shoulder peak at 991 eV was suggested previously by Li *et al.* as interstitial Zn present as defect sites in the ZnO lattice.[33] These changes between metallic and oxidised states are thus much clearer in the Auger spectra than in the core level spectra and reflect changes in the M (valence) levels of Zn.

As mentioned above, due to Zn is a highly oxidisable species, the only cases for these samples where we could definitely identify the presence of the Zn metal was for the 5% Pd : Zn (1 : 5)/TiO_2 and 5% Pd : Zn (1 : 10)/Al_2O_3 and these are shown in Fig. 4i. The equivalent core level Zn $2p_{3/2}$ XPS is shown in Fig. 4(ii), revealing a similar binding energy of ~1022 eV for these samples. This is not surprising since Zn 2p is less sensitive towards small ZnO reduction changes with only a small possible binding energy shift between Zn metal and ZnO.[34,35] Significant shifts in Zn metal and ZnO positions can, however, be expected between the metal and oxide in the Auger LMM. The Auger indicates that most of the Zn in the structure exists as Zn^{2+} which most likely to be ZnO,[36] but there is a shoulder peak at 992 eV, together with the fingerprint for the presence of the metal at 996 eV.[32] Besides that, the changes of shoulder peak in the Zn Auger profile could be either due to the oxygen vacancies in ZnO or from metallic Zn.[37,38] Similar observations are seen on Cu/ZnO/Al_2O_3 catalysts, which show increasing intensity of the 996 eV peak when the catalysts are reduced in H_2.[37] Studies by Deroubaix and Marcus on Zn $L_3M_{24}M_{24}$ Auger electron spectra of CuZn alloy show a 0.3 eV shift towards higher kinetic energy from Zn metal which is less than the shift observed for ZnO

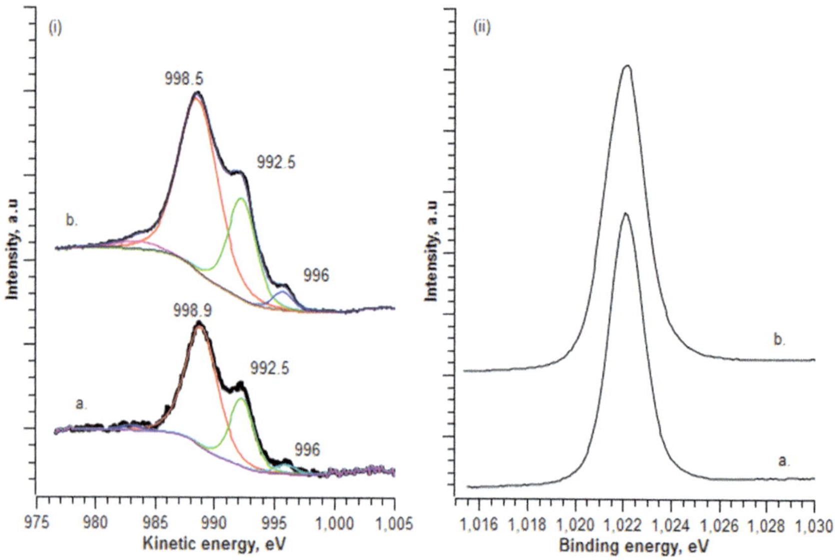

Fig. 4 (i) XPS Zn LMM Auger electron spectra and (ii) XPS Zn $2p_{3/2}$ spectra of (a) 5% Pd : Zn (1 : 5)/TiO_2 and (b) 5% Pd : Zn (1 : 10)/Al_2O_3. All the catalysts were calcined in air at 500 °C for 16 h, followed by reduction in 5% H_2/Ar at 400 °C.

– 3.9 eV.[32] We can conclude here that the 996 eV peak is from the metallic PdZn alloy whose presence is confirmed by the XRD shown above. It is very likely that a significant fraction of the Zn at the surface of these alloy nanoparticles is oxidised during transfer into the XP spectrometer, and hence the large signal seen for ZnO. It is interesting to note that though this is the case, there is little evidence for Pd oxidation, presumably due to protection from Zn/ZnO at the surface. Indeed, maybe it is protected by a covering layer.

H_2-TPR analysis. The interactions between Pd, Zn and the oxide support were probed using H_2-TPR analysis (Fig. 5). The catalysts were oxidised in air at 500 °C prior to the TPR experiment. The H_2-TPR analysis was also compared to the 5% Pd/TiO_2 and 5% Pd/Al_2O_3 catalyst. The profile of 5% Pd/TiO_2 shows a negative peak at 107 °C due to decomposition of the β-Pd hydride species.[39] However, 5% Pd/Al_2O_3 only shows a reduction peak of PdO at ~150 °C. The β-Pd hydride species is formed through hydrogen adsorption or diffusion in the Pd^0 crystallite.[40] This is a common feature of Pd/TiO_2 and the decomposition β-Pd hydride proves that the reduction of PdO occurred at temperature below this, probably before the temperature ramp. It has been reported that PdO reduces at 20–50 °C and can occur during the initial flushing of the system.[39]

The reduction of 5% Pd/ZnO and 5% Pd : Zn (1 : 10)/TiO_2 catalysts started at a similar temperature of ~50 °C. The peak for 5% PdZn (1 : 10)/TiO_2 however centred at a much lower temperature (~90 °C) than 5% Pd/ZnO. In contrast, 5% Pd/ZnO shows two peaks centred at 90 and 140 °C, which suggest the presence of two different Pd species with different environments and interaction with the ZnO support. 5% Pd : Zn (1 : 10)/TiO_2 also shows a reduction peak at ~90 °C, significantly lower than the 5% Pd/ZnO and 5% PdZn (1 : 10)/Al_2O_3 catalyst. It also loses its negative β-Pd hydride as shown in the 5% Pd/TiO_2 catalyst. These suggest

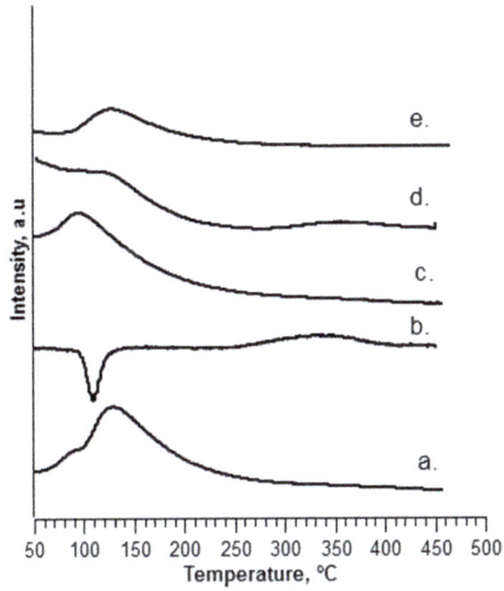

Fig. 5 H_2 TPR analysis of (a) 5% Pd/ZnO; (b) 5% Pd/TiO_2; (c) 5% Pd : Zn (1 : 10)/TiO_2; (d) 5% Pd/Al_2O_3, and (e) 5% Pd : Zn (1 : 10)/Al_2O_3. All the catalysts were calcined in air at 500 °C before measurement.

interaction between Zn and Pd on the TiO_2 catalyst which may influence the catalytic performance. 5% Pd : Zn (1 : 10)/TiO_2 shows the most activity for CO_2 hydrogenation, which suggests that Pd on this catalyst was in a close proximity with defect ZnO sites.

TEM analysis. Analysis of particle size distribution was carried out using TEM measurements. Fig. 6a shows the TEM image of 5% Pd/ZnO that was only reduced

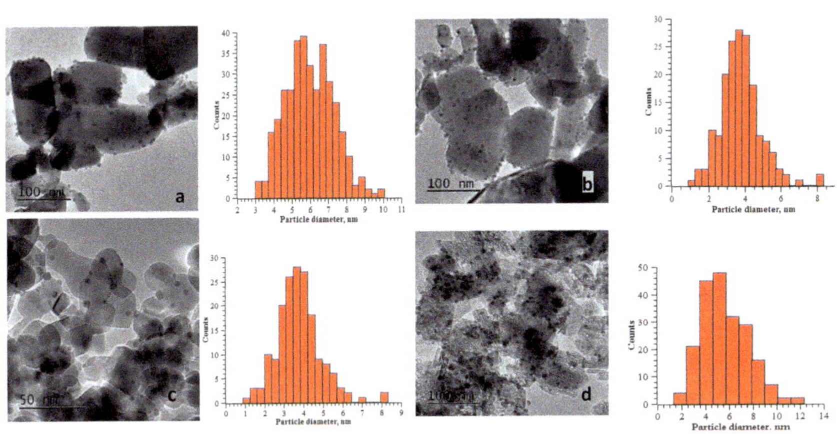

Fig. 6 TEM analysis and particle size distribution of (a) 5% Pd/ZnO reduced in 5% H_2/Ar at 400 °C, (b) 5% Pd/ZnO that was calcined at 500 °C, followed by reduction in 5% H_2/Ar at 400 °C, (c) 5% Pd : Zn (1 : 10)/TiO_2 that was calcined at 500 °C, followed by reduction in 5% H_2/Ar at 400 °C and (d) 5% Pd : Zn (1 : 10)/Al_2O_3 that was calcined at 500 °C, followed by reduction in 5% H_2/Ar at 400 °C.

in H_2 at 400 °C with the histogram of PdZn diameter distribution. The average particle diameter is ~6.1 nm. For 5% Pd/ZnO catalyst that was calcined at 500 °C followed by reduction at 400 °C, the average PdZn diameter obtained was ~5.7 nm (Fig. 6b). So these catalysts appear to be rather thermally stable. Since the specific surface area of ZnO support that was used is ~11 $m^2 g^{-1}$, the dispersion of metal nanoparticles is limited. TiO_2 and Al_2O_3 possess surface areas of 52 $m^2 g^{-1}$ and 100 $m^2 g^{-1}$, respectively, therefore better Pd and Zn nanoparticle distribution should be achieved. TEM images and particle size distribution analysis of the 5% Pd : Zn (1 : 10)/TiO_2 catalyst are shown in Fig. 6c. The catalysts were calcined at 500 °C, followed by reduction at 400 °C. The average diameter is 3.9 nm, which is significantly smaller than for the 5% Pd/ZnO catalyst. Again these particles appear to be thermally stable. Fig. 6d reveals the distribution of PdZn alloy on 5% Pd : Zn (1 : 10)/Al_2O_3 with the average particle diameter of 4.9 nm.

TPD studies of methanol on catalyst. In an attempt to understand the influence of the PdZn alloy on methanol synthesis, we carried out methanol decomposition studies on 5% Pd/ZnO and 5% Pd/TiO_2 catalysts. The 5% Pd/ZnO catalyst was reduced at 400 °C to form the PdZn alloy. All the catalysts were treated with H_2 at 400 °C before dosing with methanol. Fig. 7a shows the TPR profiles of methanol decomposition on 5% Pd/ZnO. The catalyst shows coincident CO_2 and H_2 desorption peaks at ~280 °C, which indicate the decomposition of formate species on the surface. This feature, however, was negligible on the 5% Pd/TiO_2 catalyst (Fig. 7b).

Catalytic activity

The activity of 5% Pd/ZnO catalysts for CO_2 hydrogenation reaction was investigated and is given in Table 1. The reaction was carried out at 20 bar of pressure

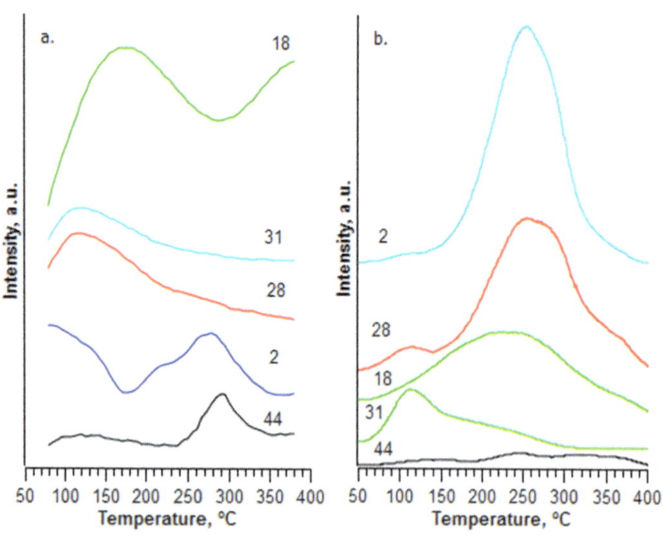

Fig. 7 TPD of (a) 5% Pd/ZnO and (b) 5% Pd/TiO_2 saturated with methanol at room temperature. 2 corresponds to H_2, 18 corresponds to H_2O, 28 corresponds to CO, 31 corresponds to methanol, and 44 corresponds to CO.

Table 1 Catalytic activity of palladium catalysts on ZnO, TiO_2 and Al_2O_3 support

Catalyst	CO_2 conv, %	CH_3OH sel, %	CO sel, %	CH_3OH productivity mmol kg_{cat}^{-1} h^{-1}
5% Pd/ZnO[a]	6.7	2	97	88.20
5% Pd/ZnO	8.2	32	68	1070
5% Pd/TiO_2	6.8	3	96	80.4
5% Pd/Al_2O_3	0.8	7	93	27.7
5% Pd : Zn (1 : 1)/TiO_2	7.6	44	55	1420
5% Pd : Zn (1 : 2)/TiO_2	9.6	38	61	1560
5% Pd : Zn (1 : 5)/TiO_2	10.1	40	59	1730
5% Pd : Zn (1 : 10)/TiO_2	6.7	53	46	1510
5% Pd : Zn (1 : 1)/Al_2O_3	2.4	4	95	38.6
5% Pd : Zn (1 : 5)/Al_2O_3	3.5	16	83	229
5% Pd : Zn (1 : 10)/Al_2O_3	8.6	19	80	635

[a] Catalyst was only reduced at 400 °C for 3 h without calcination. All the reactions were carried out at 250 °C and 20 bar. 0.5 g of catalyst was used. Data given were measured at 3 hours of reaction.

and at 250 °C reaction temperature. The catalytic activity of 5% Pd/ZnO that was only reduced at 400 °C shows ~6.7% conversion of CO_2 to CO at three hours of reaction with ~97% of CO selectivity. The catalyst activity exhibits significant improvement for methanol production when treated in static air for 16 hours at 500 °C before H_2 reduction at 400 °C. The CO_2 conversion increases to 8.7% and methanol selectivity to 31% to give 1070 mmol kg_{cat}^{-1} h^{-1} of methanol productivity.

Since it has been established that air-calcination of the catalyst to remove the carbon precursor is crucial for high methanol production, a similar annealing procedure was applied to all the catalysts. Table 1 lists the catalytic activity of the catalysts after 3 h of CO_2 hydrogenation reaction at 250 °C and 20 bar. Catalytic activity of the catalyst was also compared to 5% Pd/TiO_2 and 5% Pd/Al_2O_3. The 5% Pd/TiO_2 catalyst gives 6.8% of CO_2 conversion with 96% selectivity towards CO. The addition of Zn significantly enhances methanol productivity as was observed on 5% Pd : Zn (1 : 1)/TiO_2 (Table 1). CO_2 conversion increased with increasing zinc content up to a Pd : Zn ratio of 1 : 5, before exhibiting slight reduction at higher Pd : Zn ratios. Similar trends can be seen on the 5% Pd/Al_2O_3 and 5% Pd : Zn (1 : 10)/Al_2O_3 catalysts, with the addition of Zn significantly improving catalyst activity and methanol yield. CO_2 conversion increases from 0.8% on 5% Pd/Al_2O_3 to 8.6% on 5% Pd : Zn (1 : 10)/Al_2O_3 with methanol yield increasing from 27.7 mmol kg_{cat}^{-1} h^{-1} to 635 mmol kg_{cat}^{-1} h^{-1}, respectively. It is clear that Zn plays the role of promoter, presumably by producing PdZn active sites to initiate methanol formation. PdZn alloy deposited on TiO_2 showed superior catalytic activity for methanol production from CO_2 hydrogenation.

Effect of reaction temperatures. Further investigation of the activity of the catalyst at lower reaction temperatures was carried out in the range of 200–250 °C using the 5% Pd : Zn (1 : 10)/TiO_2 catalyst. The thermodynamic rules of CO_2 hydrogenation state that increased methanol productivity requires reaction at a lower temperature or a higher pressure. It was shown that the selectivity of methanol is higher (~94%) when the reaction was carried out at 200 °C. At this temperature the reaction has 4.2% CO_2 conversion. The conversion slowly improves with temperatures up to 250 °C. Methanol productivity is hampered at high reaction temperatures when the rWGS reaction to form CO is favoured. The 5% Pd : Zn (1 : 10)/TiO_2 catalyst achieved optimum catalytic activity to methanol at 210–220 °C to give the highest methanol yield of ~1700 mmol kg_{cat}^{-1} h^{-1}.

Discussion

PdZn alloy catalysts with narrow particle size distributions were synthesised using the CVI method. We have modified the work of Forde *et al.*[26] and Sivakumar *et al.*,[24] which showed that the CVI method can generate nanoparticles with narrow diameters due to the intimate contact between the precursor and support during the sublimation–deposition step. It is likely that the bulky structure of the acetylacetonate ion creates an isolated metal unit on the oxide support. As a result, sintering during calcination is reduced and small nanoparticles are produced. Detailed analysis using TEM shows the presence of PdZn nanoparticles with diameters between 1 and 3 nm on 5% Pd : Zn (1 : 10)/TiO_2. The 5% Pd : Zn (1 : 10)/TiO_2 catalyst was shown to retain these small nanoparticles despite calcination at 500 °C followed by reduction at 400 °C. It has been reported that catalysts with only of a few metal clusters consisting of 1–10 atoms can dominate the catalytic activity regardless of how big the mean diameter of the metal particles is.[41,42] Studies on Pd metal clusters for CO oxidation reactions showed that clusters containing only 2 Pd atoms increase the CO oxidation dramatically from a monoatomic Pd, but showed a significant drop in activity for a 7 Pd atom catalyst.[43] Studies by Rogers *et al.* on Au nanoparticle catalysts highlighted that the average diameter of metal particles alone does not correlate with catalytic activity; instead, the presence of isolated Au clusters with 1–5 atoms plays a significant role in the catalytic activity for glycerol oxidation.[41] These highlighted that it is important to make very small alloy particles and CVI is one method to achieve this. The presence of the smallest PdZn alloy with a diameter between 1 and 3 nm could be the most active PdZn alloy in the particle distribution (Table 2 and Fig. 6c).

XPS C 1s analysis and catalytic studies revealed the importance of evacuating most of the carbon from the catalyst structure in order to obtain a high activity

Table 2 Catalytic activity of 5% Pd : Zn (1 : 10)/TiO_2 at a temperature range of 200 °C to 250 °C[a]

Reaction temperatures, °C	CO_2 conv, %	CH_3OH sel, %	CO sel, %	CH_3OH productivity mmol kg_{cat}^{-1} h^{-1}	CO productivity mmol kg_{cat}^{-1} h^{-1}
200	4.2	94	5	1620	94.2
210	5.4	80	19	1777	425
220	5.2	80	19	1700	423
230	5.1	78	22	1630	459
240	5.8	63	36	1500	871
250	6.9	53	46	1510	1300

[a] All the reactions were carried out at 20 bar using 0.5 g of catalyst and were pre-treated in 5% H_2/Ar at 400 °C prior to reaction. Data given were measured at 2 h of reaction.

catalyst for methanol production. In these studies, we introduced Zn during the synthesis in an attempt to form PdZn alloy nanoparticles. Introducing Zn with molar ratios of 1Pd : 1Zn and upwards on to the Pd/TiO_2 catalyst significantly enhanced methanol productivity from 80.4 mmol kg_{cat}^{-1} h^{-1} to 1730 mmol kg_{cat}^{-1} h^{-1}, which is also more active than the 5% Pd/ZnO catalyst. The nature of the active sites in the Pd catalyst for CO_2 hydrogenation provides an insight into the mechanism of the reaction. The Pd catalyst appears to transform into a PdZn alloy under a reductive atmosphere. The PdZn alloy can exist in different compositions with a broader stability region,[31] with unique alteration of Pd electronic charges. Rehybridisation of Pd (4d)–Pd (5s, 5p) and a charge transfer between Pd (4d) → Zn (4p) during the formation of Pd–Zn bonds increase the binding energy of the core and valence levels of Pd.[30,44] This simultaneously reduces the binding energy of the core and valence levels of Zn. Pd on TiO_2 and Al_2O_3, for which alloying is unlikely at these treatment temperatures, has been shown to be unreactive for CO_2 hydrogenation to methanol. The insertion of Zn into the Pd lattice creates the PdZn alloy, which improves catalyst activity significantly, regardless of the type of metal oxide support used. We believe that the electronic structure alteration of the PdZn bimetallic alloy is responsible for its catalytic performance. TPD studies of methanol on 5% Pd/ZnO catalyst (Fig. 7a) show the presence of a formate intermediate on the surface by the coincident evolution of CO_2 and H_2 at about 280 °C, whereas without formation of the alloy on 5% Pd/TiO_2 (Fig. 7b) this is not evident. It has been shown in the past that formate is the pivotal intermediate which is central to efficient methanol synthesis.[30,45–47] In turn this implies that changes in the Pd electron density in PdZn prevent initial C=O bond breakage to make CO and enables the production and stabilisation of the formate intermediate from CO_2 hydrogenation.

The electron perturbation in the PdZn alloy exhibits similar electron density of states as metallic Cu.[48] Studies by Neyman *et al.* indicate that the surface reactivity of PdZn for methanol decomposition was similar to a Cu surface.[44] Formate decomposition on Pd revealed that formate is very unstable on the Pd surface, decomposing at low temperatures of about −50 °C.[49] In contrast, Cu stabilises formate up to 200 °C before decomposition to CO_2 and H_2.[50] All the evidence here suggests that the PdZn alloys which are formed improve CO_2 conversion and methanol selectivity. Although the mechanism of the reaction is not fully understood, we believe that the PdZn alloy is able to generate and stabilise the formate intermediate species during the CO_2 hydrogenation reaction.

Conclusions

A series of PdZn alloy catalysts were synthesised using chemical vapour impregnation (CVI) to investigate their activity in the CO_2 hydrogenation reaction. Pd on ZnO, TiO_2 and Al_2O_3 has low methanol productivity, though the productivity of the first of these increased enormously upon pre-reduction at high temperature due to the formation of a PdZn alloy. Hence we then produced a series of catalysts in which Pd and Zn were deposited by CVI. The CVI method produced highly dispersed PdZn particles with a mean diameter of 3.9 nm when deposited on TiO_2. The bulky structure of the acetylacetonate ion is believed to be responsible for creating highly isolated Pd and Zn metal species that in turn form the PdZn alloy nanoparticles with a narrow size distribution. Catalytic activity tests revealed the importance of the PdZn alloy phase for stabilizing the formate intermediate and hence a giving high activity for methanol formation from the CO_2 hydrogenation reaction.

Acknowledgements

The authors would like to acknowledge UK Catalysis Hub and the EPSRC for research funding to H. Bahruji through grant EP/I038748/1, W. Jones through EPSRC grant EP/K014714/1, J. Hayward through EC H2020 Spire grant 637016 "MefCO_2".

Notes and references

1 G. A. Olah, G. K. S. Prakash and A. Goeppert, *J. Am. Chem. Soc.*, 2011, **133**, 12881–12898.
2 S. Lee, *Methanol Synthesis Technology*, Taylor & Francis, 1989.
3 M. Behrens, *Angew. Chem., Int. Ed.*, 2014, **53**, 12022–12024.
4 I. Kasatkin, P. Kurr, B. Kniep, A. Trunschke and R. Schlögl, *Angew. Chem.*, 2007, **119**, 7465–7468.
5 E. I. Solomon, P. M. Jones and J. A. May, *Chem. Rev.*, 1993, **93**, 2623–2644.
6 M. Kurtz, N. Bauer, C. Büscher, H. Wilmer, O. Hinrichsen, R. Becker, S. Rabe, K. Merz, M. Driess, R. A. Fischer and M. Muhler, *Catal. Lett.*, 2004, **92**, 49–52.
7 M. Behrens, F. Studt, I. Kasatkin, S. Kühl, M. Hävecker, F. Abild-Pedersen, S. Zander, F. Girgsdies, P. Kurr, B.-L. Kniep, M. Tovar, R. W. Fischer, J. K. Nørskov and R. Schlögl, *Science*, 2012, **336**, 893–897.
8 O. Martin, C. Mondelli, D. Curulla-Ferré, C. Drouilly, R. Hauert and J. Pérez-Ramírez, *ACS Catal.*, 2015, **5**, 5607–5616.
9 J. T. Sun, I. S. Metcalfe and M. Sahibzada, *Ind. Eng. Chem. Res.*, 1999, **38**, 3868–3872.
10 N. Iwasa, M. Yoshikawa, W. Nomura and M. Arai, *Appl. Catal., A*, 2005, **292**, 215–222.
11 N. Iwasa, S. Masuda, N. Ogawa and N. Takezawa, *Appl. Catal., A*, 1995, **125**, 145–157.
12 H. Bahruji, M. Bowker, G. Hutchings, N. Dimitratos, P. Wells, E. Gibson, W. Jones, C. Brookes, D. Morgan and G. Lalev, *J. Catal.*, 2016, **343**, 133–146.
13 C.-H. Kim, J. Lee and D. L. Trimm, *Top. Catal.*, 2003, **22**, 319–324.
14 X.-L. Liang, X. Dong, G.-D. Lin and H.-B. Zhang, *Appl. Catal., B*, 2009, **88**, 315–322.
15 H. Zhang, J. Sun, V. L. Dagle, B. Halevi, A. K. Datye and Y. Wang, *ACS Catal.*, 2014, **4**, 2379–2386.
16 L. Fan and K. Fujimoto, *J. Catal.*, 1994, **150**, 217–220.
17 S. E. Collins, J. J. Delgado, C. Mira, J. J. Calvino, S. Bernal, D. L. Chiavassa, M. A. Baltanás and A. L. Bonivardi, *J. Catal.*, 2012, **292**, 90–98.
18 T. Fujitani, M. Saito, Y. Kanai, T. Watanabe, J. Nakamura and T. Uchijima, *Appl. Catal., A*, 1995, **125**, L199–L202.
19 J. Qu, X. Zhou, F. Xu, X.-Q. Gong and S. C. E. Tsang, *J. Phys. Chem. C*, 2014, **118**, 24452–24466.
20 Z. Zsoldos, A. Sarkany and L. Guczi, *J. Catal.*, 1994, **145**, 235–238.

21 M. Behrens, F. Studt, I. Kasatkin, S. Kühl, M. Hävecker, F. Abild-Pedersen, S. Zander, F. Girgsdies, P. Kurr, B.-L. Kniep, M. Tovar, R. W. Fischer, J. K. Nørskov and R. Schlögl, *Science*, 2012, **336**, 893–897.
22 F. Studt, M. Behrens, E. L. Kunkes, N. Thomas, S. Zander, A. Tarasov, J. Schumann, E. Frei, J. B. Varley, F. Abild-Pedersen, J. K. Nørskov and R. Schlögl, *ChemCatChem*, 2015, **7**, 1105–1111.
23 B. Sakakini, J. Tabatabaei, M. J. Watson, K. C. Waugh and F. W. Zemicael, *Faraday Discuss.*, 1996, **105**, 369–376.
24 P. Sivakumar, R. Ishak and V. Tricoli, *Electrochim. Acta*, 2005, **50**, 3312–3319.
25 V. Dal Santo, C. Mondelli, V. De Grandi, A. Gallo, S. Recchia, L. Sordelli and R. Psaro, *Appl. Catal., A*, 2008, **346**, 126–133.
26 M. M. Forde, L. Kesavan, M. I. bin Saiman, Q. He, N. Dimitratos, J. A. Lopez-Sanchez, R. L. Jenkins, S. H. Taylor, C. J. Kiely and G. J. Hutchings, *ACS Nano*, 2014, **8**, 957–969.
27 http://www.casaxps.com/.
28 http://maud.radiographema.com/.
29 P. S. Wehner, G. C. Tustin and B. L. Gustafson, *J. Catal.*, 1984, **88**, 246–248.
30 J. A. Rodriguez, *J. Phys. Chem.*, 1994, **98**, 5758–5764.
31 J. Vizdal, A. Kroupa, J. Popovic and A. Zemanova, *Adv. Eng. Mater.*, 2006, **8**, 164–176.
32 G. Deroubaix and P. Marcus, *Surf. Interface Anal.*, 1992, **18**, 39–46.
33 W. Li, L. Fang, G. Qin, H. Ruan, H. Zhang, C. Kong, L. Ye, P. Zhang and F. Wu, *J. Appl. Phys.*, 2015, **117**, 145301.
34 M. L. Cubeiro and J. L. G. Fierro, *Appl. Catal., A*, 1998, **168**, 307–322.
35 A. Bayer, K. Flechtner, R. Denecke, H.-P. Steinrück, K. M. Neyman and N. Rösch, *Surf. Sci.*, 2006, **600**, 78–94.
36 L. S. Dake, D. R. Baer and J. M. Zachara, *Surf. Interface Anal.*, 1989, **14**, 71–75.
37 S. Kuld, C. Conradsen, P. G. Moses, I. Chorkendorff and J. Sehested, *Angew. Chem., Int. Ed.*, 2014, **53**, 5941–5945.
38 V. Schott, H. Oberhofer, A. Birkner, M. Xu, Y. Wang, M. Muhler, K. Reuter and C. Wöll, *Angew. Chem., Int. Ed.*, 2013, **52**, 11925–11929.
39 N. S. Babu, N. Lingaiah, N. Pasha, J. V. Kumar and P. S. S. Prasad, *Catal. Today*, 2009, **141**, 120–124.
40 V. A. d. l. P. O'Shea, M. C. Álvarez-Galván, J. L. G. Fierro and P. L. Arias, *Appl. Catal., B*, 2005, **57**, 191–199.
41 S. M. Rogers, C. R. A. Catlow, C. E. Chan-Thaw, D. Gianolio, E. K. Gibson, A. L. Gould, N. Jian, A. J. Logsdail, R. E. Palmer, L. Prati, N. Dimitratos, A. Villa and P. P. Wells, *ACS Catal.*, 2015, **5**, 4377–4384.
42 A. A. Herzing, C. J. Kiely, A. F. Carley, P. Landon and G. J. Hutchings, *Science*, 2008, **321**, 1331–1335.
43 E. C. Tyo and S. Vajda, *Nat. Nanotechnol.*, 2015, **10**, 577–588.
44 K. M. Neyman, K. H. Lim, Z.-X. Chen, L. V. Moskaleva, A. Bayer, A. Reindl, D. Borgmann, R. Denecke, H.-P. Steinruck and N. Rosch, *Phys. Chem. Chem. Phys.*, 2007, **9**, 3470–3482.
45 M. Bowker, H. Houghton and K. C. Waugh, *J. Chem. Soc., Faraday Trans. 1*, 1981, **77**, 3023–3036.
46 M. Bowker, *Vacuum*, 1983, **33**, 669–685.
47 M. Bowker, R. A. Hadden, H. Houghton, J. N. K. Hyland and K. C. Waugh, *J. Catal.*, 1988, **109**, 263–273.

48 A. Pang Tsai, S. Kameoka and Y. Ishii, *J. Phys. Soc. Jpn.*, 2004, **73**, 3270–3273.
49 N. Aas, Y. Li and M. Bowker, *J. Phys.: Condens. Matter*, 1991, **3**, S281.
50 M. Bowker and R. J. Madix, *Surf. Sci.*, 1981, **102**, 542–565.

Faraday Discussions

PAPER

Selective hydrogenation of CO on Fe_3S_4\{111\}: a computational study

Alberto Roldan* and Nora H. de Leeuw

Received 24th October 2016, Accepted 8th November 2016
DOI: 10.1039/c6fd00224b

Fischer–Tropsch (FT) synthesis has been a recursive method to form valuable molecules from syngas. Metal surfaces have been extensively studied as FT catalysts; among them, iron presents several phases under reaction conditions, oxides and carbides, as active sites for the FT and reverse water gas shift reaction. We present CO reduction on an iron sulfide phase with spinel structure, Fe_3S_4, also considering the pathways where C–O dissociates leaving CH_x species on the surface, which may feed longer aliphatic chains *via* the FT process. We analysed the thermodynamic and kinetic availability of each step leading to O and OH species co-adsorbed on the surface as well as the formation of H_2O from the hydrogenation of the alcohol group in the molecule. This detailed analysis led to energy profiles on both active sites of the surface, and we conclude that this Fe_3S_4 surface is highly selective towards the formation of methanol, in full agreement with experimental results. These findings point out that the C–C bond formation on greigite takes place through a hydroxycarbene FT mechanism.

Introduction

Fischer–Tropsch (FT) synthesis has been the subject of intense investigation for many decades as it gives access to industrially important chemicals from simple inorganic molecules (CO and H_2).[1–3] It is considered one of the most attractive routes to convert syngas into liquid fuels and chemicals, which consists of mainly paraffins and olefins, and low levels of oxygenates including alcohols, aldehydes, ketones and acids.[4] Traditionally, four types of chain growth mechanisms have been proposed: the carbene mechanism,[5] the hydoxycarbene mechanism,[6] the CO-insertion mechanism,[7] and the CH_2-insertion mechanism.[8]

The mechanism for methanol synthesis on metal catalysts has been extensively investigated with both experimental[9–11] and theoretical methods,[11–13] in part because it is considered to be an important alternative fuel for future transportation needs, either for methanol fuel cells or for the on-board generation of H_2 for proton-exchange membrane fuel cells.[14,15] Comparing the formation enthalpies of methanol and methane, fully reduced compounds, it is obvious that

School of Chemistry, Cardiff University, Main Building, Park Place, CF10 3AT, Cardiff, UK. E-mail: roldanmartineza@cardiff.ac.uk

the latter is thermodynamically more favourable and therefore should hinder the formation of methanol. Methane is generated through the C–O bond fission, which may take place in adsorbed intermediate species (CH_xO and CH_xOH) somewhere along the CO hydrogenation pathway.[13,16] Indeed, minor CH_x species have been detected in methanol decomposition on metal surfaces, particularly at high pressures,[17] while they may desorb rather than react under ultra-high vacuum conditions.[18] Among the catalysts containing metals, iron-based catalysts have considerable merits due to their potential activity for the water–gas shift (WGS) reaction as well as their high FT activity and low cost.[3,19] The catalytic performance of these catalysts have been examined drawing the conclusion that several carbide and oxide phases are present under working conditions, among them Fe_3O_4.[20–25]

A sulfide isomorphic to Fe_3O_4, greigite, has recently showed low turnover to electro-reduce CO_2 to small molecules, methanol among them.[26] However, the presence of S on FT catalysts has been considered as a poison, which raises the queries of the actual mechanism and the role of sulfur in the sulfide catalysts. In this work, we account for the lack of CH_4 in the experimental analysis and provide evidence of the selectivity of Fe_3S_4 between hydrogenation pathways from CO to CH_3OH and CH_x on the surface, for this is crucial to complete and understand the intricate reaction networks that can be intercrossed at any stage of the reaction.

Computational details

Periodic plane-wave DFT calculations were performed using the Vienna *ab initio* simulation package (VASP),[27,28] the Perdew–Wang 91 functional[29] within the spin interpolation formula of Vosko *et al.*,[30] and a kinetic energy of 600 eV to expand the plane-waves of the Kohn–Sham valence states.[31] The inner electrons were represented by the projector-augmented wave (PAW) pseudopotentials considering also non-spherical contributions from the gradient corrections.[32] All the calculations include the long-range dispersion correction approach by Grimme,[33] which is an improvement on pure DFT when considering large polarizable atoms.[34–39] We have used the global scaling factor parameter optimized for PBE, ($s_6 = 0.75$). The optimisation thresholds were 10^{-5} eV and 0.03 eV Å^{-1} for the electronic and ionic relaxation, respectively. The Brillouin zone was sampled by a Γ-centred k-point mesh generated through a Monkhorst–Pack grid of $5 \times 5 \times 1$ k-points, which ensures electronic and ionic convergence.[40] In order to improve the convergence of the Brillouin-zone integrations, partial occupancies were determined using the tetrahedron method with Blöch corrections smearing with a set width for all calculations of 0.02 eV.

The greigite unit cell consists of eight Fe_3S_4 subunits with a cubic lattice parameter of ~9.8 Å (ref. 41 and 42) which is close to the calculated parameter resulting from the cell optimization (9.671 Å). The inverse thio-spinel arrangement is reflected by the formula AB_2S_4, where there are two possible locations for the Fe ions: tetrahedral sites (A), filled by Fe^{3+} ions, and octahedral sites (B), where both Fe^{3+} and Fe^{2+} ions co-exist.[43–49] The initial magnetic moment of the slab was described by a high spin distribution in both types of Fe with antiparallel orientation as reported previously,[50,51] which results in a ferromagnetic material.[43,44,46–48,52–54] For an accurate treatment of the electron correlation in the localized d-Fe orbital, we have used the U approximation[55,56] ($U_{\text{eff}} = 1$ eV),[49,54,57] which improves the description of localized states in this type of systems where standard LDA and GGA functionals fail.[58] The orbital spin-splitting in the valence region results in localized outermost 3d-electrons and in ordered magnetism[45,59,60] in good agreement with experimental evidence.[50,51] We have prepared the Fe_3S_4\{111\} surface as a slab model by cutting the bulk structure with the METADISE code,[61] which considers periodicity in the plane direction and provides the atomic layer stacking resulting in non-dipolar reconstructions.[62] The slab contains 56 atoms (24 Fe and 32 S) per unit cell exposing an area of 81.0 Å^2 and a thickness of sufficient size to relax the two uppermost layers (four Fe_3S_4 units), keeping the bulk structure frozen at the bottom. We added a vacuum width of 12 Å between periodic slabs, big enough to avoid interaction between periodic images. The molecules in the gas phase were relaxed in an asymmetric box of 15 × 16 × 17 Å^3.

We employed the improved dimer method (IDM), to find the transition states.[63,64] We defined the adsorption energy as the difference between the isolated species and the combined system. The activation energy (E_A) of a certain step is the energy required to surmount the potential barrier characteristic of the transition state. We have defined this energy barrier as the difference between the initial state and transition state energy for the forward process. The reaction energy (E_R) of each step is calculated as the total energy difference between the final state (product(s)) and the initial state (reactant(s)).

Results and discussion

Greigite (Fe_3S_4) is an inverse spinel-structured material whose particles produced in hydrothermal synthesis expose \{001\} and \{111\} surfaces.[26] We focused on the active \{111\} surface and, due to the presence of non-equivalent Fe ions, it has two distinct terminations depending on the relative Fe_A position with respect to the uppermost sulfur layer. The most stable termination contains two adsorption sites, Fe_A and Fe_B, originally occupying tetrahedral and octahedral bulk positions.[65] The top S layer has a bulk-like structure consisting of rows along the [0−11] direction with rhombohedral packing.

We have explored the adsorption of reactants and products on Fe_3S_4\{111\}. While CO adsorbs on Fe_A and Fe_B with a binding energy of −0.68 and −1.07 eV respectively, the adsorption of CH_3OH is 20% weaker and CH_4 does not chemisorb. Hence, all the CH_4 molecules produced from the CO reduction will desorb from the surface leading to an overall exothermic process driven by the production of H_2O. The co-adsorption of CO and four H ad-atoms is also higher in energy than adsorbed CH_3OH ($\Delta E = 0.6$ eV), indicating a favourable process.

We have represented the optimised structures of CO and CH_3OH in Fig. 1 and displayed the C–O and Fe–C distances. The C–O distances in the adsorbed states on Fe_A and Fe_B are 1–2% longer than in isolated CO and CH_3OH molecules respectively. The molecules are placed practically on top of the metal sites at a distance according to their binding energy: the distances on Fe_A are longer than on Fe_B, indicating a weaker bond. The elongation of the C–O bond is also reflected in the $\nu_{C\text{-}O}$ stretching mode, which changes by ∼100 and ∼50 cm^{-1} for CO and CH_3OH respectively compared to the gas phase molecules.

While the redox property of the surface determines the capacity to transfer electrons to the adsorbates, the acid–base character controls the gain and loss of

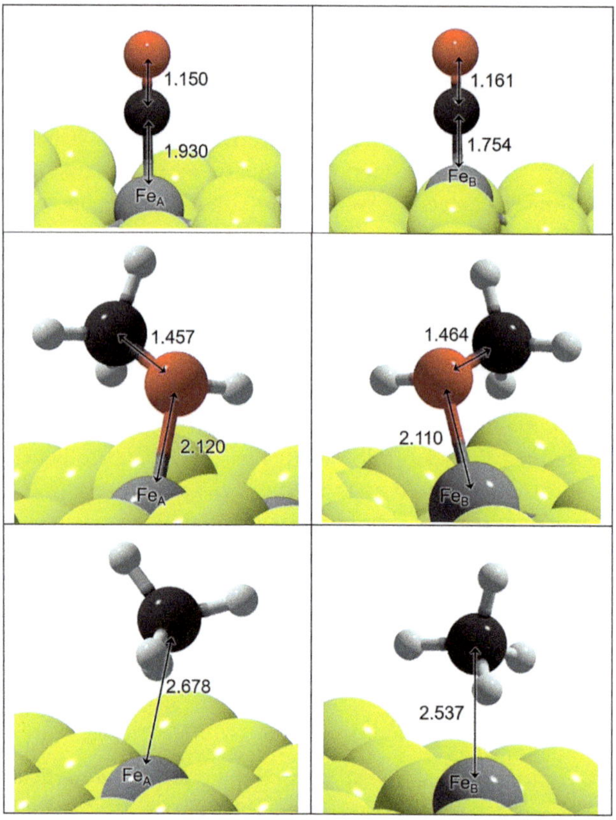

Fig. 1 Schematic representation of CO, CH_3OH and CH_4 on Fe_A and Fe_B sites of $Fe_3S_4\{111\}$. Inset distances are in Å. Colour scheme: light-grey is Fe, yellow is S, red is O, white is H and dark-grey is C.

Table 1 Relative energies (ΔE), distances (d) and C–O stretching vibrational modes (ν_{C-O}) of the different CH_xOH and CH_xO intermediates on Fe_A and Fe_B sites of the $Fe_3S_4\{111\}$ surface. The energies are relative to adsorbed CO, taking into account the corresponding number of H ad-atoms

Label	Adsorbate	Site	ΔE (eV)	d_{C-O} (Å)	d_{Fe-C} (Å)	ν_{C-O} (cm^{-1})
a1	CO	Fe_A	0.00	1.157	1.864	2049
a2	HCO		0.55	1.201	1.940	1747
a3	HCOH		0.79	1.319	1.822	1273
a4	H_2CO		1.30	1.443	2.006	895
a5	H_2COH		0.20	1.388	2.006	1131
a6	H_3CO		0.01	1.446	3.550	962
a7	H_3COH		−0.55	1.457	3.132	962
b1	CO	Fe_B	0.00	1.163	1.749	2026
b2	HCO		0.63	1.178	1.763	1940
b3	HCOH		0.89	1.330	1.781	1265
b4	H_2CO		0.07	1.131	2.054	1288
b5	H_2COH		0.55	1.396	1.997	1119
b6	H_3CO		−0.16	1.406	3.017	1085
b7	H_3COH		−0.57	1.464	2.981	944

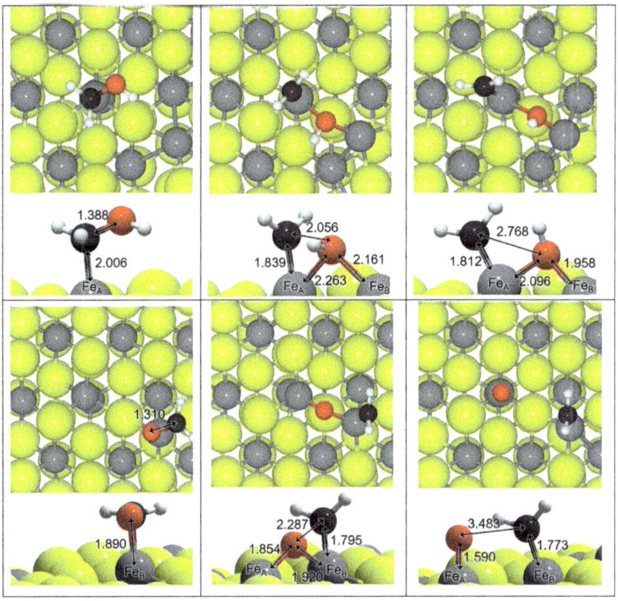

Fig. 2 Top and side views of C–O dissociation on the Fe_3S_4{111} surface: H_2COH on Fe_A and H_2CO on Fe_B. Reactants are on the right, the transition states are in the centre and the products are on the left. Colour scheme: light-grey is Fe, yellow is S, red is O, white is H and dark-grey is C. All distances in Å.

hydrogens and therefore the yields of hydrogenation products. We have explored the co-adsorption and reaction of CO and H ad-atoms on the Fe_3S_4{111} surface leading to competition between hydrogenation and C–O fission.

The adsorption of these species on transition metal surfaces has been previously investigated leading to a wide range of data. Using a similar methodology to the one we employed here, CO adsorption is reported between −0.23 and −3.21 eV (corresponding to Cu and Pd respectively),[13] while on oxides such as Fe_3O_4{111} the adsorption energies are −1.09 and −1.94 eV on Fe_A and Fe_B

Table 2 Activation (E_A) and reaction (E_R) energies for breaking the surface species C–O bond on Fe_A and Fe_B sites of the Fe_3S_4{111} surface

Label	Site	Reactants → products	E_A (eV)	E_R (eV)
R1	Fe_A	HCO → CH + O	3.91	2.08
R2		HCOH → CH + OH	2.79	1.56
R3		H_2CO → CH_2 + O	1.27	−0.55
R4		H_2COH → CH_2 + OH	1.84	1.50
R5		H_3CO → CH_3 + O	2.46	0.48
R6		H_3COH → CH_3 + OH	1.87	−0.19
R7	Fe_B	HCO → CH + O	4.34	3.48
R8		HCOH → CH + OH	1.72	1.69
R9		H_2CO → CH_2 + O	2.97	2.56
R10		H_2COH → CH_2 + OH	2.17	1.02
R11		H_3CO → CH_3 + O	2.52	1.72
R12		H_3COH → CH_3 + OH	2.32	0.69

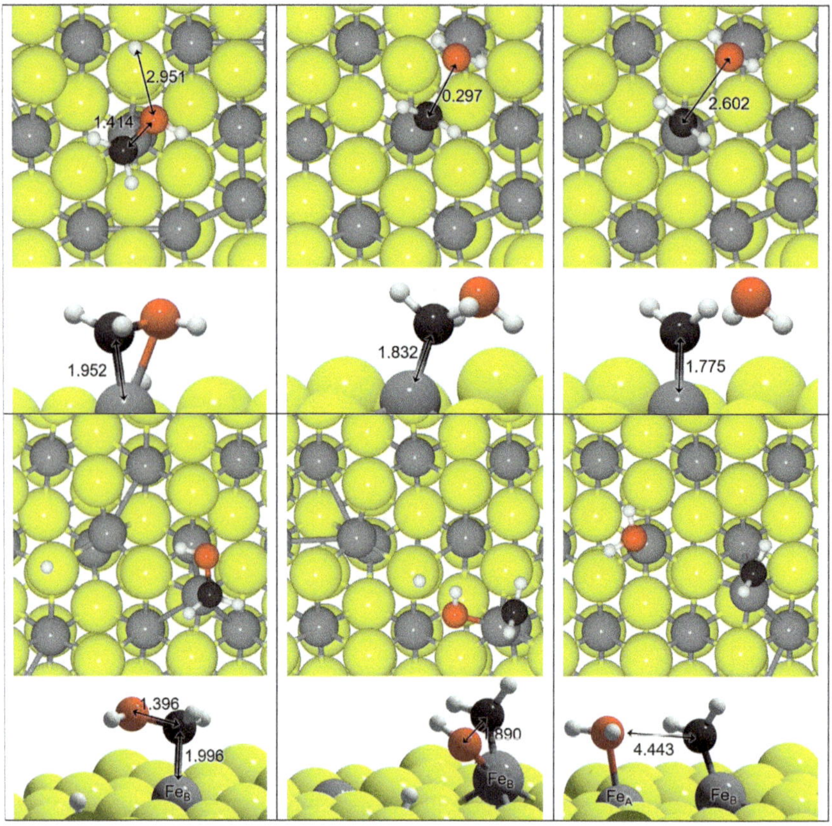

Fig. 3 Top and side views of H_2O driven C–O dissociation on the $Fe_3S_4\{111\}$ surface: H_2COH on Fe_A and on Fe_B sites. Reactants are on the right, transition states are in the centre and products are on the left. Colour scheme: light-grey is Fe, yellow is S, red is O, white is H and dark-grey is C. All distances in Å.

respectively,[66] which are stronger than on the sulfide counterpart. These reports also indicated a C–O stretching red-shift, in agreement with our calculations, except that on pure metals it is considerably lower (~1900 cm^{-1}).[12] Methanol adsorbs on top of fcc metals with an ideal surface releasing ~0.5 eV, while methoxy was found to be unstable on Pt{111}.[67,68] On $Fe_3O_4\{111\}$, the adsorption of CH_3OH is weaker than of CO, which is similar to what we found on the thio-

Table 3 Activation (E_A) and reaction (E_R) energies for the reaction $H_xCOH \rightarrow H_xC + H_2O$ on Fe_A and Fe_B sites of the $Fe_3S_4\{111\}$ surface

	Site	Reactants → products	E_A (eV)	E_R (eV)
R13	Fe_A	$HCOH + H \rightarrow CH + H_2O$	1.71	0.77
R14		$H_2COH + H \rightarrow CH_2 + H_2O$	0.93	0.72
R15		$H_3COH + H \rightarrow CH_3 + H_2O$	1.75	−0.22
R16	Fe_B	$HCOH + H \rightarrow CH + H_2O$	2.46	2.06
R17		$H_2COH + H \rightarrow CH_2 + H_2O$	1.75	0.63
R18		$H_3COH + H \rightarrow CH_3 + H_2O$	1.98	−0.15

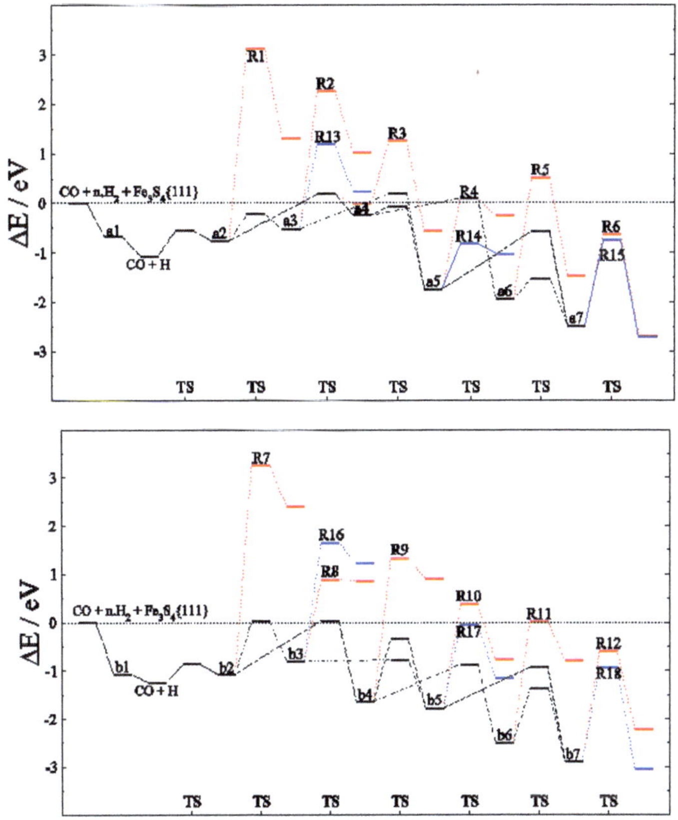

Fig. 4 Energy profile for the reduction mechanism of CO to CH_3OH (black line) on Fe_A (top) and Fe_B (bottom) sites of $Fe_3S_4\{111\}$ considering the direct scission of C–OH_x bond (red line) and its dissociation assisted by an H ad-atom leading to H_2O (blue line). Intermediates and C–O dissociation reactions are labelled according to Tables 1–3.

spinel, $Fe_3S_4\{111\}$, where the E_B difference between CH_3OH and CO is ~0.1 eV.[17] This is due to the fact that Fe_3O_4 is reluctant to modify its electronic structure while the sulfide has a milder ionic character.[50]

Adsorption of H_xCOH and H_xCO surface species ($x = 0$–3)

We have adsorbed the species involved in the generation of CH_3OH and characterised the most stable configurations (see Table 1). We found progressive changes between the different species along the reduction process: (i) the C–O distance increases by 0.3 Å, which on average is only 0.06 Å shorter in H_xCO than in the H_xCOH species (~1.392 Å), (ii) the Fe–C distance also increases by >1 Å, and (iii) the C–O stretching vibrational mode shifts by ~1000 cm^{-1}. All these changes are related to the change in carbon orbital hybridisation from sp^1 to sp^3. It is noticed that the saturation of the C by hydrogens makes the molecule bind to the surface through the O(H). This implies charge transfer from the surface and makes the distances and frequencies lay slightly out from the trend line, as in the case of H_2CO ($q = -0.5e$) and more drastically in the Fe–C distance for the methoxy bonded through the oxygen ($q = -0.35e$). The relative stability of the intermediates is expressed in Table 1 by the energy difference between the intermediate in question, and the CO with the corresponding number of hydrogens on the $Fe_3S_4\{111\}$ surface. It shows that although the formation of CH_3OH is thermodynamically favourable, there are less favourable intermediates making the process less likely, especially on Fe_A.

Although the CH_3OH formation from CO is still a matter of discussion, it is well documented on fcc metals. Cu showed good catalytic performance to reduce CO though most of the reaction intermediates (those in Table 1) adsorb weakly or endothermically.[69] This is not the case for heavier metals, of which the general trend is that adsorption weakens in line with the C-orbitals hybridisation,[13] *i.e.* the greater hybridisation the weaker the adsorption. The geometrical changes taking place on the different species, such as the increase in bond length and red-shift of the C–O stretching vibrational mode, are also common features.[12]

Dissociation of H_xCOH and H_xCO species ($x = 0$–3)

We have studied the C–O scission of each intermediate species lying on Fe_A and Fe_B taking into account that the products may be displaced to a nearby Fe site; for instance, the C–O splitting of H_2COH adsorbed on Fe_A, leads to CH_2 remaining on Fe_A while OH binds both Fe_A and Fe_B, see Fig. 2. Once the C–O bond is completely dissociated, the CH_x species lie an average of 0.4 Å closer to Fe than before dissociation. This short distance is also a consequence of the electrophilic character of the co-adsorbed O and OH groups, which withdraw electrons from nearby atoms; a co-adsorbed O ad-atom makes the Fe–C distance 0.09 Å shorter than a co-adsorbed hydroxyl on Fe_A, while it remains practically invariable on Fe_B ($\Delta d_{Fe-C} = 0.01$ Å).

Table 2 summarises the activation and reaction energies for the direct C–O scission. On Fe_A sites, the reaction energy (E_R) related to C–O bond breaking is (i) endothermic ($E_R > 1$ eV) for a2, a3 and a5 species, (ii) slightly endothermic ($0 < E_R > 1$ eV) for methoxy (a6), and (iii) exothermic ($E_R < 0$ eV) for formaldehyde (a4) and methanol (a7). Nevertheless, the high energy barriers indicate that the process is kinetically hindered and the C–O scission is unlikely; the lowest E_A is for H_2CO but still above 1 eV. The electronic and geometrical differences between sites make Fe_B behave differently: the process is endothermic for all the intermediates and E_R is lower than 1 eV only for H_3COH, while all the energy barriers are well above 1.5 eV.

The energy data in Table 2 shows that breaking the C–O bond is unfavourable under mild conditions. Although the activation and reaction energies have been found to be much smaller on pure metallic systems such as Pd{111},[13] they follow a similar trend, showing the C–O splitting to be more favourable on HCOH and H_2COH intermediates.

We have explored an alternative mechanism for C–O bond breaking driven by H_2O generation, which is schematically represented in Fig. 3. Table 3 contains the activation and reaction energies for that process on Fe_A and Fe_B, respectively. The energy trends show decreasing values as the number of hydrogens in the molecule is increased, *i.e.* the C-orbital hybridisation changes from sp^1 to sp^3. In spite of this, the process is endothermic and kinetically unfavourable.

Among the different reaction mechanisms explored, only R3 and R6 are exothermic. Despite this, they are kinetically disabled and therefore the presence of CH_x groups should be very scarce on $Fe_3S_4\{111\}$. Fig. 4 shows the energy profile for the reduction of CO to methanol (black lines) and the different paths studied describing the C–O dissociations for the direct and H_2O driven pathways (red and blue lines, respectively). Most of these pathways are hindered by the high energy barrier, however, we noticed that for $H_xCO(H)$ species with $x > 2$, the dissociation of the C–O bond requires the same or less energy than some steps undertaken during the CO reduction. Therefore, if the system has enough energy to overcome the barrier of producing $H_xCO(H)$ intermediates, the C–O dissociation is accessible and some pathways become competitive, *e.g.* R4, R6, R11, R12, R14, R15, R17 and R18.

The reaction network for CH_3OH formation on Fe_A may take place through H_xCO and H_xCOH intermediates; nonetheless, the path for the formation of a6 from a4 requires more energy to overcome the transition energy than that leading to a5. However, R14 is more kinetically feasible leading to $CH_x + H_2O$, although the process is thermodynamically unfavourable. Thus, a5 (H_2COH) is crucial for the production of methanol, the desorption of which is >1 eV below the H_3C–OH scission energy barriers. On the other hand, the C–O dissociations on Fe_B have a higher energy barrier: the stabilisation of b4 means that both H_3CO and H_3COH will be formed, the latter desorbing from the catalyst prior to dissociation. These results are in full agreement with the fact that CH_4 molecules were not detected during the experiments using gas-chromatography and ^{1}H-NMR techniques.[26]

Conclusions

We have studied systematically the competitive reactions of CO hydrogenation and C–O dissociation on both adsorption sites, Fe_A and Fe_B, of $Fe_3S_4\{111\}$. All the calculations were performed using DFT+U, including the long-range dispersion forces. We found a correlation of the C–O bond strength with the C orbital hybridization according to the bond distances and vibrational analysis, which is similar to the trends reported on metals and metal oxides. The examination of relative energies between the reactants and intermediates showed that although the reactions leading to the formation of methanol and methane are thermodynamically favorable, driven by the high energy of the H_2 molecules (and the formation of H_2O), it is necessary to overcome a reaction barrier. There are intermediates of higher energy, HCOH and H_2CO, that determine the formation of products, especially on Fe_A. The lower energy reaction pathway leads exclusively to methanol through HCOH on Fe_A while both the HCOH and H_2CO intermediates are of competing stability in the reaction mechanism on Fe_B. Thus, the C–O scission is generally an endothermic or kinetically disabled process under mild conditions on $Fe_3S_4\{111\}$; for this reason, this surface presents a selective CO reduction towards a low level of oxygenated molecules while iron oxides may lead to totally reduced molecules.

Acknowledgements

We acknowledge the Engineering & Physical Sciences Research Council (grant EP/K035355/1 and EP/K035355/2) for funding. This work made use of the ARCHER facilities, the UK's national high-performance computing service, which is provided by UoE HPCx Ltd at the University of Edinburgh, Cray Inc. and NAG Ltd, funded by the Office of Science and Technology through EPSRC's High End Computing Programme and provided *via* our membership of the HPC Materials Chemistry Consortium (EPSRC EP/L000202). The authors also acknowledge the use of the UCL Legion High Performance Computing Facility and the Advanced Research Computing @ Cardiff (ARCCA) at Cardiff University, and associated support services, in the completion of this work. All data created during this research is openly available from the University of Cardiff Research Portal at http://doi.org/10.17035/d.2016.0011347963.

References

1 C. K. Rofer-DePoorter, *Chem. Rev.*, 1981, **81**, 447–474.
2 J. J. C. Geerlings, J. H. Wilson, G. J. Kramer, H. P. C. E. Kuipers, A. Hoek and H. M. Huisman, *Appl. Catal., A*, 1999, **186**, 27–40.
3 M. E. Dry, *Catal. Today*, 2002, **71**, 227–241.
4 A. Y. Khodakov, W. Chu and P. Fongarland, *Chem. Rev.*, 2007, **107**, 1692–1744.
5 F. Fischer and H. Tropsch, *Ber. Dtsch. Chem. Ges.*, 1926, **59**, 830–831.
6 J. T. Kummer and P. H. Emmett, *J. Am. Chem. Soc.*, 1953, **75**, 5177–5183.
7 A. Y. Krylova, *Solid Fuel Chem.*, 2014, **48**, 22–35.
8 P. M. Maitlis, R. Quyoum, H. C. Long and M. L. Turner, *Appl. Catal., A*, 1999, **186**, 363–374.
9 I. A. Fisher and A. T. Bell, *J. Catal.*, 1997, **172**, 222–237.
10 J. Nerlov and I. Chorkendorff, *J. Catal.*, 1999, **181**, 271–279.
11 M. Behrens, F. Studt, I. Kasatkin, S. Kühl, M. Hävecker, F. Abild-Pedersen, S. Zander, F. Girgsdies, P. Kurr, B.-L. Kniep, M. Tovar, R. W. Fischer, J. K. Nørskov and R. Schlögl, *Science*, 2012, **336**, 893.
12 L. C. Grabow and M. Mavrikakis, *ACS Catal.*, 2011, **1**, 365–384.
13 R. García-Muelas, Q. Li and N. López, *ACS Catal.*, 2015, **5**, 1027–1036.
14 A. K. Avcı, Z. İ. Önsan and D. L. Trimm, *Top. Catal.*, 2003, **22**, 359–367.
15 D. R. Palo, R. A. Dagle and J. D. Holladay, *Chem. Rev.*, 2007, **107**, 3992–4021.
16 S. Lin, J. Ma, X. Ye, D. Xie and H. Guo, *J. Phys. Chem. C*, 2013, **117**, 14667–14676.
17 M. Baumer, J. Libuda, K. M. Neyman, N. Rosch, G. Rupprechter and H.-J. Freund, *Phys. Chem. Chem. Phys.*, 2007, **9**, 3541–3558.
18 S. K. Desai, M. Neurock and K. Kourtakis, *J. Phys. Chem. B*, 2002, **106**, 2559–2568.
19 D. Santos-Carballal, A. Roldan, R. Grau-Crespo and N. H. de Leeuw, *Phys. Chem. Chem. Phys.*, 2014, **16**, 21082–21097.
20 M. D. Shroff, D. S. Kalakkad, K. E. Coulter, S. D. Kohler, M. S. Harrington, N. B. Jackson, A. G. Sault and A. K. Datye, *J. Catal.*, 1995, **156**, 185–207.
21 R. J. O'Brien, L. Xu, R. L. Spicer and B. H. Davis, *Energy Fuels*, 1996, **10**, 921–926.
22 Q.-L. Hao, F.-X. Liu, H. Wang, J. Chang, C.-H. Zhang, L. Bai, H.-W. Xiang, Y.-W. Li, F. Yi and B.-F. Xu, *J. Mol. Catal. A: Chem.*, 2007, **261**, 104–111.
23 M. Ding, Y. Yang, J. Xu, Z. Tao, H. Wang, H. Wang, H. Xiang and Y. Li, *Appl. Catal., A*, 2008, **345**, 176–184.
24 M. Luo, H. Hamdeh and B. H. Davis, *Catal. Today*, 2009, **140**, 127–134.

25 L. A. Cano, M. V. Cagnoli, J. F. Bengoa, A. M. Alvarez and S. G. Marchetti, *J. Catal.*, 2011, **278**, 310–320.

26 A. Roldan, N. Hollingsworth, A. Roffey, H. U. Islam, J. B. Goodall, C. R. Catlow, J. A. Darr, W. Bras, G. Sankar, K. B. Holt, G. Hogarth and N. H. de Leeuw, *Chem. Commun.*, 2015, **51**, 7501–7504.

27 G. Kresse and J. Hafner, *Phys. Rev. B: Condens. Matter Mater. Phys.*, 1993, **47**, 558.

28 G. Kresse and J. Furthmüller, *Comput. Mater. Sci.*, 1996, **6**, 15.

29 J. P. Perdew, J. A. Chevary, S. H. Vosko, K. A. Jackson, M. R. Pederson, D. J. Singh and C. Fiolhais, *Phys. Rev. B: Condens. Matter Mater. Phys.*, 1992, **46**, 6671.

30 S. H. Vosko, L. Wilk and M. Nusair, *Can. J. Phys.*, 1980, **58**, 1200–1211.

31 N. D. Mermin, *Phys. Rev.*, 1965, **137**, 1441–1443.

32 G. Kresse and D. Joubert, *Phys. Rev. B: Condens. Matter Mater. Phys.*, 1999, **59**, 1758.

33 S. Grimme, *J. Comput. Chem.*, 2006, **27**, 1787–1799.

34 S. Irrera, A. Roldan, G. Portalone and N. H. De Leeuw, *J. Phys. Chem. C*, 2013, **117**, 3949–3957.

35 N. Y. Dzade, A. Roldan and N. H. de Leeuw, *J. Chem. Phys.*, 2013, **139**, 124708.

36 S. S. Tafreshi, A. Roldan, N. Y. Dzade and N. H. de Leeuw, *Surf. Sci.*, 2014, **622**, 1–8.

37 S. Haider, A. Roldan and N. H. de Leeuw, *J. Phys. Chem. C*, 2013, **118**, 1958–1967.

38 N. Dzade, A. Roldan and N. de Leeuw, *Minerals*, 2014, **4**, 89–115.

39 F. Zhang, J. D. Gale, B. P. Uberuaga, C. R. Stanek and N. A. Marks, *Phys. Rev. B: Condens. Matter Mater. Phys.*, 2013, **88**, 054112–054117.

40 H. J. Monkhorst and J. D. Pack, *Phys. Rev. B: Condens. Matter Mater. Phys.*, 1976, **13**, 5188–5192.

41 J. M. D. Coey, M. R. Spender and A. H. Morrish, *Solid State Commun.*, 1970, **8**, 1605–1608.

42 M. R. Spender, J. M. D. Coey and A. H. Morrish, *Can. J. Phys.*, 1972, **50**, 2313–2326.

43 D. J. Vaughan and J. A. Tossell, *Phys. Chem. Miner.*, 1983, **9**, 253–262.

44 D. J. Vaughan and J. A. Tossell, *Am. Mineral.*, 1981, **66**, 1250–1253.

45 D. J. Vaughan and J. R. Craig, *Am. Mineral.*, 1985, **70**, 1036–1043.

46 K. K. Surerus, M. C. Kennedy, H. Beinert and E. Munck, *Proc. Natl. Acad. Sci. U. S. A.*, 1989, **86**, 9846–9850.

47 M. J. Dekkers, H. F. Passier and M. A. A. Schoonen, *Geophys. J. Int.*, 2000, **141**, 809–819.

48 M. J. Dekkers and M. A. A. Schoonen, *Geophys. J. Int.*, 1996, **126**, 360–368.

49 L. Chang, B. D. Rainford, J. R. Stewart, C. Ritter, A. P. Roberts, Y. Tang and Q. W. Chen, *J. Geophys. Res.: Solid Earth*, 2009, **114**, 1–10.

50 A. Roldan, D. Santos-Carballal and N. H. de Leeuw, *J. Chem. Phys.*, 2013, **138**, 204712–204716.

51 D. Santos-Carballal, A. Roldan, R. Grau-Crespo and N. H. de Leeuw, *Phys. Rev. B: Condens. Matter Mater. Phys.*, 2015, **91**, 195106.

52 D. J. Vaughan and M. S. Ridout, *J. Inorg. Nucl. Chem.*, 1971, **33**, 741–746.

53 M. Braga, S. K. Lie, C. A. Taft and W. A. Lester, *Phys. Rev. B: Condens. Matter Mater. Phys.*, 1988, **38**, 10837–10851.

54 L. Chang, A. P. Roberts, Y. Tang, B. D. Rainford, A. R. Muxworthy and Q. Chen, *J. Geophys. Res.*, 2008, **113**, 1–16.
55 V. I. Anisimov, M. A. Korotin, J. Zaanen and O. K. Andersen, *Phys. Rev. Lett.*, 1992, **68**, 345–348.
56 S. L. Dudarev, G. A. Botton, S. Y. Savrasov, C. J. Humphreys and A. P. Sutton, *Phys. Rev. B: Condens. Matter Mater. Phys.*, 1998, **57**, 1505–1509.
57 A. J. Devey, R. Grau-Crespo and N. H. de Leeuw, *Phys. Rev. B: Condens. Matter Mater. Phys.*, 2009, **79**, 195126–195133.
58 I. D. R. Moreira, F. Illas and R. L. Martin, *Phys. Rev. B: Condens. Matter Mater. Phys.*, 2002, **65**, 155102–155116.
59 D. J. Vaughan and J. A. Tossell, *Am. Mineral.*, 1981, **66**, 1250–1253.
60 D. Rickard and G. W. Luther, *Chem. Rev.*, 2007, **107**, 514–562.
61 G. W. Watson, E. T. Kelsey, N. H. deLeeuw, D. J. Harris and S. C. Parker, *J. Chem. Soc., Faraday Trans.*, 1996, **92**, 433–438.
62 P. W. Tasker, *J. Phys. C: Solid State Phys.*, 1979, **12**, 4977–4984.
63 G. Henkelman and H. Jonsson, *J. Chem. Phys.*, 1999, **111**, 7010–7022.
64 A. Heyden, A. T. Bell and F. J. Keil, *J. Chem. Phys.*, 2005, **123**, 224101–224115.
65 A. Roldan and N. H. de Leeuw, *Proc. R. Soc. London, Ser. A*, 2016, 472.
66 G. S. Parkinson, *Surf. Sci. Rep.*, 2016, **71**, 272–365.
67 B. A. Sexton, *Surf. Sci.*, 1981, **102**, 271–281.
68 E. M. Karp, T. L. Silbaugh, M. C. Crowe and C. T. Campbell, *J. Am. Chem. Soc.*, 2012, **134**, 20388–20395.
69 A. A. Peterson, F. Abild-Pedersen, F. Studt, J. Rossmeisl and J. K. Norskov, *Energy Environ. Sci.*, 2010, **3**, 1311–1315.

Faraday Discussions

PAPER

In situ TEM observation of the Boudouard reaction: multi-layered graphene formation from CO on cobalt nanoparticles at atmospheric pressure

G. Marien Bremmer,[a] Eirini Zacharaki,[b] Anja O. Sjåstad,[b] Violeta Navarro,[c] Joost W. M. Frenken[d] and Patricia J. Kooyman*[e]

Received 5th September 2016, Accepted 6th October 2016

DOI: 10.1039/c6fd00185h

Using a MEMS nanoreactor in combination with a specially designed *in situ* Transmission Electron Microscope (TEM) holder and gas supply system, we imaged the formation of multiple layers of graphene encapsulating a cobalt nanoparticle, at 1 bar CO : N_2 (1 : 1) and 500 °C. The cobalt nanoparticle was imaged live in a TEM during the Boudouard reaction. The *in situ*/operando TEM studies give insight into the behaviour of the catalyst at the nanometer-scale, under industrially relevant conditions. When switching from Fischer–Tropsch syngas conditions (CO : H_2 : N_2 1 : 2 : 3 at 1 bar) to CO-rich conditions (CO : N_2 1 : 1 at 1 bar), we observed the formation of multi-layered graphene on Co nanoparticles at 500 °C. Due to the high temperature, the surface of the Co nanoparticles facilitated the Boudouard reaction, causing CO dissociation and the formation of layers of graphene. After the formation of the first patches of graphene at the surface of the nanoparticle, more and more layers grew over the course of about 40 minutes. In its final state, around 10 layers of carbon capped the nanoparticle. During this process, the carbon shell caused mechanical stress in the nanoparticle, inducing permanent deformation.

1. Introduction

Novel techniques and equipment to allow *in situ* and operando research on catalytic samples are currently under development. Using these tools, it is

[a]Huygens-Kamerlingh Onnes Laboratory, Leiden University, Niels Bohrweg 2, Leiden, 2333 CA, The Netherlands

[b]Centre for Materials Science and Nanotechnology, Department of Chemistry, University of Oslo, PO Box 1033, N-0315 Oslo, Norway

[c]TNO, Stieltjesweg 1, Delft, 2628 CK, The Netherlands

[d]Advanced Research Center for Nanolithography, Science Park 110, Amsterdam, 1098 XG, The Netherlands

[e]Department of Chemical Engineering, University of Cape Town, Private Bag X3, Rondebosch, 7701 Cape Town, South Africa. E-mail: patricia.kooyman@uct.ac.za

possible to investigate a catalyst under industrially relevant working conditions: high temperatures and high gas pressures. The so-called temperature and pressure gaps between traditional catalysis research and the industrial applications are thereby bridged, improving the relevance of the results obtained *via in situ* research.

Current demand for clean, renewable fuels has instigated interest in the well-known catalytic Fischer–Tropsch (FT) reaction, which produces hydrocarbons from synthesis gas (syngas, CO and H_2).[1] One of the catalysts frequently used for this reaction is cobalt, often in the form of supported nanoparticles.[2] Both in academia as well as industry, efforts are being undertaken to investigate this catalytic system, in order to understand the fundamental processes involved and eventually improve the performance of the catalyst.

Several research groups have used *in situ* techniques to explore the FT syngas reaction on cobalt, *e.g.* scanning tunnelling microscopy (STM),[3–5] X-ray diffraction (XRD),[6] and X-ray absorption spectroscopy (XAS),[7,8] yielding meaningful insight into the working state of the catalyst surface under industrially relevant conditions. However, the use of *in situ* transmission electron microscopy (TEM) has not yet been reported. As has been shown for other catalytic systems, such as CO oxidation on platinum nanoparticles,[9] *in situ* TEM experiments can reveal valuable information of the behaviour of the catalyst nanoparticles at the atomic scale.

To investigate the behaviour of FT catalysts at elevated pressures and temperatures in a TEM, we used a combination of a MEMS nanoreactor loaded with Co nanoparticles, inserted in a specially designed *in situ* TEM sample holder. The sample holder allowed us to heat the sample (up to 660 °C) while using a gas supply system to flow a mixture of reactant gases over the sample at ambient pressure. A residual gas analyser (RGA) monitored the product gas stream leaving the outlet of the nanoreactor.

2. Experimental

2.1 Cobalt nanoparticles

ε-Cobalt nanoparticles (NPs) were synthesized under inert conditions *via* hot injection methodology using standard procedures.[10] Oleic acid (OA, $C_{18}H_{34}O_2$, ≥99%, 0.07 g) in 1,2-dichlorobenzene (DCB, $C_6H_4Cl_2$, 99%, anhydrous, 15 mL) was heated to 170 °C under stirring, and 0.52 g of dicobalt octacarbonyl ($Co_2(CO)_8$, ≥90% Co) dissolved in 3 mL of DCB was rapidly injected. The black colloidal suspension, formed upon injection, was aged at 170 °C for 30 min before quenching with 10 mL of DCB. The NPs were flocculated with excess 2-propanol (C_3H_8O, 99.5%, anhydrous) and isolated by centrifugation. After discarding the supernatant, the NP precipitate was purified by three repetitive cycles of washing in 2-propanol and subsequent centrifugation. The ε-Co NPs dispersed in hexane (C_6H_{14}, 95%, anhydrous) were stored in a closed container in a glovebox to prevent oxidation. All reagents were supplied by Aldrich, and used without further purification.

For reference TEM imaging, a few drops of nanoparticle suspension were placed on a 400 mesh copper TEM grid coated with lacey carbon film, covered with a continuous, ultrathin (<3 nm) carbon film (Ted Pella Inc.). The solvent was left to evaporate inside the glovebox, by heating the TEM grid on a hotplate overnight at 100 °C.

2.2 Nanoreactors

In situ TEM experiments were conducted using the latest version of nanoreactors (Else Kooi Laboratory).[11] These micro-electromechanical (MEMS) devices are fully integrated on a single silicon die.[12] The nanoreactors, shown in Fig. 1, consist of a 4.5 μm deep, etched channel with an in- and outlet allowing sample loading and gas flow feeding. For heating, a platinum wire is integrated in the top part of the gas channel. This wire allows simultaneous heating and temperature readout *via* a calibrated resistance measurement. The construction of the channel is such that it is possible to go up to 14 bar and 660 °C. A set of 41 electron-transparent windows, consisting of 15 nm-thin amorphous SiN films, is embedded in the centre of the channel. The thin and chemically inert amorphous silicon nitride film also covers all internal surfaces of the nanoreactor channel. The channel is supported by pillars, to increase stiffness and reduce bulging effects when filled with gas. After the preparation of the channel, the holes used for deposition and etching are plugged to ensure a fully leak tight device.

Cobalt nanoparticles were loaded into the nanoreactor by drop-casting the suspension containing the sample on the inlet.[13] Capillary forces drew the liquid through the channel, depositing nanoparticles on all surfaces, including the electron-transparent windows. The loading was done in a glovebox, under Ar atmosphere. Subsequently, the loaded nanoreactors were stored in an airtight vessel, which was taken out of the glovebox. The vessel was gently heated to ~80 °C on a hotplate for two days, while a turbomolecular pump was connected to pump down the vessel in order to evaporate the solvent.

2.3 *In situ* TEM holder

The custom built, dedicated *in situ* TEM holder used for the experiments contains a pair of capillaries for gas feed and exhaust. The nanoreactors were placed and pressed on top of two Viton O-rings to ensure a leak-tight connection between the nanoreactor gas-channel and the capillaries. For controlling and reading out the

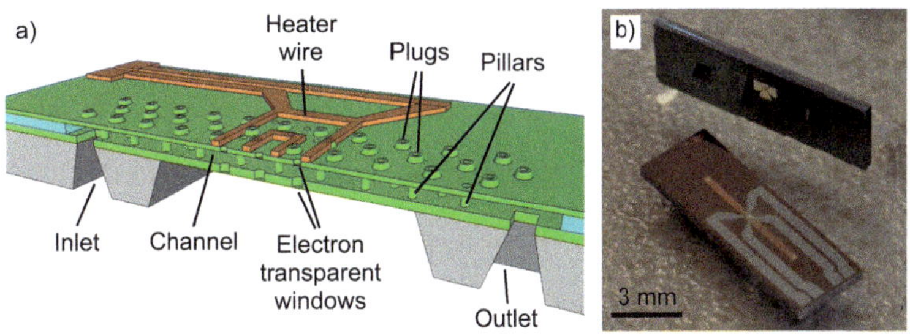

Fig. 1 (a) Schematic cross-section drawing of the nanoreactor channel, showing all relevant components. For clarity, this sketch is not drawn to scale.[10] (b) Photograph of two identical nanoreactors: the top one, viewed sideways, shows the bottom side with the channel inlet and outlet, and the bottom one, placed flat, shows the top side with the gas channel and heater wire.

temperature of the nanoreactor, four pins (two for control, two for read-out) were placed in contact with the platinum heating wire. Custom-built software digitally controlled and monitored the temperature of the sample.[14]

2.4 Gas supply system & residual gas analysis

Reactant gases were supplied by a computer controlled gas flow system (Leiden Probe Microscopy BV).[15] A schematic overview of the system is shown in Fig. 2.

Reactant gases (Praxair, purity: O_2 – 6.0; H_2 – 6.0; CO – 4.7) were introduced into the system *via* standard pressure reducing valves. The flow of each gas was controlled by Mass Flow Controllers (MFCs, Bronkhorst) and the desired mixture was obtained *via* a rotating mixing valve. Carrier gas (Linde Gas, N_2 purity: 6.0) was added to the gas mixture to create a typical total flow of $\sim$5 mL$_n$ min^{-1}. The pressure in the main gas line was regulated by a Back Pressure Controller (BPC, Bronkhorst) 'TEM holder', and the exhaust of the system was pumped by a diaphragm pump. To direct a portion of the gas mixture to the *in situ* holder, a T-piece was placed before the BPC, connected to the inlet of the holder. The capillaries directly connected to the *in situ* TEM holder were made of PEEK (poly ether ether ketone, a non-conducting material) to ensure electrical insulation of the holder when inserted into the TEM CompuStage. Vibration isolation was implemented by clamping the PEEK capillaries in-between two heavy metal slabs, reducing the mechanical noise introduced by the pumps running continuously. The exhaust capillary of the *in situ* TEM holder was directly connected to a turbomolecular pump. This setup established a pressure difference between the inlet and the exhaust of the holder, resulting in gas flow through the narrow nanoreactor channel. The pressure inside the nanoreactor channel was defined as half

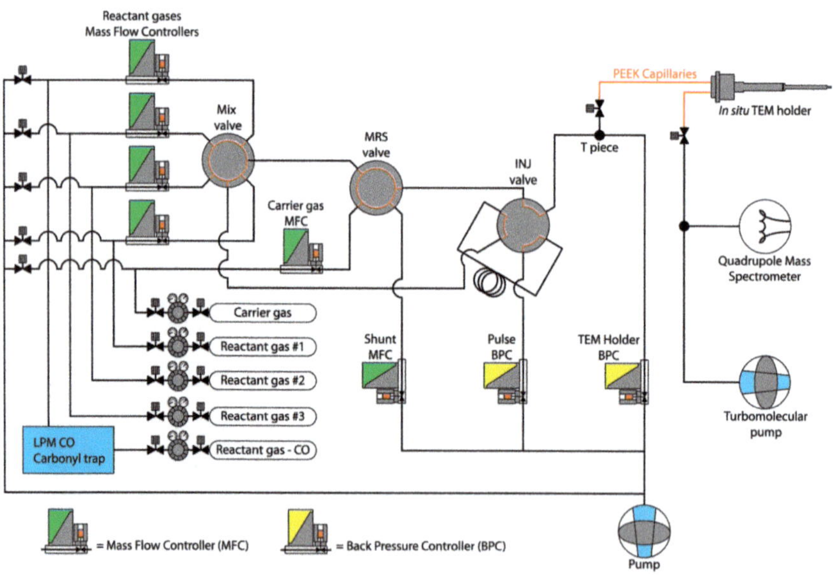

Fig. 2 Schematic overview of the gas supply system, showing the computer operated Mass Flow Controllers (MFCs), Back Pressure Controllers (BPCs), and rotating mixing valves.

of the pressure that was established by the 'TEM holder BPC' at the T-piece of the inlet side of the holder, due to the symmetry of the holder and the nanoreactor. For residual gas analysis, a quadrupole mass spectrometer (QMS, Pfeiffer Vacuum QMA-200) was connected to the exhaust line. Before entering the gas system, CO gas was led through a 'carbonyl trap' consisting of a copper tube heated to 325 °C and filled with copper-shot. This was done to decompose and remove any metal carbonyls from the gas, as nickel tetracarbonyl ($Ni(CO)_4$) and iron pentacarbonyl ($Fe(CO)_5$) can be formed in the gas when CO is stored in or passed through steel containers at high pressure (*e.g.* during the production of the gas).[16–18] Using Spacetime software[19] the data profiles from the gas supply system, QMS and nanoreactor temperature were synchronised and analysed.

2.5 Transmission electron microscopy

The TEM used for the *in situ* experiments was a C_S-corrected FEI Titan[3] 80-300 operated at 300 kV. Images were captured using a Digital Electronic camera (DE-12) with 6.0 μm pixel size and 4096 × 3072 pixels. To minimize beam effects on the sample, the electron intensity was kept below 500 electrons per Å^2 per s during the *in situ* experiments. An exposure time of 1 s was used for image acquisition.

Reference images of the cobalt nanoparticles under vacuum were acquired on an FEI monochromated Tecnai F20ST/STEM electron microscope, equipped with a Gatan Ultrascan CCD camera (4k × 4k). This TEM was operated at 200 keV.

Resulting images were analysed using Digital Micrograph 3 and Fiji software.

3. Results and discussion

3.1 Nanoparticle preparation

Cobalt nanoparticles were initially imaged on a carbon grid to obtain reference data. Fig. 3 shows a representative micrograph of a free-standing nanoparticle on carbon film. The lattice fringes with a *d*-spacing of 2.01 Å Fig. 3a, correspond to the (221) lattice plane of ε-Co (space group $P4_132$).[20] The FFT shows the two bright spots corresponding to the (221) lattice planes.

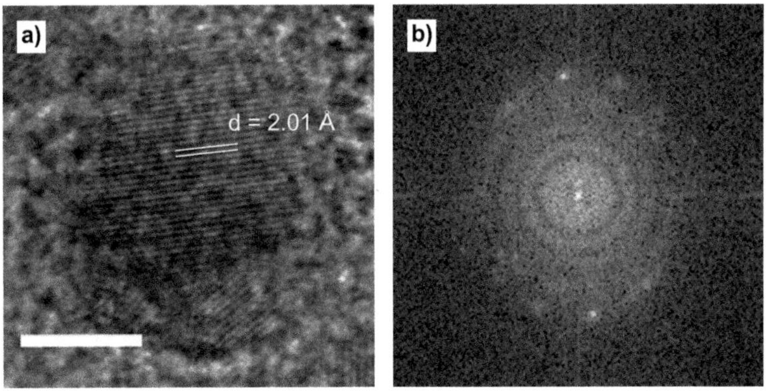

Fig. 3 (a) ε-Co nanoparticle on carbon film, showing the lattice fringes corresponding to the (221) lattice plane. (b) FFT of the image. Scale bar is 5 nm.

For the *in situ* experiments, a nanoreactor was loaded with a suspension containing cobalt nanoparticles. Prior to the Boudouard reaction, the sample was exposed to various conditions in order to conduct different experiments. The gas feed profile, QMS analyser data and temperature course of the whole experiment are shown in Fig. 4.

Initially, residual organic oleic acid (OA) molecules were removed *in situ* by heating the cobalt nanoparticles under a flow of N_2, at 1 bar and 300 °C, for 90 minutes. A micrograph of a nanoparticle on the amorphous SiN window of the nanoreactor, obtained after the heating procedure, is shown in Fig. 5.

To reduce any oxides that might have formed (although we did not observe any oxide in the TEM images), and to further reduce the amount of OA, the nanoreactor was then flushed with 5% H_2 in N_2. While flushing, the temperature was increased from 300 °C to 500 °C in 30 minutes, using increments of 50 °C, and was kept at 500 °C for the rest of the experiment. During this treatment, the

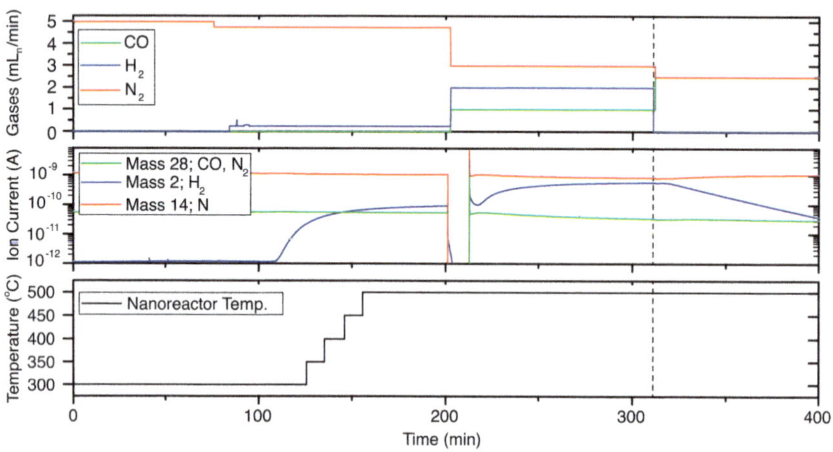

Fig. 4 Gas feed profile (top), QMS analyser data (middle), and nanoreactor temperature course (bottom) during the experiment. $t = 0$ is indicated by the vertical dashed line.

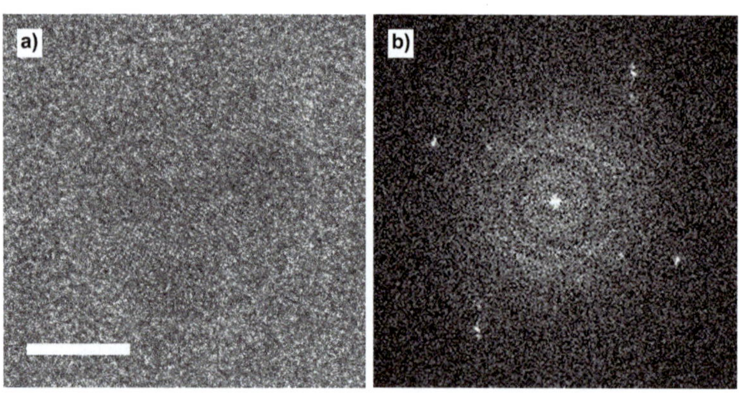

Fig. 5 (a) Cobalt nanoparticle inside the nanoreactor, at 1 bar N_2 and 300 °C. Scale bar is 5 nm. (b) FFT of the image.

nanoparticles appeared to stay in the metallic state, meaning that no oxides or carbon had formed, however we cannot exclude residues of OA remaining on the nanoparticle surface. The ε-Co phase transforms irreversibly during annealing in a non-oxidative environment to hexagonal close packing (hcp) at ~300 °C, and cubic close packing (ccp) at ~500 °C.[21]

After this reduction step, the metallic nanoparticles were exposed to Fischer–Tropsch syngas conditions. First, the valve to the TEM holder was closed and a mixture of $CO : H_2 : N_2$ (1 : 2 : 3) at 1 bar was prepared in the lines of the gas system. The closing of the valve is visible in Fig. 4b by the sudden dip in the QMS signal, at 200 min. When the reactant mixture was ready, the TEM holder valve was opened again and the gas flow entered the nanoreactor. These conditions were maintained for 100 minutes.

Finally, the feed of H_2 was removed and the gas feed mixture was adjusted to $CO : N_2$ (1 : 1) at 1 bar, while the sample temperature was kept at 500 °C. For clarity and ease of further discussion, the moment of closing the H_2 feed is defined as $t = 0$, as shown by the vertical dashed line in Fig. 4. The residual H_2 keeps flowing through the nanoreactor, until after six minutes the H_2 signal in the QMS chamber starts to decrease (Fig. 4b).

3.2 Carbon deposition

Four minutes after closing the H_2 flow, the first deposition of carbon appeared at the surface of the nanoparticle, as shown in Fig. 6. On the surface of the clean nanoparticle imaged at $t = 4:02$ minutes (Fig. 6a), the first carbon patch appeared at the top left of the nanoparticle at $t = 4:10$ (Fig. 6b, indicated by the white arrow). Then, three seconds later at $t = 4:13$, a second patch of carbon appeared on the surface at the top right of the nanoparticle (Fig. 6c, indicated by the white arrow). The carbon patches continued to grow laterally, and at $t = 4:47$ the patches joined each other and formed a continuous layer, showing a light grey fringe on top of the nanoparticle (Fig. 6d). After the first layer was formed, a second layer started to form, again at the top right (Fig. 6e). The layer continued to grow and a full second carbon layer was visible at $t = 20:48$ (Fig. 6f). At $t = 24:03$ a total of seven layers of carbon had formed on top of the particle (Fig. 6g). The layers continued to grow laterally, thereby capping the nanoparticle, as were visible at 28 : 53 (Fig. 6h). During the next 10 minutes, more carbon layers formed (Fig. 6i) until the nanoparticle reached its final state at $t = 39:10$ (Fig. 6j), covered in about 10 layers of carbon. After the formation of the last carbon layer, the nanoparticle was monitored for about 10 more minutes, but no additional carbon was deposited afterwards.

The first carbon patches were formed at two sides of the nanoparticle surface (Fig. 6b and c, white arrows). The deposition of carbon could originate from either the dissociation of adsorbed CO molecules (eqn (1)) or *via* the Boudouard disproportionation reaction (eqn (2)).[22,23]

$$CO_{ads} \rightarrow C_s + O_{ads} \tag{1}$$

$$2CO_{ads} \rightarrow CO_{2g} + C_s \tag{2}$$

Both reactions yield solid carbon at metal surfaces, but CO adsorption followed by dissociation occurs at temperatures up to around 400 °C, while the

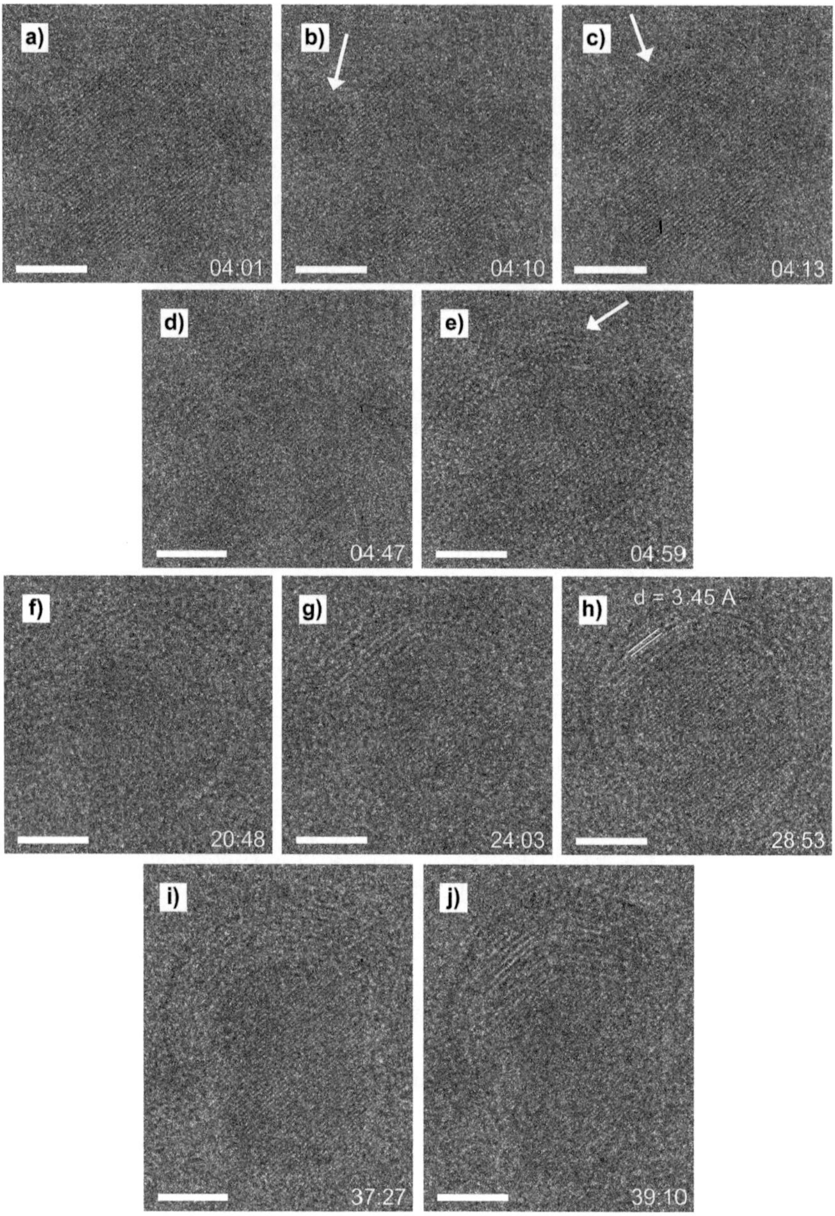

Fig. 6 Carbon deposition process on a cobalt nanoparticle at 500 °C and 1 bar CO : N_2 (1 : 1). An initially fresh nanoparticle (a) was covered by patches of carbon ((b and c), indicated by the white arrows) that then grew laterally and formed a first layer (d). A second layer formed ((e), indicated by the white arrow) and capped the nanoparticle (f). More carbon layers grew (g–i) until after 39 minutes the nanoparticle ended up it its final state (j), covered by around 10 layers of carbon. On each micrograph, the time since shutting off the H_2 feed is indicated in minutes:seconds. Scale bar is 4 nm.

Boudouard reaction is facilitated at cobalt surfaces above 427 °C.[22,23] Also, CO dissociation on cobalt leads to the formation of a single layer of carbidic carbon, while the Boudouard reaction can form multiple layers of graphitic carbon,[22] which is what we observed in our experiment. The carbon–carbon interlayer distance of the carbon layers formed on our nanoparticles is 3.45 Å, Fig. 6h, indicating that graphitic carbon was indeed formed.[24] We thus conclude that we have visualised the Boudouard reaction. Further evidence for assigning the Boudouard reaction as the cause for the observed carbon deposition could be embedded in the QMS signal, as the CO_2 signal should rise as an effect of the formation of CO_2 and C from CO. However, due to high background signals it was not possible during this experiment to draw any conclusions based on the QMS data.

The process of the formation of a new carbon layer, as shown in Fig. 7, was captured from $t = 28:15$. At this moment, there were nine carbon layers capping the cobalt nanoparticle. The next second, at 28 : 16, it was visible that the ensemble of nine layers was pushed up by a few Å. A new fringe appeared at the interlayer between the cobalt surface and the packet of carbon layers, indicating the formation of a new carbon layer. The new fringe is indicated by the black arrow in Fig. 7b. To facilitate interpretation of the weak contrast, a white arrow was drawn perpendicular to the visible *d*-spacing on the nanoparticle and the packet of carbon layers at the exact same positions in Fig. 7a and b. Contrast line profiles were obtained for both images, shown in Fig. 7c.

By identifying the cobalt fringes (Fig. 7c, blue region, *d*-spacing 2.07 Å) and the carbon fringes (Fig. 7c, grey region, *d*-spacing 3.45 Å), it is clearly visible that the additional peak displaces the earlier formed carbon layers away from the cobalt nanoparticle surface, and the grey area expands by one lattice spacing of the carbon.

3.3 Effect on the nanoparticles

During the live observation of the growing layers, the nanoparticle was changing morphology. The initially spherical nanoparticle (Fig. 6a) became elongated at the

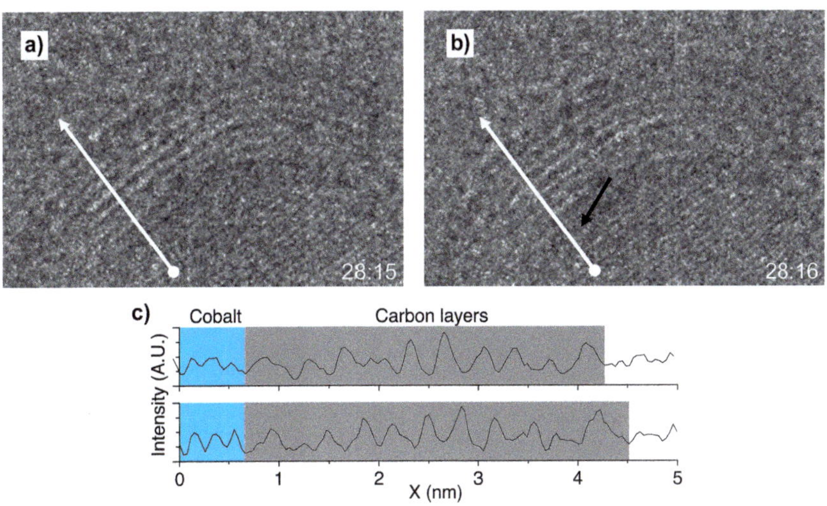

Fig. 7 Growth of a new carbon layer, indicated by the black arrows, interlayered between the cobalt surface and the set of carbon layers capping the nanoparticle, before (a) and after (b) the formation. Profiles of the white lines are shown in (c). Top profile: image (a); bottom profile: image (b). White line length is 5 nm.

end of the carbon deposition process (Fig. 6j). When the nanoparticle was covered with six carbon layers, the internal crystal structure was also temporarily altered, as shown in Fig. 8. At $t = 25:26$, the nanoparticle was still spherical (Fig. 8a), showing a uniform crystal structure. During the following ten seconds, the crystal structure shifted around and eventually a twin boundary was formed at $t = 25:35$ (Fig. 8b) in the centre of the nanoparticle. The crystal structure in the bottom right part of the nanoparticle continued showing mobility, and a second twin boundary appeared clearly at $t = 25:57$ (Fig. 8c). This situation was maintained for about 90 seconds. During that time the twin boundaries continuously shifted up and down several atomic rows. The top of the particle also showed a twin boundary for a few seconds, which was too mobile to be imaged clearly. After this period of twinning, the crystal structure of the nanoparticle became uniform again at $t = 27:34$ (Fig. 8d), which lasted until the end of the experiment.

The presence of twin boundaries is related to the amount of stress inside the nanoparticle, as the formation of boundaries leads to the release of this stress[25] and is known to occur in metallic nanoparticles.[26–28] Metallic nanoparticles

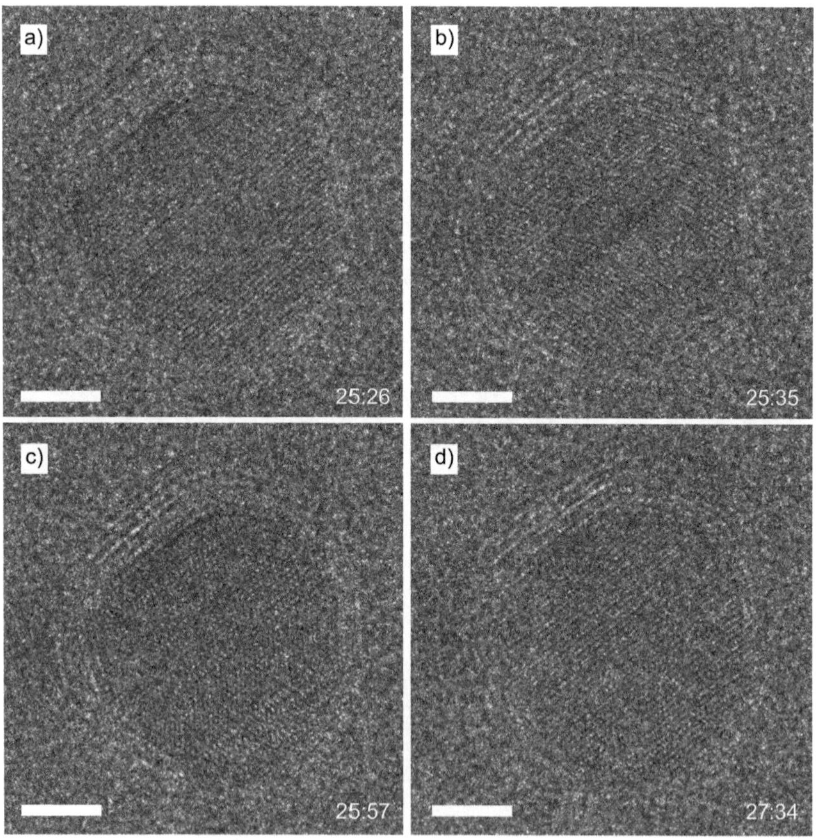

Fig. 8 Cobalt nanoparticle, capped with six layers of carbon, initially showing a uniform crystal structure (a). A first twin boundary was formed through the centre of the nanoparticle (b), followed by a second twin boundary at the bottom right (c). After around 90 seconds of continuous mobility, the crystal structure relaxed into the final, uniform state (d). Scale bar is 3 nm.

(*e.g.* Au, Co) encaged by carbon shells will eventually deform when heated to a temperature above 300 °C and irradiated with a TEM electron beam, as the electron beam displaces carbon atoms in the carbon shells, causing the shells to shrink.[29–32] Similar deformation effects have been observed when carbon nanotubes were formed *via* CH_4 decomposition on the surface of Ni nanoparticles.[33] Our experiment showed a combination of both effects, as we initially started with uniform, crystalline nanoparticles, not covered by any graphite layers. Eventually, the growing carbon layers were capping just the top half of the nanoparticle. Due to continuous electron irradiation, the carbon must have been contracting and thereby compressing the capped particle. The formed strain was relieved by the formation of the twin boundaries. The initially spherical nanoparticle then deformed into an elongated shape, fitting inside the carbon cap and thereby releasing the twin boundaries to return to a uniform crystal structure.

3.4 Carbon morphologies

After continuously imaging the nanoparticle that showed the formation of the layers of carbon, other nanoparticles were imaged. Various other nanoparticles were also covered with layers of carbon, but not all of them. Around one third of all nanoparticles that were present on electron-transparent windows were encapsulated or covered with carbon. All carbon covered nanoparticles, independent of their size, supported carbon formations that consisted of around 7 to 14 layers of carbon. No trends could be established linking nanoparticle size to the number of layers. Different morphologies of carbon structures were visible: next to encapsulated nanoparticles also multi-walled carbon nanotube formations had been formed Fig. 9a, while other nanoparticles supported spherical structures Fig. 9b that did not cap the supporting nanoparticle as tightly as the previously imaged assembly Fig. 6. As the formation of the carbon layers shown in Fig. 9 occurred before any TEM imaging was done, we conclude that the carbon deposition is not induced by the TEM electron beam.

Several of the imaged particles were stable under the electron beam. However, other particles showed shape deformation processes and formation of twin boundaries while being imaged. An example of a nanoparticle that was being

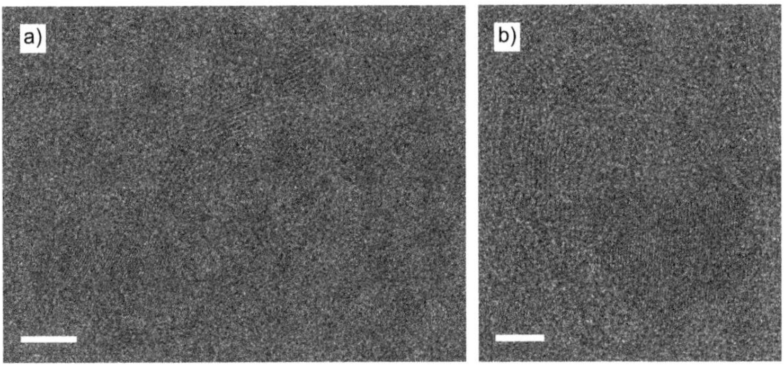

Fig. 9 Cobalt nanoparticles supporting various shapes of carbon layers. (a) A multi-walled carbon nanotube, and (b) a spherical carbon formation loosely capping the supporting cobalt nanoparticle. Scale bar is 4 nm.

deformed while imaged, after being encapsulated without electron beam irradiation, is shown in Fig. 10. When the nanoparticle was initially imaged (Fig. 10a), the shape was already heavily elongated at $t = 0$ (when TEM imaging started). After half a minute (Fig. 10b), the elongated particle started to slowly thin out in the centre, as the carbon shape around the particle was squeezing the nanoparticle. This process slowly continued (Fig. 10c) until finally at $t = 1:19$, the nanoparticle split into two separate particles (Fig. 10d). The carbon assembly

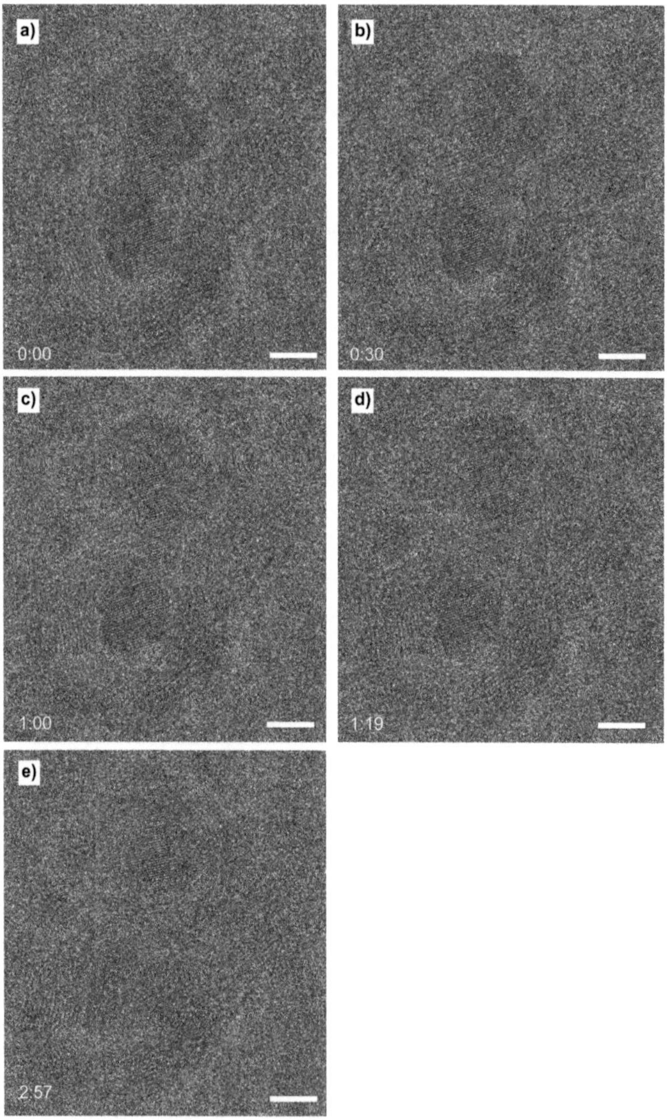

Fig. 10 Cobalt nanoparticle encapsulated and deformed by an assembly of carbon layers (a). The nanoparticle was slowly squeezed by the surrounding carbon layers (b and c), leading to the splitting of the nanoparticle into two isolated particles (d). After further deformation of the carbon assembly surrounding the particles, the system reached the final, stable state (e). Scale bar is 4 nm.

continued to deform, thereby moving the two nanoparticles around. After around three minutes the system reached its final state (Fig. 10e).

4. Conclusions and outlook

We have shown the *in situ* formation of layers of graphitic carbon on free-standing cobalt nanoparticles, at 500 °C and 1 bar CO : N_2 (1 : 1). The nanoparticles were imaged using TEM, with a specially designed *in situ* TEM holder and MEMS nanoreactor devices. Reactant gases were added to the nanoreactor *via* a gas supply system, while a QMS analysed the product gas stream. Due to the high temperature and presence of CO gas, the Boudouard reaction led to the deposition of carbon and consequent formation of graphitic carbon on the cobalt nanoparticles.

During a 40 minute tracking period, a cobalt nanoparticle was covered and capped by layers of carbon, formed one by one at the interface of the nanoparticle surface and the carbon. The layers of carbon strained the nanoparticle. As a consequence, twin boundaries appeared until the nanoparticle was permanently deformed. Several other nanoparticles were imaged afterwards showed similar deformation behaviour.

Further measurements could provide additional evidence for the Boudouard reaction by revealing the formation of CO_2 in the QMS data. The QMS data could also provide an indication whether all OA is fully removed from the surface of the nanoparticles, or if there is still OA present on the surface, however this is a topic for upcoming investigations.

Acknowledgements

This work was supported by The Netherlands Organization for Scientific Research (NWO/OCW) as part of the Frontiers of NanoScience (NanoFront) program. The research leading to these results has received funding from the European Union Seventh Framework Programme under Grant Agreement 312483 – ESTEEM2 (Integrated Infrastructure Initiative-I3). V. N. acknowledges NWO and the Technology Foundation (STW) for financial support as part of the Veni research programme (project no. 11920). We thank Prof. Dr H. W. Zandbergen and Dr Ir. F. D. Tichelaar (TU Delft) for their support in using the TEM hardware. We thank G. Pandraud and B. Morana (EKL) for supplying the nanoreactors.

References

1 A. Y. Khodakov, W. Chu and P. Fongarland, Advances in the Development of Novel Cobalt Fischer–Tropsch Catalysts for Synthesis of Long-Chain Hydrocarbons and Clean Fuels, *Chem. Rev.*, 2007, **107**(5), 1692–1744, DOI: 10.1021/cr050972v.

2 G. L. Bezemer, J. H. Bitter, H. P. C. E. Kuipers, H. Oosterbeek, J. E. Holewijn, X. Xu, F. Kapteijn, A. J. van Dillen and K. P. de Jong, Cobalt Particle Size Effects in the Fischer–Tropsch Reaction Studied with Carbon Nanofiber Supported Catalysts, *J. Am. Chem. Soc.*, 2006, **128**(12), 3956–3964, DOI: 10.1021/ja058282w.

3 V. Navarro, M. A. van Spronsen and J. W. M. Frenken, *In Situ* Observation of Self-Assembled Hydrocarbon Fischer–Tropsch Products on a Cobalt Catalyst, *Nat. Chem.*, 2016, **8**(10), 929–934, DOI: 10.1038/nchem.2613.

4 A. Banerjee, V. Navarro, J. W. M. Frenken, A. P. van Bavel, H. P. C. E. Kuipers and M. Saeys, Shape and Size of Cobalt Nanoislands Formed Spontaneously on Cobalt Terraces During Fischer–Tropsch Synthesis, *J. Phys. Chem. Lett.*, 2016, **7**(11), 1996–2001, DOI: 10.1021/acs.jpclett.6b00555.

5 B. Boeller, M. Ehrensperger and J. Wintterlin, *In Situ* Scanning Tunneling Microscopy of the Dissociation of CO on Co(0001), *ACS Catal.*, 2015, **5**(11), 6802–6806, DOI: 10.1021/acscatal.5b01684.

6 O. Ducreux, B. Rebours, J. Lynch, M. Roy-Auberger and D. Bazin, Microstructure of Supported Cobalt Fischer–Tropsch Catalysts, *Oil Gas Sci. Technol.*, 2009, **64**(1), 49–62, DOI: 10.2516/ogst:2008039.

7 S. Li, G. D. Meitzner and E. Iglesia, Structure and Site Evolution of Iron Oxide Catalyst Precursors During the Fischer–Tropsch Synthesis, *J. Phys. Chem. B*, 2001, **105**(24), 5743–5750, DOI: 10.1021/jp010288u.

8 A. K. Tuxen, S. Carenco, M. Chintapalli, C.-H. Chuang, C. Escudero, E. Pach, P. Jiang, F. Borondics, B. Beberwyck, A. P. Alivisatos, *et al.*, Size-Dependent Dissociation of Carbon Monoxide on Cobalt Nanoparticles, *J. Am. Chem. Soc.*, 2013, **135**(6), 2273–2278, DOI: 10.1021/ja3105889.

9 S. B. Vendelbo, C. F. Elkjær, H. Falsig, I. Puspitasari, P. Dona, L. Mele, B. Morana, B. J. Nelissen, R. van Rijn, J. F. Creemer, *et al.*, Visualization of Oscillatory Behaviour of Pt Nanoparticles Catalysing CO Oxidation, *Nat. Mater.*, 2014, **13**(9), 884–890, DOI: 10.1038/nmat4033.

10 E. Zacharaki, M. Kalyva, H. Fjellvåg and A. O. Sjåstad, Burst Nucleation by Hot Injection for Size Controlled Synthesis of ε-Cobalt Nanoparticles, *Chem. Cent. J.*, 2016, **10**(1), 10, DOI: 10.1186/s13065-016-0156-1.

11 Website Else Kooi Laboratory, http://ekl.tudelft.nl/EKL/Home.php, Last consulted: 05-Sept-2016.

12 J. F. Creemer, F. Santagata and B. Morana, An All-in-One Nanoreactor for High-Resolution Microscopy on Nanomaterials at High Pressures, *J. Microelectromech. Syst.*, 2011, 1103–1106, DOI: 10.1109/MEMSYS.2011.5734622.

13 I. Puspitasari, P. Saputra, B. Morana, L. Mele, F. Santagata, J. F. Creemer, F. Kapteijn and P. Kooyman, Particle Imaging and Flow Visualization of *In Situ* TEM Nanoreactors, *Microsc. Microanal.*, 2012, **18**(S2), 740–741, DOI: 10.1017/S1431927612005557.

14 J. F. Creemer, S. Helveg, P. J. Kooyman, A. M. Molenbroek, H. Zandbergen and P. M. Sarro, A MEMS Reactor for Atomic-Scale Microscopy of Nanomaterials Under Industrially Relevant Conditions, *J. Microelectromech. Syst.*, 2010, **19**(2), 254–264, DOI: 10.1109/JMEMS.2010.2041190.

15 C. T. Herbschleb, P. C. van der Tuijn, S. B. Roobol, V. Navarro, J. W. Bakker, Q. Liu, D. Stoltz, M. E. Cañas-Ventura, G. Verdoes, M. A. van Spronsen, *et al.*, The Reactor STM: Atomically Resolved Scanning Tunneling Microscopy Under High-Pressure, High-Temperature Catalytic Reaction Conditions, *Rev. Sci. Instrum.*, 2014, **85**(8), 083703, DOI: 10.1063/1.4891811.

16 C. E. H. Bawn, The Kinetics of the Decomposition of Nickel Carbonyl, *Trans. Faraday Soc.*, 1935, **31**, 440–446.

17 W. M. Goldberger and D. F. Othmer, Kinetics of Nickel Carbonyl Formation, *Ind. Eng. Chem. Process Des. Dev.*, 1963, **2**(3), 202–209, DOI: 10.1021/i260007a006.

18 T. C. Williams and C. R. Shaddix, Contamination of Carbon Monoxide with Metal Carbonyls: Implications for Combustion Research, *Combust. Sci. Technol.*, 2007, **179**(6), 1225–1230, DOI: 10.1080/00102200601057279.

19 S. Roobol, The Structure of a Working Catalyst: From Flat Surfaces to Nanoparticles, PhD thesis, Leiden University, 2014, ISBN: 9789085932048.

20 D. P. Dinega and M. G. Bawendi, A Solution-Phase Chemical Approach to a New Crystal Structure of Cobalt, *Angew. Chem., Int. Ed. Engl.*, 1999, **38**(12), 1788–1791, DOI: 10.1002/(SICI)1521-3773(19990614)38.

21 H. T. Yang, C. M. Shen, Y. G. Wang, Y. K. Su, T. H. Yang and H. J. Gao, Stable Cobalt Nanoparticles Passivated with Oleic Acid and Triphenylphosphine, *Nanotechnology*, 2004, **15**(1), 70–74, DOI: 10.1088/0957-4484/15/1/014.

22 J. Nakamura, I. Toyoshima and K. Tanaka, Formation of Carbidic and Graphitic Carbon From CO on Polycrystalline Cobalt, *Surf. Sci.*, 1988, **201**(1–2), 185–194, DOI: 10.1016/0039-6028(88)90605-X.

23 F. L. Hutson, D. E. Ramaker and B. E. Koel, Spectroscopic Evidence for Carbon–Carbon Bonding in "Carbidic" Layers on Metals, *Surf. Sci.*, 1991, **248**(1–2), 104–118, DOI: 10.1016/0039-6028(91)90065-Z.

24 O. V. Kharissova and B. I. Kharisov, Variations of Interlayer Spacing in Carbon Nanotubes, *RSC Adv.*, 2014, **4**(58), 30807–30815, DOI: 10.1039/C4RA04201H.

25 A. Mayoral, H. Barron, R. Estrada-Salas, A. Vazquez-Duran and M. José-Yacamán, Nanoparticle Stability From the Nano to the Meso Interval, *Nanoscale*, 2010, **2**(3), 335–342, DOI: 10.1039/B9NR00287A.

26 H. Hofmeister, Fivefold Twinned Nanoparticles, *Encyclopedia of Nanoscience and Nanotechnology*, American Scientific Publishers, 2004, ISBN: 1-58883-001-2.

27 A. S. Barnard, H. Konishi and H. F. Xu, Morphology Mapping of Platinum Catalysts Over the Entire Nanoscale, *Catal. Sci. Technol.*, 2011, **1**(8), 1440–1448, DOI: 10.1039/c1cy00238d.

28 J. L. Elechiguerra, J. Reyes-Gasga and M. J. Yacaman, The Role of Twinning in Shape Evolution of Anisotropic Noble Metal Nanostructures, *J. Mater. Chem.*, 2006, **16**(40), 3906–3919, DOI: 10.1039/b607128g.

29 L. Sun, J. A. Rodriguez-Manzo and F. Banhart, Elastic Deformation of Nanometer-Sized Metal Crystals in Graphitic Shells, *Appl. Phys. Lett.*, 2006, **89**(26), 263104, DOI: 10.1063/1.2403898.

30 L. Sun, A. V. Krasheninnikov, T. Ahlgren, K. Nordlund and F. Banhart, Plastic Deformation of Single Nanometer-Sized Crystals, *Phys. Rev. Lett.*, 2008, **101**(15), 156101, DOI: 10.1103/PhysRevLett.101.156101.

31 F. Banhart, J. X. Li and A. V. Krasheninnikov, Carbon Nanotubes Under Electron Irradiation: Stability of the Tubes and Their Action as Pipes for Atom Transport, *Phys. Rev. B: Condens. Matter Mater. Phys.*, 2005, **71**(24), 241408, DOI: 10.1103/PhysRevB.71.241408.

32 A. V. Krasheninnikov, P. O. Lehtinen, A. S. Foster and R. M. Nieminen, Bending the Rules: Contrasting Vacancy Energetics and Migration in Graphite and Carbon Nanotubes, *Chem. Phys. Lett.*, 2006, **418**(1–3), 132–136, DOI: 10.1016/j.cplett.2005.10.106.

33 S. Helveg, C. López-Cartes, J. Sehested, P. L. Hansen, B. S. Clausen, J. R. Rostrup-Nielsen, F. Abild-Pedersen and J. K. Nørskov, Atomic-ScaleImaging of Carbon Nanofibre Growth, *Nature*, 2004, **427**(6973), 426–429, DOI: 10.1038/nature02278.

Faraday Discussions

ROYAL SOCIETY OF CHEMISTRY

DISCUSSIONS

Designing new catalysts for synthetic fuels: general discussion

Katherine Holt, Linda Jewell, Hans Niemantsverdriet, Lebohang Macheli, Mzamo Shozi, Graham Hutchings, Tim Wezendonk, Michael Bowker, Richard Catlow, Sarah Adam, Patricia J. Kooyman, Emiel Hensen, James Hayward, Neil Coville, Kees-Jan Weststrate, Nico Fischer, Alberto Roldan, Evjeniy Redekop, Eric van Steen, Holger Friedrich, Dong-Hau Kuo, Harinarayana Bandaru, Letisha Deeplal, Aleksandar Zivkovic, Kershen Naiker, Chelsea Tucker, Thabiso Perfect Oscar Mkhwanazi, Detlef Bahnemann, Matthew Neurock, Anna Petersen, Roy Peter Forbes, Tumelo Phaahlamohlaka, Hans Schulz, David Lennon, Unni Olsbye, Moritz Wolf, Simon Kondrat, Paul Collier, Xiaohui Sun, Ying Zheng, Avelino Corma, Enrique Iglesia, Thulani Nyathi, Ding Ma, Marien Bremmer, Michael Claeys and Freek Kapteijn

DOI: 10.1039/c7fd90011b

Nico Fischer opened discussion of the paper by Ding Ma: After the ethylene treatment of your model catalyst, do you observe any cobalt carbide formation?

Ding Ma answered: No, we did not observe the formation of cobalt carbide in its XRD pattern.

Hans Niemantsverdriet commented: I fully agree with you that it is interesting to study the effect of carbon on the FTS reaction. However, I am a little concerned about the conditions you have chosen. Ethylene is a quite reactive molecule, and exposing a catalyst to it at 450 °C sounds like brute force to me. The ethylene decomposes and forms graphite on the support, and perhaps some of the carbon might be incorporated in the cobalt, which explains why the surface becomes more hydrogenating. My suggestion would be to extend these initial studies with reduced catalysts pretreated in ethylene at much lower temperatures, e.g. room temperature and upwards. If you do it systematically, you may be able to populate the surface in a subtle way with hydrocarbon species, and knowing their effect on FTS activity and selectivity would be very interesting. Surface science studies may give you some guidance on what species can be expected from ethylene.

Ding Ma answered: Yes, you are right. It is a pity that we are not able to clearly see what happens to those nanoparticles and which sites are blocked or modulated under different treatments. We surely need more work to address these issues in the future, possibly with the help of surface science as well as high pressure *in situ* TEM. But again, I want to say that soft promoters, such as in this case, surface carbon, is something that has been overlooked in the past and is worth re-visiting now.

Neil Coville asked: Co is a catalyst that is used to make carbon nanotubes. Do you see any carbon fibre (or nanotube) growth?

Ding Ma answered: Yes, we did see the formation of lots of carbon nanotubes (CNTs) or carbon nanofibers (CNFs) on those Co catalysts, especially when the particles of Co were the right size. In the TPH profiles, the hydrogenation peaks of those CNTs or CNFs appeared at a relatively high temperature, while those at low temperature corresponded to surface carbon deposition (or surface carbide, which is not able to be seen by XRD) and polymeric carbon.

Eric van Steen asked: Your DFT model presumes that carbon is present as monoatomic carbon species present on the surface. It is well known that on Co(111), C–C chains are more stable than atomic carbon.[1] Do you have experimental proof that you generate atomic carbon upon treating the catalyst in a C_2H_4/H_2 mixture at elevated temperature?

1 J. C. W. Swart, I. M. Ciobîcă, R. A. van Santen and E. van Steen, *J. Phys. Chem. C*, 2008, **112**, 12899–12904.

Ding Ma replied: No direct evidence, but normally people believe that the hydrogenation peak at very low temperature belongs to carbon atoms on the metal surface. In the current case, we just used a simplified model to simulate the modulation effect of surface carbon on Co-based FTS.

Paul Collier commented: Characterising carbon species in catalysts is difficult and neutron techniques such as those available at the STFC ISIS neutron and muon facility in the UK could help with this (*e.g.* INS). Does the carbon that you put on top of the catalyst have any catalytic activity itself? Also, carbon is soluble in some metals – is the reverse also true? Could the metals be drawn out of the surface by the carbon to form new clusters on the surface?

Ding Ma answered: Carbon itself could be active in certain catalytic reactions;[1] however, it can not dissociate CO. I believe it depends on their reactivity, some of the surface carbon could be reactive in the FTS reaction, but it surely needs the electronic modulation from the subsurface metal catalysts. You are right about your second question, it is possible. For example, normally people think that Co catalysts can break the C–O bond in CO, so normally we get hydrocarbons from pure Co-based FTS catalyst. However, under certain treatments (carbonization), part of the Co in a Co catalyst will transform into Co carbide, forming a metallic/carbide interface. Although CO could adsorb on the Co carbide, in some cases, it can not break over Co carbide. And therefore, on the interface of metallic Co and Co carbide, the formation of oxygenates was observed. This is to say, the role of carbon on the metal catalyst needs more eyes on it.

1 P. Tang, G. Hu, M. Li and D. Ma, *ACS Catal.*, 2016, **6**, 6948–6958.

Graham Hutchings asked: Following on from the discussion concerning the role of the carbon that has been laid down from ethene, does any of the carbon from the ethene end up in the products of the CO hydrogenation reaction? A relatively facile experiment would be to use $^{13}C_2H_2$ and ^{12}CO and look at the initial products. It is known that the selectivity of FT catalysts changes significantly over the first few hours of the reaction. Do you still observe the effects with longer reaction times?

Ding Ma answered: I have done more research on iron catalysts for the FTS. In the situation of iron catalysts, the carbon from the surface of iron carbide could surely participate in the formation of hydrocarbons. An example of this has been demonstrated by Hans (Niemantsverdriet) and Jose (Gracia) in their famous paper.[1] Recently we have proved that on a pure phase Fe_5C_2 catalyst, the carbon on the surface of the carbide can participate in C–C bond formation of hydrocarbons. If you begin with $Fe_5{}^{12}C_2$ and react with ^{13}CO, you can get $^{13}CH_3$–$^{12}CH_3$ ethane. I am not clear about whether this holds for a Co catalyst system, but normally people believe that the active phase of Co catalysts is metallic carbon. We need more experiments to prove whether this surface carbon can participate in chain propagation, as in the iron catalyst's case.

1 J. M. Gracia, F. F. Prinsloo and J. W. Niemantsverdriet, *Catal. Lett.*, 2009, **133**, 257–261.

Matthew Neurock remarked: This is an interesting study. The nature of the active surface will control how elementary steps are carried out. As such it is difficult to draw many mechanistic conclusions without a better understanding of the nature of the carbon on the surface. The theoretical calculations that were discussed compared energetics for methane formation and carbon coupling on a carbon covered surface with that of a bare metal surface. This, however, is not a fair comparison in that the metal surfaces under Fischer–Tropsch conditions are not bare but instead are highly covered with CO. If you were to carry out the same calculations on a CO-covered surface, you would not see the same differences between the two reactions. I would expect the barriers for methane formation to be closer to those on the C-covered surface. The calculations should be carried out on surfaces more reflective of what is present under reaction conditions.

Ding Ma responded: You are right. The calculation part of this work is somewhat too simple, and more work needs to be done on this system. I will discuss this with my DFT collaborator.

Nico Fischer asked: Do you have a measurement of the absolute amount of carbon deposited during the different presentments? How dense is the carbon layer; could it be that it protects the metal crystallites from oxidation? Are you sure that oxide species form during passivation?

Ding Ma responded: Yes, for the ethene treated sample at high temperature (450 °C), the amount of carbon could be as high as 20–25%. This means that besides the surface carbon on Co particles, there must be some fiber- or tube-like nano carbon structures in the treated catalyst. For a Co-H-C catalyst, the amount of carbon on the surface of Co is much smaller. Under TEM, you can see onion-like carbon structures on Co particles. Some of them will be oxidized during the passivation process.

Letisha Deeplal communicated: Could TGA-MS be used as a tool to quantify and identify carbon species present?

Ding Ma communicated in response: Yes, but it is best with temperature programmed oxidation or hydrogenation methods. Also, if it is possible, you could check the type and amount of residue carbon during this TGA-MS process by XPS or other carbon-sensitive methods.

Avelino Corma opened discussion of the paper by Freek Kapteijn: You make MOFs and then decompose them to obtain final C–C bonds. Why use a MOF and not just use cobalt or iron terephthalate? What is the difference, and do you have any idea why there is a difference? If you want to use the same organic that you are using for the MOFs, can you do it without having to synthesize the MOFs?

Freek Kapteijn replied: This is an excellent question. We have not considered this, being focused on MOFs, but this is a suggestion worth trying. In our study we noticed that the non-porous MOF gave poor FTS performance results after pyrolysis, so we have to do experiments to see the results.

Graham Hutchings asked: Is this a method of introducing carbon into the Fe catalyst that is similar to that described by Ma and co-workers? In their case they form the nanoparticles and then add the carbon, and in your case you start with an intimate mixing of the carbon and the metal. It then takes time for the carbon layer to form. Is it possible to comment on this?

Freek Kapteijn replied: Considering the Mössbauer results, we see that after pyrolysis and pre-reduction with hydrogen we mainly have metallic iron surrounded by a carbon shell. Under FTS conditions the iron is transformed into the active carbidic phase. So carbon introduction does not occur directly during pyrolysis, only to a small extent. Also, without pre-reduction the iron is not yet carbidic, but more FeO, also surrounded by carbon.

Neil Coville remarked: The effect seen on collapsing the MOF is interesting. Might it be possible to create massive ligands, similar to that used to construct the MOF, to give the same effect as a MOF? The only difference may be that the collapse could lead to a different type of cluster species.

Hans Niemantsverdriet commented: In Fig. 5 of your paper (DOI: 10.1039/C6FD00198J), you show very favourable stable behaviour of your catalysts. Though some particles take time to activate, after this they are surprisingly stable for some 100 h and more. You mentioned that the particles appear surrounded by a shell, which probably stabilises them against sintering. Your results remind me of work by Vannice, Delgass and myself in the 80s[1,2] on small iron particles on a very high surface area carbon support. Magnetic behaviour (absence of coupling *etc.*) indicated that these particles were effectively separated from each other in the voids between the graphitic plains of the support. These iron/iron carbide particles were small (a few nanometers only) but showed very stable FTS behaviour as well, although the selectivity wasn't too interesting. Our conclusion at the time was that because the particles were small, deactivation by carbon overlayers was avoided, as these graphitic structures would not fit well on small, curved particles. We did not consider sintering in those days.

Looking at your work, it appears to me that enclosing iron catalyst particles in *e.g.* a carbon shell, which has a similar effect to isolating them in the voids of a super-high surface area carbon, is the main reason that these particles are stabilised, as coalescence and Ostwald ripening are essentially blocked in this way. I would be interested in your views.

1 H.-J. Jung, P. L. Walker Jr. and M. A. Vannice, *J. Catal.*, 1982, **75**, 416–422.
2 J. W. Niemantsverdriet, A. M. van der Kraan, W. N. Delgass and M. A. Vannice, *J. Phys. Chem.*, 1985, **89**, 67–72.

Freek Kapteijn responded: This is indeed the interpretation we have for the stable performance. The carbon shells prevent the agglomeration, whatever the detailed mechanism is.

Patricia J. Kooyman remarked: For iron carbides prepared at low pressure in *in situ* TEM experiments, we did not observe the graphite-like shell. This shell might very well be formed only upon cooling down. For Fe-based FT catalysts characterised without exposure to air, some carbon shells were observed for some carbide particles, but upon passivation or exposure to air they grow significantly and change morphology. At the moment, we cannot be sure that the carbon shell is present during the FT process. This may be different for different carbide phases, but the *ex situ* (even after passivation) presence of a graphite-like shell does not prove it is present during FT.[1,2]

1 S. Janbroers, P. A. Crozier, H. W. Zandbergen and P. J. Kooyman, *Appl. Catal., B*, 2011, **102**, 521–527.
2 S. Janbroers, J. N. Louwen, H. W. Zandbergen and P. J. Kooyman, *J. Catal.*, 2009, **268**, 235–242.

Freek Kapteijn replied: Thanks for your comment. After pyrolysis we usually passivate the samples with diluted oxygen, so that could be similar to your oxidation treatment. Note however that we also did *in situ* Mössbauer spectroscopy, where the different iron phases could be distinguished during the different stages of treatment, although it does not give information about the carbon shell.

Hans Schulz commented: With these newly prepared iron@carbon Fischer–Tropsch catalysts, K-promoted and K-unpromoted – seeing the essential role of potassium in iron catalyst performance – could you agree with the view of potassium enhancing CO-dissociation to produce "carbon-on-site"? Carbon-on-site will have 3 options: (1) to react with iron to form the FT-active carbide, (2) to react with hydrogen to form a C1-species for FT- chain growth and (3) to react with carbon to form a deactivating carbon phase. The degree of potassium promotion must be adjusted to hydrogen availability accordingly.

The right hand side of Fig. 1 shows the kinetic scheme of potassium controlling FT performance on iron, whereas the left hand side of Fig. 1 presents the activity

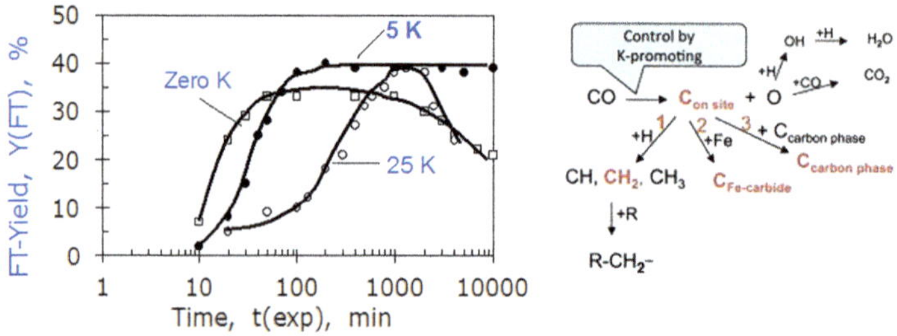

Fig. 1 Image adapted from H. Schulz, *Catal. Today*, 2014, **228**, 113–122.

time dependence in FT-synthesis with 3 differently K-promoted catalysts, indicating unfavorable too low (zero K) and too high (25 K) potassium promotion – with only the right degree of promotion (5 K) leading to high stable activity.

Evjeniy Redekop asked: Encapsulating carbon layers are hypothesized to prevent the sintering of iron oxide particles. However, these layers may also introduce transport limitations to and from the catalytic sites. This could be inferred, for example, from the inability of the K precursor to penetrate towards iron oxide on some of the samples, thus hindering K-promotion. Could these encapsulating carbon layers also affect the transport of reactants and/or reagents? Could they potentially induce a confinement effect on FT reactions?

Freek Kapteijn responded: Theoretically, yes; these shells could introduce transportation hindrance of reactants and/or products. However, the FTY we determine are of the same level or higher than those reported for *e.g.* Fe catalysts on carbon fibres, systems that are highly accessible, so mass transport limitations don't seem to be present.

James Hayward said: Your materials seem to contain a wide range of possible promoters and poisons. Do you know where in your materials they are located? Are they, for example, mostly in or on the iron, or in the carbon shell?

Freek Kapteijn replied: We show some EDX analyses of samples in the paper. How dispersed they are depends on the impurities. Some are, others seem more clustered.

Roy Peter Forbes enquired: In your paper, you stated that the size of the Fe nanoparticles obtained from the pyrolysis of the Fe-MOFs is not a function of Fe loading. Instead it was stated that the Fe nanoparticle size appears to be related to the structure and porosity of the original MOF precursor and that a trend towards smaller Fe nanoparticles was observed with increasing surface area and pore volume. With this in mind, could you please comment on varying the heating rate during the pyrolysis of these Fe-MOFs, and the effect that this may have on the resulting Fe nanoparticle size?

Freek Kapteijn responded: The heating rate also has an effect on the final particle size, as does the dwell time at the final temperature. For example, with copper MOFs, this is most clearly observed. Here the final temperature seems to have a stronger influence than the heating rate.

Linda Jewell commented: You mentioned that the carbon layer on the catalyst is porous. Have you tried to control the porosity of the carbon or vary it systematically?

Freek Kapteijn replied: No, we haven't done this. The activity is already high, the FTY is of similar or higher levels as for very accessible Fe-catalysts (*e.g.* on carbon fibres), so there was no reason to investigate this. We did however do this for the Co systems that had a low FTS activity, but we were unsuccessful in restoring high activity in that case.

Xiaohui Sun opened discussion of the paper by Michael Claeys: The new synthesis method enables the preparation of highly dispersed and optimized cobalt nanoparticles on a SiO_2 support. However, the cobalt loading is only 2.5 wt %. My question is what will the cobalt dispersion will be like if you increase the cobalt loading above 5 wt%?

Michael Claeys responded: The cobalt loading of our catalysts was 0.5 wt% and this very low loading was chosen on purpose to minimize the sintering propensity of our model catalysts. Sintering could indeed be completely avoided in our study, which was crucial in order to allow for straightforward conclusions regarding sintering. It should be noted that sintering and oxidation often go hand in hand. I would also like to highlight that these measurements were only possible due to the high sensitivity of the magnetic setup for the detection of metallic cobalt. Conventional techniques, such as laboratory-based XRD, would not have been suitable for these studies.

Although we do not have data for this exact system, it is reasonable to assume that a loading above 5 wt% would have led to severe sintering, even during reduction.

Graham Hutchings addressed Michael Claeys and Moritz Wolf: In the "surfactant-free" preparation method, the benzyl alcohol can be acting as a stabiliser for the small nanoparticles; what happens if the benzyl alcohol is replaced by ethanol: do you still make small nanoparticles?

Moritz Wolf replied: Yes, benzyl alcohol does act as a stabiliser and also remains attached to the surface of the synthesised nanoparticles. From thermogravimetric experiments we know that, depending on the size, 5–15 wt% of the washed nanoparticles seem to be organic, *i.e.* they are capped by benzyl alcohol and/or benzyl alcohol originated species. However, this layer is very thin, especially if compared to bulky, classical surfactants such as long chained oleic acid. It further allows an easy dispersion of the nanoparticles in ethanol, which is crucial for the present study and prevents major agglomeration while anchoring the nanoparticles on the support.

We have previously replaced benzyl alcohol by ethanol according to Dong *et al.*[1] Small sizes of Co_3O_4 with narrow size distributions are possible with ethanol. They even replaced benzyl alcohol by ethanol/water mixtures or pure water to increase the crystallite size. Of course, the synthesis needs to be done in autoclaves to overcome the low boiling points of the solvents. The size of the autoclave limits the amount of nanoparticles prepared *via* this synthesis and the absence of surfactants may limit the dispersibility of the nanoparticles. However, this modified synthesis allows the preparation of dispersible crystallites from 3.5 nm to up to 70 nm.

1 Y. Dong, K. He, L. Yin and A. Zhang, *Nanotechnology*, 2007, **18**, 435602–435607.

Chelsea Tucker asked: In your paper, it was mentioned that the catalyst was re-reduced after reactor testing in order to qualitatively assess the level of oxidation to cobalt oxide *versus* the amount of cobalt silicate produced. What method has been or will be utilised to quantitatively measure the extent of cobalt conversion to cobalt oxide and cobalt silicate respectively?

Michael Claeys answered: As shown in Fig. 13 of our manuscript (DOI: 10.1039/C6FD00200E), there seem to be two oxidation events, a fast one (believed to be due to cobalt silicate formation) and a slow one (believed to be due to the formation of CoO). We have attempted to verify this *via* post-run TPR experiments (Fig. 14, DOI: 10.1039/C6FD00200E), and these indeed indicate that some cobalt is re-reducible at the relatively mild reduction conditions that were applied (*i.e.* the same conditions that were used for the reduction of the catalyst precursors), and this re-reduced amount of cobalt approximately equals that which was "lost" *via* slow loss in magnetization during our model experiments (see Table 6, DOI: 10.1039/C6FD00200E). The amount of cobalt which we didn't recover here would then be indicative of hard-to-reduce cobalt silicate that has formed. This is obviously an indirect and somewhat uncertain way of distinguishing and quantifying the two cobalt phases. Ultimately, it is our intention to also conduct XAFS measurements in the near future.

Emiel Hensen said: Your results clearly show that smaller Co particles oxidize faster than larger ones. You have referred to the thermodynamic stability as a function of size. I would argue that perhaps there are also kinetic differences, with smaller particles having more defect sites that are more prone to oxidation. The graphical depiction of this process in your slides suggests that oxidation starts at the perimeter with the silica support. Can you indicate why on a non-reducible support such effects would be relevant?

Michael Claeys replied: I fully agree that defect sites on smaller crystallites will contribute to their oxidation. In a way, this is also implied in the thermodynamic calculations that we refer to.[1] In our slides, we merely indicate that silicate formation occurs at the perimeter with the silica support, while the oxidation to CoO is not limited to this area. Regarding the silicate formation, our data strongly suggest that this directly proceeds from the metallic cobalt phase and its interaction with the silica support, under the influence of the high water to hydrogen ratios that were applied. Note that this is also in agreement with thermodynamics (see Fig. 1, DOI: 10.1039/C6FD00200E), where the silicate formation can occur at ratios of around 20. In our study we even went beyond these conditions. We are currently busy doing similar studies with carbon supported model catalysts, so as to be able rule out the effects of formation of metal support species.

1 E. van Steen, M. Claeys, M. E. Dry, J. van de Loosdrecht, E. L. Viljoen and J. L. Visagie, *J. Phys. Chem. B*, 2005, **109**, 3575–3577.

Neil Coville asked: In the paper it is mentioned that hydroxyl groups affect magnetisation; could you elaborate on this? Are you suggesting that Co is oxidised by the OH groups *i.e.*, that a Co···OH interaction is being detected? What about deuteration? Will this effect be detected?

Michael Claeys answered: We have indeed shown *via* quantum chemical calculations that adsorbed species can affect the magnetization of metallic crystallites.[1] While with OH and O species an increase of magnetization is predicted, carbon-bearing species are expected to decrease magnetization. Indeed, upon the first introduction of water we experienced increases of magnetization (see Fig. 10 and 11, DOI: 10.1039/C6FD00200E), which clearly points towards OH and O species being present on the surface. This effect should obviously be more pronounced for smaller crystallites and we can confirm this with our experimental data, which show a stronger relative initial increase of magnetization for the smaller particles that were tested. I can not comment on deuteration. For this, additional calculations and/or measurements would be required. Hydrogen adsorbed on cobalt surfaces has a very weak effect on magnetization.

1 M. Claeys, M. E. Dry, E. van Steen, P. J. van Berge, S. Booyens, R. Crous, P. van Helden, J. Labuschagne, D. J. Moodley and A. M. Saib, *ACS Catal.*, 2015, **5**, 841–852.

James Hayward enquired: When your catalyst with the largest starting particles, CAT C, was reduced, your magnetic measurements implied that 24% of the particles were over the superparamagnetic point, which you give as 15–20 nm. Given that their initial size was 7 nm, this is a very large degree of sintering for particles which were shown to be very stable. Have you done any post-reduction particle size analysis, such as *via* TEM or XRD, to show evidence of these very large particles?

Moritz Wolf responded: Aside from the *in situ* magnetic measurements, we have not done post-reduction analysis. However, analysis of the magnetic measurements indicates a constant crystallite size throughout the experiments in all samples (see constant values for γ and d_V in Table S1 of the Supplementary Information for DOI: 10.1039/C6FD00200E). Post-run analysis *via* X-ray diffraction requires a synchrotron-based radiation due to the low cobalt loading of 0.5 wt %, but we have done crystallite size analysis *via* transmission electron microscopy (TEM) of the spent and passivated samples (Fig. S12). As the crystallite size did not change significantly during the experiments, the sizes in the spent samples can be assumed to be similar to the ones upon reduction. The largest identified crystallite size in CAT C, the sample comprising the largest cobalt crystallites, was 12 nm and over 80 wt% of the measured nanoparticles are smaller than 9 nm. The magnetic measurement after re-reduction of this sample results in approx. 80 wt% of cobalt larger than the critical diameter. From our previous experience, fcc-cobalt nanoparticles just below 10 nm show superparamagnetic behaviour, while larger particles are non-superparamagnetic and lead to hysteresis behaviour when measuring the sample magnetization as a function of the external field strength. This estimation fits quite well with the post-run TEM analysis, but is indeed in discrepancy with published critical diameters for superparamagnetism of 15–20 nm. This value is also temperature dependent. We are currently busy establishing the exact threshold value using experimental and theoretical approaches/techniques and initial results indicate that this value is indeed lower than previously published ones.

Hans Niemantsverdriet said: Let's consider a cobalt catalyst without support, or perhaps a carbon support. According to your work, the reason that the small particles oxidise is that the O atoms from CO dissociation are difficult to remove, and not because they are directly sensitive to water. That is a nice and understandable result. Would you agree that this is most likely the reason why smaller cobalt particles exhibit less FTS activity?

Michael Claeys responded: Our results indeed strongly indicate that oxidation is more pronounced for small crystallites and that this oxidation is due to the O atoms which originate from the CO dissociation. I would therefore agree that this mechanism can account for the lower activity of small crystallites in FT synthesis. However, as also shown in our paper (DOI: 10.1039/C6FD00200E), it should be noted that the removal of this oxygen strongly depends on the concentration of water or synthesis gas conversion levels, respectively. In previous work with differently sized cobalt crystallites,[1] which was conducted under "dry" conditions (very low conversion, and no water feeding) we also obtained a crystallite size dependency with respect to activity. Under these conditions, we did not observe oxidation, as also monitored *in situ*,[2] but still a strong structure sensitivity was found with smaller crystallites displaying lower activity. This may be due to geometrical effects, as also discussed earlier. There may therefore be different causes and contributions for structure sensitivity that are at play, depending on reaction conditions.

1 N. Fischer, E. van Steen and M. Claeys, *J. Catal.*, 2013, **299**, 67–80.
2 N. Fischer, B. Clapham, T. Feltes, E. van Steen and M. Claeys, *Angew. Chem. Int. Ed.*, 2014, **53**, 1342–1345.

Letisha Deeplal communicated: A suggestion that perhaps static chemisorption can be used to determine reducability, metal dispersion and compare crystallite sizes. In addition, CO chemisorption experiments at different temperatures could provide insight into the reaction kinetics and heats of adsorption for the CO conversion studies.

Michael Claeys communicated in reply: CO chemisorption is certainly an excellent method to determine the available cobalt surface area and average crystallite sizes. It should be noted that the magnetic technique applied here gives us access to the degree of reduction and, at least for small (superparamagnetic) particles, information on crystallite size and even size distributions from which metal surface areas can be estimated. Of course, our technique is a bulk technique and we do not have "direct" access to the metal surface. Heats of adsorption could indeed be measured using chemisorption or TPD measurements but it is often difficult to transfer these findings to the working FT system where the catalyst surface is covered with various species that ultimately may affect CO adsorption.

Kershen Naiker communicated: I wanted to find out if you considered using chemisorption to look at the effect of crystallite size? Would you prefer smaller or larger crystals for your catalysis?

Michael Claeys communicated in response: See my answer above regarding chemisorption measurements. As for the preferred crystallite size, our studies and work done by others clearly show that small crystallites display inferior activity and selectivity in the FT synthesis. The work presented here gives some insight as to why this is the case, *i.e.* preferred oxidation of small crystallites. It is therefore mandatory to use a crystallite size which is large enough to not oxidise at the targeted FT conditions of a process, yet it should be as small as possible so as to provide the largest possible cobalt utilization or per gram cobalt time yield, respectively.

Obviously, there are also other deactivation mechanisms that need to be considered, some of which are also crystallite size dependent, for example sintering. But this was not the topic of this lecture as sintering could be excluded *via* the choice of the model system used here.

Nico Fischer opened a general discussion of Ding Ma's, Freek Kapteijn's and Michael Claeys' papers: The preparation of catalyst *via* the decomposition of a well designed precursor could be a very interesting method for a number of catalysts. Could different MOF structures and compositions be synthesized in order to incorporate potential promoters in direct contact with the active material? Also, you mentioned that Co-based MOFs did not yield good catalysts, have you got any conclusive explanation regarding this observation? Does the C play a role?

Freek Kapteijn responded: In the case of Co we think the carbon layers are hardly porous, so the activity of the catalysts is very low in this case. The high methane selectivity is also striking, up to 50%.

Simon Kondrat remarked: The exact role of impurities (alkali metals, transition metals and halides) in the role of catalysis is important to control and understand. Often we cannot characterise these impurities, due to low concentrations, or find direct evidence of their effect on the catalyst system. Can the levels of these impurities be controlled in MOF formation? Could this method be a way of controlling concentrations and thereby understanding their effect on catalysis?

Freek Kapteijn replied: We have used commercial or semi-commercial samples where we had no control over the impurity level. But it is indeed possible to use self-made MOFs and add the promoters to the MOF or after pyrolysis. We showed this in our previous work.[1]

1 T. A. Wezendonk, V. P. Santos, M. A. Nasalevich, Q. S. E. Warringa, A. I. Dugulan, A. Chojecki, A. C. J. Koeken, M. Ruitenbeek, G. Meima, H.-U. Islam, G. Sankar, M. Makkee, F. Kapteijn and J. Gascon, *ACS Catal.*, 2016, **6**, 3236–3247.

Michael Bowker said: If I can make a general point on the activity dependence on particle size – particle size dependence of activity is not the easiest to measure, due to possible changes in active site availability on the surface. Promoters can also be poisons of active sites when present at too high a level[1] and as sintering occurs, so the fractional coverage by promoters may well also increase. In your case, as size increases, you see a decrease in activity. Do you feel that that may be due to increasing over-promotion as the particle size increases, and so that the intrinsic activity might be fairly constant?

1 M. Bowker, *Appl. Catal.*, 1988, **45**, 115–130.

Freek Kapteijn answered: If we make the MOF ourselves and assuming this is more pure than the commercial samples, the activity is still inversely proportional to the particle size, so the activity seems to correlate with the specific particle surface. I agree with you that an optimal promoter content should then be adjusted to the particle size. This has not been investigated systematically.

Paul Collier remarked: It is possible to introduce defects into MOFs. Would you expect this to have any chemical or topographical effect on the finished Fe-C catalyst? In VPO catalysts there is a close relationship between the location of the atoms in the precursor phase and those in the final active catalyst even after different stages of activation. Would you anticipate this to be the case with MOF precursors in your case?

Freek Kapteijn answered: I expect that the pyrolysis treatment is relatively so harsh that defects in the MOF structure will not be reflected in the pyrolysis product.

Richard Catlow commented: Neutron spectroscopy has, I think, an important role to play in our field. As shown in David Lennon's paper (DOI: 10.1039/C6FD00195E), it is able to give unique data on systems which are inaccessible to optical spectroscopy. In particular, its use could make a significant contribution to some of the problems addressed in the meeting concerning FT catalysis.

Nico Fischer responded: I agree that neutron spectroscopy should play a much larger role in our characterization mix and might be of special interest for the FT community. The major obstacle to overcome will certainly be the low temporal resolution and low measuring temperatures making "operando" type measurements currently impossible. It therefore requires a very good understanding of the quenching process so that the obtained observations can be extrapolated to the working catalyst environment.

Paul Collier said: I agree completely and hope too that neutron techniques find wider application in catalysis. In a recent review[1] Stewart Parker and myself outlined the strengths of neutron techniques and how they can be applied to study problems in catalysis.

1 S. F. Parker and P. Collier, *Johnson Matthey Technol. Rev.*, 2016, **60**, 132–144.

Eric van Steen asked: The initial activity for the MOF-derived catalysts change dramatically during the first hours on line. Can this be attributed to phase changes within the catalysts, and if so why is the rate of the phase change dependent on the catalyst precursor?

Tim Wezendonk answered: In the fixed-temperature pyrolysis process of obtaining the catalysts, the MOFs undergo a similar degree of carbonisation through decarboxylation. The use of different MOFs, however, generates catalysts with varying amounts of iron, carbon and oxygen, both on the surface and in the bulk. In addition, the iron nanoparticle size of the catalysts varies significantly, from 3.6 nm up to 6.8 nm on average. The varying activation behaviour of K-promoted samples is best described by the latter feature, as the catalysts with smallest iron nanoparticles achieve steady-state conversion under 20 h, whereas the K-Fe@C-MIL68 sample with 6.8 nm iron crystallites suggests activation over 80 h time-on-stream. Spent catalyst analysis and *in situ* Mössbauer characterisation reveal the formation of surface Hägg carbide species growing atop of the oxide core, allowing for an explanation of the diffusion-reaction-like activation curves. In addition to the nanoparticle size, we speculate that the above-mentioned change in catalyst formulation, in terms of ratios of iron oxide to carbons, can influence activation times as well: in a follow-up experiment, we confirmed that incomplete reduction of the small iron nanoparticles leads to much longer carburisation times over 100 h. This suggests that the resulting interaction of iron with the carbon framework is determined by the initial structure of the MOF.

Freek Kapteijn added: Depending on the pyrolysis conditions and the pretreatment before using the catalyst in the FTS experiment, the state of the iron differs quite a lot. *In situ* Mössbauer spectroscopic analysis has clearly shown this. Before FTS the iron is not yet in its final carbidic phase, and during FTS operation this process gradually occurs, the rate clearly depending on the particle size and oxidation state. If hydrogen pretreatment has occurred the conversion is faster than without the reduction step. In the specific case of this study, Mössbauer was not performed on the samples so one can only speculate about the different rates of conversion.

Graham Hutchings enquired: You find that the MOF structure influences the catalysis observed. If you use a mixture of MOFs do you observe intermediate behaviour, *i.e.* is it a linear behaviour, or does one structure dominate?

Freek Kapteijn replied: I have no idea. This has not been tried.

David Lennon commented: A follow-on from Richard Catlow's comments on the ability of inelastic neutron scattering (INS) to characterise carbonaceous materials – yes, I agree, INS is finding an increasing role in being able to characterise carbonaceous materials, but only when there is a reasonable degree of hydrogen associated with the carbonaceous matrix. In this way, the INS spectrum is dominated by hydrogenous vibrational modes that can reflect the local environment; the resulting vibrational spectrum can be distinctive and diagnostic of the material under examination. Stewart Parker and Peter Albers have pioneered the application of INS spectroscopy to characterise a variety of carbonaceous materials; ref. 1 (below) provides an example of their recent work.

1 P. W. Albers, W. Weber, K. Möbus, S. D. Wieland and S. F. Parker, *Carbon*, 2016, **109**, 239–245.

Katherine Holt opened discussion of Thulani Nyathi's paper: Once your catalyst is reduced, can it be regenerated by oxidation?

Thulani Nyathi responded: We heated the reactor up to 350 °C under the CO-PrOx gas and then cooled it down to 50 °C. Now when we cooled down, we saw a partial re-oxidation of the metallic Co (the degree of the re-oxidation also varied for the different crystallite sizes) between 200 and 250 °C. We saw mostly CoO during *in situ* PXRD. Co_3O_4 may have been formed too but this may either be limited to the surface or could just be a very small amount that is not detectable by PXRD. However when we cool down, CO oxidation activity is regained (but not fully) before decreasing below 200 °C. This strongly suggests that we probably did form some surface Co^{3+} species as these are believed to be responsible for the CO oxidation.

Michael Bowker questioned: Do you understand why the catalyst doesn't fully reduce? The XRD shows Cu metal and the catalytic behaviour looks metallic (methanation), yet it is not apparently 100% reduced. Could it be a core–shell system with an oxide core and metal shell, or are some particles of different sizes just too difficult to reduce?

Thulani Nyathi replied: The reduction of Co_3O_4 in CO-PrOx is not the same as what you would expect for its reduction in pure H_2.[1] During CO-PrOx you have CO, O_2, H_2, CO_2 (product of CO oxidation), H_2O (product of H_2 oxidation and CO hydrogenation) and CH_4 (product of CO hydrogenation). We haven't exactly studied the role that each gas plays in the reduction of Co_3O_4 (whether it is enhancing or preventing reduction). However, we do suspect that the presence of O_2, H_2O and to some extent CO_2, may be the reason why the catalysts do not fully reduce. The exact nature of the nanoparticles above 250 °C (upon heating) is still unknown. We proposed a core–shell system based on the changes in the product selectivity *i.e.*, CH_4 being formed more than CO_2 even when the reduction of the catalysts was not 100%. Upon decreasing the reaction temperature, we saw a partial re-oxidation of the metallic Co as well as a decrease in CH_4 and increase in CO_2. This could mean we were forming an oxide outer-shell, with a metallic inner-shell and an oxide core. We not entirely sure about the stability of these proposed structures under the conditions we used. So an alternative could be that nanoparticles with a certain size are undergoing the reduction to metallic Co and of those that reduced, a certain fraction is re-oxidised upon cooling. Even this alternative idea may need to be further investigated experimentally and/or computationally.

1 L. Lukashuk, K. Föttinger, E. Kolar, C. Rameshan, D. Teschner, M. Hävecker, A. Knop-Gericke, N. Yigit, H. Li and E. McDermott, *J. Catal.*, 2016, **344**, 1–15.

Neil Coville asked: Is it possible you are not seeing cobalt aluminates because they are amorphous?

Thulani Nyathi responded: When preparing the supported catalysts, we firstly prepared the Co_3O_4 nanoparticles using the reverse micelle technique and calcined them at 200 °C. These were then mixed with the Al_2O_3 support in distilled water and dried afterwards. This procedure was followed in order to minimise the interaction between the support and the nanoparticles and as a result, also minimise the effect the support may have on the reduction and catalytic performance of Co_3O_4. We therefore do not expect any strong interactions in the form of cobalt aluminates in this Co_3O_4/Al_2O_3, at least not in the temperature range (50–350 °C) we used for the catalyst testing.

Unni Olsbye enquired: Data in your paper (DOI: 10.1039/C6FD00217J) show that cobalt oxide is the desired active phase. If you add water or CO_2 with your feed, could it balance the concentration of this phase? Would it affect the activity of the system in other ways?

Thulani Nyathi answered: True, our data (as well as data in the literature) strongly suggests that Co_3O_4 is the active phase for CO oxidation. However, one should also remember that there is H_2 present in very high amounts during CO-PrOx which eventually out-competes CO for the reactant O_2 at elevated temperatures. So say the catalyst never reduced *i.e.*, Co_3O_4 stayed intact, the CO conversion could still decrease at elevated temperatures because of a loss in selectivity (*i.e.*, more O_2 being used to form H_2O). In fact, this loss in selectivity is observed for noble metal catalysts even though they obviously do not undergo any phase transformation since they are normally used in the reduced state.

The addition of CO_2 and H_2O is going to be part of a study that has been recently started in our research group. Although there isn't much literature on how these two gases would affect Co_3O_4 reduction during CO-PrOx, we do believe that they will have some inhibiting effect (especially H_2O). On the other hand, results in the literature show that the presence of H_2O and CO_2 decrease the catalytic activity because they adsorb and block access to the active sites. This is believed to be the case for both metal- and metal oxide-based catalysts. So at the moment, it is still not clear if these gases can somehow be used to strike a balance between catalyst phase stability and activity but we aim to investigate this with the use of multiple *in situ* techniques.

Holger Friedrich said: You clearly show that the phase transformations, and with that the really undesirable catalysis (like methanation), take place well after oxygen starvation is reached. Have you considered adding a bit more oxygen to the system? This may let you run the reaction at a lower temperature to get complete CO oxidation and may also reduce the degree of hydrogen oxidation to water, which is a higher temperature reaction.

Thulani Nyathi responded: Yes, we have considered adding more O_2 as this would indeed allow us to convert more CO to CO_2 at lower temperatures. However, this forms part of a newly began study. We are also interested in how varying the concentration of O_2 would affect the onset temperature for reduction (and degree of reduction) as we have seen that as soon as 100% O_2 conversion is reached, the catalyst starts to reduce. The addition of different concentrations of H_2O and CO_2 will also be done for similar reasons.

Holger Friedrich followed up: You postulated core–shell structured crystallites that possibly were unstable in the reaction atmosphere when cooling. Could you not maybe stop the oxygen and CO flows at 350 °C and then cool the catalyst down under nitrogen and perform an *ex situ* characterisation? Maybe unpack the reactor tube in a glove box and even prepare the samples for analysis under inert gas?

Thulani Nyathi replied: Yes, one could have done that or perhaps do this in the future. This could certainly help in confirming whether we do form particles with a core–shell structure at the different temperatures where reduction and re-oxidation take place. We are also in the process of performing quasi-*in situ* XPS where we pre-treat the sample at the temperature where we know that reduction or re-oxidation has taken place so that we can see how the surface looks in comparison to the bulk structure (as determined by PXRD and by the magnetometer).

Thabiso Perfect Oscar Mkhwanazi communicated: Have you obtained any results that suggest the formation of an interface or layer of Co_3O_4/Al_2O_3 strong metal–support interactions with respect to the crystallite size of Co_3O_4?

Thulani Nyathi communicated in response: No I haven't. H_2-TPR did not suggest the presence of very strong metal support interactions. We did see from the TPR profiles that the smaller crystallites were slightly harder to reduce but we believe that this is because they are mostly located inside the pores of the Al_2O_3 support (d_{pore} = 11.5 nm). So due to their location, they interact slightly stronger with the support and therefore fully reduce a little later. Similar catalysts have been prepared in our research group in the past (but used in the Fischer-Tropsch synthesis) and there hasn't been any indication of the presence of strong metal–support interactions. This usually becomes the case if the Al_2O_3 support is contacted with the precursor $Co(NO_3)_2 \cdot 6H_2O$ in solution and thereafter calcined. In contrast, we physically mixed Al_2O_3 with suspended Co_3O_4 nanoparticles.

Katherine Holt opened discussion of the paper by Simon Kondrat: Incorporation of more Zn into the malachite structure results in higher catalytic activity. Is this because Zn is the active catalyst or is it the effect it has on the neighbouring copper?

Simon Kondrat replied: It is generally considered that the active component of a methanol synthesis catalyst is metallic copper, with a strong correlation between Cu surface area and activity.[1] Indeed, polycrystalline Cu and Cu(110) have been shown to be active.[2] However, there has been ongoing debate on the effect of Zn/ZnO on the activity of Cu/ZnO catalysts.[3] A number of recent papers have shown that strong Cu–ZnO interaction or even Cu decorated with reduced Zn improve activity significantly.[4–6] Therefore, the incorporation of zinc into malachite or other hydroxycarbonate phases, such as georgeite, is thought to improve Cu–Zn interaction and increase turnover rates.

1 G. C. Chinchen, P. J. Denny, J. R. Jennings, M. S. Spencer and K. C. Waugh, *Appl. Catal.*, 1988, **36**, 1–65.

2 J. Yoshihara and C. T. Campbell, *J. Catal.*, 1996, **161**, 776–782.
3 M. S. Spencer, *Top. Catal.*, 1999, **8**, 259–266.
4 M. Behrens, F. Studt, I. Kasatkin, S. Kühl, M. Hävecker, F. Abild-Pedersen, S. Zander, F. Girgsdies, P. Kurr, B.-L. Kniep, M. Tovar, R. W. Fischer, J. K. Nørskov and R. Schlögl, *Science*, 2012, **336**, 893–897.
5 S. Kuld, M. Thorhauge, H. Falsig, C. F. Elkjær, S. Helveg, I. Chorkendorff and J. Sehested, *Science*, 2016, **352**, 969–974.
6 S. A. Kondrat, P. J. Smith, P. P. Wells, P. A. Chater, J. H. Carter, D. J. Morgan, E. M. Fiordaliso, J. B. Wagner, T. E. Davies, L. Lu, J. K. Bartley, S. H. Taylor, M. S. Spencer, C. J. Kiely, G. J. Kelly, C. W. Park, M. J. Rosseinsky and G. J. Hutchings, *Nature*, 2016, **531**, 83–87.

Neil Coville asked: I was interested to see your comments in the paper on the exothermic nature of the nitrate decomposition and its impact on the growth of catalyst crystallites. This it seems can be established by TGA experiments. Is there a way of controlling this decomposition reaction? Could Na nitrite work as an alternative?

Simon Kondrat answered: The control of metal nitrate decomposition and its effect of metal dispersion has been studied and several different strategies have been suggested. Firstly, the most common method is to wash out as much of the nitrate impurity as possible. Alternatively, controlling the heating ramp rate or the introduction of isotherms to moderate nitrate decomposition rate has been proposed.[1] De Jong and co-workers have shown that control of space velocity, with high flow rates, prevents metal agglomeration.[2] Interestingly, the addition of small amounts of NO to the heat treatment gas has, for other catalyst systems, been shown to scavenge radicals produced from nitrate decomposition and therefore improve metal dispersion.[3]

With regard to the use of metal nitride salts, this is an interesting idea that I have not encountered before and could be an interesting avenue to study.

1 J. van de Loosdrecht, S. Barradas, E. A. Caricato, N. G. Ngwenya, P. S. Nkwanyana, M. A. S. Rawat, B. H. Sigwebela, P. J. van Berge and J. L. Visagie, *Top. Catal.*, 2003, **26**, 121–127.
2 G. Prieto, K. P. de Jong and P. E. de Jongh, *Catal. Today*, 2013, **215**, 142–151.
3 M. Wolters, H. Daly, A. Goguet, F. C. Meunier, C. Hardacre, J. H. Bitter, P. E. de Jongh and K. P. de Jong, *J. Phys. Chem. C*, 2010, **114**, 7839–7845.

Holger Friedrich enquired: In the paper you state that it is clear that the presence of sodium zinc carbonate will result in the segregation of Zn and Cu in the precursor material. Can you explain why or how this segregation happens?

Simon Kondrat responded: The mechanism for sodium zinc carbonate formation is indeed an intriguing question. The phase itself readily forms in the presence of aqueous solutions of sodium bicarbonate and zinc salts.[1] In an *in situ* study of the conventional co-precipitation process for the formation of the zincian malachite catalyst precursor, sodium zinc carbonate is seen in the ageing process (at pH > 7) and is attributed as a reservoir for Zn during the re-crystallisation process.[2]

In regard to the current study we must consider if phase segregate sodium zinc carbonate is present from the point of precipitation. In addition, how does the phase evolve when the high sodium contain sample crystallites. Given that this would be a solid state reaction it is not clear how this would occur. We could perform an *in situ* XRD study of the initial precipitate to observe the ageing process.

1 T. E. Gier, X. H. Bu, S. L. Wang and G. D. Stucky, *J. Am. Chem. Soc.*, 1996, **118**, 3039–3040.
2 S. Zander, B. Seidlhofer and M. Behrens, *Dalton Trans.*, 2012, **41**, 13413.

James Hayward said: You mentioned that when sodium carbonate is used to precipitate the hydroxycarbonates, there is an effect on particle size from the precipitant if not sufficiently washed, and that this is also seen when ammonium carbonate is used. Your materials still contain carbonate, but not as a precipitant. Do you see this effect with your materials, and can you say whether the effect is related to the carbonate, the sodium/ammonium, or the basicity of the precipitation?

Simon Kondrat replied: In simplistic terms, the decomposition of carbonates is generally beneficial for the final catalyst surface area and retention of metal dispersion, due to its endothermic decomposition. In the case of co-precipitation we must consider the precipitating agent and also the metal salt counter ion. Conventionally the metal salts are metal nitrates. During precipitation the metal precipitate is often a carbonate, leaving the sodium cation and nitrate anion to form a salt. It is this nitrate salt that is detrimental during precursor heat treatment.

Michael Bowker asked: You have shown that it is important to get Zn into the malachite phase before calcination. What is the mechanism of this which makes a good catalyst? Is it a way of isolating the Cu nucleation centres more, and hence reducing the aggregation of Cu during the calcination process?

Simon Kondrat answered: The mixing of copper and zinc in the catalyst precursor phase is important for several reasons. The first, as you suggest, is the segregation of CuO nucleation sites and reduction of CuO aggregation during heat treatment. Secondly, good mixing of copper and zinc results in increased interaction between Cu and ZnO in the final catalyst, which is reported to facilitate partial ZnO reduction and the formation of Zn decorated Cu particles.

The specific case of zincian malachite and its evolution into a good Cu/ZnO catalysts is covered in previous work by Behrens.[1] In our paper (DOI: 10.1039/C6FD00202A), and previous work,[2] we focus on the preparation of the amorphous hydroxycarbonate phase zincian georgeite, which exists as the precursor phase to zincian malachite in the conventional catalyst preparation method. We show that, if prepared in a controlled manner, zincian georgeite can be a precursor to a highly active catalyst. The reason for this is the same as stated above; good mixing of Cu and Zn produces small Cu particles with a high degree of Cu–ZnO interaction. The reason that unaged (*i.e.* georgeite type precursors) have previously been considered unsuitable is that the preparation has not been well controlled and consequently segregated copper and zinc species have formed.

1 M. Behrens, *J. Catal.*, 2009, **267**, 24–29.
2 S. A. Kondrat, P. J. Smith, P. P. Wells, P. A. Chater, J. H. Carter, D. J. Morgan, E. M. Fiordaliso, J. B. Wagner, T. E. Davies, L. Lu, J. K. Bartley, S. H. Taylor, M. S. Spencer, C. J. Kiely, G. J. Kelly, C. W. Park, M. J. Rosseinsky and G. J. Hutchings, *Nature*, 2016, **531**, 83–87.

Nico Fischer said: In Fig. 7e of your paper (DOI: 10.1039/C6FD00202A), you report on the change in Cu and ZnO crystallite size with TOS. Under "dry" conditions ZnO remains rather stable but the Cu crystallite size increases significantly within the first minutes or so. Is this process and the extent of it understood? Does it correlate with the incorporation of Zn species in the Cu surface?

Simon Kondrat responded: The very slow crystallite growth under "dry" methanol synthesis has previously been observed and fits with our observations. However, the sudden increase in Cu crystallite size in the first 15 min has not been previously reported. It is possible that this effect is associated with our zincian georgeite derived catalysts and requires comparison with more conventionally prepared catalysts. The small crystallite size (7 nm) with high Cu loadings of 66 wt% in the zincian georgeite derived catalysts could explain this sudden growth. A thorough study of Cu surrounded by various Zn environments and even without ZnO would be required to understand this fully.

Mzamo Shozi questioned: How do you think the activity would change if you used Ga^{3+} instead of Na^{+}?

Simon Kondrat answered: While Na^{+} is considered a poison, there are reports of Ga^{3+} promoting catalytic activity.[1] The promotion is attributed to incorporation of Ga into the ZnO lattice, and it acting as a dopant to promote the n-type semiconductivity of ZnO by enhancing defect site concentration. It would be interesting to study the stability of Ga or Al promoted ZnO in similar *in situ* XRD studies reported in the paper.

1 M. Behrens, S. Zander, P. Kurr, N. Jacobsen, J. Senker, G. Koch, T. Ressler, R. W. Fischer and R. Schlögl, *J. Am. Chem. Soc.*, 2013, **135**, 6061–6068.

Richard Catlow commented: ZnO is very probably playing a substantial role in the chemistry. It is an n-type semiconductor with trapped electron states on the surface which can activate CO_2.[1]

1 G. Dutta, A. A. Sokol, C. R. A. Catlow, T. W. Keal and P. Sherwood, *ChemPhysChem*, 2012, **13**, 3453–3456.

Simon Kondrat added: I entirely agree with you that the activation of CO_2 by ZnO is likely to play an important role in the chemistry. It is also interesting to consider if other metal oxides can activate CO_2 without forming strongly bound carbonate species. The work of Rodriguez and co-workers on the inverse catalyst Cu-CeO_x shows that there might be potential in reduced Ce_2O_3.[1] Producing practical experiments based on these studies would require new innovative design techniques.

1 J. A. Rodriguez, P. Liu, D. J. Stacchiola, S. D. Senanayake, M. G. White and J. G. Chen, *ACS Catal.*, 2015, 5, 6696–6706.

Letisha Deeplal communicated: It has been debated in methanol synthesis that for copper containing catalysts, the activity may not be purely based on the reduced copper activity but rather a Cu^{+}–Cu^{0} interaction or a Cu^{+}–ZnO interaction. Have you looked into this possibility in your catalyst system, and whether the sodium plays a role in the Cu^{+}/Cu^{0} ratio which in turn would affect the catalytic activity? CO chemisorption can be used to quantify the Cu^{+} species present, and in addition to the N_2O chemisorption data you can form a relative ratio between the species and correlate this to the activity.

Simon Kondrat communicated in response: I am aware of the active site postulated by Klier and co-workers involving Cu^+ within the ZnO lattice. Generally this theory has been overturned in favour of metallic Cu or Cu decorated with Zn.[1] However, this is a good suggestion to look at the amount of Cu^+ incorporated into the ZnO structure.

1 K. C. Waugh, *Catal. Lett.*, 2012, **142**, 1153–1166.

Letisha Deeplal communicated: Have you looked into the catalyst stability over a long term study, to determine if there is any copper sintering or morphological changes of the copper particles?

Simon Kondrat communicated in reply: In our previous publication we studied zincian georgeite derived catalysts for 110 h for the water gas shift and an accelerated ageing process for the methanol synthesis reaction (80 h cycling between 190 and 250 °C).[1] Clear deactivation was observed under methanol synthesis conditions, while no significant deactivation was observed with the water gas shift reaction. However, copper surface area analysis of the catalyst after 40 h time on-line showed a substantial reduction in copper surface area.

While it is clear that copper sintering is a deactivation mechanism, other important aspects are changes in the ZnO particle size and the ZnO interaction with copper. Our *in situ* XRD study, which simulates water gas shift conditions and also high conversion conditions in methanol synthesis, showed a growth in ZnO particle size in the presence of water. Interestingly, this corresponds with a recent study of a deactivation study over 148 days which shows that changes in ZnO are the most noticeable.[2] Deactivation is due to both Cu sintering and a loss of Cu–ZnO interaction.

1 S. A. Kondrat, P. J. Smith, P. P. Wells, P. A. Chater, J. H. Carter, D. J. Morgan, E. M. Fiordaliso, J. B. Wagner, T. E. Davies, L. Lu, J. K. Bartley, S. H. Taylor, M. S. Spencer, C. J. Kiely, G. J. Kelly, C. W. Park, M. J. Rosseinsky and G. J. Hutchings, *Nature*, 2016, **531**, 83–87.
2 T. Lunkenbein, F. Girgsdies, T. Kandemir, N. Thomas, M. Behrens, R. Schlögl and E. Frei, *Angew. Chem. Int. Ed.*, 2016, **55**, 12708–12712.

Letisha Deeplal communicated: A suggestion could be to look into strong electrostatic interactions for the catalyst synthesis, by looking at work done by Feltes and Regalbuto,[1] where they were able to control the adsorption of Mn onto the active sites only and not on the support by varying the pH during the synthesis step.

1 T. E. Feltes, Y. Zhao, R. F. Kile, R. J. Meyer and J. R. Regalbuto, *ChemCatChem*, 2010, **2**, 1065–1068.

Simon Kondrat communicated in response: Many thanks. This is a very interesting suggestion that we will follow up on.

Eric van Steen opened a general discussion of Thulani Nyathi's and Simon Kondrat's papers: The conversion of methanol over copper has recently been described as a size dependent reaction. Your study seems to indicate that the crystallite size is not the dominant factor in the activity. Could the difference in the feed used be the origin of the difference in the observation?

Simon Kondrat responded: This relates to the recent work from Van den Berg *et al.* where it is claimed that activity diminishes when particle size drops below 8 nm.[1] It is important to note that the presence of ZnO has a more pronounced effect on activity. Therefore, you are also correct that conditions will have a substantial effect on catalyst structure and activity. Specifically, the CO/CO_2 ratio will have an effect on structure and activity.

1 R. van den Berg, G. Prieto, G. Korpershoek, L. I. van der Wal, A. J. van Bunningen, S. Lægsgaard-Jørgensen, P. E. de Jongh and K. P. de Jong, *Nat. Comm.*, 2016, 7, 13057.

Neil Coville said: The distribution curves for the particle sizes are quite wide as you get to larger sizes, as is typically observed. Is there an expected variation in activity with size that will affect your average activity data? Is this taken care of by counting the total number of active surface sites from chemisorption or other techniques?

Thulani Nyathi replied: When looking at the relative deviation (calculated by dividing the standard deviation by the average size), we actually do not see a significant difference between the five samples in terms of their relative spread around the average size. This relative deviation varies between 18 and 27%. Therefore from this, one could safely assume that the average activities we reported in the paper are not greatly affected by the apparent broadening of the size distributions with increasing average size.

Nico Fischer remarked: The current hypothesis of a working MeOH catalyst is that Zn decorates the surface of Cu crystallites. Could an observed crystallite size dependency (more edges and kinks for the smaller crystallites) therefore be correlated with a higher number of low coordinate Cu surface atoms which might bind Zn strongly, *i.e.* Zn covers the edge and kink sites?

Simon Kondrat answered: The dominant Cu site for methanol synthesis, with regard to low or high coordinate sites, is not entirely clear. Campbell and co-workers demonstrated reasonable reaction rates on Cu(110),[1] while Behrens and co-workers modelled the reaction on low coordinate Cu(211) and compared to Cu(111), showing that intermediates were more stable on the low coordination Cu.[2] The latter also noted that the decoration of these step sites improves activity. With regard to the strength of Zn binding to edge Cu sites, Kuld *et al.* state that Zn coverage would be greater at Cu(532) corner sites and Cu(211) than high coordinate Cu(111), as you suggested.[3] However, perhaps counter-intuitively, they modelled that smaller Cu particle size resulted in lower Zn surface coverage. Interestingly, from this it was predicted that activity would decrease with Cu particle size as Zn surface coverage decreased. Van den Berg *et al.* later illustrated the Cu particle size effect experimentally, but for catalysts both with and without

ZnO.[4] The rationale for smaller particle size being detrimental for activity was the loss of specific step-edge sites, or that the high number of low coordinate Cu sites at small particle size results in formate inhibition.

1 J. Yoshihara and C. T. Campbell, *J. Catal.*, 1996, **161**, 776–782.
2 M. Behrens, F. Studt, I. Kasatkin, S. Kühl, M. Hävecker, F. Abild-Pedersen, S. Zander, F. Girgsdies, P. Kurr, B.-L. Kniep, M. Tovar, R. W. Fischer, J. K. Nørskov and R. Schlögl, *Science*, 2012, **336**, 893–897.
3 S. Kuld, M. Thorhauge, H. Falsig, C. F. Elkjær, S. Helveg, I. Chorkendorff and J. Sehested, *Science*, 2016, **352**, 969–974.
4 R. van den Berg, G. Prieto, G. Korpershoek, L. I. van der Wal, A. J. van Bunningen, S. Lægsgaard-Jørgensen, P. E. de Jongh and K. P. de Jong, *Nat. Comm.*, 2016, **7**, 13057.

Dong-Hau Kuo commented: Inorganic catalysts, *e.g.* oxides, sulfides *etc.*, can be functionalized to have vacancy defects and n-type (electron) or p-type (hole) conducting behavior by adjusting their composition and being processed at low temperature. Conventional oxide catalysts have to be activated above 200 °C for thermoconversion. However, functionalized and active inorganic catalysts can be used even at room temperature. On the other hand, nano metal catalysts, although they are active, have no hole conducting behavior for oxidation reactions. Therefore, inorganic catalysts can be designed to have multiple functions.

Ying Zheng addressed Dong-Hau Kuo: Please elaborate – are the metal catalysts easily oxidised to become functional?

Dong-Hau Kuo responded: A recent paper pointed out that the metal oxide formed between metal and oxide contributed towards the function of catalytic conversion.[1] Metal nanoparticles can be easily oxidized or dynamically oxidized/reduced in sequence in a CO_2/CO/H_2 system. The *in situ* formed oxide, due to its non-stoichiometry, can have semiconducting behavior or become functionalized.

1 C. Tisseraud, C. Comminges, S. Pronier, Y. Pouilloux and A. Le Valant, *J. Catal.*, 2016, **343**, 106–114.

Mzamo Shozi opened discussion of the paper by Michael Bowker: Have you looked at PdZn/ZnO as a catalyst?

Michael Bowker answered: Yes, we recently reported on this and such catalysts are very good for the synthesis.[1] In that paper we reported a maximum selectivity and conversion of 60% and 11% respectively at 250 °C, 20 bar. This is only achieved after appropriate reduction to form the PdZn alloy nanoparticles.

1 H. Bahruji, M. Bowker, G. Hutchings, N. Dimitratos, P. Wells, E. Gibson, W. Jones, C. Brookes, D. Morgan and G. Lalev, *J. Catal.*, 2016, **343**, 133–146.

Tumelo Phaahlamohlaka asked: In the paper (DOI: 10.1039/C6FD00189K), the role of the catalyst support was not strongly emphasized as one of the parameters that influence the PdZn nanoparticles' activity and selectivity. Does the reducibility of the supports have any significant effect on the catalysts, especially given the different activity of the PdZn bimetallic nanoparticles on alumina and titania?

If you pre-synthesise your PdZn bimetallic nanoparticles of the same size and load them on a titania or alumina support, will the nanoparticles give different catalytic behavior, since in your paper the catalyst dispersion on titania was suggested to be the main factor responsible for the high activity and good selectivity?

Michael Bowker replied: The reducibility of the support is important. Using ZnO as a support for Pd results, after reduction, in the formation of the PdZn alloy. It may be that other reducible supports could have a similar effect, though we haven't so far made such good catalysts with Pd on TiO_2 for instance, where the SMSI effect is known to be of relevance to catalytic reactivity.

PdZn on alumina is a poorer catalyst, even though the surface area is higher. It is probably important to have the interaction of the Pd-Zn active phase with ZnO in the right conformation, that is, in close interaction together. Why exactly the alumina supported material is worse in performance is hard to define, but it could be because there is separation between the PdZn and ZnO functions, due to the wider separation of nucleation points on the higher area surface. Of course, that means there may be room for improvement of those materials.

Hans Niemantsverdriet noted: To explain the favourable behaviour of PdZn alloys, you showed a UPS spectrum from the literature, and you said, somewhat provocatively, that the PdZn alloy looks like Cu in the UPS. This sounds like a somewhat bold statement from rigid band theory. Nowadays the view is that metals retain much of their electronic character in alloys. I noticed that the Pd d-band was still visible in the UPS. Could it be that the UPS represents a PdZn system with a surface enriched in Zn, such that the Pd signal is attenuated? The actually active system may then be a Zn or ZnO phase with a low concentration of Pd exposed to the surface. That sounds more likely to me than an alloy with an overall electronic structure similar to copper.

Michael Bowker answered: PdZn is "like" Cu in the sense that the d-band is shifted well below the Fermi level, and the sp band has low intensity at the Fermi level. This and the reactivity is somewhat similar – it subtends the formation of a formate intermediate of similar stability to that on Cu, and it can make methanol rather effectively from CO_2 and H_2. On the other hand I tend to agree with you that individual metal character tends to be retained, which somewhat goes against the popular so-called "d-band theory". So it may be in this case that it is Cu like because the pivotal intermediate, the formate, is bonded both to Zn and Pd atoms; that is, the ensemble effect is important. This requires more work to identify the surface structure and composition of such nanoparticles. There is a very nice paper from Kozlov *et al.*, which shows that PdZn, in contrast to PdCu, PdAu and PdAg, tends to show no segregation of either component and is quite homogeneous in the bulk of the nanoparticle.[1]

1 S. M. Kozlov, G. Kovács, R. Ferrando and K. M. Neyman, *Chem. Sci.*, 2015, **6**, 3868–3880.

Enrique Iglesia enquired: The surface chemistry of formate species and the need for CO_2 to form them during methanol synthesis has been well-studied in work first by ICI and then by others. Such studies have implicated formate hydrogenolysis as a kinetically-relevant step. Given the phenomenological analogy that you have made between PdZn and Cu in CO_2 hydrogenation, have you looked as to how PdZn behaves in the formation of surface formates and their reactions with H_2 – for example, by forming formates through interactions with formic acid and then measuring the rate of hydrogenolysis using H_2 to form methanol isothermally as a function of coverage (which changes as the formate reacts)?

Michael Bowker replied: We have done some TPD, which shows a similar stability of the formate intermediate to that on Cu. It has been observed that formic acid, and even sometimes methanol itself can be formed in TPD, which is coincident with the formate decomposition. This then implies that self-hydrogenation to formic acid or methanol from formate is relatively facile.

Your suggestion regarding the investigation of the hydrogenation of formic acid itself is an excellent one, but we have not so far followed that path. It could probably be followed at low pressure since there is a molar decrease upon hydrogenation to methanol and water.

Chelsea Tucker asked: In your paper (DOI: 10.1039/C6FD00189K) it is mentioned that pre-reducing the Pd/ZnO catalysts at 400 °C resulted in a higher methanol productivity. Since the PdZn alloy can be formed under H_2 at 200 °C, what is the reason for the higher methanol productivity after higher temperature pre-reduction?

Michael Bowker responded: It is difficult and slow to make the alloy by reduction at 200 °C. As shown in our earlier paper using other methods to make these catalysts,[1] if the catalyst is not pre-reduced, then there is high selectivity to CO at 200 °C, since the catalyst is mainly in the monometallic Pd form, but as the temperature of reaction goes up, so does selectivity, due to the beginnings of formation of the alloy. Pre-reduction at 400 °C ensures completion of this process, and subsequent running shows selectivity at 200 °C of 80+%, whereas without pre-reduction it was ~20%.

1 H. Bahruji, M. Bowker, G. Hutchings, N. Dimitratos, P. Wells, E. Gibson, W. Jones, C. Brookes, D. Morgan and G. Lalev, *J. Catal.*, 2016, **343**, 133–146.

Eric van Steen queried: How stable are PdZn alloys in environments containing water (present during CO_2 hydrogenation)?

Michael Bowker replied: They appear to be stable over the periods we have measured, which is 16 hours. We are currently engaged in *in situ* XRD measurements to check for any structural/particle size changes when different levels of water are added to the feed.

Nico Fischer said: In Fig. 1 of your paper (DOI: 10.1039/C6FD00189K) you show the XRD patterns of model catalysts with different Zn : Pd ratios. Samples d and e both have a 1 : 10 ratio varying only in the support material chosen (TiO_2 and Al_2O_3 respectively). You generally assume full degree of alloying of the Pd, however, the intensity ratios of ZnO and PdZn are very different in samples d and e. Have you attempted to quantify the phase composition from the XRD patterns? What explanation do you have for this observation?

Michael Bowker answered: From the XRD we only see the beta 1 : 1 alloy, which is the structure with the widest range of stability in the PdZn phase diagram under the temperature ranges we have used. The difference in the XRD between d and e is simply different experiments at different times. The absolute intensities shouldn't be taken too seriously, but the relative ones are significant *e.g.* the size of PdZn relative to ZnO, for instance.

Emiel Hensen questioned: The context of CO_2 reduction with renewable H_2 that you sketched likely requires not only a new catalyst but also novel operation windows. Given that electrolysis of water is expensive and use must be made of intermittently available cheap electricity, the CO_2 conversion process must be scaled down – one approach would be to simplify the process into a once-through approach (skipping recycles). Then a catalyst active at lower temperatures than the current generation of methanol synthesis catalysts would be useful as equilibrium is shifted. Such catalysts should also be able to cope with increased water levels during CO_2 reduction. Perhaps such a catalyst would also need to be operated intermittently. The concept of Pd-Zn is interesting but likely not active enough. Can you comment on the requirements for a catalyst for the process you described?

Michael Bowker responded: If this could be achieved (very low temperature operation) that would be wonderful and may allow once-through operation as you suggest due to the higher equilibrium yield. There may then be a problem of course with heat removal, though this might be easier if it were a liquid phase process.

While our catalyst with Pd is of interest from a scientific point of view, it is not desirable either economically or environmentally to use a precious metal, so we should aim to avoid this, and we are currently working to try to make Ni-Zn catalysts. It is likely though that they would suffer from methanation problems if any traces of monometallic Ni were left. We should note that Ni-Ga has been claimed as a possible low temperature methanol catalyst,[1] though experiments have only been reported at atmospheric pressure, where the equilibrium yield is extremely low.

1 F. Studt, I. Sharafutdinov, F. Abild-Pedersen, C. F. Elkjær, J. S. Hummelshøj, S. Dahl, I. Chorkendorff and J. K. Nørskov, *Nature Chem.*, 2014, **6**, 320–324.

Neil Coville enquired: In this method I assume that the Pd acetate is vaporised. I would imagine that there are some deposits on the Schlenk tube as well? Indeed, I might expect that there could be some losses unless the whole tube is submerged in the oil bath as the catalyst would collect at the top of the tube?

Michael Bowker answered: Indeed, it is acetylacetonate and it vaporises and appears (from TEM *etc.*) to deposit fairly homogeneously over the catalyst (which has the main surface area) and little on the tube itself. This appears not to be the case when we try to make Cu catalysts, there is a great deal of deposition at the top of the tube, and we tend to get under-doping of the surface with Cu.

Simon Kondrat said: Palladium acetate is a very volatile salt and a very good precursor salt for vapour deposition. Its low sublimation temperature of ~200 °C has previously been shown to facilitate its dispersion over supports prior to its thermal decomposition.[1,2] Simple mixing of the palladium acetate with a support and heat treatment in an inert atmosphere produces high metal dispersion without the need of a vacuum. We even showed that alloys (in this case AuPd) can be produced by this mixing-decomposition method.[2] Perhaps the simple mixing and thermal treatment of palladium and zinc acetate with titania would be effective?

1 Y. Lin, K. A. Watson, M. J. Fallbach, S. Ghose, J. G. Smith, D. M. Delozier, W. Cao, R. E. Crooks and J. W. Connell, *ACS Nano*, 2009, **3**, 871–884.
2 S. A. Kondrat, G. Shaw, S. J. Freakley, Q. He, J. Hampton, J. K. Edwards, P. J. Miedziak, T. E. Davies, A. F. Carley, S. H. Taylor, C. J. Kiely and G. J. Hutchings, *Chem. Sci.*, 2012, **3**, 2965–2971.

Michael Bowker responded: That is a good suggestion, thank you. I have to say I'm very surprised that Pd acetate is that stable, since on bulk Pd substrates it decomposes near ambient temperature. However, that is in the presence of bulk metal rather than as an organometallic.

Sarah Adam commented: In your paper (DOI: 10.1039/C6FD00189K), which begins by mentioning the efforts taken to address the rising level of carbon dioxide in the atmosphere because of the effects of global warming, it is written that "Significant amounts of energy are required for the activation and conversion [of CO_2] into value-added products...". How does the conversion of carbon dioxide compare to the indirect output of CO_2 as a result of the large energy requirement for this conversion process? Would this energy need to be from renewable sources for this to be a viable process for the reduction of atmospheric CO_2 concentrations?

Michael Bowker answered: The energy can come from a variety of sources of course, but in the current climate it is important that it derives from renewable energy, especially that from solar energy sources such as photovoltaic, or even better, wind.

Note, however, that in the case here the CO_2 is merely a carrier for hydrogen which itself is the stored energy (from solar energy *via* electrolysis for instance). Converting it to methanol densifies it and makes it more transportable. It is likely that the biggest use will be to regenerate energy directly (by combustion, re-releasing the CO_2) or by releasing the H_2 again at the point-of-use. Note that if solar energy is utilised to regenerate the H_2 by photocatalysis, then an extra 50% of hydrogen can be liberated from water ($CH_3OH + H_2O + hv \rightarrow CO_2 + 3H_2$).[1]

1 M. Bowker, *Green Chem.*, 2011, **13**, 2235–2246.

Unni Olsbye remarked: The identification of a formate intermediate may suggest that CO and methanol are formed *via* separate reaction paths, and that the product selectivity is independent of conversion. Is this the case, and if not, what are the optimum methanol yields obtained at each temperature? Have you observed other products than CO and methanol at higher conversions than those shown in the paper?

Michael Bowker answered: Indeed, the RWGS may occur through a different intermediate, I think COOH has been proposed as the intermediate in the shift reaction. The product selectivity is not independent of conversion, though conversion doesn't change enormously in the range shown in Table 2 of our paper (DOI: 10.1039/C6FD00189K). This is due to decreasing equilibrium conversion to methanol with increasing temperature combined with increasing yield of RWGS. Generally, selectivity towards methanol is very high for the PdZn catalysts at low temperature, as seen in Table 2.

We have not seen other products, and conversion in such single-pass reactors is limited by equilibrium, but some methane is seen from the Pd-only (unreduced) catalysts.

Katherine Holt opened discussion of Alberto Roldan's paper: Firstly, a comment on the experimental data. Although it is true that methane was not detected as a product, this may be because solution NMR was used and methane would be a gas phase product. However, this does not change the conclusions from your paper. Secondly, a question: the reaction produces water. Does this absorb or become incorporated into the FeS surface and hence change the catalytic activity?

Alberto Roldan answered: It is true that we did not analyse the gas phase but our calculations have shown quite clearly that carbon oxides are intermediates in C–C formation, which suggests very strongly that CH_4 is not formed.

Water and OH groups would indeed adsorb on the surface saturating some dangling bonds, but we need more calculations and experiments to analyse the surface after the reaction.

Neil Coville enquired: Do you see any H_2S being evolved in surface reactions?

Alberto Roldan replied: We observed H_2 evolution at negative potentials below 0.6 V due to water reduction. We did not identify H_2S in solution.

Eric van Steen commented: Your data seem to indicate that some acetic acid is formed as well. What reaction pathway do you propose for the formation of C2 oxygenates on these systems?

Alberto Roldan responded: In the present paper (DOI: 10.1039/C6FD00224B), we proved that the C–O bond will not dissociate leading to CH_x species on the surface that would undertake the accepted CH_2 addition. Instead, we suggested C–C formation *via* $H_xCO(H)$ species adsorbed on both Fe_A and Fe_B sites.

Enrique Iglesia said: MoS_2 catalysts are by far the most studied sulfide system for CO hydrogenation because of their unique ability to form oxygenates. In this case, most if not all studies have proposed the formation of vacancies by the removal of some S atoms and their involvement in the binding and activation of

CO and H_2 with the formation of C and O atoms at such vacancies. These systems appear to also involve some carbonylation function (as in the case of Rh-based catalysts that also form oxygenates) to form methanol and then react it with CO to form acetic acid that subsequently undergoes hydrogenolysis to form ethanol. In your system, you consider stoichiometric Fe sulfides as the surface in your calculations. Have you explored the involvement of surfaces with S vacancies and their involvement and reactivity in such reactions?

Alberto Roldan answered: We have investigated the formation of S vacancies due to H_2S evolution, however, the energy barriers are above 2 eV independently of the H coverage. We have also studied CO_2 adsorption on S vacancies resulting in a similar interaction energy to that on pristine $Fe_3S_4\{111\}$ surfaces. Nevertheless, the stability of reaction intermediates may vary due to the electrons delocalised on the vacancy, therefore, further study is needed.

Matthew Neurock asked: Can you elaborate more on how the hydrogenation proceeds in your calculations? From your figures, it looks like you are using two hydrogens that are bound to the sulfur sites on the surface to carry out hydrogenation. This would indicate that you are transferring two protons to the adsorbed CO* and H_xCO* intermediates as the H that is bound to the S is protonic in nature. This would seems very unlikely. There must be electron transfer for the reaction to proceed. That would either occur *via* a change in the oxidation state of the Fe, resulting in a homolytic radical like transfer of the hydrogens or *via* the heterolytic transfer of a proton (from the S–H) and a hydride that is bound to the metal.

Alberto Roldan replied: On a naked surface, hydrogen adsorbs on the surface placing its electron in the S–H bond while the oxidation state of the Fe sites remains invariable. Once the H is transferred to the C-intermediate, the electron is donated through the S to the Fe and to the C-intermediate. Thus, the metal oxidation state is Fe(III), as shown by previous experiments.[1]

1 A. Roldan, N. Hollingsworth, A. Roffey, H.-U. Islam, J. B. M. Goodall, C. R. A. Catlow, J. A. Darr, W. Bras, G. Sankar, K. B. Holt, G. Hogarth and N. H. de Leeuw, *Chem. Commun.*, 2015, **51**, 7501–7504.

Patricia Kooyman enquired: Is the sulphide stable under the reaction conditions?

Alberto Roldan answered: The experiments were carried out at room temperature and pressure for a week and the catalysts were kept on the bench under normal conditions. Products were formed regularly, indicating that the sulfides are stable under these conditions. The catalyst samples were prepared using oleylamine, which coated the particles, thus acting as a capping agent and increasing their stability.

Detlef Bahnemann said: You are differentiating between Fe_A and Fe_B sites; I assume these are Fe(III) and Fe(II) sites, respectively? Does the oxidation state of these iron sites remain unchanged during the reaction? Or, rather, are they involved in electron transfer processes?

Alberto Roldan responded: Fe_A is Fe(III) and occupies a Td position in the bulk while Fe_B is Fe(III, II) and occupies a Oh position. Once at the surface, Fe_A and Fe_B are both (III) but their position is slightly different and they behave differently.

Mzamo Shozi communicated: Is there a possibility of the formation of H_2S and subsequent poisoning by it?

Alberto Roldan communicated in response: We have analysed the formation of H_2S but this process is not favourable kinetically (barrier ~2 eV).

Aleksandar Zivkovic communicated: You mentioned in your paper that you had to employ the Hubbard U parameter to accurately describe the strongly correlated d-Fe electrons of your system. A value of 1 eV was chosen to be the best representative. According to which property did you determine this particular value?

Alberto Roldan communicated in response: We followed previous approaches (see citations within our manuscript) that adjust the U value to reproduce the geometric and electronic structure of Fe_3S_4, where $U_{eff} = 1$ eV was shown to be sufficient to provide accurate results.

Anna Petersen opened discussion of the paper by Marien Bremmer: The synthesis of metal nanoparticles typically requires the use of surfactants or capping agents, which bind to the metal surface to stabilize the small crystallites. There is a concern that those capping agents will block the metal surface rendering it ineffective for catalysis. In your article (DOI: 10.1039/C6FD00185H) you mention applying a heat treatment in N_2 and H_2 to the particles prior to the catalytic testing in order to remove the capping agents. Have you attempted any catalytic testing without removing the capping agents? This might be an interesting approach to study whether the capping agents actually poison the metal surface, or whether they are removed during the catalytic reaction. Furthermore, did you observe any changes after removing the capping agents, such as changes in particle morphology or formation of a carbonaceous residue?

Marien Bremmer replied: We have not tested the catalytic activity of our nanoparticles without a prior treatment to remove the capping agents, yet it is indeed an interesting approach to obtain more insight in the role of the surfactant. During the treatment we did not observe any changes of the particle in terms of morphology, structure or size. Also, carbonaceous residue was not formed on or around the nanoparticles, or on the windows of the nanoreactor.

Emiel Hensen enquired: You reported the formation of carbon layers on Co nanoparticles upon exposure to CO at 500 °C. Was any attempt made to verify whether this also occurs at temperatures close to the FT conditions (*i.e.*, below 300 °C)?

Marien Bremmer answered: Due to experimental problems that occur in between room temperature and 275 °C, unfortunately we cannot successfully perform *in situ* experiments below 300 °C at this stage. We have performed an experiment at 300 °C, following the same approach of switching off the H_2 feed. During that experiment we did not observe the formation of the carbon layers.

Neil Coville asked: The mechanism for carbon growth: carbon has to diffuse through the different layers, so do you see a variation in diffusion? Do you see the rates of layer formation vary/drop? Any reason why it stopped at 12–14 layers?

Marien Bremmer replied: We do not observe a significant drop in the rate at which the layers are formed. The time it takes before a new layer is visible does vary: after the first layer, it takes over 13 minutes for the second layer to be visible. Then, for the next 10 layers, it takes between 1 and 5 minutes per layer to form. There is no evident relationship between the formation of two subsequent layers in terms of deposition rate. This behaviour might indicate that CO adsorbs on an uncovered area of the Co nanoparticle surface, and that atomic carbon is formed there. Then, the carbon diffuses over (through?) the surface of the nanoparticle, to assemble into a (new) layer of carbon, or link to an already existing layer of carbon, making it grow laterally. When the carbon layers grow laterally, eventually the surface of the nanoparticle will be covered completely, explaining why no new layers are formed after 12–14 layers.

Other nanoparticles (e.g. Fig. 9a, DOI: 10.1039/C6FD00185H) show "multi-walled nanotube-like" structures that have grown on the surface of the nanoparticle but have disconnected from the surface and grown away. This indicates that, if there is still uncovered surface available, the layers can continue to grow. This can lead to the formation of structures that can reach a size of 10 times that of the nanoparticle that is supporting it. This being said, there is still more work to be done to understand the growth mechanism.

Nico Fischer said: My first question relates to the proposed removal of the surfactant. Have you ever analysed the reactor outlet gas for traces of decomposition/hydrogenation products supporting the removal of the oleic acid?

Secondly, the graphene layers shown here seem to be sheets encompassing large surface areas of the crystallites resulting in curved surfaces. If these graphene sheets consist of continuous sheets as you seem to suggest and not of small platelets arranged in sheets, would the additional strain during the formation of additional sheets at the metal surface not lead to cracks in the outermost C layers?

Marien Bremmer responded: When the surfactant molecules come off the nanoparticle surface it should indeed be possible to detect these molecules using mass spectroscopy. The gas outlet of our nanoreactor is flowing through a Quadrupole Mass Spectrometer (QMS), which can also detect (fragments of) hydrocarbons and carbon oxide species. During the experiments, we did not observe any hydrocarbons in the QMS signal.

However, this is probably due to the very low amount of sample that is loaded in our nanoreactor (while an even smaller amount is present on the part of the nanoreactor that is actually heated). In combination with the relatively high flow of gas through the nanoreactor this results in an extremely low concentration of surfactant-derived molecules. One solution for this could be to work with a very low gas flow, but this has a negative effect on the time resolution of the QMS signal. Another possible explanation is for the evaporated molecules to be "trapped" on colder surfaces (*i.e.* the gas lines downstream) before they even reach the QMS chamber.

Due to the two-dimensional projection of the TEM images, distinguishing between continuous sheets or collections of small platelets may not be straightforward. However, in view of the sharpness of the lines in the TEM images it seems likely that the graphene is present as continuous sheets. A loose collection of small platelets, when projected parallel to the long axes of the platelets, would show a fuzzy edge due to the amount of material protruding from the flattest possible projection.

Tumelo Phaahlamohlaka questioned: The Boudouard reaction is considered to be a non-catalyzed decomposition of CO to form carbon, which would generally be amorphous carbon. In your paper the deposited carbon is highly graphitic as shown in your Fig. 6 (DOI: 10.1039/C6FD00185H). Such growth is common in the synthesis of carbon nanotubes by the proposed vapor liquid solid or vapor solid solid mechanism (VLS/VSS) in which the carbon source decomposes on the nanoparticle surface and nucleates to form graphitized carbon. With this in mind, I wonder if the carbon dissociation in eqn (1) ($CO_{ads} \rightarrow C_s + O_{ads}$) can be completely ruled out, or maybe the two reactions might be competing in your current study.

Marien Bremmer answered: According to the literature[1,2] the dissociative reaction of eqn (1) ($CO_{ads} \rightarrow C_s + O_{ads}$) will yield a single layer of carbon, while the Boudouard reaction will yield multiple layers of carbon. However, the dissociation of CO may be followed by another CO reaction with the O_{ads} to recreate free sites on the catalyst surface, effectively being the start of the Boudouard reaction.

1 J. Nakamura, I. Toyoshima and K. Tanaka, *Surf. Sci.*, 1988, **201**, 185–194.
2 F. L. Hutson, D. E. Ramaker and B. E. Koel, *Surf. Sci.*, 1991, **248**, 104–118.

Enrique Iglesia commented: If we can make an analogy with the mechanism of formation of carbon filaments or onion-like structures on Ni (during reactions of CH_4, CO_2, H_2O, CO and H_2 in methane reforming contexts), I would expect that the structures formed are determined by the prevalent chemical potential of carbon, set by the chemical reactions of the carburizing molecules. A threshold chemical potential is required to overcome the thermodynamic barriers for nucleation. Near the threshold, carbon nucleates infrequently and each particle acquires a single nucleating structure that grows rapidly and directionally to form a filament. At high chemical potentials, carbon structures nucleate at multiple points in each particle and cannot achieve a preferential direction of growth, thus forming onion-like structures. Do you see similar trends in your studies, and are you able to set and locally control the carbon chemical potential (set by the ratio of $(CO)^2$ to (CO_2) pressures in units of bar) as the carbon is formed?

Marien Bremmer answered: Indeed, the chemical potential would play an important role in the nucleation and growth of the carbonaceous structures on the metal nanoparticles. However, we have not analysed/calculated the chemical potential during our experiments.

Although our set-up allows to change the CO/CO_2 ratio easily and fairly quickly (the response time after switching off the H_2 was 6 min), for the current set of experiments we studied only a single set of conditions. Therefore, at this moment we cannot comment on the effects of changing the chemical potential of the system. However, these kind of experiments are certainly possible and might be interesting for future research.

The ratio of the different reactant gases is set in gas lines before the *in situ* sample holder and nanoreactor. The catalyst nanoparticles that we image during our experiments are at the very centre of the nanoreactor. When we tune the gas composition this affects the whole nanoreactor, thus we can control the gas ratio for the catalytic sample as a whole. A more local approach to control the gas ratio is not possible at this moment.

Tim Wezendonk asked: At the start of the FTS regime, around 90 minutes before time zero, there seems to be a decrease in the otherwise constant hydrogen evolution from the MS. Have you witnessed any change in the morphology of the Co particle or lattice fringes of the crystal plane? The same question holds for time zero, where the Boudouard reaction starts forming carbon overlayers.

Marien Bremmer replied: No, neither during the observed dip in the H_2 MS signal nor during the start of the Boudouard reaction signal did we observe any change in the nanoparticle morphology or lattice fringes.

Moritz Wolf said: You report a phase transformation of the as-prepared ε-Co to α-Co (hcp) at 300 °C and to β-Co (fcc or ccp) at 500 °C during the removal of the surfactant from the surface of your Co nanoparticles *via* annealing. Have you actually seen these phase transformations *via in situ* TEM? You cite Yang *et al.*[1] who also synthesised ε-Co and annealed the nanoparticles in 95% inert gas and 5% H_2, but they only analysed the nanoparticles after annealing at 500 °C. Hence, they only observed the phase transformation to β-Co without knowing the actual transition temperature and previous phase transformations. Where does the information of the phase transformation of ε-Co to α-Co at 300 °C come from? Nevertheless, your assumptions are in good agreement with observations from Kitakami *et al.*[2] for the annealing of 30 nm Co nanoparticles: starting with a ratio of α-Co/β-Co of 3.5, they observed a maximum concentration of α-Co at 300 °C, while pure β-Co was present at 500 °C. However, they also showed that Co nanoparticles, prepared *via* neutron sputtering at ambient temperature, below 20 nm consist of pure β-Co, while the concentration of α-Co increased significantly with size. Fischer *et al.*[3] (see references 28–30 therein for further publications on temperature induced phase transformations of Co) confirmed that β-Co is the main allotrope for Co nanoparticles below 20 nm and at high temperatures *via in situ* X-ray diffraction in reduction studies of alumina-supported Co nanoparticles. Reduction of several sizes (3–10 nm) of Co_3O_4 to metallic Co in H_2 at 450 °C resulted in single-phase β-Co. In an older publication, Lee *et al.*[4] reported the low-temperature (<400 °C) allotrope α-Co in bulk cobalt, while β-Co is the high-temperature phase at temperatures exceeding 450 °C. Can you please provide experimental proof for the mentioned phase transformations at given temperatures? Are you subsequently using the generic term "metallic cobalt" because you do not know the actual phase after annealing?

1 H. T. Yang, C. M. Shen, Y. G. Wang, Y. K. Su, T. H. Yang and H. J. Gao, *Nanotechnology*, 2004, **15**, 70–74.
2 O. Kitakami, H. Sato, Y. Shimada, F. Sato and M. Tanaka, *Phys. Rev. B*, 1997, **56**, 13849–13854.
3 N. Fischer, E. van Steen and M. Claeys, *Catal. Today*, 2011, **171**, 174–179.
4 B. W. Lee, R. Alsenz, A. Ignatiev and M. A. van Hove, *Phys. Rev. B*, 1978, **17**, 1510–1520.

Marien Bremmer answered: During our experiments and our preliminary analysis, we did not observe changes in the phase of the nanoparticle. There are several publications that report that the metastable ε-Co transforms into hcp Co at approximately 300 °C, which subsequently transforms to ccp at approximately 500 °C. Yang's work (ref. 21 in our paper, DOI: 10.1039/C6FD00185H) does not include the first phase transformation, but the following paper does document this information.[1] We can confirm, based on observations in our laboratory by means of powder X-ray diffraction, that ε-Co converts to hcp Co at approximately 300 °C, and that at 500 °C the main stacking is ccp. We use the term "metallic cobalt" as we could not document the exact atomic arrangement in the observed Co nanoparticle during the *in situ* TEM experiment. ε-Co, hcp and ccp are all metals.

1 M. Green, *Chem. Commun.*, 2005, 3002–3011.

Kees-Jan Weststrate remarked: You mention in your presentation that experiments suffer from carbon deposition due to electron beam-induced decomposition of hydrocarbon contaminants when working below 300 °C. Could these "contaminants" in fact be FT products? And in general, under "real" FTS conditions one certainly expects to produce a large quantity of volatile hydrocarbon products, so "contaminants" would be inevitable under real conditions. Would that make experiments under working conditions practically impossible or do you think there are ways to avoid/minimize carbon deposition because of electron beam decomposition of FTS products?

Marien Bremmer responded: We don't exclude the possibility that FT products are cracked under the electron beam and cause contamination on our nanoreactor windows. However, the contamination that we observe is always present, even without FT catalysts and syngas conditions, so it is hard to say whether FT products contribute to the problem or not. Whether TEM experiments at real FTS conditions would be practically possible or not is hard to say at this stage. It might be possible if the amount of hydrocarbons produced is not too high.

One way to minimise the deposition of carbon due to cracking of hydrocarbons is to minimise the intensity of the electron beam (*i.e.* using low-dose imaging). However, this comes at a significant loss of image quality, while the signal-to-noise contrast is already poor due to the amorphous SiN windows. This could be compensated for by improving the TEM hardware, *e.g.* by using a more sensitive camera, or by image post-processing to reduce noise.

Alberto Roldan opened a general discussion of the papers by Michael Bowker, Alberto Roldan and Marien Bremmer: The PdZn nanoparticles showed a similar electronic structure to Cu, which is known to activate CO_2. Do you think that the electronic structure band alignment of other metals and alloys will provide insights into the particles' reactivity towards the activation of CO_2 prior to experimental testing?

Michael Bowker replied: It may indeed do so. If the d-band is filled and hence shifted down below the Fermi level, then it may become a methanol synthesis catalyst. It would be good to make a range of catalysts for which the d-band is shifted by different amounts from the Fermi level – does it correlate with methanol synthesis activity? However, we don't use Ag or Au for methanol synthesis – perhaps the d-band is shifted too far away from the Fermi level in those cases. Note also the answer to Prof Niemantsverdriet's comment – we shouldn't put all our focus on the d-band position – the LDOS and individual chemical elemental character are also important. They dictate the stability of the pivotal intermediate, the formate.

Nico Fischer addressed Marien Bremmer: The effect of the electron beam on the crystal structure and strain thereof was mentioned during your presentation and in the associated paper. How big is the effect of the electron beam and would you say it is commonly underestimated in TEM analysis of catalyst samples?

Marien Bremmer responded: In this specific case, the effect of the electron beam on the nanoparticle is significant but indirect, as the deformation and structural reconstruction are mainly due to the contracting effect of the carbon shells.[1]

It is hard to quantify the influence of the electron beam since its effects are difficult to monitor, as we cannot obtain images without electrons, but it can never be fully excluded. In general, scientists working with *in situ* TEM on catalysis are very much aware of electron beam effects. The beam effect is almost always explicitly acknowledged and a common approach to deal with the negative consequences is to minimise the intensity and/or total dose. However, in general *ex situ* TEM characterisation of catalysts electron beam effects are sometimes not recognised or even ignored.

1 L. Sun, J. A. Rodriguez-Manzo and F. Banhart, *Appl. Phys. Lett.*, 2006, **89**, 263104.

Harinarayana Bandaru asked Michael Bowker: I noticed that your paper mainly focused on PdZn catalysts. Which Pd : Zn ratio is acting as a Cu catalyst? Do you have any experiments with Cu catalysts for comparison?

Also, in Fig. 5 of your paper (DOI: 10.1039/C6FD00189K) you observed a negative peak in the H_2 TPR profile for the 5% Pd/TiO_2 catalyst. The reduction temperature was ~50 °C for PdO. In contrast, there were no negative peaks observed for 5% Pd/Al_2O_3 and 5% Pd/ZnO_2 catalysts: could this suggest the presence of β-Pd hydride species?

Michael Bowker responded: The important PdZn phase here appears to be the beta 1 : 1 tetragonal phase, the only form we have been able to identify by XRD. In the bulk phase diagram this is present over a range of compositions from ~ 40% to 60% Zn content. It is assumed that the negative peak for Pd/TiO_2 is indeed due to hydride presence, which is common for Pd metal, but why it is not there for the other catalysts is unclear. We can only assume that this particular material is more reducible than the others, so it is already reduced in the H_2 flow at ambient temperature before the heating begins.

Neil Coville noted: XPS (and the Auger Zn peak) was used to identify the presence of Zn in the alloy. Depth profiling should also remove any ZnO layer and confirm this. Did you do this? This would also give some sense of alloy composition below the oxidized surface.

Michael Bowker answered: I agree that this could be a useful method, but we have not done this yet. Note that in answer to another question I have cited a paper which indicates that the composition of the beta PdZn alloy, unlike many others, should be rather isotropic; that is, with no preferential segregation of one component.

Katherine Holt asked: In using TPD to study the absorbed formate do you ever see evidence of the non-decomposed formate coming off the surface?

Michael Bowker replied: The major products of such TPD are CO_2 and H_2. However, if you look at a number of formate decomposition profiles, it is often the case that other products are seen to evolve coincidently with the main products, but at a very low level. See for instance the early work on ZnO [1] and also from Cu, where formic acid is seen at the level of ~1% of the CO_2 desorption.[2] This indicates that self-hydrogenation of the formate intermediate can occur during the TPD process. In the case of ZnO powder, this can go as far as the desorption of methanol with the formate decomposition.[1]

1 M. Bowker, H. Houghton and K.C. Waugh, *J. Chem. Soc., Faraday Trans. 1*, 1981, **77**, 3023–3036.
2. M. Bowker, E. Rowbotham, F. M. Leibsle and S. Haq, *Surf. Sci.*, 1996, **349**, 97–110.

Lebohang Macheli addressed Marien Bremmer: Does your surfactant form a chemical bond with the cobalt nanoparticles or is it just a layer of surfactant around the cobalt nanoparticle? If there is a chemical bond, other than the effect of blockage of active sites by the surfactant do you think there could be any electronic effect due to bond formation between the OA and the cobalt?

Marien Bremmer answered: The oleic acid (OA) surfactant bonds to the surface metal atoms via covalent bonding.[1,2] Due to the covalent bonding there is indeed an electronic effect between the (surface) cobalt atoms and the OA molecules: electrons of the surface atoms are shared with the OA oxygen atoms. This might translate into an effect on the catalyst performance (activity/selectivity). However, whether this effect exists and what the effect would be, is currently unknown. In principle, the pretreatment procedure should have removed the OA completely.

1 E. Zacharaki, M. Kalyva, H. Fjellvåg and A. O. Sjåstad, *Chem. Cent. J.*, 2016, **10**, 10.
2 A. Weddemann, I. Ennen, A. Regtmeier, C. Albon, A. Wolff, K. Eckstädt, N. Mill, M. K.-H. Peter, J. Mattay, C. Plattner, N. Sewald and A. Hütten, *Beilstein J. Nanotechnol.*, 2010, **1**, 75–93.

Matthew Neurock said to Michael Bowker: This is very interesting work. You indicate that PdZn takes on the properties of Cu and as such is active for carrying out the reduction of CO_2. Have you considered other Pd alloys such as PdAu or PdSn that might have similar properties, or other Cu alloys that might lead to improved catalytic activity?

Michael Bowker replied: We have indeed looked at other alloys. In fact, the original aim of our project was to make alloys and we were diverted by the kind of auto-alloying which we observed to make PdZn, which came as some surprise to us. PdAu shows some promise with high selectivity, but lower activity, while we were not successful in making useful Sn catalysts, they only showed high RWGS. However, I think there is a lot of room in this area for alternative synthesis strategies and other alloys. Of course, it is important to move away from precious metals, so NiZn might make a good alternative, provided the methanation activity of Ni can be suppressed. Cu alloys may have some benefit over Cu alone especially if water tolerance is improved.

Richard Catlow commented: I would like to emphasise the point I made in the discussion of Simon Kondrat's paper. As you have commented, there is strong evidence that the oxide support plays a role in the chemistry; and it is significant that the most active systems you identify involve supports such as CeO_2 and ZnO that are n-type semiconductors. These oxides have trapped electron states on their surfaces which are effective at activating CO_2; so we can envisage that the oxides activate the CO_2 by electron donation and the activated molecules are the hydrogenated at the oxide–metal interface.

Michael Bowker replied: I agree that this is a strong possibility. The bent CO_2 intermediate may straddle the interface between the metal and oxide. Also, it may be that one oxygen may fill anion vacancies which may exist in the oxide at the periphery of the metal nanoparticle.

Faraday Discussions

PAPER

Opportunities in upgrading biomass crudes†

Y. Mathieu,[a] L. Sauvanaud,[a] L. Humphreys,[b] W. Rowlands,[b] T. Maschmeyer[c] and A. Corma*[a]

Received 7th October 2016, Accepted 25th October 2016
DOI: 10.1039/c6fd00208k

An unconventional crude from biomass (biocrude) has been processed to yield a hydrocarbon stream that is not only fully processable in conventional refineries but is already close to the specification of commercial fuels such as transportation diesel. The upgrading of biocrude was carried out with a combination of hydrotreatment and catalytic cracking, yielding middle distillate as the main product.

1. Introduction

As the reserves of conventional crude are declining, new resources are being developed to fulfil the increasing demand for transportation fuels. A number of synthetic crudes can be obtained from unconventional sources, such as extra-heavy crude oils or lignocellulosic biomass. These can be processed further in existing or upgraded refinery processes, taking advantage of the huge oil processing and distribution infrastructure available already. However, these crudes introduce new problems due to their composition.

Beyond most important environmental considerations, biofuels made from abundant lignocellulose (wood, grasses, *etc.*) represent an opportunity to lower the dependence on oil. It has been reported that in the United States alone 1 billion tonnes per year of lignocellulosic biomass would be available for its transformation into fuels.[1] Yet economical transformation of lignocellulose into liquid fuels that can compete with petroleum-based products has remained elusive. Cellulosic ethanol costs remain high due to the difficulty of deconstructing cellulose, either by chemical treatment or enzymatically. Also, the yield of alcohol is constrained since not all the biomass is used. Gasification followed by Fischer–Tropsch chemistry has been demonstrated to yield high quality product, but the process efficiency is about 50% and the associated capital costs

[a]*Instituto de Tecnología Química (UPV-CSIC), Universitat Politècnica de València, Avda. dels Tarongers, s/n 46020 Valencia, Spain. E-mail: acorma@itq.upv.es*
[b]*Licella Pty. Ltd., Level 7 140 Arthur St, North Sydney, NSW 2060, Australia*
[c]*School of Chemistry, The University of Sydney, NSW 2006, Australia*

† Electronic supplementary information (ESI) available. See DOI: 10.1039/c6fd00208k

remain very high.[2] Moreover, with smaller scale units adapted to the biomass supply network, these costs increase even further. Fast pyrolysis is an efficient and cheap process to liquefy the biomass, but the quality of the liquid product obtained is too low for straightforward application to transportation fuel. Upgrading through hydrotreatment has proved to be challenging due to the high oxygen content, and the process requires harsh operating conditions.[3] Another simple technology that treats the whole of the lignocellulose is thermal hydrotreatment. In an aqueous medium when water is brought near its critical point (330–380 °C, 200–300 bar), the lignocellulose breaks apart yielding a liquid and some gas (essentially CO_2 with only very minor amounts of H_2, CO and CH_4). Compared to pyrolysis oil, this liquid is of higher molecular weight and stable with a substantially lower oxygen content that makes upgrading much easier.[4] As a result, the energetic content of the oil phase from hydrothermal treatment was reported to be in the range of 30–35 MJ kg^{-1} for a wide range of organic feeds[5] as compared to the energetic content of pyrolysis oil, which is often in the range of 15–20 MJ kg^{-1} depending on the water content.[6]

In the present manuscript the upgrading of these unconventional (bio)crudes has been studied using processes that are well-known to the refiner or that can be directly derived from such processes. In this way it should be possible to treat the biocrude from hydrothermal treatment in the existing units of oil refineries, preferably mixed with the regular oil being processed.

2. Biocrude from hydrothermal liquefaction

2.1. Processing scheme and preparation of the biocrude supplied for the study

A multistep process was implemented to convert pine wood chips into fuels. The complete processing scheme is presented in Fig. 1. We will describe in this section the steps that allowed the production of the biocrude, which was supplied by Licella, and was further used in this study to explore upgrading options (catalytic cracking and hydrotreatment).

In the first step, pine wood chips were processed hydrothermally in the presence of a catalyst, yielding a raw biocrude. One of the hurdles of such process development is the low productivity associated with batch treatment. Thus, a reactor allowing continuous treatment at 10 000 tonnes per year nameplate capacity was developed.[7] The lignocellulose (*e.g.* pine wood chips) is reduced in

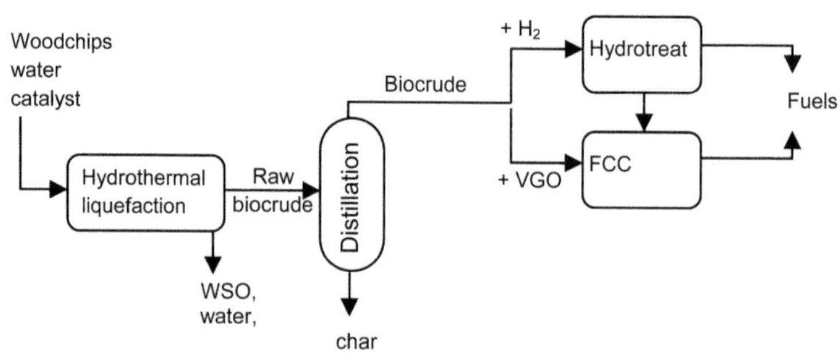

Fig. 1 Two step process for the preparation of biocrude and biocrude upgrading options.

particle size to suit the type of slurry pump being used. In these experiments the feedstock was reduced to about 500 microns in diameter; however, particle diameters of up to 2 mm have successfully been employed with the same pump and reactor configuration. The lignocellulose is suspended in water, and raised to the operating reaction temperature by a combination of electrical heating, heat exchangers and supercritical steam injection. The slurry is pumped into a vertically oriented, serpentine tubular reactor.[8] Operating conditions are close to the critical point of water, optimally sub-critical in the 330–350 °C temperature range and at 200–250 bar.[9] The residence time in the reactor is typically around 20–30 minutes for woody biomass. At this stage the processing stream is an oil–water emulsion that contains a small amount of solid material, *i.e.* mostly ash from the feedstock. In the next step the stream is partially cooled and then directly flashed at the output of the high-pressure hydrothermal reactor, splitting the emulsion and ensuring a rapid and efficient separation of oily products and water phase. Catalyst and a number of water-soluble organics are removed with the water phase. These include products such as small ketones, acids and phenols that may be recovered from the aqueous stream for further valorization. On a dry basis the oily phase recovered may represent about 30–35 wt% of the dry biomass fed to the hydrothermal reactor. This now already dewatered oil is further distilled in order to remove the heaviest part, constituted of a material that tends to transform into char when exposed to temperatures above 350 °C. This heavy fraction represents 20–30% of the raw biocrude, depending on the hydrothermal processing conditions.

Due to its characteristics, the resulting biocrude is not straightforwardly suitable for use as transportation fuel, so a final upgrading step is necessary. For that purpose we have studied two options. The simplest one is to mix the biocrude with Vacuum Gas Oil (VGO) and crack the mixture in a fluid catalytic cracking unit (Fig. 1). A second option considers first hydrotreating the biocrude, which yields a number of distillate range products and then cracking the remaining heavy fraction in a fluidized catalytic cracking (FCC) unit.

2.2. Experimental

2.2.1. Biocrude characterization. The biocrude obtained after flash separation and distillation was extensively characterized. Elemental analysis (CHNS, EuroEA, Eurovector) showed a nitrogen content of 0.2%, no detectable sulphur (<0.1%) and a hydrogen to carbon (H/C) molar ratio of 1.36 (Table S1†). We assumed that biocrude distillate did not contain ash, so the oxygen content was determined as the difference from the CHNS analysis. By this method, an oxygen content of around 12% was calculated. This product quality is much better than common pyrolysis oil, where oxygen contents as high as 40% with a H/C molar ratio close to 1.0 are commonly found. The biocrude oxygen content and H/C ratio are indeed not too far from those of the triglycerides present in plant oils or animal fats that are currently commercially hydrogenated. Simulated Distillation (SIMDIS) of this oily product following ASTM-2887 guidelines was also performed. Although SIMDIS data of oxygen-containing streams cannot be compared directly with those of pure hydrocarbon streams, since the removal of oxygen will result in a shift to lower boiling temperatures, it nevertheless gives an upper limit of the boiling point ranges in the sample. In our case this indicates that the biocrude is at least similar to a petroleum-based diesel stream, with 60 wt% of the sample having a boiling point between 216 and 359 °C (diesel-range), 20 wt% below (gasoline-range) and 20 wt% above (bottoms-range) these limits. Consistent with the above discussion, it has to be noted that compared to a petroleum-based vacuum gas oil, the biocrude's heaviest components, as detected by chromatography, are significantly lighter than the heavy end tail detected in the VGO SIMDIS chromatogram (Fig. S1†).

2.2.2. Catalysts characterization. A commercial NiMo catalyst supported on alumina was used as a hydrotreating catalyst and its general properties are summarized in Table 1. Surface area properties were obtained through nitrogen adsorption on a Micromeritics ASAP2420 instrument. Ni and Mo content was obtained by an Inductively Coupled Plasma (ICP) technique on a Varian 715-ES instrument. Unit cell size and zeolite content were determined from a powder X-ray diffraction pattern (CubixPro, PANalytical). Zeolite content is given as compared with a CBV100 zeolite sample taken as a reference. The hydrotreatment catalyst was used without further treatment and sulphided *in situ* during the start-up of the experimental runs. The catalytic cracking catalyst was a commercial grade zeolite catalyst containing 3 wt% rare earth. It was steam deactivated at 816 °C for 4 hours with 100% steam and its main properties are summarized in Table 2.

2.2.3. Hydrotreating and cracking units. The hydrotreating unit (Fig. S2†) comprised a feeding tank, an HPLC pump, a gas system (H_2, H_2S/H_2 for pretreatment, N_2 purge), a fixed bed reactor, a liquid collector, pressure control through a back-pressure-regulator (BPR) valve and a gas exhaust. The feed tank was gently heated (60 °C) and stirred to maintain the biocrude distillate at low viscosity. 4 grams of catalyst were mixed with silicon carbide (SiC) to adjust the total bed volume to 8 ml (approx. 6 grams). An additional 1 ml SiC was added on the top of the bed. The catalyst was crushed and sieved into 0.2 to 0.8 mm particles to avoid diffusion limitations. Once loaded, the catalyst was sulphided at 400 °C with a stream of 10% H_2S in H_2 (120 ml min^{-1}) for at least 12 hours.

Then, the reactor was set to operating temperature (350 °C) and the system was pressurized with hydrogen at 40 bar. Feed injection started once the pressure was stabilized. The hydrogen feed rate was adjusted to 100 N ml min^{-1}, which represents 13.3 wt% of the total feed. As this is a large excess when compared to the amount of hydrogen consumed for gas oil hydro-desulphurisation (HDS) (below 1 wt%) or even hydrodeoxygenation (HDO) of the biocrude distillate (3–6 wt% range), it was done to ensure that the hydrogen partial pressure remains high at every point in the reactor. The weight hourly space velocity was maintained at 0.5 h^{-1} during the whole run.

Mass balances were performed at regular intervals. The liquids recovered were weighed and for each liquid sample at least one gas sample was analyzed online.

Table 1 Hydrotreatment catalyst properties

Catalyst	NiMo
BET (m^2 g^{-1})	200
Pore volume (cm^3 g^{-1})	0.41
NiO (wt%)	3
MoO_3 (wt%)	23

Table 2 Catalytic cracking catalyst main properties after steam deactivation

Rare earth content (wt%)	3.0
BET (m^2 g^{-1})	144
Micropore volume	0.06
UCS (Å)	24.30
Zeolite content (DRX, wt%)	11

Gas samples were taken downstream of the BPR valve. CO, CO_2 and C_1–C_6 gas concentrations were determined. The coke yield was considered very low when estimated from the mass balance (below 1%) and was confirmed by the analysis of coke on the catalyst. An aqueous phase formed a well-defined layer at the bottom of the liquids, and could be easily extracted by pipetting. The remaining water content of the oily phase was determined by Karl Fischer titration after extraction of the aqueous phase from the liquids and was determined to be negligible. The remaining organic phase was analyzed by SIMDIS (ASTM-D-2887). Three cuts were determined from the boiling point curve to help the discussion. The gasoline-range fraction grouped the components with a boiling point below 216 °C (as determined by SIMDIS) and the diesel-range fraction consisted of components with a boiling point between 216 and 359 °C. Bottoms grouped the components boiling above 359 °C. These cuts and the discussions referred to them, especially in the catalytic cracking part, must be understood strictly as SIMDIS cuts and not as commercial fuels. Indeed, the SIMDIS of a commercial road diesel may show some gasoline and bottoms fractions. In catalytic cracking, diesel and bottoms fractions are often designed as Light Cycle Oil (LCO) and Heavy Cycle Oil (HCO) due to their aromatic nature. Some of the liquid products were also analyzed through GC × GC-MS technique, using a system configuration known as reverse GC × GC. Here, the first column (INNOWAX, 30 m × 0.25 mm × 0.25 μm) is polar, while the second one (DB5, 5 m × 0.25 mm × 0.25 μm) is apolar. A flow modulator was used, with a modulation period of 4.5 s. The oven temperature was maintained at 50 °C for 5 minutes, then ramped at a 2 °C min^{-1} ramp up to 250 °C, followed by a plateau for 60 minutes. The second oven followed the first oven program with a 10 °C offset. The FID temperature was 300 °C with the acquisition frequency 100 Hz. The MS acquisition frequency was set at 14 spectra per s in a mass range of 40–360 amu.

Cracking was performed in a fixed-bed micro activity test unit modified from ASTM-3907.[10] The operating conditions were: 500 °C reaction temperature, 30 seconds time-on-stream and a catalyst-to-oil ratio in the range of 2 to 5. These conditions were elected to maximize the middle distillates yield. Liquid products were recovered in glass receivers and gases were recovered in a water burette or a gas bag when biocrude was present in the feed, as CO/CO_2 and some of the oxygenated products would have been lost in the water. Pressure during the test remained close to atmospheric. After the reaction, a 15 minute stripping process with nitrogen at 50 ml min^{-1} was performed. The catalyst was regenerated with air at 550 °C for 3 hours. Gas and liquid samples were analyzed as described in the former paragraph. Nitrogen was used as an internal standard to determine the total volume of gas recovered when using a gas bag. C_5^+ components and oxygenates in the gas were fully quantified and their weight was added to the gasoline fraction in the mass balance. When biocrude was present in the feed, experiments were duplicated and the water was extracted from the liquid product using methanol. The extract was then analyzed by Karl Fischer titration to determine the water content. The coke yield was determined by combustion and integration of the CO_2 signal in flue gas measured by infrared spectroscopy (IR). The conversion was calculated as the sum of gas, gasoline, diesel and coke products. The mass balances performed were 95% or higher in all series.

2.3. Results and discussion

2.3.1. Direct catalytic cracking. Direct cracking of the biocrude was performed, using either a pure feed or a blend with a standard Vacuum Gas Oil (VGO) containing 10 or 30 wt% of biocrude. This was then compared to the results achieved by cracking the VGO alone. Conversion in catalytic cracking is often defined by convention as the sum of gas, gasoline and coke products. However, in the present case, owing to the unconventional nature of the crude treated, we have preferred to define the conversion as the sum of gases, gasoline, diesel and coke products, sometimes referred to as total conversion in conventional catalytic cracking. It should also be taken into consideration that diesel-range products are quite interesting at present. As can be seen in Fig. 2, (total) conversion did not vary much with the incorporation of increasing amounts of biocrude in the VGO. Dry gas selectivity increased with the presence of biocrude, while LPG selectivity decreased. Gasoline selectivity decreases and diesel (Light Cycle Oil, LCO) selectivity increases with the increase of biocrude in the feed mix, with such an amplitude that by the traditional definition, the conversion as the sum of gas, gasoline and coke would have decreased largely when biocrude is added. The result could then give the misleading impression that the biocrude is very refractory to convert. Instead, the amount of (unwanted) bottoms remaining after cracking varies only slightly when biocrude is added. Finally, coke selectivity increases with the presence of biocrude, as does coke-on-catalyst. This last point will strongly limit the direct processability of biocrude in a Fluid Catalytic Cracking Unit (FCCU), as coke yield is determined by the thermal balance of the unit. Then an important increase in coke yield resulting from the incorporation of biocrude into the VGO feed will increase the temperature in the regenerator and consequently result in a reduction of the catalyst to oil ratio in the FCCU, thus indirectly decreasing the conversion of the processed mixture. As a comparison, untreated pyrolysis oil also tends to produce increased coke yields.[11] Coke yield increase is less pronounced when the oil is hydrotreated or obtained through catalytic pyrolysis.[12–14] All these oils tend to yield as much as or more gasoline than the VGO used as standard. This is a consequence of the pyrolysis oil composition, that includes highly oxygenated small molecules that end as monoaromatics with few alkyl carbons, especially when ZSM-5 is used in catalyst formulation.[15] By the contrary, the present biocrude tends to form multi-ring aromatics that end in the diesel fraction as explained below.

The propylene to propane ratio increases with high levels of biocrude in VGO, as does the isobutene to iso-butane ratio. This implies that hydrogen transfer reactions are suppressed with higher levels of biocrude, which is a direct consequence of higher coke-on-catalyst values. Also, the higher amounts of poly-aromatics generated in the cracking of the biocrude play a role in catalyst deactivation.[16] The

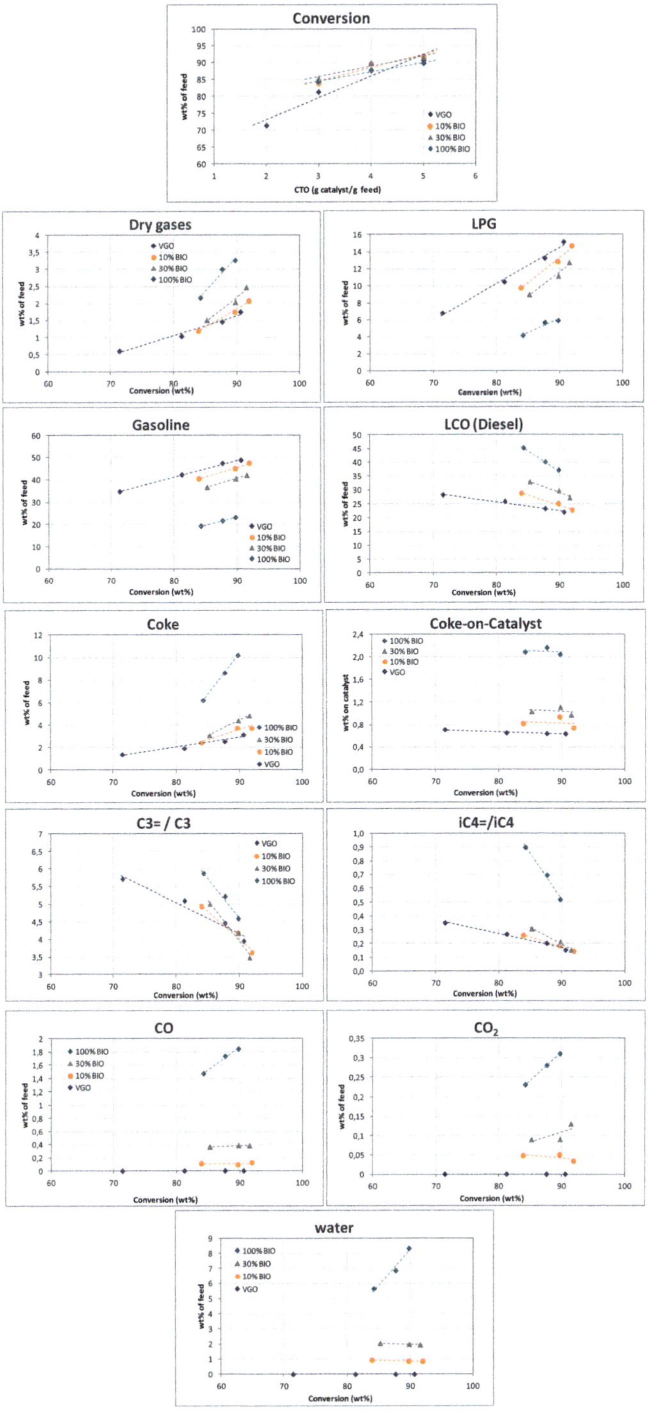

Fig. 2 Catalytic cracking of biocrude, VGO and their mixture, at 500 °C, TOS = 30 s and catalyst-to-oil ratio in the 2–5 range. Conversion, selectivity towards main products, coke-on-catalyst and olefins to paraffins ratios of interest are shown.

analysis of the diesel fraction, the yield of which increased sharply with biocrude addition, revealed that it is very aromatic in nature (Table 3). Also, when biocrude was co-fed a significant amount of oxygenated compounds, essentially phenols and its derivates, were detected in the diesel range. This is in accordance with the composition of the biocrude, where NMR analysis revealed that oxygen is bound essentially in furanic and aromatic form.[4a] The aromaticity of the diesel fraction explains why such high yields of diesel could be maintained even at the higher catalyst-to-oil ratio (giving higher conversion), as these kind of aromatics crack only to a very limited extent to yield gasoline and gaseous fragments.

Independent of the biocrude content of the feed, oxygen was converted mainly into water, with a yield in the 40–60% range. CO and CO_2 were also produced, totalling less than 10 wt% of the feed oxygen, with more CO than CO_2. This indicates that there are few acids to decarboxylate in the feed. In fact, the oxygen balance implies that 30 to 50 wt% of the feed oxygen was still retained in the liquid products or the coke. The amount of oxygenates found in the diesel products is thus consistent with the former observations.

Therefore, the direct catalytic cracking of the biocrude, while being a straightforward and very cheap option, will then be limited to low amounts in co-processing with VGO, below 10 wt%. Nevertheless, an FCC unit in which 2–5 wt % of biocrude will be introduced together with a conventional feed coming from oil could be acceptable from the point of view of the FCCU stability. Note that a 5% share in a FCC unit (FCCU) of 30 000 bpd capacity (roughly 1.5 million tons per year) still represents around 200 tons per day of biocrude to be treated.

2.3.2. Hydroprocessing followed by bottoms cracking. In an alternative approach the biocrude was hydrotreated with a commercial desulphurisation–deoxygenation catalyst in order to improve its quality. While this catalyst is primarily optimized for use with petroleum feedstocks, it also presents an appreciable activity towards deoxygenation. The hydrotreating operating conditions were soft, with a total pressure of 40 bar, a temperature of 350 °C and a WHSV of 0.5 h^{-1}. Similar hydrotreating conditions have been employed successfully with feeds such as plant oils or tall oil.[17,18]

Hydrotreating results are reported in Table 4. In spite of using a treatment pressure in the lower range compared to standard diesel hydrotreatment, deoxygenation of the biocrude was nearly complete, with a recovered water yield very similar to the oxygen content in the feed. Elemental analysis of the recovered oil (*i.e.* separated from the water phase) confirmed that the oxygen content was below 1 wt%. Other biocrudes from hydrothermal treatment were also hydroprocessed to full deoxygenation, but under much more severe operating conditions.[19,20]

Table 3 Diesel yield and composition, CTO = 5, increasing amount of distilled biocrude in VGO

% biocrude in VGO	0	10	30	100
Diesel, wt%	22.0	22.9	26.9	37.6
Composition, wt%				
Saturates	15.9	11.7	9.3	2.1
Mono-aromatics	21.8	20.5	17.6	14.5
Poly-aromatics	62.1	63.9	63.1	53.8
Oxygenates	0.2	3.9	10.1	29.7

Table 4 Operating conditions, yields and hydrocarbon product characteristics for the hydroprocessing of biocrude

Operating conditions	
Catalyst	NiMo
P (bars)	40
WHSV (h^{-1})	0.5
Reactor temperature (°C)	350
Yields, wt%	
Gas	1.9
Liquids: ibp-216 °C	28.9
216–359 °C	49.8
359 °C-fbp	8.1
Water	11.2
Hydroprocessed oil properties	
Density (g cm^{-3})	0.916
Saturates (wt%)	42.0
Mono-aromatics (wt%)	44.1
PAH & oxygenates (wt%)	13.9

The amount of CO and CO_2 in the gas remained very low, and contained less than 5 wt% of the oxygen in the feed. Oxygen was thus removed essentially through a hydrogenation–dehydration mechanism. Gas yield was significantly higher than what is commonly found for hydrocarbon processing, in which gas yield is generally below 1 wt%. The obtained oil contains more gasoline and fewer bottoms than the feed. In particular, the higher boiling point fraction (approximately 20 wt% in the feed boiling above 359 °C) was reduced by half in the oil produced. The heavy fraction may be reduced further by higher pressure processing. The density of the oil product was found to be high compared with the specifications for a road diesel, in spite of a significant light fraction in the product (30% gasoline-range products). The composition determined by GC × GC for the hydroprocessed oil showed a significant amount of aromatics and polyaromatics (PAH). Some oxygenated compounds, essentially substituted phenols, were also detected and totalled approximately 1.5 wt% of the oil sample with respect to the feed. The oxygen content of the oil resulting from the presence of these components, in the 0.2 to 0.4 wt% range, is in accordance with elemental analysis observations. These compounds were lumped with PAH in Table 4. The alkanes and naphthenes fraction (saturates) represented more than 40% of the oil, the density of which remained surprisingly high in view of this composition. Indeed, the GC × GC chromatogram revealed that the saturate fraction included a large amount of poly-cyclic naphthenic products of the decalin family, which have a substantially higher density than the corresponding paraffins with the same carbon number (decane density is 0.730 g cm^{-3} while the density of decalin – decahydronaphthalene – is 0.896 g cm^{-3}). Furthermore, the aromatic fraction contained numerous isomers of the tetralin family, which are also denser than the corresponding decalins. It may then be interesting to further treat these streams with a dearomatization/ring opening catalyst to lower product density. Little pre-treatment would be needed as the hydroprocessed product is sulphur and nitrogen free and has only very little, if any, oxygen remaining.

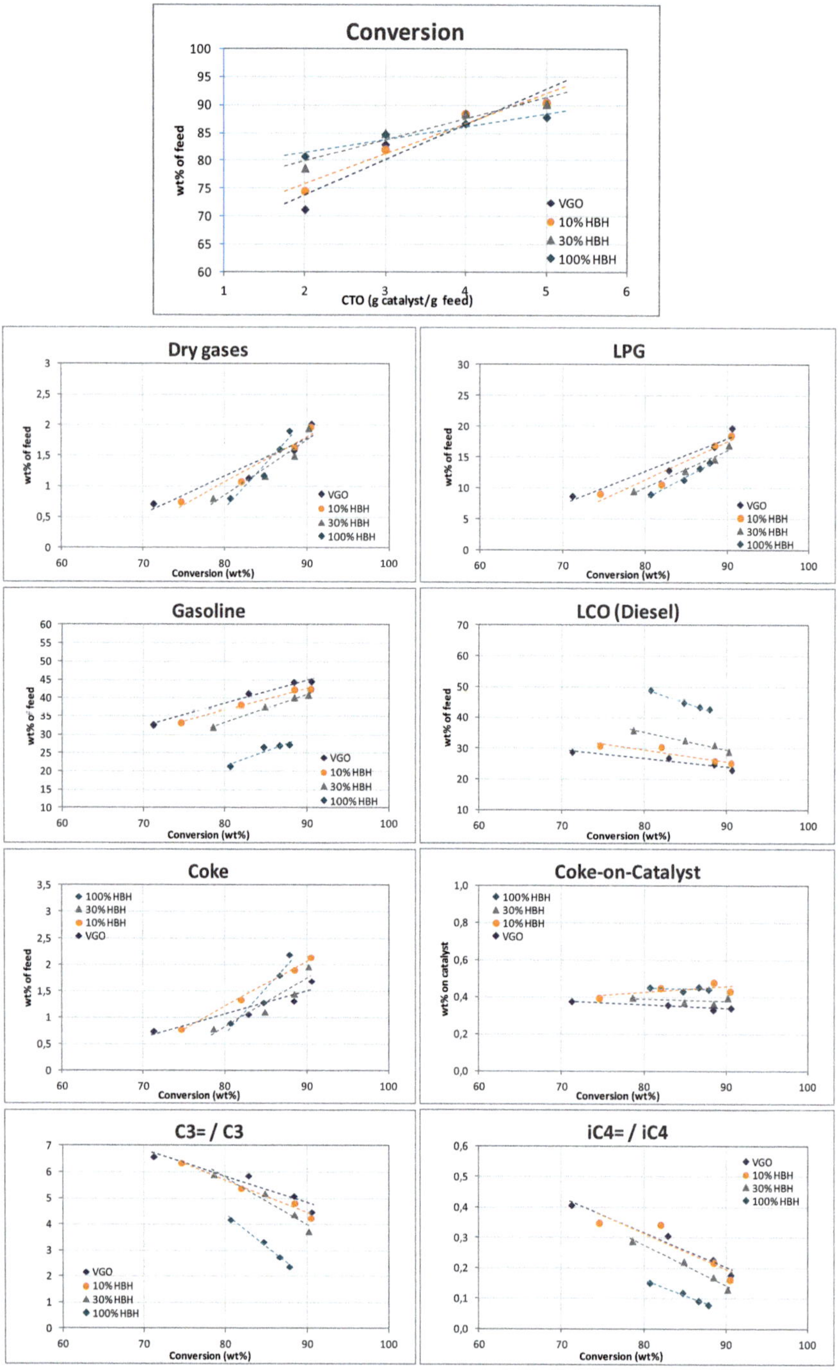

Fig. 3 Catalytic cracking of hydrotreated biocrude heavies, VGO and their mixtures at 500 °C, TOS = 30 s and catalyst-to-oil ration in the 2–5 range. Conversion, selectivity towards main products, coke-on-catalyst and olefins to paraffins ratios of interest.

The oil product from hydroprocessing at 40 bar was further fractionated using equipment and operating conditions similar to those described in the ASTM-D-1160 method. The operation yielded a gasoline-range fraction (30 wt%), a diesel-range fraction (54 wt%) and a heavy fraction (16 wt%). Note that the cut point between diesel-range and heavy fraction was lowered at 300 °C in this distillation, so that the yield of heavies increased compared to a standard cut point (340–360 °C) used in refining operations. This Hydrotreated Biocrude Heavies (HBH) fraction had a density of 0.995 g cm^{-3}, and was further processed by catalytic cracking. The same protocol as in Section 2.3.1 was used; that is, mixing with a VGO at 10 and 30 wt% levels as well as comparing with 100% HBH cracking.

Following the same reasoning as applied to the original biocrude, we report here the selectivity calculated for a conversion defined as the sum of gases, gasoline, LCO (diesel) and coke yields (Fig. 3). The (total) conversion of HBH offers some similarity with that of the biocrude in the sense that the conversion does not change much with the incorporation of a biomass-based stream into VGO, and that the selectivity largely shifts from gasoline to diesel range products. Yet the large difference lies in dry gas and coke selectivity, which are now similar for VGO and HBH. Thus, contrary to biocrude, HBH could be processed even at high blending levels in VGO, with no penalty in conversion. LPG selectivity decreased slightly with the incorporation of HBH, and the olefinicity of the LPG decreased, indicating that hydrogen transfer took place with HBH to a greater extent. This is due to the nature of HBH, which contains a significant amount of hydro-aromatics. Indeed, these compounds could be detected through GC × GC-MS analysis, as is shown in Fig. S3.† Apart from some linear paraffins, the main components of the highlighted groups of peaks in this figure included structures with 3 fused rings of the anthracene or phenanthrene family with varying degrees of saturation, from saturated 3 ring naphthenes to tri-aromatics. 3-Ring saturated systems may crack yielding gas and gasoline components, but hydro-aromatics will be prone to hydrogen transfer, yielding poly-aromatics with 2 or 3 fused aromatic rings that belong to the LCO fraction.[21] The composition of diesel fraction in the cracked liquid shown in Table 5, determined by GC × GC, shows very clearly an increase of the poly-aromatics when the amount of biocrude in the feed is increased. The poly-aromatics formed after hydrogen transfer will accumulate mainly in the LCO fraction, which partially explains the higher selectivity towards this fraction at the expense of gasoline and gases. As the oxygen content in HBH was very low, no CO/CO_2 or water could be detected in the products. Traces of phenolic and benzofuranic compounds could be observed in the diesel fraction, but always with a total amount below 0.5 wt% and they were grouped with the poly-aromatics fraction.

Table 5 Diesel yield and composition with increasing amounts of biocrude. Catalytic cracking of HBH, CTO = 4, 500 °C, TOS = 30 s

% HBH in feed	0	10	30	100
Diesel, wt%	24.7	25.9	31.0	43.2
Composition, wt%				
Saturates	19.0	16.7	13.2	4.3
Mono-aromatics	24.2	23.0	21.3	17.0
Poly-aromatics	56.8	60.3	65.6	78.7

Taking into account the amount of oxygen in phenol-like compounds, we can assume that the cracked diesel has a content of oxygen below 0.1 wt%.

3. Conclusions

Biocrudes obtained from lignocellulose through a hydrothermal treatment have been upgraded so that they can be further refined using existing infrastructure (drop-in replacement). Amounts of biocrude below 10 wt% could be directly fed to the catalytic cracker. A better option for higher blends of biocrude is processing through soft or mild hydroprocessing, yielding a diesel-range stream free of sulphur and very low in oxygen that can readily be refined with the straight-run gas oil, or may be directly used as a diesel pool additive. The heavy ends from hydrotreatment may also be sent to an FCC unit.

Acknowledgements

The authors thank Licella for material and financial support, as well as providing the biocrude used for the hydrotreating experiments. Licella gratefully acknowledges support from the Australian Government in the form of funding as part of the Advanced Biofuels Investment Readiness Program, received through the Australian Renewable Energy Agency (ARENA). Financial support by the Spanish Government-MINECO through program "Severo Ochoa" (SEV 2012-0267), CTQ2015-70126-R (MINECO/FEDER), and by the Generalitat Valenciana through the Prometeo program (PROMETEOII/2013/011) is also acknowledged.

Notes and references

1 U.S. Department of Energy. 2016. 2016 Billion-Ton Report: Advancing Domestic Resources for a Thriving Bioeconomy, Volume 1: Economic Availability of Feedstocks. M. H. Langholtz, B. J. Stokes, and L. M. Eaton (Leads), ORNL/TM-2016/160. Oak Ridge National Laboratory, Oak Ridge, TN. 448pp.

2 P. M. Maitlis and A. de Klerk, *Greener Fischer-Tropsch Processes for Fuels and Feedstocks*, Wiley, 2013, ch. 16.

3 F. M. Mercader, M. J. Groeneveld, S. R. A. Kersten, C. Geantet, G. Toussaint, N. W. J. Way, C. J. Schaverien and K. J. A. Hogendoorn, *Energy Environ. Sci.*, 2011, **4**(3), 985–997.

4 (*a*) F. Goudriaan and D. G. R. Peferoen, *Chem. Eng. Sci.*, 1990, **45**(8), 2729–2734; (*b*) A. A. Peterson, F. Vogel, R. P. Lachance, M. Fröling, M. J. Antal and J. W. Tester, *Energy Environ. Sci.*, 2008, **1**, 32–65.

5 S. A. Toor, L. Rosenthal and A. Rudolf, *Energy*, 2011, **36**, 2328–2342.

6 A. Oasmaa and S. Czernik, *Energy Fuels*, 1999, **13**, 914–921.

7 http://www.licella.com.au/commercial-demonstration-plant/.

8 L. J. Humphreys, WO Pat., 2011/032020(A1), 2011.

9 T. Maschmeyer and L. J. Humphreys, WO Pat., 2011/123897(A1), 2011.

10 A. Corma, F. Melo and J. Prieto, WO Pat., 9012317(A1), 1990.

11 A. V. Bridgwater, *Appl. Catal., A*, 1994, **116**, 5–47.

12 F. de Miguel Mercader, M. J. Groeneveld, S. R. A. Kersten, N. W. J. Way, C. J. Schaverien and J. A. Hogendoorn, *Appl. Catal., B*, 2010, **96**, 57–66.

13 C. Wang, M. Li and Y. Fang, *Ind. Eng. Chem. Res.*, 2016, **55**(12), 3525–3534.
14 G. Fogassy, N. Thegarid, Y. Schuurman and C. Mirodatos, *Green Chem.*, 2012, **14**, 1367–1371.
15 P. S. Rezaei, H. Shafaghat and W. M. A. W. Daud, *Appl. Catal., A*, 2014, **469**, 490–511.
16 R. Hughes, G. J. Hutchings, C. L. Koon, B. McGhee, C. E. Snape and D. Yu, *Appl. Catal., A*, 1996, **144**(1–2), 269–279.
17 G. W. Huber, P. O'Connor and A. Corma, *Appl. Catal., A*, 2007, **329**, 120–129.
18 J. M. Anthonykutty, K. M. Van Geem, R. De Bruycker, J. Linnekoski, A. Laitinen, J. Rasanen, A. Harlin and J. Lehtonen, *Ind. Eng. Chem. Res.*, 2013, **52**(30), 10114–10125.
19 H. P. Ruyter and J. H. J. Annee, *US Pat.*, no 4,670,613, 1987.
20 S. Jones, Y. Zu, D. Anderson, R. Allen, D. Elliot, A. Schmidt, K. Albrecht, T. Hart, M. Butcher, C. Drennan, L. Snowden-Swan, R. Davis and C. Kinchin, *PNNL report 23227*, March 2014.
21 (*a*) A. Corma, V. González-Alfaro and A. V. Orchillés, *J. Catal.*, 2001, **200**, 34–44; (*b*) A. Corma and F. J. Ortega, *J. Catal.*, 2005, **233**, 257–265.

Faraday Discussions

PAPER

Fischer–Tropsch synthesis. Evaluation of an aluminum small channel reactor

D. E. Sparks,[a] S. Vallee,[b] Zhijun Jia,†[b] W. D. Shafer[a] and B. H. Davis*[a]

Received 15th August 2016, Accepted 10th October 2016
DOI: 10.1039/c6fd00179c

Fischer–Tropsch synthesis was conducted in a small channel compact heat exchange reactor that was constructed of aluminum. While limited to lower temperature–pressure regions of the Fischer–Tropsch synthesis, the reactor could be operated in an isothermal mode with nearly a constant temperature along the length of the channel. The results obtained with the compact heat exchange reactor were similar to those obtained in the isothermal continuous stirred tank reactor, with respect to both activity and selectivity. Following a planned or unplanned shutdown, the reactor could be restarted to produce essentially the same catalytic activity and selectivity as before the shutdown.

1. Introduction

The reactor is essential for making use of a catalyst. While the reactor can never improve the activity or selectivity of a catalyst, it can certainly decrease either or both of these properties for a catalyst. Thus, there is a need to match the catalyst and reactor to obtain the optimum activity and product selectivity.[1]

The Fischer–Tropsch reaction is very exothermic. The early German work utilized fixed-bed reactors and they made efforts to control the magnitude of the exotherm.[2] One approach was to use concentric tubes with cooling water passing through the inner tube and in the space outside the outer tube (Fig. 1). The catalyst thickness was about 10 mm, a scale comparable to the size of the channels in today's microchannel reactor.[3] For the atmospheric pressure reactors, the reactors were a rectangular sheet-steel box containing a bundle of horizontal water-cooling tubes interlaced by vertical steel sheets placed 7.4 mm apart. Thus, the Germans were operating at the heat transfer scale of today's microchannel reactors in their commercial operations in the 1940s. In addition, they had standardized the design and reactor production. Each reactor was sized to

[a]*Center for Applied Energy Research, University of Kentucky, 2540 Research Park Drive, Lexington, KY 40511, USA. E-mail: burtron.davis@uky.edu*
[b]*Chart Energy & Chemicals, Inc., 2191 Ward Avenue, La Crosse, WI 54601, USA*
† Current address: CompRex, LLC, 1740 Eisenhower Drive, De Pere, WI 54115, USA.

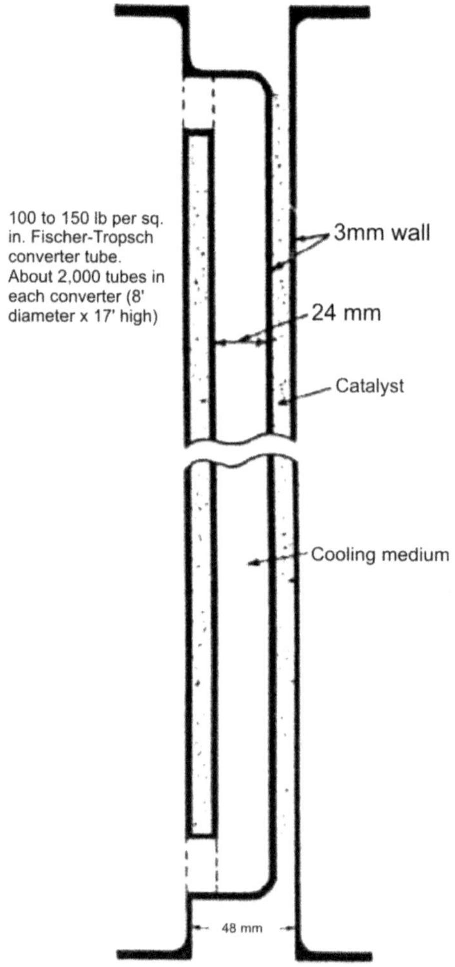

Fig. 1 A section of middle-pressure Ruhrchemie synthine reactor tube.

produce about 15 barrels per day. For each plant they employed a sufficient number of the 15 barrel per day reactors to meet the design capacity.

The introduction of auto exhaust catalysts led to the introduction of monolith or honeycomb type reactors; further development of these reactors led to smaller, higher density tubes that were wash-coated with a thin catalyst layer. Eventually the size of the tubes reached the size of today's microchannel reactor but the catalyst was only present as a thin coating on or within a thin layer on the wall of the tubes so that the weight of catalyst per reactor volume was low, as was product production.

Gas phase microreactors have been utilized for decades. For example, Emmett and co-workers[4] utilized a microreactor coupled with a gas chromatograph just three years after the development of the gas chromatograph was announced. However, it appears that microreactors with embedded high performance heat exchangers capable of handling liquids were only made in the early 1990s.[5] These were quickly utilized for the Fischer–Tropsch reaction by Battelle workers and this became the basis for a spin-off company, Velocys. Progress in the work using microchannel reactors has advanced to the point that commercialization is now considered possible.[6–8]

At least two materials have been considered for the construction of small channel reactors: stainless steel and aluminum. Aluminum, having a much higher thermal conductivity, gives superior thermal performance and a more isothermal temperature profile for the same channel geometry. As the operating temperature of the aluminum reactor increases the operating pressure is usually lowered. This limits the operating temperature–pressure range for aluminum much more than for stainless steel. However, the construction with aluminum requires a brazing time that is only a fraction (0.15–0.2) of the time that is required for stainless steel; thus the construction with aluminum is much faster and less costly. Chart Energy & Chemicals, Inc. currently manufactures small channel heat exchangers using aluminum that is sized up to $1.22 \times 1.22 \times 8.54$ m ($4 \times 4 \times 28$ feet) and can provide the heat treatment needed for these units. A unit of this size when filled with a commercial cobalt catalyst should be capable of producing about 500 barrels per day of Fischer–Tropsch products. Thus, it was desirable to test the same catalyst in a small channel reactor constructed of aluminum as well as in a conventional continuous stirred tank reactor (CSTR), and that is the purpose of the work described in this manuscript.

2. Experimental

The testing of a Fischer–Tropsch catalyst in a CSTR in our lab has been described previously.[9] Briefly, the catalyst was reduced in a flow of 25% H_2/75% He at 623 K overnight in a fixed bed reactor. After reduction it was cooled to room temperature and isolated under hydrogen, connected to the CSTR through a fitting and then 8.9 g of Süd Chemie (now Clariant) T2857 (120–170 mesh) catalyst was transferred by over-pressuring with H_2, and blown into a CSTR that contains molten Polywax 3000, a polyethylene polymer with an average carbon number of 200, or other solvent. The CSTR was closed, flushed three times with H_2, and the catalyst was reduced at 543 K overnight in flowing H_2. The syngas ($H_2/CO = 2.0$) was delivered to the slurry *via* a sparger tube located below an impeller blade turning at 750 rpm. The effluent exited the reactor and passed sequentially through two traps maintained at 333 and 273 K. Nonvolatile products were drained from the reactor through a metallic filter to a collection vessel maintained at 473 K.

A schematic of the small channel compact heat exchange reactor is shown in Fig. 2. An oil temperature control unit (A) can provide up to 68 liters per minute of TrueTherm heat transfer oil at temperatures ranging from ambient to 560 K, with a typical deviation from the set point of 1 K. The catalyst and heat exchange channels are shown in Fig. 3; the catalyst channel is about 2×4 mm and extends the length of the reactor. Mass flow controllers are used to control the feed rates of hydrogen and carbon monoxide as well as any other gas that is added. Two sets of product traps are utilized; one set contains 1 liter hot (423 K) and cold (273 K) traps used to collect samples for GC analysis or products produced with low feed gas flows. During the stabilization period following operating changes or during the major period of operation before collecting samples for analysis, the 3.8 liter hot and cold traps are used. A back pressure regulator is used to maintain reactor

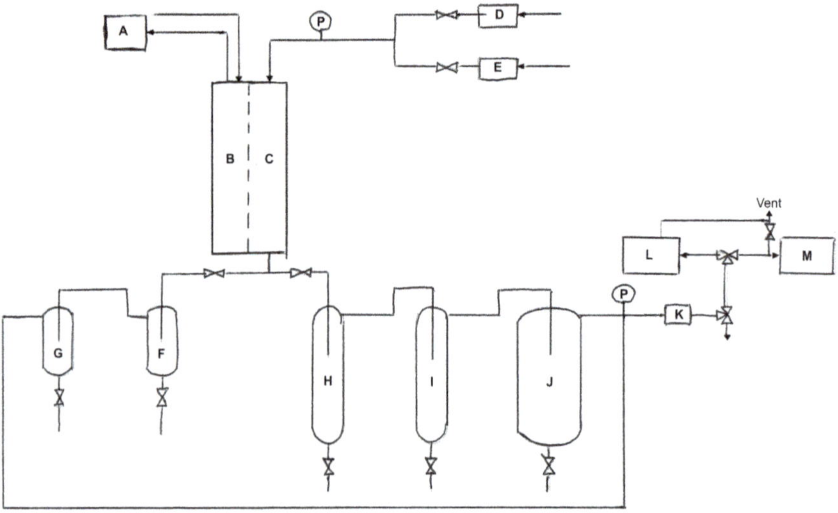

Fig. 2 Schematic for the CAER Chart Energy micro-channel reactor. (A) AEC Model TCOU-1 portable temperature control unit. (B) The heat exchange side of the Chart Energy micro-channel reactor. (C) The process side of the reactor. (D) Brooks Model 5850E mass flow control (MFC) valve for metering hydrogen. (E) Brooks 5850E MFC for metering carbon monoxide. (F) One liter warm trap, maintained at 150 °C by an electrical heat tape controlled by the LabView process control program. (G) One liter cold trap, maintained at −1 °C by a chilled water bath. (H) One gallon (3.8 L) warm trap, maintained at 150 °C. (I) One gallon cold trap. (J) A second cold trap that can be added to the product recovery train as needed. (K) Swagelok 0–500 psig back pressure regulator (BPR). (L) American Meter Company Model 200A dry test meter (OTM). (M) Agilent Model 3000A Micro GC. (P) 0–600 psig pressure gauges.

pressure and to decrease to atmospheric pressure. A dry test meter is utilized to measure the flow rate of the gas that passes through the 273 K cold trap. A picture of the installed reactor is shown in Fig. 4.

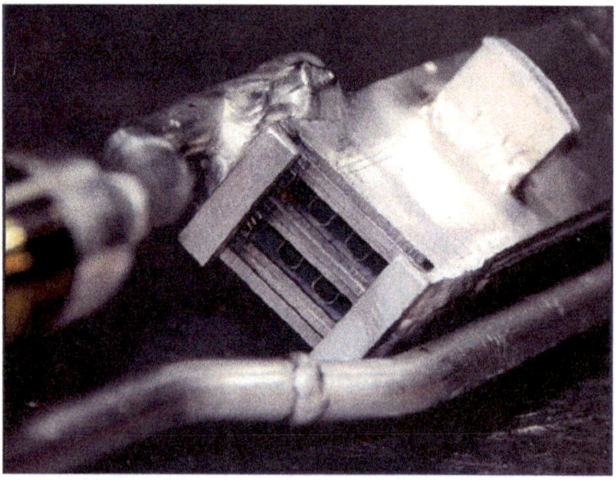

Fig. 3 Picture of the reactor showing the small channels for the catalyst and cooling fluid.

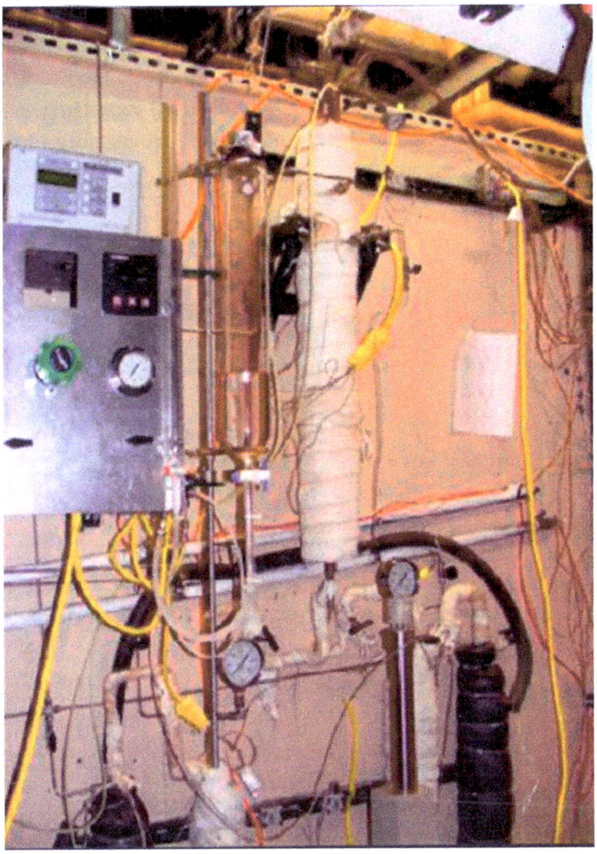

Fig. 4 Picture of the reactor represented by the schematic shown in Fig. 2.

The exit gas may be diverted to an Agilent Model 3000A Micro GC to determine H_2, CO, N_2 or other inert gases, and C_1–C_4 hydrocarbons. The water and oil layers from the hot and cold traps are separated and the water layers are combined to be analyzed for dissolved oxygenates by GC analysis. The hydrocarbon layer is analyzed using a standard GC and a high temperature GC.[10]

The catalysts utilized in this study were supplied by a commercial source and those prepared in our labs. A brief description will be presented when the results obtained using them are given in the Results section. Prior to use the catalyst was reduced in flowing hydrogen (25% H_2 and 75% He) overnight at 623 K at atmospheric pressure. It was then cooled to reaction temperature, the gas switched to pure hydrogen, and cooled to room temperature or about 373 K. While increasing the temperature to the reaction temperature, the feed was switched to a syngas mixture.

3. Results and discussion

The selection of catalyst particle size is an important factor in optimizing the productivity of the fixed bed compact heat exchange reactor. As smaller physical particles are used, the impact of diffusion effects decreases and this is desirable since diffusion limitations increase the fraction of the products that are undesirable light gases. On the other hand, the pressure drop across the catalyst bed becomes larger as the particle size decreases. Thus, there must be a trade-off between diffusion effects and pressure drop effects. Our preliminary work indicates that a physical particle size of about 100 microns is the optimum trade-off for these two factors.

3.1. Decline due to physical factors

A feature that may be encountered during slurry phase synthesis is a rapid decline in conversion during the early portion of the run (Fig. 5), and a severe example is shown by data generated by Syntroleum workers using syngas generated in a commercial coal gasifier.[11] While this data exhibits a greater decrease in activity than is usually observed, it is a feature of many slurry phase runs. However, with the compact heat exchange reactor, this initial decline in activity is usually not observed (Fig. 6). The products produced in this run had an alpha value of 0.78 with about 90% of the product, on a molar basis, exiting the reactor in the gas phase. These products combined with the moles of water produced and the unconverted syngas indicate that the moles of gases exiting the reactor greatly exceeds the moles of nonvolatile products that must leave the reactor in the liquid phase. Our earlier work has shown that most, if not all, of this initial drop in activity is due to the accumulation of the heavy solvent and product molecules in the interior of the catalyst particle. This is demonstrated in Fig. 7.[12] Here, batches of the same catalyst were treated the same, except in one run the solvent was Polywax 500 (average carbon 35) and the other was Polywax 3000 (average carbon number 200). The catalyst exhibited a constant activity during the 350 hour run in Polywax 500 but the activity decreases by nearly 2/3rd during the run in Polywax 3000; thus, accumulation of liquid occurred within the particle to cover active

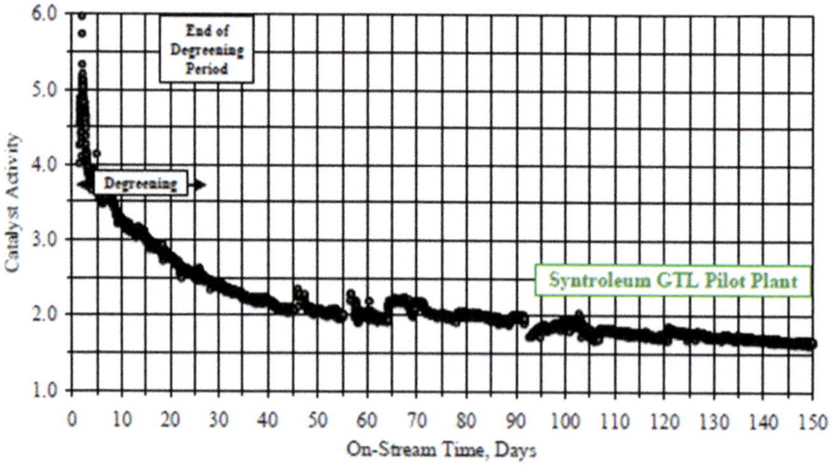

Fig. 5 Catalyst activity during the initial period of Fischer–Tropsch synthesis in a slurry reactor, using syngas from coal liquefaction at Eastman Chemicals by Syntroleum operators.[11]

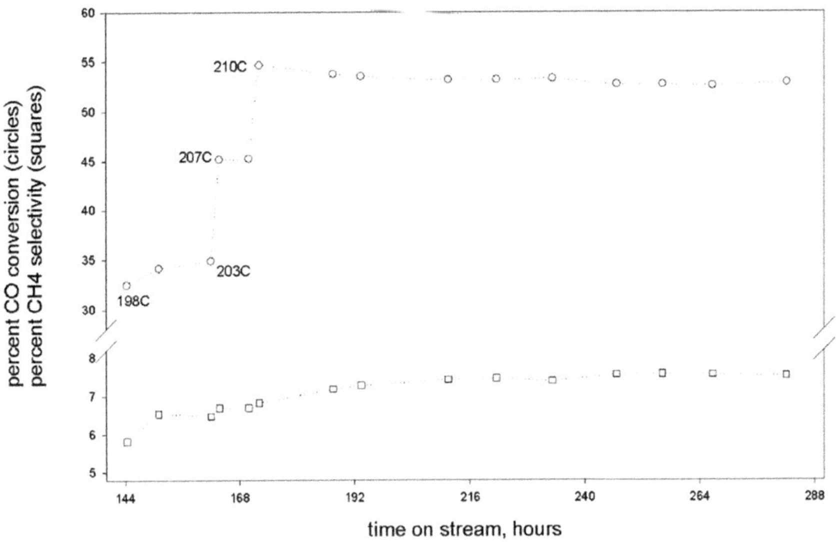

Fig. 6 Initial start-up conversion is the compact heat exchange reactor.

sites with the higher boiling solvent. The large gas flow in the compact heat exchange reactor eliminates, or greatly reduces, the impact of non-volatile product accumulation in the interior of the catalyst particles. This is a very important advantage for utilizing the fixed bed compact heat exchange reactor.

Catalyst life using a fixed bed reactor is a major factor since the time off-stream dramatically reduces productivity. In addition, many operational problems are introduced to affect regeneration. Both of these factors are detrimental to operating expenses for the process. A run was made with a laboratory catalyst, provided by a catalyst company, that lasted more than 2200 h (92 days) with a decline in conversion of less than 1% per week (Fig. 8). During a 7 week period of

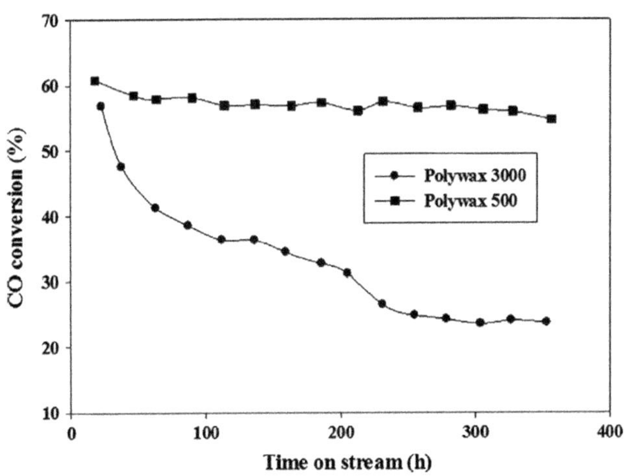

Fig. 7 Effect of the start-up solvent on CO conversion for 20–63 μm size catalyst (reaction conditions: $T = 220$ °C; $P = 20$ atm; SV-3 slph $g_{catalyst}^{-1}$; $H_2/CO = 2$).[12]

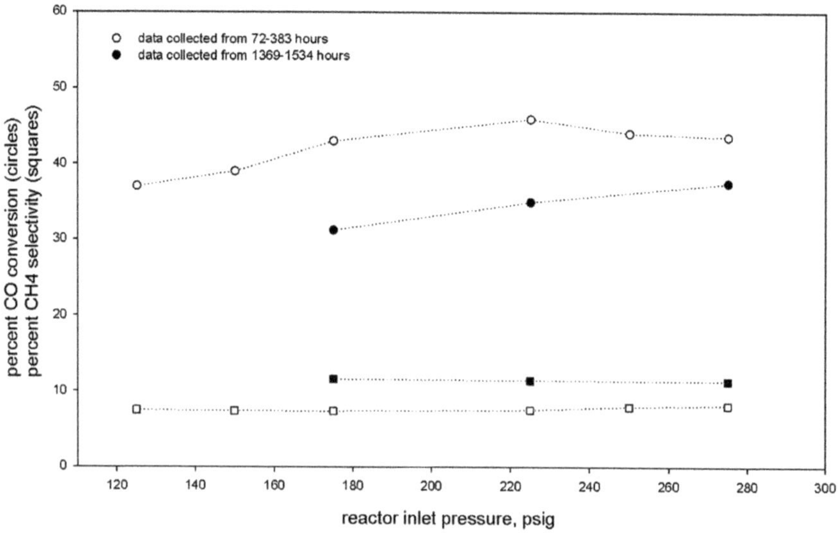

Fig. 8 CO conversion activity and methane selectivity following two regenerations in hydrogen (SV = 3, H_2/CO = 2.0, 487 K, 11.9 atm gauge).

testing in another run, the catalyst lost activity at about 2.0% CO conversion per week, somewhat higher than measured with this catalyst in other runs, but still a slow deactivation rate. The methane selectivity shows a small increase from about 9 mol% to about 11 mol%. These are impressive numbers despite the fact that the catalyst particle size was not at the optimum for selectivity.

3.2. Catalyst activation

Two *in situ* catalyst reactivations were attempted with a test catalyst. After being on-line for 690 h, the CO flow was terminated but the hydrogen flow continued as during the run at reaction temperature until methane could not be detected in the exit gas. The pressure was decreased from 11.9 to 5.1 atm gauge and the hot oil temperature increased to 518 K. The reactor was held at this temperature for 23 h before increasing the oil temperature to 528 K with a further decrease in hydrogen pressure to 3.4 atm gauge. The reactor was held at these conditions for 7 h and then cooled to 453 K to begin a normal startup. After running the regenerated catalyst for about 400 h, a second catalyst reactivation was attempted using the same procedure, except the step at 518 K was omitted. The two runs with the regenerated catalyst were the same as the initial portion of the run prior to reactivating, except that Durasyn 164 was added during the initial portion of the run but was not used for the startup following the two regenerations.

The data shown in Fig. 9 make it clear that the reactivation in a hydrogen flow restored the initial CO conversion to a high value, but that in these two runs there was a decline in CO conversion to approach a stable conversion after 20–40 h on-stream. Thus, while the reactivations were reproducible, their activity *vs.* time curves did not duplicate that of the data obtained during the initial period of the run (Fig. 9). For the initial period of the run, a flow of Durasyn was utilized so that the catalyst particles should have some liquid present; thus, the initial conversion

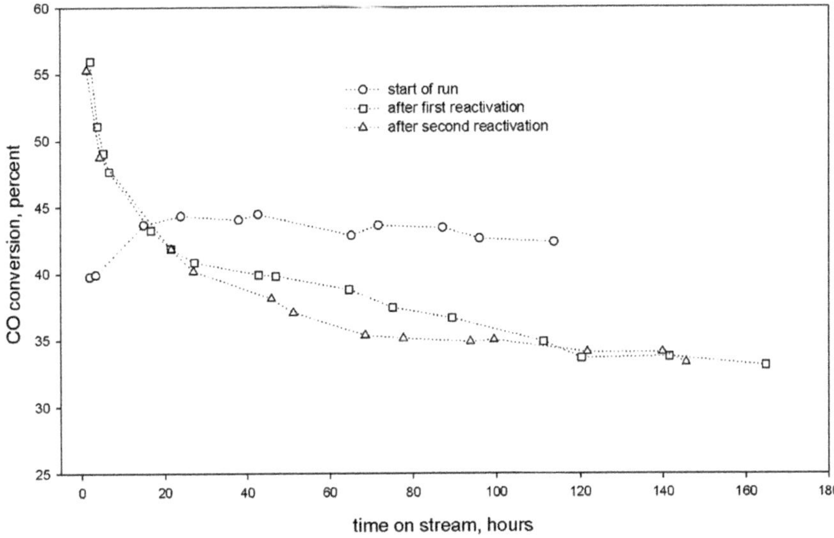

Fig. 9 CO conversion for the initial part of the run and after the 1st and 2nd reactivations in hydrogen, compared from the start of the run period (SV = 2.98, $H_2/CO = 2.0$, $T = 488$ K).

was low and increased slightly during the run to reach a steady level. The reactivations also had a detrimental impact on the methane selectivity which increased from about 7 mol% to about 14 mol% (Fig. 10). The start-up following the two reactivations were conducted without a flow of Durasyn. In both of these runs the initial activity was much higher than during the first initial period, but it

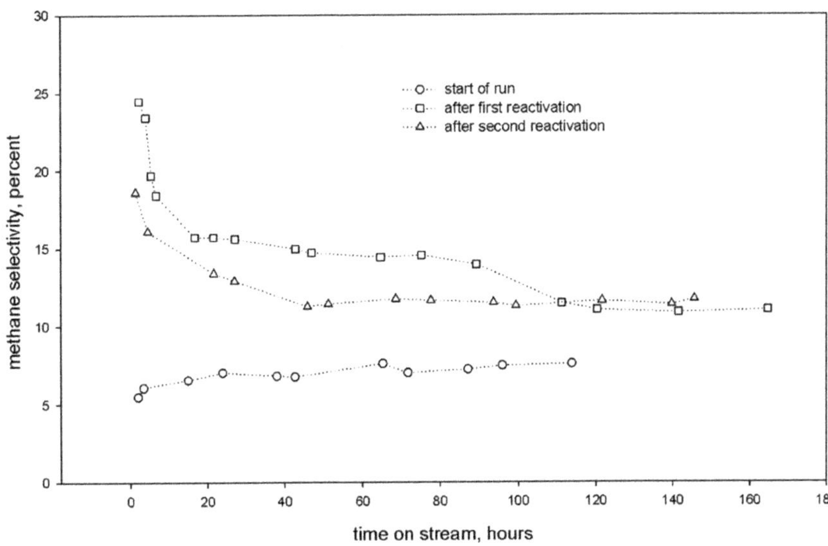

Fig. 10 Methane selectivity for the initial part of the run and after the 1st and 2nd reactivations in hydrogen, compared from the start of the run period (SV = 2.98, $H_2/CO = 2.0$, $T = 488$ K).

declined rapidly during a 20–40 hour period. In addition, sudden, high-temperature spikes were observed in both runs following reactivation, but not during the initial portion when the catalyst was started up in a flow of Durasyn. More details of these temperature excursions are presented in a following section.

3.3. Shut-down

During these exploratory studies we encountered intended and unintended shut-downs. The response to an intended shut-down because of the need to interrupt the electrical power to the lab for a period of time involved slowly cooling the reactor to room temperature and decreasing to atmospheric pressure before discounting the gas flow. After two days the run was re-started by slowly increasing the pressure and temperature under a syngas flow that was the same as used during the run (3 L h^{-1} $g_{cat}{}^{-1}$). As shown in Fig. 11, both the CO conversion and methane selectivity were the same, following the shut-down and start-up as they were before this operation.

An unintended power loss was experienced during a run that was made at different pressures (Fig. 12). Following the power loss, the syngas feed was terminated and the reactor quickly cooled to room temperature. The reactor was allowed to remain under the syngas blanket for about a day before reheating to the reaction temperature in syngas for two hours. As shown in Fig. 12, neither the CO conversion nor the methane selectivity were impacted by this unintended power loss and unscheduled reactor shut-down.

3.4. Activation energy

In many runs with cobalt–alumina catalysts in a CSTR reactor, we obtain a value for the activation energy in the range of 95 ± 10 kJ (Table 1).[13] This range of activation values has also been reported by other workers. The activation energy

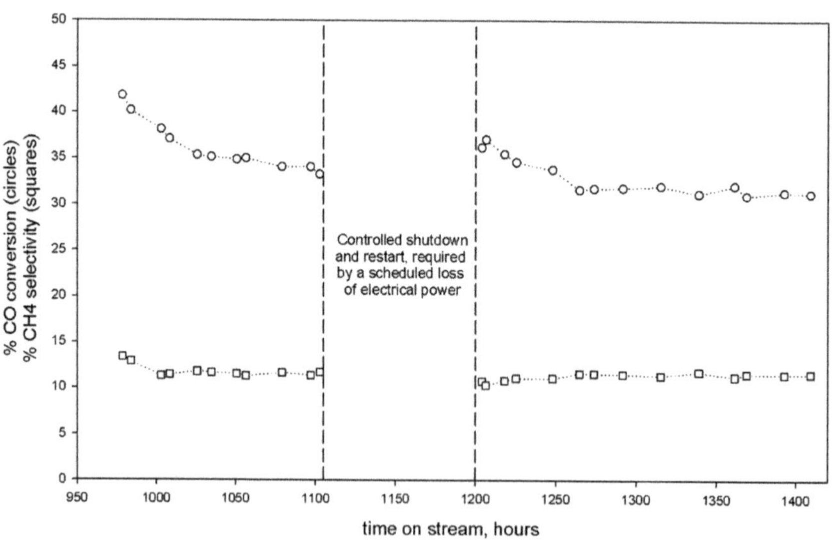

Fig. 11 CO conversion and methane selectivity before and following a scheduled shut-down.

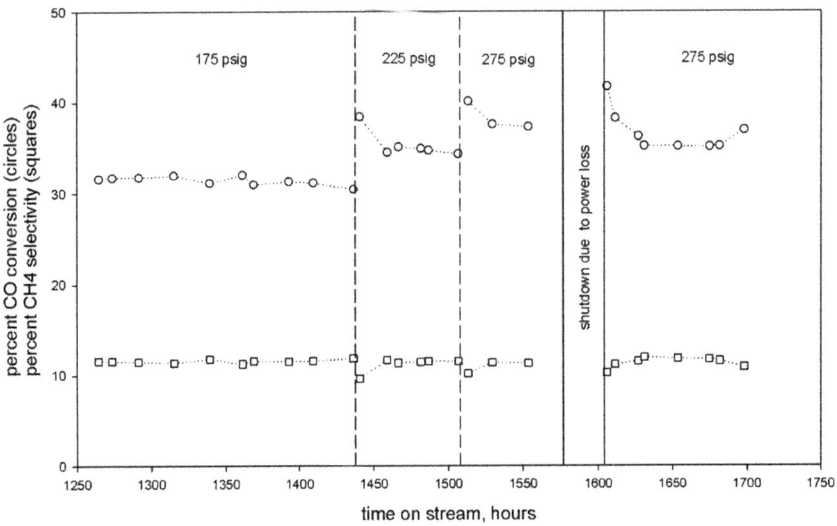

Fig. 12 CO conversion and methane selectivity before and following an unintended power loss.

was determined for three runs in the small channel compact heat exchange reactor and the values obtained were: $E_a = 83.3$, 88.2 and 103 kJ mol^{-1}. The lowest value was obtained with a larger catalyst particle size and it is probably impacted by diffusion. Even so, it is apparent that the activation energies obtained using the small channel fixed bed compact heat exchange reactor yielded activation energies that are similar to those obtained in a CSTR, where diffusional limitations are small, if present at all (Fig. 13).

3.5. Comparison with CSTR

A portion of the catalyst batch was ground to 100 microns or smaller. A run was conducted in a CSTR with the same catalyst activation and reaction conditions as utilized with another batch of the catalyst in the small channel compact heat exchange reactor. Fig. 14 shows data for the conversion of CO during two runs in the small channel compact heat exchange reactor and in a CSTR using the same reaction conditions and catalyst batch. The start-up solvent for the CSTR run was a decene trimer so that the T-shape of the trimer should limit its diffusion into smaller catalyst pores. This was so that we could obtain constant conversions with time that resemble those for the Polywax 500 run shown in Fig. 7, and we did. Furthermore, in the CSTR all catalyst particles experience the same reactant concentration and temperature so that all catalyst particles should be equally productive. The conversion data obtained during two separate runs with the portions of the same catalyst batch show that the activation and run data are very reproducible in the small channel compact heat exchange reactor. In addition, the conversion data obtained in the two types of reactors agree very well. This is consistent with the measured nearly constant reaction temperature along the entire length of the fixed bed reactor so that all catalyst particles are equally active; only the concentrations of the reactants will vary along the catalyst in the fixed

Table 1 Activation energies (E_a) of different Co catalysts[13]

Catalyst	Support	Pt/Co loading, wt%	Calcination/pretreatment	Activation energy E_a, kJ mol^{-1}
NASA 2A	Narrow pore Condea Vista SBA 200 γ-Al_2O_3	0/25	NO calcined at 350 °C for 4 h/350 °C, H_2/He = 1 : 3, 1 NL min^{-1}, 1 atm for 10 h	88
NASA 2B	Narrow pore Condea Vista SBA 200 γ-Al_2O_3	0/25	Air calcined at 350 °C for 4 h/350 °C, H_2/He = 1 : 3, 1 NL min^{-1}, 1 atm for 10 h	87
NASA 2BX	Condea Catalox SBA-150 γ-Al_2O_3	0/25	Air calcined at 350 °C for 4 h/350 °C, H_2/He = 1 : 3, 1 NL min^{-1}, 1 atm for 10 h	104
NASA 3	Medium pore Puralox HP 14 γ-Al_2O_3	0/25	Air calcined at 350 °C for 4 h/350 °C, H_2/He = 1 : 3, 1 NL min^{-1}, 1 atm for 10 h	78
NASA 4	Wide pore Degussa γ-Al_2O_3	0/25	Air calcined at 350 °C for 4 h/350 °C, H_2/He = 1 : 3, 1 NL min^{-1}, 1 atm for 10 h	77
NASA 10	Narrow pore Condea Vista SBA 200 γ-Al_2O_3	0.5/15	Air calcined at 350 °C for 4 h/350 °C, H_2/He = 1 : 3, 1 NL min^{-1}, 1 atm for 10 h	96
NASA 11	Narrow pore Condea Vista SBA 200 γ-Al_2O_3	0/15	Air calcined at 350 °C for 4 h/350 °C, H_2/He = 1 : 3, 1 NL min^{-1}, 1 atm for 10 h	111
NASA 6	PQ-SiO_2 CS-2133	0/15	Air calcined at 350 °C for 4 h/350 °C, H_2/He = 1 : 3, 1 NL min^{-1}, 1 atm for 10 h	86
NASA 8	PQ-SiO_2 CS-2133	0/25	Air calcined at 350 °C for 4 h/350 °C, H_2/He = 1 : 3, 1 NL min^{-1}, 1 atm for 10 h	94

bed reactor. Thus, we would expect a similar conversion and that is the experimental result that was obtained.

The methane selectivities obtained using the two reactors are similar but the values obtained at the higher space velocities for the CSTR seem to be too high (Fig. 15). Even so, it appears that the conversion and selectivities obtained using the two types of reactors are essentially the same and this is what is expected if the fixed bed reactor was operating under isothermal conditions.

3.6. Temperature excursions

Upon restart of the reactor with the freshly re-activated catalyst and without liquid present in the reactor, some temperature excursions were observed. This demonstrates the importance of using a start-up procedure that includes liquid injection until an adequate amount of liquid products are formed. The temperature excursions had a significant impact on conversion and selectivity. The steady state operation of the small channel reactor may be interrupted for a brief (usually minute or less) temperature excursion. The temperature excursion is measured in the channel containing the catalyst but not in the channel containing the coolant. Furthermore, this temperature excursion usually occurs only

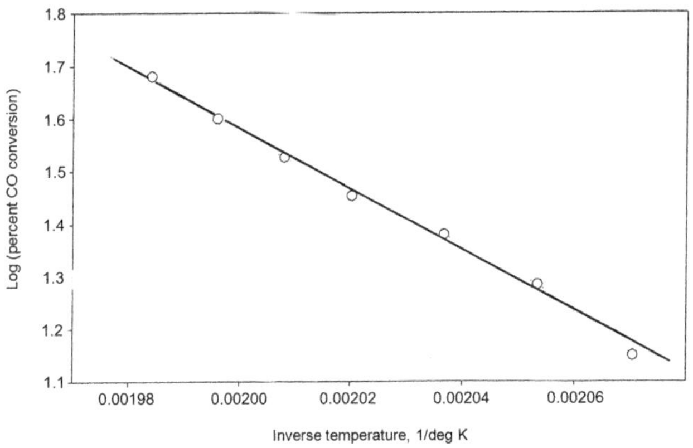

Fig. 13 Example of the activation energy plot for data obtained in a small channel compact heat exchange reactor using a Co–alumina catalyst.

near the top of the catalyst containing channel. As indicated above, during the start-up following the first regeneration, in the absence of liquid flow, three brief exotherms occurred at the top of the channel containing the catalyst (Fig. 16, left). During the start-up following the second hydrogenation reactivation, an exotherm was observed but this time it appeared in the middle of the catalyst filled channel (Fig. 16, right). This is likely because the catalyst at the top of the filled channel was already partially deactivated from the previous thermal excursions. As indicated above, the initial CO conversion was much higher than was obtained during the initial period but declined to a lower steady state CO conversion following both hydrogen reactivations.

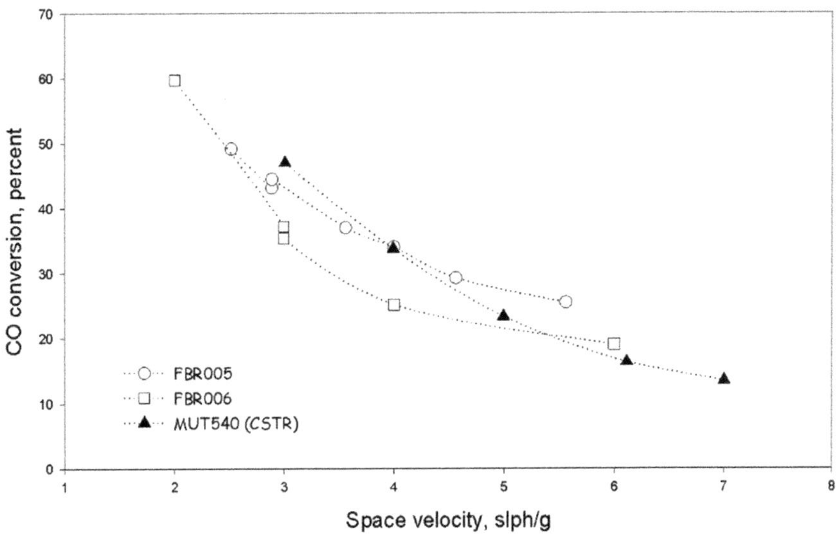

Fig. 14 CO conversion using a small channel fixed bed compact heat exchange reactor and a CSTR (487 K, 11.9 atm gauge).

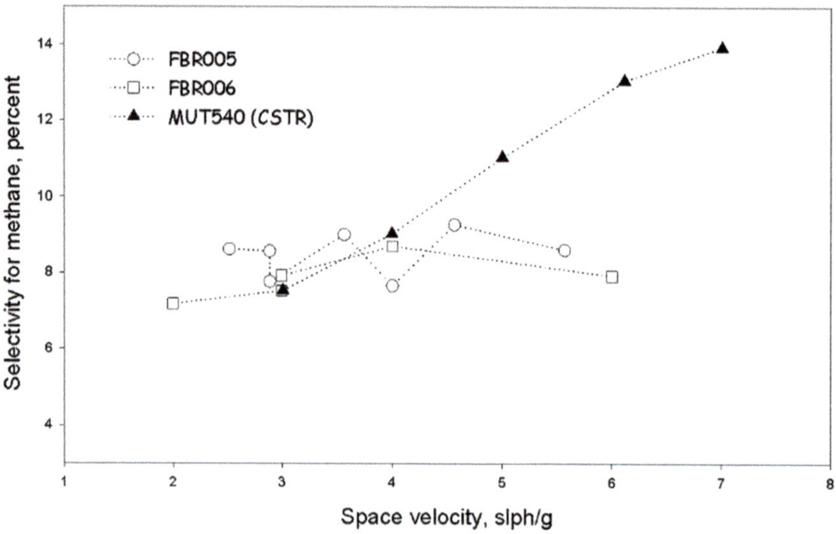

Fig. 15 Methane selectivity for runs conducted in a CSTR and in the small channel compact heat exchange reactor for the run with conversions shown in Fig. 14.

Another such excursion is illustrated in Fig. 17. This temperature excursion occurred during day 18 of the run as the temperature was being increased from 494 to 498 K. The temperature excursion was not experienced in the cooling side oil nor at thermocouples located further down the catalyst channel of the reactor. This temperature exotherm was not observed by the operators; it was apparent only when the temperature profiles were examined the next day. The lower two thermocouples showed the expected 2–4 K degree increase during the temperature change operation. However, the thermocouple in the upper bed increased to 505 K five minutes into the temperature increase and then in the sixth minute had increased to 609 K, but returned to 503 within a minute before decreasing to the set temperature of 498 K within the next eight minutes. The exotherm impacted the CO conversion. Two days (day 16) prior to the temperature increase, the space velocity was returned to the value where the CO conversion had been measured on day 5 and, within experimental error, the CO conversions

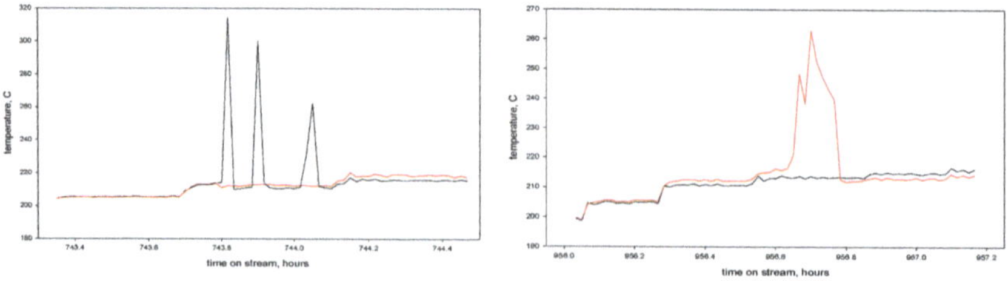

Fig. 16 (Left) Temperature excursions at the top of the catalyst filled channel, but at the middle or bottom of the catalyst bed, following start-up after the first regeneration in hydrogen. (Right) Temperature excursions at the middle of the catalyst filled channel, but at the middle or bottom of the catalyst bed, following start-up after the second regeneration in hydrogen.

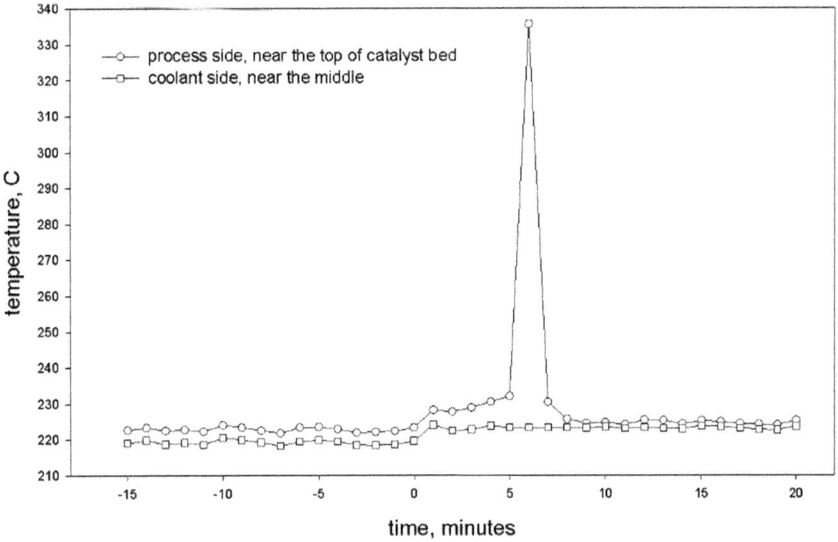

Fig. 17 Temperature measurements during the temperature ramp as the temperature was increased from 494 to 498 K (SV = 5.57 L h^{-1} g_{cat}^{-1}, H_2/CO = 2.0).

were the same (*ca.* 47%) (Fig. 18). Following the excursion, the CO conversion at the same space velocity was only about 31%. Thus, the temperature excursion caused a CO conversion decrease of about 34%. As shown in Fig. 19, the temperature excursion also caused an increase in methane selectivity from about 8 mol% to about 12 mol%.

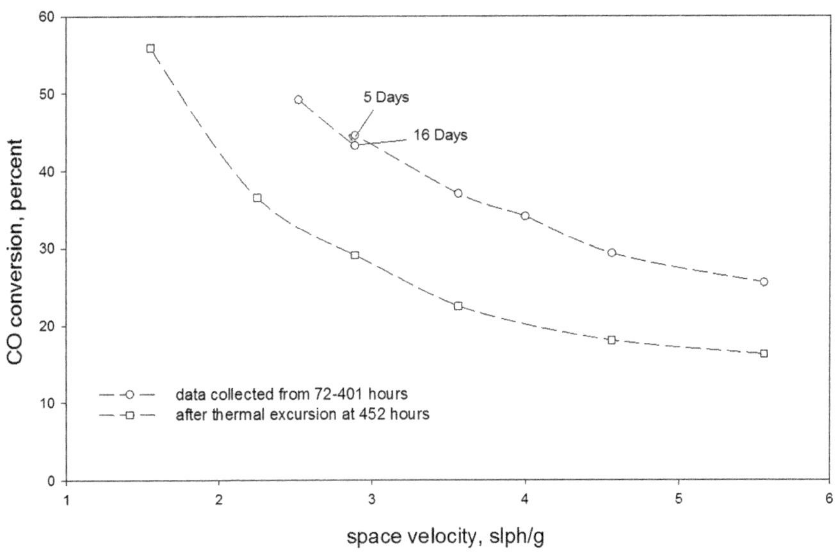

Fig. 18 CO conversion at various space velocities prior to and following the temperature excursion illustrated in this figure (T = 487 K, P = 11.9 atm gauge, H_2/CO = 2.0).

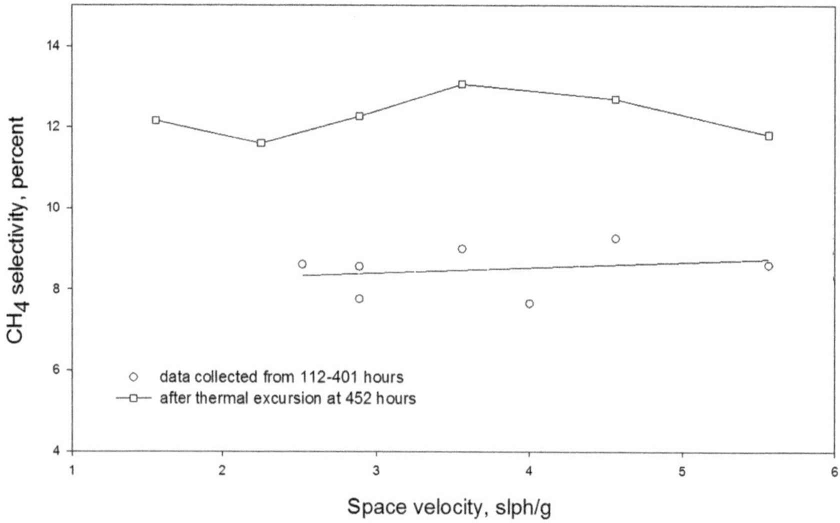

Fig. 19 Methane selectivity (mol%) prior to and following the temperature excursion shown in Fig. 17.

4. Summary

Because of its superior heat conductivity and lower manufacturing cost, an aluminum small channel compact heat exchange reactor has advantages over one prepared from stainless steel, despite the fact that the stainless steel reactor can be operated at higher temperature–pressure regions. The data obtained using the compact heat exchange reactor with a semi-commercial Co–alumina catalyst were essentially the same as those obtained with the same catalyst utilized in an isothermal CSTR, showing that the small channel reactor operated nearly isothermally along its length. In addition, a similar selectivity was obtained for the synthesis in the small channel reactor and the CSTR. Experiments indicate that the small channel reactor can be restarted following an intended or unintended shut-down and the activity and selectivity obtained following the restart is essentially the same as before the shut-down. The cause of sudden, short temperature excursions during the operation of the compact heat exchange reactor was likely due to the absence of a liquid phase during the restart procedure after catalyst reactivation.

References

1 B. H. Davis, Matching Catalyst to the Feedstock and the Reactor, *AIChE Meeting*, San Francisco, CA, 3–8 November, 2013.

2 B. H. Davis, *Top. Catal.*, 2005, **32**, 143–168.

3 C. M. Cawley and C. C. Hall in *Reviews of Petroleum Technology*, ed. F. H. Garner, E. B. Evans and G. Sele, The Inst. of Petroleum, London, 1947, vol. 7, p. 294.

4 R. J. Kokes, H. Tobin Jr and P. H. Emmett, *J. Am. Chem. Soc.*, 1955, **77**, 5860–5863; W. K. Hall and P. H. Emmett, *J. Am. Chem. Soc.*, 1957, **79**, 2091–2093.

5 K. Schubert, J. Brandner, M. Fichtner, G. Linder, U. Schygulla and A. Wenka, *Microscale Thermophys. Eng.*, 2001, **5**(1), 17–39.

6 H. J. Robota, L. Richard, S. Doskmukh and S. LeViness, *Fischer-Tropsch synthesis in a microchannel reactor: the influence of Co/SiO$_2$ catalyst structure on Fischer-Tropsch synthesis performance, International Petroleum Technology Conference, 19–22 January*, Doha, Qatar, 2014.

7 R. J. Saxton, K. Jothimurugesan and C. L. Kibby, *US Pat.* 20140343175, 2014.

8 C. Kibby, K. Jothimurugesan, T. Das, H. S. Lacheen, T. Rea and R. J. Saxon, *Catal. Today*, 2013, **215**, 131–141.

9 J. K. Neathery, R. L. Spicer, D. E. Sparks and B. H. Davis, *Stud. Surf. Sci. Catal.*, 2001, **139**, 407–414.

10 M. C. Riberio, G. Jacobs, R. Pendyala, B. H. Davis, D. C. Cronauer, A. J. Kropf and C. L. Marshall, *J. Phys. Chem.*, 2011, **115**, 4783–4792.

11 Fischer-Tropsch catalyst test on coal-derived synthesis gas, http://www.netl.doe.gov/File%20Library/Research/Coal/energy%20systems/gasification/gasifipedia/fischer-tropsch-catalyst-test-coal-derived-synthesis-gas.pdf.

12 V. R. R. Pendyala, G. Jacobs, W. Ma, J. L. S. Klettlinger, C. H. Yen and B. H. Davis, *Chem. Eng. J.*, 2014, **249**, 279–284.

13 W. Ma, M. K. Gnanamani, G. Jacobs, J. Yen and B. H. Davis, *Fischer-Tropsch synthesis on cobalt-alumina catalysts*, NASA Contractors Meeting, Cleveland, Ohio, January 2013.

Faraday Discussions

PAPER

Time- and space-resolved study of the methanol to hydrocarbons (MTH) reaction – influence of zeolite topology on axial deactivation patterns†

Daniel Rojo-Gama,[ab] Samaneh Etemadi,[a] Eliot Kirby,[a] Karl Petter Lillerud,[a] Pablo Beato,*[b] Stian Svelle[a] and Unni Olsbye*[a]

Received 8th September 2016, Accepted 10th October 2016
DOI: 10.1039/c6fd00187d

Zeolites representing seven different topologies were subjected to life-time assessment studies as methanol to hydrocarbons (MTH) catalysts at 400 °C, $P(\text{MeOH}) = 13$ kPa and $P(\text{tot}) = 100$ kPa. The following topologies were studied: ZSM-22 (TON), ZSM-23 (MTT), IM-5 (IMF), ITQ-13 (ITH), ZSM-5 (MFI), mordenite (MOR) and beta (BEA). Two experimental approaches were used. In the first approach, each catalyst was tested at three different contact times, all giving 100% initial conversion. The life-time before conversion decreased to 50% at each contact time was measured and used to calculate critical contact times (*i.e.* the contact time needed to launch the autocatalytic MTH reaction) and deactivation rates. It was found that the critical contact time is strongly correlated with pore size: the smaller the pore size, the longer the critical contact time. The second experimental approach consisted of testing the catalysts in a double tube reactor with 100% initial conversion, and quenching the reaction after 4 consecutive times on stream, representing full, partial, and zero conversion. After quenching, the catalyst bed was divided into four segments, which were individually characterised for coke content (temperature-programmed oxidation) and specific surface area (N_2 adsorption). The axial deactivation pattern was found to depend on pore size. With increasing pore size, the main source of coke formation changed from methanol conversion (1D 10-ring structures), to partly methanol, partly product conversion (3D 10-ring structures) and finally mainly product conversion (3D 12-ring structure). As a result, the methanol conversion capacity changed little with contact time for ZSM-5, while it increased with increasing contact time for the catalysts with smaller pore sizes, and decreased with increasing contact time for pore sizes larger than ZSM-5.

[a]Center for Materials Science and Nanotechnology (SMN), Department of Chemistry, University of Oslo, P.O. Box 1033, Blindern, N-0315 Oslo, Norway. E-mail: unni.olsbye@kjemi.uio.no
[b]Haldor Topsøe A/S, Haldor Topsøes Allé 1, 2800 Kgs. Lyngby, Denmark. E-mail: pabb@topsoe.dk

† Electronic supplementary information (ESI) available. See DOI: 10.1039/c6fd00187d

1. Introduction

The conversion of methanol to hydrocarbons (MTH) provides an efficient route for the production of valuable chemicals and liquid fuels from low cost carbon feedstocks such as natural gas, coal or biomass.[1,2] The MTH reaction is catalyzed by zeolites, crystalline aluminosilicates possessing strong Brønsted acidity dispersed within a microporous network, leading to shape selective catalysis.[3] The hydrocarbons formed in the MTH reaction consist of a mixture of mainly light olefins and aromatic species, and some paraffins[4] resulting from an autocatalytic complex reaction network.[2] The autocatalytic nature of the MTH reaction involves the methylation and subsequent cracking/dealkylation of adsorbed alkenes and arenes as the main reaction steps.[5,6]

In addition to the targeted products, the accumulation of coke residues in the micropores and on the outer surface of the zeolite triggers a decrease in the catalyst activity with time on stream.[7–9] The catalytic performance with respect to product selectivity and deactivation depends on the zeolite topology and operating conditions of the MTH reaction.[2,7,10] Deactivation as a result of coke deposition is reversible, and coke residues can be removed by oxidative regeneration at 300–600 °C leading to a recovery of the catalytic activity.

Great effort has been made in recent years to gain knowledge about deactivation phenomena in the MTH reaction.[7,9,11–18] Most studies have been carried out *ex situ*, analyzing the post-mortem coke content and accompanying loss of accessible surface area in the entire catalyst bed. However, when an autocatalytic reaction like MTH is carried out in a fixed bed plug flow reactor, an axial deactivation gradient is expected. Under conditions where full initial conversion is achieved, the catalyst bed can be divided into three sequential reaction zones. The feed inlet segment is defined as the initiation zone. In this first zone, partial conversion of methanol to dimethyl ether (DME), as well as a slow formation and accumulation of autocatalytic species (alkenes and arenes) take place. When sufficient autocatalytic species have been produced, the autocatalytic reaction dominates product formation, and this next zone is called the autocatalytic zone. Finally, in the product zone, located towards the bed outlet, oxygenates are fully consumed and only hydrocarbon interconversion reactions take place.[7,19]

A few studies have taken into account the axial mode of zeolite deactivation when the MTH reaction is carried out in a fixed bed plug flow reactor. Haw and Marcus proposed a "burning cigar model" to describe the deactivation and inhomogeneous aging of SAPO-34, in which a layer of active catalyst moves along the bed with increasing time on stream, leaving the deactivated catalyst in its wake.[20] This model was later confirmed by Wragg *et al.*, who used operando time- and space-resolved X-ray diffraction to monitor the deactivation of SAPO-34 as an MTH catalyst. Moreover, Wragg *et al.* allocated the initial increase in methanol conversion with time on stream over this particular catalyst to the slow accumulation of autocatalytic species in the initiation zone, which gradually transformed it into an autocatalytic zone.[21] In another contribution, Schulz studied the MTH reaction over ZSM-5 at 475 °C, also in a plug flow reactor, and visually observed that the spent catalyst was inhomogeneously deactivated.[9] The top part, which was heavily deactivated due to coke located on the outer surface of the catalyst particles, was followed by a grey reaction zone, whereas a less deactivated zone with a light blue color was observed towards the end of the bed. This color was allocated to "coke seeds", formed by secondary reactions of the MTH products.

Recently, Bleken *et al.*[14] investigated the deactivation of a microporous ZSM-5 catalyst as well as its desilicated analogue by dividing the partially deactivated catalyst bed into different segments that were subsequently characterized. They observed that desilication treatment not only improved the catalyst lifetime, but also altered the axial deactivation pattern. For the microporous parent sample, the layers located near the entrance of the catalyst bed were most severely deactivated. However, in the desilicated, meso-/micro-porous sample, the axial deactivation pattern was more homogenous and the layers at the bottom of the bed were slightly more deactivated than the preceding layers. Notably, this observation demonstrated that the axial deactivation pattern observed in previous contributions is not globally applicable to all MTH catalysts. This observation served as an inspiration for the current study.

A similar approach to the one used by Bleken *et al.* was presented by Luo *et al.*[22] who investigated the spatial and temporal evolution of confined species in SAPO-34. The authors concluded that methanol and olefins led to the formation of intermediate species, which were subsequently transformed into coke.

Very recently, another approach was applied by Müller *et al.*[13] who compared the deactivation rate of a ZSM-5 catalyst in two different reactors; a plug-flow reactor (PFR) and a continuously stirred tank reactor (CSTR). Notably, the catalyst deactivated substantially faster in the PFR, leading the authors to suggest that coke formation from methanol alone is a major source of catalyst deactivation. In an earlier contribution from the same group (the Lercher group), Sun *et al.* observed a dramatic increase in catalyst conversion capacity when butanol was co-fed with methanol over ZSM-5.[23] Together, these observations led Lercher *et al.* to revisit and support an early proposal by Hutchings *et al.*,[24] *i.e.* that coke formation in the MTH reaction is preceded by hydrogen transfer between two methanol molecules, leading to the formation of formaldehyde and methane (reaction (1)). Formaldehyde is subsequently converted to coke species. Furthermore, Lercher *et al.* suggested that Rx. (1) is inhibited by the presence of hydrocarbons.

$$2CH_3OH \rightarrow CH_4 + CH_2O \tag{1}$$

Other possible coke forming mechanisms which are supported by experimental results in the literature are reactions between methanol and hydrocarbon pool species, such as polymethyl benzene, leading *e.g.* to the hepta-methyl benzenium cation which can subsequently rearrange into a naphtalene precursor (reaction (2))[25,26]

$$\begin{aligned} C_6(CH_3)_6 + CH_3OH + H^+ &\rightarrow C_6(CH_3)_7{}^+ + H_2O \rightarrow \\ &\rightarrow C_{10}H_5(CH_3)_3 + 6\text{“H”} + H_2O + H^+ \end{aligned} \tag{2}$$

as well as reactions between hydrocarbon pool species leading again to the formation of polycyclic coke precursors, as exemplified by reaction (3)[20,27]

$$C_6H_6 + 2C_3H_6 \rightarrow\rightarrow C_{10}H_6(CH_3)_2 + 6\text{“H”} \tag{3}$$

All three classes of reactions are expected to contribute to coke formation in the MTH reaction. However, it is of interest to study to which extent the relative importance of these three reaction classes depends on catalyst topology.

Among the mathematical models put forward to simulate MTH catalyst deactivation, the models of Janssens[11] and Janssens *et al.*[12] are of particular relevance for this study. Both models consider deactivation simply as a loss of effective contact time with time on stream. The Janssens model[11] was initially developed for the commercially used ZSM-5 catalyst, and the approach resulted from the observation that the variation in product yield over ZSM-5, in particular an increase in ethene and propene yield before methanol breakthrough, was similar to the evolution of the products at an increasing space velocity.[28] A preassumption of this model is that deactivation is non-selective, *i.e.* the individual product yields obtained at a given conversion level are the same in a fresh catalyst as in a partially deactivated catalyst. This preassumption has been validated for ZSM-5,[29] ZSM-22, ZSM-23, EU-1[30] and SAPO-5[31] catalysts. A further assumption of Janssens' model is that the deactivation rate is first order in methanol and proportional to methanol conversion. These assumptions mean that coke formation from products alone is negligible (an assumption which was supported by the experimental evidence on ZSM-5 (*vide supra*)) and furthermore, that the selectivity for coke formation is constant from zero to full methanol conversion. Janssens' model was successfully applied to correlate the deactivation of a large set of ZSM-5 catalysts to their initial activity and deactivation rate.[32] However, a limitation of Janssens' first model is that it does not take into account the autocatalytic nature of the MTH reaction and hence, the more complex correlation between applied contact time and methanol conversion capacity for catalysts with a non-negligible initiation zone. This limitation was addressed by the autocatalytic deactivation model which was subsequently developed by Janssens *et al.*[12] based on the observation that the conversion capacity of ZSM-22 as an MTH catalyst increases linearly with an increase in the applied contact time, τ_0, *i.e.* with the ratio between the catalyst weight and total gas flow (W/F).

In the autocatalytic deactivation model, the rate of methanol conversion is expressed as:

$$\mathrm{d}M/\mathrm{d}\tau = -k_1 M - k_2 MP \tag{4}$$

where: M is methanol concentration and P is product concentration. Such a simplified expression is warranted by the stable steady-state product distribution reported for many MTH catalysts in the 0–80% conversion range.[33] Mathematical manipulation led to the following simplified correlation between the catalyst lifetime, $t_{1/2}$, and the deactivation coefficient (a), leading to a decrease in the conversion from 100% to 50%:

$$t_{1/2} = (\tau_0 - \tau_{\mathrm{crit}})/a \tag{5}$$

where: τ_0 is the applied contact time (W/F) and τ_{crit} is the critical contact time, *i.e.* the contact time needed to build up sufficient autocatalytic pool species to launch the autocatalytic reaction. Excellent correlation between eqn (5) and the catalyst conversion capacity was found for a ZSM-22 catalyst.[12]

The aim of this work was to elucidate the influence of zeolite topology on the axial mode of deactivation during the MTH reaction. The topologies studied comprise 1D 10-ring structures (ZSM-22, ZSM-23), 3D 10-ring structures (ZSM-5, IM-5, ITQ-13), a 1D 12-ring zeolite (mordenite) and a 3D 12-ring structure (beta). Two approaches were used: first, the methanol conversion capacity of each topology was studied at different contact times. The autocatalytic deactivation model (*vide supra*) was applied to all materials, in order to elucidate critical contact times and deactivation rates. Furthermore, a selection of the materials was subjected to life-time performance studies in a double tube reactor, where segments of the catalyst bed could be characterized separately after selected times on stream. The study revealed that the influence of hydrocarbon products as coke precursors increases with increasing pore size. As such, the catalyst conversion capacity increases with increasing contact time for smaller-pore 10-ring zeolites such as H-ZSM-22 and H-ZSM-23, in which coke is mainly formed from methanol, whereas it decreases with increasing contact time for larger 12-ring pore zeolites, such as beta zeolite. The study further suggested a correlation between pore size and critical contact time.

2. Experimental

2.1 Catalysts

Eight zeolite samples were tested in the first part of the study, where catalyst deactivation was studied as a function of applied contact time, according to the autocatalytic deactivation model. Most samples were obtained from commercial producers; ZSM-22 (Zeolyst), ZSM-23 (Zeolyst), mordenite-14386 (Ventron), ZSM-5 (Pentasil and MFI-27; both from Süd-Chemie) and beta (VALFOR CP806; Zeolyst); while IM-5 and ITQ-13 were synthesized under hydrothermal conditions as reported elsewhere.[34,35]

For the spatio-temporal investigation, the following seven zeolite samples were used: ZSM-5 (MFI-27; Süd-Chemie), beta (VALFOR CP-806, CP-7119 and CP-814E; all from Zeolyst) and mordenite (CBV-21A, Zeolyst) were obtained from commercial producers, while ZSM-22 and ZSM-23 were synthesized in-house following a previously reported experimental procedure.[36,37]

All materials were ion-exchanged with a 1 M NH_4NO_3 solution 3 × 2 h in a 75 °C water bath, followed by calcination in static air at 550 °C for 5 hours to obtain the protonated form.

2.2 Catalyst characterization

The crystallinity and purity of the fresh materials were determined using powder X-ray diffraction, collected on a Siemens Bruker D5000 using Bragg-Brentano geometry and $CuK_{\alpha1}$ radiation (λ = 1.5406 Å). The experimental results were analyzed by full-profile Le Bail refinement.

The elemental composition of the zeolites was obtained using X-ray micro-analysis (SEM/EDX).

N_2 adsorption isotherms were measured at −196 °C, using a Belsorp-mini II instrument. Prior to each measurement, fresh samples were outgassed under vacuum for 1 h at 80 °C and for 4 h at 300 °C. Spent catalysts were outgassed for 4 hours, 1 h at 80 °C and 3 h at 180 °C. Specific surface area was calculated using the BET equation[38] in the range of validity for microporous materials.[39]

FT-IR measurements were performed on fresh catalysts on a Bruker Vertex 80 with an MCT detector operating in transmission mode. Before the measurement, samples were pressed into a self-supporting wafer and were pre-treated under vacuum at 150 °C for 1 h, at 300 °C for 1 h and at 450 °C for 1 h. For CO adsorption, samples were cooled to −196 °C. After complete adsorption of CO, spectra were recorded during CO desorption.

The total amount of coke in the deactivated catalysts was obtained using TGA on a Rheometric Scientific STA 1500 instrument. Approximately 15 mg of sample was heated to 800 °C using a heating rate of 5 °C min^{-1} in 25 mL min^{-1} of synthetic air. Then, the final hold time at this temperature was 40 minutes. The coke content was normalized to the amount of zeolite used in each experiment.

The coke species retained in the catalyst after methanol conversion were analyzed by the dissolution–extraction technique.[40] Approximately 15 mg of the deactivated catalyst was dissolved in 1 mL of 15% HF for 60 minutes, inside a small capped Teflon vial. Subsequently, the soluble fraction of the coke was extracted with 1 mL of CH_2Cl_2 and analyzed with an Agilent 6890N/5793 MSD GC-MS equipped with an HP-5MS column (60 m, 0.25 mm i.d., stationary phase thickness 0.25 μm). *ortho*-Chloro-toluene was used as the internal standard to normalize all chromatograms. The NIST98 database was used for the identification of species.

2.3 Catalytic tests

Two reactor setups were used for the catalytic tests. To investigate the influence of contact time on catalyst deactivation, a U-shaped fixed-bed quartz reactor (8 mm i.d.) and 50 mg of catalyst (250–420 μm) was used.

For the spatio-temporal study, a double-tube straight quartz reactor (9 mm i.d. and 11 mm o.d.) was chosen. 200 mg of catalyst (250–420 μm) was poured in the internal reactor using quartz wool as a support for the catalyst bed. After thermal quenching at increasing times on stream, the catalyst bed was subsequently divided into 4 layers, being numbered consecutively from the entrance to the outlet of the catalyst bed.

Regardless of the reactor setup, catalytic experiments were carried out at ambient pressure and 400 °C. Prior to the catalytic tests, the reactor was heated to 550 °C under He flow. At this temperature, catalysts were calcined for 1 h in a flow of pure O_2. Then, the temperature was lowered to 400 °C, the actual reaction temperature, using He. MeOH (BDH Laboratory, purity > 99.8%) was fed to the reactor by flowing a stream of He through a methanol evaporator kept at 293 °C (P_{MeOH} = 13 kPa). The contact times (τ = catalyst mass/total gas flow [g h mol^{-1}]) were modified by adjusting the flow of He accordingly, keeping the amount of catalyst constant in each test. The reaction effluent was analyzed online using an Agilent 6890 gas chromatograph equipped with a flame ionization detector and an Agilent J&W capillary column (length 60 m, 0.530 mm i.d., stationary phase thickness of 20 μm). Conversion, selectivity and product yield were calculated on C_1 basis as described in literature.[41]

3. Results

All catalytic tests reported in this study were performed at contact times leading to 100% initial conversion. Representative conversion *versus* time on stream plots for the zeolite topologies which were subjected to spatio-temporal studies are shown in Fig. 1. It should be noted that the applied contact time, τ_0, required to obtain full initial conversion is different for each catalyst, and that the data in Fig. 1 were therefore obtained at different applied contact times (*vide infra*).

Fig. 2 shows the change of methanol conversion capacity (in g MeOH per g catalyst) for each of the materials tested at different applied contact times. The data in Fig. 2 were obtained by integrating the area under the conversion *versus* time on stream curves until complete deactivation. The catalyst lifetime to 50% MeOH conversion, $t_{1/2}$, is plotted *versus* the applied contact time, τ_0, for each catalyst in Fig. 3. As a general comment, the results in Fig. 3 show that $t_{1/2}$ depends linearly on the applied contact time for all zeolites, as predicted by the autocatalytic deactivation model.[12] Using this model, the deactivation constant (a) and the critical contact time (τ_{crit}) were derived from the data in Fig. 3, using eqn (5). The slope of the $t_{1/2}$ *versus* τ_0 line equals $1/(a)$ and the critical contact time is obtained by extrapolating the line to $t_{1/2} = 0$. Results of a and τ_{crit} are presented in Table 1.

In the following sections, data obtained for each zeolite topology will be presented separately, followed by an overall discussion of the results, finally leading to the conclusion of the study. Characterization data for the zeolite samples tested in this study can be found in Table 2. For some topologies, several materials have been tested. In some cases, the use of several samples was necessitated by the sparse amount of each sample, while in other cases, several samples were tested and compared in order to elucidate whether the observed axial deactivation pattern is a general feature of the zeolite topology studied, or specific for one sample. In each case, focus will be set on the axial deactivation profiles observed. Differences in the overall activity and deactivation rates between materials

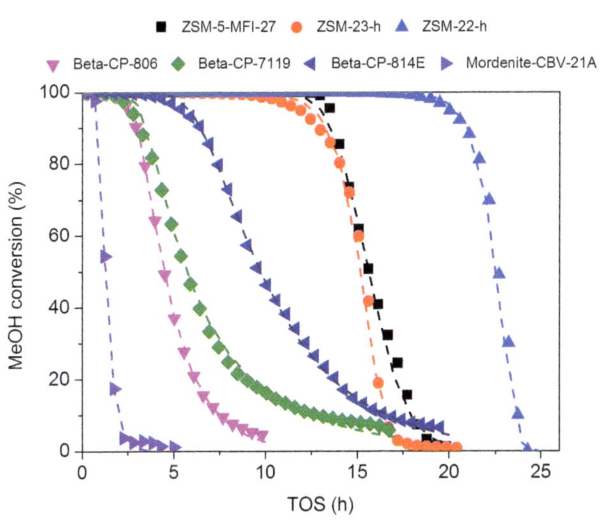

Fig. 1 Methanol conversion *versus* time-on-stream at 400 °C over the various catalysts tested. Experimentally measured data are represented by symbols while simulated deactivation curves are represented by dashed lines. $\tau_0 = 4$ g h^{-1} mol^{-1} was applied over ZSM-22-h and ZSM-23-h whereas for the rest of the catalysts $\tau_0 = 2$ g h mol^{-1} was used.

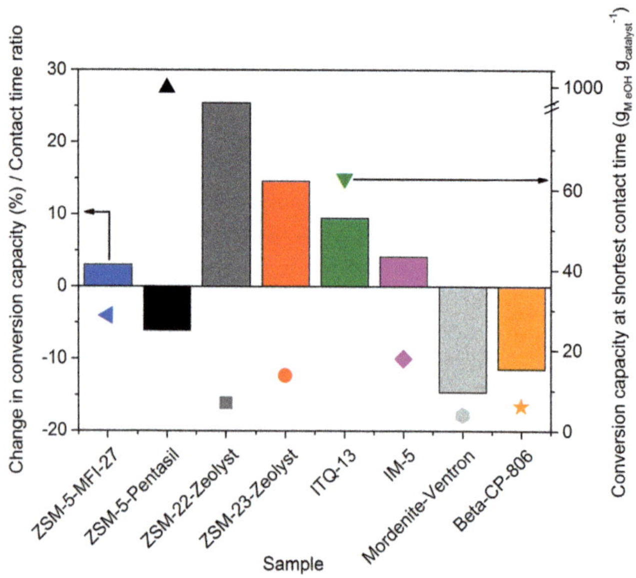

Fig. 2 Normalized variation of the methanol conversion capacity with respect to the increase in contact time on the different zeolites investigated (left axis). Methanol conversion capacity (g MeOH/g catalyst) of the different zeolites investigated at the shortest contact time (τ_0) (right axis).

belonging to the same topology, but with different morphology, acid site density or various lattice defects, have been the subject of previous reports and will not be discussed here.

3.1 Deactivation of ZSM-5 (MFI topology)

ZSM-5 is used in industrial MTG and MTP processes, and has been studied extensively over the past 40 years. It is therefore a natural starting point for this study. Two ZSM-5 catalysts were studied here, ZSM-5-pentasil and ZSM-5-MFI-27. Only ZSM-5-MFI-27 was tested until full deactivation.

As a first remark, the methanol conversion *versus* time on stream curve obtained over ZSM-5-MFI-27 (Fig. 1) displays an inverse sigmoidal shape with a rapid

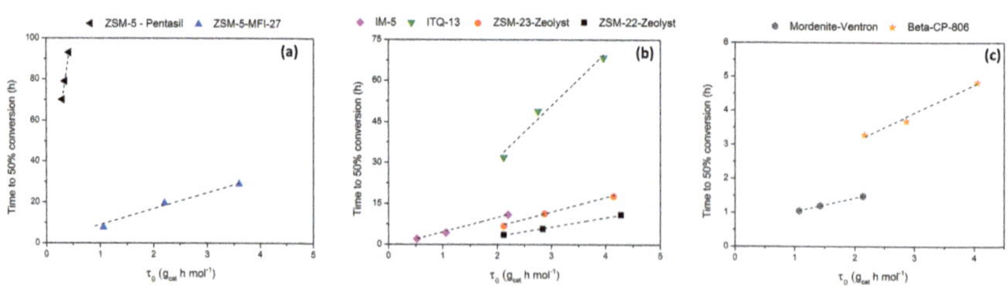

Fig. 3 Catalyst lifetime to 50% MeOH conversion ($t_{0.5}$) as a function of the applied contact time, τ_0 (weight to flow ratio W/F) on (a) ZSM-5-pentasil and ZSM-5-MFI-27, (b) IM-5, ITQ-13, ZSM-22-Zeolyst and ZSM-23-Zeolyst and (c) mordenite-Ventron and beta CP-806. The dotted lines are added as a guide for the eye.

Table 1 Calculated values of the critical contact times (τ_{crit}) and deactivation coefficients (a) using the autocatalytic deactivation model[a]

Sample	τ_{crit} (g h mol^{-1})	a (g mol^{-1})
ZSM-5-pentasil	0	0.004
ZSM-5-MFI-27	0	0.13
IM-5	0.14	0.19
ITQ-13	0.36	0.052
ZSM-22-Zeolyst	1.1	0.29
ZSM-23-Zeolyst	0.76	0.24
Mordenite-Ventron	*	2.5
Beta-CP806	*	1.2

[a] *Model not applicable, see bulk text.

decline in conversion after the breakthrough of methanol, followed by a flattening of the curve at less than 40% conversion. Numerous previous studies of ZSM-5 catalysts have shown the same conversion *versus* time-on-stream behaviour (see *e.g.* ref. 4, 12, 14, 16 and 29), which is typical for systems where the reactant is involved in coke formation.[2,42] Fig. S.1 (ESI†) shows the product yield of ZSM-5-MFI-27 with respect to time on stream. In agreement with previous studies, the product stream consists mainly of C_2–C_5 alkenes, as well as arenes and paraffins.[29,34]

The methanol conversion capacity ($g_{MeOH}\ g_{cat}^{-1}$) of the two ZSM-5 samples is only moderately influenced by the applied contact time (Fig. 2). For ZSM-5-MFI-27, a slight increase in conversion capacity with longer contact time is observed,

Table 2 Summary of the characterization results of the fresh catalysts

Sample	Si/Al ratio	BET surface area (m^2 g^{-1})	External area (m^2 g^{-1})	Total pore volume (cm^3 g^{-1})	Micropore volume[a] (cm^3 g^{-1})	Particle size (μm)	$\Delta\nu$ ((OH) cm^{-1}) upon CO adsorption
IM-5	15	482	118	0.81	0.15	0.2–1	313
ITQ-13	50	461	40	0.39	0.17	0.2–1	305
ZSM-22-Zeolyst	45	217	20	0.14	0.08	2–4	315
ZSM-23-Zeolyst	23	159	56	0.66	0.05	0.2–1	320
Mordenite-Ventron	11	573	44	0.40	0.22	0.2–1	310
ZSM-5-pentasil	45	422	113	0.61	0.14	0.05–0.1	313
ZSM-5-MFI-27	15	414	5	0.18	0.17	1–4	306
Beta-CP806	17	677	188	0.99	0.19	<0.1	311
Beta-CP-814E	14	653	248	1.05	0.16	<0.1	—
Beta-CP-7119	13	657	26	0.37	0.26	0.5	—
Mordenite-CBV-21A	11	525	27	0.26	0.19	0.1–0.3	—
ZSM-22-h	41	268	55	0.49	0.08	0.5–1	—
ZSM-23-h	28	222	45	0.26	0.07	0.5–1	—

[a] *t*-Plot.

in line with the results reported by Bleken.[29] In that study, a 10% increase in methanol conversion capacity was observed upon a ten-fold increase in contact time over a ZSM-5 catalyst. For ZSM-5-pentasil, a minor decrease in the methanol conversion capacity is observed with increasing contact time (Fig. 2). This discrepancy could be due to an extrapolation uncertainty because the cumulative conversion for ZSM-5-pentasil was extrapolated when the conversion of methanol was still at 10%.

As shown in Fig. 3, the $t_{1/2}$ values for ZSM-5-pentasil obtained at τ_0 between 0.30 and 0.43 g h mol^{-1} are considerably higher than those of ZSM-5-MFI-27, obtained at τ_0 between 1.1 and 3.6 g h mol^{-1}. Variations in the catalyst lifetime between the two ZSM-5 samples are attributed to the differences in their Si/Al ratio and crystal size (Table 2), in accordance with previous studies.[43,44]

In line with previously reported results over ZSM-5,[12] the $t_{1/2}$ *versus* τ_0 curves for ZSM-5-pentasil and ZSM-5-MFI-27 cross the *x*-axis very close to the origin (Fig. 3). In other words, τ_{crit} can be considered negligible and consequently, the autocatalytic reaction dominates in ZSM-5 regardless of the applied contact time.

The value of the deactivation constant (*a*) (see Table 1) in ZSM-5-MFI-27 far exceeds that of ZSM-5-pentasil, in line with the higher conversion capacity of ZSM-5-pentasil obtained by integrating the conversion *versus* time on stream curves (Fig. 2). Indeed, the deactivation constant in the latter material is the lowest among all the tested zeolites (*vide infra*), reflecting its superior catalyst stability.

For the spatio-temporal study, four individual tests were carried out using ZSM-5-MFI-27 at a constant $\tau_0 = 2$ g h mol^{-1}. The reaction was quenched after 5, 10, 15 and 20 hours on stream, when the methanol conversion was 100%, 100%, 80% and 0%, respectively (Fig. 1). Photographs of the catalyst bed after each test are shown in Fig. 4, top. A color gradient was observed along the catalyst bed, in particular after 5 and 10 hours on stream, *i.e.* before methanol breakthrough (Fig. 1).

The dark first layers of the catalyst bed display evidence of a more deactivated region, whereas the light grey color of the lower part of the bed suggests that this region is deactivated to a lesser extent.

The BET surface area and the total amount of coke determined by TG analysis in the four catalyst layers obtained after each time on stream are shown in Fig. 4, bottom. Concerning the spatial distribution of coke along the bed, after the 5 and 10 hour experiments, thermogravimetric analysis unequivocally shows that there is a substantial decrease in the total amount of oxidizable coke species when moving from the entrance towards the outlet of the catalyst bed. Besides, the BET area is strongly reduced in the top (entrance) layer owing to the amount of coke deposited in this fraction, while it is gradually less reduced in the bottom layers of the bed.

During the first 10 hours of reaction, the product distribution (Fig. S.1 ESI†) is independent of coke deposition, in agreement with previous studies,[11,29,45] and only after the breakthrough of methanol, the product distribution as well as the evolution of coke and BET area change. The large gradient in coke content observed during the first 10 hours of reaction was reduced for the 15 h experiment, where still a minor difference between the top and bottom layers is observed due to the residual activity of the bottom-most catalyst layer. However in the completely deactivated sample (20 h on stream), the differences in the coke

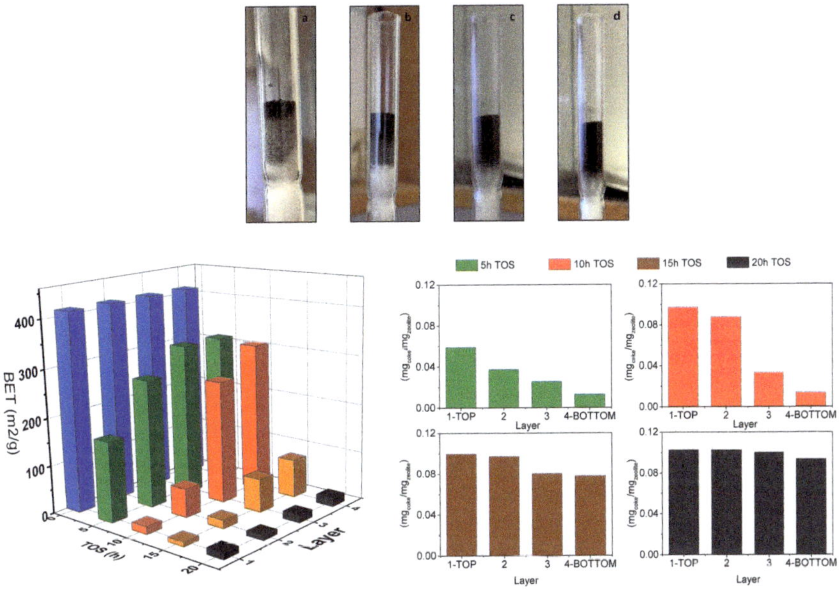

Fig. 4 Top: ZSM-5-MFI-27 deactivated at $t = 5$, 10, 15, and 20 hours (a, b, c and d, respectively) in the 9 mm reactor, $T = 400$ °C, $\tau = 2$ g h mol^{-1}. Bottom-left: BET area on partially deactivated catalyst layers of ZSM-5. Bottom-right: Normalized amounts of total coke in the different fractions of ZSM-5-MFI-27.

amount and BET area along the catalyst bed are negligible. These data are in line with previous studies of ZSM-5 catalysts.[9,13,14,19,46]

Different from the previously reported results,[29] no white layer was observed at the entrance of the catalyst bed. The first layer shows the most profound deactivation. Such discrepancy can be attributed to the different experimental conditions used. Bleken[29] used shorter contact times ($\tau = 0.05$–0.48 g h mol^{-1}). With such low contact times, even for ZSM-5, an initiation zone for the build-up of the autocatalytic species (and coke) was observed.

In order to get further insight into the distribution of coke along the bed, soluble coke species trapped in the pores were analyzed following the Guisnet protocol[40] in the four segments of ZSM-5-MFI-27. The data are reported in Fig. S.8 in the ESI.†

The distribution of soluble coke species agrees well with the TG data. Soluble species were only detected in the top layers of the catalyst bed. For the 5 and 10 hour experiments, in the bottom-most layers, no retained soluble species were extracted. It is only after the breakthrough of methanol that soluble coke species are identified in these layers. This result further confirms that the deactivation front is moving downwards along the bed and, although insoluble coke was observed in the lower fractions of the bed, it is only the fractions where methanol is present that contain soluble coke species. Finally, it should be noted that no changes were observed in the species identified at increasing times on stream. This can be interpreted as an accumulation of the same type of deactivating species throughout the bed, and/or to a steady-state situation of formation and further conversion of such species to insoluble coke.

The results obtained in this section show full agreement between the two catalysts and test methods used, and with previous studies of ZSM-5 catalysts, *i.e.* that methanol is a main source of catalyst deactivation in this topology,[29] either *via* formaldehyde (reaction (1)) or by reaction between methanol and hydrocarbon products (reaction (2)), with a smaller contribution of coke formation from products alone (reaction (3)).

3.2 Deactivation of ZSM-22 (TON topology)

Two ZSM-22 materials were tested in this contribution: the commercial ZSM-22-Zeolyst was used in studies of catalyst deactivation and methanol conversion capacity *versus* applied contact time, while the home-made ZSM-22-h was used in the spatio-temporal study. The conversion *versus* time on stream curve for ZSM-22-h is shown in Fig. 1 and has a similar shape to that observed for ZSM-5, with rapid deactivation after methanol breakthrough, albeit much less flattening of the curve at low conversion. Furthermore, differently from ZSM-5, the conversion capacity of ZSM-22-Zeolyst increases significantly with an increase in contact time (Fig. 2). Such an increase has previously been reported for ZSM-22 by Teketel *et al.*[15] and Janssens *et al.*[12]

The product yield obtained over ZSM-22-h is shown in Fig. S.2 (ESI†). In agreement with previous studies of ZSM-22,[10,15] an alkene rich product spectrum is obtained with near complete absence of aromatics.

Plots of $t_{1/2}$ *versus* contact time for ZSM-22-Zeolyst are shown in Fig. 3 (middle panel). Extrapolation of the curve to $t_{1/2} = 0$ displays a substantial initiation zone in these experiments, the critical contact time being 1.1 g h mol^{-1}, *i.e.* 25–50% of the applied contact time, τ_0, in these experiments. This observation complements the higher methanol conversion capacity of this material with increasing contact times, obtained by integrating the conversion *versus* time on stream curves until full deactivation (Fig. 2). Thereby, the applicability of the autocatalytic deactivation model for this catalyst topology is confirmed.

The deactivation constant ($a = 0.292$ g mol^{-1}, see Table 1) in ZSM-22-Zeolyst reflects its limited catalyst stability in the MTH process, in line with previous studies where its conversion capacity was found to be comparable to that of the commercial MTO catalyst, SAPO-34.[15]

The deactivation in ZSM-22-h was spatially and temporally investigated at 400 °C and at a constant contact time ($\tau_0 = 4$ g h^{-1} mol^{-1}). Fig. 5 shows the evolution of coke and BET area for the catalyst layers obtained at increasing reaction times. The MTH reaction over ZSM-22-h was quenched at 5, 15, 21 and 24 hours on stream, corresponding to 100%, 100%, 90% and 0% methanol conversion (Fig. 1).

The deactivation pattern obtained in ZSM-22-h is comparable to that previously shown for ZSM-5. During the early stages of reaction, the catalyst layer closest to the reactor entrance suffers from a higher degree of deactivation *i.e.* larger reduction in BET area and higher accumulation of coke, compared to the subsequent catalyst layers. At longer reaction times, particularly after the methanol breakthrough, the differences in terms of coke content and BET surface area are minimised among the layers of ZSM-22-h, suggesting that the catalyst bed is deactivated to a similar extent.

It is important to note that the ZSM-22 (and ZSM-23, *vide infra*) catalysts used for the contact time influence tests are different from the samples used in the

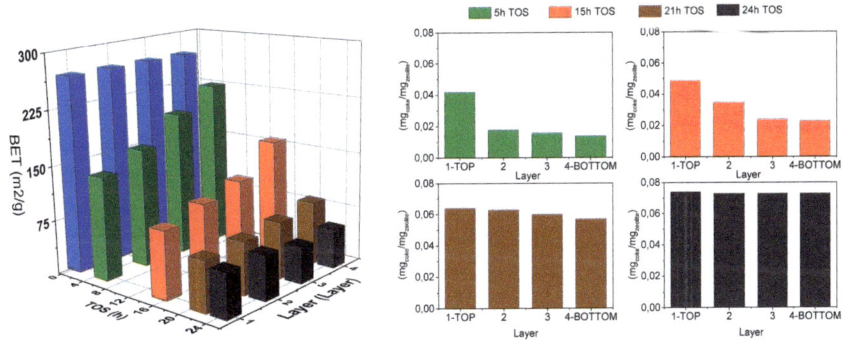

Fig. 5 Evolution of BET surface area (left) and total coke content obtained by TG (right) on ZSM-22-h at increasing reaction times.

spatiotemporal investigation, due to limited amounts of each material. In the former case commercial samples from Zeolyst were used, but in the latter, samples were synthesized in house. The catalyst lifetime of the in-house synthesized ZSM-22 shows a longer lifetime than previously reported results on ZSM-22.[15,47,48] The improvement in the catalyst lifetime of this particular sample warrants further studies.

Since the initiation zone fraction of the ZSM-22-h catalyst bed is unknown, it is of particular relevance for this study that del Campo *et al.* recently studied the deactivation of ZSM-22-Zeolyst using space- and time-resolved operando X-ray diffraction with the same contact time and slightly lower temperature (375 °C) compared to the current study.[48] Rapid deactivation (methanol breakthrough), starting immediately after the onset of the experiment, was observed. Furthermore, the unit cell expansion in the ZSM-22-Zeolyst due to the formation of coke species was invariant throughout the bed, for the duration of the test. In the same study, an acid–base treated ZSM-22-Zeolyst catalyst, with improved methanol conversion capacity compared to the parent material, was tested. Over that catalyst, the cell expansion was more rapid in the first layers of the catalyst bed compared to the subsequent catalyst layers, suggesting more rapid deactivation of the first layer of the catalyst bed, in agreement with the results obtained for ZSM-22-h in this study.

Together, current and previous studies of ZSM-22 suggest that methanol is a main source of coke formation in this topology, with substantially lower contribution of coke formation from product–product reactions. Further evidence of such a conclusion is found in a recent study by Teketel *et al.*, who observed an increasing methanol conversion capacity over ZSM-22 when co-feeding methanol with propanol or butanol. The co-feeding of these higher alcohols (which dehydrate under reaction conditions into the corresponding alkenes) led to an increase in the methanol conversion capacity by 4- or 5-fold when co-feeding propanol and butanol, respectively, in comparison with a pure methanol feed.[49]

3.3 Deactivation of ZSM-23 (MTT topology)

The ZSM-23 topology was tested at applied contact times, τ_0, between 2–4 g h mol^{-1}, using two samples; ZSM-23-Zeolyst (contact time variation) and ZSM-23-h (spatio-temporal study). Characterization data for the two samples are shown in Table 2. The conversion of methanol *versus* time on stream to complete deactivation with a contact time $\tau_0 = 4$ g h mol^{-1} is shown for the ZSM-23-h sample in Fig. 1. Its deactivation pattern was very similar to that of ZSM-22, with rapid deactivation after methanol breakthrough, and little flattening at low conversion. Product distribution *versus* time on stream and methanol conversion plots are shown in Fig. S.3† and displays similarities to the product distribution obtained over ZSM-22, with propene and C_{5+} alkenes as main products and without noticeable amounts of aromatic compounds, as it has been previously reported.[30]

An increase in the contact time resulted in a higher methanol conversion capacity for the ZSM-23-Zeolyst, as it was also observed for the ZSM-22-Zeolyst (Fig. 2). Such an increase has also previously been reported for this topology.[30]

A plot of $t_{1/2}$ *versus* applied contact time, τ_0, for this topology showed a substantial critical contact time before onset of the autocatalytic reaction (Fig. 3), *i.e.*; $\tau_{crit} = 0.76$ g h mol^{-1} with a deactivation constant $a = 0.242$ g mol^{-1} (Table 1). In accordance with the evolution of the conversion capacity at increasing contact times (Fig. 2), τ_{crit} was lower for ZSM-23-Zeolyst than for ZSM-22-Zeolyst. According to the autocatalytic deactivation model, such a result suggests that the initiation zone of this catalyst is non negligible, different from ZSM-5 and similar to ZSM-22, and furthermore, that the first zone of the catalyst bed has negligible activity in the autocatalytic MTH reaction.

For the spatiotemporal study of ZSM-23-h deactivation, a constant contact time $\tau_0 = 4$ g h mol^{-1} was used. The evolution of the BET area and coke content on the partially spent layers of the catalyst is shown in Fig. 6. A similar deactivation pattern to that obtained in ZSM-5 and ZSM-22 was observed for ZSM-23-h. For this material, the MTH reaction was quenched at the very beginning of the reaction after only 20 minutes, and after 5, 10, and 19 hours, corresponding to 100%, 100%, 98% and 0% methanol conversion, respectively (Fig. 1).

Even for the shortest experiment (20 minutes on stream), the topmost layer displays a darker color in comparison to the rest of the bed. Regarding the evolution of the BET surface area and the amount of coke, the first layers show a larger decrease of the BET area and a higher amount of coke than the layers located towards the outlet of the bed.

It is worth highlighting the rapid accumulation of coke in the first layer of the ZSM-23 catalyst bed. After only 20 minutes under reaction conditions, the total coke content in this topmost layer reaches 78% of the coke detected in the completely deactivated sample. The combination of a substantial initiation zone and a rapid deactivation of the first catalyst layer even at the shortest contact time could suggest that methanol is the main source of coke formation, and that the initial selectivity of methanol conversion *via* formaldehyde (reaction (1)) favors coke formation over the formation of autocatalytic species (*vide infra*).

3.4 Deactivation of IM-5 and ITQ-13 (IMT and ITH topology)

IM-5 and ITQ-13 were subjected to deactivation *versus* applied contact time studies, but not to spatio-temporal investigations. Both materials displayed

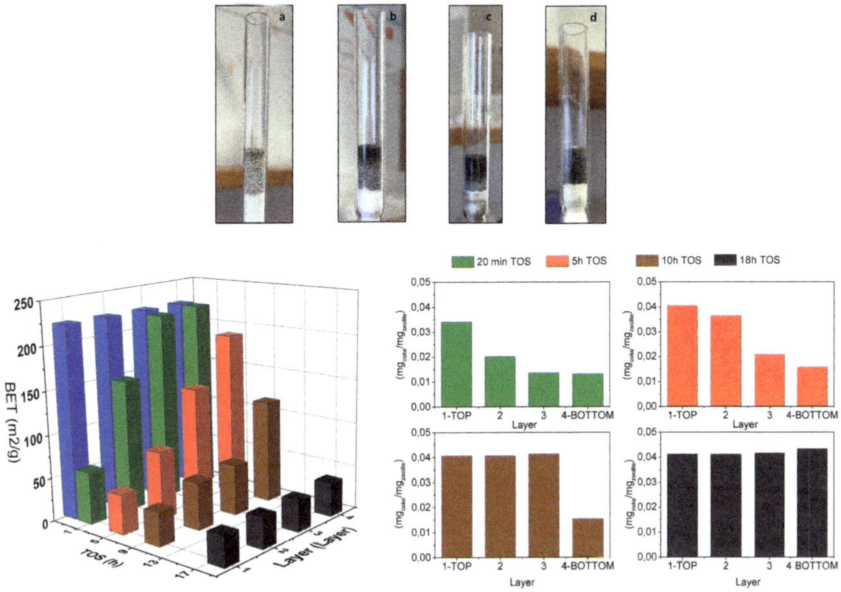

Fig. 6 Top: Photos of ZSM-23-h after quenching the reaction at (a) 20 minutes, (b) 5, (c) 10 and (d) 19 hours on stream, $T = 400$ °C, $\tau_0 = 4$ g h^{-1} mol^{-1}. Bottom-left: BET area on the partially deactivated catalyst layers of ZSM-23. Bottom-right: Normalized total coke content in ZSM-23.

a similar response to contact time variations as the 1D 10-ring topologies ZSM-22 and -23 with an increase in the methanol conversion capacity with increasing applied contact time (Fig. 2), matched by a significant critical contact time (0.14 g h mol^{-1} for IM-5, 0.36 g h mol^{-1} for ITQ-13; Fig. 3 and Table 1) which was, however, substantially shorter than for the ZSM-22 and -23 (Table 1). The deactivation constant of IM-5 (0.186 g mol^{-1}) was slightly lower than for ZSM-22 and -23, while it was much lower for ITQ-13 (0.052 g mol^{-1}, Table 1). Overall, the deactivation of these two topologies was intermediate between the 3D 10-ring ZSM-5 topology, and the more narrow-pored 1D 10-ring ZSM-22 and -23 topologies, in agreement with their intermediate pore dimensionality and size.

3.5 Deactivation of beta zeolite (BEA topology)

To the best of our knowledge, no spatio-temporal study of 12-ring zeolites for the MTH reaction has been published before. Therefore, three samples with BEA topology were subjected to spatio-temporal studies in this contribution: beta-CP-806, beta-CP-7119 and beta-CP-814E, while one sample, beta-CP-806, was subjected to applied contact time variation studies. Conversion *versus* time on stream curves for the three samples are shown in Fig. 1. Their shape is similar to one another, but is markedly different from those of the 10-ring topologies, with a rather fast breakthrough of methanol, followed by a deactivation curve for which the slope decreases with the progress of catalyst deactivation. The product selectivity over beta-CP-806 (Fig. S.4†) is dominated by aromatic compounds, followed by butenes and branched C_{5+} species, in agreement with previous studies.[50,51]

Distinct differences between 12-ring and 10-ring topologies are also observed in the tests with varying applied contact times. A decrease in the conversion capacity with increasing contact time is observed for beta-CP-806 (Fig. 2), which is opposite to the trend observed in the medium pore size zeolites (except for ZSM-5). Furthermore, in spite of the linear dependency of $t_{1/2}$ on τ_0 in the experiments performed at different contact times (Fig. 3), results of the deactivation constant (a) and τ_{crit} (Table 1) are markedly different from those observed for the medium pore size zeolites. Interestingly, a negative value is obtained for the critical contact time over beta-CP-806. This result lacks a physical meaning, since according to the autocatalytic deactivation model, it would mean that even without feeding methanol the autocatalytic reaction is dominating. While this observation evidences the shortcoming of the autocatalytic model for the beta topology, the observation of a decrease in the methanol conversion capacity with increasing applied contact time might be explained by an increase in secondary reactions between the hydrocarbon products in the large pore beta topology (Rx. (3)), leading ultimately to the experimentally observed faster coking, and as a consequence a decrease in methanol conversion capacity with a larger product zone (*i.e.* increasing contact time). It is natural to correlate the enhanced influence of product–product reactions leading to coke formation in this topology, to the large pore size of 12-ring zeolites, which facilitates the diffusion of large product molecules into the zeolite pores, even in the product zone of the catalyst bed.

The deactivation in beta-CP-806 was also investigated spatially and temporally resolved, using the same reaction conditions as for ZSM-5-MFI-27 (400 °C and $\tau_0 = 2$ g h^{-1} mol^{-1}). The MTH reaction over beta-CP-806 was quenched at 1, 3, 5 and 10 hours on stream, corresponding to 100%, 90%, 35% and 3% methanol conversion (Fig. 1).

The obtained deactivation pattern along the catalyst bed of beta-CP-806 was very different from that observed in medium pore size zeolites.

Fig. 7 shows the total coke content and the BET surface area of the partially deactivated layers of beta-CP-806. The first remarkable difference with respect to the 10-ring zeolites is the visual aspect of the catalyst bed. Even for the shortest reaction time (1 hour on stream and full methanol conversion), the catalyst bed is completely black without any axial gradient in color along the bed. Moreover, no significant difference in the reduction of BET area compared to the fresh sample were observed between the catalyst layers, again suggesting that they are equally deactivated (Fig. 7, blue bars). With increasing time on stream the degree of reduction of BET area compared to the parent material increases; and again, the deactivation seems to be homogeneous along the bed. Thermogravimetric analyses also show that coke is evenly distributed throughout the segments of catalysts in contrast to the clear gradients observed in ZSM-5, ZSM-22 and ZSM-23.

In order to get further insight into the distribution of coke along the bed, soluble coke species trapped in the pores were analyzed following the Guisnet protocol[40] in the deactivated segments of beta-CP-806. Fig. S.9 in the ESI† shows the chromatograms of the soluble coke species extracted in beta-CP-806. Clearly, soluble species are homogeneously distributed along the catalyst layers, even

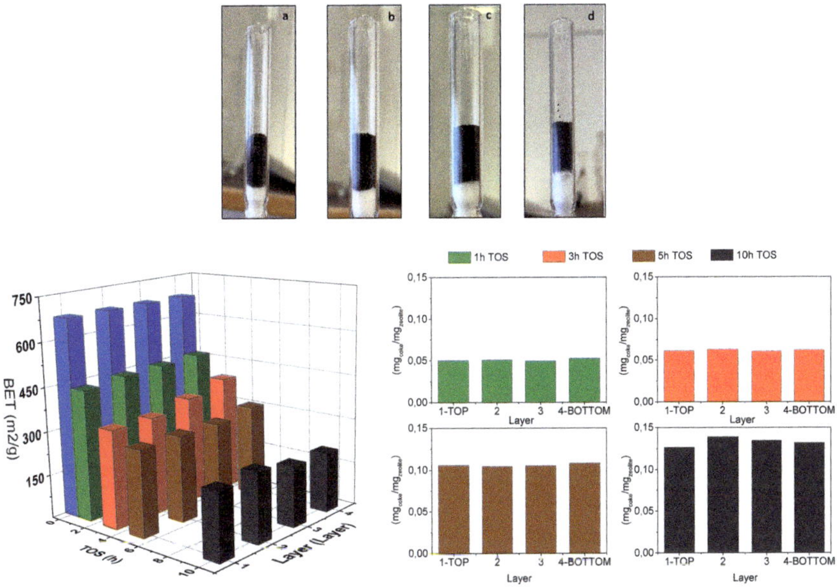

Fig. 7 Top: Photos of beta CP-806 after quenching the reaction at (a) 1 h, (b) 3 h, (c) 5 h and (d) 10 hours on stream. Bottom: Evolution of BET area (left) and normalized total coke content (right) in the layers of beta-CP-806 at increasing times on stream.

before the breakthrough of methanol, again suggesting that hydrocarbon products are strongly involved in coke formation. Finally, at increasing TOS, a decrease in the amount of soluble species is observed, together with a shift towards heavier and more thermodynamically favored species, in line with previous studies carried out over beta zeolites.[25] These results again confirm the different deactivation patterns of 10- and 12-ring topologies. Moreover, the amount of soluble species detected in the large beta-CP-806 is significantly lower than that in ZSM-5-MFI (note the differences in scale of Fig. S.8 and S.9†), suggesting rapid conversion of these species to insoluble coke species in the more spacious BEA pores.

Another difference between beta-CP-806 and the 10-ring topologies is that the product yield over beta-CP-806 is not independent of the coke content before the breakthrough of methanol (Fig. S.5 in the ESI†). Instead, the C_{5+} fraction increases at the expense of the C_4 products with increasing TOS, and reaches a maximum before the breakthrough of methanol.

In order to investigate whether the observed deactivation pattern is a general feature of the BEA topology, two additional beta samples were studied. Both beta-CP-814E (Fig. 8) and beta-CP-7119 (Fig. S.6†) show a very similar axial deactivation pattern to that observed in beta-CP-806: a uniform decrease in the BET surface area accompanied by an even accumulation of coke is observed for the different segments of the catalyst bed quenched at the same time on stream.

These results demonstrate that the observed axial deactivation pattern is not affected by the selection of the sample. Rather, it is a general pattern for samples with BEA topology.

3.6 Deactivation in mordenite (MOR topology)

Two mordenite samples were studied; mordenite-CBV-21A was used in the spatio-temporal studies, while mordenite-Ventron was used in the applied contact time variation studies. The conversion *versus* time on stream plot at 400 °C and $\tau_0 = 2$ g h mol^{-1} of mordenite-CBV-21A is shown in Fig. 1. A very rapid deactivation is observed, followed by a flattening of the curve close to full deactivation. The product yield is shown in Fig. S.7† and it shows a high initial yield of lower olefins with noticeable amounts of methane in the effluent at short reaction times, in line with previously reported results.[52]

A decrease in methanol conversion capacity with increasing contact time was observed for mordenite-Ventron, similar to the beta-CP-806 sample (Fig. 2). Furthermore, a linear correlation between applied contact time and $t_{1/2}$ was observed (Fig. 3), enabling the calculation of critical contact time and deactivation rate by the autocatalytic model. However, as for beta-CP-806, a negative critical contact time was obtained, indicating the failure of the autocatalytic deactivation model to correctly describe the deactivation of the MOR topology. Again, we hypothesize that the origin of these observations is due to a dominance of secondary reactions between hydrocarbon products that transform into coke, compared to coke originating from methanol.

Concentrating next on the spatio-temporal study of mordenite-CBV-21A (Fig. 9), the reaction was quenched after 20, 35, and 70 minutes, and 5 hours on stream, corresponding to 100%, 100%, 55% and 0% methanol conversion, respectively. The axial pattern is similar to what is observed for the beta samples: the coke content and BET surface area of each catalyst layer is similar throughout the bed at all times on stream investigated.

4. Discussion

The zeolite topologies studied in this contribution span a wide range of MTH catalysts, from a 1D 10-ring topology with high C_{5+} alkene selectivity, where alkenes are the autocatalytic engines and even monocyclic aromatic compounds are retained as inactive coke species in the pores (ZSM-22)[15,30] *via* a 3D 10-ring

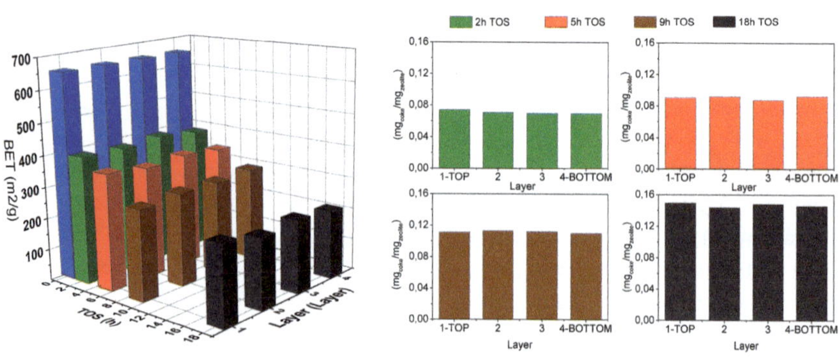

Fig. 8 Evolution of BET surface area (left) and coke content (right) on the partially deactivated layers of beta CP-814E

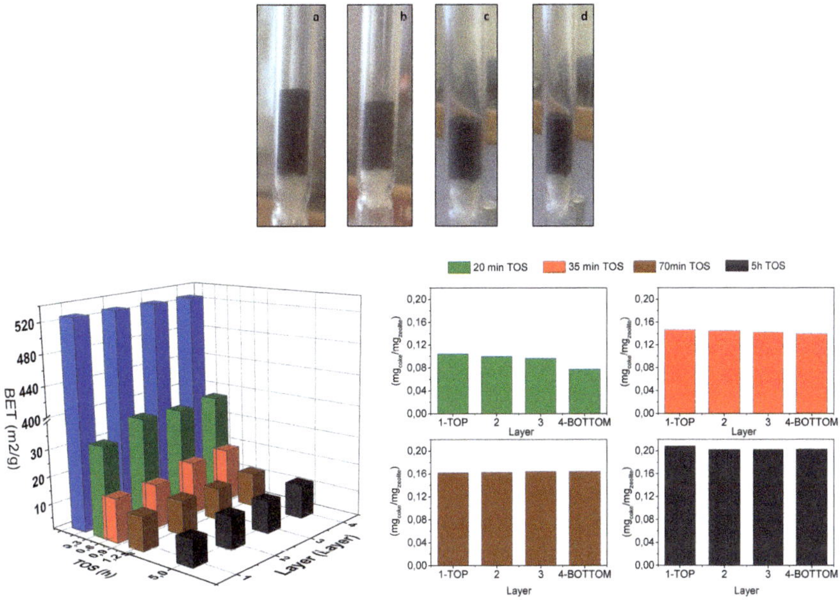

Fig. 9 Top: Visual evidence of the catalyst bed of mordenite CBV-21-A after quenching the reaction at (a) 20 min, (b) 35 min, (c) 70 min and (d) 5 hours on stream. Bottom: BET surface area (left) and coke content (right) on the partially deactivated layers of CBV-21-A at increasing TOS.

topology where alkenes and arenes work together as autocatalytic engines and where the effluent products are distributed between light alkenes, alkanes and arenes up to C_{10} (ZSM-5),[5,29] to a 3D 12-ring topology where light alkanes, alkenes and arenes up to C_{12} are the main products and where arenes together with alkenes are the dominating autocatalytic engines (beta).[50,51]

Among the 7 topologies studied, only two (ZSM-5 and ZSM-22) have previously been subjected to systematic studies of axial catalyst deactivation patterns. Hence, the systematic study reported herein fills the gap in pore size and dimensionality between those two topologies, and expands the parameter space to large-pore 12-ring zeolites.

4.1 Influence of catalyst topology on the origin and impact of coke formation

As a first general remark to the results reported in Section 3, current and previous studies of several samples belonging to the same topology demonstrate that the observed axial deactivation pattern is a general feature of each topology, and not limited to one particular sample. Considering next the origin of coke formation, *i.e.* methanol *versus* hydrocarbon products, the impact of hydrocarbons alone as coke precursors can quite straightforwardly be assessed from the spatio-temporal studies, in particular the coke amounts formed in the product zone compared to that formed in the initiation and autocatalytic zones. As such, a clear correlation is observed between zeolite topology and the axial deactivation pattern (Fig. 4–9): the larger the pore size, the more important the influence of coke formed from the conversion of hydrocarbon products. Conversely, the smaller the pore size, the more important the formation of coke from methanol. As mentioned above, the distinction between 10-ring and 12-ring topologies with respect to coke formation from hydrocarbon products alone is likely related to the facile diffusion of product molecules in and out of large-pore zeolites.

Considering next the coke formation from methanol alone, *i.e.* *via* formaldehyde (reaction (1)), compared to coke formation from methanol and hydrocarbons (reaction (2)), the correlation is less straightforward, in particular for topologies such as ZSM-5, beta, and mordenite, where no critical contact time is observed in the contact time variation experiments (Fig. 3). However, the ensemble of test data obtained for 10-ring zeolites with more narrow pore size than ZSM-5 show a substantial initiation zone, combined with coke formation in the top layer of the catalyst bed (Fig. 2, 5 and 6), and therefore strongly suggest that reaction (1) contributes to coke formation. One additional parameter may help to further clarify the relative impact of reactions (1) and (2): methane formation. If coke is formed exclusively *via* formaldehyde (Rx. (1)), then the cumulative mass of methane formed in the experiment should be similar to the cumulative mass of coke. If coke is formed from reactions between methanol and hydrocarbons (which could likely be secondary products from Rx. (1)), then the hydrogen abstraction required to subsequently form (poly-)aromatic and graphitic species could proceed by hydrogen transfer either to methanol (thereby forming methane) or to a hydrocarbon such as an alkene (thereby forming the corresponding alkane). The commonly reported correlation between the hydrogen transfer index (HTI), (*i.e.* the alkane fraction in aliphatic products), and the aromatics fraction in the effluent could suggest that hydrogen transfer to another hydrocarbon is the most common mechanism of hydrogen abstraction.[33] In this respect, it is of interest to note that for the ZSM-23 catalyst, where rapid coke accumulation was observed in the first catalyst layer during the first 20 minutes on stream, the total amount of CH_4 formed during those first 20 minutes was 0.5 mg CH_4 per $g_{catalyst}$, whereas the coke content in the first layer of ZSM-23 was 34 $mg_{coke}\ g_{catalyst}{}^{-1}$, *i.e.* close to two orders of magnitude higher. This result clearly suggests that reaction (1) mainly serves as an initiation reaction, which subsequently leads to coke formation by interactions between methanol and hydrocarbons.

Considering finally the overall performance of the various topologies with respect to conversion capacities (Fig. 2) and deactivation rates (Table 3), the ensemble of data should be considered with care due to the different Si/Al ratios and crystal sizes of the individual samples (Table 2). However, as a general trend, it is interesting to note that the 12-ring topologies have limited conversion capacity due to their high deactivation rates, while the conversion capacity of 10-ring topologies is equally related to their substantial critical contact time.

4.2 A simplified model of catalyst deactivation

Spatio-temporal studies of catalyst deactivation may be experimentally challenging for large sample libraries. It is therefore of interest to elucidate whether

Table 3 Simulated rate constants of the reactions (6)–(8)

Sample	k_1	k_2	k_3
ZSM-5-MFI-27	19 315	63	26
ZSM-23-h	1420	98	0
ZSM-22-h	2368	108	0
Beta-CP-806	3202	41	72
Beta-CP-7119	3575	17	68
Beta-CP814-E	4526	21	42
Mordenite-CBV-21A	3166	72	87

the coke origin in a given test of an MTH catalyst may be hypothesized by visual inspection of the shape of the conversion *versus* time on stream curve. If so, one could easily assess whether or not the deactivation rate of various samples could be compared by applying the autocatalytic deactivation model (which is limited to tests where methanol is a main coke precursor (*vide infra*)).

Olsbye[7] recently suggested a simple model to distinguish between coke formation from reactants and products. In that model, all reaction rates are assumed to be first order in the reacting molecule. Methanol, A, is the reactant, B is the effluent product, C is the coke and N is the number of active sites.

$$\mathrm{A} \xrightarrow{k_1} \mathrm{B} \quad -\mathrm{d}[\mathrm{A}]/\mathrm{d}t = (k_1 + k_2)[\mathrm{A}]N \tag{6}$$

$$\mathrm{A} \xrightarrow{k_2} \mathrm{C} \quad \mathrm{d}[\mathrm{C}]_i/\mathrm{d}t = (k_2[\mathrm{A}] + k_3[\mathrm{B}])N \tag{7}$$

$$\mathrm{B} \xrightarrow{k_3} \mathrm{C} \quad \mathrm{d}[\mathrm{B}]/\mathrm{d}t = (k_1[\mathrm{A}] - k_3[\mathrm{B}])N \tag{8}$$

In the current contribution, this simple model has been used to determine the rate constants k_1, k_2 and k_3 for each of the tests reported in Fig. 1, by regression analysis. The reactor was modelled as a series of 21 segments (9 in the case of MOR), where each segment was considered a CSTR reactor. The concentrations of A, B and C were integrated along the catalyst bed for each time on stream. Furthermore, the total coke content, C, and change in number of active sites, N, in each segment (i) was integrated with time on stream (j).

$$N_{i,j} = N_{i,j-1} - C_{i,j} \tag{9}$$

Fig. 1 depicts the experimental (symbols) and the simulated (dashed line) conversion curves *versus* time on stream over the samples used in the spatio-temporal study. The simulated rate constants are reported in Table 3.

As a first comment to the data in Table 3, the rate constant of the main reaction (k_1) clearly surpasses the value k_2 and k_3 in all the catalysts, as expected. Furthermore, the relative values of the two latter constants correlate with the shape of the deactivation curves, as follows.

First, the experimental MeOH conversion curves obtained for ZSM-5, ZSM-22 and ZSM-23 take an inverse S form with a rapid drop in conversion after the breakthrough of methanol. In the three medium-pore size zeolites, the rate constant corresponding to the formation of coke from the reactants (k_2) is larger than k_3, assigned to the formation of coke from the products of the reaction. In particular, for ZSM-22 and ZSM-23, where the flattening of the conversion *versus* time on stream curve is less pronounced than for ZSM-5, the model provides a value equal to 0 for k_3. This can be interpreted as if coke in these rather narrow-pore 10-ring zeolites originates exclusively from reactions involving methanol. In ZSM-5, the value of k_2 is more than two-fold of the value of k_3, suggesting that methanol is a main source of coke formation also in this case, however, with a significant contribution of coke formed from hydrocarbon products.

Second, the three beta zeolites show a deactivation curve for which the slope decreases with time on stream, as the product concentration decreases. In these 3D-12 ring zeolites, the value of k_3 is at least two-fold that of k_2, suggesting that coke is formed to a larger extent from the products of reaction.

Finally, mordenite-CBV-21A shows an intermediate situation where a rapid decline in conversion is observed after methanol breakthrough but where the slope of the curve decreases at low conversion. The calculated rate constants indicate that the formation of coke is due to both reactants and products but the latter contribute somewhat more to the formation of coke.

In conclusion, a clear correlation is observed between this simple model and the results reported in the spatio-temporal studies, suggesting that the shape of conversion *versus* time on stream curves is a valuable indicator of coke formation patterns in the catalyst bed.

4.3 Influence of catalyst topology on critical contact time

Finally, another particularly interesting observation of this study is the increase in critical contact time, *i.e.* the contact time needed to launch the autocatalytic reaction, with a decrease in pore size for the 10-ring zeolites. The critical contact time is particularly long for the 1D 10-ring materials, ZSM-22 and -23, shorter for IM-5 and ITQ-13, and negligible for ZSM-5, beta, and mordenite. In all cases, substantial coke formation is observed in the first layer of the catalyst bed. This last observation suggests that methanol conversion is significant in the first layer of all topologies. However, in some topologies it leads mainly to coke formation and only to a small fraction of hydrocarbon pool species, while in others it leads to substantial formation of hydrocarbon pool species and effluent products, alongside with coke.

We suspect that the reason for this difference is the distinct role played by polymethylated benzene (PMB) molecules in the different topologies. In ZSM-22, PMB are inert coke molecules which are trapped in the pores.[15,30] In ZSM-5 and beta zeolite, however, PMB are hydrocarbon pool engines, and are also part of the effluent products[5,29,50,51]. In the intermediate structures between ZSM-22 and ZSM-5, IM-5 and ITQ-13, PMB are eluted as effluent products, but their possible role as hydrocarbon pool engines is yet to be revealed.

To elucidate the influence of PMB as effluent products, and PMB as hydrocarbon pool engines, on the observed critical contact time, two plots were made. In the first graph, (Fig. 10), the critical contact time is plotted *versus* effluent selectivity to aromatic products at 50% conversion. We considered this selectivity the most reliable measure of pore size under reaction conditions. In the

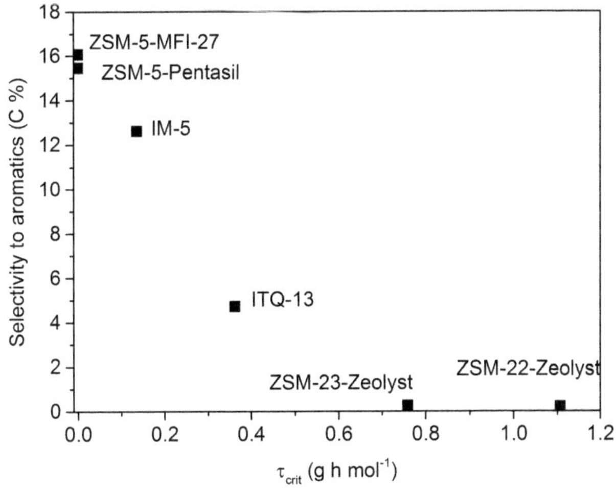

Fig. 10 Critical contact time (τ_{crit}) *versus* selectivity to aromatic products over medium pore size zeolites at 50% MeOH conversion.

second graph, the critical contact time is plotted *versus* the size of the largest sphere that can be inserted in the structure,[53] as a measure of cavity size (Fig. 11).

The plots show an inverse correlation between aromatics selectivity in the effluent and the critical contact time, whereas the correlation between the maximum cavity size and critical contact time is less evident. Based on this analysis, the ability of aromatic products to diffuse out of the zeolite crystals seems to be of primary importance for the rapid initiation of the autocatalytic MTH reaction.

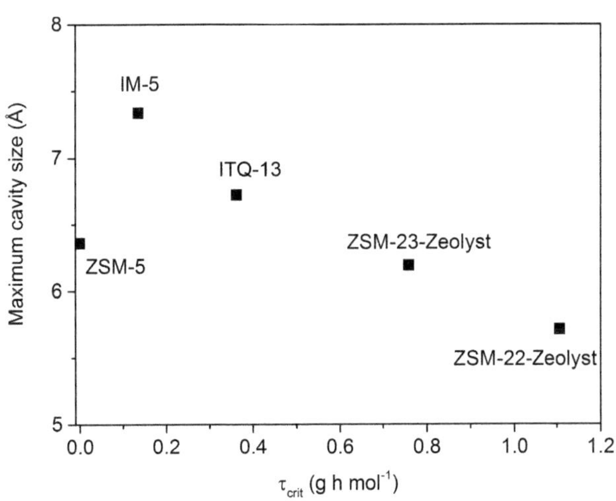

Fig. 11 Critical contact time (τ_{crit}) *versus* cavity size, measured as the largest sphere that can be inserted in the zeolite structure.

5. Conclusions

In this work we have examined the influence of zeolite topology on the axial mode of deactivation during the conversion of methanol to hydrocarbons. First, the autocatalytic deactivation model was used as a tool to determine the critical contact time and deactivation constant on seven different zeolites. The model was successfully applied in medium pore size topologies (ZSM-5, ZSM-22, ZSM-23, ITQ-13 and IM-5), yielding critical contact times and deactivation rates of each catalyst. In particular, it was found that the critical contact time increased with a decrease in pore size. Furthermore, the methanol conversion capacity of ZSM-5 varied little with contact time. However, the conversion capacity of the larger 12-ring topologies, mordenite and beta, decreased with an increase in contact time; in contrast to the 10-ring zeolites with pore sizes smaller than ZSM-5, for which the conversion capacity increased with an increase in contact time.

Furthermore, the deactivation on a selection of medium and large pore size zeolites has been investigated spatially and temporally. First, the catalyst bed was divided from top to bottom on increasingly deactivated samples. Secondly, the different layers of catalysts were characterized using nitrogen adsorption, thermogravimetric analysis and a dissolution–extraction protocol. The results revealed that zeolite topology affects not only catalyst lifetime and product distribution, but also modifies the axial mode of catalyst deactivation. Whereas the deactivation over medium pore size zeolites (ZSM-5, ZSM-22 and ZSM-23) occurs gradually from the entrance towards the outlet of the catalyst bed with gradients in the total coke content and evolution of BET area, in the case of beta and mordenite, the accumulation of coke and reduction of BET area in the larger pore size counterparts does not substantially vary along the catalyst bed layers, resulting in uniform deactivation throughout the catalyst bed.

Overall, this study has improved the understanding of the deactivation in the MTH process. Results show that the axial deactivation pattern and critical contact time change with zeolite topology, a concept which had not been addressed before.

Acknowledgements

This work is financially supported by the European Industrial Doctorates project "Zeomorph" (Grant Agreement No. 606965), as part of the Marie Curie actions (No. FP7-PEOPLE- 2013-ITN-EID). K. A. Lukaszuk and A. Molino are kindly acknowledged for the synthesis of ZSM-22-h and ZSM-23-h. Ms Sharmala Aravinthan is also acknowledged for her assistance with TG and N_2 adsorption experiments.

Notes and references

1 M. Stöcker, *Microporous Mesoporous Mater.*, 1999, **29**, 3–48.
2 U. Olsbye, S. Svelle, M. Bjørgen, P. Beato, T. V. W. Janssens, F. Joensen, S. Bordiga and K. P. Lillerud, *Angew. Chem., Int. Ed.*, 2012, **51**, 5810–5831.
3 B. Smit and T. L. M. Maesen, *Nature*, 2008, **451**, 671–678.
4 C. D. Chang and A. J. Silvestri, *J. Catal.*, 1977, **47**, 249–259.

5 M. Bjørgen, S. Svelle, F. Joensen, J. Nerlov, S. Kolboe, F. Bonino, L. Palumbo, S. Bordiga and U. Olsbye, *J. Catal.*, 2007, **249**, 195–207.
6 S. Svelle, F. Joensen, J. Nerlov, U. Olsbye, K. P. Lillerud, S. Kolboe and M. Bjørgen, *J. Am. Chem. Soc.*, 2006, **128**, 14770–14771.
7 U. Olsbye, S. Svelle, K. P. Lillerud, Z. H. Wei, Y. Y. Chen, J. F. Li, J. G. Wang and W. B. Fan, *Chem. Soc. Rev.*, 2015, **44**, 7155–7176.
8 M. Guisnet, L. Costa and F. R. Ribeiro, *J. Mol. Catal. A: Chem.*, 2009, **305**, 69–83.
9 H. Schulz, *Catal. Today*, 2010, **154**, 183–194.
10 S. Teketel, U. Olsbye, K.-P. Lillerud, P. Beato and S. Svelle, *Microporous Mesoporous Mater.*, 2010, **136**, 33–41.
11 T. V. W. Janssens, *J. Catal.*, 2009, **264**, 130–137.
12 T. V. W. Janssens, S. Svelle and U. Olsbye, *J. Catal.*, 2013, **308**, 122–130.
13 S. Müller, Y. Liu, M. Vishnuvarthan, X. Sun, A. C. van Veen, G. L. Haller, M. Sanchez-Sanchez and J. A. Lercher, *J. Catal.*, 2015, **325**, 48–59.
14 F. L. Bleken, K. Barbera, F. Bonino, U. Olsbye, K. P. Lillerud, S. Bordiga, P. Beato, T. V. W. Janssens and S. Svelle, *J. Catal.*, 2013, **307**, 62–73.
15 S. Teketel, S. Svelle, K.-P. Lillerud and U. Olsbye, *ChemCatChem*, 2009, **1**, 78–81.
16 F. Schmidt, C. Hoffmann, F. Giordanino, S. Bordiga, P. Simon, W. Carrillo-Cabrera and S. Kaskel, *J. Catal.*, 2013, **307**, 238–245.
17 D. M. Bibby, N. B. Milestone, J. E. Patterson and L. P. Aldridge, *J. Catal.*, 1986, **97**, 493–502.
18 P. L. Benito, A. G. Gayubo, A. T. Aguayo, M. Olazar and J. Bilbao, *Ind. Eng. Chem. Res.*, 1996, **35**, 3991–3998.
19 M. Kaarsholm, F. Joensen, J. Nerlov, R. Cenni, J. Chaouki and G. S. Patience, *Chem. Eng. Sci.*, 2007, **62**, 5527–5532.
20 J. Haw and D. Marcus, *Top. Catal.*, 2005, **34**, 41–48.
21 D. S. Wragg, M. G. O'Brien, F. L. Bleken, M. Di Michiel, U. Olsbye and H. Fjellvåg, *Angew.Chem., Int. Ed.*, 2012, **51**, 7956–7959.
22 M. Luo, H. Zang, B. Hu, B. Wang and G. Mao, *RSC Adv.*, 2016, **6**, 17651–17658.
23 X. Sun, S. Mueller, Y. Liu, H. Shi, G. L. Haller, M. Sanchez-Sanchez, A. C. van Veen and J. A. Lercher, *J. Catal.*, 2014, **317**, 185–197.
24 G. J. Hutchings, F. Gottschalk and R. Hunter, *Ind. Eng. Chem. Res.*, 1987, **26**, 635–637.
25 M. Bjørgen, U. Olsbye and S. Kolboe, *J. Catal.*, 2003, **215**, 30–44.
26 M. Bjørgen, U. Olsbye, S. Svelle and S. Kolboe, *Catal. Lett.*, 2004, **93**, 37–40.
27 M. Bjørgen, U. Olsbye, D. Petersen and S. Kolboe, *J. Catal.*, 2004, **221**, 1–10.
28 C. D. Chang, *Catal. Rev.*, 1984, **26**, 323–345.
29 F. L. Bleken, T. V. W. Janssens, S. Svelle and U. Olsbye, *Microporous Mesoporous Mater.*, 2012, **164**, 190–198.
30 S. Teketel, W. Skistad, S. Benard, U. Olsbye, K. P. Lillerud, P. Beato and S. Svelle, *ACS Catal.*, 2012, **2**, 26–37.
31 M. Westgård Erichsen, S. Svelle and U. Olsbye, *J. Catal.*, 2013, **298**, 94–101.
32 K. Barbera, F. Bonino, S. Bordiga, T. V. W. Janssens and P. Beato, *J. Catal.*, 2011, **280**, 196–205.
33 S. Teketel, M. Westgard Erichsen, F. Lonstad Bleken, S. Svelle, K. Petter Lillerud and U. Olsbye, in *Catalysis: Volume 26*, The Royal Society of Chemistry, 2014, vol. 26, pp. 179–217.

34 F. Bleken, W. Skistad, K. Barbera, M. Kustova, S. Bordiga, P. Beato, K. P. Lillerud, S. Svelle and U. Olsbye, *Phys. Chem. Chem. Phys.*, 2011, **13**, 2539–2549.

35 R. Castañeda, A. Corma, V. Fornés, J. Martínez-Triguero and S. Valencia, *J. Catal.*, 2006, **238**, 79–87.

36 K. Möller and T. Bein, *Microporous Mesoporous Mater.*, 2011, **143**, 253–262.

37 S. Ernst, J. Weitkamp, J. A. Martens and P. A. Jacobs, *Appl. Catal.*, 1989, **48**, 137–148.

38 S. Brunauer, P. H. Emmett and E. Teller, *J. Am. Chem. Soc.*, 1938, **60**, 309–319.

39 J. Rouquerol, P. Llewellyn and F. Rouquerol, in *Studies in Surface Science and Catalysis*, Elsevier, 2007, vol. 160, pp. 49–56.

40 P. Magnoux, P. Roger, C. Canaff, V. Fouche, N. S. Gnep and M. Guisnet, *Stud. Surf. Sci. Catal.*, 1987, **34**, 317–330.

41 B.-T. Bleken, D. Wragg, B. Arstad, A. Gunnæs, J. Mouzon, S. Helveg, L. Lundegaard, P. Beato, S. Bordiga, U. Olsbye, S. Svelle and K. Lillerud, *Top. Catal.*, 2013, **56**, 558–566.

42 F. J. Keil, *Microporous Mesoporous Mater.*, 1999, **29**, 49–66.

43 M. Sugimoto, H. Katsuno, K. Takatsu and N. Kawata, *Zeolites*, 1987, **7**, 503–507.

44 C. D. Chang, C. T. W. Chu and R. F. Socha, *J. Catal.*, 1984, **86**, 289–296.

45 D. Chen, H. P. Rebo, K. Moljord and A. Holmen, *Ind. Eng. Chem. Res.*, 1997, **36**, 3473–3479.

46 F. L. Bleken, K. Barbera, F. Bonino, U. Olsbye, K. P. Lillerud, S. Bordiga, P. Beato, T. V. W. Janssens and S. Svelle, *J. Catal.*, 2013, **307**, 62–73.

47 S. Teketel, L. F. Lundegaard, W. Skistad, S. M. Chavan, U. Olsbye, K. P. Lillerud, P. Beato and S. Svelle, *J. Catal.*, 2015, **327**, 22–32.

48 P. del Campo, W. A. Slawinski, R. Henry, M. W. Erichsen, S. Svelle, P. Beato, D. Wragg and U. Olsbye, *Surf. Sci.*, 2016, **648**, 141–149.

49 S. Teketel, U. Olsbye, K. P. Lillerud, P. Beato and S. Svelle, *Appl. Catal., A*, 2015, **494**, 68–76.

50 M. Bjørgen, S. Akyalcin, U. Olsbye, S. Benard, S. Kolboe and S. Svelle, *J. Catal.*, 2010, **275**, 170–180.

51 M. Bjørgen, F. Joensen, K.-P. Lillerud, U. Olsbye and S. Svelle, *Catal. Today*, 2009, **142**, 90–97.

52 J. W. Park, S. J. Kim, M. Seo, S. Y. Kim, Y. Sugi and G. Seo, *Appl. Catal., A*, 2008, **349**, 76–85.

53 C. Baerlocher, L. B. McCusker and D. H. Olson, *Atlas of zeolite framework types*, Elsevier, 2007.

Faraday Discussions

PAPER

An assessment of hydrocarbon species in the methanol-to-hydrocarbon reaction over a ZSM-5 catalyst

Suwardiyanto,[a] Russell F. Howe,[ab] Emma K. Gibson,[b] C. Richard A. Catlow,[bcd] Ali Hameed,[e] James McGregor,[be] Paul Collier,[f] Stewart F. Parker[bg] and David Lennon*[bh]

Received 19th September 2016, Accepted 12th October 2016
DOI: 10.1039/c6fd00195e

A ZSM-5 catalyst is examined in relation to the methanol-to-hydrocarbon (MTH) reaction as a function of reaction temperature and time-on-stream. The reaction profile is characterised using in-line mass spectrometry. Furthermore, the material contained within a catch-pot downstream from the reactor is analysed using gas chromatography-mass spectrometry. For a fixed methanol feed, reaction conditions are selected to define various stages of the reaction coordinate: (i) initial methanol adsorption at a sub-optimum reaction temperature (1 h at 200 °C); (ii) initial stages of reaction at an optimised reaction temperature (1 h at 350 °C); (iii) steady-state operation at an optimised reaction temperature (3 days at 350 °C); and (iv) accelerated ageing (3 days at 400 °C). Post-reaction, the catalyst samples are analysed *ex situ* by a combination of temperature-programmed oxidation (TPO) and spectroscopically by electron paramagnetic resonance (EPR), diffuse-reflectance infrared and inelastic neutron scattering (INS) spectroscopies. The TPO measurements provide an indication of the degree of 'coking' experienced by each sample. The EPR measurements detect aromatic radical cations. The IR and INS measurements reveal the presence of retained hydrocarbonaceous species, the nature of which are discussed in terms of the well-developed 'hydrocarbon pool' mechanism. This combination of experimental evidence, uniquely applied to this reaction system, establishes the importance of retained hydrocarbonaceous species in effecting the product distribution of this economically relevant reaction system.

[a]*Department of Chemistry, University of Aberdeen, Aberdeen, AB24 3UE, UK*
[b]*UK Catalysis Hub, Research Complex at Harwell, STFC Rutherford Appleton Laboratory, Chilton, Oxon, OX11 0FA, UK*
[c]*Department of Chemistry, University College London, 20 Gordon St., London WC1 HOAJ, UK*
[d]*Cardiff Catalysis Institute, School of Chemistry, Cardiff University, CF10 3AT, UK*
[e]*Department of Chemical and Biological Engineering, University of Sheffield, Sheffield, S1 3JD, UK*
[f]*Johnson Matthey Technology Centre, Blounts Court, Sonning Common, Reading, RG4 9NH, UK*
[g]*ISIS Facility, STFC Rutherford Appleton Laboratory, Chilton, Oxon, OX11 0QX, UK*
[h]*School of Chemistry, University of Glasgow, Joseph Black Building, Glasgow, G12 8QQ, UK. E-mail: David.Lennon@glasgow.ac.uk*

1. Introduction

As oil reserves continue to diminish, the methanol to olefin (MTO) reaction and related methanol to gasoline (MTG) reactions over zeolitic catalysts such as ZSM-5 are anticipated to play an increasing role in the landscape of large-scale chemical and fuel manufacturing operations. Although the original feedstock for these processes is methanol derived from natural gas, future feedstocks may also include methanol derived from coal and/or biomass. Looking further ahead, renewable sources such as the direct oxidation of methane produced by aerobic digestion could additionally contribute to and sustain the viability of this intricate reaction system.

Following the industrial innovation in methanol-to-hydrocarbon (MTH) chemistry that commenced in the 1980s, a vast amount of literature has subsequently materialised, examining all aspects of the associated process chemistry. A representative sample of the contemporary literature follows. Hemelsoet *et al.* looked at the MTO reaction over ZSM-5 and SAPO-34, concentrating on mechanistic aspects of the process; in particular, they examined the well-established concept of a hydrocarbon pool *via* a combination of kinetic measurements supplemented by theoretical simulations.[1] C. Wang *et al.* have also concentrated on the hydrocarbon pool chemistry connected with the ZSM-5/MTO reaction, using solid-state NMR spectroscopy to highlight the role of a variety of polymethylbenzenes as reactive intermediates.[2] S. Wang *et al.* have used DFT calculations to probe the MTO process over a range of zeolites and report that the interaction between the proton transfer reagent (methanol or water) and the zeolitic framework are key factors in the olefin elimination step.[3] Sun *et al.* have addressed issues connected with the two distinct catalytic cycles that operate for MTO chemistry over ZSM-5; namely, an olefin homologation/cracking route and the formation of the afore-mentioned hydrocarbon pool.[4] *Via* a detailed kinetic analysis, these works show the importance of a hydrogen transfer pathway involving chemisorbed methanol intermediates, which has consequences for sustaining catalyst lifetime under MTO process conditions.[4] More recently, Olsbye and co-workers have comprehensively reviewed matters connected with the MTH process over protonated zeotype catalysts.[5] This includes an appreciation that the autocatalytic feature of the process involves a role for hydrocarbons as active species but, additionally, as deactivating species. They explicitly consider the topic of catalyst deactivation by (hydro-)carbon residues, which leads on to the concept of coking. The time evolution of the (hydro-)carbon deposits at the micro- and meso-scales is also considered.[5] Such investigations employ an armoury of analytical techniques to interrogate the range of active species associated with the product formation pathways as well as with the catalyst deactivation pathways. In addition to the conventional micro-reactor technology, these include probes such as solid-state NMR spectroscopy, infrared spectroscopy, electron paramagnetic resonance (EPR) spectroscopy, thermogravimetric analysis, temperature-programmed oxidation (TPO), Raman scattering, confocal fluorescence microscopy and UV-visible micro-spectroscopy.[5]

One technique that, up until now, has been little used to examine MTO/MTH reactions over zeolites is inelastic neutron scattering (INS). Increasingly, INS is being applied to investigate heterogeneous catalysts used in a variety of economically relevant reaction systems, such as methyl chloride synthesis,[6] methanol synthesis,[7] methane reforming,[8] Fischer–Tropsch synthesis,[9,10] *etc.* The application of inelastic neutron spectrometers to investigate issues in molecular spectroscopy, including aspects of heterogeneous catalysis, has been reviewed by Parker and co-workers.[11]

In 2016 O'Malley *et al.* used a variety of neutron scattering methods, including INS, to investigate room temperature methoxylation in zeolites.[12] In the same year, Howe *et al.* reported a preliminary investigation applying INS to examine the methanol-to-gasoline reaction over a ZSM-5 catalyst.[13] This latter work demonstrated how INS could be used to obtain the vibrational spectrum of hydrocarbonaceous entities retained by the catalyst during the MTH transformation process at elevated reaction temperatures. Compared to infrared spectroscopy, INS conveys several advantages. Namely, (i) a wide range of vibrational frequencies (600–4500 cm^{-1}) is accessible without obscuration from zeolite framework modes, (ii) the method can be applied to technical catalysts with no requirement for sample preparation and (iii) no restriction on the extent of coke formation is evident. This latter attribute in particular suggests that the methodology adopted by Howe *et al.* could be used to simultaneously inspect several facets of the MTH process; specifically, the formation of higher molecular weight hydrocarbons and coke precursors.[13]

Recognising that INS can supplement the already wide range of investigations on MTH chemistry over zeolitic materials, the present study extends the preliminary ZSM-5/MTH investigation of Howe and co-workers[13] by using INS to discover if the hydrocarbonaceous species retained by the zeolite is sensitive to reaction conditions. To this end, and for a fixed methanol feed, the following four sets of reaction conditions are selected to define various stages of the reaction coordinate: (i) initial methanol adsorption at a sub-optimum reaction temperature (1 h at 200 °C), (ii) initial stages of reaction at an optimised reaction temperature (1 h at 350 °C), (iii) steady-state operation at an optimised reaction temperature (3 days at 350 °C) and (iii) accelerated ageing (3 days at 400 °C). INS spectra are presented for catalyst samples that have experienced each of these four reaction treatments. In order to contextualise the INS measurements, a multi-technique approach is adopted. Firstly, reaction profiles are characterised by in-line mass spectrometry. Further, products contained within a 'catch-pot' downstream from the reactor are analysed by gas chromatography-mass spectrometry. Post-reaction, the catalyst samples are analysed *ex situ* by a combination of temperature-programmed oxidation (TPO), electron paramagnetic resonance (EPR) spectroscopy and diffuse-reflectance infrared Fourier transform spectroscopy (DRIFTS). These additional measurements facilitate comparisons to literature work as highlighted by, amongst others, Olsbye and co-workers.[5] This is the first time that this combination of experimental techniques have been applied to the methanol to hydrocarbon reaction over a ZSM-5 catalyst. In order to ensure a balance between sensitivity and resolution,[11] two direct geometry INS spectrometers are employed to better discern the nature of the hydrocarbonaceous material retained at the catalyst surface.

2. Experimental

2.1 Catalyst characterisation

The ZSM-5 catalyst was a commercial grade catalyst supplied by Johnson Matthey (Intercat IC16926, lot no. 16296.02261). As received, it possessed a Si : Al ratio of ~30, although it also contained some residual template (tetrapropyl ammonium bromide) at ~0.7 wt%. The BET surface area was 371 $m^2\ g^{-1}$. Chemical analysis using XRF revealed the following composition (wt%): SiO_2 (95), Al_2O_3 (4.96), Fe_2O_3 (0.03), TiO_2 (0.02), CaO (0), Na_2O (0), P_2O_5 (0). In order to remove the residual template, the catalyst was calcined in air at 500 °C for 12 h prior to use. Inspection of the sample post-calcination using infrared spectroscopy showed complete removal of the template material. It is noted that this ZSM-5 catalyst differs from the material examined by Howe and co-workers in their preliminary report[13] and that used by O'Malley *et al.*[12]

2.2 Reaction testing

Reaction testing was performed on a catalyst test system designed to activate and react heterogeneous catalyst samples of the sample sizes typically required for INS measurement (10–20 g). The gas manifold, control system and heating arrangements are comprehensively described elsewhere.[14] A new design of stainless steel reactor was used to contain the catalyst samples, Fig. 1. A photograph of the reactor is presented on the left hand side of Fig. 1, whilst the right hand side shows a schematic drawing of the reactor. The internal diameter of the reactor is 35 mm and its length is 60 mm. For a comparable charge, this arrangement minimises the catalyst bed length with respect to previously used designs,[15] thereby minimising the possibility of concentration gradients across the length of the bed and ensuring greater homogeneity of sample composition throughout the sample volume.

The reactor was charged with the calcined catalyst powder (~12.0 g) and dried using the following procedure: whilst continually maintaining a flow of helium over the catalyst (1000 sccm, CK Gas, >99.0%), the sample was heated to 350 °C at a heating rate of 10 °C min^{-1} and then maintained at that temperature for 2 h. The heating was stopped and the sample was allowed to cool to ambient temperature in a continuous flow of He. The reactor was then isolated and transferred to an argon-filled glove box (MBraun UniLab MB-20-G, $[H_2O] < 1$ ppm, $[O_2] < 2$ ppm) for loading of the catalyst into an aluminium INS cell that is sealed *via* an indium wire gasket.[14] The sample was then transferred to the INS spectrometer for spectral acquisition. All INS measurements were undertaken at ~20 K. The resulting spectrum provided the vibrational spectrum of the clean, activated catalyst.

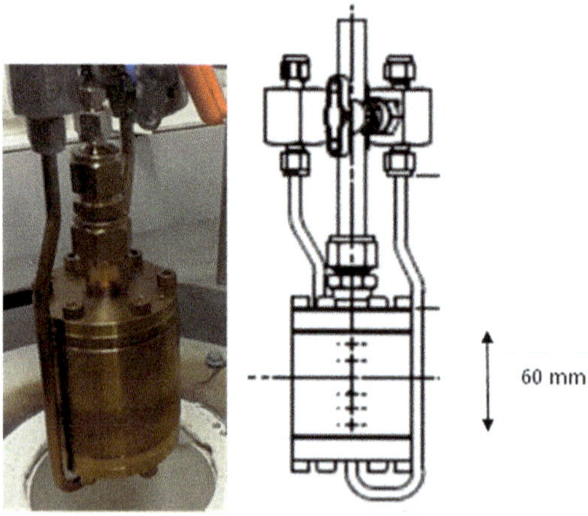

Fig. 1 Photograph (left) and schematic (right) of the stainless steel reactor.

For samples to be reacted with methanol at different temperatures and for varying times-on-stream, a clean activated catalyst sample was prepared in the same manner as described above. However, the dehydration stage was followed by dosing of the catalyst with methanol vapour by means of a bubbler arrangement whilst the catalyst was maintained at a specified reaction temperature (200, 350 or 400 °C) for a specified period of time (1 h or 3 days). Methanol (Aldrich, 99.9%) was entrained in flowing helium (150 ml min^{-1}) and the methanol exposure corresponded to a feed rate of 0.2 g methanol $g_{(cat)}^{-1}$ h^{-1}. The eluting stream was analysed using in-line mass spectrometry (Hiden Analytical, HPR-20), with the spectrometer connected to the reactor exit line *via* a differentially-pumped, heated quartz capillary. Relative intensities of various masses were normalized against the signal of the helium carrier gas. The apparatus is equipped with a 'catch-pot' downstream of the catalyst that collected liquid products. Material collected there was subsequently analysed using GC-MS (Shimadzu QP2010SE). The column employed was a DB-1MS capillary column (60 m length, 0.25 mm diameter, 0.25 µm thickness). The analytical conditions were as follows: initial oven temperature was 40 °C, held for 2 min, followed by heating at 10 °C min^{-1} to 150 °C, and then held for 3 min. The injector and detector temperatures were 250 °C.

After the designated time-on-stream, the methanol flow and heating were stopped, and the reactor cooled whilst maintaining the helium flow. The reactor was then isolated and transferred to the glovebox, where the catalyst was loaded into an Al INS cell using the procedure outlined above. During extraction of the reacted catalyst from the stainless steel reactor, a proportion of the sample was set aside for subsequent *ex situ* analysis. The samples for temperature-programmed oxidation and infrared spectroscopic measurements were stored in glass vials under Ar atmosphere, whilst samples for subsequent electron paramagnetic analysis were transferred to quartz EPR tubes and sealed *via* a Youngs tap (PTFE seal) before removal from the glove box.

2.3 Inelastic neutron scattering

Two spectrometers were utilised in this study. Due to the relative low quantities of hydrogenous species retained by the catalyst as evidenced by previous measurements,[13] the MERLIN spectrometer was used to study all the prepared catalyst samples. This is a direct geometry inelastic spectrometer that affords superior sensitivity compared to other instruments in the Target Station-1 suite of the ISIS Facility of the Rutherford Appleton Laboratory.[16] To maximise instrument sensitivity, the low resolution S-chopper package was utilised with incident energies (and frequencies) of 5244 cm^{-1} (600 Hz), 2017 cm^{-1} (400 Hz) and 1210 cm^{-1} (300 Hz).

Certain samples were selected to be run on the MAPS spectrometer, which possesses superior resolution compared to MERLIN.[16] Given the complexity of the molecules associated with surface reactions in the MTH process,[5] favourable resolution is required to discern individual molecular components. Here, the higher resolution A-chopper package was used with incident energies (and frequencies) of 5244 cm^{-1} (600 Hz), 2017 cm^{-1} (400 Hz) and 807 cm^{-1} (200 Hz). All INS spectra displayed (Fig. 9–13) are integrated over the momentum transfer range $0 \leq Q \leq 12$ Å^{-1}.[10]

2.4 Other analytical procedures

Temperature-programmed oxidation (TPO) was conducted on the catalysts post-reaction using a Micromeritics Chemisorb 2720 instrument equipped with a thermal conductivity detector. Upon removal of the sample that was contained within a glass vial from the glove box, the sealed vial was retained for approximately 3 weeks before the TPO measurements. Approximately 5 mg of catalyst was used in each case. The sample was first purged with helium (25 ml min^{-1}) before being heated in 5% O_2/He (25 ml min^{-1}) at a rate of 10 °C min^{-1} to a final temperature of 800 °C. This final temperature was then maintained for 30 min to ensure complete combustion of any carbonaceous species.

EPR spectra were measured with a JEOL FA200 spectrometer operating at 9.1 GHz at room temperature or 80 K with an Oxford Instruments ESR900 cryostat using liquid nitrogen. Spectra were recorded at low power (0.1 mW) and modulation (0.1 mT) to avoid line broadening. Methanol adsorption experiments were conducted in EPR cells fitted with high vacuum stopcocks. The zeolite was dehydrated at 350 °C on a turbo-pumped vacuum line (base pressure 10^{-6} mbar), exposed to methanol vapour (100 mbar) at room temperature, then heated to various temperatures. Concentrations of radicals were estimated by comparing doubly integrated signal intensities with that of a weak pitch standard (Bruker, 10^{13} spins per cm). Simulation of the spectrum was done using the Symphonia (Bruker) software.

Diffuse Reflectance Infrared Fourier Transform Spectroscopy (DRIFTS) spectra were collected with an Agilent Carey 680 FTIR spectrometer, taking 64 scans with a resolution of 4 cm^{-1} using a liquid nitrogen cooled MCT detector. Upon removal of the sample sealed within a glass vial from the glove box, the vial was retained for approximately 1 day before the DRIFTS measurements were undertaken. Transfer of the used catalyst samples to the DRIFTS apparatus was performed in the presence of a continuous flow of helium. Spectra were collected at every 5 °C whilst the sample was ramped to 375 °C (10 °C min^{-1}) under a 30 ml min^{-1} flow of helium.

3. Results and discussion

Five samples will be considered in this study. Table 1 defines sample history in terms of exposure to methanol at the standard rate of 0.2 g methanol $g_{(cat)}{}^{-1}$ h^{-1}. Sample 1 represents the clean, activated catalyst; this is used as a background in some of the INS measurements. Schulz has shown that there exists a marked temperature dependence in ZSM-5/methanol reactivity, with reaction below 300 °C leading to a rapid loss of activity and significant hydrocarbon deposition

Table 1 Sample descriptors and sample treatments for the five ZSM-5 samples examined

Sample no.	Sample treatment	Reaction temperature/°C	Duration
1	ZSM5 + He	350	1 h
2	ZSM5 + methanol	200	1 h
3	ZSM5 + methanol	350	1 h
4	ZSM5 + methanol	350	3 d
5	ZSM5 + methanol	400	3 d

compared to tests employed at high reaction temperatures.[17] Sample 2 is therefore intended to mimic this phase of reactivity. A reaction temperature of 350 °C is selected as being a representative reaction temperature. Sample 3 considers the case for short reaction times, whilst sample 4 represents longer reaction times. Coking is a fundamental component of MTH chemistry over ZSM-5,[5] and the elevated reaction temperature of 400 °C (sample 5) is thought to expose the catalyst to a degree of accelerated ageing. Collectively, this selection of samples provides information on whether the hydrocarbonaceous retained material reported in the preliminary INS study[13] is sensitive to reaction conditions.

3.1 In-line mass spectrometry and catch-pot analysis

Fig. 2 presents the reaction profile for sample 3, for various stages of the 3 day duration of sustained methanol conversion at 350 °C. Frame (a) corresponds to the initial catalyst conditioning period, frame (b) represents 1 day on-stream, frame (c) 2 days on stream and frame (d) 3 days on-stream. The methanol consumption is approximately constant throughout the whole reaction sequence, although a slight increase in conversion is observed in the data for the third day compared to that observed during the first two days of operation. Whereas the product distribution is broadly consistent after day 1, dramatic swings in product concentrations are observed during the initial period of reaction. Most important is the rise and subsequent fall of dimethyl ether over the ~2 h period of time presented in Fig. 2(a). It rises almost immediately on introduction of methanol, remains roughly constant for less than an hour, thereafter it drops to a low level that is subsequently maintained for the following 3 days. This profile is indicative

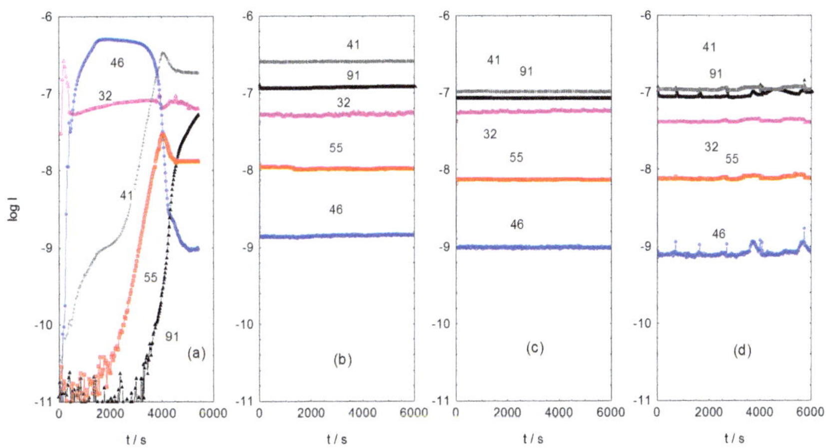

Fig. 2 Mass spectral analysis of the effluent from the reactor at 350 °C over a 3 day period of continuous reaction as a function of time on stream: (a) initial introduction of methanol; (b) after 1 day on stream; (c) after 2 days on stream; and (d) after 3 days on stream. Each frame displays the mass spectrometer output over a 1.7 h period. The numbers on the plots correspond to atomic mass numbers: 91 amu corresponds to the tropylium ion that signifies the presence of aromatic species; 55 amu corresponds to butene; 46 amu corresponds to dimethyl ether; 41 amu corresponds to propene; and 32 amu corresponds to methanol.

of dimethyl ether (DME) being an intermediate species, most likely formed from the dimerization of chemisorbed methoxy groups.[18]

During the period when DME production is greatest, a rapid increase in propene production is observed that is subsequently sustained throughout the 3 day sequence. In contrast, as DME formation falls, butene production increases and then achieves a nearly constant rate of formation thereafter. At the point where the DME signal starts to significantly decline, the aromatic signal, as signified by the tropylium ion, suddenly rises from baseline levels. This promptly levels out and achieves sustained production. At 3 days, which represents steady-state operation, propene and aromatic entities constitute the major products. The concomitant fall in DME production and rise in aromatic formation suggests that DME could be a precursor to aromatic products. The coincidence of aromatics and olefins in the product stream is consistent with well documented MTH chemistry.[4,5] It is noted that for all sets of reactions investigated here (Table 1, samples 2–5), mass spectrometric analysis of the product stream showed no obvious evidence of deactivation during the course of the reactions.

For the catch-pot analysis, two distinct phases were observed: an aqueous phase and an organic phase. The aqueous phase was essentially entirely water. Fig. 3 presents the chromatogram for the organic phase and shows a number of substituted aromatics to be present. The simplest aromatic compound observed, and in relatively low quantities, was toluene. The importance of methylation as a surface mediated reaction is demonstrated by the presence of dimethylbenzenes, trimethylbenzenes and tetramethylbenzenes in relatively significant quantities. However, in addition, although to a lesser degree, polycyclic aromatic hydrocarbons are also produced, as evidenced by the presence of a number of substituted naphthalenes. No higher molecular weight fractions were detected in the liquor. As noted above, when commenting on vapour phase products detectable by mass spectrometry, MTH activity corresponds to the

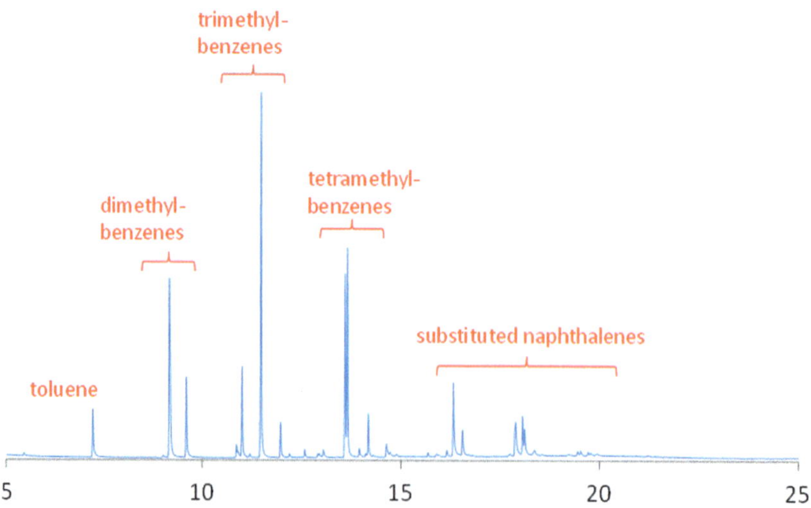

Fig. 3 Chromatograph for the organic phase taken from the catch-pot on steady-state operation at 350 °C.

formation of olefins and aromatic species *via* parallel, coincident pathways.[4,5] The main thrust of the present investigation is the application of INS to obtain new information on the nature of the hydrocarbon pool that is frequently connected to the MTH process (see Section 3.4). Although the product analysis here is qualitative, it is informative in terms of identifying relevant trends within the product distribution. A more detailed quantitative study of the different aromatic compounds would be helpful for refining mechanistic insight but this was beyond the scope of the present study.

3.2 Temperature-programmed oxidation (TPO)

The post-reaction TPO profiles for samples 2–5 are shown in Fig. 4. Reaction with methanol at a temperature below the idealised reaction temperature of 350 °C, *i.e.* 200 °C, (Fig. 4(a)) shows little carbon deposition to have occurred over the 1 h reaction period. A weak, relatively narrow signal with a peak maximum at about 280 °C is seen. This is suggestive of homogeneous hydrocarbon species that are weakly bound to the catalyst surface. Increasing the temperature to 350 °C for 1 h causes a modest increase in carbon retention with desorption occurring over a wider temperature range, *viz.* 280–620 °C, with a peak maximum at ~590 °C. This indicates that the enhanced reaction temperature has induced greater carbonaceous retention and that this carbon is more strongly bound to the catalyst, and is hence likely to be more ordered, than was the case for the lower temperature reaction. Table 2 presents the quantities of retained carbon for each sample.

It appears that the processes contributing to Fig. 4(b) are cumulative, as an extended run time of 3 days leads to a dramatic increase in the quantity of retained carbon, Fig. 4(c). The profile is less skewed than that observed in Fig. 4(b) and displays a distinct maximum at 610 °C. The even distribution of the peak at this T_{max} indicates the retained material to be reasonably well ordered and strongly bound to the catalyst. Fig. 4(d) shows the TPO plot for the aged sample

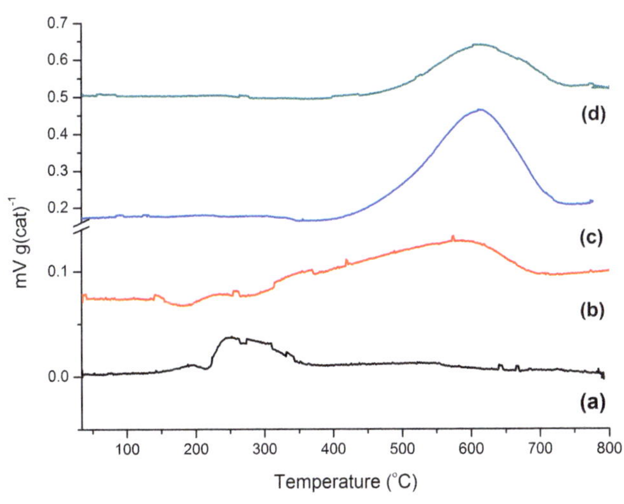

Fig. 4 Temperature-programmed oxidation profiles for the ZSM-5 catalyst post reaction: (a) 1 h at 200 °C; (b) 1 h at 350 °C; (c) 3 days at 350 °C; and (d) 3 days at 400 °C.

Table 2 Carbon retention (wt%) for samples 2–5 as determined by *ex situ* temperature-programmed oxidation

Sample no.	Sample treatment	C/wt%
2	ZSM5 + methanol, 200 °C, 1 h	2.7
3	ZSM5 + methanol, 350 °C, 1 h	2.6
4	ZSM5 + methanol, 350 °C, 3 d	7.1
5	ZSM5 + methanol, 400 °C, 3 d	4.9

(3 days at 400 °C). Although the peak maximum coincides with Fig. 4(c), it is noticeably smaller in intensity. Intuitively, this seems incorrect, as it indicates a lower degree of retention of strongly bound carbonaceous material, where one might have expected coke forming reactions to dominate. However, it is noted that Schulz has previously reported a similar trend for ZSM-5/methanol, where lower coke deposition was seen at 380 °C compared to that observed at 290 °C.[17] This pattern also meshes with the comments of Sun *et al.*, who highlight how reaction temperature can disproportionally influence the production rates of the two product streams: the aromatics based cycle dominates at low reaction temperatures but the olefin based cycle is favoured at higher reaction temperatures.[4] With respect to the disproportionate intensities observed in Fig. 4(c) and (d), it is thought that the higher operational temperature has disfavoured a coke forming pathway.

One final point for this section concerns the degree of coking. Although the TPO profiles in Fig. 4 for samples 4 and 5 show substantial amounts of carbon being present, it is noted that visual inspection of these samples post-reaction revealed the catalysts to exhibit a light grey colouration; they were most certainly not black in colour. This observation indicates the catalysts to be only mildly coked, consistent with the quantities displayed in Table 2. Bibby and co-workers report that the coke loading at which hydrocarbon production stops varies with zeolite particle size but is within the range of 14–18 wt%.[19]

3.3 Electron paramagnetic resonance (EPR) spectroscopy

A clean, activated sample of ZSM-5 was EPR silent. Fig. 5 shows the EPR spectra recorded at room temperature from used catalyst samples removed from the reactor and loaded into EPR tubes under argon. After reaction at 350 °C for 1 hour, a weak narrow line at $g = 2.003$ was observed with a line width of 1.2 mT (Fig. 5(a)). Amplification of this signal (Fig. 5(b)) revealed some indication of poorly resolved hyperfine splitting. After reaction at 350 °C for 3 days, a similar signal was measured with a 20 fold increase in intensity (Fig. 5(c)). Consistent with the TPO data (Section 3.2), the signal obtained after reaction at 400 °C for 3 days (Fig. 5(d)) had lower intensity and a significantly reduced line width (1.0 mT). The concentrations of paramagnetic species in the zeolite in each of these samples were estimated by comparison of the doubly integrated signals with a weak pitch standard. The maximum signal intensity (3 days reaction at 350 °C) corresponded to a concentration of $\sim 9 \times 10^{18}$ spins per gram, or 0.08 spins per unit cell of the zeolite.

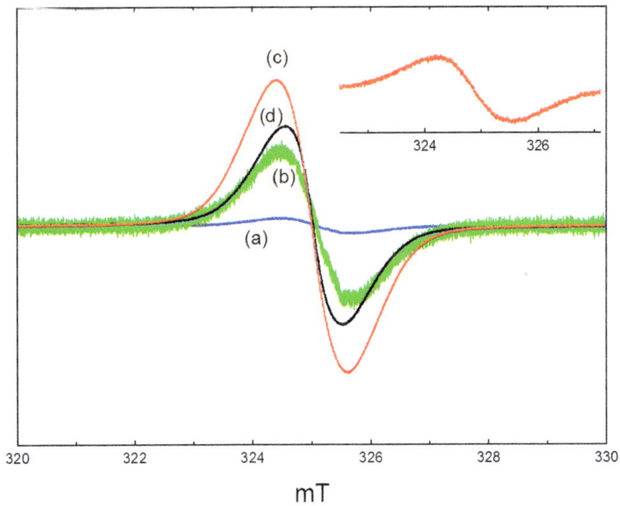

Fig. 5 EPR spectra of the ZSM-5 catalyst reacted at (a) 350 °C for 1 hour; (b) 10× expansion of (a); (c) 350 °C for 3 days; (d) 400 °C for 3 days. The inset shows the spectrum obtained by heating ZSM-5 in methanol vapour at 300 °C for 1 hour, then evacuating at 50 °C.

To explore the origins of these signals, further studies were carried out in which samples of zeolite catalyst were dehydrated in a high vacuum cell and then exposed to a static atmosphere of methanol vapour (~100 mbar) for 1 hour, sealed and transferred to the EPR spectrometer. Exposure to static methanol vapour at high temperatures gave similar signals to those seen from the used catalysts. For example, the inset in Fig. 5 shows the spectrum obtained from a zeolite exposed to methanol at 300 °C for 1 hour, then evacuated at 50 °C prior to measurement. This has the same line width as that seen from the catalysts used at 350 °C. This signal was not removed on subsequent evacuation up to 300 °C.

A notably different spectrum was obtained when the zeolite was exposed to methanol at 250 °C. No signals were detected when such a sample was evacuated at room temperature prior to measurement. Evacuation at 100 °C or 150 °C gave a signal showing a well resolved hyperfine structure, as illustrated in Fig. 6. At least 9 lines can be resolved in the room temperature spectrum, with a hyperfine coupling constant of 0.67 mT. When measured at 80 K, the signal is broadened and the apparent hyperfine splitting increases to 1.0 mT. The integrated intensity of the signal corresponded to $\sim 5 \times 10^{15}$ spins per gram of zeolite. Evacuation at higher temperatures removed this signal completely.

There are several previous reports of EPR signals being generated when methanol is reacted over zeolite catalysts. Kim *et al.* observed an 11 line signal with a splitting of 0.64 mT when methanol was reacted over HSAPO-34 and its aluminosilicate analogue HSSZ-13 at 350 °C for between 30 and 180 minutes on stream.[20] They attributed this signal to the hexamethylbenzene radical cation. The hyperfine splitting is closely similar to that seen for the 13-line spectrum of UV irradiated hexamethylbenzene in sulphuric acid.[21] A closely related spectrum was seen when hexamethylbenzene was adsorbed in H-mordenite.[20] At longer reaction times the hyperfine structure collapsed and a single symmetrical line was

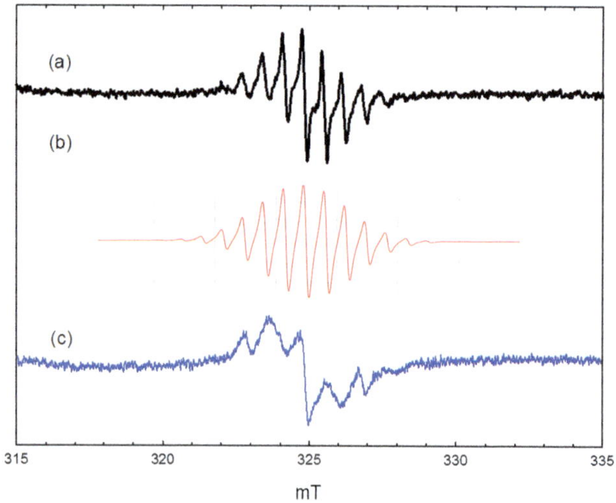

Fig. 6 EPR spectra recorded following exposure of ZSM-5 to methanol vapour at 250 °C for 30 min and evacuation at 150 °C for 30 min. (a) Measured at 25 °C; (b) simulation with parameters described in text; (c) measured at −193 °C.

observed. This was extremely narrow in the case of HSAPO-34 (~0.3 mT) but considerably broader in the case of HSSZ-13 (~1.3 mT). These authors also report that a single symmetric line was observed when methanol reacted over HZSM-5 at 350 °C for all times on stream, although no spectra were shown.

Spectra very similar to those shown in Fig. 5 were obtained by Madeira *et al.* from HZSM-5 catalysts used in the conversion of ethanol to hydrocarbons at 350 °C.[22] They report a single symmetric line with a line width of 1.0 to 1.4 mT, rising to a maximum intensity after 16 hours on stream corresponding to 3.5 micromol per gram, or ~1 radical for 50 unit cells of zeolite. Although no hyperfine splitting was resolved in the CW EPR signals, pulsed HYSCORE measurements showed the presence of several proton hyperfine couplings between 0.3 and 0.6 mT.

Spencer *et al.* report partially resolved signals from mordenite zeolite catalysts exposed to methanol for different times on stream at various reaction temperatures.[23] The signals were best resolved at shorter reaction times and the best match between observed and simulated EPR spectra was obtained for the 2,3,6,7-tetramethylnaphthalenium radical cation. At longer reaction times a single symmetrical line with a width of ~1.6 mT was observed.

Jang *et al.* describe EPR signals obtained when methanol is reacted over a phosphorus modified HZSM-5 zeolite.[24] At short reaction times (*e.g.* 5 minutes at 350 °C) they observe a partially resolved hyperfine splitting which they attribute to the 1,2,4,5-tetramethylbenzenium cation but, as in the earlier studies, the resolution of the signal was lost at longer reaction times.

It is clear from the above studies and from earlier work on aromatic molecules adsorbed in zeolites that aromatic radical cations are readily formed in acid zeolites.[25] The signals we observe from used catalysts (Fig. 5) can be readily assigned to such species, given the methylated aromatic molecules detected in the product stream (Fig. 3). In the absence of resolved hyperfine splitting, it is impossible to identify specific radicals. The absence of hyperfine splitting has been attributed by various authors to exchange interactions between radical cations. We note however that the spin concentrations measured indicate a low number of radicals per unit cell of the zeolite. The spin exchange narrowing observed suggests therefore that the radicals are not uniformly distributed throughout the zeolite pores but rather heterogeneously distributed in close proximity to each other. A further important observation is that the EPR signal from the used 400 °C catalyst sample has a significantly narrower line width than those reacted at 350 °C, even though the spin concentration is lower. This suggests a greater degree of condensation of aromatic rings in the radical cations formed at this temperature.

More surprising is the observation of well resolved proton hyperfine splitting in the signal shown in Fig. 6. The signal is identical to that reported by Kim *et al.* and assigned to the hexamethylbenzenium cation.[20] This radical, with 18 equivalent methyl protons, is expected to give a 19-line hyperfine pattern, but the intensities of the outermost lines are too weak to be detected. The intensity distribution of the inner lines matches well with that in the calculated simulation (Fig. 6(b)) using a single isotropic coupling constant of 0.67 mT (and $g = 2.003$). The observed spectrum cannot be fitted to simulations for benzenium cations containing fewer methyl groups, where two or more different proton coupling constants are expected from methyl and aromatic ring protons. Hexamethylbenzene is too large to escape from the pores of ZSM-5 and is not detected at all in the reaction products. This work shows, however, that traces of hexamethylbenzene must be formed within the zeolite at temperatures as low as 250 °C when HZSM-5 is heated in a static methanol atmosphere. The radical cation is only detected on subsequent outgassing, and it is not stable to heating above 150 °C in vacuum. We think therefore it is unlikely to play any role in the formation of the hydrocarbon pool; this is borne out also by its low concentration. The change in the appearance of the signal on cooling to 80 K (Fig. 6(c)) suggests a significant anisotropy in the proton hyperfine coupling, which is not normally seen for methyl groups in aromatic radical cations. The observed splitting of 1.0 mT would then be the parallel component, implying a perpendicular component of ~0.5 mT, not resolved in the spectrum. For comparison, the parallel and perpendicular methyl proton coupling constants in the toluene and xylene radical cations in frozen solution differ by only ~0.1 mT.[26] Further work is required to clarify the nature and structure of the radical in the zeolite environment.

3.4 Diffuse-reflectance infrared spectroscopy (DRIFTS)

Infrared spectra were recorded from the used catalyst samples by diffuse reflectance following temperature-programmed flushing in flowing helium up to the temperature at which they had been reacted. Spectra were recorded over the wavenumber range 4000 to 1000 cm^{-1}, but the region below 1700 cm^{-1} was found to be obscured by strong bands arising from the dispersive component occurring in the reflected beam from lower frequency absorption bands, due to zeolite framework stretching modes. The overtones of the framework bands at ~1900 and 2000 cm^{-1} were used to normalise band intensities. Fig. 7 shows spectra in the ν(OH) region from the fresh catalyst dehydrated at 350 °C and from the four used catalysts, in each case recorded at the temperature at which they had been reacted.

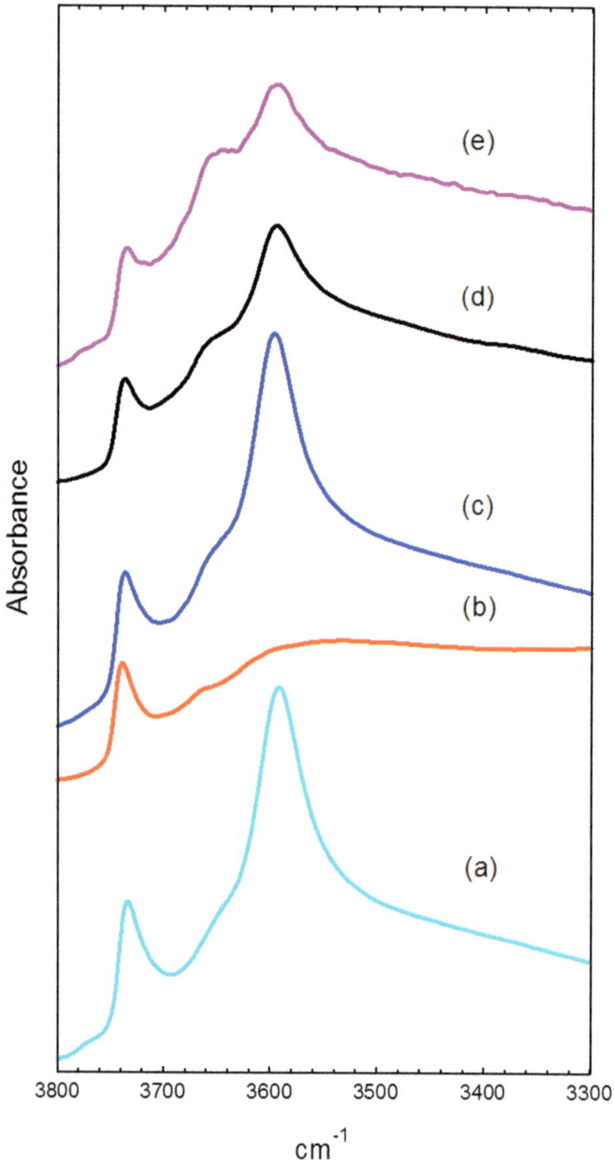

Fig. 7 Infrared spectra in the ν(O–H) region of (a) fresh catalyst; (b) catalyst after reaction at 200 °C for 1 h; (c) catalyst after reaction at 350 °C for 1 h; (d) catalyst after reaction at 350 °C for 3 days; (e) catalyst after reaction at 400 °C for 3 days. All spectra were recorded in flowing He at the temperature indicated.

The fresh catalyst shows three bands in this region: a strong band at 3595 cm^{-1} due to the Brønsted acid hydroxyl groups,[27] a weaker band at 3740 cm^{-1} due to silanol groups on the external surface of the zeolite[28] and a weak shoulder at ~3650 cm^{-1}, which has been previously assigned to AlOH groups associated with extra-framework aluminium.[29,30] The intensity of the 3595 cm^{-1} band and the relatively small contribution from the AlOH band confirms that the HZSM-5 catalyst contains a high concentration of Brønsted acid sites.

After reaction with methanol at 200 °C for 1 hour, the spectrum (measured at 200 °C) shows almost complete removal of the Brønsted OH band (Fig. 7(b)). At lower frequencies (not shown) this was replaced by an intense broad triplet of bands at ~3000 cm^{-1} (underlying the CH stretching bands discussed below), ~2300 cm^{-1}, and ~1700 cm^{-1}. This pattern is characteristic of strongly hydrogen bonded methanol and has been extensively discussed in the literature.[31] The so-called ABC triplet arises from resonant interactions between the ν(OH) mode of the zeolite and the overtone of the bending mode of the perturbed methanol hydroxyl.

The spectrum measured after reaction at 350 °C for 1 hour and flushing in helium at the same temperature gives a ν(OH) spectrum comparable to that of the fresh catalyst (Fig. 7(c)). After reaction at 350 °C for 3 days however, both the Brønsted OH band and the external SiOH band are reduced in intensity, and the AlOH band at 3650 cm^{-1} becomes more prominent (Fig. 7(d)). After 3 days of reaction at 400 °C the AlOH band grows further, and the Brønsted OH band is further reduced (Fig. 7(e)).

The loss of intensity in the Brønsted OH band at longer reaction times at 350 °C can have two causes. Firstly, methylation of the surface by methanol replaces OH groups with methoxy groups (see below).[32] Secondly, Brønsted OH groups can also be lost through hydrolysis by steam produced as a product under the reaction conditions[30,33] and the relative increase in intensity of the AlOH band at 3650 cm^{-1} indicates that some hydrolysis is also occurring here.

Fig. 8 shows the difference spectra in the CH stretching region of the used catalysts (after subtraction of the spectrum of the fresh catalyst). The catalyst reacted at 200 °C (Fig. 8(a)) shows an envelope of at least 4 bands at 3003, 2970, 2955, and 2848 cm^{-1} superimposed on the A hydroxyl component of the H-bonded methanol triplet. Three of these can be attributed to the CH stretching vibrations of hydrogen bonded methanol (the degeneracy of the asymmetric CH stretching mode is lifted in the adsorbed state). The low temperature adsorption of methanol in HZSM-5 was studied in detail by Mirth *et al.*[34] They attributed the 3 CH stretching bands observed at 2993, 2958 and 2856 cm^{-1} to hydrogen bonded protonated methanol, although as discussed more recently,[35] the methanol is unlikely to be protonated. Campbell *et al.* report CH frequencies of 3007, 2957 and 2853 cm^{-1} for hydrogen bonded methanol.[35] The remaining CH band in Fig. 8(a) at 2970 cm^{-1} is close to that assigned by several groups to the asymmetric stretch of surface methoxy groups.[29,32,33,35,36] The corresponding symmetric stretch at about 2868 cm^{-1} is hidden by the stronger 2848 cm^{-1} symmetric stretch of hydrogen bonded methanol. We conclude that the catalyst reacted at 200 °C for 1 hour contains both hydrogen bonded methanol and methoxy groups.

The infrared spectrum of the catalyst reacted at 350 °C for 1 hour contains no evidence of hydrogen bonded methanol. The Brønsted OH band is close to its original intensity, and the CH stretching region (Fig. 8(b)) shows, in addition to small contributions from methoxy groups at 2970 and 2870 cm^{-1}, a weak peak at 2920 cm^{-1}. The 2920 cm^{-1} band persists and becomes more intense in the catalyst reacted at 350 °C for 3 days (Fig. 8(c)). This spectrum also contains a stronger contribution from the methoxy groups (2970 and 2870 cm^{-1}) and an intense, broader band at 3120 cm^{-1}. Palumbo *et al.*, in their study of HZSM-5 heated in static methanol vapour at elevated temperatures, attribute a band at ~3100 cm^{-1} to aromatic CH stretching vibrations of adsorbed methylaromatic species.[37] They supported this assignment through the observation of bands in the C–C stretching

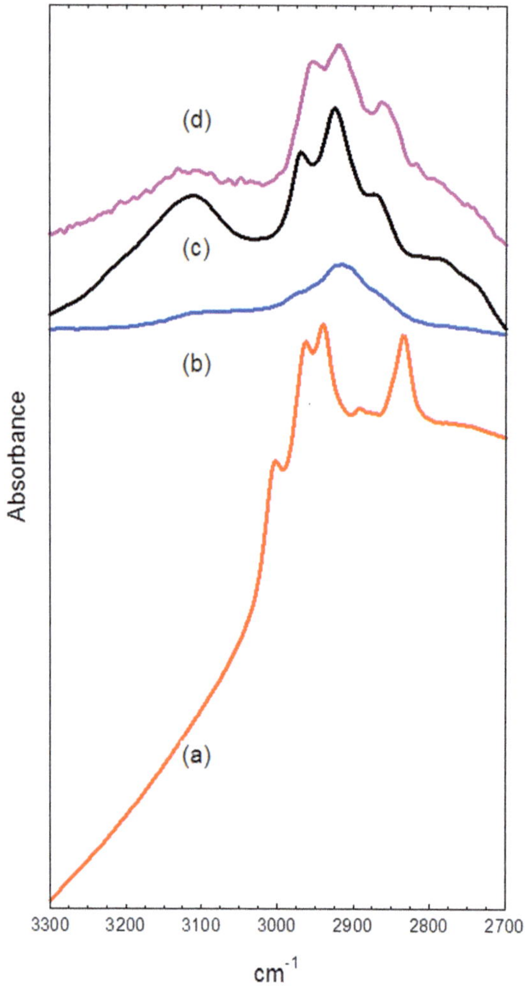

Fig. 8 Infrared difference spectra in the ν(C–H) region for the catalyst following reaction with methanol at (a) 200 °C for 1 h; (b) 350 °C for 1 h; (c) 350 °C for 3 days; (d) 400 °C for 3 days. Spectra measured in flowing He at the temperature indicated.

region (1500–1700 cm^{-1}), which are not accessible in the diffuse reflectance spectra measured in the present work. The dominant band at ~2920 cm^{-1} is assigned to methyl groups bonded to aromatic rings; for example durene (1,2,4,5-tetramethylbenzene) has a major CH stretching band at 2930 cm^{-1}.[38]

The spectrum in the CH stretching region of the catalyst reacted at 400 °C for three days (Fig. 8(d)) shows some interesting differences from that of the 350 °C catalyst. The aromatic CH stretching band at 3120 cm^{-1} is greatly reduced in intensity, and there are some shifts in the aliphatic CH stretching profile below 3000 cm^{-1}. In particular, the 2970 and 2870 cm^{-1} bands of surface methoxy groups are lost, the 2920 cm^{-1} remains, and a new band appears at 2860 cm^{-1}. Detailed discussion of these bands is not warranted, given the complexity of the system, but the reduction in the aromatic CH stretching band does imply a greater degree of methylation of the aromatic rings present in the catalyst at this temperature.

3.5 Inelastic neutron scattering (INS)

As mentioned in Section 2.3, spectra are presented using two direct geometry spectrometers: MERLIN emphasizes sensitivity and MAPS offers superior resolution.

3.5.1 Sample 1. Fig. 9 presents the INS spectra of the clean, activated ZSM-5. Fig. 9(a) shows the region of the vibrational spectrum corresponding to (O–H) and (C–H) stretching modes, with only a single peak observed at 3515 cm^{-1} that is assigned to internal hydroxyl groups. The external silanol group population detectable by IR spectroscopy (Fig. 7(a)) is not discernible in Fig. 9(a). Lower energy transitions are evident in Fig. 9(b). A broad peak in the MERLIN spectrum at 1140 cm^{-1} is assigned to an (O–H) bend; in the MAPS spectrum this is further resolved to peaks at 1179 and 1124 cm^{-1}. Weaker peaks are seen at 819 and 436 cm^{-1}, that are also thought to be hydroxyl deformations, possibly from a second type of hydroxyl group, as seen in the IR spectrum.

3.5.2 Sample 2. The INS spectrum for the 1 h 200 °C sample (sample 2) is shown in Fig. 10. Two intense peaks are seen in the higher energy scan, a ν(C–H)

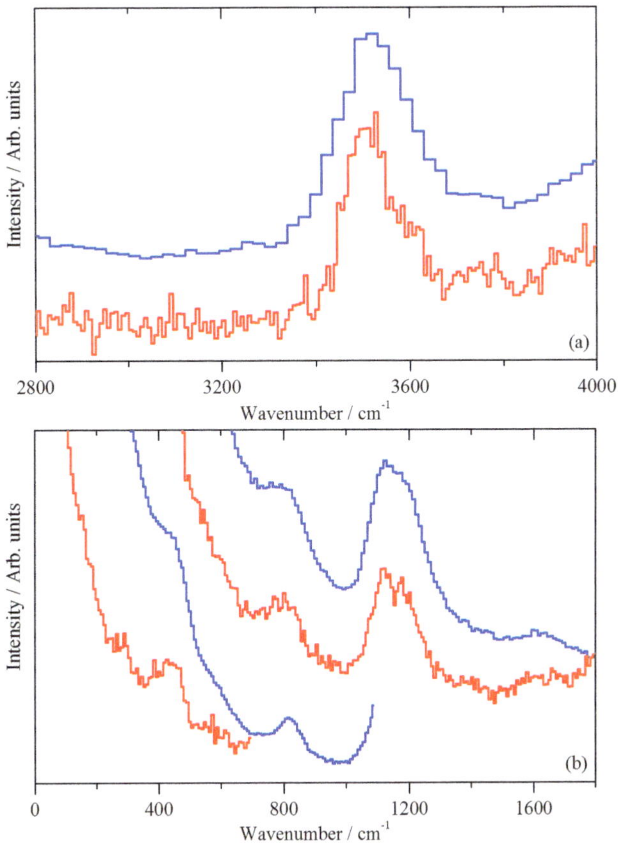

Fig. 9 INS spectra (MERLIN (blue) and MAPS (red) spectrometers) of clean, activated ZSM-5 (sample 1). (a) Spectral region 4000–2800 cm^{-1}, recorded with an incident energy of 5244 cm^{-1}; (b) spectral region 1800–0 cm^{-1}, recorded with incident energies of 2017 and 807 cm^{-1}.

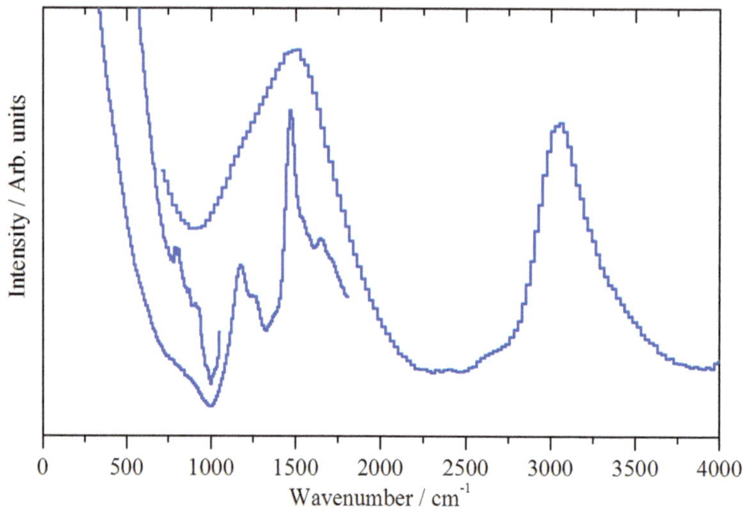

Fig. 10 INS spectra (MERLIN spectrometer, 4500–0 cm^{-1}) of the ZSM-5 catalyst after 1 h of reaction at 200 °C (sample 2). Spectra were recorded with incident energies of 5244, 2017 and 1210 cm^{-1}.

mode at 3065 cm^{-1} and a δ(C–H) mode at 1478 cm^{-1}. The infrared spectrum (Section 3.3) indicates that the sample comprises a mixture of methoxy and hydrogen-bonded methanol. The INS spectrum is consistent with this analysis. The zeolite O–H stretch at 3515 cm^{-1} has downshifted and comprises part of the long, broad tail above 3200 cm^{-1} that also includes the methanol O–H stretch. The methanol C–O–H bending mode contributes to the band at ~800 cm^{-1} that is also assigned to a framework hydroxyl deformation. The methoxy and methanol C–H stretch, bend, and rock modes are close in energy and are not resolvable on this instrument, thus both species contribute to the modes at 3065, 1470 and 1180 cm^{-1} modes.[12] The adsorbed methoxy species is formed by reaction with the Brønsted sites, eqn (1), with the water produced being swept out away from the reaction zone by the He carrier gas.

$$M\text{–}O\text{–}H + HOCH_3 \rightarrow M\text{–}O\text{–}CH_3 + H_2O. \quad (1)$$

The medium energy scan resolves a methyl rocking mode at 1172 cm^{-1} from the more intense δ(C–H) mode.[39] In fact, the methyl rock appears to be split into two components: a band head at 1172 cm^{-1} and a high energy shoulder at 1247 cm^{-1}. This splitting could indicate that either the local symmetry is highly distorted from C_{3v}, or there are two distinct methoxy populations present, perhaps originating from different hydroxyls. It is noted that Fig. 10 shows no evidence for a methyl torsion that is typically located at about 80 cm^{-1}.[39] However, the resolution of MERLIN is insufficient to resolve this from the elastic line with E_i = 1210 cm^{-1}, thus its absence in Fig. 10 does not negate the presence of methyl groups.

3.5.3 Sample 3. The INS spectrum of sample 3 (1 h, 350 °C) is shown in Fig. 11 that includes the presentation of a difference spectrum, a useful means of accounting for what hydrogenous modes have been retained by the catalyst on

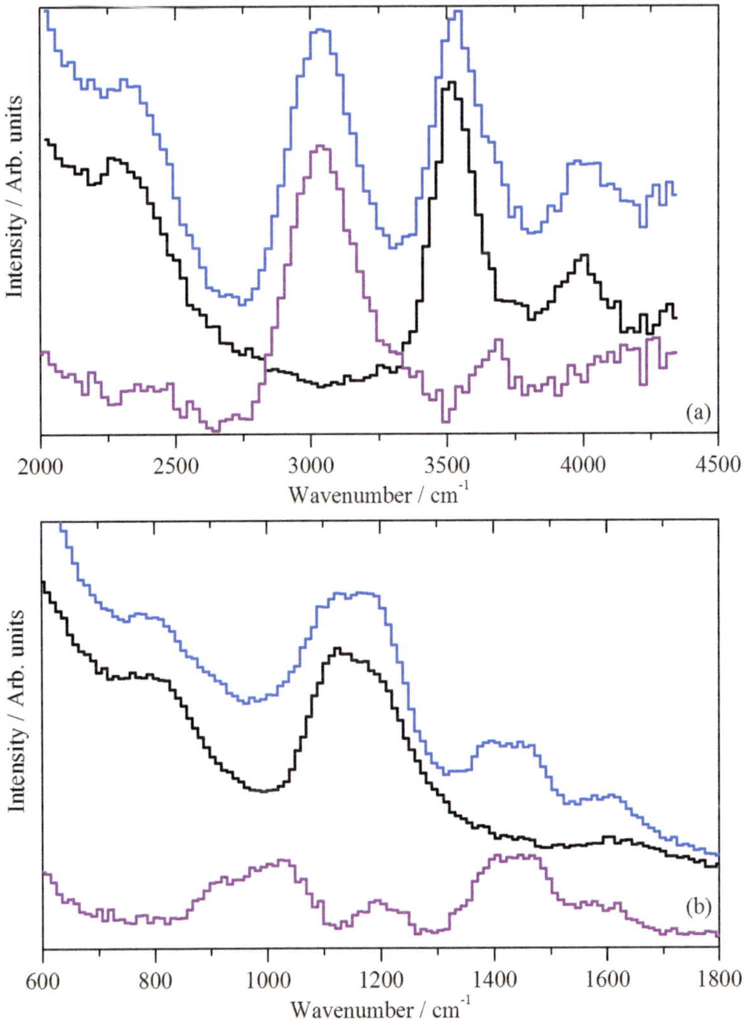

Fig. 11 INS spectra (MERLIN spectrometer) of the ZSM-5 catalyst after 1 h of reaction at 350 °C (sample 3): (a) spectral region 4500–2000 cm^{-1}, recorded with an incident energy of 5244 cm^{-1}; (b) spectral region 1800–600 cm^{-1}, recorded with an incident energy of 2017 cm^{-1}. In both (a) and (b) the blue plot represents the post-reaction spectrum, whereas the purple plot represents the difference spectrum [*i.e.* post-reaction spectrum (sample 3) – spectrum of the clean catalyst (sample 1, black)].

reaction. There is a small increase in a ν(O–H) mode at 3690 cm^{-1} and a most noticeable increase in a ν(C–H) mode centred at 3038 cm^{-1}. The intensity of this latter feature indicates a significant degree of hydrocarbon retention even for a short period of time-on-stream. The band head energy indicates this hydrocarbonaceous moiety to be olefinic/aromatic in nature, *i.e.* C(sp^2)H. Compared to the IR spectrum (Fig. 8(b)), the INS spectrum exhibits different intensities for ν(C–H). In both IR and INS spectroscopies, the intensity is directly proportional to the number of oscillators present, however, in the IR the intensity is weighted by the extinction coefficient, which is a molecular property, so it is different for aliphatic C–H, aromatic C–H and hydroxyl O–H vibrations. In contrast, for INS the weighting is by the amplitude of vibration, which in the harmonic approximation is about the same for both C–H and O–H vibrations, so that the intensities directly reflect the number of oscillators present. Further work is required to better understand the differences between the vibrational spectra of these materials obtained *via* optical and neutron scattering methods.

Fig. 11(b) presents the deformation region of the spectrum. A small increase in the δ(O–H) mode at 1150 cm^{-1} is seen, alongside broad bands centred at about 1450 and 980 cm^{-1}. The former is possibly a δ(C–H) mode associated with a population of aliphatic species that could be associated with the low energy tail of the intense feature seen in Fig. 11(a) at 3008 cm^{-1}. The 980 cm^{-1} feature is thought to represent an olefinic wag.[40]

3.5.4 Sample 4. The spectrum for sample 4 (3 days, 350 °C) is presented in Fig. 12. For the stretching region, Fig. 12(a), the ν(O–H) feature is observed at 3603 cm^{-1}. In the (C–H) stretching region, peaks at 2998 and 3090 cm^{-1} are seen, indicating a co-existence of sp^2 and sp^3 hybridised C–H bonds, which is consistent with the IR data (Fig. 8(c)). The INS relative intensities of the C(sp^3)–H and

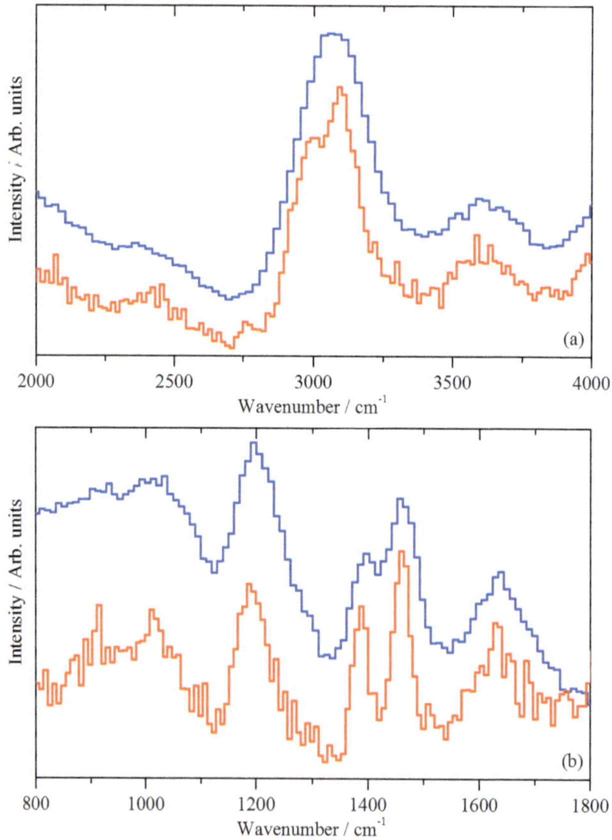

Fig. 12 INS spectra (MERLIN (blue) and MAPS (red) spectrometers) of the ZSM-5 catalyst after 3 days of reaction at 350 °C (sample 4): (a) spectral region 4000–2000 cm^{-1}, recorded with an incident energy of 5244 cm^{-1}; (b) spectral region 1800–800 cm^{-1}, recorded with an incident energy of 2017 cm^{-1}.

$C(sp^2)$–H modes are approximately equal, implying equal numbers of each. It is noted that this is inconsistent with the presence of only polymethylated aromatics, which should have a larger number of aliphatic C–H moieties.

In the deformation region, Fig. 12(b), there are peaks at 1639, 1460, 1390, and 1190, and a broad feature at 800–1100 cm^{-1} (with a maximum at 1000 cm^{-1}). Although the 1190 cm^{-1} band coincides with a hydroxyl bending mode as observed in Fig. 9(b), given the intensity of the ν(C–H) mode in Fig. 12(a), the 1190 cm^{-1} peak is assigned to a C–H mode. The obvious assignment is that the 1460, 1390, and 1190 cm^{-1} modes are the methyl asymmetric bend, symmetric bend, and rock, respectively. This assignment would require a strong methyl torsion mode below 250 cm^{-1}, but further spectral acquisition is required to discern the presence or absence of this diagnostically significant mode. It is also unusual to see the asymmetric and symmetric bending modes resolved (1460 and 1390 cm^{-1}); for chemisorbed methoxy, only a single broad peak is observed.[39] The simplest (but certainly not totally satisfactory) assignment is that all the peaks are due to coupled aromatic C–C stretch and C–H bend modes. This possibility requires further investigation.

3.5.5 **Sample 5.** The spectrum for the 3 days at 400 °C sample is presented in Fig. 13. The ν(O–H) mode is seen at 3665 cm^{-1}; this represents a shift to a higher wavenumber upon extent of reaction, possibly indicating a reduced degree of hydrogen bonding. A strong ν(C–H) feature is present at 3094 cm^{-1}, with the peak maximum implying a predominance of $C(sp^2)$H species; the only indication for the presence of $C(sp^3)$H moieties is that the peak is asymmetric to the low energy side.

In the deformation region, Fig. 13 shows some commonality with the 3 days at 350 °C spectrum (Fig. 12(b), (C–H) deformation modes at ~1456 cm^{-1} and methyl rock at ~1195 cm^{-1}), except for the presence of three new features at 1090, 939 and 782 cm^{-1}. These are thought to arise from aryl C–H wag vibrations. This could include a contribution from naphthalenic species.[40] As discussed with sample 4, the splitting of the δ(C–H) mode is thought to indicate the presence of at least two species.

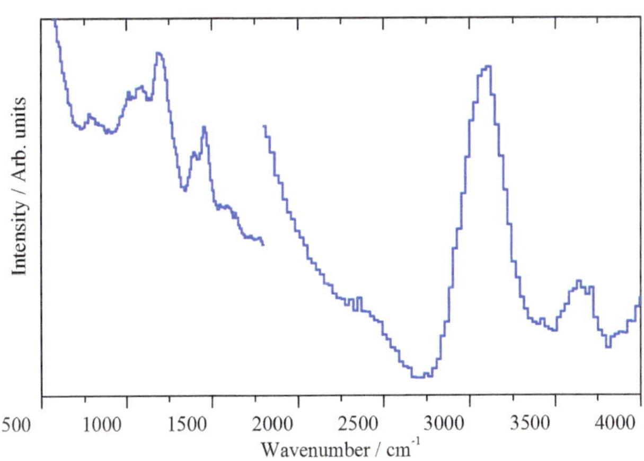

Fig. 13 INS spectra (MERLIN spectrometer) of the ZSM-5 catalyst after 3 days of reaction at 400 °C (sample 5) for the spectral region 4000–500 cm^{-1}, recorded with incident energies of 5244 and 2017 cm^{-1}.

The original intention of examining a sample after 3 days at 400 °C was to induce coking reactions. However, despite exhibiting a reasonable TPO profile albeit with a lower coke content than at 350 °C (Fig. 4, Table 2), visual inspection of the sample post-reaction indicates only mild coking at most (Section 3.2). Indeed, as reported in Section 3.1, no deactivation was observed *via* in-line mass spectrometry. Instead, the IR (Fig. 8(d)) and INS (Fig. 13) data suggest that the progression from 350 to 400 °C has caused some modification to the form of the hydrocarbonaceous material retained by the catalyst. It is possible that this modification is connected to the different emphasis on reaction pathways as a function of temperature, *i.e.* favouring the olefinic pathway over the aromatic one, as considered by Sun and co-workers.[4] Certainly, in order to access a substantial coking component, it will be necessary to examine more forcing conditions in future experiments.

3.5.6 Retained hydrocarbonaceous moieties. Considering the INS spectra together, a low reaction temperature (200 °C) has led to the formation of hydrogen bonded methanol and methoxy species, reasonable intermediates for MTH chemistry.[5] The spectra for the higher temperature runs are not so readily assignable and appear to be composed of a variety of hydrocarbonaceous species, possibly over a range of sites. These spectra are the vibrational fingerprint of the hydrocarbon pool and INS is able to establish some modification of these hydrocarbonaceous species as a function of reaction temperature and time-on-stream.

Although INS provides a welcome new handle on the retained hydrogenous population that are engaged in MTH chemistry over ZSM-5, thus far, there is insufficient data to define a precise role for how a hydrocarbon pool simultaneously facilitates the formation of olefin and aromatic products. Nonetheless, there are clearly opportunities for using INS to characterise the nature and

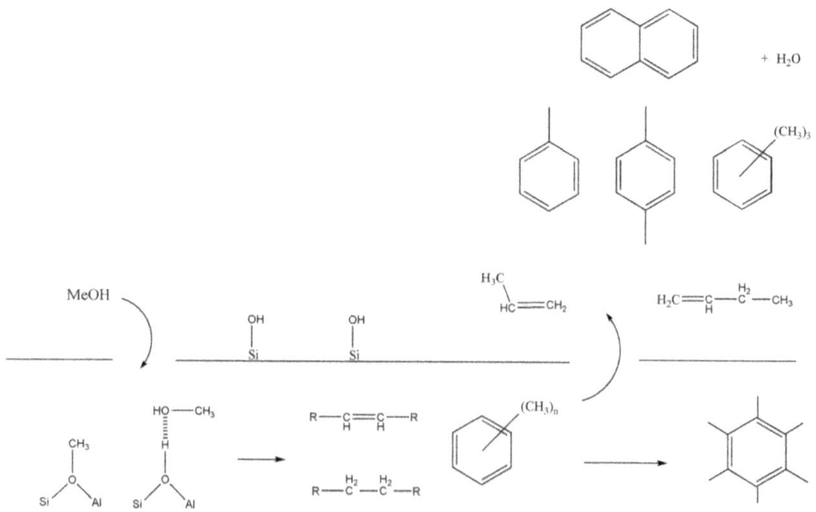

Fig. 14 A schematic representation of the principal chemical species identified as part of this multi-technique investigation of the reaction of methanol over ZSM-5 at elevated temperatures. The 'tube' is intended to indicate the internal surface of the zeolite. The aliphatic, olefinic and aromatic species in the middle of the tube are representative examples of the retained hydrocarbonaceous material revealed in the INS spectra.

quantity of the retained hydrocarbonaceous material with respect to defined process operations. For example, previously, the role of "coking" was considered a key process but Fig. 10–13 provide clear evidence that the 'coke' has a substantial hydrogen component, indicating that future studies should more correctly consider a role for hydrocarbonaceous entities. Olsbye *et al.*'s highlighting of the importance of "(hydro-)carbon residues" similarly endorses this awareness.[5] Fig. 14 attempts to illustrate the interconnectivity between the species observed in this study.

With reference to Fig. 14, it is proposed that INS is a most appropriate probe to be added to the armoury of techniques used to explore the intricacies of MTH chemistry.

4. Concluding remarks

A multi-technique approach using a combination of reaction testing (mass spectrometry and GC-MS), TPO, EPR, IR and INS has been used to examine the methanol-to-hydrocarbon reaction using a commercial grade ZSM-5 catalyst over a range of conditions, varying from constrained reaction at 200 °C, steady-state operation at 350 °C, and an attempt at accelerated ageing *via* reaction at 400 °C. The following conclusions can be drawn.

• The reaction profile for the reaction at 350 °C indicates that methanol dosing leads initially to the formation of dimethyl ether. It is assumed that this forms from the dimerization of chemisorbed methoxy species. Dimethyl ether appears to be a precursor to the simultaneous production of olefinic and aromatic products. The aromatics are typified by methylated benzenes and naphthalenes.

• TPO measurements indicate that an extended time-on-stream at 350 °C leads to a significant population of retained carbonaceous material that is strongly bound to the catalyst.

• EPR measurements indicate that continued reaction leads to an accumulation of undefined aromatic radical cations of low spin density.

• Infrared measurements have provided information on catalyst hydroxyl groups and the retention of hydrocarbonaceous species as the reaction proceeds, confirming the co-existence of sp^2 and sp^3 hybridised C–H entities for reaction at elevated temperatures.

• INS measurements provide vital information on how hydrogen partitions within the catalyst matrix as the reaction proceeds. For reaction at 200 °C, the co-existence of hydrogen bonded methanol and methoxy species, as evidenced by IR, is confirmed. The INS spectrum of the catalyst after 3 days of operation at 350 °C indicates the presence of sp^2 and sp^3 hybridised C–H species. However, no discrete molecular assignment for the retained hydrocarbonaceous species is possible as yet, indicating the complexity of the composition of the hydrocarbon pool. Increasing the reaction temperature from 350 to 400 °C modifies the form of the retained hydrocarbon to possibly include napthalenic entities.

Acknowledgements

The UK Catalysis Hub is kindly thanked for resources and support provided *via* our membership of the UK Catalysis Hub Consortium and funded by EPSRC (grants EP/I038748/1, EP/I019693/1, EP/K014706/1, EP/K014668/1, EP/K014854/1,

EP/K014714/1 and EP/M013219/1). Project support from the EPSRC *via* grant EP/K503903/1 (University of Glasgow) is also acknowledged. The STFC Rutherford Appleton Laboratory is thanked for access to neutron beam facilities. Johnson Matthey plc is thanked for the provision of the ZSM-5 catalyst and its characterisation. A. Hameed acknowledges the support of the Iraqi Ministry of Higher Education and Scientific Research and the Chemical Engineering Department, University of Technology, Baghdad, Iraq.

References

1 K. Hemelsoet, J. Mynsbrugge, K. De Wispelaere, M. Waroquier and V. Speybroeck, *ChemPhysChem*, 2013, **14**, 1526.

2 C. Wang, Y. Chu, A. Zheng, J. Xu, Q. Wang, P. Gao, G. Qi, Y. Gong and F. Deng, *Chem. – Eur. J.*, 2014, **20**, 12432.

3 S. Wang, Y. Chen, Z. Wei, Z. Qin, J. Chen, H. Ma, M. Dong, J. Li, W. Fan and J. Wang, *J. Phys. Chem. A*, 2014, **118**, 8901.

4 X. Sun, S. Mueller, Y. Liu, H. Shi, G. L. Haller, M. Sanchez-Sanchez, A. C. van Veen and J. A. Lercher, *J. Catal.*, 2014, **317**, 185.

5 U. Olsbye, S. Svelle, K. P. Lillerud, Z. H. Wei, Y. Y. Chen, J. F. Li, J. G. Wang and W. B. Fan, *Chem. Soc. Rev.*, 2015, **44**, 7155.

6 S. F. Parker and D. Lennon, *Acc. Chem. Res.*, 2014, **47**, 1220.

7 T. Kandemir, M. Friedrich, S. F. Parker, F. Studt, D. Lennon, R. Schlögl and M. Behrens, *Phys. Chem. Chem. Phys.*, 2016, **18**, 17253.

8 A. R. McFarlane, I. P. Silverwood, R. Warringham, E. L. Norris, R. M. Ormerod, C. D. Frost, S. F. Parker and D. Lennon, *RSC Adv.*, 2013, **3**, 16577.

9 N. G. Hamilton, I. P. Silverwood, R. Warringham, J. Kapitán, L. Hecht, P. B. Webb, R. P. Tooze, S. F. Parker and D. Lennon, *Angew. Chem., Int. Ed.*, 2013, **52**, 5608.

10 R. Warringham, A. R. McFarlane, D. A. MacLaren, P. B. Webb, R. P. Tooze, J. Taylor, R. A. Ewings, S. F. Parker and D. Lennon, *J. Chem. Phys.*, 2015, **143**, 174703.

11 S. F. Parker, D. Lennon and P. W. Albers, *Appl. Spectrosc.*, 2011, **65**, 1325.

12 A. J. O'Malley, S. F. Parker, A. Chutia, M. R. Farrow, I. P. Silverwood, V. Garcia-Sakai and C. R. A. Catlow, *Chem. Commun.*, 2016, **52**, 2897.

13 R. F. Howe, J. McGregor, S. F. Parker, P. Collier and D. Lennon, *Catal. Lett.*, 2016, **146**, 1242.

14 R. Warringham, D. Bellaire, S. F. Parker, J. Taylor, R. Ewings, C. M. Goodway, M. Kibble, S. R. Wakefield, M. Jura, M. P. Dudman, R. P. Tooze, P. B. Webb and D. Lennon, *J. Phys.: Conf. Ser.*, 2014, **554**, 012005.

15 I. P. Silverwood, N. G. Hamilton, A. McFarlane, R. M. Ormerod, T. Guidi, J. Bones, M. P. Dudman, C. M. Goodway, M. Kibble, S. F. Parker and D. Lennon, *Rev. Sci. Instrum.*, 2011, **82**, 034101.

16 D. Lennon, R. Warringham, T. Guidi and S. F. Parker, *Chem. Phys.*, 2013, **427**, 49.

17 H. Schulz, *Catal. Today*, 2010, **154**, 183.

18 A. R. McInroy, D. T. Lundie, J. M. Winfield, C. Dudman, P. Jones and D. Lennon, *Langmuir*, 2005, **21**, 11092.

19 D. M. Bibby, N. B. Milestone, J. E. Patterson and L. P. Aldridge, *J. Catal.*, 1986, **97**, 493.

20 S. J. Kim, H. G. Jang, J. K. Lee, H. K. Min, S. B. Hong and G. Seo, *Chem. Commun.*, 2011, **47**, 9498.
21 M. K. Carter and G. Vincow, *J. Chem. Phys.*, 1967, **47**, 302.
22 F. F. Madeira, H. Vezin, N. S. Gnep, P. Magnoux, S. Maury and N. Cadran, *ACS Catal.*, 2011, **1**, 417.
23 N. A. Spencer, E. J. Ditzel, J. S. J. Hargreaves and S. Sproules, *J. Mater. Chem. A*, 2016, **4**, 7036.
24 H. Jang, H. Min, S. Hong and G. Seo, *J. Catal.*, 2013, **299**, 240.
25 X. Liu, K. Iu, J. K. Thomas, H. He and J. Klinowski, *J. Am. Chem. Soc.*, 1994, **116**, 11811.
26 R. M. Kadam, Y. Itagaki, R. Erickson and A. Lund, *J. Phys. Chem. A*, 1999, **103**, 1480.
27 P. A. Jacobs and R. von Ballmoos, *J. Phys. Chem.*, 1982, **86**, 3050.
28 J. C. Vedrine, A. Auroux and G. Coudurier, *ACS Symp. Ser.*, 1984, **248**, 253.
29 L. Kubelkova, J. Novakova and K. Nedomova, *J. Catal.*, 1990, **124**, 441.
30 S. M. Campbell, D. M. Bibby, J. M. Coddington, R. F. Howe and R. H. Meinhold, *J. Catal.*, 1996, **161**, 338.
31 E. L. Meier, R. A. van Santen and A. P. J. Jansen, *J. Phys. Chem.*, 1996, **100**, 9282.
32 T. R. Forester and R. F. Howe, *J. Am. Chem. Soc.*, 1987, **109**, 3076.
33 S. M. Campbell, D. M. Bibby, J. M. Coddington and R. F. Howe, *J. Catal.*, 1996, **161**, 350.
34 G. Mirth, J. A. Lercher, M. W. Anderson and J. Klinowski, *J. Chem. Soc., Faraday Trans.*, 1990, **86**, 3039.
35 S. M. Campbell, X. Z. Jiang and R. F. Howe, *Microporous Mesoporous Mater.*, 1999, **29**, 91.
36 H. Yamazaki, H. Shima, H. Imai, T. Yokoi, T. Tatsumi and J. Kondo, *Angew. Chem., Int. Ed.*, 2011, **50**, 1853.
37 L. Palumbo, F. Bonino, P. Beato, M. Bjorgen, A. Zecchina and S. Bordiga, *J. Phys. Chem. C*, 2008, **112**, 9710.
38 http://webbook.nist.gov/chemistry, accessed 14th September 2016.
39 A. R. McInroy, D. T. Lundie, J. M. Winfield, C. Dudman, P. Jones, S. F. Parker, J. W. Taylor and D. Lennon, *Phys. Chem. Chem. Phys.*, 2005, **7**, 3093.
40 D. Lin-Vien, N. B. Colthup, W. G. Fateley and J. G. Grasselli, in *The Handbook of Infrared and Raman Characteristic Frequencies of Organic Molecules*, Academic Press, San Diego, 1991, p. 286.

Faraday Discussions

DISCUSSIONS

Hydrocarbon conversion in the production of synthetic fuels: general discussion

Linda Jewell, Graham Hutchings, Michael Bowker, Emiel Hensen, Michael Claeys, James Hayward, Nico Fischer, Stephen McCord, Eric van Steen, Chelsea Tucker, Matthew Neurock, Hans Schulz, Unni Olsbye, Ding Ma, Paul Collier, Ying Zheng, Avelino Corma, Freek Kapteijn, Irina Yarulina, Siphamandla Hadebe, Enrique Iglesia, David Lennon and Burtron Davis

DOI: 10.1039/c7fd90012k

Stephen McCord opened a general discussion of the paper by Avelino Corma: A key phase of your process is the pyrolysis of biomass. Did you investigate the impact of varying the heating rate applied to the biomass during pyrolysis? It has been previously reported that high heating rates have resulted in the mass fraction of gaseous products being increased. This increased volatility could result in changes to the C : H ratios in the remaining biomass crude, was this something you encountered?

If you did investigate the heating rate, did this have any notable effects on the biomass crude composition, or on the catalyst lifespan? If you did not investigate this, would you predict this to have any significant effect on performance?

Avelino Corma answered: In our case it was a hydrothermal treatment process. Nevertheless, in the case of biomass pyrolysis, heating rate, contact time and the presence of a catalyst are key variables to control the total amount of gases, hydrocarbons and the C : H : O in the biomass crude.

Ying Zheng asked: Normally looking at cracking hydrocarbons rather than aromatics, how would you foresee the catalytic cracking of the aromatics-dominated biofuels? Is catalytic cracking an effective approach to upgrade the biofuels?

Avelino Corma answered: For me, hydrotreating will give a higher yield of valuable liquid fuel, as the biocrude is rather hydrogen deficient compared to the gas oil. Fluid catalytic cracking (FCC) is an easier and less expensive process but, if available, hydrotreating would be my preferred option for the hydrothermal bio oils, as happens with pyrolysis oils. Catalytic cracking may, however, be considered for the production of aromatics for chemicals.

Hans Schulz remarked: Referring to the biocrude hydrotreating step, my question concerns catalyst consumption and used catalyst disposal in relation to common petroleum refining operation. Is it more or less difficult? And are there any specific aspects? Eventual higher content of unsaturated compounds (olefins?) in the feed – is this being obtained in hydrothermal liquefaction?

Avelino Corma answered: We performed hydrotreating for 50 h, and it is not a sufficient amount of time to properly check catalyst deactivation or contaminant build up on the catalyst, such as metals (*eg.* Na, Fe), from ppm amounts in the feed. We did note an increase of coke on the catalyst recovered after the run, compared to a similar operation run with straight run gas oil at 70 bar, a very common desulphurization operation in refineries. Thus, we may expect a higher replacement rate with this type of NiMo catalyst for biocrude hydroprocessing at low pressure. However, we do not expect to run biocrude hydrotreating commercially with a standard NiMo catalyst, but with an adapted catalyst for oxygenated feeds.

Emiel Hensen continued: The very interesting approach you described converts biomass to liquid transportation fuels using a combination of novel technology (hydrothermal treatment) and downstream processing in a refinery setting. About 17–20 wt% of the biomass ends up in fuels. To me, these numbers seem very realistic; it however implies that the impact of biomass for fuel production will be limited in the future. How much would be the GHG-savings for such a case? How realistic is it to use the biocrude as a co-feedstock in a refinery setting?

Because of the concern about the biomass to fuels route, biomass to chemicals is attracting more attention from the academic community, even though the challenges are likely to be even greater than obtaining fuels from biomass. My prediction however is that in a CO_2-constrained world with cheap electricity and cheap oil available (due to the Paris agreement), we would use electricity for transport, some liquid fuels for heavy duty transport, and use crude oil predominantly to produce chemicals. So the role of biomass will be small. Perhaps there is an opportunity for catalysis to develop efficient routes from oil to chemicals. Could you comment on this view?

Avelino Corma answered: This technology works with lignocellulose residues and uses the energy generated by the gaseous products and coke generated.

Co-processing 3 wt% in the fluid catalytic cracking process has been seen to be applicable. For a medium sized, 30 000 barrels per day fluid catalytic cracking unit (FCCU), this already represents nearly 1000 barrels per day of biocrude, that is in the range of a world-class biorefinery. In the case of hydrotreating, the amount of oxygen of the biocrude generated is almost the same as for the vegetable oils processed commercially today. I agree with your view on electricity for transport, together with diesel for heavy duty. I would add the needs for jet fuel, which is unlikely to be displaced, but in the long term maybe by hydrogen. There is also a significant societal pressure to shift towards biofuels in this area. In the case presented, the fuel introduced is in the diesel range, but a large kerosene fraction that we estimate in the 40–50 wt% range may also be distilled from the hydroprocessed fuel; the remaining being directed to gasoline and diesel pools.

Nevertheless, I also agree that a longer term oil should be used for chemicals, as well as biomass with already specific building block molecules.

Nico Fischer queried: Although not at the core of the current study, do you believe that there is potential for catalyst development for the conversion step from biomass to biocrude, currently only yielding 30–35% by mass relative to dry biomass?

Avelino Corma answered: With direct transformation of lignocellulosic feedstock, it will be difficult to get more than 30–35 wt% of a largely deoxygenated biocrude (say 10–15 wt% oxygen, compared to 50 wt% in the dry biomass). This is because of the C : H : O balance in the biomass and also the need to expel the oxygen within the process in the form of water or, preferentially, CO_2, in order to keep hydrogen in the biocrude product, unless introducing a large amount of H_2 and minimizing CO_2 production. Processes such as hydropyrolysis remain in the development stage.

Ding Ma asked: For pyrolysis, during hydrocracking you need lots of hydrogen to remove oxygen. Is it possible to use H from CH_4, *i.e.* co-refining of biomass and methane, is this possible?

Avelino Corma replied: I guess that in this case you would have to use the hydrogen formed in methane reforming for hydrotreating. Then, I find difficult to conciliate the two processes in a single step as they yield best results at very different temperature and pressure windows.

Matthew Neurock enquired: This is very nice work. Can you please elaborate more on the catalyst that was used to carry out the catalytic cracking of the resulting biocrude mixtures and what properties of the catalysts are important?

Avelino Corma replied: The catalyst is not different from a conventional FCC catalyst. Notice that in the referred case it is considered that up to a maximum of 5 wt% of oil is combined with a conventional FCC feed to be treated in the FCC unit, thus the catalyst will be tailored focusing on proper conversion of the gasoil fraction. If some adjustment would be possible, a catalyst with low hydrogen transfer would probably benefit the conversion of the biocrude, limiting the formation of aromatics.

Hans Schulz said: In general, I think the proposed process is excellently and realistically designed for the problem of oil recovery from biomass. My question concerns the right (minimum) size of the plant.

I would imagine that the size should be sufficiently large for practically/economically processing the individual flows of feed and intermediates, and for these to be available for continuous operation all year round.

Avelino Corma responded: The hydrothermal plant can be designed as large as desired owing to its modular structure. The larger it is, the cheaper the processing will be. The key logistical problem in biomass processing is getting enough of it from a small distance away (a few dozen miles). A very large scale operation may be able to gather one million tons of dry biomass per year. With a process efficiency of 21 vol% (19 wt%), this gives an output of approximately 4,000 barrels per day of hydroprocessed oil (60 million gallons per year).

Siphamandla Hadebe remarked: 1) Did you notice catalyst deactivation when you introduced the biomass? It has been shown in the literature that hydrotreating of biomass esters may follow the decarboxylation pathway and yield CO/CO_2 gases. A decrease in catalyst activity may be noted, which is associated with competitive adsorption of CO or CO_2 instead of hydrotreating the sulphur or nitrogen containing compounds contained in the feed being processed. I would like to find out the type of catalysts that were utilized and would also like to know if catalyst deactivation was noted?

2) How much maximum biomass can you have until you get deactivation? Is there a certain threshold amount that can be co-processed without compromising catalyst activity?

Avelino Corma answered: We obtained fairly stable yields during our 50 h of testing after initial stabilization. This is, however, a short time to check for long term deactivation. We did note an increase of the coke level on the catalyst recovered after the run with biocrude, compared to a similar operation with straight run gas oil (SRGO) run at 70 bar, as well as some sulfur loss from the catalyst. Some CO_2 could be detected in the produced gas, but no CO. We thus expect the standard NiMo catalyst we used to deactivate faster when processing biocrude at lower pressures than in traditional desulpurization operation. Note that specialized catalysts for dealing with oxygenated feeds exist on the market, and are expected to have a much improved deactivation pattern for these feeds compared to the standard NiMo catalyst.

Linda Jewell opened the discussion of the paper by Burtron Davis: What is the catalyst particle size that you used and how does the catalyst particle size compare to the dimensions of the channels of the reactor? How did you load the catalyst into the reactor?

Burtron Davis answered: We have operated with a catalyst particle size of about 100 microns. Since the small channels are in the order of 2 to 4 mm, the catalyst particle is only 0.05 to 0.025 times the size of the channel. With the laboratory scale reactor, we pour the catalyst powder slowly, while the reactor is subjected to vibration. A proprietary loading device is available from the manufacturer of the reactor.

Michael Claeys asked: How long was your reactor and what were the pressure drops that you experienced with the 100 micron particles that you used? Did the pressure drop increase with the formation of liquid products during the FT synthesis?

You mentioned the advantages and disadvantages of your reactor system, which uses aluminium as the main material. What are the maximum operating conditions (temperature and pressure) that could be applied with the set-up?

Burtron Davis responded: Only on one occasion did we experience a large pressure drop across the reactor bed and we have not been able to provide an answer for this. Normally when operating at about 15 atm we experience a pressure drop of 0.3 atm or less. At the gas flow rate we normally encounter with our catalysts we do not experience a significant pressure increase when operating with normal FTS conditions. The temperature-pressure limitations allow one to operate at the low FTS conditions for Co and Fe catalysts since it is the pressure difference across the Al reactor wall that is important; thus, if the pressure difference between the cooling and the reactor channels is limited to 3–5 atm, all low FTS conditions can be used.

Chelsea Tucker asked: Can small channel reactors be implemented in a modular arrangement? This would mean capacity could be increased over time to meet increasing demands, and would decrease the initial capital investment required for large-scale plants.

Burtron Davis replied: It would appear that a major advantage of these reactors is the ability to manufacture off-site and easily ship over normal highways using standard transportation means. As indicated in the oral responses, this is the policy the Germans used in the period between 1930–1950 when they manufactured reactors with a capacity of about 15 bbl/day and then installed a number of reactors of this size so they could obtain the production desired for a plant.

Nico Fischer remarked: You discuss the formation of an exotherm in your small channel reactor. As a result, do you see catalyst deactivation at this point and does the exotherm move with TOS?

Burtron Davis responded: We have encountered both short and more severe exotherms. Usually a small, rapid temperature excursion does not cause a measurable modification of the catalytic activity. If the exotherm is for a few minutes then we notice a deactivation of the catalytic activity.

Hans Schulz remarked: The provided evidence about decreasing conversion (activity) in the FT-synthesis on cobalt because of large hydrocarbon molecules accumulating in the catalyst particle pores appears as of significant theoretical and technical relevance, not commonly acknowledged. It appears to explain our common opposite observation of initially increasing activity in FT-synthesis on cobalt when applying powdered catalyst particles ($\gg$100 nm), these adhering on the surface of coarse (0.2–0.3 mm) fused silica spheres in a fixed bed of catalyst.[1] Could you agree with this view?

1 H. Schulz, *Catal. Today*, 2014, **228**, 113–122.

Burtron Davis answered: The early conversion history can be impacted by the heavy hydrocarbons and these effects should be most notable during the early period of the reaction as you indicate.

Michael Bowker queried: 1) Aluminium is a soft metal, do you anticipate any problems with the corrosion of the reactor, especially if any acids are made?

2) Al isn't easy to work with as a material due to its high friction coefficient – are all fittings/connections with steel?

Burtron Davis replied: 1) No corrosion problems have been noted to date with up to three months of continuous operation with a cobalt catalyst. The amount of acids produced with the cobalt catalyst is much less than with an iron catalyst.

2) With the reactor structure of an aluminum reactor core in a stainless steel shell, all external connections are with stainless steel. Inside the shell the aluminum and steel portions are connected using explosion bonded transition joints.

Nico Fischer remarked: Under reaction conditions you report a maximum operating temperature of 270–280 °C. What is the maximum temperature at atmospheric pressure and how do you activate your catalyst?

Burtron Davis responded: Normally the FT catalyst is activated at 400 °C or lower; thus, at atmospheric pressure the temperature limitations due to using an aluminum reactor should not be a problem.

Linda Jewell asked: You mentioned that it is much quicker to make the aluminium reactor than a stainless steel reactor. Besides the time taken to make the reactor, is there any reason why you have chosen to work with an aluminium reactor rather than a stainless steel reactor?

Burtron Davis replied: The higher thermal conductivity of aluminum allows for better heat transfer, which allows a better approach to isothermal operations. Using the current technology, the size of the plate fin reactor structures that can be brazed in a single block is up to 30 times larger for aluminum than for steel.

Freek Kapteijn opened a general discussion of the paper by Unni Olsbye: In your presentation you mention making a distinction between the 'hydrocarbon pool' and 'coke'. Can you make that distinction? Is it possible to determine the hydrocarbon pool species and coke separately?

Unni Olsbye responded: The main distinction between hydrocarbon pool species and coke is that the hydrocarbon pool species are active intermediates in product-forming cycles, while coke is not. As such, transient $^{12}C/^{13}C$ feed experiments may be used to distinguish the two types of species, since incorporation of the labelled species will mainly proceed in the pool species.

Irina Yarulina asked: Have you ever analyzed different fractions of the catalytic bed using SS NMR for spent catalysts and/or by dissolution extraction experiments. What would be the nature of the retained species in different parts of the catalytic bed? As a comment, we observe the same axial deactivation of the catalytic bed for ZSM-5 with a high amount of acid sites. However, for zeolites with high Si/Al ratio (>1000) and also for zeolites modified with the alkaline-earth metals, a different picture is observed. These catalysts seem to be deactivated from bottom to top rather from top to bottom. (Deactivation profile – White-grey-dark, then white) This might imply that deactivation for zeolites with a low number of acid sites is due to the interconversion of the products occurring at the bottom of the catalytic bed.

Unni Olsbye responded: We have not used SS NMR for fractional studies of the fixed bed flow reactor, but have indeed used dissolution extraction experiments. Such data are *e.g.* given in the SI of this paper for a H-ZSM-5 catalyst and a H-Beta catalyst. The nature of the species observed in the top, middle and lower fractions of the bed depends on the catalyst as well as the reaction conditions. In order to elucidate whether the deactivation progresses from top to bottom or starts mainly in the middle or the bottom of the bed, it is useful to perform several experiments of the same catalyst under the same conditions, but with different test durations, and compare the spacial deactivation pattern for several partially deactivated catalyst beds.

Hans Schulz remarked: In MTH on zeolites, retained organic matter (commonly termed 'coke') must be recognized as a product. Coke selectivity must be determined and time resolved, together with detailed analysis of volatile compounds, as essential for revealing the mechanisms of product formation and catalyst deactivation. Also, a time resolved quantitative mass balance is needed. Trying to interpret the observed coke on used catalysts mechanistically is problematic. Evidently, such coke on/in H-ZSM-5 after use in MTH at 400–450 °C consists of deactivating external coke, a carbon rich organic phase covering the zeolite particle surface, and included organic aromatic compounds (mainly alkylated benzene molecules). The entrapped molecules in the pores are not the reason for deactivation but are retained in the pores because of deactivation (through surface coverage).

How do these issues fit into your view of MTH-fundamentals?

Unni Olsbye responded: I fully agree with your research strategy description, and with the global picture of species being trapped inside the zeolite due to pore blocking by coke species on, or close to, the outer crystal surface. However, the initiation of coke-forming reactions, and its dependence on reaction temperature as well as catalyst characteristics, are still not fully understood and deserve further attention.

Nico Fischer opened a general discussion of the paper by David Lennon: Inelastic neutron scattering is an exciting technique which is currently not heavily used in heterogeneous catalysis research. Could you please provide us with a brief summary of its applications and opportunities, also regarding resolution, temporal resolution and applicability for working catalysts?

David Lennon responded: Parker and co-workers have recently reviewed the general technique of inelastic neutron scattering (INS) as applied to investigations in molecular spectroscopy and heterogeneous catalysis.[1-4]. One factor critical to the successful acquisition of the vibrational spectrum of active heterogeneous catalysts is careful control of the sample environment; two recent articles outline developments in that area.[5,6] In order to provide a perspective on the range of catalytic systems suited to investigation by INS spectroscopy, and some of the constraints encountered in those studies, it is useful to consider some recent examples of work undertaken at the ISIS facility of the Rutherford Appleton Laboratory (UK). Parker and Lennon have reviewed how INS has been able to inform an improved understanding of the hydrochlorination of methanol over alumina based catalysts.[7] That work contributed to the development of a new generation methyl chloride synthesis catalyst.[8] Other work, looking at the reforming of methane over alumina-supported nickel catalysts, illustrated how INS could be used to quantify the hydrogen content within the catalyst matrix post-reaction.[9,10] Of particular interest to this meeting, is the application of INS to investigate iron based Fischer-Tropsch catalysts. Notably, catalyst samples extracted from a commercial scale coal-to-liquids Fischer–Tropsch synthesis (FTS) facility exhibit a distinct INS spectrum, which hints at a role for a hydrocarbonaceous overlayer in the reaction chemistry.[11] Follow-on studies have used the CO hydrogenation reaction to investigate the surface chemistry of the iron oxide/iron carbide/CO/H_2 reaction system.[12,13] Most recently, the FTS studies have utilised the capability of a modern chopper-based INS spectrometer to assign a magnetic transition seen in the INS spectrum of a high specification FTS pre-catalyst.[14] Broadening the range of systems further, other studies have used INS to probe different routes to methanol over industrial grade copper based catalysts,[15] as well as investigations for adsorbate interactions relevant to the synthesis of methyl methacrylate monomer.[16] Concerning the matter of resolution, temporal resolution and applicability for working catalysts, the following points are briefly noted. Spectral resolution is dependent on the type of spectrometer. On a direct geometry instrument, the resolution is a function of the incident energy, Ei, at the largest energy transfer and degrades with decreasing energy transfer; typical resolution values would be in the range 1–5% Ei.[1] On an indirect geometry time-of-flight spectrometer, the resolution of the spectrometer is $\sim$1% of the energy transfer.[1] Thus, INS conveys inferior spectral resolution compared to optical techniques, such as infrared spectroscopy. Furthermore, as INS is a relatively insensitive technique, sample acquisition times are relatively long. For example, considering the MTH study presented at this meeting (DOI: 10.1039/C6FD00195E), which utilised approximately 12 g of ZSM-5 for each reaction, a scan period of approximately 6 h was required to obtain a spectrum of reasonable signal : noise ratio. Finally, regarding the applicability for working catalysts, the examples presented above indicate that INS can provide new information not otherwise readily accessible by conventional analytical techniques. At this time, operando spectroscopy represents a considerable challenge for INS, although certain chemical systems are more amenable than others.[17] In summary, INS is probably at its most powerful when combined with a range of other characterisation techniques, including reaction testing.

1 S. F. Parker, D. Lennon and P. W. Albers, *Appl. Spectrosc.*, 2011, **65**, 1325–1341.
2 S. F. Parker, A. J. Ramirez-Cuesta, P. W. Albers and D. Lennon, *J. Phys.: Conf. Ser.*, 2014, **554**, 012004.
3 S. F. Parker and D. Lennon, *J. Phys.: Conf. Ser.*, 2016, **746**, 012066.
4 S. F. Parker and P. Collier, *Johnson Matthey Technol. Rev.*, 2016, **60**, 132–144.
5 I. P. Silverwood, N. G. Hamilton, A. McFarlane, R. M. Ormerod, T. Guidi, J. Bones, M. P. Dudman, C. M. Goodway, M. Kibble, S. F. Parker and D. Lennon, *Rev. Sci. Instrum.*, 2011, **82**, 034101.
6 R. Warringham, D. Bellaire, S. F. Parker, J. Taylor, R. Ewings, C. M. Goodway, M. Kibble, S. R. Wakefield, M. Jura, M. P. Dudman, R. P. Tooze, P. B. Webb and D. Lennon, *J. Phys.: Conf. Ser.*, 2014, **554**, 012005.
7 S. F. Parker and D. Lennon, *Acc. Chem. Res.*, 2014, **47**, 1220–1227.

8 A. R. McInroy, J. M. Winfield, C. C. Dudman, P. Jones and D. Lennon, *Faraday Discuss.*, 2016, **188**, 467–479.
9 I. P. Silverwood, N. G. Hamilton, C. J. Laycock, J. Z. Staniforth, R. M. Ormerod, C. D. Frost, S. F. Parker and D. Lennon, *Phys. Chem. Chem. Phys.*, 2010, **12**, 3102–3107.
10 A. R. McFarlane, I. P. Silverwood, R. Warringham, E. L. Norris, R. M. Ormerod, C. D. Frost, S. F. Parker and D. Lennon, *RSC Adv.*, 2013, **3**, 16577–16589.
11 N. G. Hamilton, I. P. Silverwood, R. Warringham, J. Kapitán, L. Hecht, P. B. Webb, R. P. Tooze, S. F. Parker and D. Lennon, *Angew. Chem., Int. Ed.*, 2013, **52**, 5608–5611.
12 N. G. Hamilton, R. Warringham, I. P. Silverwood, J. Kapitán, L. Hecht, P. B. Webb, R. P. Tooze, W. Zhou, C. D. Frost, S. F. Parker and D. Lennon, *J. Catal.*, 2014, **312**, 221–231.
13 R. Warringham, N. G. Hamilton, I. P. Silverwood, C. How, P. B. Webb, R. P. Tooze, W. Zhou, C. D. Frost, S. F. Parker and D. Lennon, *Appl. Catal., A*, 2015, **489**, 209–217.
14 R. Warringham, A. R. McFarlane, D. A. MacLaren, P. B. Webb, R. P. Tooze, J. Taylor, R. A. Ewings, S. F. Parker and D. Lennon, *J. Chem. Phys.*, 2015, **143**, 174703.
15 T. Kandemir, M. Friedrich, S. F. Parker, F. Studt, D. Lennon, R. Schlögl and M. Behrens, *Phys. Chem. Chem. Phys.*, 2016, **18**, 17253–17258.
16 A. R. McFarlane, H. Geller, I. P. Silverwood, R. I. Cooper, D. J. Watkin, S. F. Parker, J. M. Winfield and D. Lennon, *Phys. Chem. Chem. Phys.*, 2016, **18**, 17210–17216.
17 S. F. Parker, *Chem. Commun.*, 2011, **47**, 1988–1990.

Emiel Hensen remarked: You have clearly shown the potential of INS for characterizing the vibrational signatures of coke and zeolite active sites. In your work, you refer to bridging hydroxyl groups in different spectra but their frequencies range from 3515–3665 cm^{-1}. One would assume that the vibrational band around 3600 cm^{-1} is due to bridging hydroxyl groups in ZSM-5. The one at 3665 cm^{-1} is usually assigned to OH connected to extraframework Al, while a broad band at 3515 cm^{-1} might be due to water perturbing the acid sites. Could you comment on the assignment and briefly explain how the scaling of the vibrational spectra is carried out?

David Lennon responded: The MERLIN spectrum of the clean, activated catalyst presented in Figure 9 (DOI: 10.1039/C6FD00195E) shows a single band centred at 3505 cm^{-1}, whereas the higher resolution MAPS spectrum reveals the additional presence of a high wavenumber shoulder at $\sim$3585 cm^{-1}. The central feature is assigned to Brønsted acid hydroxyl groups and the shoulder is thought to represent a population of AlOH groups from extra framework Al. On reaction, typically a single υ(O–H) feature is seen centred between 3603 (Fig. 12) and 3665 cm^{-1} (Fig. 13, DOI: 10.1039/C6FD00195E). Throughout all of the spectra presented, there is no obvious presence of retained water, which would be additionally signified by a deformational mode at about 1640 cm^{-1} and a librational mode at about 550 cm^{-1}.[1] Accepting that INS does not possess the resolution accessible by IR spectroscopy, nevertheless, there remain some noticeable differences in the IR and INS spectra of the catalyst υ(O–H) modes. For example, the silanol groups on the external zeolite surface seen at 3740 cm^{-1} in the IR spectrum (Fig.7, DOI: 10.1039/C6FD00195E) are not evident in the comparable INS spectrum (*e.g.* Fig. 9, DOI: 10.1039/C6FD00195E). These matters are currently the topic of further investigation. The frequency scaling of the INS spectra relies on the time-of-flight technique. Parker and co-workers have recently reviewed the technique and operational characteristics of INS as applied to molecular spectroscopy, including aspects of investigations involving heterogeneous catalysts.[2] The authors shared similar concerns to the questioner regarding a possible variability in the energy scale but, on checking the spectral profiles with reference standards, the reported peak positions were proven to be correct.

1 A. R. McInroy, D. T. Lundie, J. M. Winfield, C. C. Dudman, P. Jones, S. F. Parker and D. Lennon, *Catal. Today*, 2006, **114**, 403–411.
2 S. F. Parker, D. Lennon and P. W. Albers, *Appl. Spectrosc.*, 2011, **65**, 1325–1341.

Unni Olsbye questioned: I find this method very interesting. However, the high catalyst amount required could make it difficult to study particular segments of the catalyst bed. Therefore, I was wondering if it would be possible to construct the reactor as a CSTR reactor, or to eventually use a prereactor, giving the desired conversion level followed by very short contact times through the catalyst bed, which will later be collected and characterised?

David Lennon replied: These are insightful comments. The necessity of having to use relatively large catalyst masses (*ca.* 10 g) in order to obtain INS spectra of sufficient signal : noise ratio can indeed cause problems with concentration gradients for certain reactions. Professor Olsbye's work elegantly shows MTH chemistry over ZSM-5 to be a reaction prone to deactivation, with spatio-temporal analysis of the catalyst clearly demonstrating differing degrees of carbon retention along the length of the catalyst bed (DOI: 10.1039/C6FD00187D). To date, our work has utilised predominantly fixed bed reactors. Although it is acknowledged that such a configuration is susceptible to problems with concentration gradients, this reactor type is compatible with current sample handling procedures adopted at the Central Facility.[1] Professor Olsbye's suggestions on the suitability of alternative reactor configurations is instructive, and is something that we will explore further as the present project proceeds.

1. R. Warringham, D. Bellaire, S. F. Parker, J. Taylor, R. Ewings, C. M. Goodway, M. Kibble, S. R. Wakefield, M. Jura, M. P. Dudman, R. P. Tooze, P. B. Webb and D. Lennon, *J. Phys. Conf. Ser.*, 2014, **554**, 012005.

Hans Schulz commented: Detailed spectra of retained organic compounds/phases in H-ZSM-5 during methanol conversion are interesting observations, contributing to the understanding of this unique process. To some extent these compounds are addressed as pool constituents. However, this is only a very limited, tentative suggestion. A dynamic view of the MTH pool mechanism with

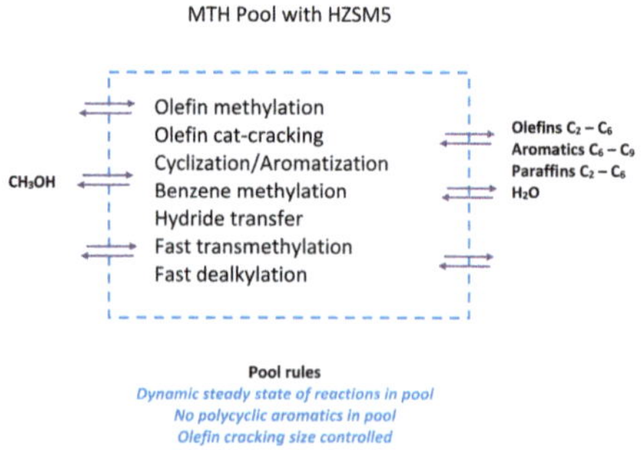

Fig. 1 A dynamic view of the MTH pool mechanism with H-ZSM-5. Adapted from M. Wei and H. Schulz, *Top. Catal.*, 2014, **57**, 683–692.

H-ZSM-5 is presented in Fig. 1. It displays the dynamic steady state of reactions and concentrations. The polyaromatic compounds are too large to be formed in the pores and can only be formed and retained on the crystallite surface. Respectively, they control deactivation, however are not pool species. This mechanistic view refers to reaction temperatures above 400 °C. At low temperatures, the alkylation of benzene rings with ethene and propene is possible and produces bigger molecules, which accumulate in the pores and provide internal deactivation. How do these conclusions fit with your observations?

David Lennon replied: I am very sympathetic to these statements, which I believe reflect the dynamic nature and degree of multiplicity of the MTH reaction system. Professor Schulz's figure nicely conveys these themes and, moreover, provides guidance on boundary conditions by the inclusion of a few 'pool rules'. I class the concepts outlined in the question as imaginative and insightful. Indeed, I am of the view that Professor Schulz's figure has connectivity to Fig. 14 of our paper (DOI: 10.1039/C6FD00195E). Figure 14 seeks to define the transitions between (i) methanol adsorption, (ii) the concept of a hydrocarbon pool existing in the internal pore structure that comprises alkanes, olefins and methylated aromatics, and (iii) the formation of a range of products of modest molecular weight ($\leq C_{10}H_6(CH_3)_2$). Furthermore, (iv) the presence of hexa-methylbenzene, as gleaned from *ex situ* EPR measurements, is decoupled from the hydrocarbon pool. Essentially, these concepts are encapsulated within Professor Schulz's figure, although it is readily acknowledged that his graphic is more specific regarding the chemical transformations taking place within the zeolite pore structure. At present, our work is insufficiently defined to address the important area of catalyst deactivation. This aspect of the catalytic process is ongoing for the project team.

Enrique Iglesia asked: How does one quench catalytic reactions and know that one has preserved what belongs there during steady-state catalysis? What is fast enough and/or cold enough given how limited our knowledge of how fast elementary steps are?

David Lennon replied: This is an important issue. Spectral acquisition of inelastic neutron scattering is typically undertaken at ~20 K in order to overcome a Debye–Waller factor in the scattering equation.[1] The approach of quenching heterogeneously catalysed reactions prior to transfer to the INS spectrometer follows the approach adopted by Wayne Goodman and co-workers.[2,3] Clearly, this is a non-ideal situation and represents a position where one is examining a catalyst sample post-reaction and not under steady-state conditions, where the real interest lies. Indeed, the authors readily acknowledge that certain reaction intermediates will not survive the transfer to the INS spectrometer and therefore the vibrational spectrum obtained *via* INS spectroscopy provides different information to, for example, an *operando* investigation utilising infrared spectroscopy. However, issues such as (i) an extended spectral range, (ii) an ability to analyse catalysts after extended times-on-stream when they may have become unobservable by optical spectroscopy, (iii) the fact that INS intensities are amenable to simulation *via* theory and are readily quantifiable, *etc.* are advantages that possess real value. Thus, accepting that INS has limitations when applied to investigating heterogeneously catalysed reaction systems, it is suggested that when the technique is combined with other spectroscopic probes plus conventional reaction testing protocols, it can provide new insight and information on hydrogeneous material retained within the catalyst matrix.

1. S. F. Parker, D. Lennon and P. W. Albers, *Appl. Spectrosc.*, 2011, **65**, 1325–1341.
2. S. Chinta, T. V. Choudhary, L. L. Daemen, J. Eckert and D. W. Goodman, *Angew. Chem., Int. Ed.*, 2002, **41**, 144.
3. C. Sivadinarayana, T. V. Choudhary, L. L. Daemen, J. Eckert and D. W. Goodman, *J. Am. Chem. Soc.*, 2004, **126**, 38.

Irina Yarulina enquired: 1) In your paper, the EPR hyperfine splitting is resolved based on the literature overview. However, I was wondering how, in general, the hyperfine splitting is resolved? Is the signal compared to the reference sample or are there other ways to assign the spectra?

2) Could you analyse other species apart from the hexamethylbenzene radical cation by EPR?

3) In DRIFTs spectra, both 2970 and 2870 cm^{-1} bands are attributed to methoxy groups. What is the difference between these two species both having methoxy groups?

David Lennon replied: 1) Figure 5 (DOI: 10.1039/C6FD00195E) presents the EPR spectra for samples 3, 4 and 5 (350 °C for 1 h, 350 °C for 3 d, and 400 °C for 3 d, respectively). In all cases a broad unresolved feature was observed. As mentioned in the paper (DOI: 10.1039/C6FD00195E), comparable spectra have been reported by other researchers for alcohols reacted over zeolites. The absence of hyperfine splitting in these spectra is attributed to the presence of exchange interactions between radical cations. In contrast, Figure 6 (DOI: 10.1039/C6FD00195E) shows that exposure of ZSM-5 to methanol vapour at 250 °C for 30 min, followed by evacuation at 150 °C for 30 mins, leads to an EPR spectrum exhibiting a well resolved hyperfine structure. Subsequent evacuation of this sample at an elevated temperature removes the signal completely. The thermal treatment that leads to the hyperfine structure seen in Figure 6 (DOI: 10.1039/C6FD00195E) that is not observable in Figure 5 (DOI: 10.1039/C6FD00195E) is thought to occur as a consequence of the desorption of hydrocarbonaceous entities that reduces the proximity of adjacent molecules, thereby minimising the degree of exchange interactions. This species is not stable when heated above 150 °C in vacuum.

With reference to the literature reported in the paper,[1] the well resolved proton hyperfine splitting seen in Figure 6(a) (DOI: 10.1039/C6FD00195E) is assigned to the hexamethylbenzenenium cation. Figure 6(b) (DOI: 10.1039/C6FD00195E) shows simulation of this entity to nicely reproduce the measured spectrum, endorsing the assignment. Other candidate molecules were considered, *e.g.* benzenium cations containing fewer methyl groups, but the simulated spectra gave inferior fits.

2. No, the only species we could unambiguously link to the observed EPR spectra was the hexamethylbenzene radical cation.

3. The 2970 and 2870 cm^{-1} bands in Figure 8(c) (DOI: 10.1039/C6FD00195E) are respectively assigned to the asymmetric and symmetric C–H stretches of

surface methoxy groups.[1]

1 A. R. McInroy, D. T. Lundie, J. M. Winfield, C. Dudman, P. Jones, S. F. Parker, J. W. Taylor and D. Lennon, *Phys .Chem. Chem. Phys.*, 2005, 7, 3093–3101.

Nico Fischer opened the general discussion, addressing David Lennon and Unni Olsbye: As you have shown, the MTH process is plagued by deactivation *via* carbon formation. Are there any ways to stabilize a working catalyst, maybe by simulating reaction conditions so that it can be analyzed in detail at a given conversion and deactivation level without a major gradient throughout the catalyst bed?

David Lennon responded: This is a very relevant comment. The central problem here is the relative insensitivity of the inelastic neutron scattering (INS) technique that nudges researchers to use relatively large catalyst masses (10–20 g). Such a catalyst charge requires larger reactors than typically encountered in laboratory testing. For a fixed bed arrangement, this leads to the possibility of concentration gradients across the length of the catalyst bed. Previous publications have also noted how the larger scale INS reactor possesses constraints in terms of accessible weight hourly space velocities compared to conventional micro-reactor testing.[1,2] In an attempt to minimise such experimental constraints, recent work has concentrated on optimising experimental arrangements, so that lower catalyst loadings (<10 g) can yield spectra of reasonable signal : noise ratios.[3] Moreover, the paper presented at this meeting (DOI: 10.1039/C6FD00195E) uses a new design of reactor that has a wider diameter than previous formats,[4] for a fixed catalyst charge; this reduces the bed length and helps minimise the scale of the problem. Preliminary infrared spectroscopic analysis of the catalyst post-reaction, sampled at different sections of the catalyst bed from the new reactor, show little evidence of sample variation as a function of sample position within the reactor. Further investigations in this area are ongoing.

1 N. G. Hamilton, R. Warringham, I. P. Silverwood, J. Kapitán, L. Hecht, P. B. Webb, R. P. Tooze, W. Zhou, C. D. Frost, S. F. Parker and D. Lennon, *J. Catal.*, 2014, **312**, 221–231.
2 R. Warringham, N. G. Hamilton, I. P. Silverwood, C. How, P. B. Webb, R. P. Tooze, W. Zhou, C. D. Frost, S. F. Parker and D. Lennon, *Appl. Catal. A*, 2015, **489**, 209–217.
3 R. Warringham, A. R. McFarlane, D. A. MacLaren, P. B. Webb, R. P. Tooze, J. Taylor, R. A. Ewings, S. F. Parker and D. Lennon, *J. Chem. Phys.*, 2015, **143**, 174703.
4 R. Warringham, D. Bellaire, S. F. Parker, J. Taylor, R. Ewings, C. M. Goodway, M. Kibble, S. R. Wakefield, M. Jura, M. P. Dudman, R. P. Tooze, P. B. Webb and D. Lennon, *J. Phys. Conf. Ser.*, 2014, **554**, 012005.

Unni Olsbye added: I think it will be difficult to simulate the reaction conditions inside the catalyst bed, since the exact composition of a fractionally converted MTH feed depends on the catalyst as well as the reaction conditions. Assuming that the goal is to study deactivation of an MTH catalyst in a fixed bed flow reactor, one option is to use the approach we reported here (DOI: 10.1039/C6FD00187D), *i.e.* to fractionate the catalyst bed after testing. Another option we have reported before, is to perform space- and time-resolved operando studies at the synchrotron.[1,2] A third possibility is to limit the single-pass conversion through the bed. This could either be done *via* a two-reactor approach, in which partial conversion is obtained in the pre-reactor, while the contact time in the second reactor (which is targeted for catalyst characterisation) is kept short; or using a single flow reactor with external recycle.

1 D. Wragg, M. G. O'Brien, F. L. Bleken, M. Di Michiel, U. Olsbye and H. Fjellvåg, *Angew. Chem., Int. Ed.*, 2012, **51**, 7956–7959.
2 P. del Campo, W. A. Slawinski, R. Henry, M. W. Erichsen, S. Svelle, P. Beato, D. Wragg and U. Olsbye, *Surf. Sci.*, 2016, **648**, 141–149.

Paul Collier asked: If you were to design the MTH catalyst from first principles, would you still end up with a zeolite or are there better catalysts that could be designed?

Unni Olsbye responded: The answer to this question is only partially known, and depends largely on the desired product. Previous studies by the Iglesia group have shown that homogeneous Lewis acid catalysts may very well catalyse the MTH reaction in liquid phase.[1] The product distribution is similar to that observed in Brønsted acidic zeolite Beta, with hexamethylbenzene and branched alkanes as major product groups. Shape selectivity, leading to more narrow product ranges, and less hydrogen transfer reactions (hence more alkenes, and less alkanes and aromatic products), is introduced by using zeolites with more narrow pore size. Use of higher reaction temperatures may further promote such product distribution, while the introduction of Lewis acid sites promotes hydrogen transfer reactions. The main pending question, which first principle calculations have not yet answered, is how to promote *e.g.* methylation and cracking reactions while hindering hydrogen transfer reactions. To conclude, zeolites are the best catalysts we know for this reaction, but I do not exclude the possibility that superior catalysts may be found in the future.

1 N. Hazari, E. Iglesia, J. A. Labinger and D. A. Simonetti, *Acc. Chem. Res.*, 2012, **45**, 653–662.

Avelino Corma replied: If we think through a carbocation mechanism, a zeolite with controlled pore dimensions, pore topology and acidity appears as a good solution to me.

Graham Hutchings commented: Just adding to the discussion concerning the question of whether non-zeolite catalysts are effective for the methanol to hydrocarbons reaction; in the 1980s there was a huge effort in the study of alternative catalysts, including many high area oxides (*e.g.* WO_3/Al_2O_3). Typically, these materials generate lots of methane in addition to hydrocarbons and operate *via* mechanisms that are different from that of ZSM-5. With ZSM-5, methane is formed in the very early stages of the reaction and also as the catalyst deactivates, but during use it does not produce any significant methane.

Avelino Corma addressed Unni Olsbye: Do you transfer hydrogen to a carbocation? Formaldehyde polymerises extremely fast under these conditions; how would you make the carbon balance in such an instance?

Unni Olsbye responded: The intermediates of the reaction are not yet fully disclosed. The selectivity for formaldehyde is low. Hence, it is probable that it reacts with a hydrocarbon to form dienes, and subsequently aromatic compounds, instead of polymerising.

Hans Schulz addressed Graham Hutchings, Avelino Corma and Unni Olsbye: Among the elemental reactions in MTH on acidic zeolites, ionic dehydrogenation plays an essential role, methane being the hydrogen-rich co-product.

At low temperatures (< 300 °C) in MTH, methane selectivity on H-ZSM-5 is initially high, as explained by reaction of CH_3^+ with hydrogen in the β-position of an olefin double bond, in order to form poly-unsaturated olefins as reactive pore constituents.

The same reaction principle, methane being formed selectively, has been observed with a highly deactivated HY-zeolite at 450 °C. As no easier reaction is possible, the methyl cation performs dehydrogenation of the coke species, which are entrapped in the super cages of the zeolite. How do these conclusions about ionic dehydrogenation fit with your basic view in the area?

Graham Hutchings replied: This description is fully consistent with the behaviour for methane formation at the start of the MTH reaction and during deactivation. There may of course be other pathways, depending on the likely sources for H transfer to the surface methoxy species and this would depend on the relative concentration of methanol and the products of MTH. At the very beginning, the methanol concentration will be very high relative to products and so H transfer from methanol may be a possible pathway.

Avelino Corma responded: I agree with the view you presented.

Unni Olsbye answered: I agree with this description, and think that methanol may also play the role of hydrogen donor, as originally proposed by Hutchings for the initial methanol–methanol reaction,[1] and very recently by Lercher and co-workers for methanol-assisted hydrogen transfer between olefins in product-rich zones of the reactor.[2]

1 G. J. Hutchings, F. Gottschalk, M. V. M. Hall and R. Hunter, *J. Chem. Soc., Faraday Trans.*, 1987, **83**, 571–583.
2 S. Müller, Y. Liu, F. M. Kirchberger, M. Tonigold, M. Sanchez-Sanchez and J. A. Lercher, *J. Am. Chem. Soc.*, 2016, **138**, 15994–16003.

Eric van Steen said: The hydrothermal liquefaction of biomass using supercritical water yields only 30–35% oil based on the amount of dry biomass fed. Theoretically, it should be possible to obtain higher oil yields (an oil selectivity of up to 80 C%). What kind of catalysts would be required to reduce char formation and increase the oil yield?

Avelino Corma responded: With a carbon content of 80% in the biocrude, and assuming a 45 wt% of carbon in the dry biomass, we get a carbon efficiency in the 55–65 wt% range. In my opinion it would be very difficult to do that without introducing large amounts of hydrogen in the biomass liquefaction step. At the end the oxygen has to be removed as much as possible as H_2O and, preferably, CO_2 so that hydrogen is conserved into the biocrude.

James Hayward addressed Avelino Corma: You use near-supercritical water to break bonds and form biocrude from lignin and cellulose. Could this technique be adapted to work for other materials? Would it be possible to generate the biocrude from, say, plastics or rubber?

Avelino Corma replied: Indeed, the process can be used to treat plastics and even organic municipal waste. The type of feed used will impact the composition of the biocrude produced. In the case of municipal solid waste (MSW), for example, a high protein content may result in a biocrude with high nitrogen content, thus more difficult to upgrade in a refinery setup. On the contrary, polyolefin plastics may give a higher quality biocrude.

Michael Claeys asked: Can you please let us have your forecast on the potential future uses of small scale reactor technology for XTL (*i.e.* Gas, Coal, Biomass and/or Waste To Liquids) applications. Will this technology play a significant role?

Burtron Davis responded: It is difficult to predict the future of FTS, let alone the reactor to be used. It would appear that the small (micro) channel reactor will find a role in smaller scale FTS operations, such as the one Velocys is using at the 1000 bbl/day plant in the US.

Nico Fischer said: Over recent years multiple companies have reported on developments in small scale FT applications. Several niches have been discussed, ranging from mobile plants on ships utilizing stored remote gas in oil fields, to the gasification of waste in landfills. Where do you see the most promising niche for this technology and which size of plant do you think will be built?

Burtron Davis answered: Your question includes the several niches where it appears that small channel reactors will be utilized for FTS.

Unni Olsbye addressed Avelino Corma: Currently, the interest in waste recycling, including polymer cracking to fuels, is increasing. If the polymer waste contains PVC, the feed will represent similar challenges to biomass, since HCl formation leads to H : C ratios below 2 in the product. Furthermore, HCl is likely to induce corrosion. To what extent do you think these processes would be suitable for polymer waste cracking?

Avelino Corma answered: If the waste polymer contains large amounts of PVC HCl is an issue to be considered. However, notice that, the hydrothermal treatment, as it is done now, works with a basic water solution that would neutralize the HCl.

Nico Fischer continued: Does the conversion of biomass to an alternative crude not compete with the extraction/synthesis of biomass based high value chemicals? What is your prediction on the feasibility of these two 'competing' approaches?

Avelino Corma answered: I believe that the amounts we are talking about are very different in both cases. My opinion is that biomass is an excellent source of chemical products already containing chemical building blocks. This is certainly a direction to follow. With respect to biomass to fuels, this is mainly depending on the policies of the countries.

Graham Hutchings addressed David Lennon: Can INS help investigate the initial stages of the methanol to hydrocarbon reaction, as this is when the hydrogen transfer reactions will be best observed?

David Lennon answered: In principle, yes. The approach one would adopt here is to prepare samples that reflect the early stages of the catalyst conditioning period. Reaction would be stopped (quenched) after fixed (short) periods of time on-stream, the catalyst sample isolated, and then transferred to the INS spectrometer for spectral acquisition. To date, we have tended to concentrate on steady-state operation for fixed reaction temperatures, but, as indicated by the question, INS could additionally be targeted to examine the hydrogen transfer reactions that may precede coke formation. With reference to the preceding paper ((DOI: 10.1039/C6FD00187D), it would be interesting to see if INS could detect the presence of formaldehyde at this early stage of the reaction coordinate.

Faraday Discussions

ROYAL SOCIETY OF CHEMISTRY

PAPER

Particulate photocatalyst sheets for Z-scheme water splitting: advantages over powder suspension and photoelectrochemical systems and future challenges

Qian Wang,[ab] Takashi Hisatomi,[ab] Masao Katayama,[ab] Tsuyoshi Takata,[ab] Tsutomu Minegishi,[ab] Akihiko Kudo,[c] Taro Yamada[ab] and Kazunari Domen*[ab]

Received 5th September 2016, Accepted 28th September 2016
DOI: 10.1039/c6fd00184j

Water splitting using semiconductor photocatalysts has been attracting growing interest as a means of solar energy based conversion of water to hydrogen, a clean and renewable fuel. Z-scheme photocatalytic water splitting based on the two-step excitation of an oxygen evolution photocatalyst (OEP) and a hydrogen evolution photocatalyst (HEP) is a promising approach toward the utilisation of visible light. In particular, a photocatalyst sheet system consisting of HEP and OEP particles embedded in a conductive layer has been recently proposed as a new means of obtaining efficient and scalable redox mediator-free Z-scheme solar water splitting. In this paper, we discuss the advantages and disadvantages of the photocatalyst sheet approach compared to conventional photocatalyst powder suspension and photoelectrochemical systems through an examination of the water splitting activity of Z-scheme systems based on $SrTiO_3$:La,Rh as the HEP and $BiVO_4$:Mo as the OEP. This photocatalyst sheet was found to split pure water much more efficiently than the powder suspension and photoelectrochemical systems, because the underlying metal layer efficiently transfers electrons from the OEP to the HEP. The photocatalyst sheet also outperformed a photoelectrochemical parallel cell during pure water splitting. The effects of H^+/OH^- concentration overpotentials and of the *IR* drop are reduced in the case of the photocatalyst sheet compared to photoelectrochemical systems, because the HEP and OEP are situated in close proximity to one another. Therefore, the photocatalyst sheet

[a]*Department of Chemical System Engineering, School of Engineering, The University of Tokyo, 7-3-1 Hongo, Bunkyo-ku, Tokyo 113-8656, Japan. E-mail: domen@chemsys.t.u-tokyo.ac.jp; Fax: +81-3-5841-8838; Tel: +81-3-5841-1148*

[b]*Japan Technological Research Association of Artificial Photosynthetic Chemical Process (ARPChem), 2-11-9 Iwamotocho, Chiyoda-ku, Tokyo 101-0032, Japan*

[c]*Department of Applied Chemistry, Tokyo University of Science, 1-3 Kagurazaka, Shinjuku-ku, Tokyo 162-8601, Japan*

design is well-suited to efficient large-scale applications. Nevertheless, it is also noted that the photocatalytic activity of these sheets drops markedly with increasing background pressure because of reverse reactions involving molecular oxygen under illumination as well as delays in gas bubble desorption. It is shown that appropriate surface modifications allow the photocatalyst sheet to maintain its water splitting activity at elevated pressure. Accordingly, we conclude that the photocatalyst sheet system is a viable option for the realisation of efficient solar fuel production.

1. Introduction

Owing to ever-growing fossil fuel consumption and concerns regarding climate change, it is imperative to develop renewable energy systems on a large scale.[1–3] Much effort in this regard has been focused on the conversion of abundant solar energy into storable chemical energy. Photocatalytic water splitting into hydrogen and oxygen has attracted considerable attention because hydrogen can be used as an energy carrier and as a feedstock for fuel cells and chemical industry.[4–7] It has been estimated that a solar-to-hydrogen energy conversion efficiency (STH) of approximately 10% and a capture area of 250 000 km^2 would provide one third of the energy needs resulting from human activities in 2050.[8] There is currently significant interest in photovoltaic (PV)-assisted photoelectrochemical (PEC) cells because these devices exhibit high STH values.[9–12] However, these high-performance systems are costly and show poor stability.[5,13,14] Indeed, a recent techno-economic analysis pointed out that the estimated levelised cost of hydrogen associated with current solar hydrogen generators is not comparable to the cost of electricity generated from fossil fuels.[1] In contrast, photocatalytic water splitting systems based on particulate semiconductors are expected to have advantages over PEC systems owing to their simplicity and cost-effectiveness.[5,15] The price of the hydrogen produced by a single bed particle suspension based on a slurry of photocatalyst powder could reach the target range of \$1 to \$2 kg^{-1} given an STH of 10%.[16] Unfortunately, the STH values achieved by water splitting using a particulate photocatalyst suspension are typically on the order of 0.1%.[17]

More than 100 different photocatalysts based on metal oxides have been reported to be active for splitting water into hydrogen and oxygen.[18,19] Nevertheless, most require ultraviolet (UV) light (λ < 400 nm) to excite the photocatalyst and drive the reaction. This is because the top of the valence band of such metal oxide semiconductors normally consists of O 2p orbitals located at a level that is approximately +3 V or more positive (*vs.* NHE at pH 0).[8,19–21] Accordingly, the band gap of the semiconductor becomes larger than 3 eV if the bottom of the conduction band is located at a potential more negative than the water reduction potential. Because nearly half of solar energy originates from visible light (400 < λ < 800 nm), a primary precondition in photocatalytic water splitting should be the utilization of visible light. Our group has demonstrated that some oxynitride photocatalysts, such as $(Ga_{1-x}Zn_x)(N_{1-x}O_x)$ and $(Zn_{1+x}Ge)(N_2O_x)$[22,23] are capable of decomposing water into H_2 and O_2 under visible light irradiation up to approximately 500 nm. The apparent quantum yield (AQY) of these materials has been increased to 5.1% at 410 nm by appropriate cocatalyst modification and post-annealing treatments.[22] Recently, a complex perovskite-type oxynitride, $LaMg_{1/3}Ta_{2/3}O_2N$, has been reported to be active in overall water splitting under irradiation up to 600 nm.[24,25] However, the number of such semiconductors that have the ability to split water under visible light irradiation is still limited.

As an alternative approach to overall water splitting by a single photocatalyst material (Fig. 1a), Z-scheme water splitting systems (Fig. 1b) based on the two-step excitation of a H_2 evolution photocatalyst (HEP) and an O_2 evolution photocatalyst (OEP) coupled by reversible redox couples have been studied, based on natural photosynthesis.[26,27] In these systems, protons on the surface of the HEP are reduced by photogenerated electrons in the conduction band whilst the reductant of the added redox couple is oxidised to the oxidant by holes in the valence band. The oxidation of water by valence band holes to produce oxygen takes place simultaneously on the surface of the OEP whilst the oxidant is reduced back to the reductant by photoexcited electrons generated in the conduction band. As a consequence, overall water splitting is achieved *via* reversible reactions of the redox mediator. The energy required to drive photocatalysts could be lower in a Z-scheme water splitting system than in a one-step excitation overall water splitting system, and so Z-scheme systems may be well-suited to the harvesting of visible light. In fact, some visible light-driven photocatalysts that do not have the ability to split water into H_2 and O_2 but are capable of evolving either H_2 or O_2 from an aqueous solution containing appropriate sacrificial reagents have been successfully applied to Z-scheme water splitting.

In Z-scheme water splitting systems, ionic electron mediators, such as $Fe^{3+/2+}$,[28–32] IO_3^-/I^-,[33–39] $[Co(phen)_3]^{3+/2+}$, $[Co(bpy)_3]^{3+/2+}$,[40] and $[SiW_{11}O_{39}Mn^{III}(H_2O)]^{5-}/[SiW_{11}O_{39}Mn^{II}(H_2O)]^{6-}$,[41] have been employed as redox couples for the electron transfer between the HEP and OEP. However, the Z-scheme water splitting reaction may also be suppressed by the redox mediators because they can readily lead to both reverse reactions and light shielding.[27,42] Therefore, the development of Z-scheme systems that do not rely on redox pairs is desirable. Some redox mediator-free systems consisting of Ru-loaded $SrTiO_3$ doped with Rh ($Ru/SrTiO_3$:Rh) or Ru-loaded $SrTiO_3$ doped with Rh and La ($Ru/SrTiO_3$:La,Rh) as HEPs and various OEPs (TiO_2, WO_3, $BiVO_4$, PRGO/$BiVO_4$, and Ta_3N_5) have been reported.[42–45] These systems decompose water into hydrogen and oxygen

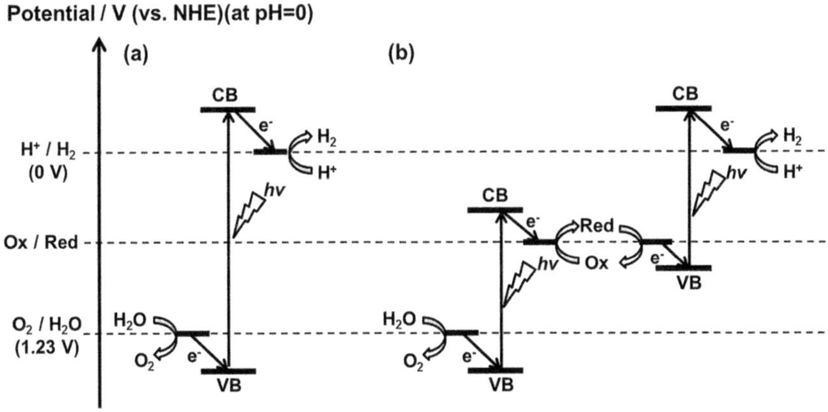

Fig. 1 Schematic energy diagrams of photocatalytic water splitting *via* (a) one-step excitation and (b) two-step excitation (Z-scheme). CB and VB stand for conduction band and valence band, respectively.

stoichiometrically under visible light without the assistance of redox couples, owing to interparticle electron transfer. The Ru/$SrTiO_3$:Rh–$BiVO_4$ combination achieves an STH of 0.12% and an AQY of 1.7% at 420 nm.[42] The incorporation of photoreduced graphene oxide (PRGO) as a solid-state electron mediator has also been reported to be effective in promoting the photocatalytic activity of this system.[45] However, the activity of these systems is strongly dependent on the pH of the reactant solution, because HEP and OEP particles need to be oppositely charged and form aggregates by electrostatic attraction so that interparticle electron transfer can occur. Therefore, the pH of the reaction solution must be adjusted between the points of zero charge of the respective photocatalysts. Effort has been devoted to the development of composites of $SrTiO_3$:Rh and $BiVO_4$ to allow intimate contact between the two kinds of materials without electrostatic interfaction.[46] However, in such cases the activity has not been enhanced drastically, possibly because chemical reactions of the two photocatalysts take place during the synthesis processes and the degree of contact between the particles is poorly controlled.

From earlier studies on redox mediator-free Z-scheme water splitting systems, it is clear that the key to improving photocatalytic activity is facilitating the interparticle electron transfer between the HEP and OEP without deteriorating their inherent properties. Recently, we designed and fabricated self-completed photocatalyst sheets based on HEP and OEP particles embedded into a conductive layer by particle transfer for Z-scheme water splitting.[47–49] In this system, the electron transfer takes place in the conductive layer, rather than *via* the aggregation of particles by attractive electrostatic forces. The HEP and OEP particles are bridged with rigid conductive layers so as to achieve efficient electron transfer and promote the activity for overall water splitting. In this work, a $SrTiO_3$:La,Rh/Au/$BiVO_4$:Mo photocatalyst sheet, in which a Au layer serves to transfer electrons, was found to exhibit an STH level of 1.1%, a record among Z-scheme pure water splitting systems, and an AQY exceeding 30% at 419 nm.[47] This photocatalyst sheet system was demonstrated to be capable of splitting water under illumination up to 520 nm. In order to extend the available wavelength and to utilise more visible light, $LaMg_{1/3}Ta_{2/3}O_2N$, having a narrower bandgap energy, has been applied as the HEP in the photocatalyst sheet system in place of $SrTiO_3$:La,Rh.[49] The resulting ($RhCrO_x$/$LaMg_{1/3}Ta_{2/3}O_2N$)/Au/rutile sheet exhibits an order of magnitude higher activity for overall water splitting than the corresponding powder suspension Z-scheme systems under the same reaction conditions. However, it has also been reported that the water splitting activity of such photocatalyst sheet systems is sensitive to the background pressure.[47] Therefore, it is necessary to understand the reaction characteristics specific to the photocatalyst sheet approach.

In this paper, we discuss the advantages and disadvantages of the photocatalyst sheet system compared to conventional powder suspensions and PEC parallel systems in Z-scheme water splitting reactions, by comparing the activity of a combination of $SrTiO_3$:La,Rh as the HEP, $BiVO_4$:Mo as the OEP, and Au as the conductor under various reaction conditions. We also discuss the effects of the incident photon flux and the origin of the drop in photocatalytic activity upon the elevation of background pressures for the photocatalyst sheet system in view of practical reaction conditions.

2. Experimental

2.1. Preparation of $SrTiO_3$:La,Rh and $BiVO_4$:Mo particles

$SrTiO_3$:La,Rh was synthesized through a two-step solid state reaction (SSR) according to the methods described in our previous paper.[43] The molar ratios of La/(La + Sr) and Rh/(Rh + Ti) were 4%. Firstly, $SrTiO_3$ was obtained by calcining a mixture of $SrCO_3$ (Kanto Chemicals, 99.9%) and rutile-type TiO_2 (rutile, Kanto Chemicals, 99.0%) powders at a molar Sr : Ti ratio of 1.03 : 1 at 1373 K for 10 h. The resulting $SrTiO_3$ was mixed with La_2O_3 (Kanto Chemicals, 99.99%), which had been freshly calcined in air at 1273 K for 12 h, and Rh_2O_3 (Kanto Chemicals, 99.9%). The mixture was ground in ethanol and then calcined at 1373 K for 6 h.

Mo-doped $BiVO_4$ was prepared *via* aqueous processes[50] at a molar Mo : V ratio of 0.05%. K_2CO_3 (Kanto Chemical, 99.5%), V_2O_5 (Wako Distilled Chemical Industries, 99.0%) and MoO_3 (Kanto Chemical, 99.5%) were mixed and then calcined in air at 723 K for 5 h to form Mo-doped $K_3V_5O_{14}$. The resulting material was added to an aqueous solution of $Bi(NO_3)_3 \cdot 5H_2O$ (Wako Distilled Chemical Industries, 99.9%) and the suspension was stirred at 343 K for 10 h. The Mo-doped $BiVO_4$ obtained in this manner was collected by filtration and washed with distilled water before drying at 343 K.

2.2. Fabrication of $SrTiO_3$:La,Rh/Au/$BiVO_4$:Mo photocatalyst sheets

$SrTiO_3$:La,Rh/Au/$BiVO_4$:Mo photocatalyst sheets were prepared by the particle transfer method,[51] as reported in our previous works.[47,48] In the initial step, a mixture of 10 mg each of $SrTiO_3$:La,Rh and $BiVO_4$:Mo powders was dispersed in 0.5 mL isopropanol and drop-cast onto a glass substrate (3 × 3 cm). After drying at room temperature, a thin Au layer (approximately 300 nm thick) was deposited onto the particles by vacuum evaporation (VFR-200M/ERH, ULVAC KIKO Inc.) at an evaporation rate of approximately 15 nm s^{-1}. The Au film, together with the particulate photocatalysts, was subsequently annealed in air at 573 K for 20 min and then transferred to a second glass plate (3 × 3 cm) with an adhesive carbon tape (Nisshin EM Co. Ltd.). The photocatalyst sheet was obtained by lifting off the primary glass plate and, in a final step, was ultrasonicated twice, for 2 min each time, in distilled water to remove excess particles accumulated on the particle layer. The samples were characterised by scanning electron microscopy (SEM; S-4700, Hitachi; JSM-7001FA, JEOL).

2.3. Fabrication of $SrTiO_3$:La,Rh/Au–$BiVO_4$:Mo/Au parallel cells

$SrTiO_3$:La,Rh/Au and $BiVO_4$:Mo/Au photoelectrodes were prepared by the particle transfer method, similar to the preparation procedures described above for the photocatalyst sheets. In this method, 20 mg of $SrTiO_3$:La,Rh or $BiVO_4$:Mo photocatalyst powder was suspended in isopropanol (0.5 mL) and ultrasonicated for 30 min. The mixture was drop-cast onto a glass substrate (3 × 3 cm) and a Au layer with a thickness of 600–800 nm was deposited under conditions similar to those used for the fabrication of the photocatalyst sheets. This process was repeated to form a Au layer with a thickness of approximately 3 μm to ensure conductivity and continuity over the electrodes while avoiding deviations in the operating pressure and temperature in the chamber and hence in the evaporation rate. The resulting specimens were annealed at 573 K for 20 min and then transferred to another glass plate (1 × 3 cm). The $SrTiO_3$:La,Rh/Au and $BiVO_4$:Mo/Au electrodes were connected using In solder (Aldrich) to form PEC parallel cells. The In was subsequently covered with epoxy (Araldite) and the exposed area after packaging was about 3.5 cm^2.

2.4. Photodeposition of the cocatalyst

Based on a prior study, Ru species were employed as a cocatalyst in all the Z-scheme systems employing $SrTiO_3$:La,Rh and $BiVO_4$:Mo.[30] These Ru species (0.03 μmol cm^{-2}) were loaded on the photocatalyst sheets and parallel cells by photodeposition using $RuCl_3 \cdot 3H_2O$ as the precursor. The photodeposition reactions were conducted in a Pyrex top-irradiation reaction vessel connected to a glass closed gas circulation system. The samples were irradiated using a Xe lamp (Lamp house, R300-3J) equipped with a cut-off filter ($\lambda > 420$ nm) until steady gas evolution rates were obtained.

2.5. Photocatalytic reactions

Z-scheme water splitting reactions were carried out in the same closed gas circulation system with overhead irradiation used for the cocatalyst deposition. The photocatalyst sheets and parallel cell samples were placed at the bottom of the reactor containing the reactant solution (40 mL). The reaction solutions were either pure water, aqueous 0.01 or 0.1 M potassium phosphate buffers (pH 6.8) or 0.1 M Na_2SO_4 (pH 6.8). H_3PO_4 (aq.) or KOH (aq.) were employed to adjust the pH of the potassium phosphate buffer solutions, whilst H_2SO_4 (aq.) or NaOH (aq.) were used for the Na_2SO_4 solution if necessary. The reactant solution was evacuated several times to degas the solution completely after which, unless otherwise noted, Ar gas (5.3 kPa) was introduced into the reaction system as the carrier gas. The reaction cell was irradiated using a 300 W Xe lamp fitted with a cut-off filter ($\lambda > 420$ nm), generating a photon flux below 540 nm of 5.2×10^{20} photons h^{-1} cm^{-2}. During irradiation, the reaction solution was maintained at 288 K using a flow of cooling water and the evolved gases were analysed by a gas chromatograph connected to the closed gas circulation system (Shimadzu GC-8A with a TCD and an MS-5A column, using Ar as the carrier gas). Photocatalytic H_2 evolution reactions were carried out in the same manner except for the use of a 10 vol% aqueous methanol solution in place of pure water. The effect of light intensity on the photocatalytic activity was determined under illumination from a 300 W Xe lamp equipped with a series of neutral-density (ND) filters.

3. Results and discussion

3.1. Comparison of the Z-scheme water splitting systems

Fig. 2 shows the time courses of the gas evolution during Z-scheme water splitting using a photocatalyst sheet. H_2 and O_2 were evolved stoichiometrically on the Ru-modified $SrTiO_3$:La,Rh/Au/$BiVO_4$:Mo photocatalyst sheet immersed in pure water (pH 6.8) under visible light illumination ($\lambda > 420$ nm), as shown in Fig. 2a. The gas production rates were two orders of magnitude higher than those obtained from the powder suspension system consisting of the $SrTiO_3$:Rh and $BiVO_4$ photocatalysts.[42] As reported previously[42–44] and as described in the Introduction, the pH of the reaction solution is a key factor affecting the interparticle electron

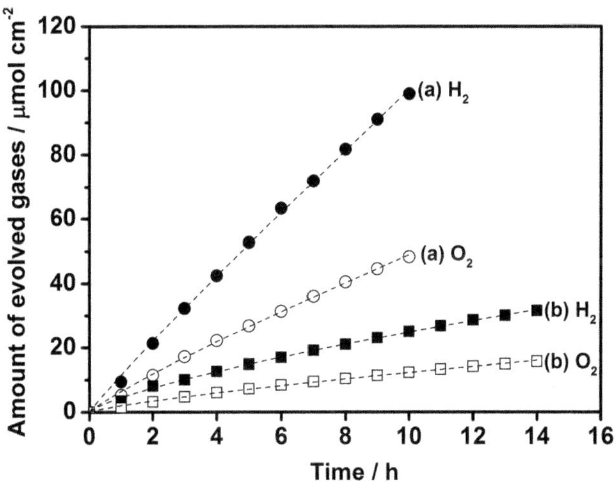

Fig. 2 Time courses of overall water splitting using a Ru-modified $SrTiO_3$:La,Rh/Au/$BiVO_4$:Mo sheet in (a) pure water (pH 6.8) (circles) and (b) an aqueous 10 mM potassium phosphate buffer at pH 6.8 (squares) under Xe lamp (300 W) illumination (λ > 420 nm).

transfer process. It has been reported that mildly acidifying the reaction solution to pH 3.5 is most effective for the $SrTiO_3$:Rh–$BiVO_4$ and $SrTiO_3$:La,Rh–$BiVO_4$ powder suspension systems.[42,48] However, the performance was still not comparable to that of the present photocatalyst sheet system operating at neutral pH. In the photocatalyst sheet system, the Au layer deposited onto the photocatalyst particles by vacuum evaporation holds the particles firmly in place irrespective of the pH. Particles that do not have intimate contact with the Au layer tend to suffer from a large grain boundary resistance and are thus eliminated through the particle transfer process.[47,48] Furthermore, the contact resistance between the photocatalyst particles and the Au layer can be reduced by a mild annealing treatment at 573 K.[47] Therefore, the photocatalyst sheet is operable in pure water and exhibits high activity regardless of the pH of the reaction solution. It is notable that, when using the photocatalyst sheet approach, the photocatalytic materials applicable to Z-scheme water splitting are not limited by the signs of the surface charges at the working pH, which is an important feature of the particle transfer method. This is in contrast to the requirements of the powder suspension system.

As can be seen from Fig. 3a, the $SrTiO_3$:La,Rh/Au–$BiVO_4$:Mo/Au parallel cell also exhibited lower water splitting activity than the photocatalyst sheet in pure water (pH 6.8). This cell did, however, generate higher activity in an aqueous solution containing Na_2SO_4 as a supporting electrolyte (Fig. 3b), and the use of a phosphate buffer led to further enhancement in the water splitting activity (Fig. 3c). The lower activity of the PEC parallel cell in pure water is attributed to the high solution resistance resulting from the low conductivity of pure water. Therefore, the use of a supporting electrolyte at a sufficient concentration is indispensable. In addition, because the H_2 and O_2 evolution reactions occur at a distance from one another, H^+ and OH^- concentration overpotentials are generated during the water splitting reaction in the parallel cell. The equilibrium

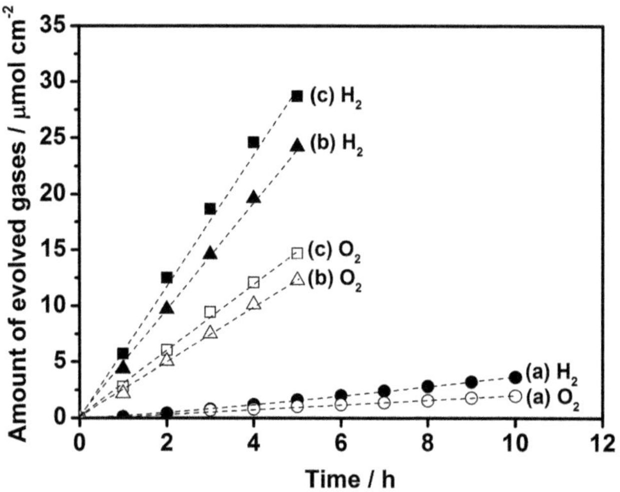

Fig. 3 Time courses of overall water splitting using a Ru-modified $SrTiO_3$:La,Rh/Au–$BiVO_4$:Mo/Au parallel cell in (a) pure water (pH 6.8) (circles), (b) an aqueous 0.1 M Na_2SO_4 solution (pH 6.8) (triangles), and (c) a 0.1 M potassium phosphate buffer (pH 6.8) (squares) under Xe lamp (300 W) illumination (λ > 420 nm).

voltage required for splitting water thus becomes greater than 1.23 V according to the Nernst equation, particularly when the activity is high. The use of a phosphate buffer effectively suppresses the pH gradient, although the photocatalytic activity of the PEC cell was still 40% lower than that of the $SrTiO_3$:La,Rh/Au/$BiVO_4$:Mo photocatalyst sheet in pure water. The high photocatalytic activity of the photocatalyst sheet is attributable to the suppression of the effects of the solution resistance and the concentration overpotentials, because the HEP and OEP are located close to one another, with spacing on the order of the particle dimensions (2 µm at most). Note that optimum particle and domain sizes are expected to exist for respective particulate photocatalyst sheet systems owing to competing effects on different scales. In individual particles, charge transfer to surface active sites is expected to be more efficient upon reducing the particle size owing to a shorter migration distance. However, recombination can be also enhanced because of lower crystallinity and a greater number of surface states. On the scale of photocatalyst sheets, when the HEP and OEP particles are smaller and located nearer to each other (*i.e.*, domain sizes are smaller), the solution resistance and the concentration overpotentials are inhibited more effectively. However, smaller particles suffer from larger grain boundary resistance as a result of aggregation as well as weaker light absorption.

Interestingly, the water splitting activity of the $SrTiO_3$:La,Rh/Au/$BiVO_4$:Mo photocatalyst sheet was lowered to 30% when the reaction was carried out in an aqueous phosphate buffer (Fig. 2b). Moreover, the water splitting activity was observed to gradually decrease over the course of the 14 hour reaction. SEM images of the sheet samples before and after the reactions (Fig. 4) show that the morphology of the photocatalyst particles was modified during water splitting in the phosphate buffer but not in pure water. It has been suggested that ionic species adsorbed on the surfaces of the photocatalyst particles and served to block the active sites.[52]

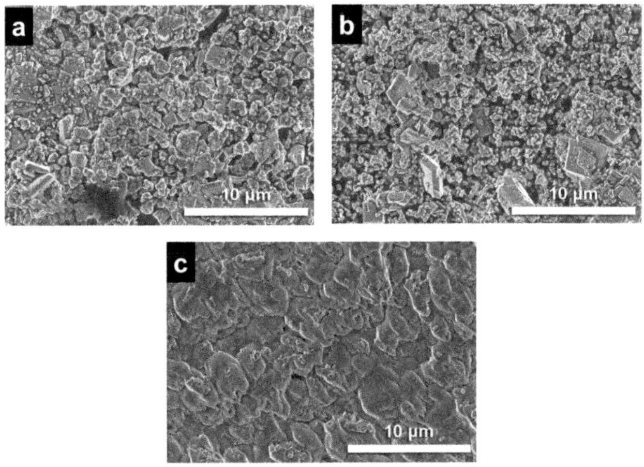

Fig. 4 SEM top-view images of $SrTiO_3$:La,Rh/Au/$BiVO_4$:Mo photocatalyst sheets (a) before, and after the water splitting reactions (b) in pure water and (c) in phosphate buffer solution.

3.2. Effect of the photon flux on the activity of the photocatalyst sheet

Fig. 5 summarises the relationship between the water splitting activity of a $SrTiO_3$:La,Rh/Au/$BiVO_4$:Mo photocatalyst sheet and the incident visible light photon flux below 540 nm. The gas evolution rate evidently increased linearly while increasing the photon flux to 1.4×10^{20} photons h^{-1} cm^{-2}. This behaviour is typical of photocatalytic reactions carried out at low light intensity.[53] Beyond 1.4×10^{20} photons h^{-1} cm^{-2}, the gas evolution rate became slightly sublinear relative to the photon flux, even though Z-scheme photocatalytic water splitting is

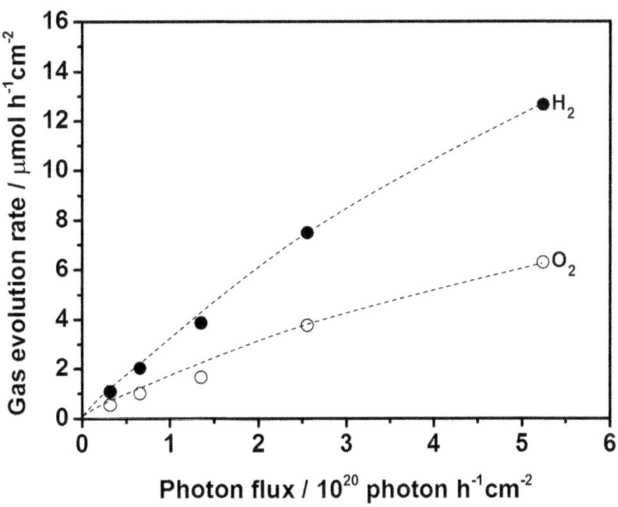

Fig. 5 Dependence of the water splitting rates over Ru-loaded $SrTiO_3$:La,Rh/Au/$BiVO_4$:-Mo sheets upon irradiated light intensity (λ < 540 nm). The reactions were carried out with a Xe lamp (300 W) equipped with a cut-off filter (λ > 420 nm) and a series of neutral density filters in pure water (pH 6.8).

a multi-electron reaction, which is likely because second-order recombination began to have an effect. It is expected that the reaction rate will become proportional to the square root of the light intensity if second-order recombination becomes the dominant process.[53] The present $SrTiO_3$:La,Rh/Au/$BiVO_4$:Mo sheet system has been found to harvest visible light up to 540 nm (ref. 48) and, in the standard AM1.5G solar spectrum, the photon flux below 540 nm is 2.0×10^{20} photons h^{-1} cm^{-2}, which falls into the transition to the sublinear region of the activity-photon flux relationship presented in Fig. 5. Therefore, the present photocatalyst sheet system may function best not when exposed to focused intense sunlight but rather when spread over a wide area under one sun light intensity, taking advantage of the scalability of the approach.

3.3. Origin of the negative pressure effect on the water splitting activity of the photocatalyst sheet

Fig. 6a shows the dependence of the water splitting activity of a Ru-loaded $SrTiO_3$:La,Rh/Au/$BiVO_4$:Mo photocatalyst sheet on the background Ar pressure. The water splitting rates were found to fall to 23 and 7% as the background pressure was elevated from 5 to 10 and 91 kPa, respectively. The strong negative effect of the background Ar pressure seen here is in agreement with our previous study.[47] In contrast, the water splitting activity of the powder suspension system was almost independent of the background pressure, as shown in Fig. 6b. These data suggest that reverse reactions, such as the recombination of H_2 and O_2 to form water and the reduction of O_2 molecules under photoexcitation, became prominent at elevated background pressures because of the unique characteristics of the photocatalyst sheet; that is, that the H_2 and O_2 are evolved in close proximity to one another. Such reverse reactions can be suppressed by surface modifications of the photocatalysts with thin oxide layers that have molecular-sieving effects such that O_2 is blocked whilst H_2 and water pass through. In fact,

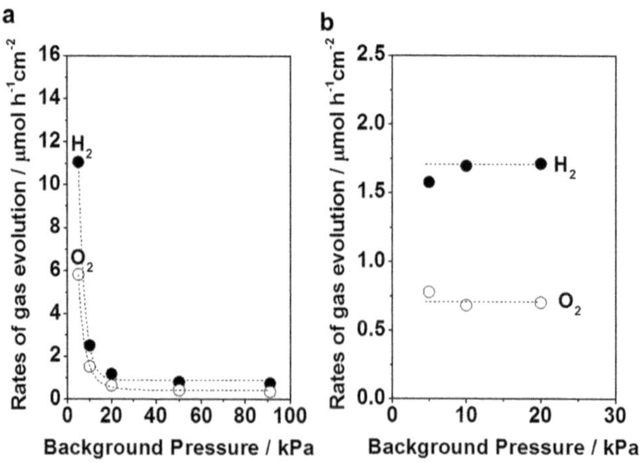

Fig. 6 Dependence of the water splitting rates over (a) Ru-loaded $SrTiO_3$:La,Rh/Au/$BiVO_4$:Mo sheets and (b) a Ru-loaded $SrTiO_3$:La,Rh–$BiVO_4$:Mo powder suspension system upon the background pressure. The reactions were carried out with Xe lamp (300 W) illumination (λ > 420 nm).

our previous work has demonstrated that a photocatalyst sheet will retain 86% of its activity even when the background pressure is increased from 5 to 20 kPa after modifications with Cr_2O_3 and amorphous TiO_2 layers.[47]

In order to discuss the origin of the negative effect of the background Ar pressure more clearly, the pressure dependence of the H_2 evolution activity of the photocatalyst sheet in an aqueous methanol solution was studied (Fig. 7). In this reaction, O_2 is not evolved because methanol is oxidised instead of water. Note that the hydrogen evolution rate was lower in aqueous methanol solution than in pure water (Fig. 6), because the carbon tape was partially dissolved by methanol. It is shown in Fig. 7 that the pressure dependence of the photocatalytic activity was much weaker in the case of the H_2 evolution reaction compared with Z-scheme water splitting (Fig. 6a). Therefore, it is confirmed that the pressure dependence is undoubtedly associated with the presence of O_2 in the system through the reverse reactions. It should also be noted that the photocatalytic activity decreased slightly with pressure even in the absence of O_2 in the system, particularly at low pressures. This suggests that the desorption of product gas bubbles may also be associated with the pressure dependence of the photocatalytic activity. Product gas bubbles tend to adhere to the $SrTiO_3$:La,Rh/Au/$BiVO_4$:Mo photocatalyst sheet surface, which is most likely because of the hydrophobicity of the material.[47] Given that the water splitting rate is constant, it is believed that the residence time of gas bubbles on the photocatalyst sheet surface becomes longer with increases in the background pressure, because the growth rate of the bubbles and thus the incremental increase in their buoyancy becomes slower. Bubbles staying on the surface may inhibit wetting of the particulate photocatalyst sheet surface and may lower the photocatalytic activity, which is similar to effects observed with $(Ga_{1-x}Zn_x)(N_{1-x}O_x)$ photocatalyst panels.[54] Modification of the photocatalyst

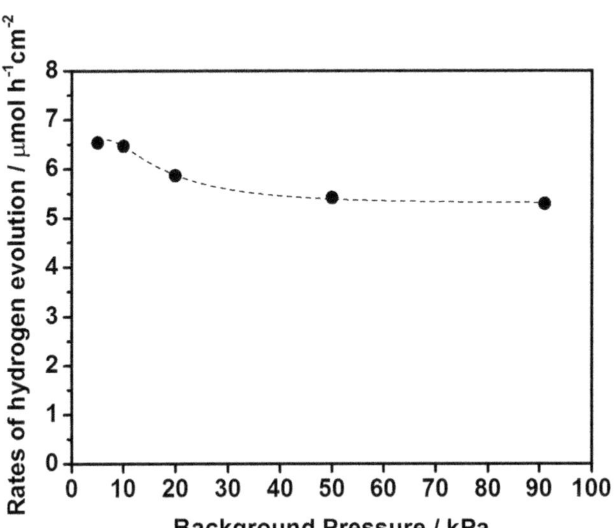

Fig. 7 Hydrogen evolution over $SrTiO_3$:La,Rh/Au/$BiVO_4$:Mo sheets from an aqueous methanol solution (10 vol%) as a function of the background pressure. The half-reactions were carried out under Xe lamp (300 W) illumination ($\lambda > 420$ nm).

particle layer with hydrophilic oxides could facilitate the desorption of bubbles and hence the penetration of water.

4. Conclusions

This work assessed the advantages of the photocatalyst sheet system over the powder suspension and PEC parallel cell systems during visible light-driven water splitting and also addressed future challenges. We demonstrated that a photocatalyst sheet exhibits higher activity in pure water than a conventional powder suspension system or PEC parallel cells in their optimal reaction solutions. The primary advantage of the photocatalyst sheet system over conventional powder suspension processes is that the electron transfer occurs through the underlying conductive layer and therefore the system can function efficiently in pure water. In addition, the solution resistance and the concentration overpotentials that are commonly observed in PEC parallel cells are inhibited in the photocatalyst sheet system because the HEP and OEP particles are located near each other. It was also confirmed that the water splitting rate increased linearly with photon flux under weak excitation. At the intensity of standard AM1.5G irradiation, the photocatalytic activity occurred at the transition to the sublinear region. Therefore, the most effective means of employing this material is believed to involve expanding the photocatalyst sheet over a large area rather than focusing the sunlight, so as to maintain the STH value.

A drastic decrease in activity with increasing background pressure was reconfirmed as the primary challenge with this photocatalyst sheet system. This is because reverse reactions involving O_2 compete with the water splitting reaction. The adhesion of gas bubbles to the hydrophobic photocatalyst sheet surface was also found to have a negative effect on the water splitting activity. These problems could be overcome through appropriate surface modifications to suppress the reverse reactions and facilitate the desorption of gas bubbles.

The photocatalyst sheet is a promising design that enables efficient and scalable pure water splitting. A variety of photocatalytic materials can be applied to photocatalyst sheets for Z-scheme water splitting because the choice of materials is not restricted by the signs of surface charges at the working pH. The potential of these photocatalyst sheet systems could be further extended by applying particulate semiconductors with narrower bandgaps.

Acknowledgements

This work was financially supported by the Artificial Photosynthesis Project of the New Energy and Industrial Technology Development Organization (NEDO) and Grants-in-Aids for Scientific Research (A) (No. 16H02417) and for Young Scientists (A) (No. 15H05494) from the Japan Society for the Promotion of Science (JSPS).

References

1 M. R. Shaner, H. A. Atwater, N. S. Lewis and E. W. McFarland, *Energy Environ. Sci.*, 2016, **9**, 2354.
2 N. S. Lewis, *Science*, 2016, **351**, aad1920.

3 R. C. Armstrong, C. Wolfram, K. P. de Jong, R. Gross, N. S. Lewis, B. Boardman, A. J. Ragauskas, K. Ehrhardt-Martinez, G. Crabtree and M. V. Ramana, *Nature Energy*, 2016, **1**, 15020.
4 K. Sivula, *J. Phys. Chem. Lett.*, 2015, **6**, 975.
5 F. E. Osterloh, *J. Phys. Chem. Lett.*, 2014, **5**, 2510.
6 J. Barber and P. D. Tran, *J. R. Soc., Interface*, 2013, **10**, 20120984.
7 M. G. Walter, E. L. Warren, J. R. McKone, S. W. Boettcher, Q. Mi, E. A. Santori and N. S. Lewis, *Chem. Rev.*, 2010, **110**, 6446.
8 K. Maeda and K. Domen, *J. Phys. Chem. Lett.*, 2010, **1**, 2655.
9 Y.-S. Chen, J. S. Manser and P. V. Kamat, *J. Am. Chem. Soc.*, 2015, **137**, 974.
10 F. F. Abdi, L. Han, A. H. M. Smets, M. Zeman, B. Dam and R. van de Krol, *Nat. Commun.*, 2013, **4**, 2195.
11 J. Brillet, J.-H. Yum, M. Cornuz, T. Hisatomi, R. Solarska, J. Augustynski, M. Graetzel and K. Sivula, *Nat. Photonics*, 2012, **6**, 824.
12 O. Khaselev and J. A. Turner, *Science*, 1998, **280**, 425.
13 J. W. Ager, M. R. Shaner, K. A. Walczak, I. D. Sharp and S. Ardo, *Energy Environ. Sci.*, 2015, **8**, 2811.
14 J. R. McKone, N. S. Lewis and H. B. Gray, *Chem. Mater.*, 2014, **26**, 407.
15 Y.-P. Yuan, L.-W. Ruan, J. Barber, S. C. J. Loo and C. Xue, *Energy Environ. Sci.*, 2014, **7**, 3934.
16 B. A. Pinaud, J. D. Benck, L. C. Seitz, A. J. Forman, Z. Chen, T. G. Deutsch, B. D. James, K. N. Baum, G. N. Baum and S. Ardo, *Energy Environ. Sci.*, 2013, **6**, 1983.
17 D. M. Fabian, S. Hu, N. Singh, F. A. Houle, T. Hisatomi, K. Domen, F. E. Osterloh and S. Ardo, *Energy Environ. Sci.*, 2015, **8**, 2825.
18 T. Hisatomi, K. Takanabe and K. Domen, *Catal. Lett.*, 2014, **145**, 95.
19 R. Abe, *J. Photochem. Photobiol., C*, 2010, **11**, 179.
20 A. Kudo and Y. Miseki, *Chem. Soc. Rev.*, 2009, **38**, 253.
21 A. Kudo, *Pure Appl. Chem.*, 2007, **79**, 1917.
22 Y. Lee, K. Teramura, M. Hara and K. Domen, *Chem. Mater.*, 2007, **19**, 2120.
23 K. Maeda, K. Teramura, D. Lu, T. Takata, N. Saito, Y. Inoue and K. Domen, *Nature*, 2006, **440**, 295.
24 T. Takata, C. Pan, M. Nakabayashi, N. Shibata and K. Domen, *J. Am. Chem. Soc.*, 2015, **137**, 9627.
25 C. Pan, T. Takata, M. Nakabayashi, T. Matsumoto, N. Shibata, Y. Ikuhara and K. Domen, *Angew. Chem., Int. Ed.*, 2015, **54**, 2955.
26 T. Hisatomi, J. Kubota and K. Domen, *Chem. Soc. Rev.*, 2014, **43**, 7520.
27 K. Maeda, *ACS Catal.*, 2013, **3**, 1486.
28 R. Niishiro, S. Tanaka and A. Kudo, *Appl. Catal., B*, 2014, **150–151**, 187.
29 H. Kato, Y. Sasaki, N. Shirakura and A. Kudo, *J. Mater. Chem. A*, 2013, **1**, 12327.
30 Y. Sasaki, A. Iwase, H. Kato and A. Kudo, *J. Catal.*, 2008, **259**, 133.
31 H. Kato, Y. Sasaki, A. Iwase and A. Kudo, *Bull. Chem. Soc. Jpn.*, 2007, **80**, 2457.
32 H. Kato, M. Hori, R. Konta, Y. Shimodaira and A. Kudo, *Chem. Lett.*, 2004, **33**, 1348.
33 K. Maeda, R. Abe and K. Domen, *J. Phys. Chem. C*, 2011, **115**, 3057.
34 R. Abe, M. Higashi and K. Domen, *ChemSusChem*, 2011, **4**, 228.
35 M. Tabata, K. Maeda, M. Higashi, D. Lu, T. Takata, R. Abe and K. Domen, *Langmuir*, 2010, **26**, 9161.

36 K. Maeda, M. Higashi, D. Lu, R. Abe and K. Domen, *J. Am. Chem. Soc.*, 2010, **132**, 5858.

37 M. Higashi, R. Abe, A. Ishikawa, T. Takata, B. Ohtani and K. Domen, *Chem. Lett.*, 2008, **37**, 138.

38 R. Abe, T. Takata, H. Sugihara and K. Domen, *Chem. Commun.*, 2005, 3829.

39 K. Sayama, K. Mukasa, R. Abe, Y. Abe and H. Arakawa, *J. Photochem. Photobiol., A*, 2002, **148**, 71.

40 Y. Sasaki, H. Kato and A. Kudo, *J. Am. Chem. Soc.*, 2013, **135**, 5441.

41 K. Tsuji, O. Tomita, M. Higashi and R. Abe, *ChemSusChem*, 2016, **9**, 2201.

42 Y. Sasaki, H. Nemoto, K. Saito and A. Kudo, *J. Phys. Chem. C*, 2009, **113**, 17536.

43 Q. Wang, T. Hisatomi, S. S. K. Ma, Y. Li and K. Domen, *Chem. Mater.*, 2014, **45**, 4144.

44 S. S. K. Ma, K. Maeda, T. Hisatomi, M. Tabata, A. Kudo and K. Domen, *Chem.–Eur. J.*, 2013, **19**, 7480.

45 A. Iwase, Y. H. Ng, Y. Ishiguro, A. Kudo and R. Amal, *J. Am. Chem. Soc.*, 2011, **133**, 11054.

46 Q. Jia, A. Iwase and A. Kudo, *Chem. Sci.*, 2014, **5**, 1513.

47 Q. Wang, T. Hisatomi, Q. Jia, H. Tokudome, M. Zhong, C. Wang, Z. Pan, T. Takata, M. Nakabayashi, N. Shibata, Y. Li, I. D. Sharp, A. Kudo, T. Yamada and K. Domen, *Nat. Mater.*, 2016, **15**, 611.

48 Q. Wang, Y. Li, T. Hisatomi, M. Nakabayashi, N. Shibata, J. Kubota and K. Domen, *J. Catal.*, 2015, **328**, 308.

49 Z. Pan, T. Hisatomi, Q. Wang, M. Nakabayashi, N. Shibata, C. Pan, T. Takata and K. Domen, *Appl. Catal., A*, 2016, **521**, 26.

50 A. Kudo, K. Omori and H. Kato, *J. Am. Chem. Soc.*, 1999, **121**, 11459.

51 T. Minegishi, N. Nishimura, J. Kubota and K. Domen, *Chem. Sci.*, 2013, **4**, 1120.

52 K. Maeda, H. Masuda and K. Domen, *Catal. Today*, 2009, **147**, 173.

53 T. Hisatomi, K. Maeda, K. Takanabe, J. Kubota and K. Domen, *J. Phys. Chem. C*, 2009, **113**, 21458.

54 A. Xiong, G. Ma, K. Maeda, T. Takata, T. Hisatomi, T. Setoyama, J. Kubota and K. Domen, *Catal. Sci. Technol.*, 2014, **4**, 325.

Faraday Discussions

PAPER

Laser-flash-photolysis-spectroscopy: a nondestructive method?

Jenny Schneider,*[a] Konstantin Nikitin,[ab] Ralf Dillert[a] and Detlef W. Bahnemann*[ab]

Received 19th September 2016, Accepted 12th October 2016
DOI: 10.1039/c6fd00193a

Herein, we report the effect of the laser illumination during the diffuse-reflectance laser-flash-photolysis measurements on the morphological and optical properties of TiO_2 powders. A grey-blue coloration of the TiO_2 nanoparticles has been observed after intense laser illumination. This is explained by the formation of nonreactive trapped electrons accompanied by the release of oxygen atoms from the TiO_2 matrix as detected by means of UV-vis and EPR spectroscopy. Moreover, in the case of the pure anatase sample a phase transition of some TiO_2 nanoparticles located in the inner region from anatase to rutile occurred. It is suggested that these structural changes in TiO_2 are caused by an energy and charge transfer to the TiO_2 lattice.

Introduction

During the last few years the demand for fundamental studies in photocatalysis became very high, especially for investigations on powdered photocatalysts due to their better comparability with the results obtained in photocatalytic tests. Laser flash photolysis spectroscopy is a widely used method to study such fundamental processes such as the formation of charge carriers and their subsequent rapid recombination and interfacial transfer kinetics.[1–3] In flash photolysis experiments the laser pulse can initiate the formation of transient species, which is usually accompanied with a change of the original optical features of the studied material. This change is spectroscopically monitored applying a flash lamp by measuring the transmittance for transparent samples or the reflectance for opaque samples, respectively, before and after the laser excitation. The optical changes induced by laser excitation in semiconductors are generally attributed to the formation of free and trapped electrons as well as of trapped holes. It is presupposed that upon illumination with the intense laser pulse no irreversible structural changes of the studied sample occur.

[a]*Leibniz University Hannover, Institute for Technical Chemistry, 30167 Hannover, Germany. E-mail: schneider@iftc.uni-hannover.de*
[b]*Saint-Petersburg State University, Laboratory "Photoactive Nanocomposite Materials", Saint-Petersburg, 198504 Russia*

TiO_2 is one of the most studied photocatalysts. In the literature different laser flash photolysis studies are presented dealing with the reaction dynamics of the charge carriers photogenerated in colloidal TiO_2 suspensions,[4–7] TiO_2 films,[8–12] and in dry TiO_2 powders.[13–16] However, most of the published transient absorption signals detected in TiO_2 do not decay to the initial value observed before the laser pulse but exhibit long-lived transients within the time scale of observation.[4,5,7,9,17,18] Serpone *et al.*[4] attributed this long-lived transient absorption to deeply trapped electrons formed *via* so-called Auger-processes. On the other hand, these irreversible changes of the optical properties might be due to irreversible alternations of the material's properties caused by changes of its stoichiometric composition or by phase transitions. However, until now no experimentally based explanation for the observed change of the optical properties of TiO_2 during the laser flash photolysis measurements has been given. Hence, we investigated the laser-induced changes of optical and morphological properties of the well-known anatase TiO_2 photocatalyst Hombikat UV100 TiO_2 by means of Raman, EPR, and UV-vis spectroscopy.

Experimental

Nanosecond diffuse reflectance laser flash photolysis spectroscopy measurements were performed using a set-up as reported in our previous paper.[19] Briefly, the excitation of the sample proceeds with an excimer laser (LPX 200) with a wavelength of 351 nm, and with a pulse duration of 15 ns. The laser intensity per pulse varied between 7 mJ cm^{-2} per pulse and 60 mJ cm^{-2} per pulse. Powder in a flat quartz cuvette has been used for all experiments. Herewith, the illumination area of the laser beam and of the analyzing light are 0.5 cm^2 and 0.196 cm^2, respectively.

Raman spectra and micrographs were recorded with a SENTERRA Raman microscope. The spectral resolution of the analyzing spectrograph was 0.5 cm^{-1}. A 532 nm laser was used as the excitation source, the power of the laser at the sample was 2.0 mW. The presented spectra were obtained at room temperature by averaging five spectra with an integration time of 10 s.

UV/vis diffuse reflectance spectra of the TiO_2 powders were recorded on a Varian Cary 100 Scan UV-vis spectrophotometer equipped with a labsphere diffuse reflectance accessory. The reflectance data were converted to $F(R)$ values according to the Kubelka–Munk theory.

EPR spectra were registered with a MiniScope MS 400 X-band EPR spectrometer at liquid nitrogen temperature. The device parameters during the measurements were as follows: microwave frequency 9.42 GHz, microwave power 5 mW, modulation frequency 100 kHz, modulation amplitude 0.15 mT.

Materials

Anatase TiO_2 Hombikat UV100 provided by Huntsman, and rutile TiO_2 provided by Crystal Global.

Results

Illumination of the TiO_2 samples studied here by intense laser pulses leads to colour changes from white to grey-blue. This effect was stable and thus no further changes were observed under ambient conditions during few months. Hence, for a better understanding of the processes occurring at the TiO_2 surface during the laser flash photolysis experiment it was important to characterize the samples before and after laser excitation. For this purpose the Hombikat UV100 powder has been analyzed by means of different spectroscopic methods presented below. To quantitatively and qualitatively understand the effect of the laser illumination the TiO_2 sample has been exposed to a higher laser intensity than that usually employed, in a typical laser flash experiment, per pulse (10×50 mJ cm^{-2} per pulse). Before its exposure to the pulsed laser irradiation the surface of the TiO_2 (UV100) sample appeared entirely white. From the micrograph of the laser illuminated TiO_2 surface, shown in Fig. 1(a), a nonhomogeneous distribution of colour centers is obvious.

Some areas exhibit dark coloration, while the majority of the sample's surface areas remained white. The Raman spectra of different sample areas are presented in Fig. 1(b). When the analyzing beam is focused on a white area, the Raman spectrum shows only anatase peaks: 149, 200, 395.5, 513 and 636.5 cm^{-1}, which can be assigned to major modes of the anatase phase as E_g, E_g, B_{1g}, and superpositions of A_{1g} and B_{1g}, and E_g, respectively.[20–23] The Raman spectrum of the dark areas exhibits additional peaks with maxima at 238, 447.5 and 610 cm^{-1}. These three bands can be attributed to the two-phonon scattering, E_g, and A_{1g} modes, of the rutile phase, respectively.[21–24] Moreover, the highest anatase peak at 149 cm^{-1}

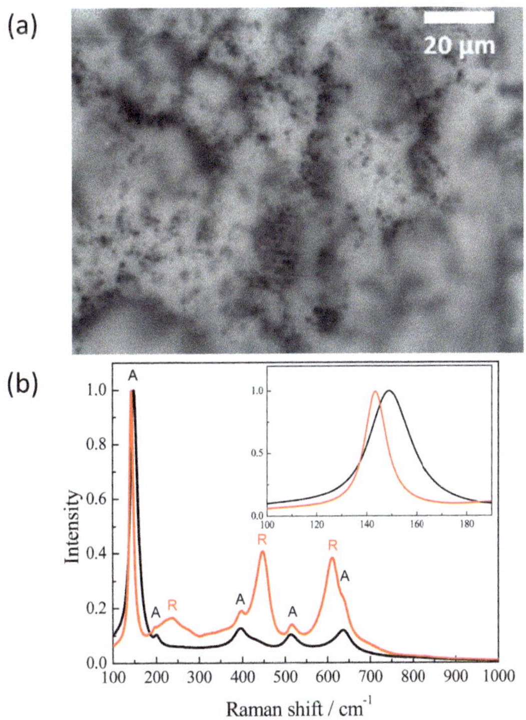

Fig. 1 (a) Micrograph of an illuminated UV100 sample surface (50× magnification). (b) Normalized Raman spectra of the illuminated UV100 sample: (black) white area and (red) dark area (exposed to the laser beam with a laser intensity of 10×50 mJ cm^{-2} per pulse).

has narrowed and shifted to 143.5 cm^{-1}. Probably, such alteration happened due to an increase in particle size,[20] while the influence of the B_{1g} mode of the rutile phase with a maximum at 143 cm^{-1} could be excluded for this interpretation because of its low line intensity.[21–23]

However, the detected anatase to rutile phase transition does not explain the dark grey-blue coloration of the UV100 TiO_2 powder after the exposure to the laser. Fig. 2(a) presents the diffuse reflectance UV-vis spectra of untreated and of illuminated TiO_2 powder, respectively. It can be seen that the absorption of the illuminated TiO_2 in the visible range increases continuously with the wavelength. This very broad visible absorption spectrum indicates the presence of long-lived Ti^{3+} species in the TiO_2 formed after the laser illumination.[3] Apparently, these trap states are energetically distributed through the entire band gap exhibiting metallic-like character. Zhu *et al.*[25] reported a wide distribution of the trapped states for electrons within the entire bandgap. The presence of the Ti^{3+} centers in the laser treated TiO_2 sample could furthermore be confirmed by means of electron paramagnetic resonance (EPR) spectroscopy. The initial uncolored sample does not show any EPR signal without UV-vis illumination, while the laser illuminated sample exhibits an intense signal as shown in Fig. 2(b). A broad peak with $g_{iso} = 1.95$ with slight asymmetry can be assigned to Ti^{3+} centers in TiO_2.[26,27]

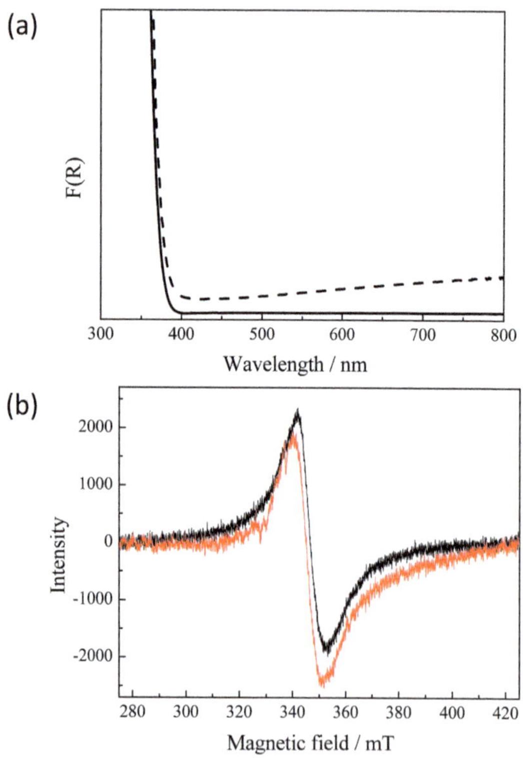

Fig. 2 (a) UV-vis diffuse reflectance spectra of (solid) untreated and (dash) illuminated UV100. (b) EPR spectra measured at 77 K of (black) illuminated UV100 TiO_2 anatase and (red) illuminated R15 TiO_2 rutile (exposed to the laser beam with laser intensity: 10 × 50 mJ cm^{-2} per pulse).

Most commonly, Ti^{3+} centers have axial symmetry and show specific signals which correspond to this anisotropy.[26–29] In our case the axial signals are not resolved because the sample temperature is still too high for this and resolved signals could only be obtained at liquid helium temperature.[26] Usually, similar EPR signals are assigned to Ti^{3+} in the rutile phase while anatase shows narrower Ti^{3+} signals with a higher *g*-value (1.98–1.99).[27–29] The formation of Ti^{3+} species in the laser-generated rutile phase could be confirmed by recording the EPR-spectra of the laser illuminated rutile phase, which exhibits the same EPR signal as laser illuminated anatase TiO_2 (see Fig. 2(b)). This also corresponds to the fact that rutile can be reduced much more easily than anatase.

Moreover, it should be noted that the Ti^{3+} centers detected here are stable in the presence of molecular oxygen as has been shown directly for Ti^{3+} centers obtained in black TiO_2 consisting of mixed anatase/rutile phases.[30] Hence, both the UV-vis spectra and the EPR results reveal that upon the laser excitation anatase/rutile TiO_2 is formed exhibiting properties similar to those of black TiO_2.

Discussion

As the Raman spectra (see Fig. 1(b)) revealed the excitation of the UV100 TiO_2 anatase powder by a 351 nm laser pulse leads to an anatase to rutile phase transition in some parts of the material. Moreover, the formation of long-lived Ti^{3+} species located in the rutile phase is detected by means of EPR- and UV-vis-spectroscopy. These results clearly prove a partial laser-induced rearrangement of the TiO_2 lattice. It is more likely that the reconstruction of the TiO_2 structure occurs at the particle surfaces rather than in the bulk, since the former exhibits a higher defect density and thus a higher reactivity.[31] However, the question arises whether the absorbed laser energy is invested into a temperature rise followed by a thermal anatase to rutile transition or whether it is directly transferred to the lattice inducing a bond breakage on the molecular level.

The laser-induced anatase to rutile phase transition has been already studied by several research groups, whereas some of them reported photoinduced thermal driven phase transitions.[32,33] Wilkinson and Willsher pointed out that materials exhibiting large optical absorption coefficients such as TiO_2 or Fe_3O_4 (*i.e.*, with $\alpha > 10^4$ cm^{-1}) are likely to experience photoinduced thermal heating.[34] Hereby the heat is homogeneously distributed over the illuminated volume resulting in a large temperature rise following the pulsed laser excitation. The temperature induced in TiO_2 by a single laser pulse can be estimated according to the following equation:[34]

$$Q_0 = \frac{E_0 \alpha}{\rho C} \tag{1}$$

Q_0: temperature (K), E_0: incident laser intensity per unit area (J cm^{-2} per pulse), C: specific heat capacity (J g^{-1} K^{-1}), ρ: density (g cm^{-3}), α: absorption coefficient (cm^{-1}).

Accordingly, the required temperature for the anatase to rutile phase transition of 680 °C can be achieved with a laser intensity of 40 mJ cm^{-2} per pulse (using values for the anatase phase), *i.e.*, $\alpha = 1.2 \times 10^4$ cm^{-1}, $C = 0.69$ J g^{-1} K^{-1}, and $\rho = 1$ g cm^{-3} (lower density value than that reported for bulk TiO_2 (3.9 g cm^{-3}) due to the powdered samples used in the experiments).[35]

However, such a photoinduced thermal heating of the illuminated volume followed by a phase transition is less probable. It contradicts the studies of Stopper and Dohrmann, who demonstrated by means of time-resolved opto-acoustic calorimetry that 88% of the released heat is dissipated in a few nano-seconds over the whole system, thus no significant temperature increase occurs.[36] Recently, Ricci *et al.*[37,38] have shown by careful analysis of the Stokes to anti-Stokes Raman peak ratio that the local temperature of the nanoparticles during the laser illumination was 370 K, thus a solely thermally driven anatase to rutile phase transition could be ruled out. Moreover, Yates *et al.*[39] have found that the photoinduced reconstruction of the TiO_2 surface was not affected by cooling the surface below 150 K.

In fact, the photoinduced desorption/adsorption of oxygen molecules on the surface of nanometer sized TiO_2 particles is reported to play a fundamental role for the non-thermal phase transition induced upon laser illumination.[37,38,40,41] Hereby, the phase transition was attributed to surface modifications resulting from a proper depletion of adsorbed oxygen on the crystal surface. The desorption of oxygen molecules as well as atoms from the TiO_2 lattice yielding oxygen vacancies leads to an enhancement of the surface chemical reactivity. Subsequently, stable chemical bonds are formed between the surfaces of neighboring anatase particles resulting in the formation of polycrystalline aggregates. The interface between the bound anatase particles, which are only present in the inner region of these agglomerated particles, provides the nucleation sites of the rutile phase.[42] The formation of the rutile phase inside of such aggregates is in agreement with our results. We, furthermore, detected long-lived trapped electrons in the rutile phase, the lower reactivity of which evinces that they are located inside the TiO_2 aggregate or cluster. The appearance of such non-reactive subsurface trapped electrons by tempering has previously been reported by Diebold *et al.*[43] The relative number of preferential nucleation sites for the anatase to rutile phase transition is found to increase as the population of oxygen vacancies at the particle surface increases.[37,38] Zhou *et al.*[31,44] reported that the rutile phase nucleates exactly at the twin boundaries generated by anatase (112) surfaces. If the direct contact between anatase particles is prevented, the phase transformation was found to be retarded or prohibited at all. For example, no phase transformation has been observed for anatase TiO_2 particles functionalized with nitric groups or with a La_2O_3 shell.[37,42]

The long-lived Ti^{3+} centers in TiO_2 detected in the present study are most likely produced by oxygen removal during the laser illumination of anatase TiO_2 (this has also been detected for the rutile phase). Desorption of one oxygen molecule leaves four additional electrons in the lattice. These electrons can either stay in the vacancy forming so-called *F*-centers, or they are transferred to Ti^{4+} yielding Ti^{3+} species. The conversion of Ti^{4+} to Ti^{3+} upon laser excitation at 355 nm has already been reported by Forsgren *et al.*[45] Concomitantly, the hydrophilicity of the surface was improved. Since we were able to detect such Ti^{3+} centers by means of EPR and UV-vis spectroscopy (see Fig. 2), this evinces that the phase transition observed in the present study occurs *via* a photoinduced oxygen desorption. However, the question arises, how the formation of the oxygen vacancies or of the rather long-lived Ti^{3+} species proceeds.

For example, Serpone *et al.*[4,46] reported that during laser flash photolysis studies long-lived Ti^{3+} centers can be produced *via* the so-called Auger recombination process. This Auger process is found to be the predominant mechanism when the number of photons exceeds the number of absorbing particles by far, which is in agreement with our experimental conditions.[47,48] In the Auger process the energy released by the recombination of an electron–hole pair is absorbed by another electron (or hole) resulting in the formation of highly energetic electrons (or holes), which can dissipate the energy either by an electron ejection from the semiconductor nanoparticle resulting in the formation of solvated electrons or by a phonon emission leading to the generation of long-lived deeply trapped electrons. The presence of solvated electrons can be ruled out here since the experiments have been performed employing dry TiO_2 powders instead of aqueous TiO_2 suspensions. However, the only experimental evidence for the formation of such Ti^{3+} species *via* an Auger process is the fact that the published transient absorption signals detected in TiO_2 do not decay to the initial value observed before the laser pulse but exhibit a long-lived transient absorption signal within the time scale of observation.[4,5,7,9,17,18,49] Moreover, the Auger recombination is normally observed in heavily doped, direct bandgap semiconductors. In direct bandgap semiconductors the recombination proceeds upon releasing photons, while in indirect semiconductors in most cases phonons are released, the energy of which is usually not high enough to excite other charge carriers.[50]

Lisachenko *et al.*[51] proposed two possible processes to explain for the photodesorption of oxygen molecules from the metal oxide surface:

(1) Photothermic processes, whereby the excitation energy is transformed into phonon energy of the solid resulting in the excitation of the vibrational mode of the surface–atom bond followed by the rupture of this bond.

(2) Photoelectronic processes, whereby the excitation energy is absorbed by the atom–surface bond thus changing the potential energy of the bond directing it towards the repulsive branch of the potential curve for the atom–surface interaction. Subsequently, the potential energy is converted into kinetic energy of the emitted particles, thus entailing the nature of the oxygen desorption. This mechanism is based on the research of Terenin, who pointed out that the emitted particle can gain the rest of the photon energy, provided that the latter exceeds the energy necessary for bond breaking.[52]

Based on Time-of-Flight (TOF) spectroscopy measurements, which allow the determination of the temperature of the desorbed molecules, Lisachenko ruled out the photothermic model.[51] In photothermic processes the average velocity of the desorbed molecules should correspond to the surface temperature. The detected kinetic temperature of the desorbed oxygen molecules, however, exceeds by far the highest surface temperature produced by the laser beam. The photodesorption of oxygen from the TiO_2 surface is therefore rather found to proceed *via* photoelectronic processes induced by electronic transitions. Hereby, the photodesorption of preadsorbed oxygen molecules (which are present on the metal oxide surface is a partially reduced form, *i.e.*, as superoxide radical anions) is caused by the capture of a photogenerated hole:[53–55]

$$h_{VB}^{+} + O_2^{-}{}_{(ads)} \rightarrow O_2(g)\uparrow \quad (2)$$

The desorption of the oxygen molecule from the TiO_2 surface lattice includes the partial neutralization of the oxygen anion:[39]

$$O^{2-} + h^+ \rightarrow O^- \tag{3}$$

The oxidation of further oxygen anions results in the transformation of two neighboring O^{2-} anions into molecular oxygen. Accordingly, two single-photon excitations are required for the formation of one oxygen vacancy:

$$2h\nu + 2O_s^- \rightarrow O_2\uparrow + 2e^- \tag{4}$$

The structural changes of the TiO_2 lattice observed in the present study can only be induced by the removal of strongly bound species. The photooxidation of lattice bound oxygen *via* photogenerated holes resulting in the release of molecular oxygen has been proposed as a mechanism for O_2 formation by Salvador[56] and for the gas phase photocatalytic oxidation of organic compounds by Pichat[57,58] and Lisachenko.[59]

The idea behind the mechanism of photoelectronic processes including the excitation energy transfer to the surface–atom bond resulting in the formation of an oxygen vacancy is also supported by other experimental techniques such as ion and electron bombardment, where the kinetic energy of the accelerated ions is transferred to the TiO_2 lattice.[60,61] Actually, Fukushima *et al.*[62] observed the anatase to rutile transition during the TiO_2 film formation by means of ion bombardment. Moreover, the laser induced phase transition of Silicon has been explained by energy transfer from an optically excited electron–hole plasma to the crystal lattice resulting in the crystal melting. This process is known as non-thermal melting.[63–65] However, the latter occurs only upon femtosecond excitation at a laser fluency of about 0.2 J cm^{-2}.

According to the above presented discussion, in the photoinduced processes studied here the absorbed light energy is transferred to the lattice inducing bond-breaking and displacement of atoms involved in the reconstructive transformation from anatase to rutile. According to Stopper and Dohrmann, 88% of the energy absorbed by colloidal TiO_2 particles after the laser pulse is dissipated in less than one nanosecond as heat over the whole system as a consequence of the charge-carrier recombination, while the remaining 12% can be stored for at least 2 μs.[36] Applying a laser pulse with an excitation wavelength of 351 nm, a total energy of 347 kJ mol^{-1} can be released as heat. According to Stopper and Dohrmann, 305 kJ mol^{-1} will be dissipated over the whole particulate system, while 42 kJ mol^{-1} can be utilized locally. Employing a laser intensity of 23 mJ cm^{-2} per pulse, one UV100 anatase nanoparticle can absorb 522 photons as estimated according to eqn (5):

$$N_p = \frac{E_0\lambda_{exc}1/6\pi d^3\alpha}{ch} \tag{5}$$

c: speed of light (m s^{-1}), *h*: Planck's constant (J s).

Assuming that every absorbed photon is utilized as much as 21 924 kJ mol^{-1} can be locally converted into bond breaking. For the cleavage of one Ti–O bond an energy of 315 kJ mol^{-1} is required. Hence, with one laser pulse of 23 mJ cm^{-2}, up to 70 Ti–O bonds per TiO_2 anatase particle can be broken. For the transition of bulk anatase to rutile it was reported that 7 out of the 24 Ti–O bonds per unit cell need to be broken leading to the cooperative displacement of both Ti and O.[37,41] It is more likely, however, that most of the Ti–O bonds are broken at the TiO_2

surface. The anatase nanoparticles of UV100 are spherical in shape, therefore the number of TiO_2 unit cells present at the surface is estimated to be 180. Hence, during each laser pulse 3 Ti–O bonds per unit cell can be broken. Subsequently, after the second laser pulse, the phase transition can be initiated.

The inhomogeneous distribution of the black spots containing the rutile phase and long-lived Ti^{3+}-centers (see Fig. 1(a)) observed here for the laser irradiated powder samples indicates different reactivities of the TiO_2 clusters at the surface. Assuming a homogeneous distribution of the photons over the illuminated TiO_2 surface all exposed TiO_2 units should be destabilized during the laser pulse. As a result bond breaking and atom displacement can occur. After the laser illumination the more stable TiO_2 units manage to return back to the initial state (this self-repair mechanism can also be supported by photo adsorption of the oxygen or water molecules), thus the initial anatase phase remains (see Fig. 1(a)). However, the more reactive TiO_2 clusters (most likely due to a higher defect density) undergo intense irreversible structural changes such as oxygen removal followed by the observed phase transition. The latter results in the reported grey-blue coloration of the TiO_2 sample. As a matter of fact, the photodeposition of metal nanoparticles on metal oxide surfaces also results in an inhomogeneous distribution of metal islands on the surface.[66]

Conclusions

In the present study the irreversible changes of TiO_2 during the diffuse reflectance laser flash photolysis experiments have been experimentally analyzed. The following laser-induced processes have been suggested to occur in anatase TiO_2 powder upon illumination with intense laser pulses: formation of electron–hole pairs, followed by energy and charge transfer to the TiO_2 lattice. The regions of the TiO_2 surface with high defect density undergo irreversible changes, such as removal of the lattice oxygen resulting in the formation of oxygen vacancies. This enhances the surface reactivity of the TiO_2 particles, thus at the interfaces of the anatase nanoparticles a phase transition to rutile occurs. The presence of Ti^{3+} centers in the illuminated powder evinces that the phase transition to rutile occurs *via* an oxygen release from the TiO_2 surface induced by the electron/energy transfer. However, the major part of the TiO_2 nanoparticles are able to conduct self-repair processes and therefore do not exhibit any significant morphological and structural changes.

These findings clearly demonstrate that the obtained transient absorption signals could be strongly influenced by such irreversible changes of the material and need to be treated with caution. This will be the topic of subsequent publications.

Furthermore, the message of the study presented here so far is that upon illumination with intense pulsed laser light structural changes of the TiO_2 photocatalyst may occur. These findings are essential for the evaluation of the underlying photocatalytic reactions. Although there are many studies which present such structural changes of the TiO_2 nanoparticles upon illumination,[32,57,58,67–71] this effect has rarely been considered for the interpretation of the photoinduced photocatalytic processes.[72] Based upon these possible reactions it is certainly highly advisable to study the role of the surface reorganization on the reactions induced upon illumination. The knowledge of these processes can most probably open new design pathways for future photocatalysts.

Acknowledgements

This work was funded in part by the German Federal Ministry of Education and Research (contract no. 13N13350, "PureBau – Untersuchung von Werkstoffsystemen für photokatalytisch hocheffiziente Baustoffe-Teilvorhaben: Oberflächenchemie der Photokatalysatoren und der Werkstoffe") and by the Project "Establishment of the Laboratory Photoactive Nanocomposite Materials" no. 14.Z50.31.0016 supported by a Mega-grant of the Government of the Russian Federation. This work was supported by the Global Research Laboratory (GRL) Programme (NRF-2014K1A1A2041044) funded by the Korea government (MSIP) through NSF. The authors thank Jinlin Nie for providing the EPR spectra.

References

1 P. V. Kamat, *Prog. React. Kinet.*, 1994, **19**, 277.
2 H. H. Mohamed, R. Dillert and D. W. Bahnemann, *Chem.–Eur. J.*, 2012, **18**, 4314.
3 J. Schneider, M. Matsuoka, M. Takeuchi, J. Zhang, Y. Horiuchi, M. Anpo and D. W. Bahnemann, *Chem. Rev.*, 2014, **114**, 9919.
4 N. Serpone, D. Lawless, R. Khairutdinov and E. Pelizzetti, *J. Phys. Chem.*, 1995, **99**, 16655.
5 D. Bahnemann, A. Henglein, J. Lilie and L. Spanhel, *J. Phys. Chem.*, 1984, **88**, 709.
6 M. Grätzel and A. J. Frank, *J. Phys. Chem.*, 1982, **86**, 2964.
7 I. A. Shkrob and M. C. Sauer, *J. Phys. Chem. B*, 2004, **108**, 12497.
8 S. A. Haque, Y. Tachibana, R. L. Willis, J. E. Moser, M. Grätzel, D. R. Klug and J. R. Durrant, *J. Phys. Chem. B*, 2000, **104**, 538.
9 T. Yoshihara, R. Katoh, A. Furube, Y. Tamaki, M. Murai, K. Hara, S. Murata, H. Arakawa and M. Tachiya, *J. Phys. Chem. B*, 2004, **108**, 3817.
10 T. Yoshihara, Y. Tamaki, A. Furube, M. Murai, K. Hara and R. Katoh, *Chem. Phys. Lett.*, 2007, **438**, 268.
11 Y. Tamaki, A. Furube, R. Katoh, M. Murai, K. Hara, H. Arakawa and M. Tachiya, *C. R. Chim.*, 2006, **9**, 268.
12 Y. Tamaki, K. Hara, R. Katoh, M. Tachiya and A. Furube, *J. Phys. Chem. C*, 2009, **113**, 11741.
13 C. J. Willsher, *J. Photochem.*, 1985, **28**, 229.
14 R. W. Kessler, G. Krabichler, S. Uhl, D. Oelkrug, W. P. Hagan, J. Hyslop and F. Wilkinson, *Opt. Acta Int. J. Opt.*, 1983, **30**, 1099.
15 A. Yamakata, J. J. M. Vequizo and H. Matsunaga, *J. Phys. Chem. C*, 2015, **119**, 24538.
16 A. Kafizas, X. Wang, S. R. Pendlebury, P. Barnes, M. Ling, C. Sotelo-Vazquez, R. Quesada-Cabrera, C. Li, I. P. Parkin and J. R. Durrant, *J. Phys. Chem. A*, 2016, **120**, 715.
17 G. Rothenberger, J. Moser, M. Gratzel, N. Serpone and D. K. Sharma, *J. Am. Chem. Soc.*, 1985, **107**, 8054.
18 D. W. Bahnemann, M. Hilgendorff and R. Memming, *J. Phys. Chem. B*, 1997, **101**, 4265.
19 J. Schneider, K. Nikitin, M. Wark, D. W. Bahnemann and R. Marschall, *Phys. Chem. Chem. Phys.*, 2016, **18**, 10719.

20 W. F. Zhang, Y. L. He, M. S. Zhang, Z. Yin and Q. Chen, *J. Phys. D: Appl. Phys.*, 2000, **33**, 912.

21 M.-S. Zhang, Z. Yin, Q. Chen, W. Xijun and J. Xiaoli, *Ferroelectrics*, 1995, **168**, 131.

22 J. Parker and R. Siegel, *J. Mater. Res.*, 1990, **5**, 1246.

23 U. Balachandran and N. Eror, *J. Solid State Chem.*, 1982, **42**, 276.

24 J. Parker and R. Siegel, *Appl. Phys. Lett.*, 1990, **57**, 943.

25 M. Zhu, Y. Mi, G. Zhu, D. Li, Y. Wang and Y. Weng, *J. Phys. Chem. C*, 2013, **117**, 18863.

26 V. Khomenko, K. Langer, H. Rager and A. Fett, *Phys. Chem. Miner.*, 1998, **25**, 338.

27 R. F. Howe and M. Grätzel, *J. Phys. Chem.*, 1985, **89**, 4495.

28 T. Berger, M. Sterrer, O. Diwald and E. Knözinger, *ChemPhysChem*, 2005, **6**, 2104.

29 J. B. Priebe, J. R. Radnik, A. J. Lennox, M.-M. Pohl, M. Karnahl, D. Hollmann, K. Grabow, U. Bentrup, H. Junge and M. Beller, *ACS Catal.*, 2015, **5**, 2137.

30 A. Naldoni, M. Allieta, S. Santangelo, M. Marelli, F. Fabbri, S. Cappelli, C. L. Bianchi, R. Psaro and V. Dal Santo, *J. Am. Chem. Soc.*, 2012, **134**, 7600.

31 Y. Zhou and K. A. Fichthorn, *J. Phys. Chem. C*, 2012, **116**, 8314.

32 H. Y. Lee, W. L. Lan, T. Y. Tseng, D. Hsu, Y. M. Chang and J. G. Lin, *Nanotechnology*, 2009, **20**, 315702.

33 H. L. Ma, J. Y. Yang, Y. Dai, Y. B. Zhang, B. Lu and G. H. Ma, *Appl. Surf. Sci.*, 2007, **253**, 7497.

34 F. Wilkinson, C. J. Willsher, S. Uhl, W. Honnen and D. Oelkrug, *J. Photochem.*, 1986, **33**, 273.

35 D. A. H. Hanaor and C. C. Sorrell, *J. Mater. Sci.*, 2011, **46**, 855.

36 K. Stopper and J. K. Dohrmann, *Z. Phys. Chem.*, 2000, **214**, 555.

37 P. C. Ricci, C. M. Carbonaro, L. Stagi, M. Salis, A. Casu, S. Enzo and F. Delogu, *J. Phys. Chem. C*, 2013, **117**, 7850.

38 P. C. Ricci, A. Casu, M. Salis, R. Corpino and A. Anedda, *J. Phys. Chem. C*, 2010, **114**, 14441.

39 S. Mezhenny, P. Maksymovych, T. L. Thompson, O. Diwald, D. Stahl, S. D. Walck and J. T. Yates Jr, *Chem. Phys. Lett.*, 2003, **369**, 152.

40 L. Stagi, C. M. Carbonaro, R. Corpino, D. Chiriu and P. C. Ricci, *Phys. Status Solidi B*, 2015, **252**, 124.

41 G. C. Vasquez, M. A. Peche-Herrero, D. Maestre, A. Gianoncelli, J. Ramarez-Castellanos, A. Cremades, J. M. Gonzalez-Calbet and J. Piqueras, *J. Phys. Chem. C*, 2015, **119**, 11965.

42 J. Zhang, M. Li, Z. Feng, J. Chen and C. Li, *J. Phys. Chem. B*, 2006, **110**, 927.

43 U. Diebold, J. Lehman, T. Mahmoud, M. Kuhn, G. Leonardelli, W. Hebenstreit, M. Schmid and P. Varga, *Surf. Sci.*, 1998, **411**, 137.

44 R. L. Penn and J. F. Banfield, *Am. Mineral.*, 1999, **84**, 871.

45 J. Forsgren, M. D. Paz, B. Leon and H. Engqvist, *J. Mater. Sci.: Mater. Med.*, 2013, **24**, 11.

46 N. Serpone, D. Lawless and R. Khairutdinov, *J. Phys. Chem.*, 1995, **99**, 16646.

47 A. W. Henglein and H. Weller, *Photochemical energy conversion*, ed. J. R. Norris and D. Meisel, Elsevier, New York, 1989, p. 161.

48 A. K. Ghosh, F. G. Wakim and R. R. Addiss, *Phys. Rev.*, 1969, **184**, 979.

49 A. Y. Ahmed, T. A. Kandiel, T. Oekermann and D. Bahnemann, *J. Phys. Chem. Lett.*, 2011, **2**, 2461.

50 F. J. Knorr, C. C. Mercado and J. L. McHale, *J. Phys. Chem. C*, 2008, **112**, 12786.
51 A. A. Lisachenko, *J. Photochem. Photobiol., A*, 2008, **196**, 127.
52 A. N. Terenin, *Uchenye zapiski LGU (rus)*, 1939, **5**, 26.
53 Z. Zhang and J. T. Yates, *J. Phys. Chem. C*, 2010, **114**, 3098.
54 J. T. Yates, *Surf. Sci.*, 2009, **603**, 1605.
55 T. L. Thompson and J. T. Yates, *J. Phys. Chem. B*, 2005, **109**, 18230.
56 P. Salvador, *Prog. Surf. Sci.*, 2011, **86**, 41.
57 H. Courbon, M. Formenti and P. Pichat, *J. Phys. Chem.*, 1977, **81**, 550.
58 H. Courbon and P. Pichat, *C. R. Seances Acad. Sci., Ser. C*, 1977, **285**, 171.
59 V. V. Titov, R. V. Milchaylov and A. A. Lisachenko, *J. Phys. Chem. C*, 2014, **118**, 21986.
60 J.-M. Pan, B. L. Maschhoff, U. Diebold and T. E. Madey, *J. Vac. Sci. Technol., A*, 1992, **10**, 2470.
61 P. J. F. M. L. Knotek, *Phys. Rev. Lett.*, 1978, **40**, 964.
62 I. W. Boyd, E. Rimini, K. Fukushima and I. Yamada, *Appl. Surf. Sci.*, 1989, **43**, 32.
63 C. V. Shank, R. Yen and C. Hirlimann, *Phys. Rev. Lett.*, 1983, **50**, 454.
64 P. Stampfli and K. Bennemann, *Phys. Rev. B: Condens. Matter Mater. Phys.*, 1994, **49**, 7299.
65 T. Zier, E. S. Zijlstra, A. Kalitsov, I. Theodonis and M. E. Garcia, *Struct. Dyn.*, 2015, **2**, 054101.
66 D. Friedmann, H. Hansing and D. Bahnemann, *Z. Phys. Chem.*, 2007, **221**, 329.
67 S. Sato, T. Kadowaki and K. Yamaguti, *J. Phys. Chem.*, 1984, **88**, 2930.
68 A. L. A. Parussulo, M. F. G. Huila, K. Araki and H. E. Toma, *Langmuir*, 2011, **27**, 9094.
69 C. Y. Wang, R. Pagel, J. K. Dohrmann and D. W. Bahnemann, *C. R. Chim.*, 2006, **9**, 761.
70 J. F. Montoya, D. W. Bahnemann, J. Peral and P. Salvador, *ChemPhysChem*, 2014, **15**, 2311.
71 J. F. Montoya, I. Ivanova, R. Dillert, D. W. Bahnemann, P. Salvador and J. Peral, *J. Phys. Chem. Lett.*, 2013, **4**, 1415.
72 C. B. Mendive, D. Hansmann, T. Bredow and D. Bahnemann, *J. Phys. Chem. C*, 2011, **115**, 19676.

Faraday Discussions

PAPER

In situ spectroscopic monitoring of CO_2 reduction at copper oxide electrode

Liying Wang, Kalyani Gupta, Josephine B. M. Goodall, Jawwad A. Darr and Katherine B. Holt*

Received 1st September 2016, Accepted 21st September 2016
DOI: 10.1039/c6fd00183a

Copper oxide modified electrodes were investigated as a function of applied electrode potential using *in situ* infrared spectroscopy and *ex situ* Raman and X-ray photoelectron spectroscopy. In deoxygenated $KHCO_3$ electrolyte bicarbonate and carbonate species were found to adsorb to the electrode during reduction and the CuO was reduced to Cu(I) or Cu(0) species. Carbonate was incorporated into the structure and the CuO starting material was not regenerated on cycling to positive potentials. In contrast, in CO_2 saturated $KHCO_3$ solution, surface adsorption of bicarbonate and carbonate was not observed and adsorption of a carbonato-species was observed with *in situ* infrared spectroscopy. This species is believed to be activated, bent CO_2. On cycling to negative potentials, larger reduction currents were observed in the presence of CO_2; however, less of the charge could be attributed to the reduction of CuO. In the presence of CO_2 CuO underwent reduction to Cu_2O and potentially Cu, with no incorporation of carbonate. Under these conditions the CuO starting material could be regenerated by cycling to positive potentials.

1 Introduction

The conversion of carbon dioxide into chemical feedstock molecules and fuels is an attractive approach to addressing carbon capture and developing sustainable energy sources.[1] Electrochemical CO_2 reduction has been well-studied over many decades, and while still far from viable industrial application, improvements in practical and theoretical understanding mean significant progress is still being made towards the technological goal.[2] All of the readily available metals,[3] many metal alloys,[4] oxides[5] and sulphides[6] have been investigated as electrodes for CO_2 reduction. The most promising and widely studied material is copper, as it is the only metal to produce hydrocarbons and appreciable quantities of $>C_1$ products during CO_2 reduction.[7] However, problems persist with poor faradaic efficiency (due to the competing H_2 evolution reaction), poor product selectivity and loss of activity over time.

Department of Chemistry, University College London, 20 Gordon St, London, WC1H 0AJ, UK. E-mail: k.b.holt@ucl.ac.uk

The use of copper oxide electrodes in this application was first reported in 1991,[8] but the field has undergone a massive resurgence over the past five years.[9–18] It has been noted by many researchers that copper electrode surface morphology and nanostructure and the presence of surface oxides plays a large role in determining product distribution and efficiency.[9–18] For this reason, the role of copper oxides and their reduction products in CO_2 reduction has been studied recently using a wide range of techniques, including X-ray diffraction,[9–12] Auger,[8,13,14] X-ray photoelectron spectroscopy (XPS),[10,11,13] temperature programmed desorption,[15,16] and X-ray absorption spectroscopy.[14,15,17] Relatedly, semiconducting copper oxides are also proposed as promising photocatalysts for CO_2 photo-reduction, hence their performance under reducing conditions is of interest.[19]

The exact nature of the active catalyst when copper oxide electrodes are used is still under question. This is because CuO undergoes reduction at the potentials employed for CO_2 reduction and hence is reduced to Cu(0). The prevailing opinion at present is that metallic Cu is the catalyst under reaction conditions, regardless of the starting material.[9–12,16–18] The most effective catalysts (in terms of overpotential, faradaic efficiency, selectivity and ability to produce higher hydrocarbons) are nanostructured copper electrodes formed from reduction of oxide-derived copper (OD-Cu).[10] These electrodes are formed through annealing of copper in air to form thick Cu_2O films and are proposed to retain metastable Cu(0) sites at grain boundaries when electrochemically reduced.[12] It is these Cu(0) sites, rather than oxides or Cu(I) sites that provide the extraordinary activity towards CO_2 and CO reduction.[10,12,16] Other studies, employing *in situ* X-ray absorption spectroscopy to monitor the Cu K-edge of different Cu(I) and Cu(II) catalyst precursors during CO_2 reduction, have likewise suggested that only metallic copper is the catalytic species and that the oxidation state of the precursor is unimportant.[17]

In contrast, a number of studies using OD-Cu electrodes for CO_2 reduction claim that Cu(I) sites and the oxygen centres of copper oxide surface structures provide CO_2 binding sites and tailored reactivity that can influence overpotential and product distribution.[8,14,15] It has been proposed that Cu_2O can promote adsorbed CO_2 activation by donating electron density,[8] that Cu(I) sites within its structure promote stronger binding to CO and are present throughout the catalytic process[15] and production of methanol as a reduction product is proportional to Cu(I) content.[14] Computational density functional theory studies of Cu_2O and CuO show that copper oxide surfaces do provide alternate and feasible binding sites for CO_2, that would lead to activation of the CO_2 molecule and a different product distribution to that seen on metallic copper.[20,21]

In this study we use cyclic voltammetry (CV), *ex situ* XPS and Raman spectroscopy and *in situ* IR spectroscopy to gain more insight into the interaction of CO_2 with CuO electrodes during electrochemical reduction. Although extensive catalytic studies have been carried out at OD-Cu, Cu_2O and CuO electrodes, most have focussed on product distribution analysis, along with some before and after electrode characterisation. Relatively few *in situ* measurements of the catalyst material have been reported.[15,17] Here we study the initial interaction of the CO_2 with the catalyst surface at low overpotentials during the first potential cycle, to determine how CO_2 binds to the CuO surface. Additionally, we determine how the presence of CO_2 in solution influences the reactivity of the CuO material and in particular how the oxidation state of the copper centres changes during potential cycling.

2 Experimental materials and procedures

2.1 CuO synthesis and characterisation

Copper(II) oxide nanoparticles were synthesised by continuous hydrothermal flow synthesis and characterised by powder X-Ray Diffraction (XRD) and Transmission Electron Microscopy (TEM).[22] The material was confirmed as CuO (tenorite) by XRD and the average particle diameter by both XRD and TEM is approximately 15 nm.

2.2 Experimental procedures

2.2.1 Cyclic voltammetry (CV). A working electrode of boron-doped diamond (BDD) modified with a layer of CuO was used, along with a counter electrode of platinum and a Ag/AgCl reference electrode. The working electrode was constructed by drop-coating CuO suspended in isopropanol onto the BDD electrode and allowing the solvent to evaporate. Aqueous electrolytes were prepared using Millipore deionised water (18 Mohm cm) and 0.5 M $KHCO_3$. Solutions were deoxygenated by bubbling through high-purity argon for 20 minutes and then maintaining a flow of argon above the electrochemical cell during measurements. Solutions were CO_2 saturated after deoxygenation with argon, by bubbling high purity CO_2 that had been through two washing stages to remove organic contaminants. All experiments were repeated three times with freshly prepared electrodes to ensure reproducibility.

2.2.2 X-ray photoelectron spectroscopy (XPS). Electrodes suitable for subsequent XPS analysis were prepared using glassy carbon rods of length *ca.* 10 cm and diameter 3 mm (Sigma Aldrich). CuO was coated onto the rods by drop-coating from isopropanol. The CuO-modified rods were stored in a vacuum desiccator overnight to ensure the removal of all of the isopropanol. Electrodes were held at different potentials in the $KHCO_3$ electrolyte, or cycled (see text), removed from the solution, dried under an argon stream and stored in glass vials before XPS analysis. XPS measurements were carried out using a Thermo Scientific K-Alpha spectrometer with Al K-alpha X-rays (1486.6 eV) and a 400 μm spot size. Three random spots were analysed per rod to ensure reproducibility. XPS spectra were peak fit using CasaXPS software with a Shirley background, making use of standard XPS databases and other literature.[23] The relative composition of the surface species was calculated from the peak areas and the relevant sensitivity factors.

2.2.3 Raman spectroscopy. CuO-modified carbon rods were prepared in an identical manner to that described above for XPS and stored in preparation for Raman analysis. A Renishaw inVia Raman Microscope with 514.5 nm laser was used, with spectra obtained from at least three random positions on the rod to ensure reproducibility. A background spectrum of the carbon rod without the CuO coating was featureless in the 100–800 cm^{-1} region shown in this study.

2.2.4 *In situ* attenuated total reflectance infrared (ATR IR) spectroscopy. An electrochemical cell was constructed above the diamond prism of the internal reflectance element of a Bruker Tensor ATR IR spectrometer. The working electrode was a BDD electrode modified with CuO, as described in Section 2.2.1. The working electrode was placed directly above the prism, with an electrode–prism spacing of approximately 15 μm. The CuO coating is in contact with both the electrode and the prism, but as the evanescent wave of the incident IR penetrates at most 2 μm above the prism, only the CuO furthest from the electrode surface is probed by the IR. Different potentials were applied to the working electrode in 0.5 M $KHCO_3$ with and without CO_2, and the IR spectra were measured at the different potentials relative to a background. Experimental conditions are described further in the text.

3 Results and discussion

3.1 Stability of CuO under potential cycling in 0.5 M $KHCO_3$ in the absence of CO_2

3.1.1 Cyclic voltammetry of CuO-modified electrode in deoxygenated $KHCO_3$. The CV of a freshly prepared CuO-modified electrode was obtained by cycling from −0.4 V to −1.2 V, then to 0.4 V and back to −0.4 V in deoxygenated 0.5 M $KHCO_3$ solution, as shown in Fig. 1. Initially currents are close to zero, indicating the presence of a passivating layer on the CuO surface, however reduction currents begin to flow at *ca.* −0.7 V (I). Reduction currents increase in magnitude negative of −1.0 V (II) and bubbles appear on the electrode surface, indicating that hydrogen evolution is taking place at this potential. No oxidation processes are noted on the reverse scan until *ca.* −0.2 V when the first of two overlapping oxidation peaks (III, IV) are observed. On cycling back to −0.4 V a reduction peak is seen at *ca.* −0.25 V (V). Previous studies[24] have assigned (I) as the reduction of Cu(II) oxide to Cu(I) oxide or Cu(0); (II) as proton reduction; (III) and (IV) as oxidation of Cu(I) or Cu(0) species to Cu(II) oxides or hydroxides and (V) as reduction of the Cu(II) products of (III) and (IV) to Cu(I) oxide. To aid with the assignment of these processes Raman spectroscopy and XPS of the CuO electrodes were carried out after application of different potentials.

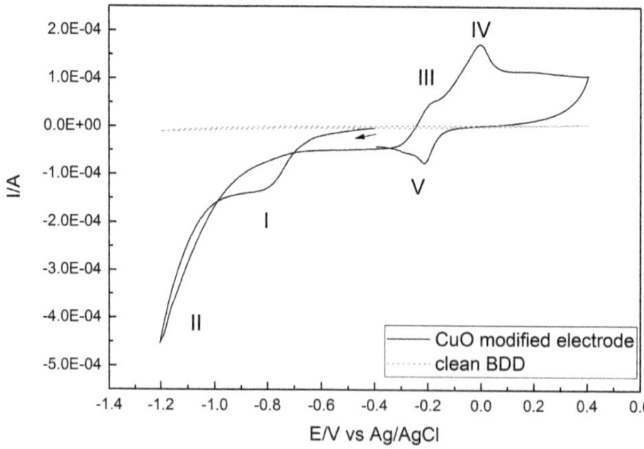

Fig. 1 CV of bare BDD (orange) and first scan of freshly CuO-modified BDD (black) electrode in Ar deoxygenated 0.5 M $KHCO_3$ electrolyte, scan rate 100 mV s^{-1}. Peak labels I–V as defined in text.

3.1.2 X-ray photoelectron spectroscopy of CuO electrodes after application of different potentials. CuO-coated carbon rod electrodes were held at potentials of −0.4 V, −0.8 V or −1.2 V for 15 min in 0.5 M $KHCO_3$ deoxygenated solution and their composition analysed *ex situ* using XPS. High resolution XPS Cu 2p spectra for the electrodes are shown in Fig. 2. The spectrum obtained for the electrode held at −0.4 V (A) is identical to that for the electrode before potential was applied (not shown). Both unused and −0.4 V treated electrodes show a Cu(II) $2p_{3/2}$ peak at

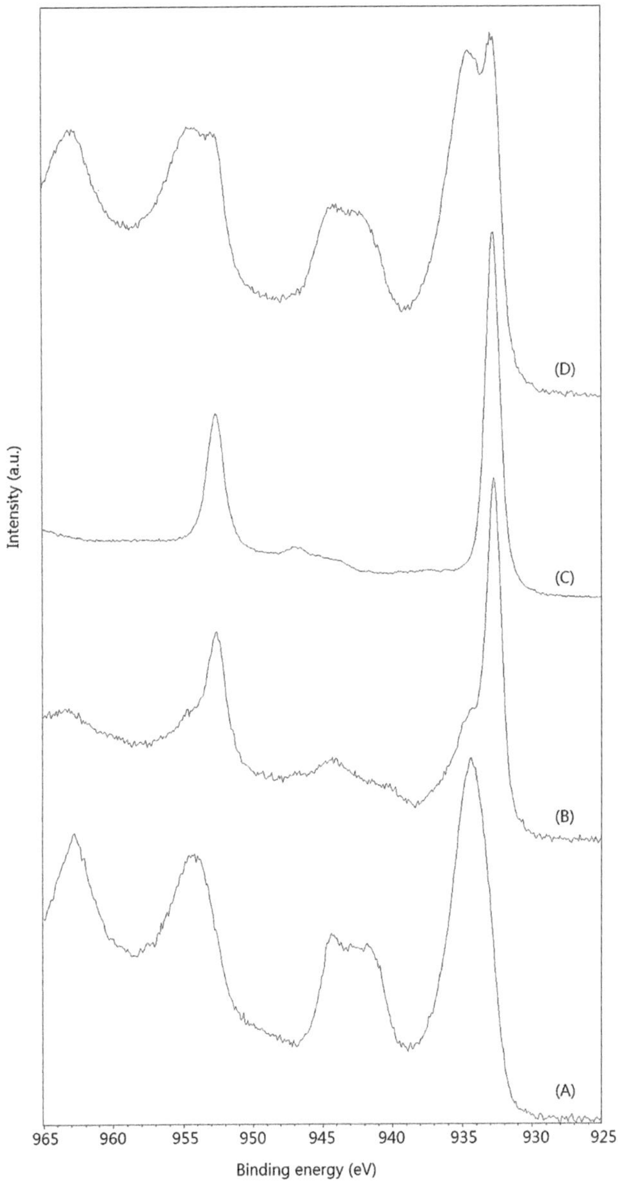

Fig. 2 High resolution Cu 2p XPS spectra for a CuO-coated carbon rod electrode after application of potential for 15 min in Ar deoxygenated 0.5 M $KHCO_3$ solution: (A) −0.4 V; (B) −0.8 V; (C) −1.2 V; (D) one full cycle from 0 V to −1.2 V to 0.4 V then back to 0 V.

935.3 eV, a Cu(II) $2p_{1/2}$ peak at 955.6 eV and well-defined Cu(II) $2p_{3/2}$ and $2p_{1/2}$ satellite peaks at 944.6 eV and 963.3 eV respectively.[25] The spectrum in both cases is consistent with only CuO being present on the surface.

After treatment at −0.8 V, the spectrum (B) shows the Cu(II) $2p_{3/2}$ and $2p_{1/2}$ peaks at 934.9 and 955.1 eV, as well as additional peaks at 932.9 eV and 952.6 eV. These new peaks can be assigned to Cu(I) or Cu(0) $2p_{3/2}$ and $2p_{1/2}$ photoelectrons.[23] The peak position and full width half maximum for Cu(0) and Cu(I) species are too similar to allow us to discriminate between the two oxidation states. The areas under the Cu(II) $2p_{3/2}$ and Cu(0)/Cu(II) $2p_{3/2}$ peaks, along with the sensitivity factor for Cu, were used to determine the relative proportion of oxidised and reduced species on the surface. After treatment at −0.8 V the proportion of Cu(II) falls from 100% to 56%, indicating that nearly half of the copper within the sampling depth of the technique is reduced. This is consistent with the assignment of reduction peak I in Fig. 1 to reduction of Cu(II)O. Treatment at −1.2 V (C) results in complete loss of the Cu(II) peaks, with the spectrum only showing the Cu(0)/Cu(I) peaks at 932.8 eV and 952.5 eV. As XPS is a surface sensitive technique, this indicates that the top several monolayers of CuO are completely reduced by treatment at −1.2 V, although the bulk may remain unchanged.

A CuO electrode was also subjected to a complete potential cycle from 0 V to −1.2 V to 0.4 V and back to 0 V. Inspection of Fig. 1 shows the electrode should undergo the reduction processes I and II and then the oxidation processes corresponding to peaks III and IV. However, as the cycle stops at 0 V, the electrode will not undergo the reduction process indicated by peak V. The Cu 2p spectrum after a complete cycle (D) showed a mixture of Cu oxidation states present on the surface, corresponding to 85% Cu(II) and 15% Cu(0)/Cu(I). This indicates that the reduced Cu species products formed during the negative-going sweep are not fully oxidised back to Cu(II) on scanning positive beyond peaks III and IV.

3.1.3 Raman spectroscopy of CuO electrodes after application of different potentials. As above, CuO-coated carbon rod electrodes were held at potentials of −0.4 V, −0.8 V or −1.2 V for 15 min in 0.5 M $KHCO_3$ deoxygenated solution and their composition analysed *ex situ* with Raman spectroscopy. As the sampling depth of Raman is greater than for XPS we expect the resulting spectra to provide information on both surface and bulk transformations. Consistent with the XPS results the CuO electrode held at −0.4 V showed identical spectral features to an unused electrode, with peaks at *ca.* 300, 350 and 630 cm^{-1} corresponding to the A_g, B_g^1 and B_g^2 modes of CuO respectively.[26]

After treatment at −0.8 V the spectrum changes significantly, consistent with the XPS and CV data which show that close to 50% of the Cu(II) content in the surface is converted to Cu(I) or Cu(0). Raman peaks are now present at *ca.* 150, 190, 220, 340, 400, 630 and 760 cm^{-1}. Definitive assignment of all the peaks is very difficult, as some are not consistent with any previous literature assignment for copper oxide or hydroxide species. Cu_2O is relatively well characterised in the literature[27] and its Raman peaks are found at *ca.* 150, 220 and 630 cm^{-1}; these are indicated by a blue asterisk on Fig. 3. The peaks at 340 and 400 cm^{-1} are tentatively assigned to Cu(II)–O modes, but are shifted from the original positions of the CuO material at 300 and 350 cm^{-1}. The peak shift and increase in intensity may result from reduction of neighbouring copper centres, or a change in bonding environment of the Cu(II) centres, perhaps due to coordination with carbonate or bicarbonate. In support of this assignment, the new peaks at 220

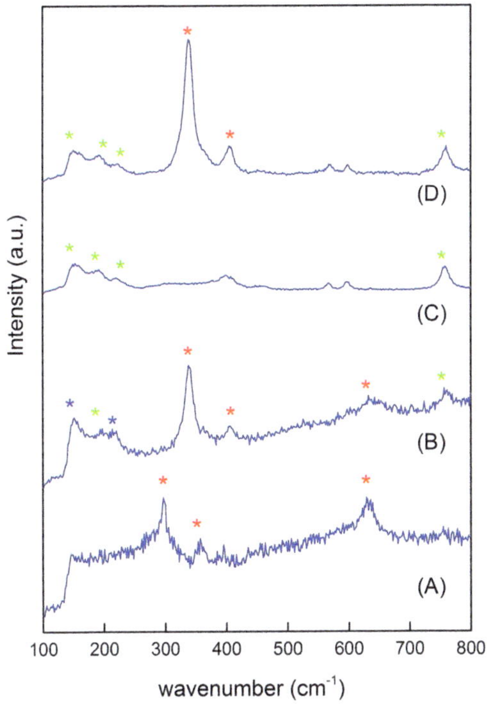

Fig. 3 Raman spectra for a CuO-coated carbon rod electrode after application of potential for 15 min in Ar deoxygenated 0.5 M $KHCO_3$ solution: (A) −0.4 V; (B) −0.8 V; (C) −1.2 V; (D) one full cycle from 0 V to −1.2 V to 0.4 V then back to 0 V. Red asterisk: CuO modes; blue asterisk: Cu_2O modes; green asterisk: Cu carbonate modes.

cm^{-1} and 760 cm^{-1} have not previously been reported for copper oxide and hydroxide species; however, Raman bands are found in similar positions for copper carbonate minerals, such as malachite.[28]

At −1.2 V the peaks at 150, 190, 220, 400 and 760 cm^{-1} remain, while the bands at 340 and 630 cm^{-1} are now absent. Additionally, a new pair of bands has emerged at 570 and 600 cm^{-1} along with a broad band at 430 cm^{-1}. The loss of the peaks at 340 cm^{-1} and 630 cm^{-1} indicate a loss of Cu(II)–O modes and this is in agreement with the XPS data that showed only Cu(I) or Cu(0) present after holding at −1.2 V. Although peaks consistent with Cu_2O remain at 150 and 220 cm^{-1}, the peak at 630 cm^{-1} is no longer present. As this peak is expected to be the strongest band for Cu_2O, its absence suggests that Cu_2O is also not present in the sample. Most of the remaining peaks show similar, but not identical positions to the copper carbonate minerals azurite and malachite; in particular the band at 760 cm^{-1} is at the position expected for carbonate modes, rather than Cu–O modes.[28]

After a complete potential cycle (D) the Raman spectrum does not closely resemble that of the CuO before potential is applied. XPS revealed that most of the copper returns to the Cu(II) oxidation state by the end of the cycle, although about 15% remains reduced. Most of the Raman peaks assigned to copper carbonate species remain (green asterisk), but the re-emergence of the peaks at 345 and 400 cm^{-1} indicate that these peaks can be more confidently be attributed to a Cu(II) species. However, the Cu(II) species is clearly not the same as the CuO starting material.

3.1.4 Chemical and structural changes to CuO electrode under cycling. From the CV, XPS and Raman data we can conclude with confidence that the reduction currents for peak I in Fig. 1 result in the partial (*ca.* 50% at the surface) reduction of CuO to a Cu(I) species at −0.8 V. Some Cu_2O is present at this potential and the remaining Cu(II) has undergone some structural or coordination change, possibly in response to reduction of neighbouring groups, or due to binding of solution bicarbonate ions. Application of −1.2 V (peak II, Fig. 1) results in further reduction, with the loss of all Cu(II) species. No Cu_2O remains at this potential, but other Cu(I) species, for example carbonate, may be present. Reduction to Cu(0) is possible and probable, but we are unable to distinguish between Cu(0) and Cu(I) with XPS and the Cu(0) is not Raman active. After one complete cycle most of the copper centres return to the Cu(II) oxidation state (85% at the surface), but the Raman spectrum shows that the resulting material has a different structure to the starting material (labelled as $Cu(I)/Cu(II)_{unknown}$ in further discussion).

Fig. 4 shows how the CV of the CuO material evolves upon continued cycling. Peaks I–V remain but increase in magnitude with increasing scan number. Beyond scan one an additional reduction peak VI is present at *ca.* −0.6 V. This can be attributed to the reduction of $Cu(I)/Cu(II)_{unknown}$ formed during the previous cycle. A possible reaction sequence for the redox events in the first cycle is therefore:

$$\text{I: } 2CuO + H_2O + 2e^- \rightarrow Cu_2O + 2OH^-$$

$$\text{II: } Cu_2O + H_2O + 2e^- \rightarrow 2Cu + 2OH^- \text{ (along with } H_2O \text{ reduction)}$$

$$\text{III, IV: Cu oxidation to } Cu(I)/Cu(II)_{unknown}$$

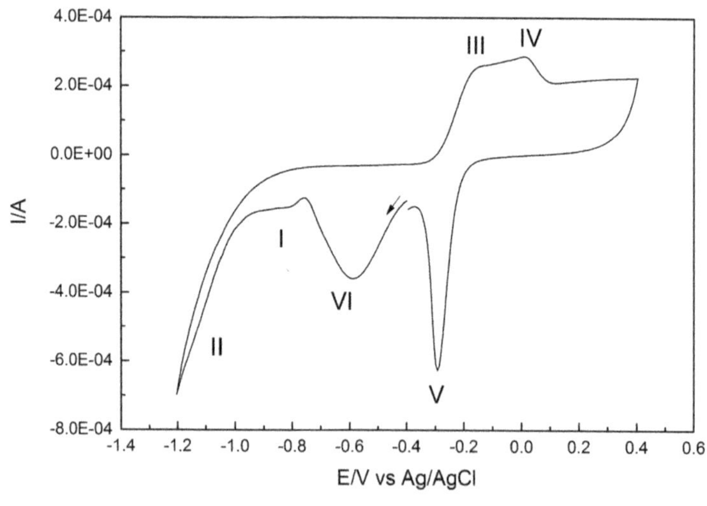

Fig. 4 Fifth CV scan of CuO-modified BDD electrode in Ar deoxygenated 0.5 M $KHCO_3$ electrolyte, scan rate 100 mV s^{-1}. Peak labels I–VI as defined in text.

Peaks V and VI are due to reduction of the Cu(I)/Cu(II)$_{unknown}$ species, which is clearly more easily reduced than the original CuO material as the reduction potentials are less negative. The current magnitude of all of the peaks increases with increased scan number indicating that a thicker layer of redox-active material is generated over repeated potential cycling. The identity of Cu(I)/Cu(II)$_{unknown}$ is not clear; however, the Raman spectrum for the product is not that of $Cu(OH)_2$ (where peaks at *ca.* 300 and 490 cm^{-1} are expected)[29] or of CuO[26] or Cu_2O.[27] The material has relatively well defined and sharp Raman peaks and its spectrum resembles copper carbonate species,[28] suggesting incorporation of electrolyte ions into the structure. The concluding message from this study is that at the potentials used for CO_2 reduction the catalyst material undergoes significant structural change and hence the starting material may not be the same as the active catalyst.

3.2 Stability of CuO under potential cycling in 0.5 M $KHCO_3$ in the presence of CO_2

The same CV, Raman and XPS investigations as described above were carried out in CO_2 saturated 0.5 M $KHCO_3$ solution. First and fifth CV scans for a CuO-modified electrode under these conditions are shown in Fig. 5. The features are broadly similar to those seen without CO_2 but in general currents are slightly higher in magnitude, reduction peaks are seen at less negative potentials, while oxidation peaks are pushed to higher positive overpotentials. Clearly the presence of CO_2 causes a change in reactivity of the CuO material, where the higher currents and decreased reduction overpotentials suggest that either the CuO material undergoes reduction more readily under these conditions, or additional reduction processes associated with the CO_2 are taking place.

The XPS data in Fig. 6a shows that the CuO actually undergoes less reduction in CO_2 saturated conditions, indicating that the enhanced reduction currents must be related to CO_2 reduction rather than reduction of the CuO. After holding

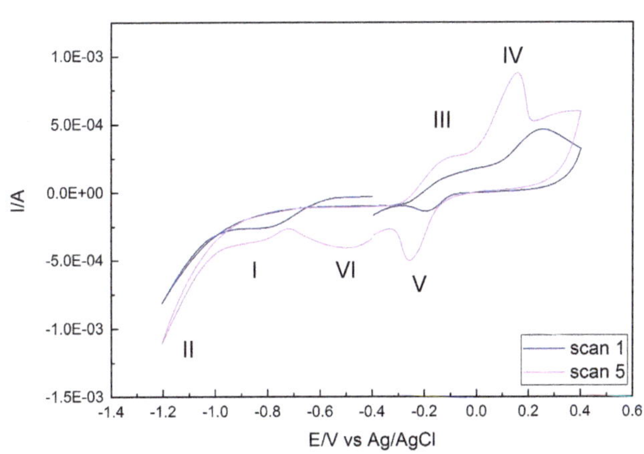

Fig. 5 First (blue) and fifth (pink) CV scan of CuO-modified BDD electrode in CO_2 saturated 0.5 M $KHCO_3$ electrolyte, scan rate 100 mV s^{-1}. Peak labels as defined in text.

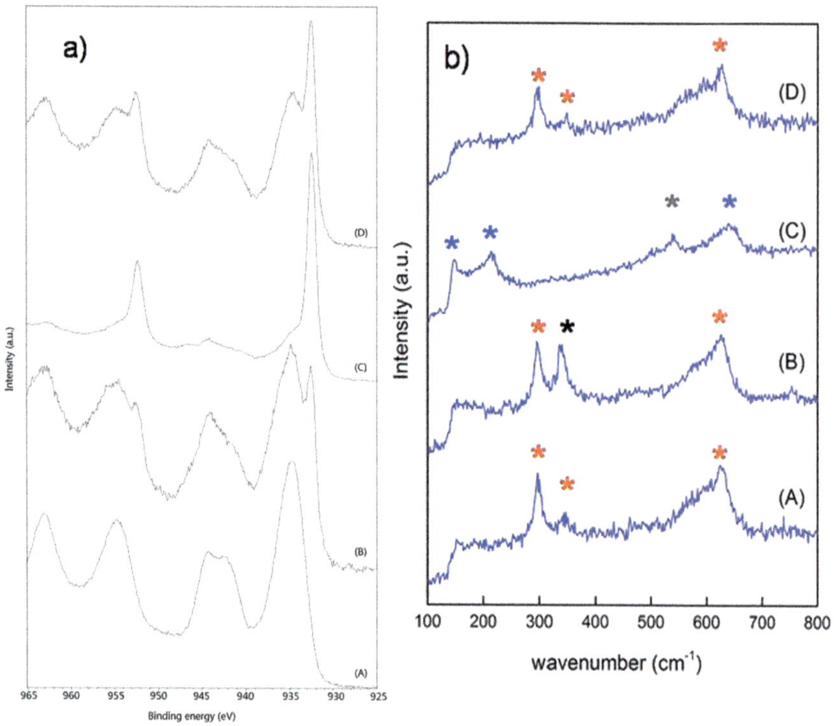

Fig. 6 (a) High resolution Cu 2p XPS spectra and (b) Raman spectra for a CuO-coated carbon rod electrode after application of potential for 15 min in CO_2-saturated 0.5 M $KHCO_3$ solution (A) −0.4 V; (B) −0.8 V; (C) −1.2 V; (D) one full cycle from 0 V to −1.2 V to 0.4 V then back to 0 V.

at −0.4 V for 15 min (A) the composition of the CuO remains unchanged but after application of −0.8 V (B) the Cu 2p spectrum shows both Cu(II) and Cu(I)/Cu(0) present in a proportion of 82% and 18% respectively. In contrast, under the same conditions in Ar-saturated solution 44% of the copper was found in the reduced form. After holding at −1.2 V more of the Cu(II) is reduced, with only about 12% remaining, but in the absence of CO_2 the XPS spectrum showed no Cu(II) remaining at this potential. After completion of one cycle a mixture of Cu(II) and Cu(I)/Cu(0) species results, as observed in Ar-saturated solution. In contrast, the proportion of Cu(II) is actually lower after cycling in CO_2 saturated solution (73% in CO_2 saturated *versus* 85% in Ar-saturated), showing that although reduction of the Cu(II) seems to be inhibited by the presence of CO_2, the oxidation of the resulting Cu(I)/Cu(0) species is likewise suppressed.

Raman spectroscopy further shows that the redox chemistry of CuO is substantially different when CO_2 is present in the solution. A comparison of Fig. 6b (with CO_2) and Fig. 3 (without CO_2) shows clearly that the product distribution in the presence of CO_2 is far less complex. After holding at −0.4 V (A) the resulting spectrum shows only the presence of CuO Raman peaks (red asterisk). Application of −0.8 V results in a new peak at *ca.* 340 cm^{-1} (black asterisk) while the peaks for CuO remain. The peak at 340 cm^{-1} was attributed previously to a Cu(II) species, where the copper centre is in a different bonding environment to the starting material. There are no peaks for Cu_2O present at this potential, while they were present at −0.8 V in the absence of CO_2 (Fig. 3). At −1.2 V peaks at 150, 220 and 640 cm^{-1} attributed to Cu_2O emerge (blue asterisk), along with an additional peak at *ca.* 540 cm^{-1} (grey asterisk). The peaks for CuO and the unknown Cu(II) species at 340 cm^{-1} are no longer present. The peak at 540 cm^{-1} is consistent with that reported for the mixed valance Cu_4O_3 species ($Cu^{II}_2Cu^{I}_2O^{II}_3$).[30] After one complete cycle (D) the spectrum is identical to that of the starting material, indicating that the reactions that resulted in Cu(I)/Cu(II)$_{unknown}$ as the main product in deoxygenated solution are greatly suppressed in the presence of CO_2. The Raman spectrum D appears to contradict the XPS data under the same conditions, as XPS shows that only 73% of the copper is returned to the Cu(II) oxidation state. However it may be that the remaining Cu(I)/Cu(0) species are Raman inactive, or are found in only a thin layer at the surface, where they would be readily detected by XPS but present in insufficient quantity to be detected by Raman.

In summary, the CuO catalyst undergoes less extreme changes in oxidation state and composition when the solution is saturated with CO_2. XPS shows that less of the Cu(II) undergoes reduction to Cu(I) or Cu(0) and Raman spectroscopy indicates reduction to Cu_2O and Cu_4O_3, with no evidence of the carbonate formation noted in the absence of CO_2. As the CV data shows enhanced currents in the presence of CO_2, this suggests that reactions additional to the reduction of the CuO catalyst must be taking place. There are two reasons why the redox behaviour of CuO could be different in CO_2 saturated solution. The first is that the pH of the solution will decrease on addition of CO_2, as more carbonic acid is formed and dissociates:

$$CO_2 + H_2O \rightarrow HCO_3^- + H^+$$

As the redox chemistry of CuO is pH dependent, the decrease in pH may change the CuO reactivity. Secondly the $CO_{2(aq)}$ molecules may preferentially bind to the CuO surface in place of the bicarbonate electrolyte ions and undergo reduction themselves while inhibiting the reduction of the CuO. *In situ* IR experiments were therefore carried out in order to identify the surface-bound species as a function of applied potential.

3.3 *In situ* IR of CuO modified electrode with applied potentials

Fig. 7 shows the 2000–800 cm^{-1} wavenumber range IR spectra for a CuO electrode at different potentials in the absence (a) and presence (b) of CO_2 in solution. The spectra are all recorded relative to a background of the CuO electrode equilibrated in the solution without a potential applied. The presented spectra are therefore difference spectra, with increases in absorbance indicating an increase in interfacial concentration of species absorbing at that wavenumber and decreases in absorbance indicating a loss of species absorbing at that wavenumber. The figure should be read from bottom to top with first −0.4 V being applied, then −0.8 V, then −1.2 V, returning to −0.8 V, −0.4 V and then 0.4 V. The ΔAbsorbance scale for (b) is approximately 5 times that for (a), so the noise due to adventitious water vapour at 1400–1800 cm^{-1} is much more pronounced in (a).

In the absence of CO_2 and with the application of −0.4 V (Fig. 7a), gains in absorbance are seen at 1635, 1362, 1012 and 845 cm^{-1} corresponding to bands for

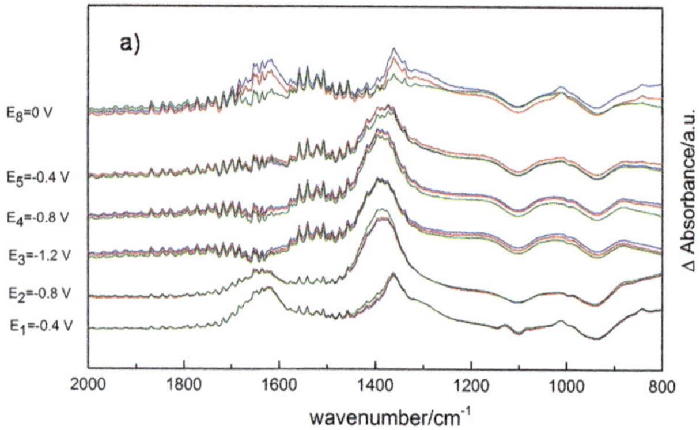

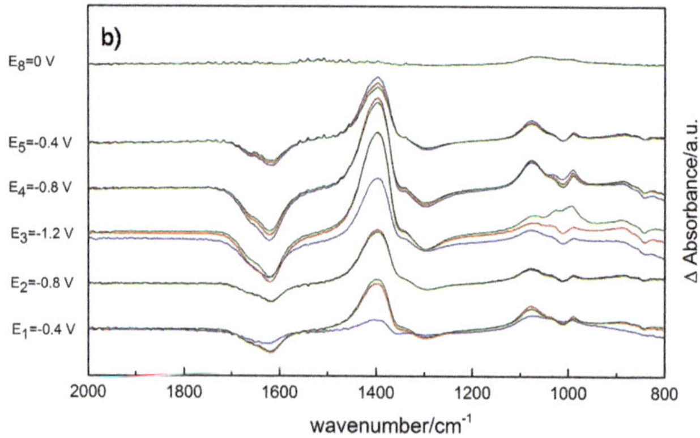

Fig. 7 ATR IR difference spectra of CuO-modified electrode under potential control (a) in argon deoxygenated 0.5 M $KHCO_3$ and (b) in CO_2 saturated 0.5 M $KHCO_3$.

bicarbonate, which are slightly shifted to higher wavenumbers than reported for the solution anion.[30] This increase in bicarbonate concentration on application of −0.4 V suggests that the applied potential encourages surface adsorption of the anion. Losses in absorption are seen at 950 cm^{-1} and 1120 cm^{-1}; these values do not correspond to carbonate or bicarbonate vibrational modes,[31] although spectral changes at similar wavenumber have been reported for CO_2 adsorption onto metal oxide surfaces and attributed to surface structure changes on adsorption.[32] On application of −0.8 V, then −1.2 V and then returning to −0.8 V the bands at 1635, 1012 and 845 cm^{-1} decrease in intensity, indicating a loss of adsorbed bicarbonate from the CuO surface. The 1362 cm^{-1} bicarbonate band is replaced by a broader band centred at *ca.* 1400 cm^{-1} that indicates the asymmetric C–O stretch of carbonate.[31] When the potential is returned to −0.4 V the bicarbonate features begin to re-emerge. At 0 V the spectrum looks similar to the initial spectrum at −0.4 V, but with an additional band at 1550 cm^{-1}, which has previously been reported for strongly coordinated bicarbonate ions on metal oxide surfaces.[32]

In the presence of CO_2 the IR difference spectra are strikingly different. The intensities of the bands are much stronger and the speciation is different. At −0.4 V a decrease in intensity is seen at the position of solution bicarbonate bands at 1617, 1320, 1010 and 840 cm^{-1}. There is a concomitant increase in bands at 1400 and 1080 and 990 cm^{-1}, consistent with surface-bound carbonate species.[32] The band at *ca.* 1080 cm^{-1} is reported to be particularly characteristic of carbonato complexes.[33] This is different from the behaviour in Ar-saturated solution, where bicarbonate was the adsorbed species and losses in metal oxide surface vibrational modes were seen at 950 and 1120 cm^{-1}. In the presence of CO_2 bicarbonate is instead lost from the interfacial region to be replaced by a different adsorbed species and the same losses in metal oxide surface vibrational modes are not observed. The differences suggest that the new adsorbed species is CO_2 in a bent configuration, similar to the structure of carbonato ligands and that it binds on different surface sites to bicarbonate. As spectral features for bicarbonate are lost at the same rate, this indicates that adsorption of the CO_2 competes and is preferential. On application of −0.8 V, −1.2 V, −0.8 V and then −0.4 V the spectral features remain similar, although their relative intensity changes, indicating that the surface carbonato species remains bound throughout the cycle at negative potentials. At 0 V the spectrum returns to the background response obtained before the application of potential, with only a band at 1080 cm^{-1} for the carbonato-species persisting.

Fig. 8 shows a schematic of the possible surface species under the different conditions. Under argon, the initial adsorbing species at −0.4 V is bicarbonate, shown in Fig. 8 as binding to the Cu centres *via* two oxygens. In pH 8 solution, hydroxide ions may be bound to the Cu centres, in which case bicarbonate would interact with the Cu–OH moieties by hydrogen bonding. Both methods of bonding would result in spectral features for adsorbed bicarbonate and disruption to surface Cu–O–(H) vibrational modes. On application of a more negative potential the bicarbonate is replaced by carbonate, possibly as a consequence of the increase in pH at the electrode interface during water reduction. Computational modelling has indicated that CO_2 binds to CuO surfaces by coordination of the central carbon atom with surface oxygen,[21] as shown in Fig. 8b. Depending on the surface structure, either one or two of the oxygen atoms of CO_2 can interact with Cu centres. Either structure leads to activation of the CO_2 and a bent configuration, consistent with the observed spectral features of the carbonato species. As indicated by the spectra, the binding sites for CO_2 adsorption in this model are different from those for bicarbonate.

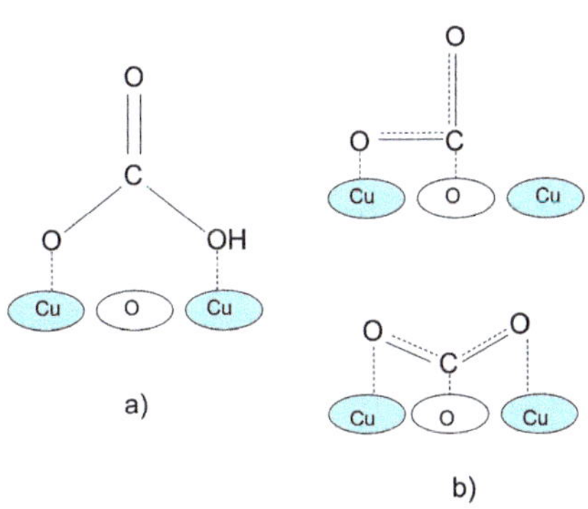

Fig. 8 Schematic of adsorbed solution species at −0.4 V in (a) argon deoxygenated 0.5 M $KHCO_3$ and (b) CO_2 saturated 0.5 M $KHCO_3$.

4 Conclusion

Catalysis studies using the same CuO nano-materials as used in this work have shown that formic acid is the main product formed at the reduction potentials studied here.[22] Formic acid is also predicted as a major product by computational modelling.[21] We have shown using *in situ* IR spectroscopy that CO_2 binds to the surface of the CuO at negative applied potentials and the spectral features are consistent with a bent, activated configuration. CO_2 is found to bind to the surface at potentials where CuO has not yet undergone reduction to Cu(I) or Cu(0) species. This shows that the CuO is able to activate CO_2 without preliminary reduction taking place. XPS and Raman analysis of the catalyst material show that reduction of the Cu content is inhibited in the presence of CO_2, despite higher reduction currents being observed in the CV. This indicates that some of the charge passed may contribute to reduction of adsorbed CO_2 rather than reduction of the catalyst.

As discussed in the introduction, the prevailing opinion is that copper oxide materials must first be reduced to copper to be effective CO_2 electrocatalysts.[9–12,16–18] This study does not contradict this, as it likely that the copper produced during reduction is a more active catalyst for CO_2 reduction than CuO. However, what we show, in support of other studies,[8,14,15] is that the CuO surface does allow for CO_2 activation at potentials positive of the reduction of CuO. Enhanced currents for CO_2 reduction are also observed at potentials where reduction of CuO is more likely to result in Cu_2O species rather than Cu(0), indicating the copper need not be fully reduced before allowing catalysis. We also show that the catalyst is relatively robust to cycling, especially in the presence of CO_2, where competitive carbonate complexation does not take place. However, the spectroscopic analysis carried out in the work investigated the first reduction cycle of pristine CuO and further studies are required to understand the catalytic performance under repeated cycling.

Acknowledgements

Liying Wang thanks the China Scholarship Council and UCL Industrial Doctoral Centre for Molecular Modelling and Materials Science (EP/G036675) for funding her PhD studentship. This work was partly funded by EPSRC Grant EP/H046313.

References

1 For a review see: G. A. Olah, G. K. S. Prakash and A. Goeppert, *J. Am. Chem. Soc.*, 2011, **133**, 12881–12898.

2 For a review see: E. E. Benson, C. P. Kubiak, A. J. Sathrum and J. M. Smieja, *Chem. Soc. Rev.*, 2009, **38**, 89–99.

3 Y. Hori, *et al.*, Electrochemical Reduction on Metal Electrodes, in *Modern Aspects of Electrochemistry*, ed. C. G. Vayenas, Springer, New York, 2008.

4 (*a*) S. Sarfraz, A. T. Garcia-Esparza, A. Jedidi, L. Cavallo and K. Takanabe, *ACS Catal.*, 2016, **6**, 2842–2881; (*b*) S. Y. Choi, S. K. Jeong, H. J. Kim, I.-H. Baek and K. T. Park, *ACS Sustainable Chem. Eng.*, 2016, **4**, 1311–1318.

5 (*a*) D. D. Zhu, J. L. Liu and S. Z. Qiao, *Adv. Mater.*, 2016, **28**, 3423–3452; (*b*) S. Lee and J. Lee, *ChemSusChem*, 2016, **9**, 333–344.

6 (*a*) K. Chan, C. Tsai, H. A. Hansen and J. K. Norskov, *ChemCatChem*, 2014, **6**, 1899–1905; (*b*) A. Roldan, N. Hollingsworth, A. Roffey, H. U. Islam, J. B. M. Goodall, C. R. A. Catlow, J. A. Darr, W. Bras, G. Sankar, K. B. Holt, G. Hogarth and N. H. de Leeuw, *Chem. Commun.*, 2015, **51**, 7501–7504.
7 (*a*) K. J. P. Schouten, Y. Kwon, C. J. M. van der Ham, Z. Qin and M. T. M. Koper, *Chem. Sci.*, 2011, **2**, 1902–1909; (*b*) A. A. Peterson, F. Abild-Pedersen, F. Studt, J. Rossmeisl and J. K. Norskov, *Energy Environ. Sci.*, 2010, **3**, 1311–1315.
8 K. W. Frese Jr, *J. Electrochem. Soc.*, 1991, **138**, 3338–3344.
9 A. Dutta, M. Rahaman, N. C. Luedi, M. Mohos and P. Broekmann, *ACS Catal.*, 2016, **6**, 3804–3814.
10 C. W. Li and M. W. Kanan, *J. Am. Chem. Soc.*, 2012, **134**, 7231–7234.
11 C. S. Chen, J. H. Wan and B. S. Yeo, *J. Phys. Chem. C*, 2015, **119**, 26875–26882.
12 C. W. Li, J. Ciston and M. W. Kanan, *Nature*, 2014, **508**, 504–507.
13 J. Xie, Y. Huang and H. Yu, *Front. Environ. Sci. Eng.*, 2015, **9**, 861–866.
14 M. Le, M. Ren, Z. Zhang, P. T. Sprunger, R. L. Kurtz and J. C. Flake, *J. Electrochem. Soc.*, 2011, **158**, E45–E49.
15 S. Lee, D. Kim and J. Lee, *Angew. Chem. Int. Ed.*, 2015, **54**, 14701–14705.
16 A. Verdaguer-Casadevall, C. W. Li, T. P. Johansson, S. B. Scott, J. T. McKeown, M. Kumar, I. E. L. Stephens, M. W. Kanan and I. Chorkendorff, *J. Am. Chem. Soc.*, 2015, **137**, 9808–9811.
17 A. Eilert, F. S. Roberts, D. Friebel and A. Nilsson, *J. Phys. Chem. Lett.*, 2016, **7**, 1466–1470.
18 W. Tang, A. A. Peterson, A. S Varela, Z. P. Jovanov, L. Bech, W. J. Durand, S. Dahl, J. K. Norskov and I. Chorkendorff, *Phys. Chem. Chem. Phys.*, 2012, **14**, 76–84.
19 C. Janaky, D. Hursan, E. Endrodi, W. Chanmanee, D. Roy, D. Liu, N. R. de Taconni, B. H. Dennis and K. Rajeshwar, *ACS Energy Lett.*, 2016, **1**, 332–338.
20 E. L. Uzanova, N. Seriani and H. Mikosch, *Phys. Chem. Chem. Phys.*, 2015, **17**, 11088–11094.
21 A. K. Mishra, A. Roldan and N. H. de Leeuw, *J. Phys. Chem. C*, 2016, **120**, 2198–2214.
22 K. Gupta, M. Bersani and J. A. Darr, *J. Mater. Chem. A*, 2016, **4**, 13786–13794.
23 (*a*) National Institute of Standards and Technology XPS Database, 15 September 2012, accessed 10 March 2016, http://srdata.nist.gov/xps/main_search_menu.aspx; (*b*) X-ray Photoelectron Spectroscopy (XPS) Reference Pages, 2015, accessed 10 March 2016, http://www.xpsfitting.com/.
24 (*a*) S. Gonzalez, M. Perez, M. Barerra, A. R. Gonzalez Elipe and R. M. Souto, *J. Phys. Chem. B*, 1998, **102**, 5483–5489; (*b*) D. Tromans and R.-H. Sun, *J. Electrochem. Soc.*, 1992, **139**, 1945–1951.
25 S. Poulson, P. M. Parlett, P. Stone and M. Bowker, *Surf. Interface Anal.*, 1998, **24**, 811–820.
26 L. Debbichi, M. C. Marco de Lucas, J. F. Pierson and P. Kruger, *J. Phys. Chem. C*, 2012, **116**, 10232–10237.
27 A. Singhal, M. R. Pai, R. Rao, K. T. Pillai, L. Lieberwirth and A. K. Tyagi, *Eur. J. Inorg. Chem.*, 2013, **2013**, 3640–3651.
28 R. L. Frost, W. N. Martens, L. Rintoul, E. Mahmutagic and J. T. Kloprogge, *J. Raman Spectrosc.*, 2002, **33**, 252–259.
29 H. Y. H. Chan, C. G. Takondis and M. J. Weaver, *J. Phys. Chem. B*, 1999, **103**, 357–365.

30 C. Toparli, A. Sarfraz and A. Erbe, *Phys. Chem. Chem. Phys.*, 2015, **17**, 31670–31679.

31 A. R. Davis and B. G. Oliver, *J. Solution Chem.*, 1972, **1**, 329–339.

32 K. D. Dobson and A. J. McQuillan, *Langmuir*, 1997, **13**, 3392–3396.

33 K. Nakamoto, *Infrared and Raman Spectra of Inorganic and Coordination Compounds*, Wiley, New York, 3rd edn, 1978.

Faraday Discussions

DISCUSSIONS

Novel photocatalysts: general discussion

Graham Hutchings, Michael Bowker, Richard Catlow, Thulani Nyathi, Ilyaas Abdullah, Michael Claeys, Neil Coville, Alberto Roldan, Pieter van Helden, Nico Fischer, Eric van Steen, Linda Jewell, Lerato Mokoloko, Detlef Bahnemann, Matthew Neurock, Freek Kapteijn, Enrique Iglesia, Philip Gibson, Katherine Holt, Kazunari Domen and Abhishek Kumar Gupta

DOI: 10.1039/c7fd90013a

Michael Bowker opened discussion of the paper by Kazunari Domen: You used a filter in the experiment in Fig. 2 (DOI: 10.1039/C6FD00184J) to pass only light at >420 nm. So what happens when you don't have the filter? If the rate is higher without the filter, why use it at all? And do you see opportunities to utilise single photon processes for pure water splitting?

Kazunari Domen replied: The filter is used because photocatalytic systems operable under visible light are studied. Even if a photocatalyst shows high apparent quantum yield under UV light, like Al-doped $SrTiO_3$, it cannot be used for solar energy conversion effectively.

To answer your last question, yes. Such systems are simpler and less expensive. This advantage may outweigh the problem of low STH values. We will continue the development of one-step excitation systems with higher apparent quantum yields.

Freek Kapteijn noted: You produce in your beautiful cell a nice but tricky mixture of H_2 and O_2, so you have to solve a separation problem. Could you construct your cell in such a way to have separate locations for hydrogen/oxygen production? What would the impact of this be on the efficiency, and what needs to be improved to go in that direction?

Kazunari Domen responded: We have considered placing HEP and OEP on opposite sides. However, problems of mass transfer and solution resistance will be critical in such a configuration, especially when the sheet is extended to a large area. In addition, photocatalyst sheets will be opaque even if transparent conducting oxides are used as a conductor layer because of scattering. It is necessary to illuminate both sides of a film immersed in water. It seems to be difficult to build such systems inexpensively.

Eric van Steen said: The results of the effect of solvent and pressure seem to indicate that in the case of the parallel plate arrangement, diffusion of ions through the liquid phase might be limiting; this does not seem to be the case for the photocatalyst sheet. The photocatalytic sheet will produce a mixture of O_2 and H_2, which will require a separate separation process. Will the efficiency of the photocatalytic sheet outweigh the need for a H_2/O_2 separation unit in commercial applications?

Kazunari Domen answered: The separation of hydrogen and oxygen is indeed a critical problem in the photocatalyst sheet approach. Thermodynamically, the minimum energy loss associated with the separation of hydrogen and oxygen is 1.1%. Therefore, the energy loss can be reduced to an acceptable level if efficient gas separation processes are available, and we believe that the efficiency and scalability of sheets can outweigh the need for gas separation. Membranes for the separation of hydrogen and oxygen are studied by specialists in the ARPChem project of New Energy and Industrial Technology Development Organization (NEDO) funded by the Ministry of Economy, Trade and Industry (METI). It is probably possible to prepare a membrane with sufficient selectivity and permeance in dry conditions. However, the performance of gas separation membranes is significantly influenced by the presence of water vapour, and appropriate solutions are under development. We believe that the photocatalyst sheet approach is feasible, but other approaches are also worthy of research because each system has its own advantages and problems. At present, nobody can tell which approach will become the most efficient as a whole in the future.

Neil Coville enquired: This looks to be a very complicated catalyst. Could you please explain how you came to the formulation historically and where you are going to go with it in the future?

Kazunari Domen explained: A one-step excitation water splitting system consists of (i) a particulate semiconductor photocatalyst that absorbs light and generates excited electrons and holes, and (ii) co-catalysts working as surface active sites for hydrogen and oxygen evolution. Sheets can be prepared simply by spreading and fixing such photocatalysts on certain substrates. In the case of sheets for Z-scheme water splitting, hydrogen evolution and oxygen evolution occur on different photocatalysts and these two photocatalysts are electrically connected by an underlying conductive layer. The working mechanism is not very complicated. The activity of particulate photocatalysts was thought to be enhanced if hydrogen evolution sites and oxygen evolution sites could be formed asymmetrically.[1] The particle transfer method was reported in 2012 as a process that enables rigid contact between particulate photocatalysts and conductive layers of arbitrary combinations.[2] The first paper on photocatalyst sheets in the present form was published in 2015.[3] Hydrogen evolution photocatalysts and oxygen evolution photocatalysts are embedded into conductive materials by particle transfer. Recently, the activity of the sheet at ambient pressure was found to be maintained when carbon was used as a conductor.[4] Therefore, the problem of backward reactions at ambient pressure has been largely overcome. The application of narrow band gap photocatalysts to sheets, the development of inexpensive and scalable reactors accommodating the photocatalyst sheet, and integration of the reactor with gas separation units are the next important challenges.

1 T. Hisatomi, T. Minegishi and K. Domen, *Bull. Chem. Soc. Jpn.*, 2012, **85**, 647–655.
2 T. Minegishi, N. Nishimura, J. Kubota and K. Domen, *Chem. Sci.*, 2013, **4**, 1120–1124.
3 Q. Wang, Y. Li, T. Hisatomi, M. Nakabayashi, N. Shibata, J. Kubota and K. Domen, *J. Catal.*, 2015, **328**, 308–315.
4 Q. Wang, T. Hisatomi, Y. Suzuki, Z. Pan, J. Seo, M. Katayama, T. Minegishi, H. Nishiyama, T. Takata, K. Seki, A. Kudo, T. Yamada and K. Domen, *J. Am. Chem. Soc.*, 2017, **139**, 1675–1683.

Neil Coville followed up: So the d-electron count is key to the catalyst?

Kazunari Domen answered: Yes. Empirically, photocatalysts that are active in overall water splitting are semiconductors consisting mainly of transition metal cations with the d^0 electronic configuration and typical metal cations with the d^{10} electronic configuration. In some cases, doping transition metal cations with d-electrons makes photocatalysts active under visible light because they induce mid-gap states. In the case of Rh-doped $SrTiO_3$, the valence state of Rh ions is an important factor. However, this is a very rare case.

Ilyaas Abdullah said: You mention that your photocatalyst sheet is operable in pure water regardless of pH; over what range did you determine that experimentally? Was corrosion of your photocatalyst sheets a problem at lower pH values? And have you considered having your OEP and HEP on opposite sides of the same film, to separate the evolution of oxygen and hydrogen?

Kazunari Domen replied: The pH values from 3 to 9 were examined for the system based on $SrTiO_3$:La,Rh and $BiVO_4$:Mo. Corrosion of $BiVO_4$ is expected at lower and higher pH values. Therefore, the performance of the present sheet at such pH values was not studied. The choice of pH value will depend on the stability of photocatalyst, cocatalyst, and conductor materials. The activity of $SrTiO_3$:La,Rh itself will improve at higher pH values.

We have considered placing HEP and OEP on opposite sides. However, problems of mass transfer and solution resistance will be critical in such a configuration, especially when the sheet is extended to a large area. In addition, a system that allows illumination of both sides of a film immersed in water will be very complicated and may not be built inexpensively.

Graham Hutchings asked: In chemical redox catalysis the surface area of the oxide can be an important design criteria; is this the case with your strontium titanate for example? In some cases we find that highly disordered or amorphous materials can be very effective. Is this the case with photocatalysis?

Kazunari Domen responded: The surface area may have an influence on the rate of reactions when adsorption of reactants is an important factor. However, we do not think that this is the case for water splitting using a photocatalyst sheet. The crystallinity is more important in most cases because photoexcited carriers have to migrate to surface active sites efficiently. We did not associate the disorder or amorphicity of $SrTiO_3$:La,Rh with the activity because the H_2 evolution site was the Ru species deposited on it.

Michael Bowker queried: I think you have worked on single photon excitation in the past, what is your maximum efficiency using single photon water splitting? How poor are they relative to these Z-scheme catalysts? And would you be happy to supply the photocatalytic community with some of your materials?

Kazunari Domen answered: The solar-to-hydrogen energy conversion efficiency (STH) of one-step excitation water splitting using $(Ga_{1-x}Zn_x)(N_{1-x}O_x)$ is 0.2%. Al-doped $SrTiO_3$ shows a higher STH value of 0.5%. However, Al-doped $SrTiO_3$ is active only under UV light ($\lambda \leq 380$ nm) and not efficient for solar energy conversion. $LaMg_{1/3}Ta_{2/3}O_2N$ is also active in overall water splitting under irradiation of up to approximately 600 nm, but the apparent quantum yield is in the order of 0.1% because of difficulty in the preparation of high-quality $LaMg_{1/3}Ta_{2/3}O_2N$. The apparent quantum efficiency tends to be low as the absorption edge wavelength of photocatalysts becomes longer. As a consequence, the STH of one-step excitation water splitting is lower than that of Z-scheme water splitting using sheets at present. Nevertheless, we will continue development of one-step excitation systems because they are simpler and less expensive. In addition, one-step excitation systems can achieve higher STH values than two-step excitation systems if the apparent quantum yield is improved to the same level.

To answer your final question: of course, yes. We have already provided our samples for and have received samples from different groups. The details should be discussed in advance, because nanoparticulate co-catalysts can be poisoned over time.

Michael Bowker opened discussion of Detlef Bahnemann's paper: You are associating the black regions you see in SEM with rutile, but rutile is white, so why do you think your rutile regions are black?

Detlef Bahnemann answered: Most certainly, rutile is white. However, following illumination with UV(A) light, electron/hole pairs are formed. While the holes are trapped at surficial oxygen atoms thus generating oxygen vacancies, the remaining electrons reduce Ti(IV) ions in the TiO_2 lattice. The Ti(III) ions formed in this process act as color centers and are responsible for the black color since they absorb light throughout the entire visible spectrum. Each of the "black" rutile nanoparticles contains several of these Ti(III) color centers. However, 4 to 5 Ti(III) ions in a particle with a diameter of 2 nm are sufficient to generate their dark coloration.

Michael Bowker continued: Do you see any evidence for a structural phase which is often characteristic of reduced titania? These are Magneli phases in which defects can cluster into planes of reduced material.

Detlef Bahnemann responded: We have not observed these phase changes, which might be due to the nanocrystalline nature of the material that has been studied in our work.

Michael Bowker commented: We should note, of course, that electron mobility is incredibly fast, while hole mobility is not.

Detlef Bahnemann replied: This is absolutely right, electrons in the conduction band of the rutile particles exhibit a very high mobility. Once they are trapped as Ti(III) ions, however, their mobility is drastically reduced. The hole mobility is low from the time of their generation.

Graham Hutchings said: During the discussion, quantum efficiency is highlighted and researchers are striving for the highest quantum yield. Why is quantum efficiency important and what would be a minimum quantum efficiency target?

Hydrogen peroxide was also mentioned as a valuable product. The hydrogen to oxygen ratio would still mean that hydrogen would be produced, but of course additional air could be added to provide the required oxygen. However, one would only make a few ppm hydrogen peroxide with current catalysts using ambient pressure hydrogen in water. Therefore the system would need to operate at high pressure and this would exacerbate the potential safety issues.

Detlef Bahnemann answered: The highest possible quantum efficiency (in the absence of any free radical chain reaction) will be unity (1.0); that is, each absorbed photon results in one chemical transformation. If we are thinking of a chemical plant that utilizes, for example, solar photons for the synthesis of fuels such as molecular hydrogen, the quantum efficiency of the overall process will determine the size of the required light absorbing unit and hence the investment costs for the factory. Currently, an overall solar energy to chemical fuel conversion efficiency of 10% is regarded as the required threshold which translates into a minimum quantum efficiency of at least 50% (0.5) over the entire visible part of the solar spectrum.

Hydrogen peroxide in a photoelectrochemical system is generated through the two-electron oxidation of water (rather than through a reductive process). Taking into account that the respective electrode is immersed in an aqueous electrolyte at ambient pressure is sufficient for its operation and therefore no special safety features will be required.

Enrique Iglesia enquired: The steps that you have discussed for the formation of dioxygen seem to lack details of the intermediate elementary steps required to accomplish the two two-electron reductions required for O_2 evolution. Could you comment on this, especially in the context of what is known about the microscopic reverse of this reaction – the capping of non-vicinal vacancies by O_2? Such steps require the sequence of the two-electron oxidations, *via* the formation of OOH species or H_2O_2 in order to avoid the need to form unstable vicinal vacancies that can accommodate the four electrons that must be transferred upon the dissociative activation of O_2.

Detlef Bahnemann replied: Various scenarios are currently discussed to account for the photocatalytic formation of molecular oxygen from the oxidation of water molecules.[1] Assuming that the photogenerated holes are trapped at different terminal surface hydroxide ions this would generate surface bound hydroxyl radicals readily yielding hydrogen peroxide as the first stable product provided that they are located at vicinal sites. Otherwise, a surface diffusion process of at least one of these surface hydroxyl radicals will be required. Alternatively, several research groups discuss the formation of $O^{\bullet-}$ radicals either at the surface or in the TiO_2 lattice.[1] Despite of their negative charge these oxygen radical anions may either also dimerize yielding a peroxo species or they may react with an electron trapped at a titanium site as a Ti^{3+} species effectively resulting in the recombination of one electron/hole pair thus evincing the micro reversibility of all free radical generation processes as you indicated. I think it is fair to assume that the subsequent two electron transfer steps finally leading to the formation of molecular oxygen are kinetically and thermodynamically relatively simple reactions as compared with the initial peroxide formation described before.

1 J. Schneider, M. Matsuoka, M. Takeuchi, J. Zhang, Y. Horiuchi, M. Anpo and D. W. Bahnemann, *Chem. Rev.*, 2014, **114**, 9919–9986.

Eric van Steen noted: The obtained black spots on the anatase are inhomogeneous. Would the inhomogeneous distribution of black spots also be obtained with well-defined, uniformly shaped anatase?

Detlef Bahnemann responded: Yes it would. The black particles are those which have first of all changed their phase from anatase to rutile since their surfaces are "protected" from a self-healing mechanism which usually fills the oxygen vacancies present at the particle surface through reactions with water molecules. Following the trapping of the electrons as Ti(III) ions, the black spots are generated which survive for a long time only when their re-oxidation by, for example, molecular oxygen is prevented. Hence, in such a well-defined system only the innermost particles will be formed and remain as black color centers.

Matthew Neurock asked: Can you please elaborate more on the formation of O_2 and its removal on the surface? Perhaps I misunderstood, but it looked like you were considering both of the oxygens may be coming from the lattice. I would think that the O_2 formation would more likely involve an oxygen in the lattice and a terminal bound OH* or H_2O.

Detlef Bahnemann responded: Indeed, both of the oxygen atoms in a photocatalytically formed O_2 molecule might originate from bridging oxygen atoms of the titania matrix.[1] However, this does not rule out the mechanism you suggested. I am pretty sure that both of these mechanisms are operating together (side by side).

1 J. F. Montoya, D. W. Bahnemann, P. Salvador and J. Peral, *Catal. Sci. Technol.*, 2017, 7, 902–910.

Nico Fischer enquired: Assuming there is an established hydrogen economy and space is available for effective photocatalytic H_2O splitting (the separation challenge of O_2 and H_2 is also solved in this scenario), what technology do you foresee dominating for hydrogen storage? Currently compression competes with adsorption, metal hydrides and chemical carriers.

Detlef Bahnemann replied: My personal favourites are chemical carriers and metal hydrides, in particular for mobile applications. A tank cooled down and/or compressed to liquify its content is most certainly not something that I would want to "carry around" in my future car.

Matthew Neurock opened discussion of the paper by Katherine Holt: You indicated that you need to go to reducing conditions and that the active catalyst may be Cu(0). Based on other electrocatalytic systems, this would appear to the be the more likely site but I was a little confused as your last figure seemed to focus on Cu(I)O sites for CO_2 adsorption. What is the known step(s) that controls the rate? Is it considered to be electron transfer to the CO_2 to form the bent CO_2 intermediate? Does this fall out of a Tafel slope analysis?

Katherine Holt responded: From our *in situ* experiments we are able to see which species are adsorbed to the electrode surface at different potentials and from the *ex situ* experiments we can observe which Cu oxidation states are present at these potentials. Taking both together it is clear that CO_2 can absorb to the surface at -0.4 V (from IR) but that at this potential CuO is not reduced (still Cu(II) according to XPS and Raman). The final figure represents this situation at -0.4 V – *i.e.* the adsorbed species on the electrode, which is still CuO. However most catalytic studies are carried out at more negative potentials, with CO_2 reduction products observed from about -0.8 V (*vs.* Ag / AgCl). So we start with the situation shown in the figure, but when we sweep the potential more negative the CuO is reduced and it is likely the surface is Cu(0). The rate controlling step for CO_2 reduction is the first electron transfer step to form the bent radical anion intermediate and yes, this can be determined and analysed using Tafel plots. Hence if you activate the molecule by adsorbing in a bent configuration prior to reduction you should lower the activation energy and hence the overpotential required for reduction.

Alberto Roldan asked: At the surface, Cu atoms are less coordinated than in the bulk of Cu_2O. Thus, which oxidation state are these atoms in? Also, as carbonate buffer solution may be responsible for the signals detected, have you tested, for instance by isotope labelling, that it is actually CO_2?

Katherine Holt answered: We would need to look more closely at the surface structure of the particles to be sure, but from XPS, which is a surface sensitive technique we only see Cu(II) species in the starting material. However we did not look too closely for evidence of trace Cu(I) sites so I could not say for sure there are none present. As for using labelled species to distinguish between adsorbed species due to CO_2 and those due to the electrolyte carbonate and bicarbonate species, that is an excellent suggestion that should be considered for future studies.

Richard Catlow noted: Reduced CuO will have a significant concentration of surface vacancies with trapped electrons (F-centres). We know for other systems, including ZnO and MgO [1] that these species are effective at CO_2 activation by electron donation and are likely to be playing a significant role here.

1 C. A. Downing, A. A. Sokol and C. R. A. Catlow, *Phys. Chem. Chem. Phys.*, 2014, **16**, 184–195.

Katherine Holt responded: This is a good point and actually one of the first papers that addressed electrochemical CO_2 reduction at copper oxide[1] proposed that the electronic properties of copper oxide and surface defect sites in particular played a significant role in CO_2 activation.

1 K. W. Frese Jr, *J. Electrochem. Soc.*, 1991, **138**, 3338–3344.

Enrique Iglesia said: The coupling of H_2O electrolysis to form hydrogen and CO_2 reduction seems to require some sort of kinetic synergy to justify the chemical complexity and the requirements for the concerted reaction. One way to achieve this is to use electrodes that do not carry out electrolysis of water because of a specific hurdle in the desorption of H_2, since this causes a high "virtual pressure" of hydrogen at the surfaces. Is in fact the search for catalytic materials guided by the specific requirement that electrodes do not desorb H_2? Are these reactions also routinely compared with those in which the hydrogen is present as a given pressure of H_2 to test whether any synergies and much higher rates are indeed achieved and that we are not just activating CO_2 electrochemically and using the H_2 desorbed during H_2O electrolysis?

Katherine Holt replied: Yes it is correct that the hydrogen evolution reaction (HER) competes directly with CO_2 reduction at many electrodes. All metal electrodes have been characterised in terms of the products of CO_2 reduction, with a large number classified as producing hydrogen as the major product. Clearly it is desirable to identify an electrode material that can provide a source of hydrogen atoms that result in preferred products of CO_2 reduction (formic acid, hydrocarbons) rather than evolving hydrogen. Most work in this area still relies somewhat on serendipity and the studies you describe (investigating effect of partial pressure of hydrogen) are not yet done as a matter of course.

Detlef Bahnemann questioned: Could you please comment on the products formed upon the reduction of carbon dioxide? I am particularly interested in the various organic reaction products formed.

Katherine Holt answered: The reported reaction products depend on the exact experimental conditions used and are a function of electrode preparation, applied potential and solution pH *etc.* The reported products will also depend on the analytical method used to determine the products *e.g.* whether gas phase or solution analysis is undertaken. For the material used in this study, analysis of the solution phase products showed that formic acid is the major product,[1] but gas phase products were not determined. In studies using the most active catalysts, oxide derived copper (OD-Cu), a wide range of products including >C1 hydrocarbons are reported, which is why these electrode materials are favoured. Another reason why CuO is of interest is that methanol has been reported as a product; this is quite unusual for electrochemical CO_2 reduction and why the oxide is proposed to play a role in product distribution.

1 K. Gupta, M. Bersani and J. A. Darr, *J. Mater. Chem. A*, 2016, **4**, 13786–13794.

Graham Hutchings commented: In your experiments you start with copper oxide and generate carbonate under steady state operation. What happens if you start with copper carbonate or predominantly carbonate?

Katherine Holt answered: The carbonate seems to form mainly when there is no CO_2 present, as it seems that CO_2 surface binding is a competitive process to bicarbonate and carbonate adsorption. It would be interesting to see if starting with copper carbonate has any influence on the CO_2 reduction activity as one might expect that the surface binding of CO_2 would be different than on copper oxides. If the activity was worse with the carbonate this would tell us that carbonate incorporation from the use of bicarbonate electrolytes deactivates electrode performance and should be avoided.

Thulani Nyathi communicated: Following up on your response to the question on which Cu oxidation state is the active species, you said if you were to do a time-resolved experiment your Cu^{2+} will eventually reduce to Cu(0) even in the presence of CO_2. You also added that with time, the reaction rate will decline. So what would be the cause of the decline in activity assuming that the reduction from Cu^{2+} to Cu(0) is not the cause?

Katherine Holt communicated in response: A decline in CO_2 reduction activity over time is noted for many electrode materials and is reported as a problem even if you start with a metallic copper electrode. One proposed reason is the presence of trace metal contaminants present in all electrolytes as many commercial sources contain some degree of trace metal content. Holding the electrode at negative potentials for a long period of time causes the reduced metal species to plate onto the electrode and deactivate it. The solution to this problem is to clean electrolytes prior to use with pre-electrolysis, which can be tedious. It has also been reported that black films, proposed as carbonaceous species have been observed to form on Cu after prolonged CO_2 reduction, although I don't think the coating has been identified and is not reproducible. It may also be that, as we observed, electrolyte species like carbonates can start to adsorb and irreversibly form surface films that may block catalytic activity. Finally, the rate of reaction becomes limited by how quickly adsorbed species and intermediates can be removed and it may be that the surface just becomes saturated with these species that don't desorb.

Michael Bowker opened a general discussion of Kazunari Domen's, Detlef Bahnemann's and Katherine Holt's papers: It has been mentioned by Prof Bahnemann regarding ways of storing H_2, that it is likely that chemical storage is best. Methanol is one option for such storage. But how do we get the H_2 back out of methanol if we need to recover it? Here, we could maybe use photocatalysis again with methanol as a hole scavenger to get the H_2 out ($CH_3OH + H_2O \rightarrow 3H_2 + CO_2$) and as a bonus, we get an extra mole of hydrogen from water. So this may be a possible future way of getting the stored hydrogen back.

Detlef Bahnemann noted: We always forget a technical problem, that is that we don't always have temperatures of 20 °C and above. In particular, in the desert we can have temperatures below 0 °C and frozen water would break most reactors and seals. So technically, I am a bit sceptical that we should use liquid water in the system – maybe water vapour would be a better choice? Maybe, in the future we should focus on gas phase reactors rather than on liquid phase reactors.

Kazunari Domen responded: We have been developing reactors that accommodate photocatalyst sheets and a water layer 1 mm or less in thickness. We believe that volume expansion when water turns into ice does not cause critical problems if the thickness (volume) of water is thin (small) and a reactor is made of soft materials such as plastic, but it is necessary to confirm this experimentally. Water vapour splitting is an interesting approach. In fact, it has been studied using particulate photocatalysts since the 1980s, and proton exchange membrane electrolysers coupled with a PV cell recently by different groups.[1–3]

1 K. Domen, S. Naito, M. Soma, T. Onishi and K. Tamaru, *J. Chem. Soc., Chem. Commun.*, 1980, 543–544.
2 F. Dionigi, P. C. K. Vesborg, T. Pedersen, O. Hansen, S. Dahl, A. Xiong, K. Maeda, K. Domen and I. Chorkendorff, *Energy Environ. Sci.*, 2011, **4**, 2937–2942.
3 S. Kumari, R. Turner White, B. Kumar and J. M. Spurgeon, *Energy Environ. Sci.*, 2016, **9**, 1725–1733.

Michael Claeys addressed Kazunari Domen: Photocatalytic water splitting yields hydrogen and oxygen, which need to be separated as discussed earlier on. We have also spoken about the potential necessity to find a "carrier" (*e.g.* an organic molecule) for easier storage and transport. What is your opinion regarding the best method for hydrogen storage?

Also, what are you doing with the oxygen which is formed in your process? Is it a target product that would be isolated and sold, or are there scenarios where this would not be economical and one would simply release the oxygen into the atmosphere? For example, photovoltaic water splitting plants which are located in remote "photon rich" areas?

Kazunari Domen replied: The toluene-methylcyclohexane (TL-MCH) system, liquid hydrogen, and methanol are possible hydrogen carriers. TL and MCH are liquids at ambient temperature and TL can be sequentially hydrogenated and dehydrogenated on a commercial level owing to recent progress in dehydrogenation catalysts. Liquid hydrogen can be transported by energy-efficient manners. Methanol can be used to produce lower olefins as raw materials of value-added chemicals. Ammonia is also very interesting as a hydrogen carrier because its oxidation products (nitrogen and water in ideal combustion) are environmentally benign and nitrogen as a raw material is available everywhere. Production of ammonia becomes energy-efficient if a large amount of renewable solar hydrogen and an ammonia synthesis catalyst operable at low temperature and pressure are available.

Michael Bowker commented: Going back to hydrogen storage molecules, it could indeed be that ammonia is used rather than methanol, though it can't be the basis for carbon chemical production. Regarding photocatalysis in the liquid phase or gas phase, it is often the case that the gas phase reaction is faster. It's not often reported that there can be a problem with the liquid phase in that agglomeration occurs for certain materials, including titania, to create agglomerates of ~1 micron and higher. These then interact inefficiently with light compared with the fundamental particles which are often around 20–50 nm. The gas phase reactions don't have that problem.

Detlef Bahnemann responded: In my opinion, we will see all kinds of fuels generated by photocatalytic systems including ammonia, molecular hydrogen, methane, and methanol. Personally, I favour the hydrogen "model" for the first step since this could utilize quite a bit of the energy infrastructure that currently already exists.

Indeed, gas phase reactions exhibit considerable advantages as compared with the condensed liquid phase. However, currently, the appropriate reactor engineering of optimal photocatalytic gas phase reactors is still lagging behind.

Freek Kapteijn addressed Kazunari Domen: What is your opinion about the preferred method of utilizing solar radiation? Either through direct photocatalytic water splitting with the cell you presented, but with the necessity to separate the H_2/O_2 mixture, or to do it indirectly using photovoltaics followed by electrolysis yielding separate H_2 and O_2. What is the best option in terms of efficiency at present and economics?

Kazunari Domen replied: It is difficult and probably unfruitful to tell which approach is the best at present. As is described in our recent introductory lecture paper,[1] photovoltaic-assisted electrochemical approaches have achieved high STH values exceeding 10% but are too expensive to be cost-competitive with existing energy resources. Photocatalyst systems may be fabricated inexpensively, with the proviso that their solar-to-hydrogen energy conversion efficiency (STH) values lag behind those of the photovoltaic-powered systems at present. Photoelectrochemical approaches fall into an intermediate category. Overall, fundamental improvements in both STH values and cost are still needed. We believe that the photocatalyst sheet approach is feasible because the efficiency and scalability can probably outweigh the need for gas separation. Of course, other approaches are also worthy of research because each system has its own advantages and problems.

1 T. Hisatomi and K. Domen, *Faraday Discuss.*, 2017, DOI: 10.1039/C6FD00221H.

Lerato Mokoloko communicated: Can fluorescence spectroscopy be used to evaluate the properties of small carbon particles on a surface? This is a known technique to evaluate the size distribution of carbon dots (<10 nm) in solution.[1]

1 Y. Liu, N. Xiao, N. Gong, H. Wang, X. Shi, W. Gu and L. Ye, *Carbon*, 2014, **68**, 258–264.

Abhishek Kumar Gupta communicated in response: Yes, fluorescence spectroscopy is a very effective tool for evaluating the properties of small carbon particles on a surface, since carbon dots are inhomogeneously broadened due to the slower relaxation of the solvent molecules around them. Time-resolved experiments show significant energy redistribution, relaxation among the emitting states, and spectral migration of fluorescence spectra on the nanosecond time scale. The excitation-dependent multicolor emission in time-integrated spectra is typically governed by the relative population of these emitting states.[1]

1 S. Khan, A. Gupta, N. C. Verma and C. K. Nandi, *Nano Lett.*, 2015, **15**, 8300–8305.

Pieter van Helden opened general discussion of the Concluding Remarks by Philip Gibson: You mentioned the importance of Fischer–Tropsch research in South Africa, as well as the number of areas of research in Catalysis for Fuels which need to be addressed. What is your view on the current balance of research activities in this field in South Africa?

Philip Gibson responded: Catalysis research in South Africa has been and still is biased towards Fischer–Tropsch research. There are three main reasons for this: firstly, the role of the Fischer–Tropsch process in the context of the South African fuel supply (>30% is synthetic fuel); secondly, the size and role of Sasol as a South African-based company in terms of setting/proposing/funding research directions in a South African context; and thirdly, the relatively small size of the catalysis research community in South Africa.

Well-resourced catalysis research groups take a long time to set up and continued effort to have a real impact on their field. It is understandably difficult to change direction at the same fast pace as the regional and global demands in catalysis are changing. The "new" global oil price scenario and the future role of renewable energy in transportation are two important factors that are challenging the future application of Fischer–Tropsch processes. There will obviously be niche applications, some of which will be coupled to regional strategic drivers like energy security, associated and stranded gas resources *etc.*

Fischer–Tropsch technology *per se* is a mature technology. Research groups should continuously assess their value proposition to future FT processes against the value of applying their hard earned skill sets on a maybe more important catalysis challenge. Fundamental understanding of processes creates tremendous scientific satisfaction but it must always contribute to the design/prediction of better catalysts in a reasonably short time frame. That is the success criteria from an industrial perspective.

Given the diminishing role of Fischer–Tropsch in the next 20–30 years in a global context and the relative maturity of the technology I would propose that the size of the academic research effort in South Africa is probably adequate.

Graham Hutchings said: I would like to address the topic of alloys in the FT reaction. Alloys for H_2O_2 were briefly mentioned in the preceding discussion. In this case, Pd when alloyed with Au or Sn is a very effective catalyst for the synthesis of hydrogen peroxide. In particular, by designing the alloy you can tune the selectivity. At present in FT synthesis there is an emphasis on supported Fe, Co and Ru FT catalysts. The question is – have we missed using alloys? Or has it been tried and they don't really work?

Philip Gibson responded: The use of alloys is an interesting concept and although not new, I believe it is coming into vogue due to our increased understanding and ability to use computational chemistry to predict experimental outcomes coupled to the fundamental understanding of metal surface structures and their dynamics. The exact role of metal promoters in the FT systems mentioned above are often overlooked. PGM metal additives are generally referred to as reduction promoters, Mn is accepted to be an olefin selectivity promoter, *etc.* I believe that the function of effective metal promoters should be explored in terms of their ability to form alloys with the base metal *i.e.* cobalt or iron. Once we understand these known examples, I am convinced we will identify further opportunities to design or influence FT catalyst selectivity through the use of non-conventional alloys. See my view in the closing remarks paper (DOI: 10.1039/C7FD00096K).

Enrique Iglesia said: Thank you for the thoughtful remarks and excellent summary of our "arguments" and discourse. This is more an observation than a comment, although I would hope that you could rebut the observation as inaccurate. As you mention risk in the context of technology development, I travelled back to my industry days and reminisced about why I left. The development of technology had become a cost instead of a benefit and a source of profit and advantage in the marketplace. The deployment of new technology had become a risk, one that was not reasonable especially for disruptive technologies. The outlets for discoveries were few in a short-sighted marketplace that valued companies quarterly and not over their history. The thinking "out-of-the-box" was fun but not profitable; why not do the thinking and the fun in a place that did not seek profits, such as academia? How much of these feelings about the past are no longer relevant, in South Africa and throughout the world?

Nico Fischer noted: You mention that the willingness to take risks with regard to implementing new developments in the industrial environment has decreased over the last decades. I am wondering, why is this? Is it really just that humanity is at a level of comfort that no significant external pressure is present as it was for example during the early 20th century when the ammonia synthesis was established in just 7 years starting from lab scale to a 600 tonnes per day plant? Surely the global economy is still/again under immense pressure? Global trends or drivers are regularly cited – still, even these do not seem to support "risky" business decisions.

Philip Gibson replied: My subjective view is that hard core technical people are no longer represented on the boards and/or decision making bodies of industrial companies to the same extent as in the previous decades. Experienced technical people developed a sixth sense, *i.e.* they developed a gut feel for technology risks and could influence decisions through their standing and "trust" in their respective organizations. Techno-economic considerations are also no longer the main drivers for the implementation of new technologies. Global markets, regional economics, environmental legacy, regional/global politics and well informed and demanding customers are all part of the decision criteria. It is very complex to get all of these right. It is getting increasingly difficult to address all the risk aspects in the available time, thus delaying decisions. Modern corporate governance also dictates that the consequences of wrong decisions are more dire than in the past.

Michael Claeys communicated: Biomass has been discussed at the meeting as an important source for future fuel production. Seeing that this is the first Faraday Discussion meeting on the African continent, I would be interested in your opinion regarding the use of biomass in a South African/African context? Issues of water scarcity and competition with food production come to mind, and may prohibit this technology or at least limit its importance in comparison to the use of fossil resources.

Philip Gibson communicated in response: There are very few regions suited for the cultivation of biomass on a sustainable basis in Southern Africa. The huge geographical area adds logistical challenges in addition to the default objection related to competition with food production. It is my understanding that a holistic energy balance simply does not make sense in a southern African context, *i.e.* relatively poor transport infrastructure and looming food insecurity. The use of biomass in niche opportunities *i.e.* corn fermentation, wood pulp into existing coal gasifiers *etc.*, will continue to be evaluated and might make sense in opportunistic specific regional areas. The east coast of South Africa and Mozambique are, to my knowledge, some of the very few areas that can be considered for large scale biomass cultivation.

Faraday Discussions

ROYAL SOCIETY OF CHEMISTRY

PAPER

Catalysis for fuels: Concluding Remarks

P. Gibson

Received 8th March 2017, Accepted 9th March 2017
DOI: 10.1039/c7fd00096k

The 2017 Faraday Discussion on the topic of "Catalysis for Fuels" was unique in the sense that it was the first Faraday Discussion to be held on the continent of Africa. "Catalysis" and "Fuels" are both topics that could be widely interpreted and the session topics proved to be a relevant spread of old and new *i.e.* Fischer–Tropsch chemistry, biomass refining, zeolite conversions and photocatalysis. Most of the papers were underpinned by fundamental studies, catalyst design approaches, reports of *in operando* characterization and detailed speciation and micro kinetic analyses. Examples of commercial application were offered under the headings of biomass conversion, Fischer–Tropsch and to some extent photocatalysis. Cognisance was given to the increasingly important role of catalytic metals in terms of scarcity, cost and environmental impact. The potential role of novel alloys in addressing some of the catalytic mechanistic challenges turned out to be one of the central themes during the discussions. The following remarks are an attempt to draw parallels between the topics under discussion and the author's subjective view of current universal questions that could/should be highlighted under the "Catalysis for Fuels" heading.

The occasion of the Faraday Discussion in Cape Town, South Africa

South Africa is unique in a global context as more than 30% of its transportation fuel demand is being met through synthetic fuel production *via* the Fischer–Tropsch process. The chemistry and catalysis associated with the Fischer–Tropsch synthesis have thus been historical areas of focus in a South African academic and corporate context. The refining of synthetic crude places unique demands on conventional *i.e.* crude oil derived refinery catalysts and processes. The typical absence of heteroatom (N & S) based compounds, the absence of aromatic compounds and the presence of oxygenated compounds are all factors that require a very specific fundamental understanding of the behaviour and application of catalysts in synthetic crude refining. Some of these characteristics are of course also present in biorefining processes which is an interface that is relevant to the present holistic view of regional and national fuel pools and specifications. Catalysis

Group Technology, Sasol South Africa (Pty) Ltd, South Africa. E-mail: philip.gibson@sasol.com

is a crucial discipline when it comes to providing the scientific and technological foundation for making cleaner, more efficient and economically viable fuels. Based on the above it is fair to say that "Catalysis for Fuels" is a fitting theme for the first RSC Faraday Discussion on the African continent.

A subjective backdrop to the theme and associated discussions

In setting the scene for the ensuing discussion one can rightly ask: so what are the innovations and trends that one can reasonably expect to feature in a "state-of-the-art" account on "Catalysis for Fuels"?

Fossil raw materials

The use of catalysis in support of the conversion of fossil fuels may appear to be relatively mature, yet there are clear challenges that require fundamental improvements in this field. The amount of crude oil refined everyday worldwide amounts to about 85–90 million barrels. This implies that even the smallest improvements in selectivity or effectiveness in the processes used translate to a significant impact. Refinery catalysts still represent $7 billion per annum (~25% of total catalyst sales) in value.[1]

Robust catalysts and processes for conversion and clean-up of ever heavier and poorer quality crudes are needed. These will require increased porosity to avoid mass transfer limitations and at the same time they will need to be more resistant to poisoning and deactivation. Alternative reactor technologies *i.e.* slurry technologies should be explored as potential solutions for heavy oil or bottom of the barrel upgrading.

The compositional differences between shale gas, NGLs and tight oil is for now a phenomena specific to the North American continent but the challenges associated with adapting refineries to cope with these multi-sourced feedstocks should help to define future "Catalysts for Fuels". Flexible catalytic systems that can accept mixtures of biomass-derived and conventional petroleum feedstock will in all likelihood become a feature of European refineries. Integrated flexible processes and catalysts for efficient carbon coupling, allowing responses to changes both in raw material and energy input, as well as product functionalization are needed.

A more detailed wish list of refinery developments, albeit from an industrial perspective, is presented below:

• New catalyst technologies (zeolite based) for hydrocracking and FCC resid upgrading to improve profitability (bottom of the barrel upgrading).

• Greener acid catalysts for alkylation (i-butane + 1-butene) and indirect alkylation (olefin dimerization) – successful operation of the first AlkyClean Solid Acid Alkylation unit (paraffin + olefin) for high octane production was presented in 2016.[2]

• Refining/petrochemical integration is a major drive for many refineries, with a focus on the following:

◦ Existing FCC units cannot keep up with propylene demand growth. This has resulted in greener FCCs with a move away from fuels to chemicals, with the prime objective to maximize propylene yield. As a result, there has been significant development in the area of zeolite catalysts to achieve a shift from fuels mode to petrochemical mode of operation.

∘ Purified aromatics can go to PX or be re-blended into gasoline. The market for aromatics and PX is to double by 2030. Further development of shape selective zeolites would be key in the future.

∘ The olefin-rich fraction can be re-cracked into propylene, aromatized into BTX, or re-blended into gasoline with no change in octane value.

∘ Catalysis for the valorization of light hydrocarbons is becoming more important – the conversion of methanol mixed with C5–C6 hydrocarbons (including pentane, hexane, and cyclohexane) over Zn/ZSM-5 catalyst to yield aromatics was reported.[2] The result confirmed that it is feasible to use the exothermic methanol conversion to enhance the endothermic transformation of some C5–C6 hydrocarbons into aromatics.

Optimised catalysts based on non-precious metals can replace noble metals (scarce and expensive) and heavy transition metals that are linked to health or environmental concerns. The role of computational chemistry in searching for alternative metals or metal alloys is highlighted later in this discussion.

Less scientific, but nevertheless a driving force for business pull, is the need for catalysts that create unique and "brandable" product properties. This is especially true for commodity fuel retailers. Consumer differentiation based on environmentally benign feedstock and/or niche product qualities is on the increase.

Bio-based fuels

A detailed understanding of the fundamentals of the various forms of catalysis, process technology, agriculture and biomass production chains is essential if we are to arrive at an economically viable and sustainable bio-based economy. Catalysis is a key enabler in those processes. The main challenges for catalysis involving biomass and renewable resources are (1) selectivity, *i.e.* how to convert the complex biomass mixture to the target molecules – either drop-in or new molecules – in as few steps as possible thus preventing intermediate separations, and (2) stability, *i.e.* how to achieve a catalyst that can handle poisons present in the biomass and that survives the conditions needed to convert biomass.

These challenges can be broken down into the following themes:[3]

• Understanding and controlling the reaction mechanisms and catalyst functions such as hydrogenation, and acid/base-functions, for operation in aqueous media.

• Understanding and controlling the catalyst selectivity (both chemoselectivity as well as regioselectivity) for over-functionalised feedstock, as this will lead to fewer by-products and thus a more efficient process in general.

• Understanding and controlling the catalyst stability to overcome typical issues with destruction, fouling, or poisoning of the catalyst.

• Activating biomass, *i.e.* making the solid biomass suitable for more facile downstream processing.

- Developing new catalytic routes to new molecules with pre-defined properties.
- Integrating the catalysis with separation technology in order to mitigate the challenges presented by high dilution, feed contaminants or abundant by-products.
- Implementing biofeedstock into downstream operations *i.e.* the co-processing of vegetable oils in hydrotreatment units.
- Developing hydrodeoxygenation (HDO) catalysts for feedstock with high oxygen content, which is a high priority.

CO_2 conversion

The conversion of carbon dioxide into energy carriers using renewable energy sources is essential in order to cope with the negative impact of increasing atmospheric CO_2. It would also solve the problem of storage of excess renewable energy. In the short term, better catalysts for electrolysers need to be developed. A sustainable solution to the energy challenge involves the development of cheap photocatalytic materials and their integration into solar-to-hydrogen devices with an efficiency of at least 20%. Together with catalytic technology for the production of chemicals from CO_2, such devices could vastly contribute to the realization of a more carbon neutral chemical industry.

Direct electrochemical CO_2 reduction will require breakthroughs in the field of electrocatalysis. A more likely scenario involves water electrolysis by renewable electrons generated by photovoltaics or direct photocatalytic water splitting to produce the H_2 needed for the reduction of CO_2.

If H_2 could be cheaply produced from solar energy, the production of synthesis gas from CO_2 would be within reach. From synthesis gas, a wide range of products can be manufactured with established catalytic technology. Another aspect of the photocatalytic reduction of water or carbon dioxide is that pure oxygen can be obtained as a by-product. This by-product stream has economic value, because the alternative route, *via* air separation, is very expensive.[3]

Bi-functional and multi-functional catalysts

Integrating multiple catalytic reactions into a single system is an important challenge across the fields of biocatalysis and homogeneous catalysis. Integration can improve the efficiency of catalytic processes and lead to entirely new functions. Man-made catalytic systems are free from the restrictions inherent to biology, permitting the use of a wider range of experimental conditions and targeting a wider range of functions. In the petrochemical industry the notion of process intensification by combining multiple catalytic functions is advanced and several processes operate based on multiple catalysts arranged in a stacked bed. Another important example of a multi catalyst device in heterogeneous catalysis is the use of spatially and temporally separated reduction catalysts and de-NO_x catalysts in vehicle exhaust systems.

The main challenges that need to be addressed in order to make more complex catalytic systems are catalyst and process compatibility – ideally between different catalysts, but at least between conditions and/or process parameters. Another challenge results from differences in the rates at which catalysts lose their activity. A workable integrated multi catalyst system will therefore have to offer the opportunity of independent catalyst activity management. When combining chemical processes side by side or within a single system, catalysis is essential as it allows independent control over the rates at which the various reactions occur. Such control is crucial if the various reactions are to be efficiently integrated.

Catalysis fundamentals

The future development of new and improved catalytic material and processes will require a shift towards rational design supported by fundamental knowledge. The integration of spectroscopic and microscopic tools, along with predictive modelling strategies, will allow for the design and synthesis of new molecules and complex materials with structural and chemical properties that are well defined and controlled over various length- and time-scales. Improved fundamental knowledge of catalysts, their functioning and manufacture, might enable the replacement of noble scarce and toxic elements by more abundant and environmentally safe metals for catalysis. A "wish list" of fundamental deliverables could include the following:[3]

• Computational methods that account for the complexity of chemical and catalytic processes on various time- and length-scales (Fig. 1).

• A transient view of catalysis that accounts for processes not in equilibrium *i.e.* not the common steady-state view.

• Precise control of morphology, composition and defect structure, leading to influence over optical, electronic and catalytic properties on multiple scales.

• Tools to translate chemical descriptors of catalysis and catalysts into predictors and design rules for new molecules and synthetic methods.

• Control and efficient generation of highly reactive catalytic intermediates.

• Precise control of selectivity in synthesis and catalysis, including regulation mechanisms to accommodate changes in reaction conditions.

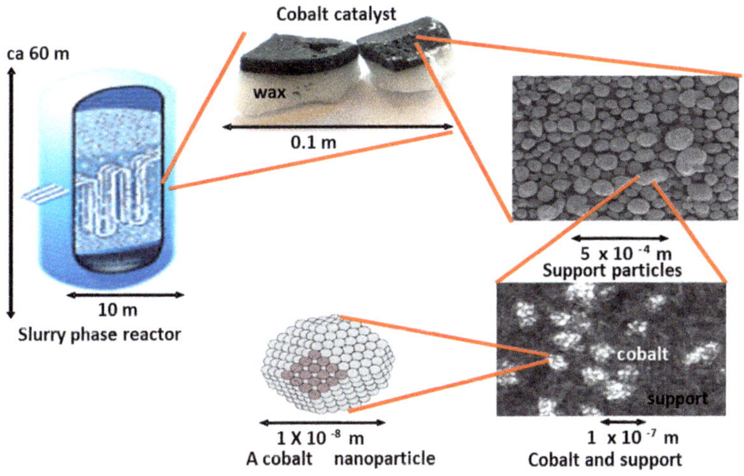

Fig. 1 Relevant length scales for the Sasol SSDP™ GTL process.

• The targeted activation of specific catalytic sites and the control of reaction speeds and mechanisms, allowing for the cost effective utilization of catalyst materials and an efficient replacement of noble metals by cheaper and more abundant materials.

Having made the assumption that many of the in vogue research elements discussed above should by necessity feature in a discussion about "Catalysis for Fuels", a brief overview of some of the scientific presentations and ensuing discussions are presented below.

Catalysis for Fuels

Some of the thoughts that stuck with me, possibly because they fit very nicely with my predetermined expectations were the apt illustration of stabilization and manipulation of transition states during the introductory lecture (Enrique Iglesia, University of California, Berkley). The use of confinement can induce thermodynamic changes of a reactive intermediate; preferential solvation can lead to the preferential stabilization of a targeted transition state. Crowded surfaces and lattice compression were also presented as fundamental means to manipulate reactive intermediates.

Comments on the debate of the Fischer–Tropsch synthesis mechanism

This particular discussion stood out, not only because of the enthusiastic and informed participation of all participants but as an example of how answers to certain scientific questions reach maturity only after many man years' worth of DFT calculations, meticulous experimental design and execution, many sleepless nights and a vision to pursue answers.

In this section, three papers (DOI: 10.1039/c6fd00205f, DOI: 10.1039/c6fd00191b, DOI: 10.1039/c6fd00197a) were presented that addressed the FT mechanism from three points of view using surface science and micro-kinetic methodologies. During the discussion a number of issues were addressed regarding the current fundamental understanding of the Fischer–Tropsch mechanism. It was clear that there have been huge advancements made in understanding this complicated reaction network and a number of aspects were noted that the delegates in the discussion agreed on. Amongst these was the fact that the FT mechanism proceeds in a high coverage regime, where the adsorbed CO plays a critical role in ensuring the appropriate kinetics landscape. Furthermore, there seems to be agreement on the nature of the FT monomer (CH*) and coupling steps. However, a number of points were heavily debated: the relation of order of the FT reaction to the rate controlling step(s), the nature of these rate and selectivity controlling steps, the extent of CO coverage under working FT conditions (*i.e.* >1 mL or restricted to about 0.67 mL), the interplay in a multisite reaction network *versus* a single site view and how that determines how you think about initiation of the reaction and the viability of diffusion of reactants on and between active sites. Despite these disagreements there seems to be meaningful progress in "solving the FT mechanism", fuelled by more holistic mechanistic assessments underpinned by quality DFT and experimental data and critical thinking.

Designing new catalysts for synthetic fuels

A theme that came to the fore during the discussion of catalyst design was that of the use of metal alloys.

The use of computational chemical models that relate steric and electronic features of molecular structures to chemical behaviour is fairly mature and is called quantitative structure-activity relationships (QSAR), or "*in silico* design and optimisation". This strategy is familiar within the pharmaceutical industry where QSAR-based screening of molecules that target specific biochemical receptors has been used to identify novel possibilities as improved bioactive compounds with specific biologic function.[4] This approach has also been applied within homogeneous catalysis.[5] Now that computational tools used to model heterogeneous systems have achieved a level of maturity (in software, hardware and methodologies), researchers are able to compute whole databases of energetic properties of heterogeneous systems. Thus QSAR-type studies can now also be performed on heterogeneous systems. This approach is similar to high-throughput screening, but where some calculated property of the system, say the adsorption of a carbon atom, is statistically linked to electronic properties of the system, say the d-band centre and dispersion.

This strategy was applied by the Nørskov group[6] where it led to the successful identification of novel catalysts for the reduction of CO_2 to methanol. This reaction is typically performed on Cu/Zn catalysts but a computational screening study that attempted to relate the adsorption energies of C and O on stepped fcc-based bimetallic compounds led to the identification of Ni/Ga systems as viable alternatives. Laboratory synthesis of these bimetallic systems resulted in novel catalysts with activities at least on par with the usual Cu/Zn systems, but with considerably lower production of CO (which is produced in competition with methanol).

At Sasol, in order to accelerate our understanding of possible catalysts that may be alternatives in the Fischer–Tropsch synthesis, we attempted a QSAR-type study where we related the adsorption energies of C and O on stepped fcc-alloys to a variety of electronic properties we could calculate for the bulk fcc-alloy system. We hoped that in this manner we would be able to perform a screening study by modelling only the bulk structure (a fast calculation compared to a full surface model and surface reactions), and thereby identify candidates for alternative Fischer–Tropsch catalysts. We did not limit ourselves to an established understanding of catalytic activity as it relates to d-band structure but instead performed a principal component analysis (PCA)[7] in an attempt to relate various electronic properties to the adsorption energies of C and O. Interestingly, the PCA validated existing knowledge of how the d-band centre and dispersion affects adsorption energies of C and O (confirming the well-known Newns-Anderson model).[8] Assuming that a Co-catalyst is the "ideal" Fischer–Tropsch catalyst, we could identify a variety of novel alloys which we believe may have competitive Fischer–Tropsch activity if they could be synthesised in an fcc-lattice. A similar approach is presently being used to identify other novel catalytic systems for the conversion of syngas.

Hydrocarbon conversion in the production of synthetic fuels

The plenary paper "Opportunities in upgrading biomass crudes" (DOI: 10.1039/c6fd00208k) clearly illustrated that the challenges associated with biomass derived crudes can be overcome through the selective activation of the biomass followed by innovative selection of conventional refinery processes. This approach makes it possible to treat biocrude derived from hydrothermal treatment in existing oil refineries, preferably integrated with regular oil feedstock. A holistic view and careful comparison of biocrude characteristics with conventional refinery feedstock can lead to the discovery of perfect conventional analogies *i.e.* vegetable oil is very similar to biocrude mixed with VGO-vacuum gas oil. The approach reported is an example of "thinking inside the box" rather than looking for radically new processes and/or catalysts.

The theme of integrating biomass and conventional refining is unique in that the driving force for innovation is not solely an economic pull or technology push but a combination of economics, technology and regional/national politics. Researchers are encouraged to take a broader view when they want to solicit support for development programs in this area.

The plenary paper discussed above was also one of the very few where classical engineering input was obvious, already in the conceptualization phase of the program. This is maybe a resource area that catalytic scientists and researchers can pro-actively target when defining their research hypotheses.

MeOH as both a potential energy carrier and chemical feedstock continues to be in vogue throughout the catalysis community. The plethora of recent successful start-ups of MTP and MTO facilities in China is evidence of the progress towards the concept of a "Methanol Economy". Paper DOI: 10.1039/c6fd00187d addressed the influence of zeolite topology on the axial mode of deactivation during the conversion of methanol to hydrocarbons. It was shown that zeolite topology not only affects catalyst lifetime and product distribution, but also modifies the axial mode of deactivation. This is a new insight and has not been addressed before.

Novel photocatalysts

Paper DOI: 10.1039/c6fd00184j described so-called *Z*-scheme photocatalytic water splitting based on the two step excitation of an oxygen evolution photocatalyst (OEP) and a hydrogen evolution photocatalyst (HEP). This is a promising approach toward the utilization of visible light. The novel approach of creating a photocatalyst sheet *i.e.* HEP and OEP particles embedded in a conductive layer was elegantly illustrated. The design of the photocatalytic sheets is also well suited for large scale applications.

Selected parts of the discussion highlighted the safety risks associated with having pure hydrogen and oxygen streams in close proximity on a large scale. Several participants pointed out that although the primary focus tends to fall on the hydrogen generation, the oxygen or in some cases also hydrogen peroxide "by-product" have equal value.

The use of laser-flash-photolysis-spectroscopy as a non-destructive method to study the morphological and optical properties of TiO_2 powders (DOI: 1039/c6fd00193a) is a response to the fast-growing demand for fundamental studies in photocatalysis, especially for investigations on powdered photocatalysts due to their better comparability with the results obtained in photocatalytic tests. The findings elegantly illustrated that the obtained transient absorption signals could be strongly influenced by irreversible changes of the material. This has implications for the evaluation of the underlying photocatalytic reactions.

Conclusion

At the start of these closing remarks, a subjective overview of topics and themes were suggested that, in the view of the author, had to be addressed during the course of any comprehensive discussion on "Catalysis for Fuels". During the discussions a scorecard was kept to tick off topics, themes and debates which could be correlated to the "hot" topic list. The meeting achieved a first class pass. Having listened to the participating authors introduce their papers and the resultant engaging discussions by all participants, this format of scientific debate is not only unique but highly successful in sparking further research.

Acknowledgements

I thank the organizers for inviting a hardened industrial catalysis practitioner to close such a prestigious event and the participants for making the first Faraday Discussion on the African continent such a success. The ability of the session chairs to keep discussions focussed and entertaining is acknowledged. The author acknowledges Sasol colleagues, Pieter van Heerden, Pieter van Helden and Jan-Albert van Den Berg for their insights into refinery catalysis and computational chemistry.

References

1 What's new in Catalysis, http://www.hydrocarbonprocessing.com/, March 2015.
2 *European Refining Technology Conference*, 2016.
3 M. Barroso, J. H. Bitter, Q. Broxterman, B. de Bruin, F. R. van Buren, U. Hanefeld, E. J. M. Hensen, S. Otto, E. T. C. Vogt and B. M. Weckhuysen, *Catalysis-Key to a Sustainable Future*, The Netherlands Institute for Catalysis Research, 2015.
4 G. Sliwoski, S. Kothiwale, J. Meiler and E. W. Lowe, Computational Methods in Drug Discovery, *Pharmacol. Rev.*, 2014, **66**, 334–395.
5 K. C. Harper, E. N. Bess and M. S. Sigman, Multidimensional steric parameters in the analysis of asymmetric catalytic reactions, *Nat. Chem.*, 2012, **4**, 366–374.
6 F. Studt, I. Sharafutdinov, F. Abild-Pedersen, C. F. Elkjær, J. S. Hummelshøj, S. Dahl, I. Chorkendorff and J. K. Nørskov, Discovery of a Ni–Ga catalyst for carbon dioxide reduction to methanol, *Nat. Chem.*, 2014, **6**, 320–324.
7 S. Pirhadi, F. Shiri and J. B. Ghasemiac, Multivariate statistical analysis methods in QSAR, *RSC Adv.*, 2015, **5**, 104635–104665.
8 A. Vojvodic, J. K. Nørskov and F. Abild-Pedersen, Electronic Structure Effects in Transition Metal Surface Chemistry, *Top. Catal.*, 2014, **57**, 25–32.

Poster titles

A selective adsorption of platinum on to Co_3O_4/SiO_2 for a synergistic promotion in the Fischer–Tropsch synthesis, **S. Govender and E. van Steen**, *University of Cape Town, South Africa*

MefCO_2: methanol for fuels from CO_2 – copper-based catalysts for green methanol synthesis, **J. Hayward, M. Bowker, S. Golunski, C. Brookes, W. Jones and P. Smith**, *Cardiff Catalysis Institute, UK*

CO_2 and H_2 interaction with nickel–yttria stabilized zirconia interface: a DFT study, **A. Essadek, A. Roldan and N. H. de Leeuw**, *University College London, UK*

C–C coupling catalysed by bio-inspired iron sulfide, **A. Roldan and N. H. de Leeuw**, *Cardiff University, UK*

Interrogating the effects of microwave irradiation on the physico-chemistries of Pd-decorated ferroalloy core–shell nanostructures for alkaline fuel cells, **K. Ozoemena, N. Mphahlele, N. Palaniyandy, K. Raju, M. Modibedi, M. Mathe and C. Billing**, *Council for Scientific and Industrial Research (CSIR), South Africa*

Synthesis and characterization of supported nanoparticles for the synthesis of biofuels, **M. Shozi, Q. Gazu and H. B. Friedrich**, *University of KwaZulu-Natal, South Africa*

Effect of varying the cation ratio in chromium molybdate catalysts for the oxidative dehydrogenation of n-octane, **H. Bandaru, S. Singh and H. B. Friedrich**, *University of KwaZulu-Natal, South Africa*

Redox properties of metal oxides, **K. Naicker, H. B. Friedrich, S. Singh and A. Mahomed**, *University of KwaZulu-Natal, South Africa*

Formulation and testing of MOF derived Fe@C catalysts for High Temperature Fischer–Tropsch synthesis (HTFTS), **L. Oar-Arteta, M. J. Valero-Romero, T. Wezendonk, X. Sun, F. Kapteijn and J. Gascon**, *Delft University of Technology, Netherlands*

Cu-based SCILL catalysts for the competitive hydrogenation of octanal in the presence of octene, **L. Deeplal and H. B. Friedrich**, *University of KwaZulu-Natal, South Africa*

Iron modified hydroxyapatities in the activation of n-octane, **D. Padayachee, V. Dasireddy, K. Bharuth-Ram, H. B. Friedrich and S. Singh**, *University of KwaZulu-Natal, South Africa*

Surface S species formed on Au/TiO_2 during the catalytic reduction of SO_2: poisons, promoters or spectators? **L. Jewell, T. Ngwenya and I. Nongwe Beas,** *Unisa, South Africa*

Multifunctional Zn(II) complexes: photophysical properties and catalytic transesterification toward biodiesel synthesis, **A. K. Gupta, P. Chullikkattil Parameswaran and A. Dhir,** *Indian Institute of Technology, Mandi, India*

Metal-organic-framework mediated cobalt/N-doped carbon hybrids as superior catalysts for the hydrogenation of nitroarenes, **X. Sun, A. Olivos Suarez, L. Oar-Arteta, E. Rozhko, D. Osadchii, A. Bavykina, F. Kapteijn and J. Gascon,** *Delft University of Technology, Netherlands*

The nature of Fe species during MOF pyrolysis and FTS, **T. Wezendonk, M. Makkee, V. P. Santos, A. Chojecki, M. Ruitenbeek, G. R. Meima, A. I. Dugulan, J. Gascon and F. Kapteijn,** *Delft University of Technology, Netherlands*

Stable nano-gold catalyst for high temperature applications, **J. Moma, P. Franklyn, D. Barrett, M. Scurrell and D. Billing,** *University of the Witwatersrand, South Africa*

Surface and shape modification of FeS nanocrystals by cysteine adsorption for heterogeneous catalytic applications, **N. Dzade, A. Roldan and N. H. de Leeuw,** *Utrecht University, Netherlands*

Photocatalytic property of visible light responsive silver nanoparticles–titania composite thin films fabricated by Molecular Precursor Method, **L. Daniel, P. Gawanas and U. Veikko,** *University of Namibia, Namibia*

Synthesis and characterization of reduced graphene oxide supported Co nanocatalysts for Fischer–Tropsch synthesis, **G. B. Sundaram, N. Fischer and M. Claeys,** *University of Cape Town, South Africa*

The interplay of Lewis and Brønsted acid sites in the catalytic performance of ZSM-5 in the methanol-to-olefins process, **I. Yarulina, M. Radersma, M. Goesten, F. Kapteijn and J. Gascon,** *Delft University of Technology, Netherlands*

Synthesis of higher alcohols on iron-based catalysts, **M. Schaller, E. Reichelt, H. Lund and M. Jahn,** *Fraunhofer IKTS, Germany*

Characterizing active sites of Fischer–Tropsch synthesis: issues from selectivity, **H. Schulz,** *Karlsruhe Institute of Technology, Engler–Bunte-Institute, Germany*

The use of CdSe based nanowires for the photoelectrocatalytic production of hydrogen gas from aqueous solutions, **I. Abdullah, M. Fawzy, M. Blumenthal and P. Levecque,** *University of Cape Town, South Africa*

Hollow carbon spheres as ideal support for hydrogen spillover studies on Fischer–Tropsch catalysts, **T. Phaahlamohlaka, L. Jewell and N. Coville,** *University of the Witwatersrand, South Africa*

Activation of alkanes to primary alcohols: "commissioning" a microbial cell factory that incorporates substrate transport, electron supply and biocatalysis, **B. White, C. J. Fenner, M. S. Smit and S. T. L. Harrison,** *University of Cape Town, South Africa*

Operando Raman spectroscopy as a probe to investigate the structure/activity relationship of iron-based Fischer–Tropsch catalysts, **H. Kotzé, N. Fischer and M. Claeys,** *University of Cape Town, South Africa*

Fe-catalysed alkene oxidation: effect of steric and electronic parameters on non-heme oxygenase activity, **A. Swarts, L. Swartzberg and C. van Sittert,** *North-West University, South Africa*

Size controlled preparation of monodisperse and highly crystalline Co_3O_4 nanoparticles in absence of surfactants, **M. Wolf, N. Fischer and M. Claeys,** *University of Cape Town, South Africa*

Reaction and interconversion of oxygenates in the Fischer–Tropsch synthesis, **J. C. Q. Fletcher, G. Govender, M. de Croon and J. Schouten,** *University of Cape Town, South Africa*

Co_3O_4 morphology in CO-PrOx reaction – an *in situ* study, **M. Khasu, M. Claeys and N. Fischer,** *Institute for Catalysis Research and c*change (DST-NRF Centre of Excellence in Catalysis), South Africa*

Metal–support interactions in cobalt–alumina model catalysts for the Fischer–Tropsch synthesis, **A. Petersen, M. Claeys and E. van Steen,** *University of Cape Town, South Africa*

Effect of metal–support interactions during CO-PrOx over Co_3O_4/Al_2O_3 catalysts: an *in situ* study, **T. Nyathi, N. Fischer, A. York and M. Claeys,** *University of Cape Town, South Africa*

Molecular dynamics modelling of ionic liquids for CO_2 capture, **Q. Parker, R. Bell and N. H. de Leeuw,** *University College London/Cardiff University, UK*

Catalysts design for the partial hydrogenation of highly unsaturated bio-fuels, **E. Olaoye, O. Oyetunji, A. Munyaneza, G. Kumar and F. Mbaiwa,** *University of Botswana, Botswana*

Growing whole cell catalyst system allows improved efficiency and performance of the catalyst CYP153A6 for end-terminal activation of the linear alkane octane, **S. T. L. Harrison, O. Olaofe, C. Fenner and M. S. Smit,** *University of Cape Town, South Africa*

Specificity and selectivity of different CYP153 alkane hydroxylases, **M. S. Smit, D. Opperman and C. Jacobs,** *University of the Free State, South Africa*

Formation of hydrocarbons and catalyst deactivation in MTH on zeolite HZSM-5, as depending on reaction time and temperature, **H. Schulz,** *Karlsruhe Institute of Technology, Engler-Bunte-Institute, Germany*

Influence of different fuels on the properties of solution-combustion synthesized palladium/ceria catalysts for low-temperature methane combustion, **T. P. O. Mkhwanazi, H. B. Friedrich, A. S. Mahomed and S. Singh,** *University of KwaZulu-Natal, South Africa*

Investigating the effect of oxidation promoters on cobalt oxide catalysts in the preferential oxidation of carbon monoxide, **E. Trenor, M. Claeys, N. Fischer and P. J. Kooyman,** *University of Cape Town, South Africa*

Synthesis of boron-doped hollow spheres with a homogeneous shell thickness and catalytic performance of Cu/B-HCS for aerobic oxidation of ethyl benzene, **I. Nongwe Beas, L. Jewell and N. Coville,** *Unisa, South Africa*

Synthesis and characterisation of spherical, monodisperse silver nanoparticles, **S. Adam and P. J. Kooyman,** *University of Cape Town, South Africa*

Primary hydrocracking of n-C_{16} in shape-selective MFI zeolites by competitive adsorption of water, **R. Brosius, P. J. Kooyman and J. C. Q. Fletcher,** *University of Cape Town, South Africa*

Detailed Fischer–Tropsch product distribution over Fe catalyst: A 2D GC study, **M. Fadlalla, H. Niemantsverdriet, N. Fischer and M. Claeys,** *University of Cape Town, South Africa*

Dry methane & ethanol reformation in a catalytic membrane shift reactor: a process for the production of high purity hydrogen and/or syngas, **S. McCord and J. McGregor,** *University of Sheffield, UK*

Surface modification of cobalt oxide with silanes for improved catalytic performance in Fischer–Tropsch catalysis, **L. Macheli and E. van Steen,** *DST-NRF Centre of Excellence in Catalysis, c*change, Centre for Catalysis Research, Department of Chemical Engineering, University of Cape Town, South Africa*

Synthesis of platinum and gold nano-sized particle supported on TiO_2, **A. Kunene and E. van Steen,** *University of Cape Town, South Africa*

Oxidation of an iron catalyst as a function of water partial pressure during Fischer–Tropsch synthesis, **R. Stegmann, M. Claeys, N. Fischer, H. Kotzé and H. Niemantsverdriet,** *University of Cape Town, South Africa*

Zinc aluminate as novel support for cobalt-based Fischer–Tropsch synthesis at high conversion, **C. Tucker and E. van Steen,** *University of Cape Town, South Africa*

Platinum nanoparticles to study metal location in shape-selective hydrocracking, **D. Nel, R. Brosius, P. J. Kooyman and J. C. Q. Fletcher,** *University of Cape Town, South Africa*

Performance of $NiFe_2O_4$–SiO_2–TiO_2 magnetic photocatalyst for the effective photocatalytic reduction of Cr(VI) in aqueous solutions, **M. O. Ojemaye, O. O. Okoh and A. I. Okoa,** *University of Fort Hare, South Africa*

Time-resolved characterisation of hierarchical diffusion and reactions in zeolites and zeotypes with Temporal Analysis of Products (TAP), **E. A. Redekop, J. S. Martinez-Espin, M. Mortén, B.-T. L. Bleken, R. Fushimi, M. Mykland and U. Olsbye,** *University of Oslo, Norway*

The Faraday Discussions Poster Prize for the best poster was jointly awarded to Mr Xiaohui Sun of Delft University of Technology, Netherlands, for his poster on metal-organic-framework mediated cobalt/N-doped carbon hybrids as superior catalysts for the hydrogenation of nitroarenes, and to Mr Moritz Wolf of University of Cape Town, South Africa, for his poster on size controlled preparation of monodisperse and highly crystalline Co_3O_4 nanoparticles in absence of surfactants.

List of participants

Mr Ilyaas Abdullah, *University of Cape Town, South Africa*
Miss Sarah Adam, *University of Cape Town, South Africa*
Dr Evans Adei, *Kwame Nkrumah University of Science and Technology, Ghana*
Mr Paul Akomeah, *Kwame Nkrumah University of Science and Technology, Ghana*
Professor Edet Archibong, *University of Namibia, Namibia*
Professor Dr Detlef Bahnemann, *Leibniz Universität Hannover, Germany*
Mr Harinarayana Bandaru, *University of KwaZulu-Natal, South Africa*
Professor Michael Bowker, *Cardiff University, United Kingdom*
Mr Marien Bremmer, *Leiden University, Netherlands*
Dr Roald Brosius, *University of Cape Town, South Africa*
Professor Richard Catlow, *University College London, United Kingdom*
Professor Michael Claeys, *University of Cape Town, South Africa*
Dr Jezreel Cloete, *University of Stellenbosch, South Africa*
Dr Paul Collier, *Johnson Matthey, United Kingdom*
Professor Avelino Corma, *Instituto de Tecnología Química (UPV-CSIC), Spain*
Dr Marie Cote, *Royal Society of Chemistry, United Kingdom*
Professor Neil Coville, *University of the Witwatersrand, South Africa*
Dr Likius Daniel, *University of Namibia, Namibia*
Dr Burtron Davis, *University of Kentucky, USA*
Professor Dr Nora de Leeuw, *Cardiff University, United Kingdom*
Mr Dominic De Oliviera, *University of Cape Town, South Africa*
Miss Letisha Deeplal, *University of KwaZulu-Natal, South Africa*
Professor Kazunari Domen, *The University of Tokyo, Japan*
Dr Nelson Dzade, *Utrecht University, Netherlands*
Mr Abdel-Aziz Essadek, *Cardiff University, United Kingdom*
Mr Mohamed Fadlalla, *University of Cape Town, South Africa*
Dr Nico Fischer, *University of Cape Town, South Africa*
Dr Jack Fletcher, *University of Cape Town, South Africa*
Dr Roy Peter Forbes, *University of the Witwatersrand, South Africa*
Professor Holger Friedrich, *University of KwaZulu-Natal, South Africa*
Mr Thobani Gambu, *University of Cape Town, South Africa*
Mr Philip Gibson, *Sasol Group Technology, South Africa*
Mr Sandeeran Govender, *University of Cape Town, South Africa*
Mr Abhishek Kumar Gupta, *Indian Institute of Technology, Mandi, India*
Dr Siphamandla Hadebe, *Sasol Group Technology, South Africa*
Ms Hilaria Hakwenye, *University of Namibia, Namibia*
Professor Susan T.L. Harrison, *University of Cape Town, South Africa*
Dr James Hayward, *Cardiff University, United Kingdom*
Professor Dr Emiel Hensen, *Technische Universiteit Eindhoven, Netherlands*
Dr Katherine Holt, *University College London, United Kingdom*
Professor Graham Hutchings, *Cardiff University, United Kingdom*
Professor Enrique Iglesia, *University of California at Berkeley, USA*
Mr Andrew Jeskins, *Royal Society of Chemistry, United Kingdom*

Professor Linda Jewell, *Unisa, South Africa*
Professor Dr Freek Kapteijn, *Delft University of Technology, Netherlands*
Mr Motlokoa Khasu, *Institute for Catalysis Research and c*change, South Africa*
Dr Simon Kondrat, *Cardiff Catalysis Institute, United Kingdom*
Professor Dr Patricia Kooyman, *University of Cape Town, South Africa*
Dr Hendrik Kotzé, *University of Cape Town, South Africa*
Dr Gopendra Kumar, *University of Botswana, Botswana*
Miss Avela Kunene, *University of Cape Town, South Africa*
Professor Dong-Hau Kuo, *National Taiwan University of Science and Technology, Chinese Taipei*
Professor David Lennon, *University of Glasgow, United Kingdom*
Mr Aubin-Maurice Liwanga-Ehumbu, *Sasol Group Technology, South Africa*
Professor Ding Ma, *Peking University, China*
Mr Lebohang Macheli, *University of Cape Town, South Africa*
Mr Wijnand Marquart, *University of Cape Town, South Africa*
Dr Foster Mbaiwa, *University of Botswana, Botswana*
Mr Stephen McCord, *University of Sheffield, United Kingdom*
Miss Mary Mensah, *Kwame Nkrumah University of Science and Technology, Ghana*
Mr Thabiso Perfect Oscar Mkhwanazi, *University of KwaZulu-Natal, South Africa*
Ms Lerato Mokoloko, *University of the Witwatersrand, South Africa*
Dr John Moma, *University of the Witwatersrand, South Africa*
Mr Muhammad Motala, *University of Cape Town, South Africa*
Dr Tshiamo Motshegwa, *University of Botswana, Botswana*
Ms Nonhlanhla Mphahlele, *CSIR, South Africa*
Dr Apollinaire Munyaneza, *University of Botswana, Botswana*
Dr Isaac Nongwe Beas, *Unisa, South Africa*
Mr Kershen Naicker, *University of KwaZulu-Natal, South Africa*
Miss Dayle Nel, *University of Cape Town, South Africa*
Professor Matthew Neurock, *University of Minnesota, USA*
Professor Dr Hans Niemantsverdriet, *SynCat@DIFFER, Syngaschem BV, Eindhoven, Netherlands*
Miss Liisa Niitsoo, *Royal Society of Chemistry, United Kingdom*
Mr Thulani Nyathi, *University of Cape Town, South Africa*
Dr Lide Oar-Arteta, *Delft University of Technology, Netherlands*
Mr Mike Ojemaye, *University of Fort Hare, South Africa*
Mr Oluwasegun Emmanuel Olaoye, *University of Botswana, Botswana*
Professor Unni Olsbye, *University of Oslo, Norway*
Professor Olayinka Oyetunji, *University of Botswana, Botswana*
Mr Drushan Padayachee, *University of KwaZulu-Natal, South Africa*
Miss Qamreen Parker, *University College London/Cardiff University, United Kingdom*
Mr Adli Peck, *University of Cape Town, South Africa*
Miss Anna Petersen, *University of Cape Town, South Africa*
Mr Tumelo Phaahlamohlaka, *University of the Witwatersrand, South Africa*
Dr Evgeniy Redekop, *University of Oslo, Norway*
Dr Erik Reichelt, *Fraunhofer IKTS, Germany*
Dr Alberto Roldan, *Cardiff University, United Kingdom*
Mr Max Schaller, *Fraunhofer IKTS, Germany*
Professor Dr Hans Schulz, *Karlsruhe Institute of Technology, Engler-Bunte-Institute, Germany*
Dr Mzamo Shozi, *University of KwaZulu-Natal, South Africa*
Professor Anja Olafsen Sjåstad, *University of Oslo, Norway*
Professor Martha Smit, *University of the Free State, South Africa*
Ms Claire Springett, *Royal Society of Chemistry, United Kingdom*

Miss Rosalind Stegman, *University of Cape Town*, *South Africa*
Mr Xiaohui Sun, *Delft University of Technology*, *Netherlands*
Dr Ganesh Babu Sundaram, *University of Cape Town*, *South Africa*
Dr Andrew Swarts, *North-West University*, *South Africa*
Dr Richard Tia, *Kwame Nkrumah University of Science and Technology*, *Ghana*
Miss Erin Trenor, *University of Cape Town*, *South Africa*
Miss Chelsea Tucker, *University of Cape Town*, *South Africa*
Professor Veikko Uahengo, *University of Namibia*, *Namibia*
Ms Tracey van Heerden, *University of Cape Town*, *South Africa*
Dr Pieter van Helden, *Sasol Group Technology*, *South Africa*
Professor Eric van Steen, *University of Cape Town*, *South Africa*
Professor Dr Manie Vosloo, *North-West University*, *South Africa*
Dr Kees-Jan Weststrate, *SynCat@DIFFER, Syngaschem BV, Eindhoven*, *Netherlands*
Mr Tim Wezendonk, *Delft University of Technology*, *Netherlands*
Ms Bronwyn White, *University of Cape Town*, *South Africa*
Mr Moritz Wolf, *University of Cape Town*, *South Africa*
Mrs Irina Yarulina, *Delft University of Technology*, *Netherlands*
Professor Dr Ying Zheng, *University of Edinburgh*, *United Kingdom*
Mr Aleksandar Zivkovic, *Cardiff University*, *United Kingdom*